Molecular Materials with Specific Interactions

CHALLENGES AND ADVANCES IN COMPUTATIONAL CHEMISTRY AND PHYSICS

Volume 4

Series Editor:

JERZY LESZCZYNSKI
Department of Chemistry, Jackson State University, U.S.A.

Molecular Materials with Specific Interactions – Modeling and Design

Edited by

W. Andrzej Sokalski

Department of Chemistry
Wrocław University of Technology,
Poland

 Springer

A C.I.P. Catalogue record for this book is available from the Library of Congress.

ISBN-10 1-4020-5371-1 (HB)
ISBN-13 978-1-4020-5371-9 (HB)
ISBN-10 1-4020-5372-X (e-book)
ISBN-13 978-1-4020-5372-6 (e-book)

Published by Springer,
P.O. Box 17, 3300 AA Dordrecht, The Netherlands.

www.springer.com

Printed on acid-free paper

CONTENTS

PREFACE

Intermolecular interactions constitute one of the major forces determining numerous specific properties and self organization of condensed matter including biological systems. There is no better illustration of their importance than the introduction in the famous textbook *The Feynman Lectures on Physics*:

If, in some cataclysm, all of scientific knowledge were to be destroyed, and only one sentence passed on to the next generations of creatures, what statement would contain the most information in the fewest words? I believe it is the atomic hypothesis that all things are made of atoms – little particles that move around in perpetual motion, attracting each other when they are a little distance apart, but repelling upon being squeezed into one another.

This statement refers directly to the concept of the potential energy function describing interaction energy as a function of the distance sufficient to model various properties of matter that are directly or indirectly related to the intermolecular forces. Recent progress in computational and experimental techniques opens the perspective for rational *de novo* design of new materials with desired properties being governed by specific interactions.

The main purpose of this volume is to present an overview of selected state-of-the-art computational methodologies and to present examples of applications. The first six chapters contain an in-depth description of several new techniques applied in modeling and designing new molecular materials.

The first chapter by Moszyński presents in a systematic and comprehensive manner the current state-of-the-art theory of intermolecular interactions. Numerous examples illustrate how theoreticians and experimentalists working in tandem may gather valuable quantitative results related to intermolecular interactions, like accurate potential functions, interaction-induced properties, spectra and collisional characteristics or dielectric, refractive or thermodynamic properties of bulk phases. On the other hand the most advanced Symmetry Adapted Perturbation Theory (SAPT) enables validation of more approximate variation-pertubation models which could be applied to the analysis of specific interactions in much larger molecular systems, for example enzyme-drug interactions discussed in Chapter VIII by Berlicki *et al.*

The second chapter by Wesołowski presents a general overview of the recent developments in Density Functional Theory (DFT), which currently constitutes the most popular tool for modeling molecular materials. Particular emphasis has been given to the systematic discussion of approximations employed in DFT and their performance for various types of molecular aggregates.

Mezoscopic and microscopic modeling methods applied in modeling biopolymers are reviewed in the third chapter by Lesyng and coworkers, allowing for a better description of biomolecular recognition processes as well as corresponding free energy changes.

The fourth chapter by Michalak and Ziegler presents an excellent introduction into DFT-based first principle molecular dynamics capable of modeling complex chemical reactions. Considerable effort has been made by the authors to explain in detail the specifics of ab initio calculations wherever they differ from conventional techniques.

A very promising methodology bridging quantum mechanics and molecular mechanics (QM/MM), allowing the mechanisms of enzymatic reactions to be analyzed in detail, has been reviewed in the fifth chapter by Mulholland and Grant.

The chapter written by Ramos and coworkers reviews various computational techniques used to study protein-protein interactions, with particular attention given to thermodynamic characteristics of mutated proteins and their interactions.

The contribution of Paneth and coworkers demonstrates how substrate-enzyme interactions could be explored using experimentally determined kinetic isotope effects and QM/MM calculations.

The remaining three chapters illustrate various applications of molecular modeling methods in exploring various properties of complex protein systems. Nakano and coworkers analyse the mechanism of [NiFe] hydrogenase involved in hydrogen production. Renugopalakrishnan studies bacteriorhodopsin which is important in information storage technologies. Filipek and coworkers concentrate on rhodopsin as being an important target for pharmacological intervention.

Intermolecular interactions define crucial characteristics of materials for hydrogen storage materials. This topic is discussed in detail in the chapter by Cheng *et al.* devoted to molecular dynamics simulations of single-walled carbon nanotubes (SWNT) with molecular hydrogen. The properties of modified SWNTs, in the contribution from Politzer *et al.*, are also analyzed from the point of view of potential applications in molecular electronics.

Molecular electronic applications are also covered in the contribution from Zhou and Hagelberg discussing interactions of various organic molecules with silicon surface, whereas Zhou *et al.* concentrate on fullerene deposition on silicon and GaAs surfaces.

The last chapter by Michalkova *et al.* presents a review of the experimental and theoretical data on nerve agent interactions with different surfaces. Particular attention is given to molecular simulations of interaction and decomposition of phosphoroorganic compounds on various metal and metal oxide clusters.

Of course the examples discussed in this volume cover only a small fraction of possible applications. We hope that with the rapid progress of computational techniques many more molecular materials will soon be the subject of rational design *in silico*.

Let me express my gratitude to all authors for their contributions which will guide readers into the exciting world of molecular modeling.

W. Andrzej Sokalski
Wrocław, Poland
June 2006

CHAPTER 1

THEORY OF INTERMOLECULAR FORCES:
AN INTRODUCTORY ACCOUNT

ROBERT MOSZYNSKI

Quantum Chemistry Laboratory
Faculty of Chemistry, Warsaw University
Pasteura 1, 02–093 Warsaw, Poland

Abstract: Modern theory of intermolecular forces is reviewed. The concept of the interaction potential is introduced within the Born-Oppenheimer separation of the electronic and nuclear motions. Various supermolecule approaches for the calculation of accurate interaction potentials are discussed. Perturbation theory of intermolecular forces is reviewed in great details. The problem of symmetry-adaptation is explained and a general symmetry-adapted perturbation theory is formulated. Convergence properties of various symmetry-adapted expansions are surveyed, and illustrated on several examples. Physical interpretation of the interaction potential in terms of the four fundamental interaction components: electrostatics, induction, dispersion, and exchange-repulsion is thoroughly exposed. Many-electron formulation of the symmetry-adapted perturbation theory in both the wave function and density functional approaches is introduced. One-center and multicenter multipole expansions neglecting the charge-overlap effects, as well as the bipolar expansion accounting for these effects are discussed. The relation of some supermolecule approaches with the perturbation theory of intermolecular forces is briefly sketched. Approximate models that can be deduced from the rigorous theory of inter-molecular forces, and applicable to the interactions of large systems are discussed. Finally, perturbation theory of nonadditive interactions in trimers and of the collision-induced electric properties of binary collisional complexes is also reviewed. The theory part is completed by an exposition of methods needed on the route from intermolecular potentials and collision-induced properties to physically measurable quantities such as the Raman spectra, rovibrational spectra, scattering cross sections, as well as thermodynamic, dielectric, and refractive properties of dilute gases. The present status of symmetry-adapted perturbation theory applied to the calculations of state-of-the-art *ab initio* potential energy surfaces and collision-induced properties is presented, and illustrated by means of applications to rovibrational spectra of Van der Waals molecules, scattering cross sections and pressure broadening coefficients, collision-induced Raman spectra of atomic gases, solvation processes, and thermodynamic, dielectric, and refractive properties of dilute gases. Theoretical results are compared with high accuracy experimental data

Keywords: intermolecular forces, Born-Oppenheimer approximation, supermolecular method, polar-ization approximation, symmetry-adapted perturbation theory, electrostatics, induction, dispersion, exchange-repulsion, multipole expansion, one-center and multicenter

1

W. A. Sokalski (ed.), Molecular Materials with Specific Interactions, 1–152.
© 2007 *Springer.*

expansions, bipolar expansion, charge overlap and damping effects, approximate models, relation between supermolecule approaches and perturbation theory, nonadditive interactions, collision-induced properties, optical spectra of Van der Waals complexes, scattering cross sections, dielectric, refractive, and thermodynamic properties of dilute gases, state-of-the-art *ab initio* potential energy surfaces, infrared spectra, integral and differential cross sections, pressure broadening coefficients, thermodynamic virial coefficients for binary complexes, high accuracy *ab initio* nonadditive potentials, radial distribution functions from first principle computer simulations, solvation processes, state-of-the-art *ab initio* Raman spectra, and dielectric and refractive properties of atomic gases, comparison with high precision experimental data

1. INTRODUCTION

The importance of intermolecular interactions in physics, chemistry, and biology does not need to be stressed. Intermolecular potentials determine the properties of non-ideal gases (pure) liquids, solutions, molecular solids, and the behavior of complex molecular ensembles encountered in biological systems. They describe the so-called non-bonded contributions, as well as the special hydrogen bonding terms, that are part of the force fields used in simulations of processes as enzyme-substrate binding, drug-receptor interactions, etc. A few examples showing important applications of intermolecular potentials include the Monte Carlo and molecular dynamics simulations of biological systems, studies of processes in the earth's atmosphere, or interstellar chemistry. One can even imagine significant technological advances from the studies of weak intermolecular forces. With a proper manipulation of the energetic and stereochemical features of the interactions between molecules one may design artificial receptor molecules capable of binding substrate species strongly and selectively, just as biological enzymes do, leading to the construction of new materials.

Let us start this review by briefly repeating that the concept of the intermolecular potential is based on the Born-Oppenheimer separation of the Schrödinger equation for the electronic and nuclear motions. The solution of the first step, the electronic structure problem, for a number of (clamped) nuclear coordinates yields the potential surface for the nuclear motions. For an individual molecule the latter are the vibrations, rotations, and translations of this molecule. For an ensemble of interacting molecules, which can be distinguished as such because they are held together by strong covalent or ionic chemical bonds, there is a hierarchy of the strong intramolecular forces which determine the internal vibrations of the molecules, and much weaker intermolecular forces which determine their relative translations and rotations. This paper is concerned with the relatively weak intermolecular interactions, Van der Waals forces and hydrogen bonding, in particular. They play a crucial role in molecular complexes which may be collisional, as in crossed molecular beam experiments, or truly bound, as occurring in high concentrations in cold supersonic nozzle beams, but also, in lower concentrations, in bulk phases.

Intermolecular potentials depend on the intermolecular degrees of freedom, i.e. on the coordinates describing the relative translations and rotations of the molecules in a complex, but also on the intramolecular coordinates describing the molecular geometries. Because of the above mentioned hierarchy, it is mostly allowed to make another adiabatic separation, namely between the intramolecular vibrations with high frequencies and the intermolecular modes with much lower frequencies. The latter can then be described with an intermolecular potential averaged over the fast intramolecular vibrations. The intermolecular motions have mostly large amplitudes, because the potential surfaces corresponding to the weak intermolecular forces are rather flat. Often, there are multiple (equivalent or non-equivalent) minima on the potential surfaces that are accessible through thermal motions of the complex or quantum mechanical tunneling. An appropriate theoretical treatment of the dynamics of such weakly bound complexes requires the knowledge of the full potential surface, not just of the second derivatives, the force constants, at the minimum of the potential, as in the harmonic oscillator model.

The interaction energy of two molecules A and B is defined in the Born-Oppenheimer approximation as the difference between the energy of the dimer E_{AB} and the energies of the monomers E_A and E_B

$$E_{int} = E_{AB} - E_A - E_B. \tag{1-1}$$

It is assumed that the internal coordinates of the monomers A and B used in the calculations of E_A and E_B are the same as within the dimer AB. Thus, the interaction energy depends on the separation R between the centers of mass of the monomers, on the Euler angles characterizing their mutual orientation, and on monomer's internal coordinates.

Present day methods and computers have evolved to a stage where it is possible to obtain accurate intermolecular potentials from *ab initio* electronic structure calculations. Basically there are two methods to do so: the supermolecule method and the symmetry-adapted perturbation theory. In the supermolecular approach the interaction energy is computed directly from Eq. (1-1) by subtracting the sum of monomer's energies from the energy of the dimer. In practical calculations, due to the use of incomplete basis sets, this method always contains the basis set superposition error resulting from a nonphysical lowering of monomer's energy in dimer's calculation due to "borrowing" of the basis set from the interacting partner. Since for all many-electron systems the errors in the total energies are much larger than the interaction energy itself, the accuracy of the computed potential depends on a cancellation of these large errors. Moreover, a supermolecular calculation does not allow any physical insight into the nature of intermolecular interactions. Methods taking into account specific nature of intermolecular interactions, i.e. the perturbation methods have actually been useful in providing an interpretation and an estimation of reliability of potential energy surfaces obtained by the supermolecular approach.

The purpose of this chapter is to review different aspects of the intermolecular interaction phenomenon. Previous review articles were restricted to some specific aspects of intermolecular forces. Refs. (1–4) concentrated on the perturbation theory of intermolecular forces, while Refs. (5–6) on the supermolecular methods. Finally, Refs. (7–10) reported expositions of the dynamics of Van der Waals molecules. It is now clear that the perturbation theory approach to intermolecular interactions provides the basic conceptual framework within which the intermolecular forces are discussed. It defines asymptotic constraints on potential energy surfaces obtained by any theoretical method, and can accurately predict the complete intermolecular potential energy surfaces for Van der Waals and hydrogen-bonded complexes. Therefore, in this review we will mostly concentrate on the perturbation theory approach, namely the symmetry-adapted perturbation theory (SAPT) for pair and three-body interactions, and for the interaction-induced properties. However, the supermolecule method, and methods needed on the route from intermolecular potentials to quantities measured in the experiments will also be discussed. All these theoretical aspects will be illustrated by high accuracy *ab initio* calculations with a special emphasis on the comparison between state-of-the-art theoretical results and high precision experimental data. Due to space limitations of the present review we will not present experimental techniques used in precise measurements of spectra, scattering cross sections, or bulk properties. We will also not discuss fitting methods used on the route from the experimental data to the interaction potentials. However, we would like to stress that very accurate potentials reproducing a wealth of the experimental data can be obtained in this way. See, for instance, Refs. (11–18) for some typical examples.

We start our review in Section 2 with a brief discussion of the separation of the electronic and nuclear motions, the so-called clamped nuclei or Born-Oppenheimer approximation. This approximation is of fundamental importance for the theory of intermolecular forces since by its very definition the concept of the interaction potential appears in this approximation. We continue in Section 3 with a short exposition of merits and drawbacks of the supermolecular method. The discussion of the simplest Rayleigh-Schrödinger perturbation expansion of the interaction energy, including an analysis of its convergence properties is presented in Section 4. The exchange effects and the symmetry adaptation problem are also discussed in this section. In this section we formulate various symmetry-adapted perturbation theories, and discuss the convergence properties of these SAPT expansions for model systems. We also discuss the relation between the symmetry-forcing procedure (weak or strong) and the convergence of the perturbation series. In Section 5 we show that the low-order polarization and exchange corrections have a simple physical interpretation and can be related to some monomer properties such as one- and two-particle density matrices, linear and quadratic response functions, etc. Section 6 is devoted to the discussion of the multipole expansion which can be viewed as the limit of SAPT for very large intermolecular separations. We discuss one-center expansions based on the global multipole moments and polarizabilities, as well as multicenter expansions based on the distributed multipole analysis. We also outline the bipolar expansion that takes into account the important charge overlap (penetration) effects. In Section 7 the many-electron implementation of SAPT is outlined. The many-electron theory is formulated both in the wave function and in the density functional approaches. Section 8 describes the relations of the many-electron SAPT method with the supermolecular approaches based on the Hartree-Fock method and the many-body perturbation theory with the Møller-Plesset partitioning of the dimer's Hamiltonian. An approximate analysis of the free energy of solvation, as defined by the selfconsistent reaction field theory, in terms of SAPT contributions is also briefly outlined. The theory reported in Sections 2–8 is exact, but it can hardly be applied to the interactions of very large molecules such as, e.g. an inhibitor molecule and an enzyme residue. Therefore, Section 9 is devoted to some simplified models of pair interactions between molecules. In particular, the relations of these models with the modern theory of intermolecular forces, as well as their merits and drawbacks, are discussed. A short exposition of the empirical force fields is also given, again in connection with the theory of intermolecular forces. In Section 10 we formulate the symmetry-adapted perturbation theory of three-body interactions, and show that the low-order contributions to nonadditive three-body potentials can be expressed through the linear and quadratic response functions, and one- and two-particle density matrices. We also discuss the physical interpretation of various three-body contributions to the interaction energy. In Section 11 we briefly sketch the application of the SAPT approach to collision-induced dipole moments and polarizabilities of Van der Waals complexes. Since the concept of the interaction energy appears in the Born-Oppenheimer approximation, interaction potentials are not directly measurable. Section 12 gives a brief overview of the methods needed on the route from intermolecular potentials and collision-induced properties to quantities measured in various experiments, such as Raman spectra, dielectric second virial coefficients, rovibrational spectra, scattering cross sections, or thermodynamic second virial coefficients. In this section we also mention simulation techniques including modern approaches based on the density functional theory. Applications of SAPT to several weakly bound binary complexes and larger clusters with a special emphasis on the comparison between state-of-the-art theoretical results and highly precise experimental data are surveyed in Section 13. In this section we also show how the present day theory can help the experimentalists to interpret their data. Finally, in Section 14 we give a brief outlook for the future of the theory of intermolecular forces.

2.   THE BORN-OPPENHEIMER APPROXIMATION

The Born-Oppenheimer separation[19−22] of the electronic and nuclear motions in molecules is probably the most important approximation ever introduced in molecular quantum mechanics, and will implicitly or explicitly be used in all subsequent sections of this chapter. The Born-Oppenheimer approximation is crucial for modern chemistry. It allows to define in a rigorous way, within the quantum mechanics, such useful chemical concepts like the structure and geometry of molecules, the molecular dipole moment, or the interaction potential. In this approximation one assumes that the electronic motions are much faster than the nuclear motions. Therefore, one can assume that slow changes in the nuclear positions do not strongly affect the physical description of the electronic motions, leading to an adiabatic separation of the electronic and nuclear problems. In the Born-Oppenheimer approximation the electronic problem is solved for an (infinite) set of nuclear coordinates. The eigenvalue of the electronic Schrödinger equation, the electronic energy as a function of these coordinates, is nothing else than the potential energy surface. This energy surface is used as the potential operator in the second step of the Born-Oppenheimer approximation, the problem of nuclear motions in a given potential.

Various derivations of the Born-Oppenheimer approximation can be found in the literature. See Refs. (23–24) for typical reviews. The applicability of this approximation has been proven on several examples, cf. the seminal works of Kolos and Wolniewicz[25–29] for various states of the hydrogen molecule. The results of Kolos and collaborators on H_2 and of other authors for other systems were reviewed in several papers, cf., e.g. Refs. (24–32). Since intermolecular forces can only be discussed for fixed geometries of the interacting monomers, the Born-Oppenheimer approximation is a natural framework for the discussion of intermolecular interactions. Therefore, in this section we will briefly review all approximations leading to to the separation of the electronic and nuclear motions, and discuss situations in which this fundamental approximation fails.

2.1.   Born-Huang Expansion of the Total Wave Function

We consider a closed-shell dimer AB with N_A electrons and n_A nuclei that can be assigned to the monomer A, and N_B electrons and n_B nuclei that can be assigned to the monomer B. The set of electronic coordinates $\mathbf{r}_i, i = 1, \ldots N_A + N_B$ will be denoted by $\{\mathbf{r}\}$, while the nuclear coordinates, $\mathbf{R}_\gamma, \gamma = 1, \ldots n_A + n_B$, will be denoted in short by $\{\mathbf{R}\}$. The coordinates of electrons and nuclei are defined in a space-fixed frame. The nuclear masses will be denoted by M_γ, and atomic units $m_e = e = \hbar = 1$ will be used throughout this chapter. The Schrödinger equation for the total wave function Ψ^{tot} can be written as,

$$\left(-\frac{1}{2} \sum_{\gamma \in AB} \frac{1}{M_\gamma} \nabla_\gamma^2 + H^{\text{el}} - \mathcal{E} \right) \Psi^{\text{tot}}(\{\mathbf{r}\}, \{\mathbf{R}\}) = 0, \qquad (1\text{-}2)$$

where H^{el} denotes the nonrelativistic Coulombic Hamiltonian describing the electronic motions for a fixed set of the nuclear coordinates. The sum of the electronic and nuclear terms in Eq. (1-2) will be called the total Hamiltonian,

$$\mathcal{H} = -\frac{1}{2} \sum_{\gamma \in AB} \frac{1}{M_\gamma} \nabla_\gamma^2 + H^{\text{el}}. \qquad (1\text{-}3)$$

The total wave function of the dimer can be represented by the Born-Huang expansion[22],

$$\Psi^{\text{tot}}(\{\mathbf{r}\}, \{\mathbf{R}\}) = \sum_k \Psi_k^{\text{el}}(\{\mathbf{r}\}; \{\mathbf{R}\})\chi_k(\{\mathbf{R}\}) \qquad (1\text{-}4)$$

where the electronic wave function Ψ_k^{el} fulfills the electronic Schrödinger equation,

$$H^{\text{el}}\Psi_k^{\text{el}} = E_k(\{\mathbf{R}\})\Psi_k^{\text{el}}, \qquad (1\text{-}5)$$

and χ_k denotes the wave function describing the nuclear motions. The electronic equation (1-5) is solved for a fixed set of nuclear coordinates $\{\mathbf{R}\}$, so the electronic wave function Ψ_k^{el} depends parametrically on $\{\mathbf{R}\}$, while the electronic energy $E_k(\{\mathbf{R}\})$ is a function of these coordinates. The index k numbers the solutions of the electronic Schrödinger equation, i.e. it labels the electronic states. Note that the electronic energy $E_k(\{\mathbf{R}\})$ for the dimer AB is nothing else than the energy E_{AB} appearing in the definition of the interaction energy, Eq. (1-1).

By inserting the expansion (1-4) into the Schrödinger equation (1-2), multiplying from the left by $\left(\Psi_l^{\text{el}}\right)^*$, and integrating over the electronic coordinates we get the following set of coupled differential equations for the nuclear wave function χ_k,

$$\left[-\frac{1}{2}\sum_{\gamma\in AB}\frac{1}{M_\gamma}\nabla_\gamma^2 + C_{kk}(\{\mathbf{R}\}) + E_k(\{\mathbf{R}\}) - \mathcal{E}_k \right]\chi_k$$

$$= \sum_{l\neq k}\left[C_{kl}(\{\mathbf{R}\}) + \sum_{\gamma\in AB}\frac{1}{2M_\gamma}\left(\nabla_\gamma\cdot\mathbf{B}_\gamma^{kl} + \mathbf{B}_\gamma^{kl}\cdot\nabla_\gamma\right) \right]\chi_l, \qquad (1\text{-}6)$$

where

$$\mathbf{B}_\gamma^{kl} = \frac{1}{M_\gamma}\langle\psi_k^{\text{el}}|\nabla_\gamma\psi_l^{\text{el}}\rangle_{\{\mathbf{r}\}}, \qquad (1\text{-}7)$$

and

$$C_{kl}(\{\mathbf{R}\}) = \langle\Psi_k^{\text{el}}| -\frac{1}{2}\sum_{\gamma\in AB}\frac{1}{M_\gamma}\nabla_\gamma^2|\Psi_l^{\text{el}}\rangle_{\{\mathbf{r}\}}. \qquad (1\text{-}8)$$

The notation $\langle\cdots|\cdots|\cdots\rangle_{\{\mathbf{r}\}}$ means that the integration is performed only over the electronic coordinates. The diagonal term $C_{kk}(\{\mathbf{R}\})$ is the so-called diagonal or adiabatic correction for the nuclear motions, while the nondiagonal terms are often referred to as the nonadiabatic terms. Note parenthetically that the vector $\mathbf{B}_\gamma^{kk}$ vanishes for real electronic wave functions Ψ_k^{el}, so terms involving $\mathbf{B}_\gamma^{kk}$ do not appear on the l.h.s. of Eq. (1-6).

Equation (1-6) shows that the electronic and nuclear motions are coupled through the nondiagonal terms $C_{kl}(\{\mathbf{R}\})$, $\nabla_\gamma\cdot\mathbf{B}_\gamma^{kl}$, and $\mathbf{B}_\gamma^{kl}\cdot\nabla_\gamma$. One should note that both the diagonal and nondiagonal terms appearing in Eq. (1-6) are inversely proportional to nuclear masses, so for heavy systems the r.h.s. of Eq. (1-6) and the diagonal correction should be negligible.

We wish to end this section by saying that Eq. (1-6) can also be obtained by minimizing the energy functional with respect to the functions χ_k,

$$\mathcal{E}[\chi_1, \chi_2, \ldots] = \langle \Psi^{\text{tot}} | \mathcal{H} \Psi^{\text{tot}} \rangle - \lambda \left(\langle \Psi^{\text{tot}} | \Psi^{\text{tot}} \rangle - 1 \right), \tag{1-9}$$

where λ is the Langrange multiplier and the trial wave function is given by Eq. (1-4). This means that Eq. (1-6) has a well defined meaning (in the sense of the variational principle) even when the Born-Huang expansion (1-4) is limited to a finite number of terms.

2.2.   Adiabatic and Born-Oppenheimer Approximations

In the adiabatic approximation[21,22] one assumes that the right hand side of Eq. (1-6) can be neglected. This approximately corresponds to the following ansatz for the total wave function,

$$\Psi^{\text{tot}}_{\text{ad},k}(\{\mathbf{r}\}, \{\mathbf{R}\}) = \Psi^{\text{el}}_k(\{\mathbf{r}\}; \{\mathbf{R}\})\chi_k(\{\mathbf{R}\}). \tag{1-10}$$

The nuclear motions, i.e. the rotations and vibrations of the molecule, are described by a Schrödinger-type equation,

$$\left(-\frac{1}{2} \sum_{\gamma \in AB} \frac{1}{M_\gamma} \nabla_\gamma^2 + C_{kk}(\{\mathbf{R}\}) + E_k(\{\mathbf{R}\}) - \mathcal{E}_k \right) \chi_k = 0. \tag{1-11}$$

Thus, in the adiabatic approximation the nuclear motions are described by a Schrödinger equation with the potential operator given by the sum of the electronic energy $E_k(\{\mathbf{R}\})$ and the diagonal (adiabatic) correction $C_{kk}(\{\mathbf{R}\})$.

If the molecule is in its ground state, i.e. Eq. (1-2) is solved for the lowest molecular state, it can be shown[33] that the expectation value of the total (electronic plus nuclear) Hamiltonian with the wave function $\Psi^{\text{tot}}_{\text{ad},k}$ of Eq. (1-10) gives the lowest energy calculated by solving Eq. (1-11),

$$\mathcal{E}_0 = \langle \Psi^{\text{tot}}_{\text{ad},0} | \mathcal{H} | \Psi^{\text{tot}}_{\text{ad},0} \rangle. \tag{1-12}$$

This means that the adiabatic energy represents the upper bound to the exact eigenvalue of the total Hamiltonian.

The Born-Oppenheimer or clamped nuclei approximation[19,20] is obtained when, in addition to using the wave function (1-10), the diagonal (adiabatic) correction is neglected in Eq. (1-11), and the potential operator in the Schrödinger equation for the nuclear motions is simply defined as the eigenvalue of the electronic Schrödinger Equation (1-5). It can be shown[34] that in the Born-Oppenheimer approximation the energy obtained by solving Eq. (1-11) with C_{kk} put equal to zero represents a lower bound to the exact eigenvalue of the total Hamiltonian. One should note that this result is of litle practical use since upper bounds are usually obtained.

2.3. Electronic Schrödinger Equation

The Schrödinger equation describing the electronic motions for a fixed set of nuclear coordinates $\{\mathbf{R}\}$ is given by Eq. (1-5). This equation is the basis for the theory of intermolecular forces, since the interaction energy is closely related to the electronic energy of the dimer, cf. Eq. (1-1). Different methods of solving Eq. (1-5) lead to different approaches to the theory of intermolecular forces, and will be discussed in great details in this chapter.

When deriving the coupled equations or the Schrödinger-type equation for the nuclear motions we assumed that the electronic and nuclear coordinates are defined in a space-fixed (laboratory) frame. Usually, in quantum chemical applications, Eq. (1-5) is solved in a body-fixed frame attached to the molecule or to the Van der Waals complex. It should be stressed, however, that the kinetic term and the Coulomb interaction terms describing the electron-nucleus attraction, and electron-electron and nucleus-nucleus repulsion are invariant with respect to the choice of the coordinate system. This means that the mathematical form of the electronic Schrödinger equation does not change when going from the space-fixed to the body-fixed coordinate frame. Therefore, Eq. (1-5) can be solved in the body-fixed frame (as it is usually done). Care should be taken, however, when using an analytical representation of the electronic energy as the potential operator in the Schrödinger equation for the nuclear motions. Here, the choice between space-fixed and body-fixed frames will lead to different equations, and the analytical expression for the potential energy surface must be written in terms of those nuclear coordinates that are used in the second step of the Born-Oppenheimer approximation, cf. Section 12.4.

2.4. Schrödinger Equation for the Nuclear Motions

The Schrödinger equation describing the nulcear motions, Eq. (1-11), contains too many variables, as the translational motion of the center-of-mass (c.o.m.) has not been separated out. When the c.o.m. motion is separated out, and when the origin of the space-fixed coordinate system is located in the c.o.m. of the nuclei, Eq. (1-11) takes the form,

$$\left(-\frac{1}{2} \sum_{\gamma=2,\,\gamma\in AB} \frac{1}{M_\gamma} \nabla_\gamma^2 + C_{kk}(\{\mathbf{R}\}) + E_k(\{\mathbf{R}\}) - \mathcal{E}_k \right) \chi_k = 0, \tag{1-13}$$

where the adiabatic correction is given by,

$$C_{kk}(\{\mathbf{R}\}) = \langle \Psi_k^{\text{el}} | -\frac{1}{2} \sum_{\gamma=2,\,\gamma\in AB} \frac{1}{M_\gamma} \nabla_\gamma^2 + \frac{1}{2M_{\text{tot}}} \left(\sum_{\gamma=2,\,\gamma\in AB} \nabla_\gamma \right)^2 - \frac{1}{2} \left(\sum_i \nabla_{\mathbf{r}_i} \right)^2 | \Psi_k^{\text{el}} \rangle_{\{\mathbf{r}\}}.$$

$$\tag{1-14}$$

Here, M_{tot} is the total mass of the nuclei. Note that when eliminating the motion of the center-of-mass we arbitrarily eliminated the first nuclear coordinate (we could eliminate any single coordinate γ). Equation (1-13) can further be transformed to the body-fixed frame. When applying the so-called Eckart conditions[35] one would get the standard Watson's Hamiltonian describing the nuclear motions in molecules[36].

Equation (1-13) or its body-fixed equivalent is of little use for Van der Waals complexes, as it discriminates one nuclear coordinate, e.g. $\gamma = 1$. Specific mathematical forms of Hamiltonians describing the nuclear motions in Van der Waals dimers have been developed (7). This point will be discussed in more details in Section 12.4. Here we only want to stress that whatever the mathematical form of the Hamiltonian is used to solve the problem of nuclear motions, the results will be the same, if the Schrödinger equation is solved exactly. However, in weakly bound complexes there is a hierarchy of motions due to the strong intramolecular forces which determine the internal vibrations of the molecules, and to much weaker intermolecular forces which determine their relative translations and rotations. This hierarchy allows to make a separation between the intramolecular vibrations with high frequencies and the intermolecular modes with much lower frequencies. Such a separation of the fast intramolecular vibrations and slow rotation-vibration-tunneling motions can be performed if a suitable form of the Hamiltonian for the nuclear motions in Van der Waals molecules is used.

2.5. Failures of the Born-Oppenheimer Approximation and the Nonadiabatic Approach

Equation (1-6) does not show when the off diagonal terms on the right hand side become important. To judge the importance of the nonadiabatic effects it is most convenient to use the perturbation theory[37-40]. The Hamiltonian $\mathcal{H}$ can be represented by the sum

$$\mathcal{H} = \mathcal{H}_0 + \mathcal{H}'. \tag{1-15}$$

The zeroth-order operator will be defined by the following eigenvalue equation:

$$\mathcal{H}_0 \Psi_{ad,k}^{tot} = \mathcal{E}_k \Psi_{ad,k}^{tot}, \tag{1-16}$$

while the perturbation operator $\mathcal{H}'$ is given by the difference $\mathcal{H} - \mathcal{H}_0$, and fulfills the condition,

$$\langle \Psi_{ad,0}^{tot} | \mathcal{H}' | \Psi_{ad,0}^{tot} \rangle = 0. \tag{1-17}$$

The energy to the first-order is given by the expectation value of $\mathcal{H}$ with $\Psi_{ad,0}^{tot}$, and is given by the lowest energy calculated by solving Eq. (1-11). Thus, the corrections due to nonadiabatic effects appear only in the second order of the perturbation theory. Using the well known expressions of the Rayleigh-Schrödinger perturbation theory,

$$\mathcal{E}_0^{(2)} = \sum_{k \neq 0} \frac{|\langle \Psi_{ad,0}^{tot} | \mathcal{H}' | \Psi_{ad,k}^{tot} \rangle|^2}{\mathcal{E}_0^{(0)} - \mathcal{E}_k^{(0)}}, \tag{1-18}$$

where $\mathcal{E}_k^{(0)}$ is the energy of the kth state of $\mathcal{H}_0$ obtained by solving Eq. (1-11), one immediately sees that $\mathcal{E}_0^{(2)}$ becomes important when the denominator in Eq. (1-18) is small, i.e. when the rovibrational levels corresponding to the ground electronic state ($k = 0$) and to an excited electronic state (with a particular value of k) are very close in energy. This occurs when the potential energy surfaces show avoided crossings or conical intersections, i.e. when the electronic states interact via the coupling terms appearing on the r.h.s. of Eq. (1-6). It should be noted that since $\mathcal{H}_0$ is only formally defined via the eigenvalue problem the matrix elements appearing in the numerator of Eq. (1-18) are not defined. It can be shown, however, that they are given by,

$$\langle \Psi_{ad,0}^{tot} | \mathcal{H}' | \Psi_{ad,k}^{tot} \rangle = \langle \Psi_{ad,0}^{tot} | - \frac{1}{2} \sum_{\gamma \in AB} \frac{1}{M_\gamma} \nabla_\gamma^2 | \Psi_{ad,k}^{tot} \rangle, \tag{1-19}$$

and can be related to the quantities $C_{kl}(\{\mathbf{R}\})$ and $\mathbf{B}_\gamma^{kl}$ by the following expression,

$$\langle \Psi_{ad,0}^{tot} | \mathcal{H}' | \Psi_{ad,k}^{tot} \rangle = \int \chi_0^\star(\mathbf{R}) C_{0k}(\{\mathbf{R}\}) \chi_k(\mathbf{R}) d\mathbf{R}$$
$$+ \int \chi_0^\star(\mathbf{R}) \sum_{\gamma \in AB} \frac{1}{2M_\gamma} \left(\nabla_\gamma \cdot \mathbf{B}_\gamma^{0k} + \mathbf{B}_\gamma^{0k} \cdot \nabla_\gamma \right) \chi_k(\mathbf{R}) d\mathbf{R}. \tag{1-20}$$

The failures of the Born-Oppenheimer separation of the electronic and nuclear motions show up in the spectra of molecules as homogeneous or heterogeneous perturbations in the spectra[41]. See, e.g. Ref. (42) for an example, a fully *ab initio* study of the spectrum of the calcium dimer in a coupled manifold of electronic states. Theoretical methods needed to describe the dynamics of molecules in nonadiabatic situations are being developed now. See Ref. (43) for a review.

3. SUPERMOLECULAR APPROACH TO INTERMOLECULAR INTERACTIONS

Supermolecular methods are based on the definition of the interaction energy, Eq. (1-1). However, the electronic Schrödinger equation can rarely be solved in a nearly exact way, so the eigenvalues appearing in Eq. (1-1) should rather be replaced by some approximations, $\widetilde{E}_{AB}$, $\widetilde{E}_A$, $\widetilde{E}_B$, and $\widetilde{E}_{int}$, to the exact ground state energies and to the exact interaction energy, respectively. These approximate energies are obtained by using a specific (approximate) method of solving the clamped-nuclei

Schrödinger equation. Usually the exact interaction energy is four to seven orders of magnitude smaller than the terms subtracted in Eq. (1-1). Since with the *ab initio* methods available at present the errors $\widetilde{E}_{AB} - E_{AB}$, $\widetilde{E}_A - E_A$, and $\widetilde{E}_B - E_B$ are always much larger than the interaction energy itself, the quantity $\widetilde{E}_{\text{int}}$ can be a good approximation to E_{int} if a cancellation of these large errors occurs. The experience gained thus far[5] suggests that this may be the case if Eq. (1-1) is applied with due care. The chosen approximate method must be size-consistent[44] and it should sufficiently account for the electron correlation, i.e. go beyond the Hartree-Fock level. Finally, the approximate monomers energies should be calculated with the full basis of the dimer, i.e. the computed interaction energy should be corrected for the basis set superposition error. When these three criteria are met by an approximate electronic structure method, then a quite accurate description of the intermolecular interaction is possible.

The size-consistency requirement is not easy to meet. Indeed, the popular variational metods selfconsistent field methods based on restricted (RASSCF) active spaces[45,46], limited configuration interaction in its single-reference (CI) and multireference (MRCI) versions[47], or the second-order perturbation theory based on the CASSCF reference function (CASPT2)[48] are not size-consistent. The complete active space selfconsistent theory (CASSCF) is size consistent when the active spaces of the dimer and of the monomers are correctly chosen. If the same active space is used in the calculations for the dimer and for the monomers the method is no longer size consistent. The limited CI expansions are usually restricted to single and double excitations, and can approximately be corrected for the size-inconsistency using a kind of correcting terms (size-consistency correc- tions) due to Pople[49], Davidson[50], and others[51]. It should be stressed here that for some difficult cases like the interactions involving open-shell monomers with spatially denerate ground states or interactions in the excited states, size-consistent methods based on the open-shell coupled-cluster ansatz in the Hilbert space[52] or in the Fock space[53] suffer from the intruder state problems[54,55], and cannot be used in practice. In such cases only size-inconsistent methods quoted above are available.

For the interaction of closed-shell molecules several size-consistent methods for solving the Schrödinger equation are available. The density functional theory (DFT), although not based on the Schrödinger equation, also gives the energies of molecules. Among the size-consistent approaches, the Møller-Plesset perturbation theory (MPPT) also known as many-body perturbation theory (MBPT)[56,57] and the coupled-cluster (CC) method[58−60] gained a lot of popularity. The experience gained thus far[5] shows that the second-order of the Møller-Plesset perturbation theory (MP2) accounts for a large part of the electron correlation, and in most cases is considered to give a qualitatively correct description of the interaction energy. See Ref. (5) for examples illustrating this point. In many cases, e.g. for hydrogen-bonded systems, the MP2 interaction energy is in a semiquantitative agreement with very accurate results obtained by more elaborate methods. The inclusion of the third order in the Møller-Plesset expansion (MP3) does not improve (and often worsens) the results, and one has to go to the full fourth order (MP4) to get an improvement. For the interaction of simple closed-shell monomers with single bonds the MP4 approach gives rather accurate results.

When a high accuracy is thought for one has to resort to nonperturbative methods based on the coupled cluster ansatz. In many cases the CCD[61,62] or CCSD[63] approaches, i.e. the coupled cluster methods restricted to double or single and double excitations, respectively, do not offer any benefits compared to the simple MP2 method. It is fully established now that most accurate results are obtained by using the so-called CCSD(T) method, i.e. the coupled-cluster method limited to single, double, and noniterative triple excitations[64,65]. See Refs. (5–6) for a detailed discussion of this point. More elaborate methods that fully account for the triple excitations[66] are computationally very expensive, and cannot be used in practice, even for small systems like the water dimer. For some systems even the quadrupole excitations are very important[67,68]. A prominent example is the dimer of carbon monooxide $(CO)_2$. For this system even the CCSD(T) method fails to reproduce the correct dipole-dipole asymptotics of the potential energy surface[68], and methods incorporating the fifth-order of the Møller-Plesset theory[69,70] are necessary.

For large systems correlated methods based on the wave function approach (i.e. based on the clamped-nuclei Schrödinger equation) cannot be applied in practice due to computational limitations. On the other hand different variants of the density functional theory with gradient corrected exchange-correlation functionals such as, e.g. B3LYP[71–73], gained a lot of popularity. These methods perform very well for the geometry optimizations of hydrogen-bonded complexes, so at first glance one would expect that they should give reliable interaction energies for such systems. It is worth noting that the DFT method is computationally much less demanding than the MP2 method. In fact, the computational effort is about the same as for a Hartree-Fock calculation, so DFT could be a good alternative to computationally expensive methods based on the Møller-Plesset theory and the coupled cluster ansatz. Unfortunately this is not the case. Extensive comparisons[74] between the supermolecule DFT and CCSD(T) results showed that DFT with any gradient corrected exchange-correlation functional fails to reproduce the correct anisotropy of the potential energy surfaces for Van der Waals and hydrogen-bonded complexes. In some cases the differences were as large as a factor of two or even more. Surprisingly, supermolecule DFT calculations failed for hydrogen-bonded complexes, even though the geometry optimization gives the correct geometries of the minima and saddle points on the potential energy surfaces. Thus, with the exchange-correlation functionals available at present, DFT cannot be considered as a viable computational tool for intermolecular potential energy surfaces.

The discussion of the preceding paragraphs clearly shows that indeed supermolecule method should be applied with great care. So we wish to end this section by saying that even though the supermolecule approach is conceptually very simple it cannot be used by simply running standard black box programs of quantum chemistry.

4.   PERTURBATION THEORY OF INTERMOLECULAR FORCES

4.1.   Rayleigh-Schrödinger Perturbation Theory

We consider the interaction of two closed-shell monomers A and B in their ground states, described by the wave functions Φ_0^A and Φ_0^B, respectively, which are eigenstates of the respective monomer's Hamiltonians H_A and H_B with the corresponding eigenvalues E_0^A and E_0^B, respectively. The Schrödinger equation for the noninteracting system AB can be written as,

$$H_0\Phi_0 = E_0\Phi_0, \tag{1-21}$$

where $H_0 = H_A + H_B$, $\Phi_0 = \Phi_0^A\Phi_0^B$, and $E_0 = E_0^A + E_0^B$. When the interaction between the monomers A and B is switched on, the Schrödinger equation for the interacting system AB is given by,

$$(H_0 + V)\Psi = E\Psi, \tag{1-22}$$

where V is the intermolecular interaction operator collecting all Coulombic interactions between the nuclei and electrons of A and the nuclei, electrons of B, E is the total energy of the dimer AB, and Ψ is the electronic wave function of the dimer. For simplicity we omitted the subscript "el" used in Section 3. We consider the interaction of two closed-shell ground state monomers, so Ψ will refer to the ground state electronic wave function of the dimer and the index k numbering the electronic state will be suppressed as well.

Equation (1-22) can conveniently be rewritten in the following form,

$$\Psi = \Phi_0 + \hat{R}_0(E_{\text{int}} - V)\Psi, \quad E_{\text{int}} = \langle\Phi_0|V\Psi\rangle, \tag{1-23}$$

where $E_{\text{int}} = E - E_0$ is the interaction energy. The operator $\hat{R}_0$ is the so-called reduced resolvent of H_0, which may be viewed as the inverse of the operator $H_0 - E_0$ in the space orthogonal to Φ_0, and is defined by the following spectral expansion,

$$\hat{R}_0 = \sum_{k \neq 0} \frac{|\Phi_k\rangle\langle\Phi_k|}{E_k - E_0}, \tag{1-24}$$

where E_k and Φ_k denote the excited state eigenvalues and eigenfunctions of H_0. Equations (1-23) can be solved by applying the Rayleigh-Schrödinger (RS) perturbation theory to the dimer wave function Ψ and to the interaction energy E_{int}. To this end we parametrize the Hamiltonian $H = H_0 + V$ with a complex parameter ζ, $H(\zeta) = H_0 + \zeta V$, where the physical value of ζ is obviously equal to one. The interaction energy E_{int} and the wave function of the dimer Ψ become functions of ζ, and can be expanded as power series in ζ,

$$\Psi(\zeta) = \Phi_0 + \sum_{n=1}^{\infty} \zeta^n \Phi_{\text{pol}}^{(n)} \quad \text{and} \quad E_{\text{int}}(\zeta) = \sum_{n=1}^{\infty} \zeta^n E_{\text{pol}}^{(n)}. \tag{1-25}$$

The individual corrections $E_{\text{pol}}^{(n)}$ and $\Phi_{\text{pol}}^{(n)}$ appearing on the r.h.s. of Eqs. (1-25) are referred, after Hirschfelder[75], to as the *n*th-order *polarization energy* and *polarization wave function*, respectively. The *n*th-order energy correction is given by

$$E_{\text{pol}}^{(n)} = \langle \Phi_0 \mid V \mid \Phi_{\text{pol}}^{(n-1)} \rangle, \tag{1-26}$$

while the polarization wave functions can be obtained from the following recursion relation,

$$\Phi_{\text{pol}}^{(n)} = -\hat{R}_0 V \Phi_{\text{pol}}^{(n-1)} + \sum_{k=1}^{n-1} E_{\text{pol}}^{(k)} \hat{R}_0 \Phi_{\text{pol}}^{(n-k)}, \tag{1-27}$$

with $\Phi_{\text{pol}}^{(0)} \equiv \Phi_0$.

Although the equations defining the polarization wave functions and energies are simple, they cannot be applied in practice to describe weak intermolecular interactions because the expansions (1-25) are either divergent or converge much too slowly for $\zeta = 1$. As shown in extensive theoretical[76-78] and numerical[79-81] studies the series (1-25), if convergent, do not converge to the physical ground state of the dimer, except for one- and two-electron dimers such as H_2^+ (Ref. 79) and H_2 (Ref. 80). For the interaction of two-electron systems the energy first reaches very quickly the average of the energies of all states including the mathematical, Pauli-forbidden, solutions of the Schrödinger equation corresponding to the same dissociation limit (the so-called Coulomb energy),

$$Q = \frac{\sum_{\nu} f_{\nu} \, {}^{\nu}E_{\text{int}}}{\sum_{\nu} f_{\nu}}, \tag{1-28}$$

where ν labels possible permutational symmetry of the states of the dimer that can be obtained from a given set of the zeroth-order monomers states, and ${}^{\nu}E_{\text{int}}$ is the corresponding interaction energy. Then the series converges very slowly to the energy of the mathematical ground state of the Hamiltonian H which corresponds to the fully symmetric, Pauli-forbidden solution of the Schrödinger equation[81]. It is worth noting here that Pauli-forbidden solutions of the Schrödinger equation appear because the exclusion principle is not obeyed by the perturbation equations of the polarization theory. The situation is even more complex when one of the monomers has more than two electrons. It was shown[77,78] by using group-theoretical arguments that the polarization expansion is divergent in this case. These theoretical findings were recently supported by numerical calculations for the ground state of LiH[82,83].

The origins of this pathological convergence pattern of the polarization expansion can most easily be explained by considering the simplest case of a hydrogen atom A interacting with a proton B at a large internuclear distance R[84]. In this case Φ_0 is the $1s_A$ hydrogenic orbital located at the proton A. Since the exact wave function Ψ is symmetric with respect to the reflection in the plane perpendicular to the internuclear axis and passing through its midpoint, the correct form of Ψ at large

R is $\Psi \approx 1s_A + 1s_B$. Obviously, $\Phi_0 = 1s_A$ is not a good approximation to Ψ. The component $1s_B$ due to the perturbation V is as large as the unperturbed function itself. Hence, the operator V cannot be considered as a small perturbation. Since the polarization wave functions for this system are all localized at the nucleus A, i.e. they decay exponentially with the distance from the nucleus A, the polarization expansion can possibly recover the $1s_B$ component of Ψ, localized on the nucleus B, only in very high orders.

For the interaction of many-electron systems the polarization expansion is strictly divergent. This divergence is due to the fact that the physical ground-state of the dimer is submerged in the Pauli-forbidden continuum of mathematical solutions of the Schrödinger equation corresponding to other than antisymmetric eigenstates of the Hamiltonian. The Pauli-forbidden continuum originates from the fact that, when the exclusion principle is not obeyed, i.e. when the antisymmetry of Ψ is not forced, the electrons assigned initially to the system A can fall into the Coulomb wells of the system B by means of the strong nucleus-electron attraction, ejecting some other electrons into the continuum. This means that if the exclusion principle is violated, the physical ground-state is a bound state submerged in the continuum of states of the same symmetry. For such a state the wave function is not normalizable, and consequently, the standard Rayleigh-Schrödinger perturbation theory cannot be convergent[85].

This convergence failure of the polarization expansion is illustrated in Table 1-1, where we report high-order results for the ground states of the helium dimer He_2 and lithium hydride LiH. An inspection of Table 1-1 shows that for the interaction of two two-electron monomers (He atoms) the polarization expansion converges to the Pauli-forbidden, mathematical ground state of the dimer[81]. This is illustrated in the second column of the table, where the convergence pattern is reported for a very small interatomic distance, $R = 1$ bohr. Around the potential minimum the polarization series also converge to the ground mathematical state, but the convergence is prohibitively slow, and the perturbation expansion effectively recovers only the Coulomb part of the interaction energy. The situation is different for the interaction of a three electron monomer with a one electron system, LiH[83]. In this case the physical ground state is submerged in the continuum of the Pauli-forbidden mathematical states, and the polarization expansion diverges. This divergence is observed for both $R = 10$ and 12 bohr. Still, the partial sums of the series first quickly reach the Coulomb energy, and then the series slowly diverge.

Although the polarization expansion cannot be used to compute the interaction energies of weakly bound complexes, it does provide the correct asymptotic expansion of the interaction energy in the following sense[84]

$$E_{\text{int}} = \sum_{n=1}^{N} E_{\text{pol}}^{(n)} + O(R^{-\kappa(N+1)}), \tag{1-29}$$

where $\kappa = 2$ if at least one of the interacting molecules has a net charge and $\kappa = 3$ if both molecules are neutral. After projection with the operator $^v\!\mathcal{A}$ corresponding

Table 1-1. Convergence of the polarization expansion for the interaction of two ground-state helium atoms at $R = 1$ and 5.6 bohr, and of the lithium and hydrogen atoms in their ground states at $R = 10$ and 12 bohr. The Coulomb energies represent 53.50% (He_2, $R = 5.6$ bohr), 73.4% (LiH, $R = 10$ bohr), and 85.53% (LiH, $R = 12$ bohr) of the energies of the fully symmetric (Pauli forbidden) states. The quantity $\delta(n)$ represents the percent error of the perturbation series through the nth-order with respect to the variational interaction energy of the Pauli forbidden state

	He_2		LiH	
	$R = 1$	$R = 5.6$	$R = 10$	$R = 12$
n		$\delta(n)$		
2	−54.22	−54.82	−49.31	−74.98
3	−38.19	−55.16	−53.21	−76.70
4	−25.37	−54.57	−56.94	−78.65
5	−17.11	−54.34	−59.63	−79.95
6	−11.07	−54.13	−61.88	−81.00
7	−6.99	−53.98	−63.72	−81.84
8	−4.14	−53.87	−65.27	−82.51
9	−2.27	−53.78	−66.58	−83.07
10	−1.06	−53.71	−67.68	−83.51
15	0.35	−53.54	−71.33	−84.81
20	0.09	−53.49	−73.80	−85.35
25	0.00	−53.47	−77.76	−85.73
30	−0.01	−53.46	−89.89	−86.34
35			−137.54	−87.85

to a given permutational symmetry ν, the polarization expansion for the wave function gives the correct[84] asymptotic expansion for the exact (unnormalized) wave function Ψ

$$\Psi = {}^{\nu}\!A\Phi_0 + \sum_{n=1}^{N} {}^{\nu}\!A\Phi_{\mathrm{pol}}^{(n)} + O(R^{-\kappa(N+1)}). \qquad (1\text{-}30)$$

The function Φ_0 or a finite sum of the polarization wave functions are bad approximations to the exact wave function Ψ. Indeed, it was shown by Jeziorski and Kolos[84] that $\|\Psi - \Phi_0\| \approx 1$. By contrast, the function ${}^{\nu}\!A\Phi_0$ is a much better approximation, since in view of Eq. (1-30) $\|\Psi - {}^{\nu}\!A\Phi_0\| = O(R^{-\kappa})$[84].

Although the function ${}^{\nu}\!A\Phi_0$ would seem a natural zeroth-order approximation for a perturbation expansion of the interaction energies, it unfortunately is not an eigenfunction of H_0. One could try to introduce a new partitioning of the Hamiltonian, $H = \widetilde{H}_0 + \widetilde{V}$, such that ${}^{\nu}\!A\Phi_0$ is an eigenfunction of $\widetilde{H}_0$, but no really successful construction of $\widetilde{H}_0$ has been reported to date. On the other hand, one could keep the natural partitioning of the Hamiltonian, $H = H_0 + V$, and modify the perturbation equations in such a way that the function ${}^{\nu}\!A\Phi_0$ can be used as the zeroth-order approximation. Such a modification leads to symmetry-adapted perturbation theories (SAPT). In the next section we review the problem of the symmetry adaptation of the perturbation expansions for the intermolecular interaction energies of weakly bound complexes, we introduce the concept of symmetry-forcing in the perturbation theory of intermolecular forces, and briefly discuss the convergence of various SAPT expansions, and its relation with the employed symmetry-forcing.

4.2.   Symmetry Adaptation

As discussed in the previous section, the perturbation expansion based on Eqs. (1-23) shows pathological convergence properties. To introduce the symmetry-adaptation into Eqs. (1-23) let us note that the exact solution of this equation possesses a definite permutational symmetry ν corresponding to an appropriate irrep of the group $S_{N_A+N_B}$. Thus, any operator enforcing the proper permutational symmetry acting on Ψ will give Ψ. This is no longer true if Eqs. (1-23) are parametrized with a complex parameter ζ. The group $S_{N_A+N_B}$ is no longer a symmetry group of the parametrized Hamiltonian $H(\zeta) = H_0 + \zeta V$, so the wave function $\Psi(\zeta)$ does not belong to an irrep of this group. The key idea of the symmetry-adaptation consists in replacing Eqs. (1-23) by equations in which Ψ is replaced by a symmetry-forcing operator acting on Ψ, parametrizing these equations with a complex parameter ζ, and expanding the parametrized equations as power series in ζ. The resulting perturbation expansions will no longer be equivalent to the polarization expansion, since the parametrized equations are equivalent to the original Schrödinger equation only for $\zeta = 1$.

To introduce the perturbation expansions corresponding to perturbation theories with different symmetry adaptation, it is useful to introduce a general concept of the interpolation function ${}^{\nu}\epsilon(\zeta)$ defined such that[86]:

$$
{}^{\nu}\epsilon(0) = 0, \qquad {}^{\nu}\epsilon(1) = {}^{\nu}E_{\text{int}}, \tag{1-31}
$$

where the index ν labels the permutational symmetry of the state of interest at $\zeta = 1$. The perturbation expansion corresponding to a given symmetry adaptation scheme is then defined as a power series expansion of ${}^{\nu}\epsilon(\zeta)$,

$$
{}^{\nu}\epsilon(\zeta) = \sum_{n=1}^{\infty} \zeta^n \, {}^{\nu}E^{(n)}, \tag{1-32}
$$

while the interaction energy ${}^{\nu}E_{\text{int}}$ is obtained by summing the expansion (1-32) at $\zeta = 1$.

Interpolation functions corresponding to various symmetry adaptation schemes defined in Refs. (84,86–88) can conveniently be written in the following general form,

$$
{}^{\nu}\epsilon(\zeta) = \zeta \frac{\langle \Phi_0 | V \, {}^{\nu}\mathcal{G} \, {}^{\nu}\Psi(\zeta) \rangle}{\langle \Phi_0 | \, {}^{\nu}\mathcal{G} \, {}^{\nu}\Psi(\zeta) \rangle}, \tag{1-33}
$$

where the wave function $^{\nu}\Psi(\zeta)$ is a solution of the equation,

$$^{\nu}\Psi(\zeta) = \Phi_0 + \hat{R}_0 \left(\zeta \frac{\langle \Phi_0 | V \, {}^{\nu}\widetilde{\mathcal{G}} \, {}^{\nu}\Psi(\zeta) \rangle}{\langle \Phi_0 | \, {}^{\nu}\widetilde{\mathcal{G}} \, {}^{\nu}\Psi(\zeta) \rangle} - \zeta V \right) \, {}^{\nu}\mathcal{F} \, {}^{\nu}\Psi(\zeta), \qquad (1\text{-}34)$$

and $^{\nu}\mathcal{G}$, $^{\nu}\widetilde{\mathcal{G}}$, and $^{\nu}\mathcal{F}$ are operators enforcing the proper permutational symmetry, such that $^{\nu}\mathcal{G}^{\mu}\Psi(\zeta = 1) = {}^{\nu}\widetilde{\mathcal{G}}^{\mu}\Psi(\zeta = 1) = {}^{\nu}\mathcal{F}^{\mu}\Psi(\zeta = 1) = \delta_{\mu\nu} \, {}^{\nu}\Psi(\zeta = 1)$. Thus, for $\zeta = 1$ Eqs. (1-33) and (1-34) are equivalent to the Schrödinger equation (1-23). One may also note here that when $^{\nu}\mathcal{G} = {}^{\nu}\widetilde{\mathcal{G}} = {}^{\nu}\mathcal{F} = 1$, i.e. when no symmetry is enforced, the Taylor expansion of Eqs. (1-33) and (1-34) defines the perturbation equations of the polarization theory, Eqs. (1-26) and (1-27). Different choices of the symmetry operators $^{\nu}\mathcal{G}$, $^{\nu}\widetilde{\mathcal{G}}$, and $^{\nu}\mathcal{F}$ lead to different symmetry-adapted perturbation expansions with various levels of mathematical sophistication, and different convergence properties. The proposed symmetry-adapted perturbation theories can be divided into two categories. In the first category, corresponding to the so called *weak symmetry forcing*[87,88], the symmetry operator appears only in the energy expressions. The perturbation equations do not contain any nonlocal symmetry operator. Only these type of theories have been applied thus far to interactions of many-electron systems. In the second class, corresponding to the *strong symmetry forcing*[87,88] the symmetry operators enter the perturbation equations and complicate significantly their solution when the interacting monomers have more than two electrons.

In the simplest symmetrized Rayleigh-Schrödinger (SRS) perturbation theory[87−89] the symmetry forcing operator appears only in the energy expression. Hence, this formalism employs weak symmetry forcing and the operators $^{\nu}\widetilde{\mathcal{G}}$, $^{\nu}\mathcal{F}$, and $^{\nu}\mathcal{G}$ are given by,

$$^{\nu}\widetilde{\mathcal{G}} = {}^{\nu}\mathcal{F} = 1, \qquad {}^{\nu}\mathcal{G} = {}^{\nu}\mathcal{A}, \qquad (1\text{-}35)$$

where $^{\nu}\mathcal{A}$ is the projection operator on the appropriate representation of the symmetric group $S_{N_A+N_B}$, where N_A and N_B denote the number of electrons in the monomer A and B, respectively. The SRS perturbation corrections to the interaction energy, $^{\nu}E_{SRS}^{(n)}$, are given by[87],

$$^{\nu}E_{SRS}^{(n)} = {}^{\nu}N_0 \left[\langle \Phi_0 | V \, {}^{\nu}\mathcal{A}\Phi_{pol}^{(n-1)} \rangle - \sum_{k=1}^{n-1} {}^{\nu}E_{SRS}^{(k)} \langle \Phi_0 | \, {}^{\nu}\mathcal{A}\Phi_{pol}^{(n-k)} \rangle \right], \qquad (1\text{-}36)$$

where $^{\nu}N_0 = \langle \Phi_0 | \, {}^{\nu}\mathcal{A}\Phi_0 \rangle^{-1}$.

Extensive numerical calculations for small systems (H_2^+ (Ref. 79,87), H_2 (Ref. 89,90) and He_2 Ref. (81) show that the convergence properties of the SRS theory are excellent. Since in most cases the polarization expansion is divergent, one can expect that for many-electron monomers the SRS expansion will not be strictly convergent. However, the experience gained thus far for large many-electron systems suggests that a second-order SRS calculation correctly accounts for all major polarization and exchange contributions to the interaction energy. In the region of the Van der Waals minimum it should be accurate to within a few percent.

Compared to the SRS theory, the perturbation scheme proposed by Murrell and Shaw[91], and independently by Musher and Amos[92] introduces only a slight complication. As shown by Jeziorski and Kolos[88] the corresponding symmetry-forcing operators are given by,

$$^{\nu}\widetilde{\mathcal{G}} = {}^{\nu}\mathcal{G} = {}^{\nu}\mathcal{A}, \quad {}^{\nu}\mathcal{F} = 1, \tag{1-37}$$

while the MSMA perturbation equations can conveniently be written as:

$$^{\nu}E_{MSMA}^{(n)} = {}^{\nu}N_0 \left[\langle \Phi_0 | V \ {}^{\nu}\mathcal{A} \ {}^{\nu}\Psi_{MSMA}^{(n-1)} \rangle - \sum_{k=1}^{n-1} {}^{\nu}E_{MSMA}^{(k)} \langle \Phi_0 | \ {}^{\nu}\mathcal{A} \ {}^{\nu}\Psi_{MSMA}^{(n-k)} \rangle \right], \tag{1-38}$$

$$^{\nu}\Psi_{MSMA}^{(n)} = -\hat{R}_0 V \ {}^{\nu}\Psi_{MSMA}^{(n-1)} + \sum_{k=1}^{n-1} {}^{\nu}E_{MSMA}^{(k)} \hat{R}_0 \ {}^{\nu}\Psi_{MSMA}^{(n-k)}, \tag{1-39}$$

with ${}^{\nu}\Psi_{MSMA}^{(0)} \equiv \Phi_0$.

As shown in Refs. (79,86,93), the MSMA theory does not introduce any improvement in the convergence properties compared to the SRS theory. In fact the MSMA theory diverges already for the lowest $^2\Sigma_u^+$ state of H_2^+, while the SRS theory remains convergent for this state[87]. This divergence can be explained by the analytical structure of the MSMA interpolation function around $\zeta = 1$. It was shown by Jeziorski, Schwalm, and Szalewicz[86] that for the H_2^+ ion the interpolation function corresponding to the MSMA theory has a real pole for a value of the parameter ζ slightly small that one. This explains the success of Padé summation techniques applied to the MSMA perturbation series[93].

It is important to stress here that in each order of the SRS and MSMA theories the energy correction can be separated into the polarization and exchange parts,

$$^{\nu}E_{SRS/MSMA}^{(n)} = E_{pol}^{(n)} + {}^{\nu}E_{exch}^{(n)}. \tag{1-40}$$

The exchange contributions ${}^{\nu}E_{exch}^{(n)}$, vanish exponentially as a function of inter-monomer distance in each order, so at large intermonomer separations the SRS and MSMA results coincide with the RS results. This means that perturbation schemes employing the weak symmetry forcing are compatible with the asymptotic expansion of the interaction energy, cf. Eq. (1-29). Note that the exchange terms result from the exchange of electrons (quantum mechanical tunneling) between the unperturbed monomers, and represent repulsive contributions to the interaction energy. Therefore, the exchange contributions are often referred to as the exchange-repulsion.

The first perturbation expansion employing the strong symmetry-forcing was introduced by Eisenschitz and London[94] as early as in 1932, and rediscovered later by Hirschfelder[95], Van der Avoird[96], and Peierls[97]. The corresponding symmetry-forcing operators are given by[88],

$$^{v}\widetilde{\mathcal{G}} = {}^{v}\mathcal{G} = {}^{v}\mathcal{F} = {}^{v}\mathcal{A}, \tag{1-41}$$

and the perturbation equations can be written as,

$$^{v}E_{\mathrm{ELHAV}}^{(n)} = {}^{v}N_0 \left[\langle \Phi_0 | V \, {}^{v}\mathcal{A} \, {}^{v}\Psi_{\mathrm{ELHAV}}^{(n-1)} \rangle - \sum_{k=1}^{n-1} {}^{v}E_{\mathrm{ELHAV}}^{(k)} \langle \Phi_0 | \, {}^{v}\mathcal{A} \, {}^{v}\Psi_{\mathrm{ELHAV}}^{(n-k)} \rangle \right], \tag{1-42}$$

$$^{v}\Psi_{\mathrm{ELHAV}}^{(n)} = -\hat{R}_0 V \, {}^{v}\mathcal{A} \, {}^{v}\Psi_{\mathrm{ELHAV}}^{(n-1)} + \sum_{k=1}^{n-1} {}^{v}E_{\mathrm{ELHAV}}^{(k)} \hat{R}_0 \, {}^{v}\mathcal{A} \, {}^{v}\Psi_{\mathrm{ELHAV}}^{(n-k)}, \tag{1-43}$$

with $^{v}\Psi_{\mathrm{ELHAV}}^{(0)} \equiv \Phi_0$.

Extensive numerical studies for model systems show that forcing the symmetry in the equations for the energy and wave function corrections in each order of the perturbation theory is very efficient and the ELHAV theory converges very fast. For the triplet state of LiH and at the interatomic distance corresponding to the Van der Waals minimum the convergence radius of the ELHAV theory is equal to 1.60[83]. It is remarkable that the symmetry-forcing procedure characteristic of the ELHAV method effectively eliminates the coupling with the Pauli-forbidden continuum in which the ground state of the LiH is submerged, and leads to a convergent SAPT series. Unfortunately, this excellent convergence of the ELHAV series is observed only in high orders of the perturbation theory. Numerical investigations and analytical solutions for the H_2^+ ion have shown that the ELHAV expansion fails to recover in the second order the well-known induction and dispersion components of the interaction energy and, consequently, the correct asymptotics of the interaction energy. This wrong asymptotic behaviour of the interaction energy persists in any finite-order and for any system. This means that the ELHAV theory is not consistent with the asymptotic conditions (1-29) and (1-30), and cannot be used in practice to compute the interaction energies.

Since the ELHAV theory shows such excellent convergence properties, but a wrong asymptotic behavior, Jeziorski and Kolos[88] proposed a perturbation scheme based on an intermediate symmetry-forcing. The corresponding symmetry-forcing operators are given by[88],

$$^{v}\widetilde{\mathcal{G}} = {}^{v}\mathcal{G} = 1, \qquad {}^{v}\mathcal{F} = {}^{v}\mathcal{A}, \tag{1-44}$$

and the perturbation equations can be written as,

$$^{v}E_{\mathrm{JK}}^{(n)} = \langle \Phi_0 | V \, {}^{v}\Psi_{\mathrm{JK}}^{(n-1)} \rangle - \sum_{k=1}^{n-1} {}^{v}E_{\mathrm{JK}}^{(k)} \langle \Phi_0 | \, {}^{v}\Psi_{\mathrm{JK}}^{(n-k)} \rangle, \tag{1-45}$$

$$\nu\Psi_{JK}^{(n)} = -\hat{R}_0 V \ ^{\nu}\mathcal{A} \ ^{\nu}\Psi_{JK}^{(n-1)} + \sum_{k=1}^{n-1} \ ^{\nu}E_{JK}^{(k)} \hat{R}_0 \ ^{\nu}\mathcal{A} \ ^{\nu}\Psi_{JK}^{(n-k)}, \tag{1-46}$$

with $\ ^{\nu}\Psi_{JK}^{(0)} \equiv \Phi_0$. Unfortunately, the intermediate symmetry-forcing only partly corrects the wrong asymptotic behavior of the perturbation series. It can be shown[87] that the JK theory through the second order is fully equivalent to the SRS theory, and thus fulfils Eqs. (1-29) and (1-30) for $N = 2$ and 1, respectively. Higher order terms in the JK expansion do not show the proper asymptotic behavior. However, the convergence properties of the JK perturbation series are as good as for the ELHAV theory[83,87]. Since for neutral systems errors introduced by the wrong asymptotic behavior of the interaction energy vanish with the intermolecular distance R like R^{-9}, the JK theory may be considered as a convergent SAPT expansion giving the correct asymptotics up to and including the R^{-8} terms.

The symmetry forcing employed in the Hirschfelder-Silbey theory[98] is more sophisticated. In this approach, one considers simultaneously all eigenfunctions of the Hamiltonian H which are asymptotically degenerate with the physical solution of the Schrödinger equation for the dimer, i.e., one has to consider the interpolation functions with symmetry labels $\mu = 1, \ldots, f$ corresponding to all irreducible representations of $S_{N_A+N_B}$ entering the induced product $[\nu_A] \otimes [\nu_B] \uparrow S_{N_A+N_B}$. This induced product is carried by the $\binom{N_A+N_B}{N_A}$ functions obtained by operating with all $(N_A + N_B)!$ permutations of the group $S_{N_A+N_B}$ on the product of two functions, one of symmetry ν_A on A and one of symmetry ν_B on B. Since in this theory all asymptotically degenerate solutions are coupled, Eqs. (1-33) and (1-34) have to be slightly modified. The wave function $\ ^{\nu}\Psi(\zeta)$ on the r.h.s. of Eqs. (1-33) and (1-34) should be replaced the vector $(\ ^1\Psi(\zeta), \ldots, \ ^f\Psi(\zeta))$, and the corresponding symmetry forcing operators are given by[88],

$$\nu\tilde{\mathcal{G}} = \ ^{\nu}\mathcal{G} = \ ^{\nu}\mathcal{F} = \ ^{\nu}\mathcal{K}, \quad \ ^{\nu}\mathcal{K}(\ ^1\Psi, \ldots, \ ^f\Psi) = \ ^{\nu}\mathcal{A} \sum_{\mu=1}^{f} \ ^{\mu}\Psi. \tag{1-47}$$

The symmetry forcing operator of the HS theory acts on a vector of f functions $\ ^{\mu}\Psi$ and gives, as the result, a single function $\ ^{\nu}\mathcal{A} \sum_{\mu=1}^{f} \ ^{\mu}\Psi$. It can be shown[88] that the perturbation equations of the Hirschfelder-Silbey theory can conveniently be rewritten as,

$$\nu E_{HS}^{(n)} = \ ^{\nu}N_0 \left[\langle \Phi_0 | V \ ^{\nu}\mathcal{A} F^{(n-1)} \rangle - \sum_{k=1}^{n-1} \ ^{\nu}E_{HS}^{(k)} \langle \Phi_0 | \ ^{\nu}\mathcal{A} F^{(n-k)} \rangle \right], \tag{1-48}$$

$$F^{(n)} = -\hat{R}_0 V F^{(n-1)} + \sum_{k=1}^{n} \sum_{\nu=1}^{f} \ ^{\nu}E_{HS}^{(k)} \hat{R}_0 \ ^{\nu}\mathcal{A} F^{(n-k)}, \tag{1-49}$$

$$\langle \Phi_0 | F^{(k)} \rangle = \delta_{k0}, \tag{1-50}$$

where $F^{(0)} \equiv \Phi_0$ and

$$F^{(n)} = \frac{1}{f} \sum_{\mu=1}^{f} {}^\mu \Psi^{(n)}. \tag{1-51}$$

In each order of the HS theory the energy correction separates into the polarization and exchange parts, and the exchange contributions vanish exponentially as a function of intermonomer distance in each order, so at large intermonomer separations the HS results coincide with the RS results. However, it is worth noting here that the Hirschfelder-Silbey[98], although compatible with the asymptotic expansions (1-29) and (1-30), is affected by the Pauli-forbidden continuum and diverges for LiH[83].

Perturbation equations of the Hirschfelder-Silbey theory can be derived by using the concept of the primitive (localized) wave function F formally defined as,

$$F = \sum_{\mu=1}^{f} c_\mu \, {}^\mu \Psi, \tag{1-52}$$

where c_μ are some coefficients (to be determined). Obviously, the physical wave function is recovered from F by projection,

$${}^\nu \Psi = {}^\nu A F. \tag{1-53}$$

It is easy to show that the primitive function F satisfies the following equation

$$(H - E_0)F = QF - \sum_{\nu}^{f} {}^\nu K \, {}^\nu A F, \tag{1-54}$$

where Q is the Coulomb part of the interaction energy, cf. Eq. (1-28), and the exchange energies ${}^\nu K$ are defined as,

$${}^\nu K = {}^\nu E_{\text{int}} - Q. \tag{1-55}$$

Equation (1-54) holds for any choice of the coefficients c_μ, so an extra condition needed to uniquely specify F must be imposed. The function F of the Hirschfelder-Silbey theory fullfils the following condition:

$$\langle \, {}^\nu A \Phi_0 | H_0 - E_0 | F \rangle = 0, \ \text{for } \nu = 1, \dots, f. \tag{1-56}$$

Thus, indeed it may be viewed as a kind of localized function, although the localization condition (1-56) cannot easily be interpreted. Other localization schemes are also possible[99-106]. For instance, one could ask that F can be obtained from Φ_0 by the action of the Bloch operator of the quasidegenerate perturbation theory. In this case one gets the so-called Bloch localization[103]. Another possible choice, the so-called Kato localization[102], requires that the distance between the unperturbed wave function and the primitive (localized) function is minimal. Perturbation equations corresponding to these various localization schemes were derived in Ref. (104) (see also Refs.(105–106) for their application in the context of the perturbation analysis of the Hartree-Fock interaction energy). Unfortunately, the convergence properties of the perturbation theories based on the Bloch and Kato localization conditions are very poor already for the HeH system, so most probably these schemes will be of little use for many-electron systems.

The convergence pattern of various symmetry-adapted perturbation theories with weak and strong symmetry forcing is illustrated in Table 1-2 for the distances of the Van der Waals minima of the $b^2\Sigma_u^+$ state of H_2^+ ion and the $a^3\Sigma^+$ state of the LiH system. An inspection of the numerical results for H_2^+ (Refs. 79,87) and LiH (Refs. 83,107) reported in this table shows that none of the symmetry-adapted perturbation schemes is perfect. The asymptotic problems with the ELHAV series in low orders are clearly seen, especially for the lowest triplet state of LiH. The convergence rate of the HS expansion does not differ much from that of the SRS and MSMA series. It seems that the SRS theory employing weak symmetry forcing, although divergent, gives fairly accurate results in low orders, and represents a good compromise between accuracy and numerical complication.

Various symmetry forcing schemes, convergence properties, and asymptotic behavior are summarized in Table 1-3. An inspection of this Table leads to quite pessimistic conclusions. All perturbation theories that are asymptotically consistent

Table 1-2. Convergence of the SRS, MSMA, HS, and ELHAV expansions for the $b^2\Sigma_u^+$ state of H_2^+ at $R = 12.5$ bohr and for the $a^3\Sigma^+$ state of LiH at $R = 11.5$ bohr. $\delta(n)$ represents the percent error of the perturbation series through the nth-order with respect to the variational interaction energy of the physical state

	H_2^+, $b^2\Sigma_u^+$, $R = 12.5$ bohr				LiH, $a^3\Sigma^+$, $R = 11.5$ bohr			
	SRS	MSMA	HS	ELHAV	SRS	MSMA	HS	ELHAV
n				$\delta(n)$				
2	−0.633	−0.633	−0.629	−32.474	−11.724	−11.724	−11.723	−80.749
3	0.277	0.288	0.287	−7.192	−12.334	−12.326	−12.334	−32.749
4	0.175	0.187	0.188	−1.618	−9.452	−9.443	−9.448	−13.099
5	0.069	0.081	0.083	−0.367	−7.838	−7.828	−7.833	−5.413
6	0.021	0.034	0.037	−0.084	−6.397	−6.387	−6.393	−2.269
7	0.000	0.013	0.016	−0.019	−5.240	−5.230	−5.234	−0.964
8	−0.009	0.004	0.007	−0.004	−4.290	−4.280	−4.284	−0.416
9	−0.013	−0.001	0.003	−0.001	−3.516	−3.505	−3.509	−0.182
10	−0.015	−0.002	0.002	−0.000	−2.884	−2.873	−2.877	−0.081
20	−0.016	−0.004	0.000	0.000	−0.398	−0.385	−0.366	0.000
30	−0.016	−−0.004	0.000	0.000	−0.205	−0.192	−3.982	0.000
40	−0.016	−0.004	0.000	0.000	−13.722	−13.715		0.000

Table 1-3. Summary of the symmetry forcing operators, convergence properties, and asymptotic correctness of various symmetry-adapted perturbation theories

	$^{\nu}\widetilde{\mathcal{G}}$	$^{\nu}\mathcal{G}$	$^{\nu}\mathcal{F}$	Forcing	Convergence	Asymptotics
RS	1	1	1	weak	no	yes
SRS	1	$^{\nu}\mathcal{A}$	1	weak	no	yes
MSMA	$^{\nu}\mathcal{A}$	$^{\nu}\mathcal{A}$	1	weak	no	yes
HS	$^{\nu}\mathcal{K}$	$^{\nu}\mathcal{K}$	$^{\nu}\mathcal{K}$	strong	no	yes
ELHAV	$^{\nu}\mathcal{A}$	$^{\nu}\mathcal{A}$	$^{\nu}\mathcal{A}$	strong	yes	no
JK	1	1	$^{\nu}\mathcal{A}$	strong	yes	no

with the polarization theory are divergent, and all convergent perturbation schemes show a wrong asymptotic behavior.

Recently, Adams[108] and Patkowski et al.[107,109] have shown that it is possible to construct a symmetry-adapted perturbation expansion that converges for the interaction of many-electron systems and is simultaneously compatible through all orders with the asymptotic expansions (1-29) and (1-30). This goal could be achieved by an appropriate regularization of the attractive singularities in the Coulomb potential. Indeed, one may think that the elimination of the singularities from the attractive nucleus-electron potential will stop the electron flow from one monomer to the other, and this should lead to a convergent perturbation theory. Patkowski et al.[107,109] split the Coulomb electron-nucleus attraction into the singular v_p and regular v_t parts. The subscripts p and t indicate that v_p and v_t are responsible for the polarization and tunnelling aspects of the interaction phenomenon. When the one-electron Coulomb terms in the intermolecular interaction operator V are replaced by the regularized functions, one obtains the regularized interaction operator V_p,

$$V_p = -\sum_{\beta \in B}\sum_{i=1}^{N_A} Z_\beta v_p(\mathbf{r}_{\beta i}) - \sum_{\alpha \in A}\sum_{j=1}^{N_B} Z_\alpha v_p(\mathbf{r}_{\alpha j})$$

$$+\sum_{i=1}^{N_A}\sum_{j=1}^{N_B} \frac{1}{r_{ij}} + \sum_{\alpha \in A}\sum_{\beta \in B} \frac{Z_\alpha Z_\beta}{r_{\alpha\beta}}, \tag{1-57}$$

where the summation over α and β runs over the nuclei of the monomer A and B, respectively, Z_γ is the charge of the nucleus γ, and the regularized potential v_p is given by,

$$v_p(\mathbf{r}) = \frac{1}{r}\left(1 - e^{-\eta r^2}\right). \tag{1-58}$$

Here, η is the so-called regularization parameter. In principle, η is an arbitrary and positive real number. Note that the full interaction operator V is recovered from V_p in the limit $\eta \to \infty$. The tunneling part of V, V_t, is given by,

$$V_t \equiv V - V_p = -\sum_{\beta \in B}\sum_{i=1}^{N_A} Z_\beta v_t(\mathbf{r}_{\beta i}) - \sum_{\alpha \in A}\sum_{j=1}^{N_B} Z_\alpha v_t(\mathbf{r}_{\alpha j}), \tag{1-59}$$

where

$$v_t(\mathbf{r}) = \frac{1}{r} - v_p(\mathbf{r}). \tag{1-60}$$

In the limit $\eta \to \infty$ V_t tends to zero. Thus, the regularization parameter η effectively introduces a partitioning of V into a nonsingular, regular part V_p, and a singular part V_t.

Since the operator V_p does not have any singularities, the SRS theory employing V_p as a perturbation is convergent and provides all long-range and the majority of the exchange contributions to the interaction energy. The regularized corrections to the wave function and to the interaction energy are given by Eqs. (1-27) and (1-36), respectively, with V replaced by V_p. For a suitable choice of the regularization parameter η the regularized SRS expansion, R-SRS, must converge since the neglect of V_t shifts the unphysical Pauli-forbidden continuum above the physical ground state. Given the fact that V_t neglected in the R-SRS treatment is a short-range potential, cf. Eqs. (1-59) and (1-60), the R-SRS expansion exhibits the correct asymptotic behavior given by Eqs. (1-29) and (1-30). The remaining small part of the interaction energy due to the singular perturbation $V_t = V - V_p$, can efficiently be recovered using the symmetry-forcing procedure characteristic of the JK or ELHAV theories. Since the V_t operator is of the short-range type and gives only exponentially vanishing contributions to the interaction energy, there are no difficulties with the wrong asymptotics of the ELHAV corrections and the ELHAV expansion converges very fast in this case. The perturbation equations of the regularized ELHAV theory denoted by R-ELHAV are given by Eqs. (1-42) and (1-43), except that the interaction operator V is replaced by V_t, the zeroth-order Hamiltonian is replaced by $H_0 + V_p$, and the reduced resolvent and the zeroth-order wave function also correspond to the Hamiltonian $H_0 + V_p$. This means that zeroth-order wave function is not the product of the wave functions of the isolated monomers A and B, but rather the converged wave function of the regularized polarization expansion. The explicit form the these equations is as follows,

$$
{}^{\nu}E^{(n)}_{\text{R-ELHAV}} = {}^{\nu}N_0 \Big[\langle \Phi_p | V_t \, {}^{\nu}\mathcal{A} \, {}^{\nu}\Psi^{(n-1)}_{\text{R-ELHAV}} \rangle
$$
$$
- \sum_{k=1}^{n-1} {}^{\nu}E^{(k)}_{\text{R-ELHAV}} \langle \Phi_p | \, {}^{\nu}\mathcal{A} \, {}^{\nu}\Psi^{(n-k)}_{\text{R-ELHAV}} \rangle \Big], \tag{1-61}
$$

$$
{}^{\nu}\Psi^{(n)}_{\text{R-ELHAV}} = -\hat{R}_p V_t \, {}^{\nu}\mathcal{A} \, {}^{\nu}\Psi^{(n-1)}_{\text{R-ELHAV}}
$$
$$
+ \sum_{k=1}^{n-1} {}^{\nu}E^{(k)}_{\text{R-ELHAV}} \hat{R}_p \, {}^{\nu}\mathcal{A} \, {}^{\nu}\Psi^{(n-k)}_{\text{R-ELHAV}}, \tag{1-62}
$$

where Φ_p is the solution of the regularized Schrödinger equation,

$$(H_0 + V_p)\Phi_p = E_p \Phi_p, \tag{1-63}$$

and $^{\nu}\Psi_{R-ELHAV}^{(0)} = \Phi_p$, The reduced resolvent operator $\hat{R}_p$ is defined in the standard way,

$$\hat{R}_p = (H_0 + V_p - E_p)^{-1}Q_p, \tag{1-64}$$

where Q_p denotes an orthogonal projection operator on the space orthogonal to Φ_p,

$$Q_p = 1 - |\Phi_p\rangle\langle\Phi_p|. \tag{1-65}$$

The regularization scheme discussed above is only one example of various possible variants, cf. Ref. (110) for a review. Other variants differ in the way the regular and singular parts are treated, and in the definitions of the regularized Coulomb potential. For instance, one could solve the Schrödinger equation of the dimer using a double perturbation theory with the regular potential V_p treated by the standard Rayleigh-Schrödinger perturbation theory, and with the singular perturbation V_t treated by a perturbation theory with the strong symmetry-forcing. Or one can devise an "all-in-one" R-SRS+R-ELHAV approach which employs Φ_0 as the zeroth-order wave function for both the R-SRS and R-ELHAV treatments. All these approaches differ a bit in the convergence pattern of the perturbation expansions, but the most important goal of the regularization procedure, namely the asymptotic correctness of the interaction energy in all orders of the perturbation theory and the convergence of the perturbation series, is achieved. This means that by introducing the concept of regularization into the theory of intermolecular forces one could finally define convergent perturbation expansions compatible with the multipole expansion in all orders. Numerical results for the singlet and triplet states of the LiH molecule show that the regularization of the intermolecular interaction operator leads indeed to convergent perturbation expansions. The regularized SAPT even provided an accurate description of the chemical bond in LiH. It remains to be seen whether such good results will be obtained for larger systems as well.

5. PHYSICAL INTERPRETATION OF THE LOW-ORDER POLARIZATION AND EXCHANGE ENERGIES

The polarization and exchange energies through the second order have an appealing physical interpretation. Except for the second-order exchange terms they can also be rigorously related to monomer properties which considerably facilitates their practical evaluation.

5.1. Electrostatic Energy

The first-order polarization energy, often referred to as electrostatic energy, is given by

$$E_{pol}^{(1)} \equiv E_{elst}^{(1)} = \langle\Phi_0^A\Phi_0^B \mid V \mid \Phi_0^A\Phi_0^B\rangle. \tag{1-66}$$

As shown in Ref. (111) the expression for $E_{elst}^{(1)}$ can be rewritten in terms of the total charge distributions $\rho_A^{tot}(\mathbf{r})$ and $\rho_B^{tot}(\mathbf{r})$ of the unperturbed monomers,

$$E_{elst}^{(1)} = \int \int \rho_A^{tot}(\mathbf{r}_1)\frac{1}{|\mathbf{r}_1 - \mathbf{r}_2|}\rho_B^{tot}(\mathbf{r}_2)d\mathbf{r}_1 d\mathbf{r}_2, \tag{1-67}$$

where the total charge distribution for monomer A is given (in atomic units) by

$$\rho_A^{tot}(\mathbf{r}) = \sum_{\alpha \in A} Z_\alpha \delta(\mathbf{r} - \mathbf{R}_\alpha) - \rho_A(\mathbf{r}). \tag{1-68}$$

The term containing Dirac's delta $Z_\alpha \delta(\mathbf{r} - \mathbf{R}_\alpha)$ represents the contribution from the positive point charge Z_α at the position $\mathbf{R}_\alpha$ of the nucleus α, and $-\rho_A(\mathbf{r})$ is the electronic charge distribution, given by the diagonal element of the first-order density matrix normalized to the number of electrons in the monomer A.

Equations (1-67) and (1-68) show that the first-order polarization energy represents the energy of the electrostatic interaction of the unperturbed monomers' charge distributions, and is referred to as the *electrostatic energy*. At large intermonomer distances R the electrostatic energy can be represented as a sum of classical electrical interactions between the permanent multipole moments of the unperturbed monomers. One should note, however, that the electrostatic energy also contains important short-range components due to the mutual penetration (damping, charge overlap) of the monomers' electron clouds. This short-range part of the electrostatic energy makes significant contribution to the stabilization energy of der Waals complexes and cannot be neglected in any accurate calculation of the potential energy surfaces for such systems.

5.2.   First-order Exchange (Heitler-London) Energy

The first-order energy in the SAPT theories is given by

$$E^{(1)} = \frac{\langle \Phi_0 | V | \mathcal{A}\Phi_0 \rangle}{\langle \Phi_0 | \mathcal{A}\Phi_0 \rangle}, \tag{1-69}$$

where $\mathcal{A}$ is the antisymmetrizer. When Φ_0 is an exact eigenfunction of H_0 this energy is identical with the so-called Heitler-London energy:

$$E_{HL}^{(1)} = \frac{\langle \mathcal{A}\Phi_0 | H - E_0 | \mathcal{A}\Phi_0 \rangle}{\langle \mathcal{A}\Phi_0 | \mathcal{A}\Phi_0 \rangle}. \tag{1-70}$$

To separate the polarization and exchange parts of $E^{(1)}$ one has to use the following decomposition of the total antisymmetrizer[112]

$$\mathcal{A} = \frac{N_A! N_B!}{(N_A + N_B)!}(1 + \mathcal{P})\mathcal{A}_A \mathcal{A}_B, \tag{1-71}$$

where $\mathcal{A}_A$ and $\mathcal{A}_B$ are the antisymmetrizers for the monomers A and B, respectively and $\mathcal{P}$ collects all permutations (with appropriate sign factors) interchanging at least one pair of electrons between the interacting monomers. By inserting Eq. (1-71) into Eq. (1-69) one finds that

$$E^{(1)} = E^{(1)}_{\mathrm{elst}} + E^{(1)}_{\mathrm{exch}}, \tag{1-72}$$

where

$$E^{(1)}_{\mathrm{exch}} = \frac{\langle \Phi_0 \mid V - E^{(1)}_{\mathrm{elst}} \mid \mathcal{P}\Phi_0 \rangle}{1 + \langle \Phi_0 \mid \mathcal{P}\Phi_0 \rangle}. \tag{1-73}$$

This expression vanishes exponentially at large R since the functions Φ^A_0 and Φ^B_0 decay exponentially with the distance from the centers of the respective monomers[113]. The $E^{(1)}_{\mathrm{exch}}$ component represents the main exchange contribution to the interaction energy. At the Van der Waals minima it usually accounts for over 90% of the total exchange effect. The interpretation of $E^{(1)}_{\mathrm{exch}}$ is very simple: it represents the effect of taking the expectation value of the full Hamiltonian with the simplest possible function $(\mathcal{A}\Phi_0)$ representing in the zeroth order the resonance tunneling of electrons between all available equivalent minima.

An accurate evaluation of $E^{(1)}_{\mathrm{exch}}$ is difficult because the multiple electron exchange operators included in $\mathcal{P}$ prevent us from expressing this quantity in terms of monomer properties. For the intermonomer distances corresponding to typical Van der Waals minima Eq. (1-73) can greatly be simplified by considering only the single exchange approximation[114,115]. Since the resulting approximate value of $E^{(1)}_{\mathrm{exch}}$ is quadratic in the intermolecular overlap densities $\rho_{lm}(\mathbf{r}) = \psi_l(\mathbf{r})\psi_m(\mathbf{r})$ (orbital ψ_l on A and ψ_m on B), it is denoted by $E^{(1)}_{\mathrm{exch}}(S^2)$,

$$E^{(1)}_{\mathrm{exch}}(S^2) = -\langle \Phi_0 \mid V - E^{(1)}_{\mathrm{elst}} \mid \mathcal{P}_1 \Phi_0 \rangle, \tag{1-74}$$

where $\mathcal{P}_1$ denotes the sum of all $N_A N_B$ transpositions of electrons between the monomers. Equation (1-74) represents a very good approximation since its error is of the fourth order in the intermonomer overlap densities. It can be shown that $E^{(1)}_{\mathrm{exch}}(S^2)$ can be expressed through the one- and two-particle density matrices of the unperturbed monomers[116]

$$E^{(1)}_{\mathrm{exch}}(S^2) = \int\int \left(\tilde{v}(\mathbf{r}_1, \mathbf{r}_2) - \frac{1}{N_A N_B} E^{(1)}_{\mathrm{elst}} \right) \rho_{\mathrm{int}}(\mathbf{q}_1, \mathbf{q}_2) d\mathbf{q}_1 d\mathbf{q}_2, \tag{1-75}$$

where

$$\rho_{\mathrm{int}}(\mathbf{q}_1, \mathbf{q}_2) = -\rho_A(\mathbf{q}_1 \mid \mathbf{q}_2)\rho_B(\mathbf{q}_2 \mid \mathbf{q}_1)$$
$$- \int\int \Gamma_A(\mathbf{q}_1\mathbf{q}_3 \mid \mathbf{q}_1\mathbf{q}_4)\Gamma_B(\mathbf{q}_2\mathbf{q}_4 \mid \mathbf{q}_2\mathbf{q}_3) d\mathbf{q}_3 d\mathbf{q}_4,$$

$$-\int \Gamma_A(\mathbf{q}_1\mathbf{q}_3 \mid \mathbf{q}_1\mathbf{q}_2)\rho_B(\mathbf{q}_2 \mid \mathbf{q}_3)d\mathbf{q}_3$$

$$-\int \rho_A(\mathbf{q}_1 \mid \mathbf{q}_3)\Gamma_B(\mathbf{q}_2\mathbf{q}_3 \mid \mathbf{q}_2\mathbf{q}_1)d\mathbf{q}_3, \tag{1-76}$$

and $\tilde{v}(\mathbf{r}_i, \mathbf{r}_j)$ is a modified interelectronic interaction potential

$$\tilde{v}(\mathbf{r}_i, \mathbf{r}_j) = r_{ij}^{-1} - N_B^{-1}\sum_{\beta \in B} Z_\beta r_{\beta i}^{-1} - N_A^{-1}\sum_{\alpha \in A} Z_\alpha r_{\alpha j}^{-1}$$

$$+ N_A^{-1}N_B^{-1}\sum_{\alpha \in A}\sum_{\beta \in B} Z_\alpha Z_\beta R_{\alpha\beta}^{-1}, \tag{1-77}$$

defined such that

$$\sum_{i \in A}\sum_{j \in B} \tilde{v}(\mathbf{r}_i, \mathbf{r}_j) = V. \tag{1-78}$$

Finally, ρ_X and Γ_X, with $X = A$ or B, are the conventional one- and two-particle density matrices for monomer X, normalized to N_X and $N_X(N_X - 1)$, respectively. In Eqs. (1-75) and (1-76) $\mathbf{q}_i = (\mathbf{r}_i, \mathbf{s}_i)$ denotes the space and spin coordinates of the ith electron. Since theoretical methods for the evaluation of the density matrices ρ_X and Γ_X for many-electron molecules are well developed, Eqs. (1-75) and (1-76) enable practical calculations of the first-order exchange energy using accurate electronic wave functions of the monomers A and B [116].

5.3.   Induction Energy

The second-order polarization energy $E_{pol}^{(2)}$ is given by

$$E_{pol}^{(2)} = -\langle \Phi_0 \mid V\hat{R}_0 V \mid \Phi_0\rangle. \tag{1-79}$$

The induction energy, $E_{ind}^{(2)}$, is obtained when the reduced resolvent $\hat{R}_0$, Eq. (1-24), is restricted to terms where one of the monomers is in the ground state and the other in the excited state. The corresponding expression is given by

$$E_{ind}^{(2)} = E_{ind}^{(2)}(A \leftarrow B) + E_{ind}^{(2)}(B \leftarrow A), \tag{1-80}$$

where

$$E_{ind}^{(2)}(A \leftarrow B) = -\langle \Phi_0^A \mid \Omega_B \hat{R}_{0A}\Omega_B \mid \Phi_0^A\rangle, \tag{1-81}$$

and a similar definition holds for $E_{ind}^{(2)}(B \leftarrow A)$. Here, Ω_B denotes the operator of the electrostatic potential generated by the unperturbed monomer B

$$\Omega_B = \sum_{i \in A} \omega_B(\mathbf{r}_i), \quad \omega_B(\mathbf{r}_i) = \int \frac{1}{r_{ij}}\rho_B^{tot}(\mathbf{r}_j)d\mathbf{r}_j. \tag{1-82}$$

The operator $\hat{R}_{0A}$ is the part of the resolvent, Eq. (1-24), in which B is in its ground state and the sum is over the excited states of A.

Equation (1-81) has the form of the second-order energy correction for the monomer A perturbed by the static field generated by the unperturbed monomer B. This field, corresponding to the potential ω_B, induces a modification $\Phi_{ind}^{(1)}(A \leftarrow B) = -\hat{R}_{0A}\Omega_B\Phi_0^A$ in the wave function of the monomer A, and the change in the energy due to this modification is equal to $E_{ind}^{(2)}(A \leftarrow B)$. Thus, the second-order induction energy results from the mutual polarization of the monomers by the static fields of their unperturbed partners. Asymptotically, at large R, $E_{ind}^{(2)}$ is fully determined by the permanent multipole moments and static multipole polarizabilities of the monomers. At finite R additional information is needed to account for the short-range, penetration part of $E_{ind}^{(2)}$. This information is contained in the short-range part of the electrostatic potentials $\omega_X(\mathbf{r})$, $X = A$ or B, and in the polarization propagators of the monomers. The polarization propagator is a molecular property, which fully describes the linear response of a molecule to an arbitrary one-electron perturbation[117,118]. It is defined for an arbitrary frequency ω by

$$\Pi_{kk'}^{ll'}(\omega) = -\langle \Phi_0^A \mid E_k^l \hat{R}_A(-\omega) E_{k'}^{l'} \mid \Phi_0^A \rangle$$
$$- \langle \Phi_0^A \mid E_{k'}^{l'} \hat{R}_A(\omega) E_k^l \mid \Phi_0^A \rangle, \tag{1-83}$$

where E_k^l is the spin-free unitary group generator (orbital replacement operator), defined by

$$E_k^l = \sum_{\sigma=1}^{2} a_{l\sigma}^\dagger a_{k\sigma} \tag{1-84}$$

and $a_{l\sigma}^\dagger$ $(a_{k\sigma})$ is a creation (annihilation) operator associated with the spinorbital $\psi_k\sigma$, $\sigma = \alpha$ or β. Further, $\hat{R}_A(\omega)$ is the frequency dependent resolvent operator defined as $\hat{R}_A(\omega) = (H_A - E_0^A + \omega)^{-1}Q_A$, and $Q_A = 1 - |\Phi_0^A\rangle\langle\Phi_0^A|$. The induction energy $E_{ind}^{(2)}(A \leftarrow B)$ is related to the polarization propagator at $\omega = 0$ by the equation[119],

$$E_{ind}^{(2)}(A \leftarrow B) = \frac{1}{2}(\omega_B)_i^k(\omega_B)_{l'}^{k'}\Pi_{kk'}^{ll'}(0), \tag{1-85}$$

where $(\omega_B)_i^k$ is the matrix element of the electrostatic potential $\omega_B(\mathbf{r})$, i.e. $(\omega_B)_i^k = \langle\psi_l|\omega_B|\psi_k\rangle$. The Einstein summation convention over repeated lower and upper indices is used in Eq. (1-85) and further on in this paper. Since the electron densities (needed to calculate ω_B) and the static propagators can be calculated as the first and second derivatives of the monomer energy with respect to appropriate perturbations, the existing quantum chemical technology[120-128] for the calculations of analytic first and second derivatives can directly be employed to study induction interactions in the region where the charge overlap effects play an important role, i.e. in the region of the Van der Waals minimum and at shorter distances.

Equation (1-85) can conveniently be rewritten in terms of the density suscepti-bility functions of the monomers. The density susceptibility function of the monomer X may be viewed as the coordinate representation of the polarization propagator, and is given by the following expression:

$$\alpha_X(\mathbf{r}, \mathbf{r}'|\omega) = \sum_{k,k'} \sum_{l,l'} \Pi_{kk'}^{ll'}(\omega) \phi_k^\star(\mathbf{r}) \phi_l(\mathbf{r}) \phi_{k'}^\star(\mathbf{r}') \phi_{l'}(\mathbf{r}'). \tag{1-86}$$

It follows directly from the Eq. (1-85) that the expression for the induction energy can be rewritten as,

$$E_{\text{ind}}^{(2)}(A \leftarrow B) = \frac{1}{2} \int \int \omega_B(\mathbf{r}) \omega_B(\mathbf{r}') \alpha_A(\mathbf{r}, \mathbf{r}'|0) d\mathbf{r} d\mathbf{r}'. \tag{1-87}$$

Although at first glance Eq. (1-87) can be taken as a simple reformulation of Eq. (1-85), such a reformulation it is not purely of academic use, since the density susceptibilities of the monomers can be expanded in terms of a single set of atomic orbitals, making of $\alpha_A(\mathbf{r}, \mathbf{r}'|\omega)$ a two index, and not a four index quantity and thus greatly simplifying calculations of the induction energy for large systems. The application of this technique is considered in more details in Section 7.

5.4.   Exchange-induction Energy

The second-order exchange energy in the SRS theory, defined as $E_{\text{exch}}^{(2)} = E_{\text{SRS}}^{(2)} - E_{\text{pol}}^{(2)}$, separates naturally into two contributions: *exchange-induction* and *exchange-dispersion* energies

$$E_{\text{exch}}^{(2)} = E_{\text{exch-ind}}^{(2)} + E_{\text{exch-disp}}^{(2)}. \tag{1-88}$$

The exchange-induction energy is an energetic effect resulting from the antisym-metrization of the induction wave function,

$$\Phi_{\text{ind}}^{(1)} = \Phi_{\text{ind}}^{(1)}(A \leftarrow B)\Phi_0^B + \Phi_0^A \Phi_{\text{ind}}^{(1)}(B \leftarrow A), \tag{1-89}$$

and can be viewed as a coupling between the induction interaction and the electron exchange. At the distances corresponding to the Van der Waals wells, it is sufficient to consider only the single-exchange part of the exchange-induction energy. Higher-order terms (in S^2) have been computed for the helium dimer and found to be negli-gible in the region of the Van der Waals minimum [129]. In this approximation $E_{\text{exch-ind}}^{(2)}$ is given by the following expression [130]

$$E_{\text{exch-ind}}^{(2)}(S^2) = -\langle \Phi_0 \mid (V - E_{\text{elst}}^{(1)})(\mathcal{P}_1 - \overline{\mathcal{P}}_1) \mid \Phi_{\text{ind}}^{(1)} \rangle, \tag{1-90}$$

where $\overline{\mathcal{P}}_1 = \langle \Phi_0|\mathcal{P}_1\Phi_0\rangle$ and $\Phi_{\text{ind}}^{(1)}$ is given by Eq. (1-89). In the repulsive part of the intermolecular potential the exchange-induction energy quenches a substantial part of the induction contribution and cannot be neglected in any quantitatively accurate calculation.

5.5. Dispersion Energy

The second-order dispersion energy $E_{\text{disp}}^{(2)}$ is defined as the difference between the second-order polarization and induction energies, $E_{\text{disp}}^{(2)} = E_{\text{pol}}^{(2)} - E_{\text{ind}}^{(2)}$. One can also use the following direct definition

$$E_{\text{disp}}^{(2)} = -\langle \Phi_0 \mid V \hat{R}_{AB} V \mid \Phi_0 \rangle, \tag{1-91}$$

where the operator $\hat{R}_{AB}$ is that part of $\hat{R}_0$, cf. Eq. (1-24), which involves only excited states on both monomers A and B. By its very definition the dispersion interaction represents a pure intermolecular correlation effect. It may be viewed as the stabilizing energetic effect of the correlations of instantaneous multipole moments of the monomers. Since the classic work of Casimir and Polder[131] we know that, asymptotically at large R, the energy of the dispersion interaction can be expressed in terms of the dynamic multipole polarizabilities of the monomers. A powerful generalization of the Casimir and Polder result has been reported in Refs. (132–137). The authors of Refs. (132–137) have shown that the complete dispersion energy, including the charge-overlap effects, can be expressed, via the Casimir-Polder type integral, in terms of the polarization propagators of the isolated monomers

$$E_{\text{disp}}^{(2)} = -\frac{1}{4\pi} v_{l_1 n_1}^{k_1 m_1} v_{l_2 n_2}^{k_2 m_2} \int_{-\infty}^{+\infty} \Pi_{k_1 k_2}^{l_1 l_2}(i\omega) \Pi_{m_1 m_2}^{n_1 n_2}(-i\omega) d\omega. \tag{1-92}$$

In the above expression we assumed that k_1, k_2, l_1, l_2 and m_1, m_2, n_1, n_2 label the orbitals of monomers A and B, respectively. We also introduced the following notation for the Coulomb integrals:

$$v_{l_1 n_1}^{k_1 m_1} = \langle \psi_{l_1}(1) \psi_{n_1}(2) | r_{12}^{-1} | \psi_{k_1}(1) \psi_{m_1}(2) \rangle. \tag{1-93}$$

Equation (1-92) is very important since in the region of the Van der Waals minimum the charge-overlap contribution to the dispersion energy is always substantial. Additionally, the powerful computational techniques, developed in the 1980's to obtain accurate polarization propagators[118] can be utilized via Eq. (1-92) in the calculations of the dispersion energies at finite distances.

Similarly as in the case of the induction energy, Eq. (1-92), can be rewritten in terms of the dynamic susceptibilities of the isolated monomers:

$$E_{\text{disp}}^{(2)} = -\frac{1}{4\pi} \int_{-\infty}^{+\infty} \int \int \int \int \frac{\alpha(\mathbf{r}_1, \mathbf{r}_2 | i\omega) \alpha(\mathbf{r}_1', \mathbf{r}_2' | i\omega)}{|\mathbf{r}_1 - \mathbf{r}_2||\mathbf{r}_1' - \mathbf{r}_2'|} d\mathbf{r}_1 d\mathbf{r}_2 d\mathbf{r}_1' d\mathbf{r}_2' d\omega. \tag{1-94}$$

The density susceptibilities of the monomers can be expanded in terms of a single set of atomic orbitals, making of $\alpha_A(\mathbf{r}, \mathbf{r}' | \omega)$ a two index quantity and thus greatly simplifying calculations of the dispersion term for large systems. See Section 7 for a more detailed discussion of this technique applied to the calculations of the dispersion energy.

5.6. Exchange-dispersion Energy

The exchange-dispersion energy $E^{(2)}_{\text{exch-disp}}$ is the energetic effect of the antisymmetrization of the dispersion wave function $\Phi^{(1)}_{\text{disp}}(A \cdots B) = \Phi^{(1)}_{\text{pol}} - \Phi^{(1)}_{\text{ind}}$, and can be interpreted as a coupling between the dispersion interaction and the electron exchange. In the single-exchange approximation $E^{(2)}_{\text{exch-disp}}$ is given by[130]

$$E^{(2)}_{\text{exch-disp}}(S^2) = -\langle \Phi_0 \mid (V - E^{(1)}_{\text{elst}})(\mathcal{P}_1 - \overline{\mathcal{P}}_1) \mid \Phi^{(1)}_{\text{disp}}(A \cdots B) \rangle. \quad (1\text{-}95)$$

The effect of multiple exchanges has been computed for the He dimer and found to be negligible in the region of the Van der Waals minimum[129]. The exchange-dispersion contribution is relatively small, quenching usually only a few percent of the dispersion energy.

5.7. Third-order Polarization and Exchange Contributions

The third-order polarization energy can written as the sum of three distinct contributions,

$$E^{(3)}_{\text{pol}} = E^{(3)}_{\text{ind}} + E^{(3)}_{\text{ind-disp}} + E^{(3)}_{\text{disp}}, \quad (1\text{-}96)$$

where the corrections with subscripts ind, ind-disp, and disp refer to the third-order induction, mixed induction-dispersion, and dispersion energies, respectively.

The third-order induction energy can be represented as a sum of two terms describing the (second-order) polarization of the monomer A by the fields of the monomer B and *vice versa*, and of one mixed term corresponding to the simultaneous polarization of both monomers by the field of their partners. It has been shown[119] that the third-order induction energy (including the charge-overlap contribution) can be expressed through the static polarization propagators and quadratic response functions of the isolated monomers,

$$E^{(3)}_{\text{ind}} = \frac{1}{6}(\omega_B)^k_i (\omega_B)^{k'}_{i'} (\omega_B)^{k''}_{i''} \Pi^{ll'l''}_{kk'k''}(0,0)$$

$$+ \frac{1}{6}(\omega_A)^m_n (\omega_A)^{m'}_{n'} (\omega_A)^{m''}_{n''} \Pi^{nn'n''}_{mm'm''}(0,0)$$

$$+ (\omega_B)^k_i (\omega_A)^m_n v^{k'm'}_{l'n'} \Pi^{ll'}_{kk'}(0) \Pi^{nn'}_{mm'}(0), \quad (1\text{-}97)$$

where $\Pi^{l_1 l_2}_{kk_1 k_2}(\omega_1, \omega_2)$ denotes the quadratic polarization propagator (quadratic response function)[138],

$$\Pi^{ll_1 l_2}_{kk_1 k_2}(\omega_1, \omega_2) = \langle \Phi^A_0 \mid E^l_k \hat{R}_A(-\omega)(E^{l_1}_{k_1} - \rho^{l_1}_{k_1}) \hat{R}_A(-\omega_2) E^{l_2}_{k_2} \Phi^A_0 \rangle$$

$$+ \langle \Phi^A_0 \mid E^{l_2}_{k_2} \hat{R}_A(\omega_2)(E^{l_1}_{k_1} - \rho^{l_1}_{k_1}) \hat{R}_A(\omega) E^l_k \Phi^A_0 \rangle$$

$$+ \langle \Phi^A_0 \mid E^{l_1}_{k_1} \hat{R}_A(\omega_1)(E^l_k - \rho^l_k) \hat{R}_A(-\omega_2) E^{l_2}_{k_2} \Phi^A_0 \rangle + (1 \leftrightarrow 2), \quad (1\text{-}98)$$

where $\omega = \omega_1 + \omega_2$,

$$\rho_k^l = \langle \Phi_0^A \mid E_k^l \Phi_0^A \rangle, \tag{1-99}$$

and $(1 \leftrightarrow 2)$ denotes three additional terms with all symbols with indices 1 and 2 interchanged (including those with k_1, l_1, k_2, and l_2).

The third-order intermolecular correlation contribution $E_{\text{pol}}^{(3)} - E_{\text{ind}}^{(3)}$ separates into two parts: the induction-dispersion energy $E_{\text{ind-disp}}^{(3)}$ and the third-order dispersion energy $E_{\text{disp}}^{(3)}$. The induction-dispersion effect results from the coupling of the induction and dispersion interactions and gives the following contribution to the interaction energy[119]

$$E_{\text{ind-disp}}^{(3)} = E_{\text{ind-disp}}^{(3)}(A) + E_{\text{ind-disp}}^{(3)}(B), \tag{1-100}$$

$$
\begin{aligned}
E_{\text{ind-disp}}^{(3)}(A) &= -\frac{1}{4\pi} (\omega_B)_\lambda^\kappa \, v_{\lambda_1 \nu_1}^{\kappa_1 \mu_1} \, v_{\lambda_2 \nu_2}^{\kappa_2 \mu_2} \\
&\quad \times \int_{-\infty}^{+\infty} \Pi_{\kappa \kappa_1 \kappa_2}^{\lambda \lambda_1 \lambda_2}(i\omega, -i\omega) \Pi_{\mu_1 \mu_2}^{\nu_1 \nu_2}(-i\omega) d\omega,
\end{aligned} \tag{1-101}
$$

where the symbols have the same meaning as in Eqs. (1-97) and (1-92). The expression for $E_{\text{ind-disp}}^{(3)}(B)$ can be obtained from Eq. (1-101) by interchanging symbols pertaining to monomers A and B. It is worth noting that the induction-dispersion contribution can be obtained in a second-order perturbation treatment if in zeroth order the monomers are fully deformed by the induction effects[139]. This means that the induction-dispersion energy is a second-order intermolecular correlation effect.

The third-order dispersion energy $E_{\text{disp}}^{(3)}$ is a true third-order intermolecular correlation term. Despite some efforts[140], this energy could not be expressed, even asymptotically, through some monomer properties. The calculations for the water dimer and the HF dimer[141] have shown, however, that even for these polar systems the contribution of the third-order dispersion energy is small (1–2 % of the total interaction energy at the equilibrium configurations). It remains to be seen if this optimistic result holds also for other complexes.

The partitioning of the third-order exchange energy into exchange-induction, exchange-induction-dispersion, and exchange-dispersion components has not been derived thus far even for pairwise additive interactions. In the second order this splitting is rather natural since the first-order polarization wave function can be written as the sum of wave functions describing the induction and dispersion interactions, $\Phi_{\text{pol}}^{(1)} = \Phi_{\text{ind}}^{(1)} + \Phi_{\text{disp}}^{(1)}(A \cdots B)$. The third-order exchange energy is solely defined by the intermolecular interaction and exchange operators, and the second-order polarization wave function $\Phi_{\text{pol}}^{(2)}$, so the splitting of $E_{\text{exch}}^{(3)}$ into the sum $E_{\text{exch-ind}}^{(3)} + E_{\text{exch-ind-disp}}^{(3)} + E_{\text{exch-disp}}^{(3)}$ is defined by the splitting of $\Phi_{\text{pol}}^{(2)}$ into components describing the induction, induction-dispersion, and dispersion interactions.

This partitioning of $\Phi_{\text{pol}}^{(2)}$ does not come out naturally, but it can be defined in such a way that[139,142]

$$\Phi_{\text{pol}}^{(2)} = \Phi_{\text{ind}}^{(2)} + \Phi_{\text{ind-disp}}^{(2)} + \Phi_{\text{disp}}^{(2)}, \tag{1-102}$$

and

$$E_{\text{ind}}^{(3)} = \langle \Phi_0 | V\Phi_{\text{ind}}^{(2)} \rangle, \quad E_{\text{ind-disp}}^{(3)} = \langle \Phi_0 | V\Phi_{\text{ind-disp}}^{(2)} \rangle, \quad E_{\text{disp}}^{(3)} = \langle \Phi_0 | V\Phi_{\text{disp}}^{(2)} \rangle. \tag{1-103}$$

Calculations of the third-order exchange contributions have not been performed thus far. The results reported in Ref. (81) for the total third-order exchange effect for the helium dimer suggest that they quench a large part of the third-order polarization contribution.

6. MULTIPOLE EXPANSION OF THE INTERACTION ENERGY

According to London's theory[143,144] the interaction energy can be represented as an asymptotic expansion in powers of R^{-1},

$$E_{\text{int}}(R, \omega_A, \omega_B, \widehat{\mathbf{R}}) \sim \sum_{n=1}^{\infty} \frac{C_n(\omega_A, \omega_B, \widehat{\mathbf{R}})}{R^n}. \tag{1-104}$$

The Van der Waals constants $C_n(\omega_A, \omega_B, \widehat{\mathbf{R}})$ depend on the Euler angles ω_A and ω_B specifying the orientation of the monomers in an arbitrary space-fixed frame, and on the polar angles $\widehat{\mathbf{R}} = (\beta, \alpha)$ determining the orientation of the intermolecular axis ($\mathbf{R}$ is assumed to join the monomer centers of mass) with respect to the same space-fixed frame.

Since the interaction energy as a function of R has an essential singularity at infinity[145] due to the exponential terms resulting from the charge overlap and exchange effects, the knowledge of the Van der Waals constants is not sufficient to reconstruct the function $E_{\text{int}}(R, \omega_A, \omega_B, \widehat{\mathbf{R}})$ at finite R, even if the series (1-104) were convergent. Actually, it appears that the series (1-104) diverge for any finite value of R. Despite this divergence, for sufficiently large distances the expansion (1-104) can approximate the exact interaction energy arbitrarily closely in the sense that[146]

$$\left| E_{\text{int}}(R, \omega_A, \omega_B, \widehat{\mathbf{R}}) - \sum_{n=1}^{N} \frac{C_n(\omega_A, \omega_B, \widehat{\mathbf{R}})}{R^n} \right| = O(R^{-N-1}). \tag{1-105}$$

Therefore, knowing the Van der Waals constants is very useful to estimate the interaction energy at large distances, and is necessary to guarantee the correct large R asymptotic behavior of the potential energy surface $E_{\text{int}}(R, \omega_A, \omega_B, \widehat{\mathbf{R}})$.

The coefficients $C_n(\omega_A, \omega_B, \widehat{\mathbf{R}})$ are uniquely defined by the interaction function $E_{\text{int}}(R, \omega_A, \omega_B, \widehat{\mathbf{R}})$ and can in principle be deduced from the following equations,

$$C_1(\omega_A, \omega_B, \widehat{\mathbf{R}}) = \lim_{R \to \infty} R E_{\text{int}}(R, \omega_A, \omega_B, \widehat{\mathbf{R}}), \tag{1-106}$$

and

$$C_n(\omega_A, \omega_B, \widehat{\mathbf{R}}) = \lim_{R \to \infty} R^n \left(E_{\text{int}}(R, \omega_A, \omega_B, \widehat{\mathbf{R}}) - \sum_{k=1}^{n-1} \frac{C_k(\omega_A, \omega_B, \widehat{\mathbf{R}})}{R^k} \right). \tag{1-107}$$

The remainder of this section will be devoted to the discussion of computational methods of Van der Waals constants without prior knowledge of the interaction energy, and to other (possibly convergent) angular expansions of the interaction energy components.

6.1. One-center Expansion

Explicit expressions for the Van der Waals constants may be obtained by invoking the well-known [147] multipole expansion of the operator V. In an arbitrary space-fixed coordinate system, this expansion can be written as

$$V = \sum_{n=1}^{\infty} \frac{V_n}{R^n}, \quad V_n = \sum_{l=0}^{n-1} V_{l,n-l-1}. \tag{1-108}$$

The operator V_{l_A, l_B} is physically interpreted as representing the interaction of the instantaneous 2^{l_A} moment with respect to center A with the instantaneous 2^{l_B} moment with respect to center B and can be expressed in terms of irreducible spherical or reducible Cartesian tensor operators of multipole moments. The operator V_{l_A, l_B} can be written as

$$V_{l_A, l_B} = X_{l_A, l_B} \sum_{m=-l_A-l_B}^{l_A+l_B} (-1)^m C_{-m}^{l_A+l_B}(\widehat{\mathbf{R}}) [\mathbf{M}_{l_A} \otimes \mathbf{M}_{l_B}]_m^{l_A+l_B}, \tag{1-109}$$

where

$$X_{l_A, l_B} = (-1)^{l_B} \left(\frac{2l_A + 2l_B}{2l_A} \right)^{1/2}, \tag{1-110}$$

and the spherical multipole moment operator is given by,

$$M_{l_X}^{m_X} = \sum_{p \in X} Z_p r_p^{l_X} C_{l_X}^{m_X}(\widehat{\mathbf{r}}_p). \tag{1-111}$$

Here, the summation index p runs over all particles, both nuclei and electrons, of the molecule X, Z_p are the charges of those particles, and $C_l^m(\hat{\mathbf{r}})$ is a spherical harmonic in the Racah normalization.[148] The irreducible tensor product of two multipole moment tensors $\mathbf{M}_{l_A} = \{M_{l_A}^{m_A}, m_A = -l_A, \ldots, +l_A\}$ and $\mathbf{M}_{l_B} = \{M_{l_B}^{m_B}, m_B = -l_B, \ldots, +l_B\}$ is defined as,

$$[\mathbf{M}_{l_A} \otimes \mathbf{M}_{l_B}]_m^l = \sum_{m_A=-l_A}^{l_A} \sum_{m_B=-l_B}^{l_B} M_{l_A}^{m_A} M_{l_B}^{m_B} \langle l_A m_A; l_B m_B \mid lm \rangle. \tag{1-112}$$

Here $\langle l_1 m_1; l_2 m_2 \mid lm \rangle$ is the Clebsch-Gordan coefficient.[148]

The spherical form of the multipole expansion is very useful if we are looking for the explicit orientational dependence of the interaction energy. However, in some applications the use the conceptually simpler Cartesian form of the operators V_{l_A, l_B} may be more convenient. Moreover, unlike the spherical derivation, the Cartesian derivation is very simple, and "can be followed by everybody who knows how to differentiate a function of x, y and z"[149]. To express the operator V_{l_A, l_B} in terms of Cartesian tensors we have to define the reducible, with respect to SO(3), tensorial components of multipole moments,

$$M_{l_X}^{\{\gamma\}} = \sum_{p \in X} Z_p r_{p, \gamma_1} r_{p, \gamma_2} \ldots r_{p, \gamma_{l_X}}, \tag{1-113}$$

where r_{p, γ_i} is the γ_ith Cartesian coordinate of the particle p, i.e. $\gamma_i = 1, 2$ or 3, so that $r_{p, \gamma_i} = x_p, y_p$, and z_p, respectively, $\{\gamma\}$ denotes the set of indices $\{\gamma_1, \gamma_2, \ldots, \gamma_{l_X}\}$ and the coordinates are measured in a space-fixed system with its origin at the center of mass of molecule X. In this notation the operator V_{l_A, l_B} can be written as

$$V_{l_A, l_B} = \sum_{\{\alpha\}} \sum_{\{\beta\}} M_{l_A}^{\{\alpha\}} \, T_{\{\alpha\}, \{\beta\}}^{[l_A + l_B]} \, M_{l_B}^{\{\beta\}}, \tag{1-114}$$

where the tensor $T_{\{\alpha\}, \{\beta\}}^{[l_A + l_B]}$ describing the orientational dependence of the interaction between the instantaneous 2^{l_A} moment on molecule A and the instantaneous 2^{l_B} moment on molecule B is given by

$$T_{\{\alpha\}, \{\beta\}}^{[l_A + l_B]} = R^{l_A + l_B + 1} \frac{(-1)^{l_A}}{l_A! l_B!} \left(\nabla_{\alpha_1} \nabla_{\alpha_2} \ldots \nabla_{\alpha_{l_A}} \nabla_{\beta_1} \nabla_{\beta_2} \ldots \nabla_{\beta_{l_B}} \right) \left(\frac{1}{R} \right), \tag{1-115}$$

and the sums run over all distinct sets $\{\alpha_1, \alpha_2, \cdots \alpha_{l_A}\}$ and $\{\beta_1, \beta_2, \cdots \beta_{l_B}\}$. Explicit expressions for the tensors $T_{\{\alpha\}, \{\beta\}}^{[l_A + l_B]}$ have been derived by Mulder et al.[150] for $l_A + l_B \leq 6$. Specific formulas applying to linear and tetrahedral molecules have been reported in Ref. (151) for $l_A + l_B \leq 7$.

Although Eqs. (1-109) and (1-114) represent the very same multipole expansion of the intermolecular interaction operator, the expressions for the transformations between the spherical and Cartesian form are quite complex. In view of the Laplace equation, the partial traces $\nabla_\gamma \nabla_\gamma \left(\frac{1}{R}\right)$ vanish. Therefore, the operators $M_{l_x}^{\{\gamma\}}$ can be expressed in terms of $M_{l_x}^{m_x}$. The formal relationship between Cartesian tensors and their irreducible spherical components has been thoroughly investigated by Coope et al.[152,153,154] and by Stone[155,156]. Stone derived[155] a general scheme of reducing a Cartesian tensor of rank n into several spherical components and investigated in detail properties of Cartesian-spherical transformation coefficients[156].

The truncated multipole expansion,

$$V^N = \sum_{n=1}^{N} \frac{V_n}{R^n} \tag{1-116}$$

can be used to define the Van der Waals constants. By applying the Rayleigh-Schrödinger perturbation theory to the Schrödinger equation with the Hamiltonian H^N

$$H^N = H_0 + V^N, \tag{1-117}$$

and using $1/R$ as the expansion parameter Ahlrichs[145] has shown that the Van der Waals constants entering the asymptotic expansion (1-104) can be computed from the following recursive formulas,

$$C_n = \sum_{k=1}^{n} \langle \Phi^{[0]} \mid V_k \Phi^{[n-k]} \rangle, \tag{1-118}$$

and

$$\Phi^{[n]} = -\sum_{k=1}^{n} \hat{R}_0 (C_k - V_k) \Phi^{[n-k]}, \tag{1-119}$$

where the superscript $[n]$ at $\Phi^{[n]}$ denotes the order in $1/R$. For simplicity we have omitted the dependence of the Van der Waals constants on the angles $(\omega_A, \omega_B, \widehat{\mathbf{R}})$ in Eqs. (1-118) and (1-119). Note that the Hamiltonian H^N has a purely continuous spectrum. Consequently, the operator V^N cannot be considered as a small perturbation and the RS perturbation theory based on the partitioning (1-117) of the Hamiltonian H^N is divergent for each R, and has only a formal sense. Ahlrichs has proved[145], however, that C_n and $\Phi^{[n]}$, as defined by Eqs. (1-118) and (1-119), exist in the sense of the Hilbert space theory.

Although a direct application of Eqs. (1-118) and (1-119) is straightforward, in practice the Van der Waals constants are obtained from the constants $C_n^{(k)}$ appearing in the asymptotic expansion of the polarization energies $E_{\text{pol}}^{(k)}$

$$E_{\text{pol}}^{(k)} \sim \sum_{n=1}^{\infty} \frac{C_n^{(k)}}{R^n}, \tag{1-120}$$

and then representing each constant C_n as a finite perturbation expansion in V,

$$C_n = \sum_{k=1}^{M} C_n^{(k)}. \tag{1-121}$$

Here M denotes the smallest integer satisfying $M > n/\kappa - 1$, where $\kappa = 3$ if both interacting molecules are neutral and $\kappa = 2$ if one molecule has a net charge. Such a procedure is legitimate since in view of Eq. (1-29), the polarization expansion of the interaction energy gives a correct asymptotic representation of the interaction energy. It can also be shown[84] that the constants $C_n^{(k)}$ can be computed from the standard equations of the polarization perturbation theory, provided that the operator V is replaced by its truncated multipole expansion (1-116) with $N \geq n$. For instance, the Van der Waals constants $C_n^{(1)}$ and $C_n^{(2)}$ are given by

$$C_n^{(1)} = \langle \Phi_0 \mid V_n \Phi_0 \rangle, \tag{1-122}$$

$$C_n^{(2)} = - \sum_{k=\kappa}^{n-\kappa} \langle \Phi_0 \mid V_k \hat{R}_0 V_{n-k} \Phi_0 \rangle. \tag{1-123}$$

Equations (1-122) and (1-123) have been applied with success to compute Van der Waals constants for quite large systems[157–160].

Although Eqs. (1-123) and (1-138) can be applied in practice to compute the Van der Waals constants $C_n^{(k)}$, these constants depend in a quite complicated way on the angles $(\omega_A, \omega_B, \widehat{\mathbf{R}})$. If these constants were computed from Eqs. (1-122) and (1-123), such calculations would have to be performed for each orientation of interacting molecules. Therefore, it is preferable to introduce the multipole expansions for the interaction energy components $E_{\text{elst}}^{(1)}$, $E_{\text{ind}}^{(2)}$, and $E_{\text{disp}}^{(2)}$, in a such a way that the whole angular dependence is separated. As shown in Ref. (161) for all intermolecular separations the kth-order polarization correction $E_{\text{pol}}^{(k)}$ can be written in terms of a complete orthogonal set of angular functions labeled by $\{\Lambda\} = \{L_A, K_A, L_B, K_B, L\}$,

$$E_{\text{pol}}^{(k)} = \sum_{\{\Lambda\}} {}^{\{\Lambda\}}\tilde{\mathcal{E}}_{\text{pol}}^{(k)}(R) A_{\{\Lambda\}}(\omega_A, \omega_B, \widehat{\mathbf{R}}), \tag{1-124}$$

$$A_{\{\Lambda\}}(\omega_A, \omega_B, \widehat{\mathbf{R}}) = (-1)^{L_A+L_B+L}$$

$$\times \sum_{M_A=-L_A}^{L_A} \sum_{M_B=-L_B}^{L_B} \sum_{M=-L}^{L} \begin{pmatrix} L_A & L_B & L \\ M_A & M_B & M \end{pmatrix}$$

$$\times \mathcal{D}_{M_A,K_A}^{(L_A)^*}(\omega_A) \mathcal{D}_{M_B,K_B}^{(L_B)^*}(\omega_B) C_M^L(\widehat{\mathbf{R}}), \tag{1-125}$$

where $\mathcal{D}_{M,K}^{(L)}(\omega)$ denotes an element of the Wigner rotation matrix[148], C_M^L is the spherical harmonics in the Racah normalization, and the expression in large parentheses is a $3j$ symbol[148]. Note that the expression for the angular function $A_{\{\Lambda\}}(\omega_A, \omega_B, \widehat{\mathbf{R}})$ reported in Ref. (162) missed the factor $(-1)^{L_A+L_B+L}$ (Ref. 163).

By using the multipole expansion, we in fact replace the exact radial expansion coefficients $^{\{\Lambda\}}\tilde{\mathcal{E}}_{\text{pol}}^{(k)}(R)$ in Eq. (1-124) by the approximate coefficients $^{\{\Lambda\}}\mathcal{E}_{\text{pol}}^{(k)}(R)$, which are power series in R^{-1}. Closed expressions for the latter have been given [149,161] in terms of the irreducible spherical tensors of multipole moments and polarizabilities.

The radial part of the electrostatic energy in the multipole approximation is given by [149,161,164,165,166]

$$^{\{\Lambda\}}\mathcal{E}_{\text{elst}}^{(1)}(R) = (-1)^{L_A}\delta_{L_A+L_B,L}$$

$$\times \left[\frac{(2L_A+2L_B+1)!}{(2L_A)!(2L_B)!}\right]^{1/2} \frac{Q_{L_A}^{K_A}Q_{L_B}^{K_B}}{R^{L_A+L_B+1}}, \tag{1-126}$$

where $Q_{L_X}^{K_X}$ denotes the spherical component of the 2^{L_X} moment of the molecule X computed in a convenient molecule-fixed coordinate system,

$$Q_{L_X}^{K_X} = \langle \Phi_0^X | \tilde{M}_{L_X}^{K_X} | \Phi_0^X \rangle, \tag{1-127}$$

and $\tilde{\mathbf{M}}_{l_X}$ is the multipole moment operator of the monomer X in the body-fixed frame. Equation (1-126) shows that the first-order polarization energy in the multipole approximation is represented by the classical electrical interaction between the permanent multipole moments of the unperturbed monomers.

The radial component of the second-order induction energy in the multipole approximation can be written as [149,161,164,165,166]

$$^{\{\Lambda\}}\mathcal{E}_{\text{ind}}^{(2)}(R) = -\frac{1}{2}\sum_{l_A=1}^{\infty}\sum_{l_A'=1}^{\infty}\sum_{l_B=0}^{\infty}\sum_{l_B'=0}^{\infty} C_{\{\Lambda\},\text{ind}-A}^{\{\Lambda\}} R^{-n}$$

$$-\frac{1}{2}\sum_{l_A=0}^{\infty}\sum_{l_A'=0}^{\infty}\sum_{l_B=1}^{\infty}\sum_{l_B'=1}^{\infty} C_{\{\Lambda\},\text{ind}-B}^{\{\Lambda\}} R^{-n}, \tag{1-128}$$

where $\{\lambda\}$ is the set of indices $\{\lambda\} = \{l_A, l_A', l_B, l_B'\}$, $n = l_A + l_A' + l_B + l_B' + 2$, and the long-range induction coefficient describing the polarization of the monomer A, $C_{\{\lambda\},\text{ind}-A}^{\{\Lambda\}}$, is given by

$$C_{\{\lambda\},\text{ind}-A}^{\{\Lambda\}} = \xi_{l_A l_A' l_B l_B'}^{L_A L_B L} \alpha_{(l_A l_A')L_A}^{K_A}(0)[\mathbf{Q}_{l_B} \otimes \mathbf{Q}_{l_B'}]_{K_B}^{L_B}. \tag{1-129}$$

The symbol $\alpha_{(l_X, l_X')L_X}^{K_X}(0)$ denotes the irreducible component of the multipole polarizability

$$\alpha_{(l_X l_X')L_X}^{K_X}(\omega) = \sum_{n\neq0}\frac{2(E_n^X - E_0^X)}{(E_n^X - E_0^X)^2 - \omega^2}$$

$$\times [\langle \Phi_0^X | \tilde{\mathbf{M}}_{l_X} | \Phi_n^X \rangle \otimes \langle \Phi_n^X | \tilde{\mathbf{M}}_{l_X'} | \Phi_0^X \rangle]_{K_X}^{L_X}, \tag{1-130}$$

and the numerical constant $\xi_{l_A l'_A l_B l'_B}^{L_A L_B L}$ is given by

$$\xi_{l_A l'_A l_B l'_B}^{L_A L_B L} = (-1)^{l_A + l'_A} \left[\frac{(2l_A + 2l_B + 1)!(2l'_A + 2l'_B + 1)!}{(2l_A)!(2l_B)!(2l'_A)!(2l'_B)!} \right]^{1/2}$$
$$\times [(2L_A + 1)(2L_B + 1)(2L + 1)]^{1/2}$$
$$\times \langle l_A + l_B, 0; l'_A + l'_B, 0 \, | \, L, 0 \rangle$$
$$\times \left\{ \begin{array}{ccc} l_A & l'_A & L_A \\ l_B & l'_B & L_B \\ l_A + l_B & l'_A + l'_B & L \end{array} \right\}, \tag{1-131}$$

where the quantity between curly braces denotes a $9j$ symbol [148]. The energies and wave functions appearing in Eq. (1-130) belong to the spectrum of the Hamiltonian H_X of the monomer X. In view of Eq. (1-129) the first term on the r.h.s. of Eq. (1-128) corresponds to the energy of the polarization of the monomer A by the permanent multipole moments of the monomer B, so the induction energy in the multipole approximation is represented by the classical interaction between permanent multipole moments of one monomer and induced multipole moments of the other.

Finally, the radial part of the dispersion energy in the multipole approximation is given by [149,161,164,165,166]

$$^{\{\Lambda\}}\mathcal{E}_{\text{disp}}^{(2)}(R) = -\sum_{l_A=1}^{\infty} \sum_{l'_A=1}^{\infty} \sum_{l_B=1}^{\infty} \sum_{l'_B=1}^{\infty} C_{\{\Lambda\},\text{disp}}^{\{\Lambda\}} R^{-n}, \tag{1-132}$$

where the long-range dispersion coefficient $C_{\{\Lambda\},\text{disp}}^{\{\Lambda\}}$ can be written as the Casimir-Polder integral [131],

$$C_{\{\Lambda\},\text{disp}}^{\{\Lambda\}} = \frac{1}{2\pi} \xi_{l_A l'_A l_B l'_B}^{L_A L_B L} \int_0^{\infty} \alpha_{(l_A l'_A) L_A}^{K_A} (i\omega) \alpha_{(l_B l'_B) L_B}^{K_B} (i\omega) d\omega. \tag{1-133}$$

Equations (1-124) and (1-133) are valid in an arbitrary space-fixed coordinate system. However, since the angular functions $A_{\{\Lambda\}}(\omega_A, \omega_B, \hat{\mathbf{R}})$ are invariant with respect to any frame rotation [162], a specific choice of the coordinate system may considerably simplify Eq. (1-125). In particular, in the body-fixed coordinate system with the z axis along the vector $\mathbf{R}$ the polar angles $\hat{\mathbf{R}} = (\beta, \alpha)$ are zero. Using the fact that $C_M^L(\hat{\mathbf{R}} = (0,0)) \equiv \delta_{M,0}$ [148], one gets,

$$\tilde{A}_{\{\Lambda\}}(\omega_A, \omega_B) = (-1)^{L_A + L_B + L}$$
$$\times \sum_{M_A = -L_A}^{L_A} \begin{pmatrix} L_A & L_B & L \\ M_A & -M_A & 0 \end{pmatrix}$$
$$\times \mathcal{D}_{M_A, K_A}^{(L_A)*} (\omega_A) \mathcal{D}_{-M_A, K_B}^{(L_B)*} (\omega_B). \tag{1-134}$$

Table 1-4. Quantum numbers and angular functions specifying the angular dependence of the interaction energy. The angle δ is given by the difference $\delta = \alpha_A - \alpha_B$. Only quantum numbers not equal to zero are shown. The entries "a", "d", and "p" in columns 1 and 2 refer to atom, diatomic, and polyatomic molecule, respectively

A	B	Quantum numbers	Angular function
a	a	—	1
a	d	L_B	$P_{L_B}(\cos\beta_B)$
a	p	$L = L_B, K_B$	$C_{K_B}^{L_B}(\beta_B, \alpha_B)$
d	d	L_A, L_B, L	$P_{L_A}(\cos\beta_A)P_{L_B}(\cos\beta_B)\exp(iM_A\delta)$
d	p	L_A, L_B, K_B, L	$P_{L_A}(\cos\beta_A)D_{-M_A,K_B}^{(L_B)*}(0,\beta_B,\gamma_B)\exp(iM_A\delta)$
p	p	L_A, K_A, L_B, K_B, L	$D_{M_A,K_A}^{(L_A)*}(0,\beta_A)D_{-M_A,K_B}^{(L_B)*}(0,\beta_B,\gamma_B)\exp(iM_A\delta)$

The summary of angular coordinates, quantum numbers, and angular functions for some specific systems is given in Table 1-4. Further simplifications can be obtained if one considers the molecular symmetry groups of the monomers. For all point groups, except for the tetrahedral and cubic groups, all symmetry operators can be constructed from the inversion I, n-fold rotation about the principal (z) axis $R_z(2\pi/n)$, and twofold rotation about the x axis $R_x(\pi)$. Therefore, to determine the components of the multipole moment and polarizability tensors that span the totally symmetric representation of the symmetry group, i.e. that are invariant under operations of the symmetry group, it is enough to determine the action of these three operators on the multipole moment and polarizability tensors[164,165]. It follows from Refs. (164–165) that the multipole moment and polarizability tensors transform under these operations according to:

$$I: \quad Q_L^M \to (-1)^L Q_L^M, \quad \alpha_{(ll')L}^M \to (-1)^{l+l'}\alpha_{(ll')L}^M, \tag{1-135}$$

$$R_z(2\pi/n): \quad Q_L^M \to e^{-2\pi iM/n}Q_L^M, \quad \alpha_{(ll')L}^M \to e^{-2\pi iM/n}\alpha_{(ll')L}^M, \tag{1-136}$$

$$R_x(\pi): \quad Q_L^M \to (-1)^L Q_L^{-M}, \quad \alpha_{(ll')L}^M \to (-1)^L \alpha_{(ll')L}^{-M}. \tag{1-137}$$

Using the transformation rules given above one can easily derive the (non-zero) components of spherical tensors that are invariant under the molecular symmetry group. This, in turn, can be used to obtain the multipole expansions of the electrostatic, induction, and dispersion energies for the interactions of specific systems, see, e.g. Refs. (167–168) for expressions applying to atom-diatom and diatom-diatom interactions. In general, the symmetry-adaptation of a tensor to the molecular symmetry group can be obtained by a reduction with respect to the full rotation-reflection group $O(3)$, followed by a subduction of the $O(3)$ irreducible representations to the point symmetry group of the molecule[169]. This symmetry-adaptation scheme has been applied with success to derive all components of the (hyper)polarizability tensors that are invariant under $D_{\infty h}$[170].

Although the spherical form of the multipole expansion is definitely superior if the orientational dependence of the electrostatic, induction, or dispersion energies is of interest, the Cartesian form[171–174] may be useful. Mutual transformations between the spherical and Cartesian forms of the multipole moment and (hyper)polarizability tensors have been derived by Gray and Lo[175]. The symmetry-adaptation of the Cartesian tensors of quadrupole, octupole, and hexadecapole moments to all 51 point groups can be found in Ref. (176) while the symmetry-adaptation of the Cartesian tensors of multipole (hyper)polarizabilities to simple point groups has been considered in Refs. (172–175).

The long-range electrostatic and induction coefficients are exclusively expressed through the multipole moments and polarizabilities of the isolated monomers, so they can routinely be computed by various quantum-chemical methods. The calculations of the long-range dispersion coefficients are somewhat more sophisticated, as they require the knowledge of the dynamic multipole polarizabilities at imaginary frequency. Nowadays this problem is solved, however, and accurate long-range dispersion coefficients can be computed. The review of all methods that can be applied to obtain such coefficients is out scope of this review. Here we only want to mention that the MBPT approach of Wormer and collaborators has been successfully applied to various Van der Waals complexes providing state-of-the-art values of the long-range dispersion coefficients[177–182].

The multipole expansion of the intermolecular interaction operator is divergent in most part of the configuration space, the region of convergence being restricted to the Cartesian product of all spheres $r_i \leqslant R/2$, $r_\alpha \leqslant R/2$, $r_j \leqslant R/2$, and $r_\beta \leqslant R/2$, for $i, \alpha \in$ A, and $j, \beta \in$ B[183], where the indices i and j refer to electrons while α and β to nuclei. This particular region corresponds to that part of the configuration space in which the electrons initially assigned to molecules A and B are "localized" on their original monomers. If the operators $V_{l_A l_B}$ are interpreted as multiplicative operators in the Hilbert space, the series (1-108) is divergent for each R.

The asymptotic expansions of the polarization corrections $E_{\text{pol}}^{(k)}$ are divergent for all values of R, although this fact has been rigorously proven for $k = 2$ and only for the H_2^+ and H_2 systems. Vigné-Maeder and Claverie[184] have shown that the multipole expansion of the electrostatic energy is convergent, although to a spurious value, if the unperturbed charge distributions of the interacting molecules are approximated by Gaussian functions. Dalgarno and Lewis[185] have shown that the multipole expansion of the second-order induction energy for the H_2^+ ion is divergent for all R,

$$E_{\text{ind}}^{(2)} \sim - \sum_{n=1}^{\infty} \frac{(2n+2)!(n+2)}{n(n+1)2^{2n+2}R^{2n+2}}. \tag{1-138}$$

Later Young[186] proved that the multipole expansion of the second-order dispersion energy for the H_2 molecule is divergent as well,

$$E_{\text{disp}}^{(2)} \sim - \sum_{l_A=1}^{\infty} \sum_{l_B=1}^{\infty} \frac{C_{l_A, l_B}^{(2)}}{R^{2l_A+2l_B+2}}, \tag{1-139}$$

where

$$C^{(2)}_{l_A,l_B} > \frac{(2l_A+2l_B)!(l_A+1)(l_B+1)}{2^{2l_A+2l_B+1}}. \tag{1-140}$$

Both expansions are rapidly divergent, and not summable by any summation techniques.

6.2. Multicenter Expansions

Since the single-center multipole expansion of the interaction energy is divergent, one could use a kind of multicenter expansion. One can hope that the multipole expansion will provide better results if multipole moments and polarizabilities localized at various points of a molecule are used instead of global multipole moments and polarizabilities. This idea forms the basis of the so-called *distributed multipole analysis* of the electrostatic, induction, and dispersion interactions between molecules [187–195].

In the following considerations we will need the multipole expansion of the operator r_{12}^{-1} as series of products of operators depending on the coordinate of the particle 1 with respect to a center a, $\mathbf{r}_1$, of the particle 2 with respect to another center b, $\mathbf{r}_2$, and on the coordinates describing the relative position of the centers a and b, $\mathbf{R}_{ab}$,

$$r_{12}^{-1} = \sum_{L_A=0}^{\infty} \sum_{L_B=0}^{\infty} (-1)^{L_B} \binom{2L_A+2L_B}{2L_A}$$

$$\times \sum_{M=-L_A-L_B}^{L_A+L_B} (-1)^M \left[\mathbf{m}_{L_A}(\mathbf{r}_1) \otimes \mathbf{m}_{L_B}(\mathbf{r}_2) \right]^M_{L_A+L_B}$$

$$\times C^{L_A+L_B}_M(\hat{\mathbf{R}}_{ab}) R^{-L_A-L_B-1}_{ab}, \tag{1-141}$$

where the one-electron multipole moment operator of the monomer X, $m^{M_X}_{L_X}$, in the laboratory system of axes, is given by the following expression,

$$m^m_l(\mathbf{r}) = r^l C^l_m(\hat{\mathbf{r}}), \qquad M^{M_A}_{L_A} = \sum_{i=1}^{N_A} m^{M_A}_{L_A}(\mathbf{r}_i). \tag{1-142}$$

The space-fixed operators defined by Eq. (1-142) can be transformed to the system of axes located at the center a,

$$\tilde{m}^{K_A}_{L_A} = \sum_{M_A=-L_A}^{L_A} m^{M_A}_{L_A} D^{(L_A)^*}_{M_A K_A}(\omega^a_A), \qquad \tilde{M}^{K_A}_{L_A} = \sum_{i=1}^{N_A} \tilde{m}^{K_A}_{L_A}(\mathbf{r}_i). \tag{1-143}$$

A formal definition of the distributed multipole moments require some partitioning of the space $\mathbb{R}^3$ into regions corresponding to the set of centers $\{a\}$ and $\{b\}$ in the molecules A and B, respectively,

$$\mathbb{R}^3 = \bigcup_{a \in A} V_a, \qquad \mathbb{R}^3 = \bigcup_{b \in B} V_b. \tag{1-144}$$

Usually, the region defined by V_a will be associated with the atom a of the molecule A. The distributed multipole moment at the site a is defined by the following expression:

$$Q_{L_A}^{K_A}(a) = \int_{V_a} \widetilde{m}_{L_A}^{K_A}(\mathbf{r}) \rho_A^{\text{tot}}(\mathbf{r}) d\mathbf{r}. \tag{1-145}$$

If we extend the integration volume V_a to the full $\mathbb{R}^3$ space we will get the multipole moment of the monomer A in the coordinate system located at the site a.

It should be stressed that the multipole moment $Q_{L_A}^{K_A}$ of the monomer A cannot be obtained by a simple summation of the distributed moments $Q_{L_A}^{K_A}(a)$, since the latter are defined with respect to local systems of axes located at the sites. However, transforming the moments $Q_{L_A}^{K_A}(a)$ to the center-of-mass of the monomer A using the translation formula [148],

$$\widetilde{Q}_{L_A}^{K_A}(a) = \sum_{l_A=0}^{L_A} \sum_{k_A=-l_A}^{l_A} \left[\binom{L_A+K_A}{l_A+k_A} \binom{L_A-K_A}{l_A-k_A} \right]^{1/2}$$

$$\times (-1)^{L_A-l_A} Q_{l_A}^{k_A}(a) t_a^{L_A-l_A} C_{K_A-k_A}^{L_A-l_A}(\widehat{\mathbf{t}}_a), \tag{1-146}$$

where $\widetilde{Q}_{L_A}^{K_A}(a)$ denotes the distributed multipole moment tensor at the center a with respect to the system of axes located in the center-of-mass of the monomer A, one can make such a summation. The center-of-mass is translated by the vector $\mathbf{t}_a = (t_a, \widehat{\mathbf{t}}_a)$ from the site a. For completness we also give the translation formula for the multipole polarizability tensor:

$$\alpha_{(ll')L}^M(\omega) = \sum_{\lambda, \lambda'} \sum_{\Lambda, \Lambda'} t^{l+l'-\lambda-\lambda'}$$

$$\times [(2l+1)(2l'+1)(2\lambda+1)(2\lambda'+1)]^{1/2}$$

$$\times \langle l-\lambda, 0; l'-\lambda', 0 \mid \Lambda, 0 \rangle$$

$$\times \left[\binom{2l}{2\lambda} \binom{2l'}{2\lambda'} \right]^{1/2} \begin{Bmatrix} l-\lambda & \lambda & l \\ l'-\lambda' & \lambda' & l' \\ \Lambda & \Lambda' & L \end{Bmatrix}$$

$$\times \sum_{\Omega=-\Lambda}^{\Lambda} \sum_{\Omega'=-\Lambda'}^{\Lambda'} C_\Omega^\Lambda(\widehat{\mathbf{t}}) \widetilde{\alpha}_{(\lambda\lambda')\Lambda'}^{\Omega'}(\omega). \tag{1-147}$$

It follows from the partitioning of $\mathbb{R}^3$ according to Eq. (1-144) that the expression for the electrostatic energy, Eq. (1-67), can be rewritten as,

$$E_{\text{elst}}^{(1)} = \sum_{a \in A} \sum_{b \in B} \int_{V_a} \int_{V_b} \frac{\rho_A^{\text{tot}}(\mathbf{r}_1) \rho_B^{\text{tot}}(\mathbf{r}_2)}{r_{12}} d\mathbf{r}_1 d\mathbf{r}_2. \tag{1-148}$$

Inserting for each pair (a, b) the multipole expansion of r_{12}^{-1} with respect to centers located at sites a and b, cf. Eq. (1-141), and using Eqs. (1-142) and (1-145) one gets the following expression for the multicenter distributed multipole expansion of the electrostatic energy:

$$E_{\text{elst}}^{(1)} \sim \sum_{\{\Lambda\}} (-1)^{L_A} \delta_{L_A + L_B, L} \left[\frac{(2L_A + 2L_B + 1)!}{(2L_A)!(2L_B)!} \right]^{1/2}$$

$$\times \sum_{a \in A} \sum_{b \in B} \frac{Q_{L_A}^{K_A}(a) Q_{L_B}^{K_B}(b)}{R_{ab}^{L_A + L_B + 1}} A_{\{\Lambda\}}(\omega_A^a, \omega_B^b, \widehat{\mathbf{R}}_{ab}), \tag{1-149}$$

where the function $A_{\{\Lambda\}}(\omega_A^a, \omega_B^b, \widehat{\mathbf{R}}_{ab})$ is given by Eq. (1-125). Note that in the case of the distributed multipole expansion the function $A_{\{\Lambda\}}$ depends on the Euler angles ω_A^a and ω_B^b of the monomers A and B with respect to the local coordinate systems located on sites a and b, respectively, while $\widehat{\mathbf{R}}_{ab}$ are the polar angles of the vector $\mathbf{R}_{ab}$ connecting the sites a and b.

The multicenter expansion of the induction energy in terms of the distributed multipole moments and polarizabilities can be obtained is a similar way starting from Eq. (1-87) rewritten as follows,

$$E_{\text{ind}}^{(2)} = \sum_{aa' \in A} \sum_{bb' \in B} \int_{V_a} \int_{V_{a'}} \int_{V_b} \int_{V_{b'}} \left(\frac{\rho_B^{\text{tot}}(\mathbf{r}_3) \rho_B^{\text{tot}}(\mathbf{r}_4) \alpha_A(\mathbf{r}_1, \mathbf{r}_2 | 0)}{r_{13} r_{24}} \right.$$

$$\left. + \frac{\rho_A^{\text{tot}}(\mathbf{r}_1) \rho_A^{\text{tot}}(\mathbf{r}_4) \alpha_B(\mathbf{r}_3, \mathbf{r}_4 | 0)}{r_{13} r_{24}} \right) d\mathbf{r}_1 d\mathbf{r}_2 d\mathbf{r}_3 d\mathbf{r}_4. \tag{1-150}$$

Inserting the multipole expansions of the operators r_{13}^{-1} and r_{24}^{-1} with respect to the pairs of sites (a, b) and (a', b'), respectively, and defining the distributed polarizability tensor,

$$\alpha_{(l_X l'_X) L_X}^{K_X}(xx' | \omega) = \int_{V_x} \int_{V_{x'}} \alpha_X(\mathbf{r}, \mathbf{r}' | \omega) \left[\widetilde{\mathbf{m}}_{l_X}(\mathbf{r}) \otimes \widetilde{\mathbf{m}}_{l'_X}(\mathbf{r}') \right]_{K_X}^{L_X} d\mathbf{r} d\mathbf{r}', \tag{1-151}$$

one gets the following expression for the multicenter expansion of the induction energy in terms of the distributed multipole moments and polarizabilities:

$$E_{\text{ind}}^{(2)} \sim - \sum_{a,a' \in A} \sum_{b,b' \in B} \Bigg(\sum_{l_A,l_A'=1}^{\infty} \sum_{l_B,l_B'=0}^{\infty} \sum_{\{\Lambda\}} \frac{C_{\{\Lambda\},\text{ind}-A}^{\{\Lambda\}}(aa'bb')}{R_{ab}^{l_A+l_B+1} R_{a'b'}^{l_A'+l_B'+1}}$$

$$+ \sum_{l_A,l_A'=0}^{\infty} \sum_{l_B,l_B'=1}^{\infty} \sum_{\{\Lambda\}} \frac{C_{\{\Lambda\},\text{ind}-B}^{\{\Lambda\}}(aa'bb')}{R_{ab}^{l_A+l_B+1} R_{a'b'}^{l_A'+l_B'+1}} \Bigg)$$

$$\times A_{\{\Lambda\}}^{\{\Lambda\}}(\omega_A^a, \omega_A^{a'}, \omega_B^b, \omega_B^{b'}, \widehat{\mathbf{R}}_{ab}, \widehat{\mathbf{R}}_{a'b'}), \tag{1-152}$$

where

$$C_{\{\Lambda\},\text{ind}-A}^{\{\Lambda\}}(aa'bb') = \alpha_{(l_A l_A')L_A}^{K_A}(aa'|0)[\mathbf{Q}_{l_B}(b) \otimes \mathbf{Q}_{l_B'}(b')]_{K_B}^{L_B}, \tag{1-153}$$

Note that the formula for $\alpha_{(l_X,l_X')L_X}^{K_X}(xx'|0)$, Eq. (1-151), can be rewritten as the following sum-over-states expression,

$$\alpha_{(l_X l_X')L_X}^{K_X}(xx'|\omega) =$$

$$\sum_{n \neq 0} \frac{2(E_n^X - E_0^X)}{(E_n^X - E_0^X)^2 - \omega^2} \Big[\langle \Phi_0^X | \widetilde{\mathbf{M}}_{l_X} | \Phi_n^X \rangle_x \otimes \langle \Phi_n^X | \widetilde{\mathbf{M}}_{l_X'} | \Phi_0^X \rangle_{x'} \Big]_{K_X}^{L_X},$$

$$\tag{1-154}$$

where $\langle \cdots | \cdots \rangle_x$ means that the integration is performed over the volume assigned to the site x, i.e. V_x. Note also that the multipole moment operators appearing in Eqs. (1-151) and (1-154), labeled by l_X and l_X', are defined with respect to systems of axes located at the site x and x', respectively. Finally, the angular function $A_{\{\Lambda\}}^{\{\Lambda\}}(\omega_A^a, \omega_A^{a'}, \omega_B^b, \omega_B^{b'}, \widehat{\mathbf{R}}_{ab}, \widehat{\mathbf{R}}_{a'b'})$ is given by,

$$A_{\{\Lambda\}}^{\{\Lambda\}}(\omega_A^a, \omega_A^{a'}, \omega_B^b, \omega_B^{b'}, \widehat{\mathbf{R}}_{ab}, \widehat{\mathbf{R}}_{a'b'})$$

$$= \sum_m \sum_{m'} \sum_{m_A} \sum_{m_B} \langle l_A m_A; l_B m_B | l_A + l_B, m \rangle$$

$$\times X_{l_A l_B}^m X_{l_A' l_B'}^{m'} \sum_{m_A'} \sum_{m_B'} \langle l_A' m_A'; l_B' m_B' | l_A' + l_B', m' \rangle$$

$$\times \sum_{k_A} \sum_{k_A'} \sum_{K_A=-L_A}^{L_A} \langle l_A k_A; l_A' k_A' | L_A K_A \rangle$$

$$\times \sum_{k_B} \sum_{k_B'} \sum_{K_B=-L_B}^{L_B} \langle l_B k_B; l_B' k_B' | L_B K_B \rangle$$

$$\times \mathcal{D}_{m_A k_A}^{(l_A)*}(\omega_A^a) \mathcal{D}_{m_A' k_A'}^{(l_A')*}(\omega_A^{a'}) \mathcal{D}_{m_B k_B}^{(l_B)*}(\omega_B^b) \mathcal{D}_{m_B' k_B'}^{(l_B')*}(\omega_B^{b'})$$

$$\times C_m^{l_A+l_B}(\widehat{\mathbf{R}}_{ab}) C_{m'}^{l_A'+l_B'}(\widehat{\mathbf{R}}_{a'b'}). \tag{1-155}$$

The multicenter expansion of the dispersion energy in terms of the distributed multipole polarizabilities can be obtained is same way starting from Eq. (1-94). The final expression reads:

$$E_{\text{disp}}^{(2)} \sim - \sum_{a,a' \in A \, b,b' \in B} \sum_{l_A,l'_A=1}^{\infty} \sum_{l_B,l'_B=1}^{\infty} \sum_{\{\Lambda\}} \frac{C_{\{\Lambda\},\text{disp}}^{\{\Lambda\}}(aa'bb')}{R_{ab}^{l_A+l_B+1} R_{a'b'}^{l'_A+l'_B+1}}$$

$$\times A_{\{\Lambda\}}^{\{\Lambda\}}(\omega_A^a, \omega_A^{a'}, \omega_B^b, \omega_B^{b'}, \widehat{\mathbf{R}}_{ab}, \widehat{\mathbf{R}}_{a'b'}), \tag{1-156}$$

where

$$C_{\{\Lambda\},\text{disp}}^{\{\Lambda\}}(aa'bb') = \frac{1}{2\pi} \int_0^\infty \alpha_{(l_A l'_A)L_A}^{K_A}(aa'|i\omega)\alpha_{(l_B l'_B)L_B}^{K_B}(bb'|i\omega)d\omega,$$

$$\tag{1-157}$$

and the angular function $A_{\{\Lambda\}}^{\{\Lambda\}}$ is given by Eq. (1-155).

It should stressed that unlike in the case of the electrostatic energy, the expressions for the long-range coefficients and the angular function $A_{\{\Lambda\}}^{\{\Lambda\}}$ defining the multicenter multipole expansions of the induction and dispersion energies are different. This difference is due to the fact that in the multicenter expansions the products of the $\mathcal{D}$ functions, $\mathcal{D}_{m_A k_A}^{(l_A)*}(\omega_A^a) \, \mathcal{D}_{m'_A k'_A}^{(l'_A)*}(\omega_A^{a'})$ and $\mathcal{D}_{m_B k_B}^{(l_B)*}(\omega_B^b) \, \mathcal{D}_{m'_B k'_B}^{(l'_B)*}(\omega_B^{b'})$, and of the spherical harmonics $C_m^{l_A+l_B}(\widehat{\mathbf{R}}_{ab})C_{m'}^{l'_A+l'_B}(\widehat{\mathbf{R}}_{a'b'})$ cannot be represented by their Clebsch-Gordan series, since the arguments of the two $\mathcal{D}$ functions and of the two C functions are different. Therefore, one cannot recouple the Clebsch-Gordan coefficients to a $9j$ symbol, and simplify Eq. (1-155) to the form of Eq. (1-125).

The weakest point of the multicenter expansions based on the distributed multipole moments and polarizabilities is the definition of regions assigned to atoms. Indeed, the region of space associated with an atom (site) a is not uniquely defined. The most natural definition comes from the Bader's atoms in molecules theory [196]. In this approach the atomic basins V_a are defined in such a way that the dividing surfaces between two atomic basins in a single molecule are the zero-flux surfaces, determined from the following condition [196],

$$\nabla\rho_A(\mathbf{r}) \cdot \mathbf{n}(\mathbf{r}) = 0, \tag{1-158}$$

where $\mathbf{n}(\mathbf{r})$ is the vector normal to the surface. According to Bader imposing the zero flux condition leads directly to the topological definition of an atom in a molecule. This is in a sense true since the electron density ρ_A has maxima at the positions corresponding to the nuclei of the constituting atoms a. This means that V_a can formally be defined as the region of space with boundaries given by the zero-flux surface. Note parenthetically that the atomic basins defined in such a way are highly non-spherical, and that the integration over V_a may be difficult to perform.

The use of the Bader's basins together with Eq. (1-145) as the basis for the distributed multipole and polarizability analysis was proposed by Angyan and collaborators[197]. Other definitions and other methods leading to a distribution of multipole moments over the sites are possible[187,188,190,191,192,193,194]. For instance, Sokalski and Poirier[191] proposed an allocation algorithm of the distributed multipole moments based on the Mulliken population analysis. In this approach, the so-called Cumulative Atomic Multipole Moment method, half of the multipoles are allocated to each of the sets of sites *a* and *b* from which the basis functions came. Other allocation algorithms are also available. We refer the reader to Refs. (190–198) for a more detailed discussion of this point. Eq. (1-151) is the most natural definition of the distributed polarizability[197]. However, other definitions are also possible[189,190,195]. Usually, distributed polarizabilities are defined using some basis set partitioning techniques, and are highly nonunique.

The multicenter multipole expansions of the electrostatic, induction, and dispersion energies are usually convergent, although the convergence rate strongly depends on the allocation algorithm used to define the distributed multipole moments and polarizabilities. Note, however, that even though the series are convergent, they do not converge to the exact electrostatic, induction, and dispersion energies, since even a convergent multicenter expansion does not account for the penetration (charge overlap) effects which decay exponentially with the distance between the molecules. Therefore, multicenter expansions provide us with that part of the interaction energy that is due to the interactions of permanent, induced, and instantaneous multipole moments. The penetration part (charge-overlap effects and damping), which is a purely quantum-mechanical effect, is not accounted for. This part of the interaction energy is non-negligible, and cannot be neglected in any accurate calculation.

6.3. Bipolar Expansion of the Interaction Energy

The electrostatic, induction, and dispersion terms can be expanded in a convergent series closely related to the multipole expansion, but fully accounting for the charge-overlap effects, the so-called bipolar expansion introduced by Buehler and Hirschfelder[199,200]. In the local coordinate systems with the origins located at the centers of masses of the monomers A and B, separated by the distance R, and with their x and y axes parallel and aligned along the z axes, the distance between two particles in space can be expressed as follows,

$$\frac{1}{r_{12}} = \sum_{l_A,l_B=0}^{\infty} \sum_{m=-l_<}^{l_<} \frac{[(2l_A+1)(2l_B+1)]^{1/2}}{4\pi}$$

$$\times B_{l_A l_B}^{|m|}(r_1, r_2, R) C_{l_A}^m(\tilde{\theta}_1, \tilde{\phi}_1) C_{l_B}^{-m}(\tilde{\theta}_2, \tilde{\phi}_2) \tag{1-159}$$

where r_i, $\tilde{\theta}_i$, $\tilde{\phi}_i$ are the polar coordinates of the *i*th particle and $l_<$ denotes the smaller value of l_A and l_B. The coordinates of particle 1 are measured in the system

A of axes, while those of the particle 2 in the system B. The expression for the function $B_{l_A l_B}^{|m|}(r_1, r_2, R)$ has a different form in four regions of space:

$$\{I: R > r_1 + r_2\}; \quad \{II: r_2 > R + r_1\};$$

$$\{III: r_1 > R + r_2\}; \quad \{IV: |r_1 - r_2| \leqslant R \leqslant r_1 + r_2\}. \tag{1-160}$$

In the first three regions the expressions for $B_{l_A l_B}^{|m|}(r_1, r_2, R)$ are simple combinatorial formulas containing l_A, l_B, and m times a product of powers of r_1, r_2, and R, while in the last region the expression contains a (finite) sum of powers of r_1, r_2, and R:

$$B_{l_A l_B}^{|m|}(r_1, r_2, R) = \begin{cases} \dfrac{(-1)^{l_A + l_B}(l_A + l_B)!}{(l_A + |m|)!(l_B + |m|)!} \dfrac{r_1^{l_A} r_2^{l_B}}{R^{l_A + l_B + 1}}, & r_1, r_2 \in I \\[2ex] \dfrac{(-1)^{l_A + l_B}(l_B - |m|)!}{(l_A - l_B)!(l_A + |m|)!} \dfrac{r_1^{l_A} r_2^{-l_B - 1}}{R^{l_A - l_B}}, & l_A \geq l_B, \ r_1, r_2 \in II \\[2ex] \dfrac{(l_A - |m|)!}{(l_A - l_B)!(l_B + |m|)!} \dfrac{r_1^{-l_A - 1} r_2^{l_B}}{R^{l_B - l_A}}, & l_B \geq l_A, \ r_1, r_2 \in III \\[2ex] \sum_{k,l} A_{l_A l_B}^{|m|}(k, l) \dfrac{r_1^{k - l_A - 1} r_2^{l - l_B - 1}}{R^{k + l - l_A - l_B - 1}}, & r_1, r_2 \in IV, \end{cases}$$

$$\tag{1-161}$$

where $A_{l_A l_B}^{|m|}(k, l)$ are numerical coefficients that can be obtained from the recursion formulas derived by Buehler and Hirschfelder[199,200].

The expansion given above is exact, except for $r_{12} = 0$. If the terms resulting from the regions II-IV are neglected, one recovers the standard multipole expansion of the interaction operator. Substituting the bipolar expansion of $1/r_{12}$ and analogous expansions for other terms of the operator V into the matrix elements $\langle \Phi_0 | V | \Phi_{pol}^{(n)} \rangle$, the bipolar expansion of a given polarization correction is obtained.

The bipolar expansion was first applied to intermolecular interactions by Koide[201] and by Linder et al.[202,203]. The electrostatic energy was considered in Ref. (204) while dispersion interactions were treated in Refs. (201–205) for two hydrogen atoms, and in Ref. (206) for many-electron atoms. Later, this work was extended to interactions of atoms with diatomics by Rosenkrantz and Krauss[207] and to arbitrary systems by Knowles and Meath[208].

For the case of molecule-molecule interactions the bipolar expansion does not introduce any simplifications compared to the exact calculations using sum-over-state expressions discussed in Section 5. However, this approach allows to judge the importance of the charge-overlap (penetration) effects neglected in the multipole approximation. Calculations performed thus far show that the overlap effects are significant in the region of the Van der Waals minimum and for smaller separations. The explicit knowledge of these effects can be used to devise functional forms of the damping functions for the electrostatic, induction, and dispersion energies, which in turn can be used in constructions of semiempirical potentials, cf. Section 9.3.

6.4.   Importance of the Charge-overlap (Damping) Effects

To illustrate the convergence of the multipole expansion with and without the damping effects, and to compare the results with the full nonexpanded calculations, in Figure 1-1 we report the dispersion energy for the He–K$^+$ ion computed in various approximations[209]. The nonexpanded results were computed from the following expression,

$$E_{\text{disp}}^{(2)} = Ae^{-\beta R} - \sum_{n=1}^{3} \frac{C_{2n}}{R^{2n}} f_n(R; \widetilde{\beta}),\tag{1-162}$$

which, except for the long range coefficients C_{2n}, was fitted to the *ab initio* SAPT points[209]. The exponential term appearing in Eq. (1-162) is the so-called spherical dispersion term, while the sum on the r.h.s. is often referred to as the damped multipole expansion. The function f_n is the damping function. The functional form of f_n is not known, but we know the limiting values of f_n,

$$\lim_{R \to 0} f_n(R; \widetilde{\beta}) = 0, \qquad \lim_{R \to \infty} f_n(R; \widetilde{\beta}) = 1.\tag{1-163}$$

It follows from the limiting values that the damping function prevents the divergence of the multipole expansion at small internuclear distances R. In many applications the damping function is taken in the Tang-Toennies form[210–212], cf. Section 9.3,

$$f_n(R; \widetilde{\beta}) = 1 - \exp(-\widetilde{\beta}R) \sum_{k=0}^{n} \frac{(\widetilde{\beta}R)^k}{k!}.\tag{1-164}$$

The results computed from Eq. (1-162) and a similar expression for the induction energy will be referred to as the nonexpanded results. In many applications one approximates the induction or dispersion terms by their damped multipole expansions, cf. Section 9.3. The latter are given by Eq. (1-162) with the parameter A set

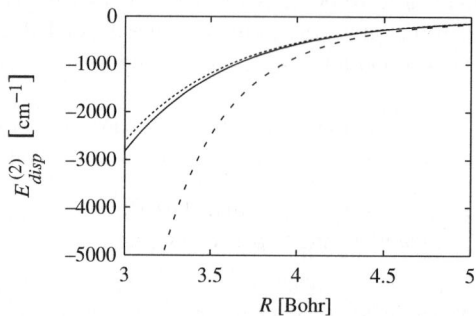

Figure 1-1. Dispersion energy $E_{\text{disp}}^{(2)}$ for the He–K$^+$ ion as a function of the internuclear distance R represented by its multipole expansion (large dashed line), damped multipole expansion (small dashed line), and by the nonexpanded results (full line)

equal to zero. Finally, the multipole-expanded results are obtained from Eq. (1-162) when puting A equal to zero and f_n equal to one.

An inspection of Figure 1-1 shows that the multipole expansion of the dispersion energy clearly diverges at small R. At larges distances, it fully recovers the nonexpanded results from *ab initio* calculations. The inclusion of the damping functions in the multipole expansion prevents the latter to diverge at small R, and except for small internuclear distances the actual values of the dispersion energy computed with the damped multipole expansion neglecting the spherical term are relatively close the nonexpanded results. This shows that the part of the charge overlap effects represented by the spherical term can be neglected in some approximate models based on one-center or multicenter multipole expansions.

7. MANY-ELECTRON FORMULATION OF THE SRS THEORY

In principle, the theory reviewed in Sections 4–6 can be applied to interactions of arbitrary systems if the full configuration interaction (FCI) wave functions of the monomers are available, and if the matrix elements of H_0 and V can be constructed in the space spanned by the products of the configuration state functions of the monomers. For the interactions of many-electron monomers the resulting perturbation equations are difficult to solve, however. A many-electron version of SAPT, which systematically treat the *intramonomer* correlation effects, offers a solution to this problem.

A general approach to the intramonomer correlation problem is known as the many-electron (or many-body) SAPT method[88,141,213–215]. In this method the zeroth-order Hamiltonian H_0 is decomposed as $H_0 = F + W$, where $F = F_A + F_B$ is the sum of the Fock operators, F_A and F_B, of monomer A and B, respectively, and W is the intramonomer correlation operator. The correlation operator can be written as $W = W_A + W_B$, where $W_X = H_X - F_X$, $X = A$ or B. The total Hamiltonian can be now be represented as $H = F + V + W$. This partitioning of H defines a double perturbation expansion of the wave function and interaction energy. In the SRS theory the wave function is obtained by expanding the parametrized Schrödinger equation as a power series in ζ and λ,

$$(F + \zeta V + \lambda W)\Psi(\zeta, \lambda) = E(\zeta, \lambda)\Psi(\zeta, \lambda), \qquad (1\text{-}165)$$

where the parameters ζ and λ are introduced to order the double perturbation expansion, and their physical value is equal to one. Note that $\Psi(0, 0) = \Phi_A^{HF}\Phi_B^{HF}$ is the product of the Hartree-Fock determinants of the unperturbed monomers, and $\Psi(0, \lambda) = \Phi_0^A(\lambda)\Phi_0^B(\lambda)$, where $\Phi_0^X(\lambda)$ is the eigenfunction of the Hamiltonian $F_X + \lambda W_X$, $X = A$ or B. The polarization energy corrections are obtained by expanding the function,

$$E(\lambda, \zeta) = \langle \Phi_0^A(\lambda)\Phi_0^B(\lambda) \mid F + \zeta V + \lambda W \mid \Psi(\lambda, \zeta)\rangle, \qquad (1\text{-}166)$$

while the SRS energy corrections are obtained by expanding

$$^{\nu}\mathcal{E}(\zeta,\lambda) = \frac{\langle \Phi_0^A(\lambda)\Phi_0^B(\lambda) \mid F + \zeta V + \lambda W \mid {}^{\nu}\mathcal{A}\Psi(\zeta,\lambda)\rangle}{\langle \Phi_0^A(\lambda)\Phi_0^B(\lambda) \mid {}^{\nu}\mathcal{A}\Psi(\zeta,\lambda)\rangle}. \tag{1-167}$$

The expansion of Eqs. (1-166) and (1-167) leads to the so-called double perturbation expansion of the interaction energy,

$$E_{\text{int}} = \sum_{n=1}^{\infty}\sum_{k=0}^{\infty}(E_{\text{pol}}^{(nk)} + E_{\text{exch}}^{(nk)}), \tag{1-168}$$

where the indices n and k denote the orders of $E_{\text{pol}}^{(nk)}$ and $E_{\text{exch}}^{(nk)}$ with respect to the intermolecular interaction and intramonomer correlation, respectively. The nth-order polarization and exchange contributions can be obtained by a direct summation of the expansion (1-168) over k. Explicit expressions for the individual corrections $E_{\text{pol}}^{(nk)}$ and $E_{\text{exch}}^{(nk)}$ can be obtained using the techniques of many-body perturbation theory and the coupled-cluster method. See Refs. (116–119, 137–141, 214–219) for the details of the derivations of open-ended expressions valid for the interaction of the monomers of arbitrary size.

For typical closed-shell systems the convergence of the expansion (1-168) in the intramonomer correlation is very satisfactory. However, in some cases the convergence in W may be poor, and one has to resort to some nonperturbative approaches. The nonperturbative treatments can most easily be devised by using Eqs. (1-67), (1-75), (1-85), and (1-92). Indeed, the monomers properties entering these expressions can be evaluated with highly correlated wave functions of the coupled-cluster singles and doubles (CCSD) method. The corresponding theoretical developments can be found in Refs. (218–219) The SAPT program[220,221] contains modules that compute the polarization and exchange corrections to interaction energy of closed-shell dimers that are correlated at the Møller-Plesset or coupled-cluster level. See Ref. (220-221) for more details.

The multipole representation of the intramonomer correlation corrections $E_{\text{elst}}^{(1n)}$, $E_{\text{ind}}^{(2n)}$, and $E_{\text{disp}}^{(2n)}$ can be obtained by using the standard Møller-Plesset expansions of the multipole moments and static and dynamic polarizabilities. For example, the relevant long-range dispersion coefficients are given by the Casimir-Polder integral (1-133) with exact polarizabilities replaced by their kth and lth-order correlation corrections in the Møller-Plesset series with $k + l = n$. Wormer et al.[177] developed a diagrammatic many-body perturbation theory of the correlation effects on the dynamic polarizabilities and a general scheme for the calculations of correlated long-range dispersion coefficients. The long-range dispersion coefficients corresponding the the sum $E_{\text{disp}}^{(20)} + E_{\text{disp}}^{(21)} + E_{\text{disp}}^{(22)}$ can now be routinely computed using the Polcor package[177,178].

Application of the conventional wave function approach in the symmetry-adapted perturbation theory (SAPT) has been shown to give very accurate description of the dispersion interaction and has provided intermolecular potentials which performed very well in numerous applications to spectroscopy and to simulation of bulk properties. The high accuracy of the conventional SAPT treatment comes, however, at the price of the very steep scaling of the computational effort which grows roughly as N^7, where N is the number of atoms. This steep scaling of the computational time with the system size makes applications to large polyatomic monomers unfeasible at present. These failures of the conventional wave function SAPT approach leaves a large class of weakly interacting polyatomic complexes without of an adequate theoretical treatment.

Jansen and Hesselmann[222−225] and independently Williams and Chabalowski[226], and Misquitta et al.[227], developed a method, referred to as SAPT(DFT), which solves the difficulty described above. In this method, the electrostatic, induction, and dispersion components of the interaction energy are obtained using the electron densities ρ_A and ρ_B from DFT calculations and the frequency-dependent density-density response functions $\alpha_A(\mathbf{r}_1, \mathbf{r}_1'|i\omega)$ and $\alpha_B(\mathbf{r}_2, \mathbf{r}_2'|i\omega)$ of the monomers computed with the time-dependent DFT techniques. It has been proved essential that the monomer exchange correlation potentials $v_{xc}(\mathbf{r})$, employed in the calculations, are asymptotically corrected at large $|\mathbf{r}|$. To reconstruct the total interaction energy, the electrostatic, induction, and dispersion energies are combined with the first and second-order exchange terms computed using the wave function expressions (1-75), (1-90), and (1-95) and the Kohn-Sham determinants of the monomers. Employing the asymptotically corrected exchange correlation potentials for monomers is also essential in this case.

The evaluation of the expressions (1-67), (1-87), and (1-94) can be simplified by using the density fitting technique[225,227]. In this approach the product $\phi_k(\mathbf{r})\phi_l(\mathbf{r})$ is expanded in terms of auxiliary atomic orbitals χ_k. In such a case the orbital expressions corresponding to Eqs. (1-67), (1-87), and (1-94) greatly simplify. For the dispersion energy Eq. (1-94) takes the form:

$$E_{\text{disp}}^{(2)} = -\frac{1}{2\pi} \int_0^\infty \text{tr}\, \mathbf{C}_A(i\omega)\, \mathbf{J}\, \mathbf{C}_B^T(i\omega)\, \mathbf{J}^T\, d\omega, \tag{1-169}$$

where $\mathbf{C}_A$ and $\mathbf{C}_B$ denote the matrices of the expansion coefficients in χ_k, $\mathbf{J}$ stands for the matrix of two-index Coulomb integrals, and $\mathbf{X}^T$ denotes the transposition of the matrix $\mathbf{X}$, The evaluation cost of Eq. (1-169) scales as $M^3 I$, where M is the number of auxiliary orbitals (proportional to the number of atoms) and I is the number of ω-integration points (of the order of 10).

The resulting SAPT(DFT) potential energy curves turn out to be very accurate in the wide range of intermolecular separations. For the benzene dimer[225,228] the results are very close to those of the much more expensive CCSD(T) treatment. For systems of the size of the benzene dimer and for the triple-zeta quality basis sets, a SAPT(DFT) calculation actually takes less time than a conventional supermolecular DFT calculation. Due to the favorable computational scaling the SAPT(DFT) approach is applicable to much larger molecules than any method used thus far for a reliable calculation of dispersion-dominated interaction potential.

It should be stressed that although the SAPT(DFT) approach is very appealing, it suffers from some drawbacks. The exact density functional theory in its time-independent and time-dependent versions gives the exact electron densities and density susceptibility functions of the monomers. This means that in principle the SAPT(DFT) method can reproduce the *exact* electrostatic, induction, and dispersion energies, if the *exact* exchange-correlation potential is known. This is not the case for the exchange terms, which are written in terms of nondiagonal elements of the density matrices and density matrix susceptibility functions. The latter quantities are not defined within the DFT formalism, and are computed from the Kohn-Sham orbitals, which, even if exact, do reproduce the exact electron density, but not the entire density matrix. Surprisingly enough, the approximation of the exact density matrix by the Kohn-Sham density matrix seems to work very well[227].

8.   RELATIONS BETWEEN THE PERTURBATION THEORY OF INTERMOLECULAR FORCES AND SUPERMOLECULAR APPROACHES

Symmetry-adapted perturbation theory is a useful tool for accurate calculations of the potential energy surfaces for weakly bound complexes, but it can also be used to interpret the results of supermolecular calculations. Numerous studies[105,106, 229–234] have been reported to interpret the supermolecule Hartree-Fock (HF) and Møller-Plesset (MP) perturbation theory interaction energies, as well as selfconsistent reaction field (SCRF) free energies of solvation in terms of physically meaningful contributions. It should be stressed that the supermolecule interaction energies suffer from basis set superposition error, and it is not *a priori* obvious that for comparison with SAPT one should take the results corrected for this error. However, theoretical arguments and numerical results presented in Ref. (81) show that a perturbation theory expansion in a finite basis can possibly converge only to the supermolecular interaction energy computed using the so-called Boys-Bernardi counterpoise correction for the basis set superposition error[235]. This means that the use of the Boys-Bernardi counterpoise correction is fully legitimate, in agreement with the conclusions reached in Refs. (236–242).

8.1.   Hartree-Fock Theory

A symmetry-adapted perturbation theory approach for the calculation of the Hartree-Fock interaction energies has been proposed by Jeziorska et al.[105] for the helium dimer, and generalized to the many-electron case in Ref. (106). The authors of Refs. (105–106) developed a basis-set independent perturbation scheme to solve the Hartree-Fock equations for the dimer, and analyzed the Hartree-Fock inter-action energy in terms of contributions related to many-electron SAPT reviewed in Section 7. Specifically, they proposed to replace the Hartree-Fock equations for the canonical orbitals of the dimer by noncanonical equations for orbitals localized on the monomers. Several localizations conditions can be exploited, but it was found advantageous[106] to employ a generalization of the localization condition used in the (many-electron) Hirschfelder-Silbey theory[98]. The perturbation expansion of the orbitals defines an expansion of the HF interaction energy, E_{int}^{HF},

$$E_{int}^{HF} = E_{elst}^{(10)} + E_{exch}^{(10)} + E_{ind,resp}^{(20)} + E_{exch-ind,resp}^{(20)}$$

$$+ E_{exch-def,resp}^{(20)} + \cdots, \qquad (1\text{-}170)$$

where the subscript "resp" appearing in the symbols of the induction and exchange-induction energies means that the orbital relaxation effects present in the electron density and response functions were taken into account. All terms appearing on the r.h.s. of Eq. (1-170) were defined in Section 7, except for the exchange-deformation energy, $E_{exch-def,resp}^{(20)}$. This contribution, specific to the Hartree-Fock theory, does not appear in the SRS theory. Thus, it can be viewed as that part of the exchange energy which cannot be recovered by perturbation theory employing weak symmetry-forcing. For systems with long-range induction interactions this contribution vanishes faster at large intermonomer distances than the exchange-induction energy itself[243,244].

Numerical results for the equilibrium geometries of the He–C_2H_2, He–CO, and Ar–HF complexes[106] are summarized in Table 1-5. Consecutive entries in this table represent the contributions appearing on the r.h.s. of Eq. (1-170). Also reported are the total nth-order approximations to E_{int}^{HF}, denoted by $E_{int}^{HF}(n)$. An inspection of Table 1-5 shows that the perturbation expansion (1-170) converges rapidly. The second-order approximation $E_{int}^{HF}(2)$ reproduces the exact Hartree-Fock results to within 3%. One may note that for some systems the exchange-deformation energy is far from negligible, so the SRS theory does not fully recover the Hartree-Fock interaction energy.

Table 1-5. Comparison of low-order approximations (in Hartree) to the Hartree-Fock interaction energies of the He–C_2H_2, He–CO, and Ar–HF complexes. The expression $(-N)$ denotes the factor 10^{-N}

	Ar–HF	He–C_2H_2	He–CO
$E_{elst}^{(10)}$	$-0.1487(-4)$	$-0.1999(-4)$	$-0.1921(-3)$
$E_{exch}^{(10)}$	$0.9345(-4)$	$0.9907(-4)$	$0.6362(-3)$
$E_{int}^{HF}(1)$	$0.7859(-4)$	$0.7911(-4)$	$0.4447(-3)$
$E_{ind,resp}^{(20)}$	$-0.1915(-4)$	$-0.5908(-5)$	$-0.2272(-3)$
$E_{exch-ind,resp}^{(20)}$	$0.3508(-5)$	$0.5691(-5)$	$0.1570(-3)$
$E_{exch-def,resp}^{(20)}$	$-0.4568(-5)$	$-0.3891(-5)$	$-0.3107(-4)$
$E_{int}^{HF}(2)$	$0.5837(-4)$	$0.7498(-4)$	$0.3427(-3)$
E_{int}^{HF}	$0.5652(-4)$	$0.7417(-4)$	$0.3326(-3)$

8.2. Møller-Plesset Theory

An analysis similar to the one above has not been performed thus far for the Møller-Plesset theories. However, Chalasinski and collaborators[229–233] proved that the polarization part of the supermolecule nth-order Møller-Plesset energy, $\Delta E_{\text{int}}^{\text{MP}n}$, $n \leqslant 4$, is given by,

$$\Delta E_{\text{int}}^{\text{MP}n} = E_{\text{elst,resp}}^{(1n)} + E_{\text{ind,resp}}^{(2n)} + E_{\text{disp}}^{(2,n-2)} + E_{\text{exch}}^{\text{MP}n} + \cdots. \tag{1-171}$$

Here $\Delta E_{\text{int}}^{\text{MP}n}$ denotes the nth-order correlation part of the supermolecule MPn interaction energy. Equation (1-171) shows that the supermolecule MP2 interaction energy correctly accounts for the leading intramonomer correlation corrections to the electrostatic and induction energies, and for the major part of the dispersion energy. This explains why it could be used with success to several Van der Waals and hydrogen-bonded complexes[5]. The physical structure of the MPn exchange terms, $E_{\text{exch}}^{\text{MP}n}$, is not well understood. A perturbation theory analysis of the MP2 equations for the pair functions in the localized representation[233] suggests that $E_{\text{exch}}^{\text{MP2}}$ accounts for the major part of the uncorrelated exchange-dispersion energy $E_{\text{exch-disp}}^{(20)}$ (the so-called K_1 term[129]), and for some parts of $E_{\text{exch}}^{(11)}$ and $E_{\text{exch}}^{(12)}$ corrections.

In order to get an idea how well a standard SAPT calculation can reproduce $\Delta E_{\text{int}}^{\text{MP2}}$ Bukowski et al.[245] analyzed the performance of the ansatz,

$$E_{\text{SAPT}}^{\text{MP2}} = E_{\text{elst,resp}}^{(12)} + E_{\text{ind,resp}}^{(22)} + E_{\text{disp}}^{(20)} + E_{\text{exch}}^{(11)} + E_{\text{exch}}^{(12)} + E_{\text{exch-disp}}^{(20)}, \tag{1-172}$$

for the helium dimer at various distances. Note that Eq. (1-172) is equivalent to Eq. (1-171) except that $E_{\text{exch}}^{\text{MP2}}$ is approximated by the sum $E_{\text{exch}}^{(11)} + E_{\text{exch}}^{(12)} + E_{\text{exch-disp}}^{(20)}$. The results are illustrated in Table 1-6. An inspection of this table shows that the agreement between the supermolecule MP2 interaction energy and the approximate MP2-SAPT results computed from Eq. (1-172) is reasonable. The authors of Ref. (245) also reported the inter- and intramonomer part of the MP2 interaction energy (denoted by $\Delta E_{\text{int}}^{\text{MP2}}(\text{inter})$ and $\Delta E_{\text{int}}^{\text{MP2}}(\text{intra})$, respectively). These terms were computed using properly localized MP2 pair functions and Hartree-Fock

Table 1-6. SAPT and localized MP2 components of $\Delta E_{\text{int}}^{\text{MP2}}$ for the He dimer. Energies are in Kelvin, and distances in bohr

Component	$R = 4.0$	$R = 5.6$	$R = 7.0$
$\Delta E_{\text{int}}^{\text{MP2}}$	-117.42	-16.003	-3.780
$E_{\text{SAPT}}^{\text{MP2}}$	-128.22	-16.250	-3.792
$\Delta E_{\text{int}}^{\text{MP2}}(\text{inter})$	-132.74	-16.683	-3.812
$E_{\text{disp}}^{(20)} + E_{\text{exch-disp}}^{(20)}$	-140.63	-16.627	-3.848
$\Delta E_{\text{int}}^{\text{MP2}}(\text{intra})$	15.32	0.681	0.032
$E_{\text{elst,resp}}^{(12)} + {}^tE_{\text{ind}}^{(22)} + E_{\text{exch}}^{(11)} + E_{\text{exch}}^{(12)}$	12.31	0.378	0.012

orbitals. See Ref. (245) for the details. The sum of ΔE_{int}^{MP2}(inter) and ΔE_{int}^{MP2}(intra) is obviously equal to ΔE_{int}^{MP2}. It is interesting to note that the agreement between the sum of the intermonomer correlation terms (dispersion and exchange-dispersion energies) and the localized intermonomer part of ΔE_{int}^{MP2} is very good. The agreement between the localized intramonomer part and the sum of the intramonomer correlation contributions is less satisfactory. Since the electrostatic and induction terms are included in ΔE_{int}^{MP2}, the level of disagreement suggests that the first-order exchange terms in MP2 and SAPT are different.

8.3. Coupled Cluster Theory

The relation between the supermolecule coupled cluster approach and the perturbation theory of intermolecular forces in even less obvious than the case of the Møller-Plesset theory, and no formal analysis has been reported in the literature thus far. Rode et al.[68] analyzed the long-range behavior of the CCSD(T) method[65], and showed that this method, although very popular and in principle accurate, may lead to wrong results for systems with the electrostatic term strongly depending on the electronic correlation, e.g. the CO dimer.

For the CO dimer the supermolecule MP4 and CCSD(T) results are very different. In particular, the location of the global and local minima, and of the saddle points changes drastically when going from the MP4 to the CCSD(T) levels[68]. On the other hand the standard many-electron SAPT approach based on the Schrödinger equation gave a good agreement with the MP4 calculations. However, this simple picture of close agreement between MP4 and standard SAPT is destroyed when the electrostatic term $E_{elst}^{(122)}$ is included. In the multipole expansion this term describes the interaction two MP2 dipole moments of the CO monomers. Since the dipole moment changes sign when going from the Hartree-Fock to the MP2 levels of theory[246], it is clear that the $E_{elst}^{(122)}$ contribution must be important. Since $E_{elst}^{(122)}$ is not included in the supermolecule MP4 interaction energy, the MP4 results cannot reliable. However, also the CCSD(T) method does not fully account for $E_{elst}^{(122)}$ and cannot be trusted in the case of the CO dimer. To explain this assertion we give a brief diagrammatic analysis of the MP5 method, of the $E_{elst}^{(122)}$ contribution to the MP5 interaction energy, and of CCSD(T). We follow the analysis of the MP5 diagrams proposed by Raghavachari et al.[247].

The diagrammatic representation of $E_{elst}^{(122)}$ reported in Figure 1-2 shows that this term is included in the supermolecule MP5 interaction energy[70]. The diagrams reported in Figure 1-2 correspond to the E_{SS}^5, $E_{ST}^5 + E_{TQ}^5(I)$, and $E_{QQ}^5(II)$ diagrams, respectively, of the MP5 theory[247,69,70]. Note that the diagrams (b) and (c) in Figure 1-2 separately do not correspond to the E_{ST}^5 and $E_{TQ}^5(I)$ terms defined in Ref. (247), but their sum corresponds to the sum $E_{ST}^5 + E_{TQ}^5(I)$. One may note that unlike those in Figure 1-2 the diagrams E_{ST}^5 and $E_{TQ}^5(I)$ of Ref. (247) have "long denominators", cf. Figure 1-3. However, as shown in Ref. (70) these "long denominators" cancel out, and the final representation of the sum $E_{ST}^5 + E_{TQ}^5(I)$ is given by diagrams with denominators that are excitation energies of the separate

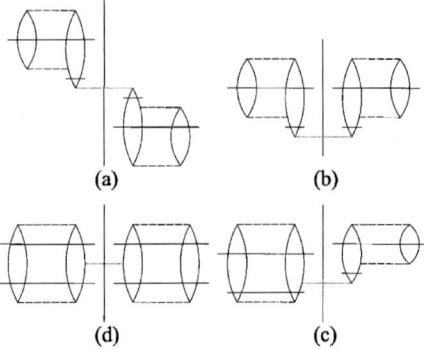

Figure 1-2. Schematic diagrammatic representation of the $E_{\text{elst}}^{(122)}$ correction (Brandow skeletons). The horizontal lines represent the denominators, while the vertical bar separates the monomers A and B. The two-electron integral corresponding to the dotted interaction line is a Coulomb integral. The dashed interaction lines represent antisymmetric two-electron integrals of the monomers. Diagram (a) is the intermolecular perturbation theory form of the MP5 contribution E_{SS}^5, diagram (d) of $E_{\text{QQ}}^5(II)$, while (b) and (c) are combinations of E_{ST}^5 and $E_{\text{TQ}}^5(I)$

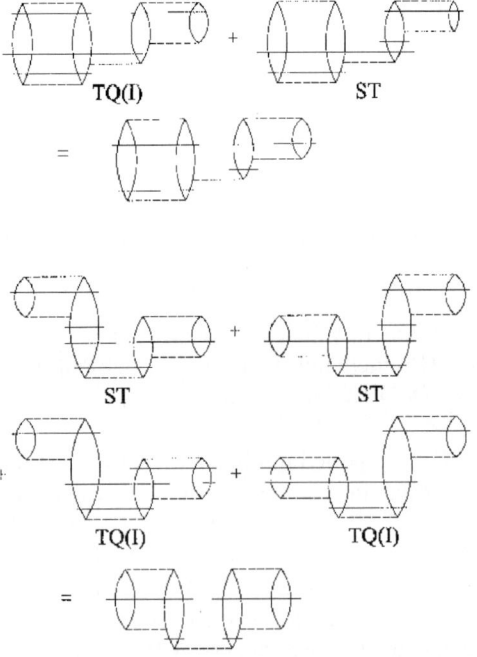

Figure 1-3. Schematic representation of the E_{ST}^5 and $E_{\text{TQ}}^5(I)$ MP5 diagrams, and illustration of the factorization of the "long denominators"

monomers, as those shown in Figure 1-2. This cancellation of the "long denominators" is diagrammatically illustrated in Figure 1-3. It is worth noting that the individual terms E_{ST}^5 and $E_{TQ}^5(I)$ of the MP5 theory of Raghavachari et al.[247] will not approach, at large distances, the corresponding diagrams of the $E_{elst}^{(122)}$ correction. Only their sum shows a correct long-range behavior.

The supermolecule CCSD(T) method[65] does not include all diagrams which, at large distances, approach the term $E_{elst}^{(122)}$. Table I of Ref. (247) shows that the $E_{QQ}^5(II)$ diagrams are completely neglected in CCSD(T) calculations, while the $E_{TQ}^5(I)$ terms are included with a coefficient of 1/2, rather than unity. Hence, in the CCSD(T) theory only half of the sum $E_{ST}^5 + E_{TQ}^5(I)$ approaches, at large distances, the corresponding contributions to $E_{elst}^{(122)}$. The remaining $\frac{1}{2}E_{ST}^5$ term does not appear by itself in the physically most important MP5 contribution to the interaction energy, namely the $E_{elst}^{(122)}$ term. In the case of the CO dimer, the omitted $E_{QQ}^5(II)$ and $\frac{1}{2}\left(E_{ST}^5 + E_{TQ}^5(I)\right)$ terms of $E_{elst}^{(122)}$ represent a substantial contribution (about 10%) to the interaction energy of the CO dimer, and seriously reduce the accuracy of the CCSD(T) results[68].

It should be stressed that the analysis presented above is general, and applies to any system. However, for the majority of Van der Waals complexes the electrostatic term $E_{elst}^{(122)}$ will not be as important as it is for the CO dimer. On the other hand, this analysis shows that any supermolecule method should be applied with great care, and an understanding of the supermolecule results in terms of contributions as defined by the symmetry-adapted perturbation theory is necessary.

8.4. Selfconsistent Reaction Field Theory

The solvent effects are often described within a semiempirical selfconsistent reaction field theory (SCRF)[248]. In this theory the free energy of solvation is obtained from a set of selfconsistent equations describing the interaction of the solute (denoted by S) with the solvent modeled by a polarizable continuum characterized by a dielectric constant ϵ. In the SCRF formalism, as developed by Rivail and collaborators[249–250] the solute-solvent system is modeled by a polarizable continuum (characterized by a dielectric constant ϵ) in which the solvent molecule is immersed within an ellipsoidal cavity[251,252]. The Hamiltonian describing the solute in the cavity is given by,

$$H = H_0 + V^{SCRF},$$ (1-173)

where H_0 is the Hamiltonian of the solute in the vacuum, and the operator V^{SCRF} describes the interaction between the permanent multipole moments of the solute, and the moments of the reaction field created by the cavity,

$$V^{SCRF} = \sum_{l=0}^{\infty} \sum_{m=-l}^{l} R_l^m M_l^m.$$ (1-174)

Here M_l^m denotes the operator of the mth spherical component of the multipole moment of order l, and the reaction field moments are given by,

$$R_l^m = \sum_{l'=0}^{\infty} \sum_{m=-l'}^{l'} f_{ll'}^{mm'} \langle \Psi_S | M_l^m | \Psi_S \rangle. \tag{1-175}$$

The numerical factors $f_{ll'}^{mm'}$ are the so-called reaction field factors, and depend on the dielectric constant ϵ of the solvent, and on the geometrical parameters of the ellipsoidal cavity. The wave function Ψ_S appearing in Eq. (1-175) is solution of the Schrödinger equation with the Hamiltonian given by Eq. (1-173). Thus, in the SCRF theory one solves a nonlinear Schrödinger equation that describes the interactions of the solute molecule with a polarizable continuum representing the solvent. The free energy of solvation is given by the expectation value,

$$\Delta G = \frac{\langle \Psi_S | H_0 + \frac{1}{2} V^{\text{SCRF}} | \Psi_S \rangle}{\langle \Psi_S | \Psi_S \rangle} - E_0, \tag{1-176}$$

where E_0 is the exact energy of the solute in the vacuum. One may note that the factor of half appearing in the expression for ΔG is a direct consequence of the fact that the solvent is assumed to be a linear dielectric.

In practice, the Schrödinger equation with the Hamiltonian of Eq. (1-173) is first solved within the self-consistent field approximation[252], leading to the so-called SCRF free energy of solvation, ΔG^{SCRF}. If the correlation corrections are included, e.g. via the MP2 approach[255,256], we get the MP2-SCRF free energy of solvation $\Delta G^{\text{MP2-SCRF}}$.

It should be stressed that the relation between the SCRF and SAPT approaches is not obvious, as the former describes the solvation energetics in terms of the free energy of solvation at a finite temperature T, while in the latter one considers the interaction energy between the molecule of the solute and all molecules of the solvent at $T = 0$ K. One should also note that in the SCRF theory the solvent is modeled by a polarizable continuum, so the SCRF Hamiltonian is semiempirical. Still, by assuming a discrete equivalent of the SCRF Hamiltonian one can get approximate relations between SAPT and SCRF at $T = 0$ K. A SAPT analysis of the free energy of solvation ΔG within the SCRF method was reported in Ref. (234). It was shown that the free energy of solvation ΔG is given by,

$$\Delta G = \sum_{i=1}^{N} \left(\frac{1}{2} E_{\text{elst}}^{(1)}(\text{SB}_i) + E_{\text{ind}}^{(2)}(\text{B}_i \leftarrow \text{S}) + \cdots \right), \tag{1-177}$$

where $E_{\text{ind}}^{(2)}(\text{B}_i \leftarrow \text{S})$ is the part of the induction energy that describes the polarization of the solvent (S), i.e. the interaction between the permanent moments of the N solute molecules ($\{\text{B}_i\}_{i=1}^{N}$) with the moments induced on the solvent by the electrostatic field of the solute. Equation (1-177) shows that the SCRF theory correctly accounts for the electrostatic and major induction effects, but neglects the

Table 1-7. Comparison of the components and global solvation energies (in kcal mol^{-1}) for the four forms of the palladium complex PdH$_3$Cl(NH$_3$)$_2$ in dichloromethane calculated by the discrete and continuum methods

	1-*mer*	2-*trans*	1-*fac*	2-*cis*
ΔG^{SCRF}	-16.8	-33.2	-11.1	-22.8
$\Delta G^{MP2-SCRF}$	-15.6	-28.3	-10.3	-19.0
ΔG^{SAPT}	-22.2	-30.8	-11.0	-22.7
E^{HF}_{int}	-13.1	-21.6	-7.7	-17.0
E^{MP2}_{int}	-20.3	-26.5	-13.6	-21.0

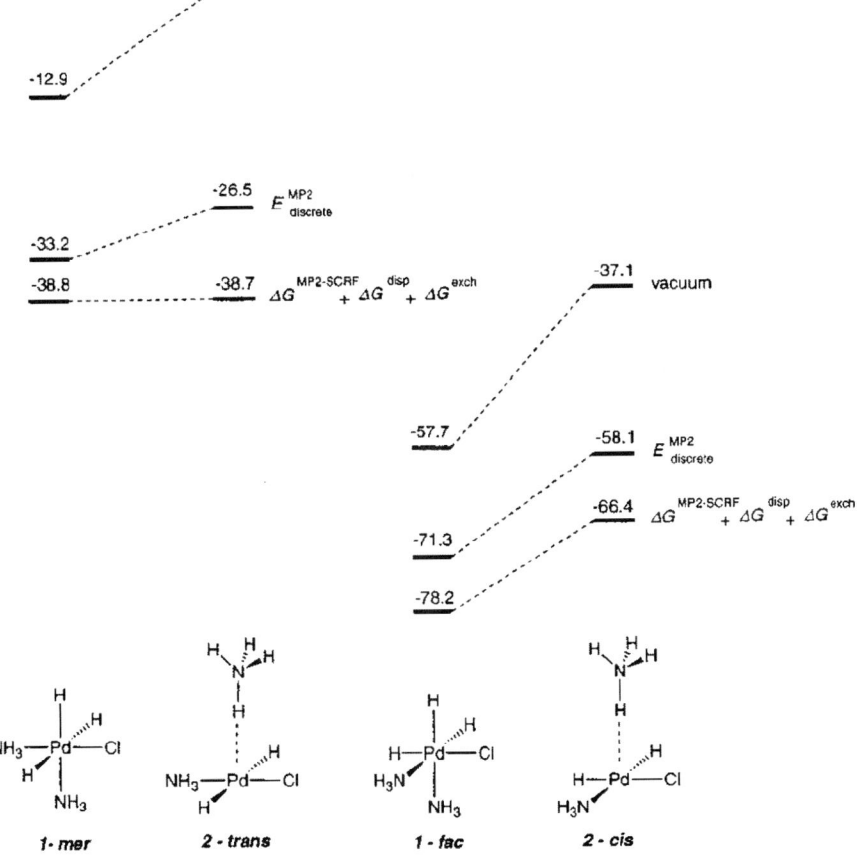

Figure 1-4. Relative stabilization energies for the four forms of the PdH$_3$Cl(NH$_3$)$_2$ complex in the CH$_2$Cl$_2$ solvent calculated by the discrete supermolecule and continuum SCRF methods at the MP2 level. The zero of energy corresponds to the 2-trans form in the vacuum

induction energy corresponding to the interaction of the permanent moments of the solvent with the moments induced on the solute by the electrostatic field of the solvent, $E_{ind}^{(2)}(S \leftarrow B_i)$. For strongly polar solvents this term may be very important. When comparing the SAPT and SCRF results care should be taken of the fact that in the SAPT approach Eq. (1-177) is evaluated at $T = 0$ K, while the SCRF calculations are done at a finite T. To get one-to-one correspondence between the two sets of calculations, the SAPT results should be Boltzmann averaged over all configurations of the solvent molecules[184,257]. It is worth noting that Eq. (1-177) can efficiently be evaluated within the multipole approximation[161] for a large number of the solvent molecules, and for a sufficient number of configurations to perform the Boltzmann average. This approach would present an *ab initio* alternative to the semiempirical SCRF-type calculations.

Ref. (234) reported a theoretical study of the solvent effects on various isomers of the palladium hydride complex $PdH_3Cl(NH_3)_2$ in dichloromethane. The influence of the solvent was investigated by discrete MP2 and SAPT, and continuum SCRF calculations. The theoretical relation between SCRF and SAPT, Eq. (1-177), was fully confirmed by the numerical results from the discrete SAPT and continuum SCRF calculations, cf. Table 1-7 and Figure 1-4. Interestingly, both the discrete MP2 and continuum SCRF models predicted the same relative stabilities for the isomers of $PdH_3Cl(NH_3)_2$ in dichloromethane. Small energetic differences between the results of the discrete and continuum calculations could be explained by the entropy effects, neglected in the discrete model.

9. APPROXIMATE MODELS FOR PAIR INTERACTION POTENTIALS

Accurate *ab initio* calculations of the interaction energy are very time consuming, and therefore restricted to relatively small systems. For the interaction of large molecules approximate models are usually used. These models are usually derived from sound theoretical approaches discussed in the preceding sections, but contain some simplifying assumptions to make the problem tractable with the present day computers. Model potentials are obtained by introducing some approximations into the equations of the perturbation theory or by introducing some empirical parameters. In the remaining of this section we will briefly discuss such approximate models with a special emphasis on their connection with the perturbation theory of intermolecular forces.

9.1. Morokuma Partitioning of the Hartree-Fock Interaction Energy

In 1971 Morokuma[258] proposed a simple partitioning of the Hartree-Fock interaction energy into some physically interpretable contributions, hopefully related to the components of the interaction energy as defined by SAPT. In this method one removes from the Fock matrix and from the energy expression the integrals (in the atomic basis) which are assumed to be unrelated to the considered type of interaction. Then, by calculating the energy with such a partial Fock matrix, one gets the contribution coming from this type of interaction. The original scheme of Morokuma[258] consists of four steps. In the first step the expectation value of the total Hamiltonian of the dimer AB is computed using the determinantal wave function of the dimer constructed from the (nonorthogonal) orbitals of the monomers A and B, and neglecting all integrals in the atomic basis that contain overlap between basis functions of A and B, i.e. that contain products of the type $\rho_{pq}^{AB}(\mathbf{r}) = \chi_p^A(\mathbf{r})\chi_q^B(\mathbf{r})$, where $\{\chi_p^X\}_{p=1}^{n_X}$ is the atomic basis of the monomer X, X=A or B. It can be shown that such a procedure gives the following approximation to the energy of the dimer, denoted by $E_{AB}^{HF}(1)$:

$$E_{AB}^{HF}(1) = E_A^{HF} + E_B^{HF} + E_{elst}^{(10)}, \tag{1-178}$$

where E_X^{HF} is the Hartree-Fock energy of the monomer X, X=A or B, calculated in the monomer basis. In the second step the integrals involving intermolecular overlap are kept in the calculations of the expectation value of the Hamiltonian. The corresponding energy of the dimer, $E_{AB}^{HF}(2)$, can be decomposed as follows:

$$E_{AB}(2) = E_A^{HF} + E_B^{HF} + E_{elst}^{(10)} + E_{exch}^{(10)} + \Delta_L^{(0)} + \Delta_M^{(0)}, \tag{1-179}$$

where the so-called zeroth-order exchange terms $\Delta_L^{(0)}$ and $\Delta_M^{(0)}$ are the Landshoff and Murrell delta terms[115]. In fact $E_{AB}^{HF}(2) - E_A^{HF} - E_B^{HF}$ is nothing else than the Heitler-London energy of Eq. (1-70) calculated with the determinantal wave functions of the monomers A and B. Eq. (1-179) follows directly from the work of Jeziorski and collaborators[115]. Note that the Murrell term decays as the fourth power of the typical overlap integral S, while the Landshoff term behaves as S^2 in the monomer basis set, and vanishes identically if the full basis of the dimer is used to compute the orbitals of the isolated monomers A and B[115]. In the third step of the Morokuma scheme[258] one repeats the calculation of step 1, except that the Hartree-Fock equations are iterated up to selfconsistency. The third step is fully equivalent to the Hartree-Hartree-Fock method proposed by Sadlej[259]. The orbitals of one monomer are distorted by the electrostatic field of the other in a selfconsistent manner, i.e. the orbital-dependent field ω_X, X=A or B, cf. Eq. (1-82), is also polarized. It follows from Ref. (259) that the energy of the dimer in the third step, $E_{AB}^{HF}(3)$, can related to some SAPT contributions as follows:

$$E_{AB}(3) = E_A^{HF} + E_B^{HF} + E_{elst}^{(10)} + E_{exch}^{(10)} + \Delta_L^{(0)} + \Delta_M^{(0)}$$
$$+ E_{ind,resp}^{(20)} + E_{ind,resp}^{(30)} + \cdots, \tag{1-180}$$

where $E_{ind,resp}^{(30)}$ represents the third-order induction energy computed from the expression (1-97) with linear and quadratic response functions from the random phase approximation, and the dots denote the fourth and higher-order induction contributions.

Equation (1-180) can be obtained from a variational calculation of the energy of the dimer with a wave function that is product of determinantal wave functions of the monomers constructed from the optimal (selfconsistent) orbitals. Unfortunately, the optimization of the orbitals of one monomer in the electrostatic field of the other does not prevent the unphysical transfer of electrons from one system to the other, since the antisymmetry of the wave function of the dimer is not preserved, and consequently, the Pauli principle is not satisfied. This may lead to some unphysical results in the short range[260]. It should also be stressed that the interaction part of Eq. (1-180) cannot be obtained, as proposed in Ref. (261), by taking the expectation value of the interaction operator V with the product of determinantal wave functions of the monomers constructed from the optimal (selfconsistent) orbitals. This would result in an overcounting of the induction terms, already in the second order.

The last step of the Morokuma scheme[258] is the calculation of the Hartree-Fock energy of the dimer with the selfconsistent orbitals of the dimer. Thus, in this step one gets the exact Hartree-Fock energy of the dimer AB. In view of Eq. (1-170) the energy of the dimer in the fourth step, $E_{AB}^{\mathrm{HF}}(4)$, can be decomposed as follows,

$$
\begin{aligned}
E_{AB}(4) = {} & E_A^{\mathrm{HF}} + E_B^{\mathrm{HF}} + E_{\mathrm{elst}}^{(10)} + E_{\mathrm{exch}}^{(10)} + \Delta_L^{(0)} + \Delta_M^{(0)} \\
& + E_{\mathrm{ind,resp}}^{(20)} + E_{\mathrm{ind,resp}}^{(30)} + E_{\mathrm{exch-ind}}^{(20)} + E_{\mathrm{exch-def}}^{(20)} + \cdots .
\end{aligned}
\tag{1-181}
$$

In the original paper of Morokuma[258] the sum of the exchange-induction and exchange-deformation contributions was ascribed to the charge transfer.

The partitioning scheme of Morokuma[258] was often criticized, because the charge transfer term, defined as the difference between the interaction energies computed in the fourth and third steps, was in some cases unusually large. This is solely due to the basis set superposition error[235]. In the original paper of Morokuma[258] and in the subsequent applications the energies of the monomers were computed in their respective bases, and not in the full basis of the dimer, so the basis set superposition error strongly affected the results. Sokalski and collaborators[262,263] suggested to perform the calculations for the monomers in the full basis of the dimer. Obviously, this idea was a cure for the Morokuma partitioning, but the calculations became more expensive than the ordinary Hartree-Fock calculations.

It follows from the description given above that the Morokuma partitioning is basis-set dependent. In particular, it depends on which atomic orbitals are considered as belonging to the molecule A and which to molecule B. Such an assignment is in general arbitrary, e.g. a basis function centered off the nuclei may cause troubles because it is difficult to decide *a priori* which molecule it belongs to. However, Eqs. (1-178) and (1-181) show that there is a clear relation between this partitioning and the SAPT approach at the Hartree-Fock level. This means that the contributions to the Hartree-Fock interaction energy as defined by Morokuma[258] can be rewritten in a basis-set independent way, and thus have a well defined meaning in the limit of a complete basis.

In 1976 Kitaura and Morokuma[264] suggested some improvements of the original partitioning[258]. Unfortunately, the improvements introduced in Ref. (264) made this partitioning much more complicated, and the relation with the SAPT approach was lost. Unlike in the original Morokuma scheme[258], the physical contributions introduced in the 1976 paper cannot be rewritten in a basis set independent way. The charge-transfer term was redefined, and does not have a well defined meaning. Finally, a new mixing interaction contribution, sometimes large, had to be added. The definitions of the contributions to the interaction energy as defined in Ref. (264) are again basis set dependent. As a consequence some contributions to the interaction energy do not have well defined limits when the basis set becomes complete. For example, the basis set saturated values of the so-called polarization component require two complete basis sets centered on A and B. In such a case, however, the charge transfer term becomes (partly) redundant and the interaction energy cannot be viewed as containing independent polarization and charge transfer components[265,266].

9.2.   Variation-Perturbation Approach

The essential problem with the partitioning of the Hartree-Fock interaction energy is related to the fact that the so-called Hartree-Fock deformation or delocalization energy,

$$E_{def}^{HF} \equiv E_{int}^{HF} - E_{HL}^{(10)}, \tag{1-182}$$

where $E_{HL}^{(10)}$ is the interaction part of the Heitler-London energy computed with single determinantal wave functions of the monomers, cannot easily be interpreted in terms of SAPT contributions. Early work of Jeziorski and Van Hemert[267] suggested that the deformation energy is mostly dominated by the second-order induction energy, but subsequent works[259,268] did not support this conclusion. It is now clear how to understand the physical meaning of the Hartree-Fock deformation energy, cf. Section 8.1, but a computationally simple scheme does not come out naturally from this work[105,106].

Sokalski and collaborators[269,270] proposed a variation-perturbation approach for the calculation of the Hartree-Fock interaction energy in terms of some physically meaningful contributions that can be applied to the interactions of large systems[271,272]. The Hartree-Fock interaction energy is decomposed into the Heitler-London and deformation contributions. The Heitler-London energy is computed in the dimer-centered basis set from the variational expression (1-70), which can easily be evaluated using orthogonalized orbitals of the isolated monomers. The electrostatic term, $E_{elst}^{(10)}$, is also at hand, so the zeroth iteration of the SCF procedure for the dimer gives $E_{HL}^{(10)}$, $E_{elst}^{(10)}$, and a Heitler-London exchange term E_{exch}^{HL}, which is a very good approximation to $E_{exch}^{(10)}$,

$$E_{exch}^{HL} = E_{exch}^{(10)} + \Delta_M^{(0)}, \tag{1-183}$$

since the Murrell's delta term decays with R as S^4. Note that the Landshoff delta term, present in Eq. (1-179), vanishes identically in the dimer-centered basis set[115].

The electrostatic term can further be decomposed into the multipole and penetration components,

$$E_{\text{elst}}^{(10)} = E_{\text{elst-mult}}^{(10)} + E_{\text{elst-pen}}^{(10)}. \tag{1-184}$$

The multipole part can efficiently be estimated from the distributed multipole analysis[191]. In this way the electrostatic penetration contribution is obtained. One may note that the accuracy of the electrostatic term can be increased by keeping the penetration part from Eq. (1-184), and replacing the Hartree-Fock distributed multipole moments by some correlated, e.g. MP2 moments. Finally, the intramonomer correlation term and the dispersion energy can be evaluated from the expression,

$$E_{\text{intra}}^{\text{MP2}} + E_{\text{disp}}^{(20)} = E_{\text{int}}^{\text{MP2}} - E_{\text{elst-mult}}^{(12)}, \tag{1-185}$$

where $E_{\text{elst-mult}}^{(12)}$ is the multipole part of the electrostatic energy calculated with the MP2 multipole moments.

Although this scheme does not give a full information about the physical components of the interaction energy, it efficiently splits them into terms decaying exponentially with R, i.e. the Heitler-London exchange and electrostatic penetration, and into terms decaying as some power of $1/R$, i.e. the electrostatic multipole term, Hartree-Fock deformation (dominated in the long range by the induction energy), and the sum of the intra and intermonomer MP2 correlation terms which are essentially dominated by the dispersion contribution.

We wish to end this section by saying that the variation-perturbation approach as discussed above, introduces a natural hierarchy of gradually more and more sophisticated models starting from the crude evaluation of the electrostatic energy in the distributed multipole approximation, and ending with the inclusion of the intramolecular and dispersion contributions at the MP2 or even more correlated level.

9.3. Tang-Toennies Model

Even more simplistic models of intermolecular potentials were proposed by Tang and Toennies[210,211,212], and by Meath and collaborators[273]. In these models, referred to as $\text{TT}^{(6,10)}$ [210] and XC[273] models, respectively, the interaction potential between two rigid molecules is given by,

$$V(R, \omega_A, \omega_B) = \mathcal{E}_{\text{rep}}(R, \omega_A, \omega_B)$$
$$- \sum_{\{\lambda\}} \frac{C_{\{\lambda\}}(\omega_A, \omega_B)}{R^n} f_n(R; \beta(\omega_A, \omega_B)), \tag{1-186}$$

where $n = l_A + l'_A + l_B + l'_B + 2$, $\mathcal{E}_{rep}$ denotes the repulsion energy, and f_n is the damping function, and is given by Eq. (1-164). In the TT model the repulsive term was taken in the Born-Mayer form [274],

$$\mathcal{E}_{rep}(R, \omega_A, \omega_B) = A(\omega_A, \omega_B) \exp(-\beta(\omega_A, \omega_B)R), \qquad (1\text{-}187)$$

while in the XC model it was assumed that the repulsion contribution can be derived from the electrostatic by using the following semi-empirical expression,

$$\mathcal{E}_{rep}(R, \omega_A, \omega_B) = [a(\omega_A, \omega_B) + b(\omega_A, \omega_B)R] E_{elst}^{(10)}, \qquad (1\text{-}188)$$

where the electrostatic term was represented by the following analytic form,

$$E_{elst}^{(10)} = -A(\omega_A, \omega_B) \exp\left(a_0(\omega_A, \omega_B)R + a_1(\omega_A, \omega_B)R^{-1} + a_2(\omega_A, \omega_B)R^{-2}\right). \qquad (1\text{-}189)$$

The angular dependence of all the parameters and of the long-range coefficients C_n is expressed through the angular functions $\tilde{A}_{\{\Lambda\}}$:

$$X(\omega_A, \omega_B) = \sum_{\{\Lambda\}} X_{\{\Lambda\}} \tilde{A}_{\{\Lambda\}}(\omega_A, \omega_B), \qquad (1\text{-}190)$$

$$C_{\{\Lambda\}}(\omega_A, \omega_B) = \sum_{\{\Lambda\}} C_{\{\Lambda\}}^{\{\Lambda\}} \tilde{A}_{\{\Lambda\}}(\omega_A, \omega_B), \qquad (1\text{-}191)$$

where X in Eq. (1-190) stands for any of the parameters A, β, a, b, a_0, a_1, a_2, etc., while the long-range coefficients $C_{\{\Lambda\}}^{\{\Lambda\}}$ are given by Eq. (1-129) for the induction term, and by Eq. (1-133) for the dispersion. Very often the long-range constants in such models are sums of the induction and dispersion long-range coefficients.

The long-range coefficients appearing in Eq. (1-186) can be obtained from *ab initio* calculations or from some semiempirical models. As far as *ab initio* calculations are concerned, the many-body perturbation theory approach developed by Wormer and collaborators [177] was successfully applied to various Van der Waals complexes providing state-of-the-art values of the long-range dispersion coefficients [177,179−18]. These coefficients, in turn, can be used to define the exact asymptotics of the dispersion energy, or may serve as an *ab initio* input to empirical potentials fitted to reproduce the high-resolution spectroscopic data, see, e.g. Refs. (177,275) Very accurate values of the isotropic $C_6^{(0)}$ dispersion coefficients can be obtained from pseudo-spectral expansions in terms of (experimental) dipole oscillator strengths [276]. These data, available now for many systems (see Refs. (276–294)) are considered to constitute benchmark values for *ab initio* calculations.

In the early works by Tang and Toennies on interatomic interactions[210,211] the repulsive term $\mathcal{E}_{\text{rep}}$ was fitted to the Hartree-Fock interaction energy. In the model extended to the atom-molecule case[212] the repulsion energy was fitted to the Heitler-London energy computed with the Hartree-Fock determinants for the isolated monomers. Nowadays the models of Eqs. (1-186) and (1-189) are mostly used in the fits of empirical potentials to some experimental data, see, e.g. Ref. (295), or to fit the *ab initio* points to the functional form of Eqs. (1-186) and (1-187), see, e.g. Ref. (296). One may note that the $TT^{(6,10)}$ model is a generalization of the simple exp-6 potential introduced by Born and Mayer[274].

9.4. Atom-Atom and Site-Site Potentials

For the interaction of large molecules the angular expansions of the potential parameters, Eqs. (1-190) and (1-191), may be slowly convergent, and the calculation of the potential may become prohibitively time consuming. Therefore, in many applications the so-called atom-atom potentials are used. The functional form of an atom-atom potential partly follows from the distributed multipole analysis[297],

$$V(R, \omega_A, \omega_B) = \sum_{a \in A} \sum_{b \in B} \left[\mathcal{E}_{\text{rep}}^{ab}(\omega_A^a, \omega_B^b) \right.$$
$$\left. - \sum_{\{\Lambda\}} \frac{C_{\{\Lambda\}}^{ab}(\omega_A^a, \omega_B^b)}{R_{ab}^n} f_n\left(R_{ab}, \omega_A^a, \omega_B^b\right) \right], \qquad (1\text{-}192)$$

where $n = l_A + l'_A + l_B + l'_B + 2$ and the long-range coefficient appearing in Eq. (1-192) is given by,

$$C_{\{\Lambda\}}^{ab}(\omega_A^a, \omega_B^b) = \sum_{\{\Lambda\}} C_{\{\Lambda\}}^{\{\Lambda\}}(ab)\tilde{A}_{\{\Lambda\}}(\omega_A^a, \omega_B^b). \qquad (1\text{-}193)$$

Similarly as in the Tang-Toennies model[210,211,212] the repulsive term is taken in the Born-Mayer form[297],

$$\mathcal{E}_{\text{rep}}^{ab}(R_{ab}, \omega_A^a, \omega_B^b) = A_{ab}(\omega_A^a, \omega_B^b) \exp\left(-\beta_{ab}(\omega_A^a, \omega_B^b)R_{ab}\right). \qquad (1\text{-}194)$$

The parameters A_{ab} and β_{ab} are expanded in terms of the functions $\tilde{A}_{\Lambda\}}$ as follows,

$$X_{ab}(\omega_A^a, \omega_B^b) = \sum_{\{\Lambda\}} X_{ab}^{\{\Lambda\}} \tilde{A}_{\{\Lambda\}}(\omega_A^a, \omega_B^b), \qquad (1\text{-}195)$$

where X_{ab} stands for A_{ab} or β_{ab}. The parameters $A_{ab}^{\{\Lambda\}}$ and $\beta_{ab}^{\{\Lambda\}}$ are usually fitted to some *ab initio* data. See Ref. (298) for a model application of this approach to the repulsive interactions in the water dimer.

Note that the angular expansion in Eqs. (1-192) and (1-193) is simplified compared to the distributed multipole expansion, cf. Eq. (1-155), so unlike in the

Tang-Toennies model, the functional form of Eq. (1-192) is not exact. Therefore, the long-range coefficients appearing in Eqs. (1-192) and (1-193) cannot be derived from any *ab initio* calculations. However, Eq. (1-192) can possibly be used to fit *ab initio* points to such an analytic form.

We wish to end this section by saying that in many cases Eq. (1-192) is even further simplified by neglecting the angular dependence of the parameters in the repulsive term and of the long-range coefficients, cf. Eqs. (1-193) and (1-195). Obviously, such a functional form, although very simple, cannot be derived from any theory. In fact, one should add "nonadditive" atom-atom terms depending on three, four, etc. centers, in order to cure the problem.

9.5. Empirical Force Fields

For the interaction of very large systems such as polypeptides, polynucleotides, or other biomolecular systems *ab initio* calculations are not feasible even at the Hartree-Fock level, and one has to resort to some empirical models of interaction potentials. Usually, these potentials depend on all intra and intermolecular degrees of freedom and are called force fields. Several analytic forms of the force fields have been designed. One of the most popular is the Amber (Assisted Model Building with Energy Refinement) package[299], where the force field V is expressed in the following way[300]:

$$V = \frac{1}{2} \sum_{\text{bonds}} K_r (r - r_{eq})^2 + \frac{1}{2} \sum_{\text{angles}} K_\theta (\theta - \theta_{eq})^2$$

$$+ \sum_{\text{dihedrals}} \frac{V_n}{2} [1 + \cos(n\phi - \gamma)]$$

$$+ \sum_{i<j} \left[\frac{A_{ij}}{r_{ij}^{12}} - \frac{B_{ij}}{r_{ij}^6} + \frac{q_i q_j}{\epsilon r_{ij}} \right]. \tag{1-196}$$

Other potential functions differ in the way the potential is partitioned into various contributions representing intra and intermolecular interactions. Cornell and collaborators[300] proposed an approach to determine the force field parameters based as much as possible on *ab initio* calculations. In this work each biomolecular system is divided into small residua, for which geometry optimization can be performed by an *ab initio* method. *Ab initio* calculations give partial atomic charges on atoms and the equilibrium geometry, i.e. the equilibrium values for the bond lengths, and planar and dihedral angles.

The first and second terms in Eq. (1-196) describe changes in the potential energy of the system associated with deformations of bond lengths and planar angles between two and three connected atoms, respectively. The summation is performed over all bonds and planar angles in the molecule. All the bonds and planar angles between adjacent atoms are described by simple harmonic potentials with parameters r_{eq}, K_r, θ_{eq} and K_θ chosen to reproduce the experimental normal mode frequencies.

The third term of Eq. (1-196) describes the energy change associated with a rotation around a single bond in a molecule. The summation is performed over all dihedral angles. The value of V_n gives the height of the barrier to torsional motions around a given bond. The value of the index n indicates the number of equivalent minima on the potential energy curve for ϕ varying from 0° to 360°. The phase γ determines the minimum value of the torsional angle. Dihedral angle parameters are calibrated to reproduce the energies of torsional motions for a set of small model compounds. Torsional energy profiles are usually derived from *ab initio* calculations with the MP2 method.

The last term in the formula (1-196) describes electrostatic and Van der Waals interactions between atoms. In the Amber force field the Van der Waals interactions are approximated by the Lennard–Jones potential with appropriate A_{ij} and B_{ij} force field parameters parametrized for monoatomic systems, i.e. $i = j$. Mixing rules are applied to obtain parameters for pairs of different atom types. Cornell et al.[300] determined the parameters of various Lenard-Jones potentials by extensive Monte Carlo simulations for a number of simple liquids containing all necessary atom types in order to reproduce densities and enthalpies of vaporization of these liquids. Finally, the energy of electrostatic interactions between non–bonded atoms is calculated using a simple classical Coulomb potential with the partial atomic charges q_i and q_j obtained, e.g. by fitting them to reproduce the electrostatic potential around the molecule.

The first two terms in Eq. (1-196) are often referred to as "hard" degrees of freedom since relatively high energies are required to cause significant deformations from the equilibrium geometries. By contrast, the last two terms are considered as "soft" degrees of freedom and a proper parametrization of these terms represents the main challenge for designing a good force field. This is due to the fact that most of the variations in the structure and relative energies of molecules result from a complex interplay between torsional and non–bonded contributions.

In the functional form of other force fields additional terms, such as the cross terms are included. These terms couple internal coordinates, e.g. changes of planar angle are coupled with adjacent bonds stretching. Cross terms are important in many cases, e.g. for a better reproduction of vibrational spectra of molecules.

Cieplak et al.[301] proposed to include an additional term in the formula of Eq. (1-196) that models the induction effects through the infinite order,

$$V_{ind} = -\frac{1}{2} \sum_i \alpha_i \mathbf{F}_i^{(0)} \cdot \mathbf{F}_i, \qquad (1\text{-}197)$$

where α_i is the isotropic polarizability of ith atom, $\mathbf{F}_i$ is the electrostatic field on the atom i due to all other charges and induced dipoles, and $\mathbf{F}_i^{(0)}$ is the electrostatic field on the ith atom due to permanent atomic charges only. Calculations of the infinite order induction terms are relatively time consuming, so in actual calculations the approach of effective charges is more frequently used. In that approach no induction terms are explicitly included in the force field formula, but the charges on atoms are adjusted so as to partially reproduce the induction effects.

It should be stressed that the functional form of the force field, Eq. (1-196) may seem to be oversimplified, and one could think of much more elaborated models. One should keep in mind, however, these force field are mostly used in Monte Carlo or molecular dynamics simulations of biological systems or condensed phases (cf. Section 12.7), and the calculation of V for a given geometry should be as fast as possible.

10. NONADDITIVE INTERACTIONS

In this section we will review the symmetry-adapted perturbation theory of pairwise nonadditive interactions in trimers. This theory was formulated in Ref. (302). We will show that pure three-body polarization and exchange components can be explicitly separated out and that the three-body polarization contributions through the third-order of perturbation theory naturally separate into terms describing the pure induction, mixed induction-dispersion, and pure dispersion interactions.

In the earlier sections of this chapter we reviewed the many-electron formulation of the symmetry-adapted perturbation theory of two-body interactions. As we saw, all physically important contributions to the potential could be identified and computed separately. We follow the same program for the three-body forces and discuss a triple perturbation theory for interactions in trimers. We show how the pure three-body effects can be separated out and give working equations for the components in terms of molecular integrals and linear and quadratic response functions. These formulas have a clear, partly classical, partly quantum mechanical interpretation. The exchange terms are also classified; for the explicit orbital formulas we refer to Ref. (302).

10.1. Supermolecular Approach

We consider the interaction of N monomers, $X_1, X_2, \ldots, X_N$. The interaction energy characterizing the interactions in the cluster is given by,

$$E_{\text{int}} = E_{X_1 X_2 \cdots X_N} - \sum_{I=1}^{N} E_{X_I}, \tag{1-198}$$

where $E_{X_1 X_2 \cdots X_N}$ is the total clamped-nuclei energy of the cluster and E_{X_I}, $I = 1, \ldots, N$, is the energy of the monomer X_I. We follow the common usage and indicate an m-body contribution to the interaction energy of an n-body cluster $(m \leq n)$ by $E_{\text{int}}(m, n)$. The interaction energy of the N-body cluster can be represented by the following many-body expansion,

$$E_{\text{int}} = E_{\text{int}}[2, N] + E_{\text{int}}[3, N] + \ldots + E_{\text{int}}[N, N]. \tag{1-199}$$

Obviously, such an expansion of the interaction energy can be defined without ambiguity only when the quantum states of all subsystems can rigorously be specified. For strongly interacting systems, such as metals or chemically bound molecules, this condition is not fulfilled and the suitability of the many-body expansion can be questioned. See Refs. (303–307) for extensive studies of nonadditive effects in clusters. The K-body contribution to the N-mer energy can be written directly in terms of the total energies of all subclusters containing up to K monomers [308],

$$E_{\text{int}}[K, N] = \sum_{I=1}^{K}(-1)^{K-I}\binom{N-I}{K-I} S_{\text{tot}}[I, N], \tag{1-200}$$

where $S_{\text{tot}}[I, N]$ is the sum of the total energies of all I-monomer subclusters of the whole N-mer.

Eq. (1-200) can be applied to compute $E_{\text{int}}[K, N]$ within the supermolecular approach [5]. However, the number of calculations that need to be performed increases rapidly with N. Moreover, there is a substantial loss of significant figures resulting from the performed subtractions, and care should taken when performing supermolecule calculations.

10.2.   Perturbation Theory of Three-Body Interactions

We consider the closed-shell systems A, B, and C. The total Hamiltonian of the trimer ABC can be written as,

$$H = H_0 + V^{AB} + V^{BC} + V^{CA}, \tag{1-201}$$

where $H_0 = H^A + H^B + H^C$ is the sum of Hamiltonians of the isolated monomers, the operator V^{XY} collects all Coulombic interactions between electrons and nuclei of monomers X and Y. The parametrized interaction energy of the trimer is defined by the asymmetric energy expression

$$E_{\text{int}}(\zeta, \eta, \chi) = \langle \Phi_0 \mid \zeta V^{AB} + \eta V^{BC} + \chi V^{CA} \mid \Psi(\zeta, \eta, \chi)\rangle, \tag{1-202}$$

where $\Phi_0 = \Phi_0^A \Phi_0^B \Phi_0^C$ is the ground state eigenvalue of the unperturbed Hamiltonian H_0, and Φ_0^X denotes the ground state wave function of the monomer X, and the parameters ζ, η, and χ have the physical value of unity. The function $\Psi(\zeta, \eta, \chi)$ is the exact solution of the Schrödinger equation of the trimer with the Hamiltonian (1-201).

The interaction energy and wave function are expanded in the usual manner as a power series in the three perturbation parameters ζ, η, and χ,

$$E_{\text{int}}(\zeta, \eta, \chi) = \sum_{i,j,k} \zeta^i \eta^j \chi^k E_{\text{pol}}^{(ijk)}, \quad \Psi(\zeta, \eta, \chi) = \sum_{i,j,k} \zeta^i \eta^j \chi^k \Phi_{\text{pol}}^{(ijk)}.$$

$$\tag{1-203}$$

Hence $E_{\text{pol}}^{(ijk)}$ and $\Phi_{\text{pol}}^{(ijk)}$ denote the polarization energy and wave function of ith-order in V^{AB}, jth-order in V^{BC}, and kth-order in V^{CA}. The energy and wave function perturbation corrections are solutions of triple Rayleigh-Schrödinger (RS) perturbation equations[88,309]; see Ref. (302) for the explicit form of these equations.

As discussed above, the polarization expansion neglects the exchange effects. We saw in the earlier sections that the simple symmetrized Rayleigh-Schrödinger (SRS) perturbation theory[87,89] shows satisfactory convergence properties (see Ref. (310) for a computational study of the convergence properties for the quartet state of H_3) and can be applied in practice to many-electron systems. The expressions for the energy corrections $E^{(ijk)}$ are obtained by expanding the function,

$$\mathcal{E}_{\text{int}}(\zeta, \eta, \chi) = \frac{\langle \Psi_0 \mid (\zeta V^{AB} + \eta V^{BC} + \chi V^{CA}) \mathcal{A} \Psi(\zeta, \eta, \chi) \rangle}{\langle \Psi_0 \mid \mathcal{A} \Psi(\zeta, \eta, \chi) \rangle}, \qquad (1\text{-}204)$$

where $\mathcal{A}$ is the full antisymmetrizer of the trimer, as a triple perturbation expansion,

$$\mathcal{E}_{\text{int}}(\zeta, \eta, \chi) = \sum_{i,j,k} \zeta^i \eta^j \chi^k E^{(ijk)}, \qquad (1\text{-}205)$$

and $E^{(000)} = 0$. Note that we consider the interaction of three closed-shell monomers, so the projection operator $''\mathcal{A}$ is simply the antisymmetrizer, and the symmetry index ν can be suppressed. Explicit expressions for $E^{(ijk)}$ in the SRS theory are reported in Ref. (302). The exchange contribution to the interaction energy in each order can now be defined by

$$E_{\text{exch}}^{(ijk)} \equiv E^{(ijk)} - E_{\text{pol}}^{(ijk)}. \qquad (1\text{-}206)$$

The expressions for $E_{\text{pol}}^{(ijk)}$ and $E_{\text{exch}}^{(ijk)}$ contain both pair-wise additive and three-body contributions to the interaction energy, and pure three-body terms have to be separated out. Thus, the polarization and exchange contributions to the interaction energy can be decomposed into $E_{\text{pol}}^{(ijk)}(2, 3)$ and $E_{\text{exch}}^{(ijk)}(2, 3)$, and $E_{\text{pol}}^{(ijk)}(3, 3)$ and $E_{\text{exch}}^{(ijk)}(3, 3)$, respectively. By definition, all terms $E_{\text{pol}}^{(ijk)}$ and $E_{\text{exch}}^{(ijk)}$ with $ij + ik + jk \neq 0$, i.e., with at least two non-zero perturbation orders, are pure three-body contributions,

$$E_{\text{pol}}^{(ijk)} = E_{\text{pol}}^{(ijk)}(3, 3), \qquad E_{\text{exch}}^{(ijk)} = E_{\text{exch}}^{(ijk)}(3, 3), \qquad ij + ik + jk \neq 0, \qquad (1\text{-}207)$$

while all contributions $E_{\text{pol}}^{(i00)}$, $E_{\text{pol}}^{(0j0)}$, and $E_{\text{pol}}^{(00k)}$ are purely additive, i.e.

$$E_{\text{pol}}^{(i00)} = E_{\text{pol}}^{(i00)}(2, 3), \quad E_{\text{pol}}^{(0j0)} = E_{\text{pol}}^{(0j0)}(2, 3), \quad E_{\text{pol}}^{(00k)} = E_{\text{pol}}^{(00k)}(2, 3). \qquad (1\text{-}208)$$

By contrast, the exchange contributions $E_{\text{exch}}^{(i00)}$, $E_{\text{exch}}^{(0j0)}$, and $E_{\text{exch}}^{(00k)}$ are not necessarily additive, since in general the antisymmetrizer will interchange the coordinates of electrons belonging to all three monomers. The two-body contribution to $E_{\text{exch}}^{(i00)}$ can formally be identified by moving in the expression for $E_{\text{exch}}^{(i00)}$ the third monomer to infinity, i.e.

$$E_{\text{exch}}^{(i00)}(2,3) = \lim_{R_{BC}\to\infty} \lim_{R_{CA}\to\infty} E_{\text{exch}}^{(i00)}\big|_{R_{AB}=\text{const.}} \cdot \tag{1-209}$$

Since the polarization wave functions $\Phi_{\text{pol}}^{(i00)}$ defining $E_{\text{exch}}^{(i00)}$ are purely additive, i.e. $\Phi_{\text{pol}}^{(i00)} = \Phi_{\text{pol}}^{(i00)}(2,3)$, the two-body term as defined by Eq. (1-209) is equal to $E_{\text{exch}}^{(i00)}$ as defined by the SRS theory of two-body interactions[89]. Thus, to extract pure three-body contribution to $E_{\text{exch}}^{(i00)}$ one has to subtract the $E_{\text{exch}}^{(i00)}$ term of the two-body SRS theory[89].

10.3. Physical Interpretation of the Polarization Effects

The electrostatic energy $E_{\text{elst}}^{(1)}$ is additive, so the first nonadditive polarization contribution is given by the second-order induction term,

$$E_{\text{pol}}^{(2)} = E_{\text{ind}}^{(110)} + E_{\text{ind}}^{(101)} + E_{\text{ind}}^{(011)}, \tag{1-210}$$

where, e.g., $E_{\text{ind}}^{(110)}$ is explicitly given by expression in terms of the linear response function of Eq. (1-83),

$$E_{\text{ind}}^{(110)} = (\omega_A)_m^n (\omega_C)_{m'}^{n'} \Pi_{nn'}^{mm'}(0). \tag{1-211}$$

As before, m and n label arbitrary orbitals on B. Similar expressions for $E_{\text{ind}}^{(101)}$, etc., can be easily found by a proper permutation of A, B, and C. The operator ω_X, X=A, B, or C, is explicitly defined by

$$\Omega_X^Y = \sum_{i=1}^{N_Y} \omega_X(\mathbf{r}_i), \tag{1-212}$$

where

$$\omega_X(\mathbf{r}_i) = v_X(\mathbf{r}_i) + \int \rho_X(\mathbf{r}_j) r_{ij}^{-1} d\mathbf{r}_j, \tag{1-213}$$

N_Y is the number of electrons of the monomer Y, $v_X(\mathbf{r}_i)$ describes the interaction of the ith electron of the monomer Y with the nuclei of the monomer X, and $\rho_X(\mathbf{r}_j)$ is the electron density of the monomer X.

Equation (1-211) clearly shows that $E_{\text{ind}}^{(110)}$ represents the three-body induction term corresponding to the interaction of the electrostatic field of the molecule C with the molecule B polarized by the electrostatic field of A. In the multipole approximation this contribution can be interpreted as the result of the interaction of permanent moments of monomer C with moments induced on monomer B by the electrostatic field of the monomer A.

Note that Eq. (1-210) represents the only second-order non-additive polarization term. The remaining non-additive polarization terms are all of third-order at least, i.e., in Eq. (1-203): $i + j + k \geq 3$. The three-body induction energy, $E_{\text{ind}}^{(3)}$, is defined as that part of $E_{\text{pol}}^{(3)}$ that can be obtained by complete neglect of the *intermonomer* correlation effects. The difference $E_{\text{pol}}^{(3)} - E_{\text{ind}}^{(3)}$ represents all intermonomer correlation effects, and separates into contributions due to pure third-order dispersion interactions and to the coupling of the second-order dispersion interaction with the induction interaction:

$$E_{\text{pol}}^{(3)} = E_{\text{disp}}^{(3)} + E_{\text{ind-disp}}^{(3)} + E_{\text{ind}}^{(3)}. \tag{1-214}$$

The dispersion nonadditivity $E_{\text{disp}}^{(3)}$ arises from the coupling of intermonomer pair correlations in subsystems XY and YZ via the intermolecular interaction operator V^{ZX}. This contribution can be expressed as a generalized Casimir-Polder formula,

$$
\begin{aligned}
E_{\text{disp}}^{(3)} = -\frac{1}{2\pi} v_{ln}^{km} v_{n'p}^{m'x} v_{p'l'}^{x'k'} \\
\times \int_{-\infty}^{+\infty} \Pi_{kk'}^{ll'}(-i\omega)\Pi_{mm'}^{nn'}(i\omega)\Pi_{xx'}^{pp'}(i\omega)d\omega.
\end{aligned}
\tag{1-215}
$$

The orbitals p and x are on C and, as before, k and l are on A and m, n on B. For the interaction of three spherically symmetric atoms the third-order dispersion nonadditivity contains the famous Axilrod-Teller-Muto triple-dipole interaction[311,312].

The induction-dispersion contribution, in turn, can be interpreted as the energy of the (second-order) dispersion interaction of the monomer X with the monomer Y deformed by the electrostatic field of the monomer Z (note that we have six such contributions). In particular, when X=A, Y=B, and Z=C the corresponding induction-dispersion contribution in terms of response functions is given by,

$$
\begin{aligned}
E_{\text{ind-disp}}^{(210)} = -\frac{1}{4\pi} (\omega_C)_n^m v_{l_1 n_1}^{k_1 m_1} v_{l_2 n_2}^{k_2 m_2} \\
\times \int_{-\infty}^{+\infty} \Pi_{k_1 k_2}^{l_1 l_2}(i\omega)\Pi_{m_1 m_2 m}^{n_1 n_2 n}(-i\omega, 0)d\omega.
\end{aligned}
\tag{1-216}
$$

Similar expressions for $E_{\text{ind-disp}}^{(120)}$, etc., can be easily found by a proper permutation of symbols pertaining to monomers A, B, and C.

The mechanism of the third-order three-body induction interactions is somewhat more complicated. It can be shown that one can distinguish three principal categories. The first mechanism is simply the interaction of permanent moments on the monomer C with the moments induced on B by the nonlinear (second-order) effect of the electrostatic potential of the monomer A plus contributions obtained by interchanging the roles of the monomers A and C. In terms of quadratic response function it takes the form,

$$E_{\text{ind}}^{(210)}(B \leftarrow A, C) = \frac{1}{2}(\omega_A)_n^m (\omega_A)_{n'}^{m'} (\omega_C)_{n''}^{m''} \Pi_{mm'm''}^{nn'n''}(0,0). \qquad (1\text{-}217)$$

Note that we have six contributions of this type corresponding to six possible permutations of the indices A, B, and C.

The second mechanism is the interaction between the multipole moments induced on A and C by the electrostatic potential of the monomer B. The induction energy component corresponding to this particular interaction will be denoted by $E_{\text{ind}}^{(111)}(A \leftarrow B; C \leftarrow B)$, and can be written as,

$$E_{\text{ind}}^{(111)}(A \leftarrow B; C \leftarrow B) = (\omega_B)_i^k (\omega_B)_p^x v_{p'l'}^{x'k'} \Pi_{kk'}^{ll'}(0) \Pi_{xx'}^{pp'}(0). \qquad (1\text{-}218)$$

Since by definition $E_{\text{ind}}^{(111)}(A \leftarrow B; C \leftarrow B) = E_{\text{ind}}^{(111)}(C \leftarrow B; A \leftarrow B)$, we have three contributions of this kind.

The third mechanism corresponds to the interaction of multipole moments induced in monomers B and C by the electrostatic potentials of monomers A and B, respectively:

$$E_{\text{ind}}^{(210)}(A \leftarrow B; B \leftarrow C) = (\omega_B)_i^k (\omega_C)_n^m v_{l'n'}^{k'm'} \Pi_{kk'}^{ll'}(0) \Pi_{mm'}^{nn'}(0). \qquad (1\text{-}219)$$

Again we have six contributions of this type corresponding to six possible permutations of the indices A, B, and C.

We wish to end this section by saying that similarly as in the two-body case, non-additive induction, induction-dispersion, and dispersion terms have well defined asymptotic behaviors from the multipole expansions of the intermolecular interaction operators. For instance, the leading term in the multipole expansion of the three-body dispersion energy for three atoms in a triangular geometry is given by the famous Axilrod-Teller-Muto formula[311,312],

$$E_{\text{disp}}^{(3)}(3,3) \sim \frac{C_9}{R_{AB}^3 R_{BC}^3 R_{CA}^3}(3\cos\theta_A \cos\theta_B \cos\theta_C + 1), \qquad (1\text{-}220)$$

where

$$C_9 = \frac{3}{\pi} \int_0^\infty \alpha_A(i\omega)\alpha_B(i\omega)\alpha_C(i\omega)d\omega. \qquad (1\text{-}221)$$

In Eq. (1-220) R_{XY} denotes the distance between the atoms X and Y, while θ_A, θ_B, and θ_C are the inner angles in the triangle ABC. General, open-ended formulas for the multipole-expanded induction, induction-dispersion, and dispersion energies through the third order are reported in Ref. (302). Specific applications to the Ar_2–HF trimer and comparison of the multipole-expanded and nonexpanded results is given in Ref. (313).

10.4.   Exchange Effects

In order to arrive at a closed expression for $E_{\text{exch}}^{(ijk)}$ the total antisymmetrizer of the trimer is approximated by

$$\mathcal{A} \approx \frac{(N_A + N_B + N_C)!}{N_A! N_B! N_C!} \mathcal{A}^A \mathcal{A}^B \mathcal{A}^C (1 + \mathcal{P}), \tag{1-222}$$

where N_X is the number of electrons of the monomer X, $\mathcal{A}^X$ is the antisymmetrizer for the monomer X, and the operator $\mathcal{P}$ collects all intermolecular two and three cycle permutations,

$$\mathcal{P} = P^{AB} + P^{BC} + P^{CA} + P^{ABC}, \tag{1-223}$$

with

$$P^{XY} = -\sum_{i \in X} \sum_{j \in Y} P_{ij} \quad \text{and} \quad P^{XYZ} = \sum_{i \in X} \sum_{j \in Y} \sum_{k \in Z} \left(P_{ijk} + P_{jik} \right). \tag{1-224}$$

The leading first-order exchange nonadditivity is given by the sum,

$$E_{\text{exch}}^{(1)} = E_{\text{exch}}^{(100)} + E_{\text{exch}}^{(010)} + E_{\text{exch}}^{(001)}, \tag{1-225}$$

where the first term is

$$E_{\text{exch}}^{(100)} = \langle \Phi_0 \mid V^{AB}(Q^{AB} - \langle Q^{AB} \rangle) \mid \Phi_0 \rangle \tag{1-226}$$

and due to the truncation of the antisymmetrizer

$$Q^{AB} = \mathcal{P} - P^{AB} \quad \text{and} \quad \langle Q^{AB} \rangle = \langle \Phi_0 \mid Q^{AB} \mid \Phi_0 \rangle. \tag{1-227}$$

The other terms of Eq. (1-225) follow by permutation of A, B and C. Similarly as in the two-body case the first-order exchange nonadditivity can be expressed through one- and two-particle density matrices of the isolated monomers[314],

$$E_{\text{exch}}^{(1)}(3, 3) = \int \Big(\bar{v}^{AB}(\mathbf{r}, \mathbf{r}') \gamma_{\text{int}}^{ABC}(\mathbf{q}, \mathbf{q}')$$

$$+ \bar{v}^{BC}(\mathbf{r}, \mathbf{r}') \gamma_{\text{int}}^{BCA}(\mathbf{q}, \mathbf{q}')$$

$$+ \bar{v}^{CA}(\mathbf{r}, \mathbf{r}') \gamma_{\text{int}}^{CAB}(\mathbf{q}, \mathbf{q}') \Big) d\mathbf{q} d\mathbf{q}' \tag{1-228}$$

where $\bar{v}^{AB}(\mathbf{r}, \mathbf{r}') = \tilde{v}^{AB}(\mathbf{r}, \mathbf{r}') - E_{\text{elst}}^{(100)}/N_A N_B$, and similar definitions hold for $\bar{v}^{BC}(\mathbf{r}, \mathbf{r}')$ and $\bar{v}^{CA}(\mathbf{r}, \mathbf{r}')$. Finally, the interaction-density matrix, $\gamma_{\text{int}}^{ABC}$, is given by

$$\gamma_{\text{int}}^{ABC}(\mathbf{q}, \mathbf{q}') = -\rho_A(\mathbf{q}|\mathbf{q}) \int \rho_B(\mathbf{q}'|\mathbf{q}'')\rho_C(\mathbf{q}''|\mathbf{q}')d\mathbf{q}''$$

$$- \rho_B(\mathbf{q}'|\mathbf{q}') \int \rho_A(\mathbf{q}|\mathbf{q}'')\rho_C(\mathbf{q}''|\mathbf{q})d\mathbf{q}''$$

$$+ \rho_A(\mathbf{q}|\mathbf{q}') \int \rho_B(\mathbf{q}'|\mathbf{q}'')\rho_C(\mathbf{q}''|\mathbf{q})d\mathbf{q}''$$

$$+ \rho_B(\mathbf{q}'|\mathbf{q}) \int \rho_A(\mathbf{q}|\mathbf{q}'')\rho_C(\mathbf{q}''|\mathbf{q}')d\mathbf{q}''$$

$$- \rho_A(\mathbf{q}|\mathbf{q}) \int \Gamma_B(\mathbf{q}'\mathbf{q}''|\mathbf{q}'\mathbf{q}''')\rho_C(\mathbf{q}'''|\mathbf{q}'')d\mathbf{q}''d\mathbf{q}'''$$

$$- \rho_B(\mathbf{q}'|\mathbf{q}') \int \Gamma_A(\mathbf{q}\mathbf{q}''|\mathbf{q}\mathbf{q}''')\rho_C(\mathbf{q}'''|\mathbf{q}'')d\mathbf{q}''d\mathbf{q}'''$$

$$+ \int \rho_A(\mathbf{q}|\mathbf{q}'')\Gamma_B(\mathbf{q}'\mathbf{q}'''|\mathbf{q}'\mathbf{q})\rho_C(\mathbf{q}''|\mathbf{q}''')d\mathbf{q}''d\mathbf{q}'''$$

$$+ \int \rho_A(\mathbf{q}|\mathbf{q}'')\Gamma_B(\mathbf{q}'\mathbf{q}''|\mathbf{q}'\mathbf{q}''')\rho_C(\mathbf{q}'''|\mathbf{q})d\mathbf{q}''d\mathbf{q}'''$$

$$+ \int \Gamma_A(\mathbf{q}\mathbf{q}''|\mathbf{q}\mathbf{q}''')\rho_B(\mathbf{q}'|\mathbf{q}'')\rho_C(\mathbf{q}'''|\mathbf{q}')d\mathbf{q}''d\mathbf{q}'''$$

$$+ \int \Gamma_A(\mathbf{q}\mathbf{q}''|\mathbf{q}\mathbf{q}')\rho_B(\mathbf{q}'|\mathbf{q}''')\rho_C(\mathbf{q}'''|\mathbf{q}'')d\mathbf{q}''d\mathbf{q}'''$$

$$+ \int \Gamma_A(\mathbf{q}\mathbf{q}''|\mathbf{q}\mathbf{q}''')\Gamma_B(\mathbf{q}'\mathbf{q}_1|\mathbf{q}'\mathbf{q}'')\rho_C(\mathbf{q}'''|\mathbf{q}_1)d\mathbf{q}''d\mathbf{q}'''d\mathbf{q}_1$$

$$+ \int \Gamma_A(\mathbf{q}\mathbf{q}''|\mathbf{q}\mathbf{q}''')\Gamma_B(\mathbf{q}'\mathbf{q}'''|\mathbf{q}'\mathbf{q}_1)\rho_C(\mathbf{q}_1|\mathbf{q}'')d\mathbf{q}''d\mathbf{q}'''d\mathbf{q}_1. \tag{1-229}$$

In order to consider higher-order exchange effects, we write the first-order wave function $\Phi_{\text{pol}}^{(100)}$ as,

$$\Phi_{\text{pol}}^{(100)} = \Phi_{\text{ind}}^{(1)}(A \leftarrow B)\Phi_0^B\Phi_0^C + -12pt\Phi_0^A\Phi_{\text{ind}}^{(1)}(B \leftarrow A)\Phi_0^C$$

$$+ \Phi_{\text{disp}}^{(1)}(A \cdots B)\Phi_0^C, \tag{1-230}$$

where $\Phi_{\text{ind}}^{(1)}(X \leftarrow Y)$ is the standard induction wave function corresponding to the polarization of the monomer X by the monomer Y, and $\Phi_{\text{disp}}^{(1)}(X \cdots Y)$ is the dispersion wave function for the pair XY[1,315].

The second-order exchange nonadditivity splits into exchange-induction, $E_{\text{exch-ind}}^{(2)}$, and exchange-dispersion, $E_{\text{exch-disp}}^{(2)}$, three-body contributions:

$$E_{\text{exch-ind}}^{(200)} = \langle \Phi_0 | (V^{AB} - \langle V^{AB}\rangle)Q^{AB}\Phi_{\text{ind}}^{(1)}(A \leftarrow B)\Phi_0^B\Phi_0^C \rangle$$

$$- \langle \Phi_0 | (V^{AB} - \langle V^{AB}\rangle)\langle Q^{AB}\rangle\Phi_{\text{ind}}^{(1)}(A \leftarrow B)\Phi_0^B\Phi_0^C \rangle$$

$$- \langle \Phi_0 | (V^{AB} - \langle V^{AB}\rangle)\langle Q^{AB}\rangle\Phi_0^A\Phi_{\text{ind}}^{(1)}(B \leftarrow A)\Phi_0^C \rangle,$$

$$+ \langle \Phi_0 | (V^{AB} - \langle V^{AB}\rangle)Q^{AB}\Phi_0^A\Phi_{\text{ind}}^{(1)}(B \leftarrow A)\Phi_0^C \rangle, \tag{1-231}$$

$$E_{\text{exch-ind}}^{(110)} = \langle \Phi_0 \mid (V^{AB} - \langle V^{AB} \rangle) P \Phi_0^A \Phi_{\text{ind}}^{(1)} (B \leftarrow C) \Phi_0^C \rangle$$

$$- \langle \Phi_0 \mid (V^{AB} - \langle V^{AB} \rangle) \langle P \rangle \Phi_0^A \Phi_{\text{ind}}^{(1)} (B \leftarrow C) \Phi_0^C \rangle$$

$$+ \langle \Phi_0 \mid (V^{AB} - \langle V^{AB} \rangle) P \Phi_0^A \Phi_0^B \Phi_{\text{ind}}^{(1)} (C \leftarrow B) \rangle$$

$$- \langle \Phi_0 \mid (V^{AB} - \langle V^{AB} \rangle) \langle P \rangle \Phi_0^A \Phi_0^B \Phi_{\text{ind}}^{(1)} (C \leftarrow B) \rangle$$

$$+ \langle \Phi_0 \mid (V^{BC} - \langle V^{BC} \rangle) P \Phi_{\text{ind}}^{(1)} (A \leftarrow B) \Phi_0^B \Phi_0^C \rangle$$

$$- \langle \Phi_0 \mid (V^{BC} - \langle V^{BC} \rangle) \langle P \rangle \Phi_{\text{ind}}^{(1)} (A \leftarrow B) \Phi_0^B \Phi_0^C \rangle$$

$$- \langle \Phi_0 \mid (V^{BC} - \langle V^{BC} \rangle) \langle P \rangle \Phi_0^A \Phi_{\text{ind}}^{(1)} (B \leftarrow A) \Phi_0^C \rangle$$

$$+ \langle \Phi_0 \mid (V^{BC} - \langle V^{BC} \rangle) P \Phi_0^A \Phi_{\text{ind}}^{(1)} (B \leftarrow A) \Phi_0^C \rangle, \qquad (1\text{-}232)$$

$$E_{\text{exch-disp}}^{(200)} = - \langle \Phi_0 \mid (V^{AB} - \langle V^{AB} \rangle) \langle Q^{AB} \rangle \Phi_{\text{disp}}^{(1)} (A \cdots B) \Phi_0^C \rangle$$

$$+ \langle \Phi_0 \mid (V^{AB} - \langle V^{AB} \rangle) Q^{AB} \Phi_{\text{disp}}^{(1)} (A \cdots B) \Phi_0^C \rangle \qquad (1\text{-}233)$$

$$E_{\text{exch-disp}}^{(110)} = \langle \Phi_0 \mid (V^{AB} - \langle V^{AB} \rangle) P \Phi_0^A \Phi_{\text{disp}}^{(1)} (B \cdots C) \rangle$$

$$- \langle \Phi_0 \mid (V^{AB} - \langle V^{AB} \rangle) \langle P \rangle \Phi_0^A \Phi_{\text{disp}}^{(1)} (B \cdots C) \rangle$$

$$- \langle \Phi_0 \mid (V^{BC} - \langle V^{BC} \rangle) \langle P \rangle \Phi_{\text{disp}}^{(1)} (A \cdots B) \Phi_0^C \rangle$$

$$+ \langle \Phi_0 \mid (V^{BC} - \langle V^{BC} \rangle) P \Phi_{\text{disp}}^{(1)} (A \cdots B) \Phi_0^C \rangle. \qquad (1\text{-}234)$$

11. SYMMETRY-ADAPTED PERTURBATION THEORY OF THE INTERACTION-INDUCED PROPERTIES

Potential energy surfaces of weakly bound dimers and trimers are the key quantities needed to compute transition frequencies in the high resolution spectra, (differential and integral) scattering cross sections or rate coefficients describing collisional processes between the molecules, or some thermodynamic properties needed to derive equations of state for condensed phases. However, some other quantities governed by weak intermolecular forces are needed to describe intensities in the spectra or, more generally, infrared and Raman spectra of unbound (collisional complexes) of two molecules, and dielectric and refractive properties of condensed phases. These are the interaction-induced (or collision-induced) dipole moments and polarizabilities.

During a collision between two molecules the intermolecular interaction leads to distortions of their charge distributions, so that a collisional complex possesses a dipole moment and polarizability in excess of the sum of these properties of the isolated molecules. These excess properties, referred to as the interaction-induced or collision-induced dipole moment and polarizability, are defined as incremental parts of the properties of the complex AB due to intermolecular interactions. So, the interaction-induced dipole moment of a pair of molecules A and B is given by the difference between the dipole moment of the complex AB and the sum of dipole moments of the noninteracting molecules A and B,

$$\Delta \mu_i = \mu_i^{AB} - \mu_i^A - \mu_i^B, \tag{1-235}$$

where μ_i^{AB} is a Cartesian component of the dipole moment of the dimer AB, and μ_i^A and μ_i^B denote components of the dipole moments of the isolated molecules A and B, respectively. Similarly, the interaction-induced polarizability $\Delta \alpha_{ij}$ of a pair of molecules A and B is defined as the excess polarizability of the collisional pair AB due to intermolecular interactions, i.e.

$$\Delta \alpha_{ij} = \alpha_{ij}^{AB} - \alpha_{ij}^A - \alpha_{ij}^B, \tag{1-236}$$

where α_{ij}^{AB} is a Cartesian component of the dimer polarizability tensor, and α_{ij}^A and α_{ij}^B denote components of the polarizability tensors of the isolated monomers A and B, respectively. Equations (1-235) and (1-236) can be conveniently rewritten by using the Hellmann-Feynman theorem,

$$\Delta \mu_i = - \left(\frac{\partial E_{int}}{\partial F_i} \right)_{\mathbf{F}=0}, \tag{1-237}$$

$$\Delta \alpha_{ij} = - \left(\frac{\partial^2 E_{int}}{\partial F_i \partial F_j} \right)_{\mathbf{F}=0}, \tag{1-238}$$

where E_{int} denotes here the interaction energy for the dimer AB in the presence of a static, uniform electric field $\mathbf{F}$. Equations (1-237) and (1-238) show that the interaction-induced dipole moment and polarizability can be obtained from standard finite field calculations, if the field-dependent interaction energy can be computed using, e.g. the symmetry-adapted perturbation theory. Subsequently the interaction-induced dipole moments and polarizabilities can be obtained from finite difference formulas.

In view of Eqs. (1-237) and (1-238), the components of the interaction-induced dipole moment and polarizability can written as,

$$\Delta \mu_i = \Delta \mu_{i,\mathrm{pol}}^{(1)} + \Delta \mu_{i,\mathrm{exch}}^{(1)} + \Delta \mu_{i,\mathrm{ind}}^{(2)} + \Delta \mu_{i,\mathrm{disp}}^{(2)} + \Delta \mu_{i,\mathrm{exch}}^{(2)} + \cdots, \tag{1-239}$$

$$\Delta \alpha_{ij} = \Delta \alpha_{ij,\mathrm{pol}}^{(1)} + \Delta \alpha_{ij,\mathrm{exch}}^{(1)} + \Delta \alpha_{ij,\mathrm{ind}}^{(2)} + \Delta \alpha_{ij,\mathrm{disp}}^{(2)} + \Delta \alpha_{ij,\mathrm{exch}}^{(2)} + \cdots,$$

$$\tag{1-240}$$

where the superscripts indicate the order in the intermolecular potential. Obviously an nth order contribution to $\Delta\mu_i$ or $\Delta\alpha_{ij}$ is obtained by differentiating once or twice the corresponding contribution of the nth order SAPT interaction energy.

Equation (1-239) relates the interaction-induced part of the dipole moment of the complex AB to the distortion of the electron density associated with the electrostatic, exchange, induction, and dispersion interactions between the monomers. The polarization contributions to the dipole moment through the second-order of perturbation theory ($\Delta\mu_{i,\text{pol}}^{(1)}$, $\Delta\mu_{i,\text{ind}}^{(2)}$, and $\Delta\mu_{i,\text{disp}}^{(2)}$) have an appealing, partly classical, partly quantum, physical interpretation. The first-order multipole-expanded polarization contribution $E_{\text{pol}}^{(1)}(\mathbf{F})$ is due to the interactions of permanent multipole moments on A with moments induced on B by the external field $\mathbf{F}$, and vice versa. The terms linear in $\mathbf{F}$ give $\Delta\mu_{i,\text{pol}}^{(1)}$. The mechanism that yields the second-order induction dipole $\Delta\mu_{i,\text{ind}}^{(2)}$ is somewhat more complicated, and one can distinguish two principal categories. The first mechanism is the interaction of a permanent multipole on monomer A with a multipole on B induced by the nonlinear (second-order) effect of both a permanent multipole on A and the external field $\mathbf{F}$ (plus a contribution obtained by interchanging the roles of the monomers A and B). The second mechanism is the interaction of a multipole moment on A, induced by a permanent multipole on B, with a moment on B induced by the external field $\mathbf{F}$, and vice versa. Again, the energy terms that are linear in $\mathbf{F}$ give the corresponding interaction induced dipoles. Finally, the dispersion term $\Delta\mu_{i,\text{disp}}^{(2)}$ represents the intermonomer correlation contribution to the dipole moment of the dimer AB. Various physical contributions to the interaction-induced polarizability can be classified analogously. The polarization, induction, and dispersion contributions to the collision-induced dipole moments and polarizabilities have well defined asymptotics coming from the multipole expansion. See Ref. (163) for general open-ended expressions valid for any collisional pairs.

It is worth noting that the contribution from the kth-order contribution to the collision-induced dipole moment, $\Delta\mu_m^{(k)}$, can be written in terms of the complete orthogonal set of angular functions labeled by $\{\Lambda\} = \{L_A, K_A, L_B, K_B, L\}$, λ and m,

$$\Delta\mu_m^{(k)} = \frac{1}{\sqrt{3}} \sum_{\{\Lambda\}} \sum_{\lambda} d_{\{\Lambda\}\lambda}^{(k)}(R) \, A_{\{\Lambda\}m}^{(\lambda)1}(\omega_A, \omega_B, \widehat{\mathbf{R}}), \tag{1-241}$$

where m label the spherical rather than Cartesian component of the collision-induced dipole moment, and the angular function $A_{\{\Lambda\}m}^{(\lambda)1}(\omega_A, \omega_B, \widehat{\mathbf{R}})$ is given by,

$$A_{\{\Lambda\}m}^{(\lambda)1}(\omega_A, \omega_B, \widehat{\mathbf{R}}) = \sum_{M_A=-L_A}^{L_A} \sum_{M_B=-L_B}^{L_B} \sum_{\mu=-\lambda}^{\lambda} \sum_{M=-L}^{L} \langle L_A M_A; L_B M_B | \lambda\mu \rangle \langle \lambda, \mu; L, M | l, m \rangle$$

$$\times D_{M_A,K_A}^{(L_A)^*}(\omega_A) D_{M_B,K_B}^{(L_B)^*}(\omega_B) C_L^M(\widehat{\mathbf{R}}). \tag{1-242}$$

In Ref. (163) the techniques introduced above were illustrated for He–He and He–H_2 collisional complexes. In Refs. (316,317) the property functions were applied in full quantum-statistical calculations of the dielectric second virial coefficient and of the polarized and depolarized Raman spectrum of the He gas. Some further applications were reported by Rizzo and collaborators[318–322].

12. FROM INTERMOLECULAR POTENTIALS AND COLLISION-INDUCED PROPERTIES TO THE MEASURED PROPERTIES OF ISOLATED COMPLEXES AND CONDENSED PHASES

In the previous sections we gave a detailed overview of the theory of intermolecular forces applied to interaction potentials and collision-induced properties. The starting point for these investigations is the electronic Schrödinger equation in the Born-Oppenheimer approximation. The solution of this equation by a supermolecule method or by symmetry-adapted perturbation theory provides us with the inter-action potential and interaction-induced dipole and polarizability functions needed to solve the nuclear motion problem—the second step in the Born-Oppenheimer approximation—and to compute the quantities measured in high-resolution spectro-scopic and scattering experiments or characterizing the condensed phases. In this section we give a brief description of the theoretical methods needed on the route from the intermolecular potentials and properties to rovibrational spectra, collision-induced Raman spectra, collision cross sections (thermodynamic and dielectric) second virial coefficients, and properties of condensed phases.

12.1. Collision-induced Raman Light Scattering in Atomic Gas

In the collision-induced Raman experiments the laser light of wavenumber ω_0 is scattered inelastically by the interacting atoms in the gas. The intensities of the depolarized, $D(\nu)$, and polarized, $P(\nu)$, scattered light are given by[323,324],

$$D(\nu) = \frac{2}{15}\omega^3\omega_0 G(\nu), \qquad P(\nu) = \omega^3\omega_0 A(\nu), \tag{1-243}$$

where ν is the frequency shift, $\omega = \omega_0 - 2\pi\nu/c$, and c is the speed of light. The frequency shift ν is negative for the Stokes and positive for the anti-Stokes bands. The spectral functions $G(\nu)$ and $A(\nu)$ can be written as,

$$G(\nu) = \frac{2hc\lambda_B^3}{(2I+1)^2}\sum_{J,J'} g_J(2J+1)$$

$$\times \int_0^\infty e^{-E/k_BT} b_{J'}^J |\langle E', J'|\gamma(R)|E, J\rangle|^2 dE, \tag{1-244}$$

$$A(\nu) = \frac{2hc\lambda_B^3}{(2I+1)^2} \sum_J g_J (2J+1)$$

$$\times \int_0^\infty e^{-E/k_B T} |\langle E', J|\alpha(R)|E, J \rangle|^2 dE, \qquad (1\text{-}245)$$

where the trace, $\alpha(R)$, and anisotropy, $\gamma(R)$, are the invariants of the tensor, and are defined by the equations:

$$\alpha = \frac{1}{3}(2\Delta\alpha_{xx} + \Delta\alpha_{zz}), \qquad \gamma = \Delta\alpha_{zz} - \Delta\alpha_{xx}. \qquad (1\text{-}246)$$

Furthermore $E' - E = h\nu$, $J' = J, J \pm 2$, h is the Planck constant, k_B is the Boltzmann constant, T denotes the temperature, $\lambda_B = (h^2/2\pi\mu k_B T)^{\frac{1}{2}}$ is the de Broglie wavelength, μ is the reduced mass of the collisional complex, and I and g_J designate the nuclear spin and nuclear spin statistical weight, respectively. The constants $b_{J'}^J$ are given by

$$b_{J'}^J = (2J'+1)\begin{pmatrix} J' & J & 2 \\ 0 & 0 & 0 \end{pmatrix}^2, \qquad (1\text{-}247)$$

and the matrix elements of the trace and anisotropy of the polarizability appearing in Eqs. (1-244) and (1-245) are defined as,

$$\langle E', J'|X(R)|E, J \rangle = \int_0^\infty \chi_{E'}^{J'*}(R)X(R)\chi_E^J(R)dR, \qquad (1\text{-}248)$$

where the scattering wave functions $\chi_E^J(R)$ are solutions of the radial Schrödinger equation describing the relative motion of the atoms in the potential $V(R)$, subject to the energy normalization condition.

12.2.   Dielectric Second Virial Coefficients of Atomic Gases

It is well known that for atomic gases at low densities the Clausius-Mossotti function can be related to the atomic polarizability via the following virial relation:

$$\frac{\epsilon - 1}{\epsilon + 2} = \frac{4\pi\alpha_0}{3}\rho + B_\epsilon(T)\rho^2 \cdots, \qquad (1\text{-}249)$$

where ϵ is the dielectric constant, α_0 is the atomic polarizability, and ρ denotes the gas number density. At higher pressures deviations from the linear dependence on ρ are observed, and they can be attributed to intermolecular interactions. Buckingham and Pople[325] have shown that the leading correction to Eq. (1-249), quadratic in the gas density, is given by $B_\epsilon(T)\rho^2$, where the second dielectric virial coefficient $B_\epsilon(T)$ is related to the interatomic potential $V(R)$ and interaction-induced polarizability trace by,

$$B_\epsilon(T) = \frac{2\pi}{3} \int_0^\infty \alpha(R) \exp(-V(R)/k_B T)R^2 dR. \qquad (1\text{-}250)$$

At very low temperatures Eq. (1-250) is no longer valid, and one has to use the exact quantum-statistical expression. The quantum equivalent of Eq. (1-250) has been developed in Ref. (317) from the general relation between the second dielectric virial coefficient and the ordinary (thermodynamic) second virial coefficient of an atomic gas in a uniform electric field[317],

$$B_\epsilon(T) = -\frac{4\pi k_B T}{3} \left(\frac{\partial^2 B_2(T; F)}{\partial F^2} \right)_{F=0}, \tag{1-251}$$

where $B_2(T; F)$ denotes the ordinary (pressure) second virial coefficient for the gas in the electric field F. Using Eq. (1-251) and an expression for $B_2(T; F)$ in terms of the field-dependent Slater sum[326], one finds the following formula for $B_\epsilon(T)$[317],

$$B_\epsilon(T) = \frac{2\pi\lambda_B^3}{3} \sum_{J=0}^{\infty} (2J+1) \left[1 + \frac{(-1)^{J+2I}}{2I+1} \right]$$

$$\times \left[\sum_n e^{-E_{nJ}/k_B T} \langle n, J | \alpha(R) | n, J \rangle \right.$$

$$\left. + \frac{2}{\pi} \int_0^\infty e^{-\hbar^2 k^2/2\mu k_B T} \langle E, J | \alpha(R) | E, J \rangle dk \right], \tag{1-252}$$

where

$$\langle n, J | \alpha(R) | n, J \rangle \equiv \int_0^\infty \chi_{nJ}^\star(R) \alpha(R) \chi_{nJ}(R) dR, \tag{1-253}$$

$|\chi_{nJ}\rangle = \psi_{nJ}(R)$ denote the bound-state eigenfunctions of the Schrödinger equation for the relative motion in the potential $V(R)$, with eigenvalues E_{nJ}, and $|E, J\rangle$ are the scattering states with energies $E = \hbar^2 k^2/2\mu$ defined above.

Equation (1-251) can also be used to derive the semiclassical expansion of the second dielectric virial coefficient. Indeed, one may hope that at intermediate temperatures an expansion of $B_\epsilon(T)$ as a power series in $\hbar^2$ will give sufficiently accurate results, making full quantum-statistical calculations unnecessary. Thus, one can approximate $B_\epsilon(T)$ as,

$$B_\epsilon(T) = B_\epsilon^{(0)}(T) + B_\epsilon^{(1)}(T) + B_\epsilon^{(2)}(T), \tag{1-254}$$

where the classical term $B_\epsilon^{(0)}(T)$ is given by the r.h.s. of Eq. (1-250), while the quantum corrections of the order $\hbar^2$ and $\hbar^4$ (denoted by $B_\epsilon^{(1)}(T)$ and $B_\epsilon^{(2)}(T)$, respectively), can be written as[316],

$$B_\epsilon^{(1)}(T) = -\frac{\pi^2 \hbar^2}{9\mu k_B^2 T^2} \int_0^\infty \exp(-V(R)/k_B T)$$

$$\times \left[\frac{\alpha(R)}{k_B T} \left(\frac{dV}{dR} \right)^2 - 2\frac{dV}{dR}\frac{d\alpha}{dR} \right] R^2 dR, \tag{1-255}$$

$$B_\epsilon^{(2)}(T) = \frac{\pi^2 \hbar^4}{180\mu^2 k_B^3 T^3} \int_0^\infty \exp(-V(R)/k_B T)$$

$$\times \left[\frac{\alpha(R)}{k_B T} f(R) + g(R) \right] R^2 dR, \tag{1-256}$$

where the functions $f(R)$ and $g(R)$ are given by[316],

$$f(R) = \left(\frac{d^2 V}{dR^2}\right)^2 + \frac{2}{R^2}\left(\frac{dV}{dR}\right)^2 + \frac{10}{9k_B T}\frac{1}{R}\left(\frac{dV}{dR}\right)^3$$

$$- \frac{5}{36k_B^2 T^2}\left(\frac{dV}{dR}\right)^4, \tag{1-257}$$

$$g(R) = -2\frac{d^2 V}{dR^2}\frac{d^2\alpha}{dR^2} - \frac{4}{R^2}\frac{dV}{dR}\frac{d\alpha}{dR} - \frac{10}{3k_B T}\frac{1}{R}\left(\frac{dV}{dR}\right)^2\frac{d\alpha}{dR}$$

$$+ \frac{5}{9k_B^2 T^2}\left(\frac{dV}{dR}\right)^3\frac{d\alpha}{dR} \tag{1-258}$$

It is worth noting that Eqs. (1-250) and (1-255)–(1-256) can be used to evaluate the full quantum correction even at low temperatures by means of the simplest Padé approximant $[1/1]$[317].

12.3. Refractive (Kerr) Second Virial Coefficients of Atomic Gases

Similarly as the trace, the anisotropy of the polarizability tensor of diatomic collisional systems can also be related to some macroscopic properties, namely to the refractive properties of atomic gases. The so-called Kerr constant, the anisotropy of the refractive index in the parallel and perpendicular directions to the external static electric field is given by,

$$\lim_{F \to 0} \frac{2(n_\parallel - n_\perp)}{27F^2} = \frac{4\pi\gamma_0}{81}\rho + B_K(T)\rho^2 + C_K(T)\rho^3 + \cdots, \tag{1-259}$$

where F is the external electric field in the z direction, $n_\parallel$ and $n_\perp$ denote the refraction coefficients in the paralell and perpendicular directions to the field $\mathbf{F}$, γ_0 is the atomic hyperpolarizability, and B_K and C_K are the second and third Kerr virial coefficients. At high temperatures the second Kerr virial coefficient can be related to the anisotropy of the collision-induced polarizability tensor[327],

$$B_K(T) \approx B_K^{(0)}(T) = \frac{2\pi}{81}\int_0^\infty \frac{(\gamma(R))^2}{5k_B T}e^{-V(R)/k_B T}R^2 dR. \tag{1-260}$$

A quantum expression and a semi-classical expansion can be derived[328], as in the case of the dielectric virial coefficient. We wish to end this short section by saying that recently Rizzo and collaborators reported a general virial expansion of various properties of atomic gases[322].

12.4.   Rovibrational Spectra of Weakly Bound Complexes

High-resolution spectroscopic experiments provide a detailed experimental information on the shape of the intermolecular potential in the attractive regions. Recent improvements in supersonic beams and new laser techniques increased dramatically the sensitivity and resolution in the near-infrared region and opened to high-precision measurements the difficult far-infrared region. The latter development made it possible to investigate directly intermolecular vibration bands which are very sensitive probes of the shape of intermolecular potentials. The new spectroscopic techniques provide a lot of accurate data on interaction potentials for atom-molecule complexes, as well as on more complicated systems such as the HF, ammonia or water dimers.

Weakly bound complexes display unusual structural and dynamical properties resulting from the shape of their intermolecular potential energy surfaces. They show large amplitude internal motions, and do not conform to the dynamics and selection rules based on the harmonic oscillator/rigid rotor models[329]. Consequently, conventional models used in the analysis of the spectroscopic data fail, and the knowledge of the full intermolecular potential and dipole (or polarizability) surfaces, and the exact quantum-mechanical calculations of the energy levels and intensities are essential to determine the assignments of the observed transitions.

Depending on the strength of the anisotropy in the interaction potential, nuclear motions in weakly bound Van der Waals complexes are usually described using a set of coordinates related to a space-fixed or body-fixed frame[7]. When the anisotropy of the potential in the region of the Van der Waals minimum is relatively weak one can expect that the molecule in the complex should behave as a nearly free rotor, i.e. that the space-fixed description is appropriate. As a consequence, the energy levels and infrared transitions can be approximately classified by the use of the case (*a*) coupling of Bratoz and Martin[330] (see Ref. (7) for reviews). Moreover, the intramolecular vibrations can, to a good approximation, be decoupled from the intermolecular modes due to their high frequency, and vibrationally averaged rotational constants can be used. For a general AB complex with molecules A and B characterized by two sets of rotational constants A_X, B_X, and C_X, X=A or B, we have the following Hamiltonian describing the nuclear motion,

$$H = T_A + T_B - \frac{\hbar^2}{2\mu R} \frac{\partial^2}{\partial R^2} R + \frac{l^2}{2\mu R^2} + V, \qquad (1\text{-}261)$$

$$T_X = A_X j_{xX}^2 + B_X j_{yX}^2 + C_X j_{zX}^2, \qquad (1\text{-}262)$$

where μ is the reduced mass of the dimer, $\mathbf{j}_X$ is the angular momentum of the molecule X, X=A or B, A_X, B_X, and C_X are the rotational constants of the monomer X, and $\mathbf{l}$ denotes the angular momentum associated with the end-over-end rotation of the complex.

The wave function $\chi^{JM}(\omega_A, \omega_B, R, \widehat{\mathbf{R}})$ can be expanded in a basis of products of radial and angular functions of the form,

$$\langle \omega_A, \omega_B, R, \widehat{\mathbf{R}} | n, j_A, k_A, j_B, k_B, j_{AB}, l; JM \rangle = N\phi_n(R)$$

$$\times \sum_{m_A=-j_A}^{j_A} \sum_{m_B=-j_B}^{j_B} \sum_{k_{AB}=-\min(J,j_{AB})}^{\min(J,j_{AB})} \mathcal{D}_{m_A k_A}^{(j_A)^*}(\omega_A) \mathcal{D}_{m_B k_B}^{(j_B)^*}(\omega_B)$$

$$\times \langle j_A m_A; j_B, m_B | j_{AB} k_{AB} \rangle$$

$$\times \sum_{m_l=-l}^{l} C_{m_l}^{l}(\widehat{\mathbf{R}}) \langle j_{AB} k_{AB}; l m_l | JM \rangle, \tag{1-263}$$

where N is the normalization constant, $\phi_n(R)$ denotes the radial function, and $\widehat{\mathbf{R}}$ stands for the spherical polar angles of $\mathbf{R}$ with respect to a space-fixed frame. Note that the functions $\mathcal{D}_{m_A k_A}^{(j_A)^*}$ and $\mathcal{D}_{m_B k_B}^{(j_B)^*}$ appearing in Eq. (1-263) describe rotations of the molecules A and B, respectively, within the complex. These rotations characterized by the quantum numbers j_A and j_B are first coupled to the total internal angular momentum j_{AB}. The overall rotation of the complex is described by the spherical harmonics $C_{m_l}^{l}$. The end-over-end angular momentum l is coupled with the total internal angular momentum j_{AB} to give the total angular momentum J. The angular basis functions have a well defined parity $p = (-1)^{j_A + j_B + l}$ with respect to the space-fixed inversion, so the full Hamiltonian, Eq. (1-261), is blocked in both p and J. Within each block various j_A, j_B, and l are mixed through the potential. The basis functions of Eq. (1-263) look quite complicated, but they simplify a lot for specific cases, cf. Table 1-8.

For an atom–diatom system the basis (1-263) reduces to the product of radial functions and Clebsch-Gordan coupled spherical harmonics,

$$[C^j(\widehat{\mathbf{r}}) \otimes C^l(\widehat{\mathbf{R}})]_M^J = \sum_{m_l, m_j} \langle j m_j; l m_l | JM \rangle C_{m_j}^{j}(\widehat{\mathbf{r}}) C_{m_l}^{l}(\widehat{\mathbf{R}}), \tag{1-264}$$

where $\widehat{\mathbf{r}}$ are the polar angles of the vector $\mathbf{r}$ pointing from one atom to the other in the diatomic molecule in the space-fixed system of axes. In the limit of vanishing

Table 1-8. Quantum numbers and space-fixed basis functions for different dimers. The entries "a", "d", and "p" in columns 1 and 2 refer to atom, diatomic, and polyatomic molecule, respectively

A	B	j_A	k_A	j_B	k_B	j_{AB}	Angular basis function
a	a	0	0	0	0	0	$C_M^J(\widehat{\mathbf{R}})$
a	d	0	0	j_B	0	j_B	$[C^{j_B}(\widehat{\mathbf{r}}_B) \otimes C^l(\widehat{\mathbf{R}})]_M^J$
a	p	0	0	j_B	k_B	j_B	$[D_{\cdot, k_B}^{(j_B)^*}(\omega_B) \otimes C^l(\widehat{\mathbf{R}})]_M^J$
d	d	j_A	0	j_B	0	j_{AB}	$[[C^{j_A}(\widehat{\mathbf{r}}_A) \otimes C^{j_B}(\widehat{\mathbf{r}}_B)]^{j_{AB}} \otimes C^l(\widehat{\mathbf{R}})]_M^J$
d	p	j_A	0	j_B	k_B	j_{AB}	$[[C^{j_A}(\widehat{\mathbf{r}}_A) \otimes D_{\cdot, k_B}^{(j_B)^*}(\omega_B)]^{j_{AB}} \otimes C^l(\widehat{\mathbf{R}})]_M^J$
p	p	j_A	k_A	j_B	k_B	j_{AB}	$[[D_{\cdot, k_A}^{(j_A)^*}(\omega_A) \otimes D_{\cdot, k_B}^{(j_B)^*}(\omega_B)]^{j_{AB}} \otimes C^l(\widehat{\mathbf{R}})]_M^J$

anisotropy the quantum numbers j, which describes the rotation of the diatom in the dimer, and l, which corresponds to the rotation of the vector $\mathbf{R}$ are good quantum numbers. The total angular momentum $\mathbf{J} = \mathbf{j} + \mathbf{l}$ is always conserved, due to the isotropy of space, but j and l are broken by the anisotropy in the potential. A degenerate (j, l)-level splits into sublevels $J = |j - l|, \ldots, j + l$ under the influence of the anisotropy. If these splittings are small, like in the Ar–H$_2$[331] and He–C$_2$H$_2$[302] cases, the states can still be labeled to a good approximation by j and l.

When the leading anisotropic term is large compared to the rotational constant of the complex, and small compared to the rotational constants of the free molecules, the energy levels and infrared transitions can approximately be classified using the case (b) coupling of Bratoz and Martin[330], i.e. the molecules in the complex should behave as hindered rotors. Choosing the *embedded* reference frame such that the vector $\mathbf{R}$ connecting the centers of mass of the molecules A and B defines the new z axis, the Hamiltonian describing the nuclear motion can be written as[332–334],

$$H = T_A + T_B - \frac{\hbar^2}{2\mu R} \frac{\partial^2}{\partial R^2} R + \frac{\mathbf{J}^2 + (\mathbf{j}_A + \mathbf{j}_B)^2 - 2(\mathbf{j}_A + \mathbf{j}_B) \cdot \mathbf{J}}{2\mu R^2} + V. \tag{1-265}$$

Here, the operator $\mathbf{J}$ is the total angular momentum operator in the space-fixed frame, and T_X, X=A and B, is defined by Eq. (1-262). Note, that the present coordinate system corresponds to the so-called "two-thirds body-fixed" system of Refs. (7–334). Therefore, the internal angular momentum operators $\mathbf{j}_A$ and $\mathbf{j}_B$, and the pseudo angular momentum operator $\bar{\mathbf{J}}$ *do not* commute, so the second term in Eq. (1-265) cannot be factorized.

The wave function describing the nuclear motion can now be expanded in a basis of products of radial and angular functions of the form,

$$\langle \omega_A, \omega_B, R, \widehat{\mathbf{R}} | n, j_A, k_A, j_B, k_B, j_{AB}, K; JM \rangle = \phi_n(R)$$

$$\times \sum_{m_A = -j_A}^{j_A} \sum_{m_B = -j_B}^{j_B} \langle j_A m_A; j_B, m_B | j_{AB} K \rangle$$

$$\times \left(\mathcal{D}_{m_A k_A}^{(j_A)*}(\omega_A) \mathcal{D}_{m_B k_B}^{(j_B)*}(\omega_B) \mathcal{D}_{MK}^{(J)*}(\alpha, \beta, 0) \right.$$

$$\left. + p(-1)^J \mathcal{D}_{m_A, -k_A}^{(j_A)*}(\omega_A) \mathcal{D}_{m_B, -k_B}^{(j_B)*}(\omega_B) \mathcal{D}_{M, -K}^{(J)*}(\alpha, \beta, 0) \right), \tag{1-266}$$

where (β, α) are the polar angles of the $\mathbf{R}$ vector in the space-fixed coordinates. The angular basis functions have a well defined parity p, and the full Hamiltonian, Eq. (1-265), is blocked in both $p(-1)^J$ and J. Within each block functions with different K are mixed through the off-diagonal Coriolis interaction, $2(j_{xA} + j_{xB})\bar{J}_x + 2(j_{yA} + j_{yB})\bar{J}_y$. For specific systems the angular part of the basis function (1-266) can be simplified using the rules given in Table 1-4 for the angular expansions of the interaction potentials.

For an atom–diatom system the basis (1-266) reduces to

$$[C_K^j(\vartheta, \varphi)\mathcal{D}_{M,K}^{(J)*}(\alpha, \beta, 0) + p(-1)^J C_{-K}^j(\vartheta, \varphi)\mathcal{D}_{M,-K}^{(J)*}(\alpha, \beta, 0)],$$

(1-267)

where ϑ and φ are the spherical angles of the $\mathbf{r}$ vector in the body-fixed coordinates. The only rigorously conserved quantum numbers are, again, the total angular momentum J and the spectroscopic parity $\sigma = p(-1)^J$. The diatom rotational quantum number j, and the projection K of $\mathbf{J}$ (or $\mathbf{j}$) onto the intermolecular axis, are only approximately conserved. This conservation is broken by off-diagonal Coriolis interaction. Since K is the projection of an angular momentum, states with $K = 0, \pm 1$, etc., are denoted as Σ, Π, etc. In addition, levels with $\sigma = +1$ and $\sigma = -1$ will be designated by superscripts e and f, respectively. For $K = 0$ only e parity states exist. The case b coupling of Bratoz and Martin[330] gives a very simple classification of the rovibrational energy levels of the complex: each monomer rotational level j is split into $j+1$ levels corresponding to any $J \geq |K|$ with $K = 0, \pm 1, \pm 2, \ldots, \pm j$. The inclusion of the Coriolis interaction introduces further splitting of the states with $|K| \neq 0$ (the so-called l-doubling) into states with e and f parity labels.

It should be stressed that for a fixed set of quantum numbers j_A, k_A, j_B, k_B, j_{AB}, J, M, and K running from $-\min(J, j_{AB})$ to $+\min(J, j_{AB})$ the basis functions of Eq. (1-266) span the same space as the basis functions of Eq. (1-263) with l running from $|J - j_{AB}|$ to $J + j_{AB}$. This means that the Hilbert spaces spaned by the basis functions (1-263) and (1-266) are isomorphic. Consequently, the final quantum states (eigenvalues and eigenvectors) will be the same in both bases. The specific choice of the mathematical form of the Hamiltonian, Eq. (1-261) or (1-265), and consequently, of the basis depends on the anisotropy of the potential energy surface.

We wish to add that Martin and Bratoz[330] also considered a case (c) corresponding to almost rigid complex. The treatment of the dynamics in the case (c) does not differ from the standard treatment of rotations and vibrations in rigid molecules with the Watson's Hamiltonian for nuclear motions.

The infrared transitions obey the following selection rules: $|\Delta p| = 1$, and $|\Delta J| = 1$ or 0. The wave functions for the initial and final states obtained by solving the Schrödinger equation with the Hamiltonian of Eq. (1-261) or (1-265) can be used to compute the infrared absorption intensities for the complex. The infrared absorption coefficient $\mathcal{J}(J'' \to J')$ for the transition $J'' \to J'$ is proportional to,

$$\frac{\exp(-E_{J''}/k_B T)}{Z(T)} (E_{J'} - E_{J''}) S(J'' \to J'),$$

(1-268)

where E_J denotes the energy of the state labeled by J, $Z(T)$ is the partition function, the line strength is given by,

$$S(J'' \rightarrow J') = \sum_{M''=-J''}^{J''} \sum_{M'=-J'}^{J'} \sum_{m=-1}^{1} |\langle \chi^{J'M'} | \mu_m | \chi^{J''M''} \rangle|^2, \qquad (1\text{-}269)$$

and μ_m is the spherical component of the dipole moment of the dimer (for the far-infrared transitions) or the spherical component of the transition dipole (for the near-infrared transitions).

12.5.   Scattering Cross Sections for Rotational Excitation

Molecular beam scattering experiments provide direct and detailed information about the repulsive part of the interaction potential between the colliding particles. With such scattering data available, detailed studies of the short and intermediate-range parts of the potential energy surfaces can readily be made, provided that accurate theoretical methods, e.g. quantum close-coupling approach, are used to describe the scattering phenomena.

The coordinate system used in the close-coupling method is the space-fixed frame. For simplicity we consider the atom–diatom scattering. The wave function $\chi_E^{JM}(R, \widehat{\mathbf{r}}, \widehat{\mathbf{R}})$ for an atom-rigid rotor system corresponding to the total energy E, total angular momentum J, and its projection M on the space-fixed z axis can be written as an expansion,

$$\chi_E^{JM}(R, \widehat{\mathbf{r}}, \widehat{\mathbf{R}}) = \sum_{j,l} \phi_{jl,E}^J(R)[C^j(\widehat{\mathbf{r}}) \otimes C^l(\widehat{\mathbf{R}})]_M^J, \qquad (1\text{-}270)$$

where the angular functions are defined by Eq. (1-264). The radial functions $\phi_{jl}^J(R)$ are solutions of the system of coupled differential equations (close-coupling equations):

$$\left[-\frac{\hbar^2}{2\mu R} \frac{\partial^2}{\partial R^2} R + \frac{\hbar^2 l(l+1)}{2\mu R^2} + V_{jl,jl}^J(R) + b_0 j(j+1) - E \right] \phi_{jl,E}^J(R) =$$
$$- \sum_{(j'l') \neq (jl)} V_{jl,j'l'}^J(R) \phi_{j'l',E}^J(R), \qquad (1\text{-}271)$$

where b_0 is the rotational constant of the diatom in its ground vibrational state, and the angular matrix elements of the potential, $V_{jl,j'l'}^J(R)$, can be written in terms of the Percival-Seaton coefficients[335]. One may note that for a given state j, the associated channels (j, l) are found from the triangular condition

$$|J - j| \leq l \leq J + j. \qquad (1\text{-}272)$$

The asymptotic form of the radial functions $\phi_{jl,E}^J(R)$ determines the matrix elements of the scattering matrix, $S_{jl,j'l'}^J$. These in turn define the state-to-state integral and differential cross sections by the usual expressions[336]. For instance, the integral cross section is given by,

$$\sigma(j'' \to j') = \frac{\pi}{(2j''+1)k_{j''}^2} \sum_{J=0}^{\infty} \sum_{l'',l'} |S^J_{j''l'',j'l'} - \delta_{j''j'}\delta_{l''l'}|^2. \tag{1-273}$$

where

$$\frac{\hbar^2 k_{j''}^2}{2\mu} = E - b_0 j''(j''+1). \tag{1-274}$$

After a proper Boltzmann averaging of the integral cross section over the kinetic energy in the center of mass system, E_{CM}, one get the rate constants,

$$k_{j''j'}(T) = \langle v\sigma(j'' \to j') \rangle = \left(\frac{8k_BT}{\pi\mu}\right)^{1/2}$$
$$\times \int_0^\infty \frac{E_{CM}}{k_BT} \sigma(j'' \to j') e^{-\frac{E_{CM}}{k_BT}} d\left(\frac{E_{CM}}{k_BT}\right) \tag{1-275}$$

where v denotes the relative velocity.

It is worth noting that the elements of the scattering matrix also define generalized cross sections which, after the Boltzmann average, give the pressure broadening coefficients of the spectral lines of the diatom in the bath of colliding atoms, as well as the quantum expressions for various transport coefficients of diatomic molecules in dilute atomic gases, such as the viscosity coefficient, diffusion coefficient, thermal conductivity, etc.[337,338].

12.6. Thermodynamic Second Virial Coefficients

It is well known that the equation of state for a gas at a temperature T is given by the virial expansion of the form,

$$\frac{p}{k_BT} = \rho + B_2(T)\rho^2 + \cdots, \tag{1-276}$$

where p denotes the pressure and $B_2(T)$ is the second virial coefficient which is related directly to the intermolecular potential. Thus, *ab initio* calculations of the potentials and, subsequently, of the virial coefficients give theoretical equation of state for dilute gases. Thermodynamic second virial coefficients also provide additional information about the accuracy of the theoretical interaction potentials, in particular about the correctness of the volume of the Van der Waals well. For the atom-linear molecule case the expressions for the classical virial coefficient, $B_2^{(0)}(T)$, and the first quantum correction, $B_2^{(1)}(T)$ were derived by Pack[339]. The first quantum correction splits into terms due to relative translational motions, $B_R^{(1)}(T)$, to molecular rotations, $B_A^{(1)}(T)$, and to the Coriolis term, $B_C^{(1)}(T)$:

$$B_2(T) = B_2^{(0)}(T) + B_{R\,2}^{(1)}(T) + B_A^{(1)}(T) + B_C^{(1)}(T), \tag{1-277}$$

where the consecutive terms on the r.h.s. of Eq. (1-277) are given by[339]:

$$B_2^{(0)}(T) = \pi N_A \int_0^\infty \int_0^\pi \left(1 - \exp\left[-\frac{V(R, \vartheta)}{k_B T}\right]\right) R^2 \sin \vartheta dR d\vartheta, \tag{1-278}$$

$$B_R^{(1)}(T) = \frac{N_A \pi \hbar^2 \beta^3}{24\mu} \int_0^\infty \int_0^\pi \exp\left[-\frac{V(R, \vartheta)}{k_B T}\right]$$
$$\times \left(\frac{\partial V}{\partial R}\right)^2 R^2 \sin \vartheta dR d\vartheta, \tag{1-279}$$

$$B_A^{(1)}(T) = -\frac{N_A \pi \beta^2 b_0}{12} \int_0^\infty \int_0^\pi \exp\left[-\frac{V(R, \vartheta)}{k_B T}\right]$$
$$\times \sum_{l=0}^\infty l(l+1) V_l(R) P_l(\cos \vartheta) R^2 \sin \vartheta dR d\vartheta, \tag{1-280}$$

$$B_C^{(1)}(T) = -\frac{N_A \pi \hbar^2 \beta^2}{24\mu} \int_0^\infty \int_0^\pi \exp\left[-\frac{V(R, \vartheta)}{k_B T}\right]$$
$$\times \sum_{l=0}^\infty l(l+1) V_l(R) P_l(\cos \vartheta) \sin \vartheta dR d\vartheta. \tag{1-281}$$

Here, $V(R, \vartheta)$ is the interaction potential depending on the distance R from the atom to the center of mass of the molecule and on the angle ϑ between the molecular axis and the line connecting the atom with the center of mass of the molecule, N_A is Avogadro's constant, and $V_l(R)$ are the radial coefficients in the expansion of the potential in Legendre polynomials P_l,

$$V(R, \vartheta) = \sum_{l=0}^\infty V_l(R) P_l(\cos \vartheta). \tag{1-282}$$

Let us end this section by saying that the expressions (1-278)–(1-281) as well as the specific form of the expansion (1-282) can be generalized to the interaction of arbitrary molecules[339,340].

12.7. Simulations of Condensed Phases

Computer simulations which model microscopic behavior of condensed phases provide important insight into various physical, chemical, and biological phenomena that cannot easily be obtained by quantum-mechanical methods. Among quantities computed using simulations methods one can cite transport properties such as diffusion and viscosity or thermal conductivity, radial distribution functions related to the X-ray or neutron scattering form factors, etc. It is not surprising that computer simulations have recently become a rapidly developing field of science[341].

Two types of computer simulation procedures are used to describe properties of condensed phases: the Monte Carlo method based on the Metropolis algorithm and the molecular dynamics approach. Both approaches consider a cell containing typically between 10^3 and 10^5 molecules interacting via a given potential which may contain non-additive terms. The cell is considered to be surrounded by identical replicas of itself on each side. The Monte Carlo method[342] is based on random changes of molecular coordinates within the cell. If the energy change produced by the shift of molecular coordinates is zero or negative with respect to the previous configuration of molecules, the new configuration is accepted. If the energy increased by the change of coordinates, the new configuration is accepted in a proportion of the cases which is given by $\exp(-\Delta U/k_B T)$, where ΔU is the energy shift. Very many changes of coordinates are made in this way, producing a series of configurations which occur with a probability proportional to the Boltzmann factor $\exp(-U_N/k_B T)$, where U_N is the total potential energy for a given configuration. The series of configurations obtained in such a way are analyzed to give static properties of the system, such as the internal energy, pressure, and radial distribution functions.

In the molecular dynamics approach[343] a basic cell and its replicas are again defined. Starting from an initial configuration with a well defined total (kinetic plus potential) energy, the motions of the molecules are followed by numerical integration of the classical equations of motion of all molecules. The forces acting on each molecule are summed, and new positions and velocities are calculated after a small time step, usually of the order of 10^{-15} s. This procedure is repeated for typically 10^3 to 10^6 time steps. At each step of the time chain the average contributions similar to those obtained in the Monte Carlo procedure can be evaluated. In addition, the partition of the total energy into kinetic and potential terms can be obtained, and the distribution of molecular velocities can be studied. The molecular dynamics method is a more powerful technique than the Monte Carlo procedure, as it yields in addition to equilibrium properties of the system also the transport properties, such as the diffusion coefficients, viscosity or thermal conductivity.

We wish to end this section by saying that now it is possible to perform molecular dynamics simulations "on the fly" without precomputing the potential energy surface. This idea was introduced by Carr and Parrinello[344,345], and is known as the Carr-Parrinello dynamics. In this approach the nuclear motions are treated classically within the molecular dynamics method, but the energy and force are precomputed for each configuration of the nuclei with a suitable version of the density functional theory. If applicable with trust, this method would unable computer simulations of condensed phases without calculations of the potential energy surfaces. Applications to the liquid water are very promising[345], so for large systems this method will become a viable alternative to classical simulations based on empirical force fields.

13.   ILLUSTRATIVE APPLICATIONS

13.1.   Pair Potentials and Modelling of Spectroscopic, Collisional, and Thermodynamic Properties of Binary Complexes

All the formulas described in Section. 7 have been implemented in the computer program SAPT2[220,221] which can routinely be used to compute intermolecular pair potentials. Some of the results, for Ar–H_2[331,346–347], He–HF[348–350], water dimer[351,352], He–CO[353,354], Ne–CO[355], He–C_2H_2[356–358], and Ne–C_2H_2[359] were already summarized in the previous review articles[7–10]. Among the results which were recently obtained, the pair potentials of He–CO_2[296] and Ar–CH_4[360–363] were very thoroughly tested with various experimental data from high-resolution spectroscopic and collisional experiments, as well as with the thermodynamic second virial coefficients[364].

13.1.1.   He–CO_2 complex

A typical feature of the potentials for weakly interacting systems is that their shape is determined by a subtle balance between the geometry dependence of the repulsive short range interactions and the, mostly attractive, long range forces. This is also the case for He–CO_2. The potential surface of He–CO_2[296] has a T-shaped equilibrium structure, with $R_e = 6.34$ bohr, $\Theta_e = 92.2°$, and $D_e = -53.49$ cm^{-1}, where R_e, Θ_e, and D_e are the equilibrium distance, equilibrium angle between the vectors pointing from the carbon atom to the helium atom, and from one oxygen atom to the other, and the binding energy, respectively. See Figure 1-5 for the shape of the potential energy surface. The total interaction energy and its components around the global and local minima are reported in Table 1-9. Table 1-9 shows that the first-order exchange energy and the dispersion energy are two major contributions to the interaction potential determining its anisotropy in the region of the global minimum, while the induction energy is less important and strongly quenched by its exchange counterpart (the exchange-induction energy). Surprisingly, the electrostatic energy, which is of purely penetrational character, is more important than the weak quadrupole-induced dipole induction interaction. Obviously, the dispersion energy favors the linear He$\cdots$O=C=O minimum in the potential energy surface. However, the short-range energy, dominated by the first-order exchange energy, behaves, to a good approximation, in a reversed manner to $E_{\text{disp}}^{(2)}$, and shows a stronger anisotropy. Thus, the position of the minimum is mainly determined by the anisotropy of the exchange-repulsion term. The barrier for moving away from this minimum to the linear OCO–He geometry ($\Theta = 0°$) is

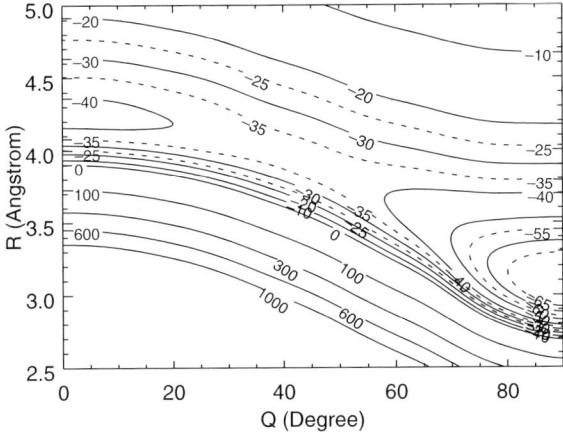

Figure 1-5. Contour plot of the SAPT potential for the He–CO$_2$ complex

relatively high, suggesting that the He atom will show a hindered rotation around the CO$_2$ rotor.

A different picture is observed for the local minimum of $\epsilon_m = -28.94$ cm^{-1} located at the linear He$\cdots$O=C=O geometry for $R_m = 8.03$ bohr. Around the local

Table 1-9. Angular dependence of the He – CO$_2$ interaction energy (in cm^{-1}) near the global ($R = 6$ bohr) and local ($R = 8$ bohr) minima

Θ	$0°$	$30°$	$60°$	$90°$
		Global minimum		
$E_{elst}^{(1)}$	-643.93	-274.03	-39.51	-7.88
$E_{exch}^{(1)}$	3066.48	1335.32	205.40	42.72
$E_{ind}^{(2)}$	-387.72	-137.13	-16.68	-4.94
$E_{exch-ind}^{(2)}$	361.63	128.80	13.69	1.81
$E_{disp}^{(2)}$	-703.23	-417.71	-144.75	-81.26
$E_{exch-disp}^{(2)}$	94.97	44.72	7.63	1.83
E_{int}^{SAPT}	1601.11	587.31	12.55	-49.53
		Local minimum		
$E_{elst}^{(1)}$	-5.93	-3.15	-0.60	-0.10
$E_{exch}^{(1)}$	35.65	19.00	3.60	0.52
$E_{ind}^{(2)}$	-2.30	-1.25	-0.31	-0.20
$E_{exch-ind}^{(2)}$	1.12	0.59	0.09	0.01
$E_{disp}^{(2)}$	-56.64	-40.19	-19.43	-13.07
$E_{exch-disp}^{(2)}$	1.49	0.76	0.14	0.03
E_{int}^{SAPT}	-27.86	-25.08	-16.68	-12.83

minimum the anisotropy of the potential is much less pronounced. Unlike the global minimum, the local minimum is mostly bound by the dispersion forces, despite nonnegligible values of the repulsive exchange components, cf. Table 1-9. Surprisingly, the short-range electrostatic energy is again much more important than the long-range induction term and substantially contributes to the well depth of the secondary minimum.

The potential energy surface for the He–CO_2 complex was used[296] in converged variational calculations to generate bound rovibrational states and the infrared spectrum corresponding to the simultaneous excitation of the ν_4 vibration and internal rotation in the CO_2 subunit within the complex. Due to the high barrier separating the minima, the complex behaves like a semirigid asymmetric top and the rovibrational energy levels can classified with the asymmetric top quantum numbers. The computed frequencies of the infrared transitions in the ν_4 band of the spectrum are in a very good agreement with the high resolution experimental data of Weida et al.[365]. This is illustrated in Figure 1-6, were we compare the theoretical spectrum with the results of high-resolution measurements[365]. Indeed, the agreement between the computed and observed transition frequencies and intensities is excellent. Since this piece of the spectrum probes the anisotropy of the potential in the region of the potential well, the level of agreement presented on Figure 1-6 suggests that the *ab initio* potential is very accurate in this region.

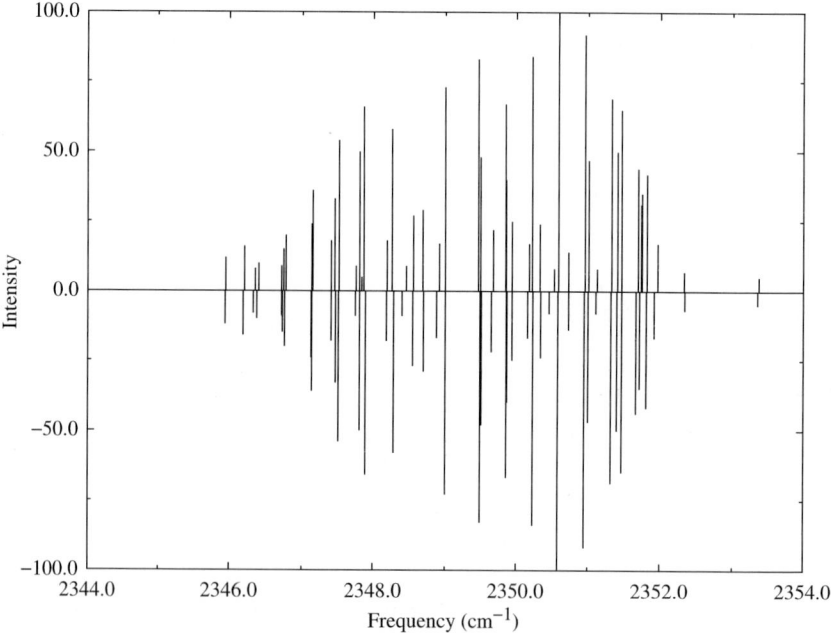

Figure 1-6. Theoretical (upper panel) and experimental (lower panel) infrared spectrum for the He–CO_2 complex

In the experimental work of Ref. (365) several weak spectroscopic transitions were observed that could not be assigned to the ν_4 fundamental band of the complex. On the basis of the integrated band intensities it was assumed that these transitions correspond to the ν_5 intermolecular bending mode, although no definite assignments could be proposed. Only the position of the first excited bending state could approximately be located at $9 \pm 2 \text{ cm}^{-1}$ above the ground state. The calculations of bound rovibrational levels reported in Ref. (296) showed a large number of levels that can be attributed to the ν_5 band. In particular, the first bending frequency was predicted at 8.805 cm^{-1}, in very good agreement with the experimental estimate quoted above. Assuming that the initial states in the unassigned portion of the experimental spectrum correspond to the ν_5 states of the complex, one could try to assign these lines, and a tentative assignment of the observed ν_5 band was proposed[296]. The agreement in the line positions is within $0.1–0.3 \text{ cm}^{-1}$. Such an accuracy can be expected from the theoretical side. The ν_4 transitions were reproduced within 0.05 cm^{-1}. Since they correspond to low-lying energy states they can be computed accurately, the interaction potential being accurate in region of the global minimum. The ν_5 lines lie much higher in energy and are sensitive to the regions of the potential surface close to the dissociation limit. Here the SAPT potential may be less accurate, so an error in the transition frequencies of the order of tenths of the wavenumber is not surprising. The theoretical intensity pattern closely follows the experimental one suggesting that the proposed assignment is correct.

As a further test of the *ab initio* potential, calculations of the pressure broadening coefficients of the R branch rotational lines of the ν_4 spectrum of CO_2 in a helium bath at various temperatures were performed[296]. [The pressure broadening coefficients reflect

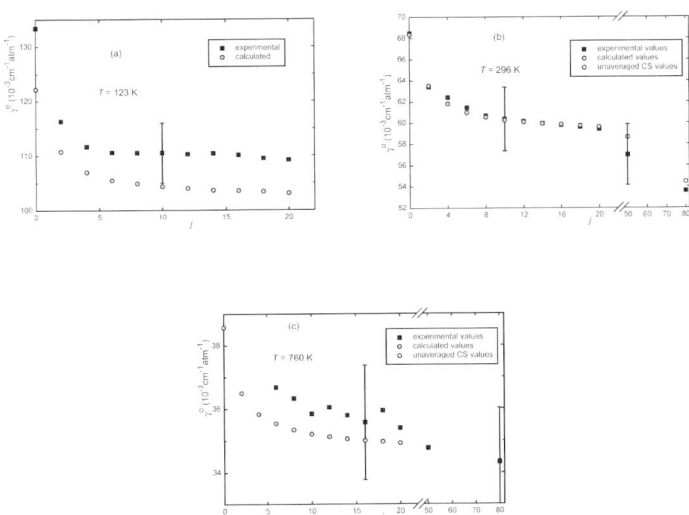

Figure 1-7. Theoretical (open circles) and experimental (full circles) pressure broadening coefficients of the CO_2 spectral lines in the helium bath at various temperatures

thebroadeningofthespectrallinesofamoleculeinabathofagasatatemperature T.] Very good agreement is found with the wealth of experimental results for various rotational states of CO_2 at different temperatures, cf. Figure 1-7. The agreement between the theory and experiment observed on Figure 1-7 suggests that the anisotropy of the *ab initio* potential in the repulsive region is correct. Finally, the potential was also tested[296] by computing the second virial coefficients at various temperatures. Again, the agreement between theory and experiment is satisfactory, cf. Figure 1-8, suggesting that the volume of the Van der Waals well is correctly reproduced by the *ab initio* potential.

All these comparisons between theory and various high precision experimental data show that the *ab initio* SAPT potential for He–CO_2 reproduces various physical properties of the complex. This suggests that the potential is accurate not only around the van der Waals well, but also its anisotropy in the attractive and repulsive regions, as well as the volume of the van der Waals well are correct.

The SAPT potential for the He–CO_2 complex was also used in the calculations of the rovibrational spectra of the He$_N$–CO_2 clusters[366]. High resolution experimental data were also reported in this paper. Comparison of the theoretical and experimental effective rotational constants B and other spectroscopic characteristics as functions of the cluster size N is shown on Figure 1-9. Again, the agreement between the theory and experiment is impressive showing that theory can describe with trust spectroscopic characteristics of small clusters He$_N$–CO_2. This especially true for the effective rotational constant and the frequency shift of the CO_2 vibration due to the solvation by the helium atoms. One may note in passing that the clusters He$_N$–CO_2 with the number of helium atoms N around 20 do not exhibit all the properties of the CO_2 molecule in the first solvation shell of the (quantum) liquid helium at very low temperatures.

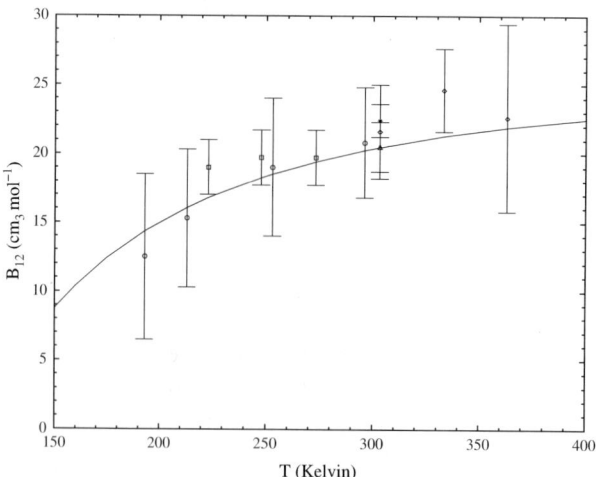

Figure 1-8. Theoretical (full line) and experimental (open circles) second virial coefficient of He–CO_2 at various temperatures

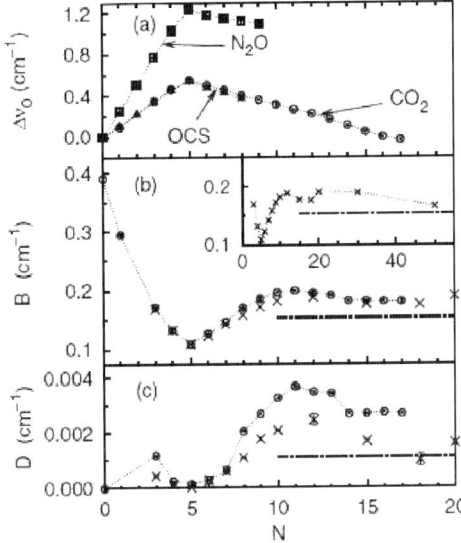

Figure 1-9. Variation of the vibrational band origin $\Delta\nu_0$, effective rotational constant B, and centrifugal distortion constant D with the size N of the cluster He_N–CO_2. Circles and dots indicate the experimental and theoretical results, respectively. The upper panel also includes for comparison the value of $\Delta\nu_0$ for clusters of He with OCS and N_2O. The inset in the middle panel shows the results calculated in an extended range of helium atoms in the cluster. The dashed lines indicate the nanodroplet results of Miller and collaborators.[367]

13.1.2. Ar–CH_4 complex

An example that we describe in somewhat more detail is the pair potential of Ar–CH_4 calculated by SAPT[360]. The position of the argon atom is described by the vector $\mathbf{R} = (R, \Theta, \Phi)$ pointing from the center-of-mass of methane to the Ar atom. A two-dimensional cut that displays most of the interesting features of this three-dimensional intermolecular potential is shown in Figure 1-10. For large R the preferred direction of approach of the Ar atom is along one of the C–H bonds ($\Theta = 55°$), as if the C–H$\cdots$Ar bond were a hydrogen bond. At shorter distance, however, also the steric repulsion is the largest for this orientation, and the deepest attractive well occurs where the short range repulsion is the weakest, i.e. for the Ar atom in between three C–H bonds ($\Theta = 125°$). In Figure 1-10 one can observe the origin of this behavior. The long range attraction is mostly caused by dispersion forces. The attraction caused by induction is much smaller at large R, but increases steeply with decreasing R, when the charge clouds of Ar and CH_4 start to overlap. This latter effect is due to penetration, i.e., incomplete screening of the nuclear charges by the electron clouds. The small R behavior is dominated by the short range repulsion, however, with the contributions of first-order exchange, exchange-induction, and exchange-dispersion in decreasing order of importance. Both the long range attraction and the short range repulsion are largest for Ar along one of

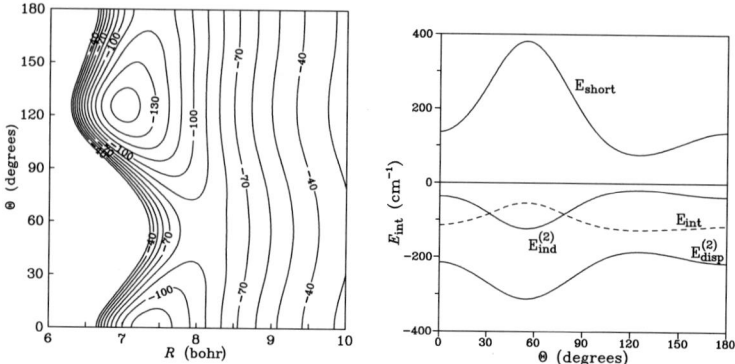

Figure 1-10. Contour plot of the SAPT potential (left panel) and the angular dependence of the potential and its components (right panel) for Ar–CH$_4$

the C–H bonds ($\Theta = 55°$), and smallest when the Ar atom approaches one of the faces of the CH$_4$ tetrahedron ($\Theta = 125°$). Since the long range R^{-n} contributions decrease less rapidly with increasing R than the exponential short range terms this explains the observed behavior, which is typical for a Van der Waals complex. Even for hydrogen bonded complexes one finds such behavior, but in that case the (first-order) electrostatic and (second-order) induction forces are more dominant, and the equilibrium geometry of the complex is often determined by these long range forces.

The same Ar–CH$_4$ potential was used in extensive close-coupling calculations for elastic and rotationally inelastic scattering at various energies. Both the total differential Ar–CH$_4$ scattering cross sections[361] and the integral state-to-state cross sections for the rotational transitions of CH$_4$ induced by collisions with Ar (360) agree well with the experimental data of Buck et al.[368] and Chapman et al.[369]. This is illustrated on Figures 1-11 and 1-12.

An inspection of these figures shows that the overall agreement between theory and experiment is very good, suggesting that the anisotropy of the potential in the repulsive region is correct. Surprisingly enough, the *ab initio* SAPT potential performs better than the empirical potential of Ref. (368) fitted to the experimental total differential scattering cross sections.

Another test to which the Ar–CH$_4$ potential was subjected, is the calculation of the second virial coefficients at various temperatures. The agreement between the theory and experiment is illustrated on Figure 1-13. An inspection of this figure show that the agreement between the measured and computed second virial coefficients is good, so similarly as in the case of He–CO$_2$ mixtures, the volume of the interaction potential well for Ar–CH$_4$ is correct.

Finally, the SAPT potential of Ref. (360) was used to generate the infrared spectrum of the Ar–CH$_4$ complex. This spectrum, in the region of the ν_3 mode of CH$_4$, was measured by Miller in 1993 and presented at the 1994 Faraday Discussion

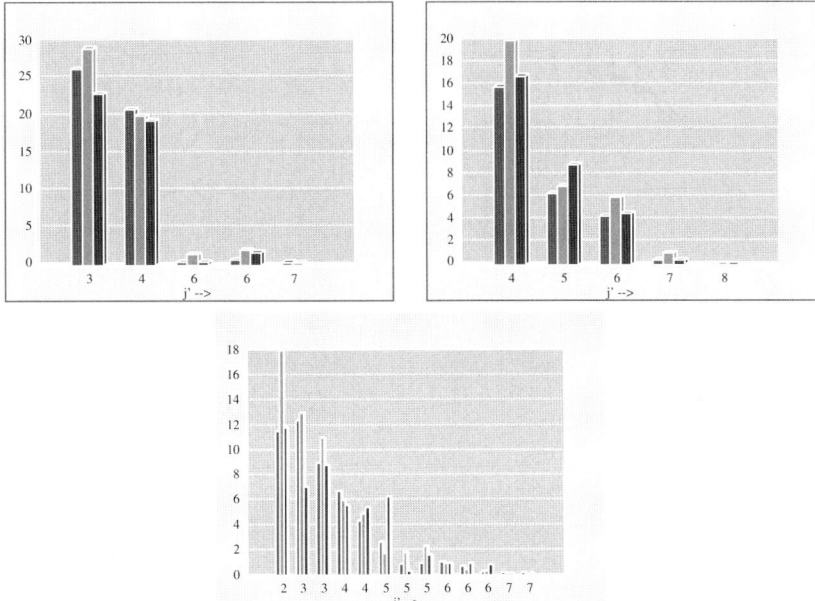

Figure 1-11. Theoretical and experimental relative integral cross sections for Ar–CH$_4$. The experimental data are displayed in red, the results displayed in blue and green were generated from the SAPT and empirical potentials, respectively. The labels *A*, *E*, and *F* refer to the total nuclear spin states of methane. The empirical potential was fitted to the total differential cross sections

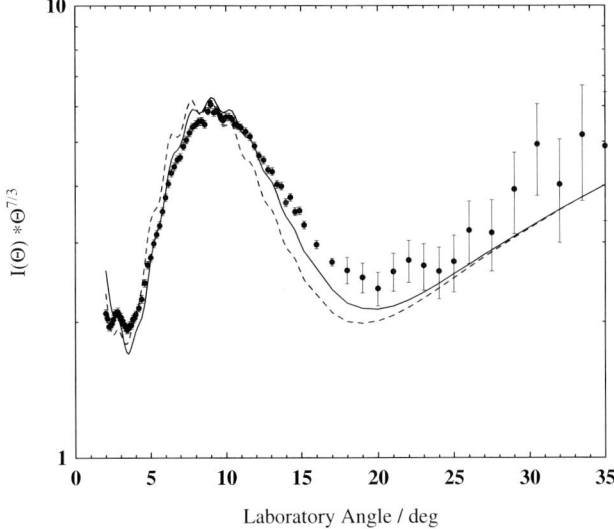

Figure 1-12. Theoretical and experimental total differential cross sections for Ar–CH$_4$. The full and dashed lines correspond to the close-coupling results generated from the SAPT potential and from the empirical potential fitted to these data, respectively

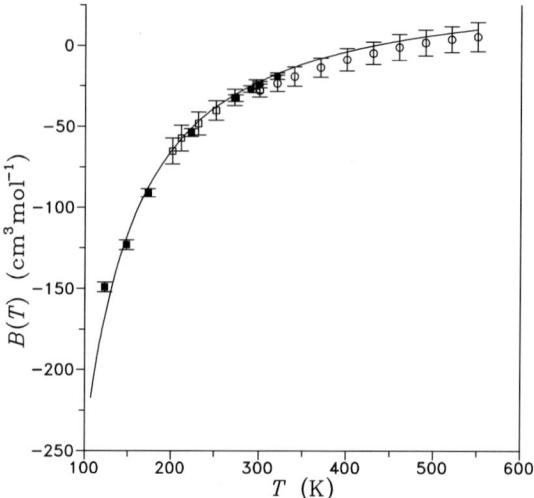

Figure 1-13. Theoretical (full line) and experimental (circles and rectangles) second virial coefficient of Ar–CH$_4$ at various temperatures

on Van der Waals molecules[370]. It shows a lot of detailed structure, with many lines more or less grouped in seven bands, over a range of 40 cm^{-1} around the band origin of the ν_3 mode (3020 cm^{-1}) which could not yet be assigned or understood, however. Recently, the bound levels of Ar–CH$_4$ were calculated from the SAPT potential. Also the quasi-bound levels of the complex with the ν_3 mode excited were calculated with the use of the same ground state potential surface, but it was explicitly taken into account that the ν_3 mode has three degenerate sublevels and the Coriolis coupling between the vibrational angular momentum of these sublevels and the (hindered) internal rotation of the CH$_4$ subunit inside the complex was included. With the use of the ν_3 transition dipole moment function—it was assumed that the weak interaction with Ar does not affect this transition dipole moment—the infrared spectrum in the region of the ν_3 mode could then be generated completely *ab initio*. This spectrum is shown in Figure 1-14, next to Miller's experimental high resolution spectrum. It is obvious that the agreement is very good and that the structure of the measured spectrum can be fully understood from the *ab initio* calculations. More details are given in two papers[362,363], where it is shown in more detailed displays of each of the seven bands that even the individual lines agree very well in most cases, and can thus be assigned.

These examples, and the previous summaries of the results for other dimers[7,8], demonstrate that the pair potentials from *ab initio* SAPT calculations are indeed accurate. Another, more global, comparison with experiment which confirms this finding was made by computations of the (pressure) second virial coefficients of all of these dimers over a wide range of temperatures[364].

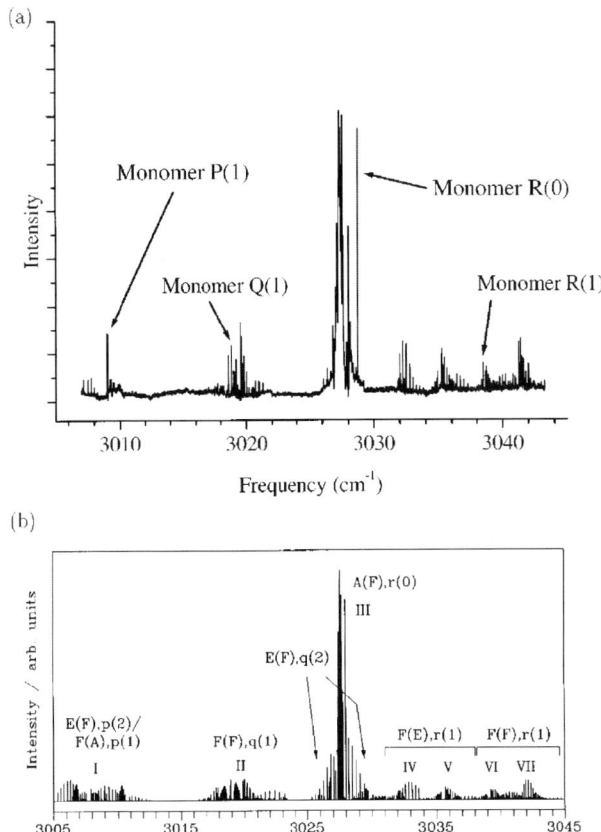

Figure 1-14. Experimental (upper panel) and theoretical (lower panel) infrared spectrum of the Ar–CH$_4$ complex

13.2. Nonadditive Interactions, Spectroscopic Signatures of Molecular Clusters, and Simulations of Condensed Phases

In this section we present some results obtained with the SAPT code for three-body interactions, SAPT3[371]. Routine applications of SAPT to three-body interactions are relatively scarce. Here we concentrate on the water clusters with a special emphasis on the simulations of the liquid water properties starting from *ab initio* SAPT potentials for pair and three-body interactions and on clusters of water with hydrogen chloride in the context of protolytic dissociation of HCl in small water clusters. Other applications of SAPT to, e.g. Ar$_2$–HF trimer can be found in Ref. (313).

13.2.1.   Water clusters and simulations of gaseous and liquid
water properties

Small water clusters $(H_2O)_n$ are hydrogen bonded complexes with each monomer acting simultaneously as a proton donor and proton acceptor. The trimer has a triangular equilibrium structure with rather strongly non-linear hydrogen bonds, the tetramer has a square planar system of hydrogen bonds, and the pentamer has a slightly puckered, pentagonal hydrogen bonded framework, cf. Figure 1-15 for a schematic representation of geometries corresponding to the global minima of these clusters.

The importance of various pairwise additive and non-additive interaction energy contributions for the equilibrium geometries of the water trimer, tetramer, and pentamer were studied in Ref. (372). See Table 1-10 for the results. The results reported in Table 1-10 show that a substantial part of the binding energy originates from the three-body contributions: 17% for the trimer, 26% for the tetramer, and 29% for the pentamer. The dominant three-body term is the second order induction energy, mainly due to the dipole induced-dipole interactions, but also the third order induction energy is important, so if one wishes to include induction effects by iteration[298,373] of the induced dipole moments and the corresponding electric fields, one should proceed with this iteration beyond the first step. The contribution of the third order induction-dispersion energy is small and, even though the dispersion energy is an important component of the pair hydrogen bonding energy, the Axilrod-Teller three-body dispersion energy is even smaller. The three-body exchange effects are substantial, however, so one cannot restrict the treatment of non-additive effects in water to the classical induction terms only.

Simulations of the liquid water properties have been the subject of many papers, see Ref. (374) for a review. Recently a two-body potential for the water dimer was computed by SAPT(DFT)[375]. Its accuracy was checked[375] by comparison with the experimental second virial coefficients at various temperatures. As shown on Figure 1-16, the agreement between the theory and experiment is excellent. Given an accurate pair potential, and three-body terms computed by SAPT[376], simulations of the radial O–O, O–H, and H–H distribution functions could be

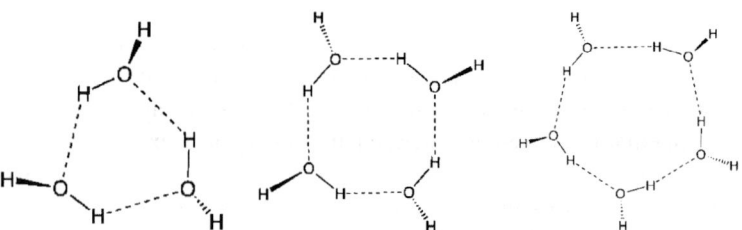

Figure 1-15. Structures of the global minima for the water trimer $(H_2O)_3$, tetramer $(H_2O)_4$, and pentamer $(H_2O)_5$

Table 1-10. Decomposition of the pair and three-body interaction energies (in kcal/mol) for the structures corresponding to the global minima of small water clusters

		Trimer	Tetramer	Pentamer
2-body	$E_{pol}^{(1)}$	-26.645	-48.963	-64.090
	$E_{ind}^{(2)}$	-12.252	-25.364	-34.146
	$E_{disp}^{(2)}$	-9.121	-16.088	-20.932
	E_{exch}	36.583	72.445	96.634
	E_{int}^{SAPT}	-11.435	-17.970	-22.534
	$E_{int}^{CCSD(T)}$	-11.624	-18.111	-22.201
3-body	$E_{ind}^{(2)}$	-1.351	-3.169	-4.551
	$E_{ind}^{(3)}$	-0.688	-1.165	-1.251
	$E_{ind-disp}^{(3)}$	-0.090	0.026	0.195
	$E_{disp}^{(3)}$	0.060	0.077	0.045
	E_{exch}	-0.345	-1.958	3.558
	E_{int}^{SAPT}	-2.414	-6.189	-9.120
	$E_{int}^{CCSD(T)}$	-2.371	-6.081	-8.978
4-body	$E_{int}^{CCSD(T)}$		-0.562	-1.220
5-body	$E_{int}^{CCSD(T)}$			-0.009
total	E_{int}^{SAPT}	-13.849	-24.159	-31.654
	$E_{int}^{CCSD(T)}$	-13.995	-24.754	-32.481

performed, and compared with the experimental data[377]. The old work of Clementi and collaborators[378,379] suggested that higher many-body (four-body, etc.) terms are important, so simulations of the radial distribution functions also included these higher terms in the many-body expansion of the total interaction potential. The results of the Monte Carlo simulations that used the SAPT(DFT) pair potential, SAPT three-body potential, and higher N-body contributions from the iterative induction calculations[298,373] are presented in Figure 1-17. An inspection of this figure shows that pair potential alone reproduces reasonably well the minima and maxima of the radial O–H and H–H distribution functions. However, it fails to reproduce the O–O radial distribution function. Adding the three-body terms makes the agreement between theory and experiment almost quantitative for the O–H and H–H radial distribution functions, and semiquantitative for the O–O distribution function. Finally, taking into account four-body and higher many-body effects one can get a quantitative agreement between the theory and experiment. The results presented in Figures 1-16 and 1-17 show that using high accuracy *ab initio* methods for the calculation of pair and nonadditive potentials it is possible now to characterize the equation of state for a dilute gas, and the structural characteristics of the liquid, as well as some thermodynamic characteristics such as the temperature dependence of the water density.

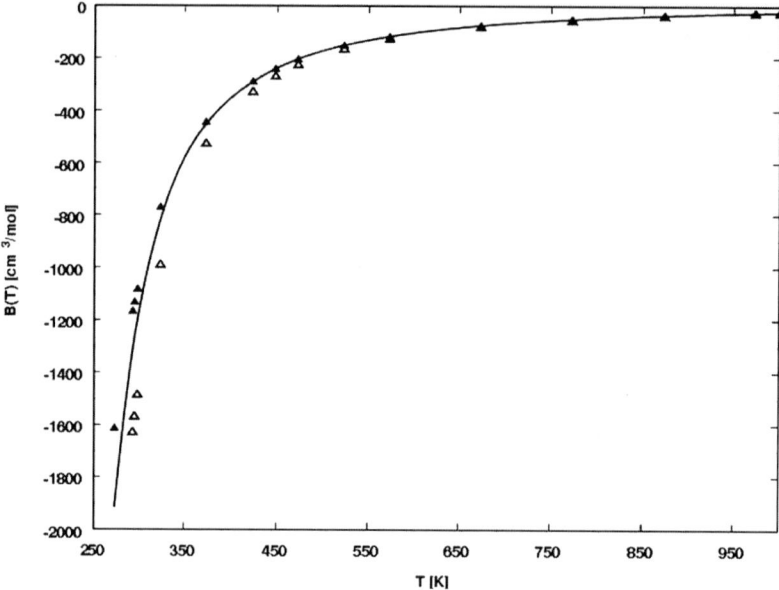

Figure 1-16. Theoretical (full line) and experimental (open triangles) second virial coefficient for gaseous water at various temperatures

13.2.2.   *Clusters of water and HCl and their spectroscopic signatures*

Theoretical and experimental studies of the interactions between water molecules and hydrogen chloride are of fundamental importance for the understanding of the production of stratospheric chlorine molecules which, in turn, take part in the catalytic ozone depletion reactions. This mainly heterogeneous atmospheric reaction begins with the adsorption of the HCl molecules on the surface of water icicles is the source of the stratospheric chlorine atoms in the polar regions[380–382]. Chlorine molecules are photolysed by solar radiation and the resultant chlorine atoms take part in the destruction of the stratospheric ozone. The study of the $(H_2O)_n HCl$ clusters is an important step towards understanding of the behavior of the HCl molecule on the ice surface[383–386].

Only the first two members of the $(H_2O)_n HCl$ series have been observed by means of high resolution spectroscopy. The rotational spectrum of the dimer H_2OHCl has been first recorded and analyzed by Legon et al.[387], and reinvestigated by Kisiel and collaborators[388]. Recently, Kisiel et al.[389,390] reported the observation of the second member of the series $(H_2O)_2 HCl$, also by rotational spectroscopy. Refs. (391–392) reported a detailed theoretical investigations of $(H_2O)_2 HCl$, and compared theoretical predictions with the experimental results from the rotational spectroscopy[389,390]. The structure, molecular properties, and the qualitative picture of the vibration-rotation-tunneling dynamics were in excellent agreement with the experimental data[390,391]. This very good agreement between theory and experiment

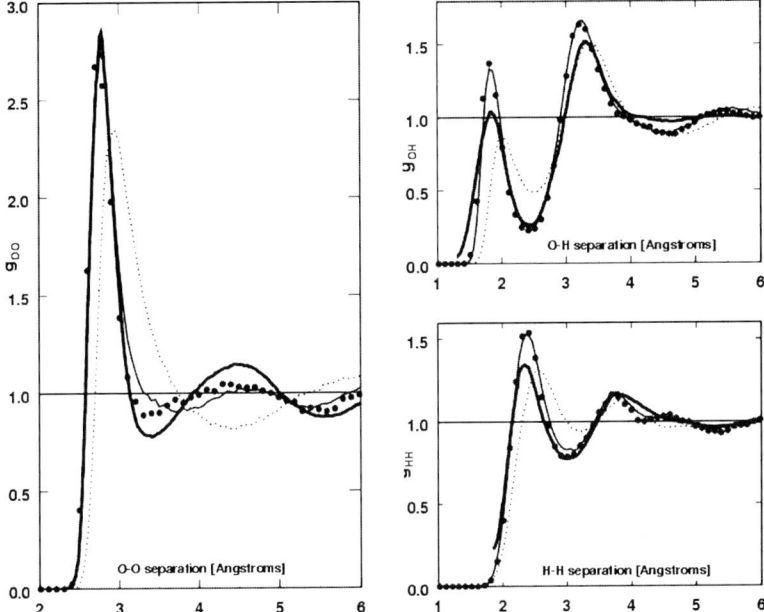

Figure 1-17. Comparison of the experimental (full circles) and theoretical O–O (left panel), O–H (upper right panel), and H–H (lower right panel) radial distribution functions computed from the pair SAPT(DFT) (dotted line), pair+three-body (thin line), and full, pair+three-body+N-body, potentials (thick solid line), respectively

is illustrated on the example of the dipole moment and nuclear quadrupole coupling constants of the trimer. Unlike the water trimer, the $(H_2O)_2HCl$ trimer has a strong dipole moment of 2.31 D. This could be guessed since the trimer was observed in the microwave experiments. Somewhat surprisingly, the dipole moment of the cluster is strongly enhanced compared to the dipole moment resulting from the addition of the three dipole moment vectors of the respective monomers. The latter quantity represents only 48% of the dipole moment of the cluster. This suggests that the mechanism leading to such a large dipole moment enhancement is governed by the polarization of the HCl molecule by the electric charge distributions on the two water monomers[163]. The best theoretical estimates[391] of the total dipole moment of $(H_2O)_2HCl$ and of the components along the principal axes of the inertia tensor of the cluster, obtained at the CCSD(T) level, agree very well with the experimental values derived from the Stark effect measurements in the microwave spectra[390], see Table 1-11 for the comparison. The theoretical total dipole moment agrees with the microwave experiment within 0.6%. The *a* component is reproduced within 0.2%, while for the small *b* component the error of the *ab initio* calculations amounts to 7%. However, due to some constraints in the interpretation of the experimental data, the *c* component was set equal to zero. Therefore, the experimental value of the *b* component should be considered as effective, and the overall agreement between

Table 1-11. The dipole moment (in D) and nuclear quadrupole coupling constants (in MHz) for the chlorine nucleus of the $(H_2O)_2HCl$ cluster

	aug-cc-pVDZ	aug-cc-pVTZ	Extrap.	Best est.	Experiment
μ_T	2.422	2.375	2.355	2.314	2.328(3)
μ_a	2.349	2.295	2.272	2.227	2.232(3)
μ_b	−0.574	−0.595	−0.604	−0.615	0.662(2)
μ_c	0.150	0.151	0.151	0.149	0.00
χ_{zz}	−43.15	−46.16	−47.43		−49.4(5)
χ_{xx}	19.94	21.23	21.77		22.8(5)
χ_{yy}	23.21	24.93	25.65		26.6(15)

the theory and experiment is satisfactory. It is worth noting that a good agreement between the theory and experiment was obtained for the components of the nuclear quadrupole coupling constants. The results are also reported in Table 1-11. Here the calculations were performed using the analytical derivatives of the second-order Møller-Plesset energy. This shows that structure predicted by the *ab initio* calculations is indeed very accurate, and that the ground-state averaging effects are not very important.

Let us discuss in more details comparisons between theory and micro-wave experiments for the $(H_2O)_2HCl$ trimer. To predict a qualitative pattern of the lines in high-resolution spectra of the $(H_2O)_2HCl$ trimer one has to consider the permutation-inversion (PI) group of the complex[393]. The full PI group of the cluster is given by the product $G = G_2 \otimes S_2 \otimes S_5$, where $G_2 = \{E, E^*\}$ is a two-element group of inversion (E denotes the identity, and E^* the space-fixed inversion), S_2 is the permutation group of two oxygen atoms, and S_5 is the permutation group for the five hydrogen atoms. The full PI group contains $2 \times 2 \times 5! = 480$ elements. Such a big group would be very impractical to classify the vibration-rotation-tunneling (VRT) levels and to label possible spectroscopic transitions. However, most of the elements of this group correspond to unfeasible operations. Indeed, out of the 5! permutations of the protons in the cluster, only the permutations (12), (34), and (12)(34) are feasible. Other permutations would correspond to the motions in the complex that break chemical bonds. Moreover, the barrier to the exchange of two water monomers in the cluster is relatively high, ≈ 8 kcal/mol, so the permutation-inversion operations related to S_2 can be neglected as well. Thus, we end up we the following PI group of feasible operations: $G_8 = \{E, E^*, (12), (34), (12)(34), (12)^*, (34)^*, (12)(34)^*\}$. One may note that the permutation-inversion operation E^* correspond to the flipping motions in the trimer, while the PI operations $(12)^*$ and $(34)^*$ govern the so-called tunneling motions. The flipping motion is accomplished by rotating one water monomer about its donor hydrogen bond, while the two bifurcation-tunneling pathways involve one monomer having its protons on the opposite sides of the hydrogen-bonded ring. The free donor proton replaces the hydrogen-bonded one, and the latter is moved to the other side of the ring. During this exchange a flipping of the free acceptor proton takes place. *Ab initio* calculations show that the barriers for the flipping and bifurcation-tunneling motions in the trimer are relatively low, of the order of 0.3 kcal/mol for the flipping and 2 kcal/mol for the bifurcation-tunneling. The VRT states of the complex can be classified according to the irreducible representations (irreps) of the G_8 group which can conveniently be represented as a direct product $G_8 = \{E, E^*\} \otimes \{E, (12), (34), (12)(34)\}$, so the classification of the VRT states can be obtained from the correlation between the states classified under $G_2 = \{E, E^*\}$ and G_8. The energy levels corresponding to the G_2 classification will be denoted by E_{A^+} and E_{A^-}. The correlation between the irreps of G_2 and G_8 shows that the states $A^\pm$ are split into quartets with symmetry labels $A^\pm_{1,2}$ and $B^\pm_{1,2}$. Thus, the bifurcation-tunneling motions introduce the splitting of the flipping tunneling states into quartets. Since the barriers corresponding to the bifurcation-tunneling motions are larger, one may expect that the splitting into quartets will be much smaller than the splitting corresponding to the flipping motions. The schematic representation of the energy levels for the nonrotating $(H_2O)_2HCl$ trimer are presented in Figure 1-18.

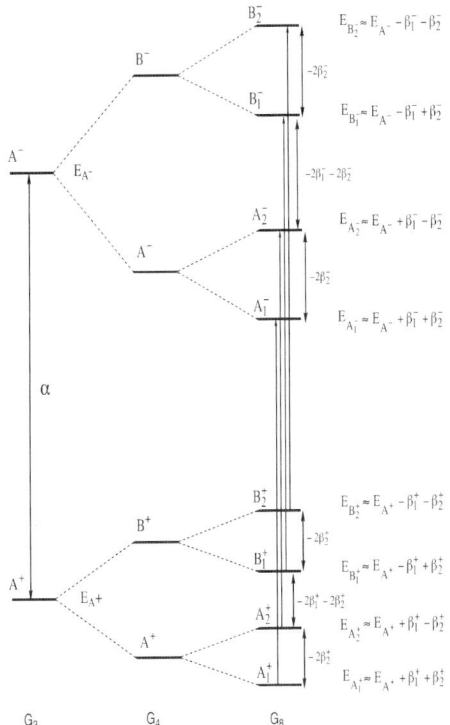

Figure 1-18. Schematic representation of the VRT levels for the non–rotating $(H_2O)_2HCl$ trimer and its $HODOH_2HCl$ isotopomer according to the PI groups G_8 and G_4

An approximate location of the energy levels in Figure 1-18 can be justified by considering the following (very general) Hamiltonian for the nuclear motion:

$$H = H^{\text{flip}} + H^{\text{bif}} + H^{\text{rot}} + H^{\text{Cor}}, \tag{1-283}$$

where the consecutive terms on the r.h.s. denote Hamiltonians describing the flipping and bifurcation-tunneling motions (the so-called internal motions), the external rotational part given by the asymmetric top Hamiltonian, and the Coriolis coupling term. To simplify the discussion let us consider the nonrotating $J = 0$ case, so that the two last terms on the r.h.s. of Eq. (1-283) can be neglected, assume that the intermolecular potential does not depend on the bifurcation-tunneling coordinates and that the eigenvalues and eigenfunctions of the Hamiltonian describing the flipping motion are known. The corresponding wave functions $\Psi_{A\pm}$ are adapted to the G_2 group. Then, the eigenvalues of H_{int} can be found by computing the expectation value of this Hamiltonian with wave functions obtained from $\{\Psi_{A+}, \Psi_{A-}\}$ by symmetry adaptation to G_8. The corresponding wave functions are obtained by applying Wigner's symmetry projection operators of the group $G_4 = \{E, (12), (34), (12)(34)\}$ on Ψ_{A+} and Ψ_{A-}. The quantity α represents the splitting of the levels exclusively due to the flipping motions. The parameters $\beta_{1,2}^{\pm}$, which define the splitting of the tunneling flipping states into quartets, are given by the following matrix elements:

$$\beta_{1,2}^{\pm} = \langle E^* \Psi_{A\pm} | H^{\text{bif}} \pi_{1,2} \Psi_{A\pm} \rangle, \quad \pi_1 = (12), \quad \pi_2 = (34). \tag{1-284}$$

Obviously, the ground state of the complex must be described by the totally symmetric (nodeless) wave function, so it is of A_1^+ symmetry. This suggests that both β_1^+ and β_2^+ are negative. It is, nevertheless, very difficult to estimate the magnitude of the parameters $E_{A\pm}$ and $\beta_{1,2}^{\pm}$ without a prior knowledge of the potential energy surface of the cluster.

Having the qualitative pattern of the VRT levels, one can turn to electric dipole allowed transitions. The allowed dipole transitions between states labeled by the irreps Γ_1 and Γ_2 of G_8 are obtained from the relation[393]:

$$\Gamma_1 \otimes \Gamma_2 \supset \Gamma^* = A_1^-, \tag{1-285}$$

where Γ^* is the antisymmetric irrep of G_8. This leads to the following dipole selection rules:

$$A_{1,2}^+ \leftrightarrow A_{1,2}^- \quad B_{1,2}^+ \leftrightarrow B_{1,2}^-. \tag{1-286}$$

Thermal relaxation between states of the same $A_{1,2}$ or $B_{1,2}$ symmetry will occur independently of the parity, so at very low temperatures the upper quartet states will be much less populated.

The knowledge of the permutation-inversion group allows us to determine the intensity pattern that should be observed in the microwave experiments. It can be obtained by considering the spin-statistical weights. These have been obtained by generating the representation spanned by all possible proton spin functions, $\Gamma_{\text{spin}}^{\text{tot}} = 18A_1 \oplus 6A_2 \oplus 6B_1 \oplus 2B_2$, and by requiring that:

$$\Gamma_{\text{spin}} \otimes \Gamma_{\text{VRT}} \supset \Gamma_{\text{int}} \equiv B_2, \tag{1-287}$$

where Γ_{spin} denotes one of the irreps entering the decomposition of $\Gamma_{\text{spin}}^{\text{tot}}$. The resulting spin-statistical weights show that the intensity pattern in the high-resolution spectra should be 9:3:3:1.

The qualitative picture of the tunneling dynamics and the large values of the computed a and b components of the dipole moment suggest that the lowest lying vibrational states of $(H_2O)_2HCl$ will arise from the vibration-rotation-tunneling motions of the two water subunits in the cluster, and are predicted to be of B_2^+, B_1^+, A_2^+, and A_1^+ symmetry with the spin-statistical weights of 18:9:9:2, respectively. Thus, the rotational spectrum should consist of transitions allowed by the a and b dipole moment components, which are further split into quartets belonging to the four low-lying vibration-rotation-tunneling states.

The qualitative picture of the spectroscopic features of the $(H_2O)_2HCl$ cluster presented above is in a perfect agreement with the microwave measurements of Kisiel and collaborators[389]. The observed rotational spectra show four states, denoted in Ref. (389) by S, S', W, and W'. On the basis of analysis it was possible to assign these states to B_2^+, B_1^+, A_2^+, and A_1^+ symmetries, cf Figure 1-19. The small and large splittings follow approximately the pattern of the levels presented in Figure 1-18, suggesting that the barriers for the internal (flipping and bifurcation-tunneling) motions are realistic. Furthermore, the spectrum of the HClHODOH$_2$ isotopomer shows splittings into doublets, again in agreement with the G_4 picture. The most intense transitions are the a-type transitions, in agreement with the large value of the a component of the dipole moment of the cluster. The c-type transitions were not observed, and this is nicely explained by the very small value of the c component of the dipole moment. Finally, the intensity pattern observed in the rotational spectrum is 4:2:2:1, while theory based on the G_8 group predicts 9:3:3:1.

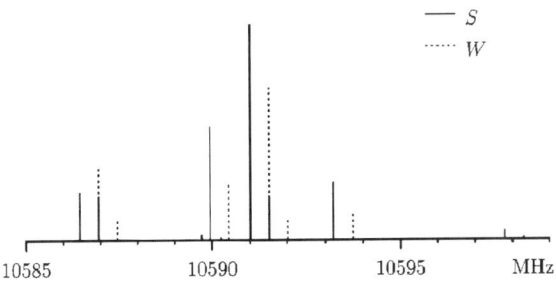

Figure 1-19. Stick diagram of the (a) and (b)-type $J''_{K''_a K''_c} = 1_{01} \rightarrow J'_{K'_a K'_c} = 2_{02}$ rotational transitions in $(H_2O)_2HCl$

Small disagreement may be due to some specific non-equilibrium properties of the supersonic expansion.

In a subsequent paper[394] quantitative predictions of the vibration-rotation-tunneling motions in the trimer were reported. Dynamical calculations were based on the SAPT pair and three-body potential energy surfaces for the flipping motions of the non-hydrogen-bonded protons in the $(H_2O)_2HCl$ trimer. The corresponding experimental spectrum has not been reported thus far, so it is difficult to judge the accuracy of these theoretical predictions.

13.3. Solvation Processes in Small Water Clusters

The proton transfer is one of the most important reactions in chemistry since it takes place in organic, inorganic, and bioorganic processes, both in the stoichiometric and catalytic regimes. Moreover, the proton transfer occurring in the $(H_2O)_2HCl$ cluster is particularly important since the formation of Cl^- from HCl interacting with water molecules seems to be a prerequisite for the Cl_2 formation responsible for the ozone depletion. Thus, in the study of the $(H_2O)_2HCl$ trimer, the proton transfer has also been investigated. It is known that the neutral forms $(H_2O)_nHCl$, and the ionic forms $(H_3O^+Cl^-)(H_2O)_{n-1}$, can coexist for n equal to 4 and 5, but the ionic form $(H_3O^+Cl^-)(H_2O)$ has not been observed thus far. In order to get an estimation of the barrier for the HCl dissociation in the $(H_2O)_2HCl$ cluster, the structure of the dissociated trimer was optimized[395] with the distances of the OH bonds in the hydronium ion kept frozen and equal to 1.04 Å 1.03 Å and 0.98 Å. The structure of the dissociated complex remains a cyclic geometry. Compared to the neutral global minimum this structure shows a global shortening of the distances between heavy atoms since the interaction between the different entities is stronger due to their ionic character. For instance, the $O \cdots O$ distance is 2.497 Å at the MP2 level compared with the corresponding distance of 2.79 Å for the global minimum.

From the energetic point of view, the ionic form lies $+11.73$ kcal/mol higher than the global minimum at the CCSD(T) level. Compared to a similar study for the water trimer[396], the values for the HCl dissociation are substantially smaller by around 20 kcal/mol than the one obtained by Siegbahn for the dissociation of the water trimer computed with the same optimization constrains. This difference can be interpreted as a crude sign of the stronger acidity of hydrogen chloride compared to water. To conclude, we should add that a transition state implying a proton transfer can also be characterized. Its structure is cyclic and corresponds to the proton transfer from one water molecule to the other, with a symmetric substructure $(H_5O_2^+)$ interacting with the chlorine anion. In the $(H_5O_2^+)$ substructure, one of the hydrogen atom is equidistant from the two water molecules of 1.22 Å. This transition state connects the (global) neutral minimum through an ionic pathway with a high barrier of $+13.76$ kcal/mol at the CCSD(T) level. The height of this barrier suggests that the dissociation process will be extremely slow, even at high temperatures.

One may ask whether the dissociation process can occur in the next hydrate of this series, namely the tetramer $(H_2O)_3HCl$. It is known from the crystallographic X-ray diffraction study that a stable crystal of $(H_2O)_3HCl$ stoichiometry exists[397], and its (rotationally unresolved) near-infrared spectrum was observed[398]. The X-ray and spectroscopic data on the crystal suggest that the HCl molecule in the crystal is fully dissociated, i.e. the stoichiometry of the tetramer should rather be written as $(H_2O)_2(H_3O)^+Cl^-$. On the other hand the near-infrared spectrum of the tetramer in the argon matrix[399] suggests that HCl is not dissociated.

The protolytic dissociation process in the tetramer was studied in Ref. (400) As could be guessed from the matrix isolation studies of the cluster, the global minimum on the potential energy surface of the trihydrate corresponds to a neutral form of the cluster. Calculations predicted a cyclic rectangular hydrogen-bonded structure with both HCl and three water monomers acting simultaneously as proton donors and acceptors. The structure of the neutral trihydrate does not agree with the structure "guessed" in Ref. (399) from the analysis of the infrared matrix isolation spectra. The authors of Ref. (399) suggested that the OH stretch at $3430\,cm^{-1}$ is too small for a water molecule forming hydrogen bonds with only two molecules, and suggested that this particular water molecule should be surrounded by two other waters and by the HCl molecule. However, the harmonic frequencies corresponding to the theoretical structure of the global minimum[400] agree quite well with the experimental frequencies of Ref. (399). In particular, the OH stretch frequency in the arrangement $ClH\cdots OH\cdots O$, is $3379\,cm^{-1}$, in a good agreement with the experimental value of $3430\,cm^{-1}$. This suggests that the analysis of the matrix isolation spectra leading to a geometry with one water molecule attached to a cyclic $(H_2O)_2HCl$ trimer[399] is not correct.

There is another minimum on the potential energy surface of the $(H_2O)_3HCl$ trihydrate corresponding to an ionic form of the cluster. The ionic minimum is characterized by a cyclic structure with the Cl^- ion hydrogen-bonded simultaneously to the two water molecules and the hydronium ion. It is interesting to note that the ionic form is less stable, as it is located $5.2\,kcal/mol$ above the neutral minimum. Therefore, the ionic structure of the trihydrate observed in the X-ray experiments [397] cannot directly be related to this ionic minimum. In fact, the structure observed in crystals must be considered as effective, taking into account the crystal field effects.

The mechanism leading from the neutral to ionic minima is presented in Figure 1-20. First, the proton is transferred from the HCl molecule to the water molecule. This step is followed by another proton transfer from the hydronium ion to the second water molecule and by a flipping motion of the free (non-hydrogen-bonded) OH bonds from the up-up-down to the down-down-up position. The latter step is responsible for the formation of the transition state. The next step is achieved by performing a flipping motion of the free OH bonds directly to the ionic minimum. Although the optimizations were performed at the B3LYP level, B3LYP energetic locations of the structures were confirmed by single-point CCSD(T) calculations, so the mechanism presented on Figure 1-20 should be reliable.

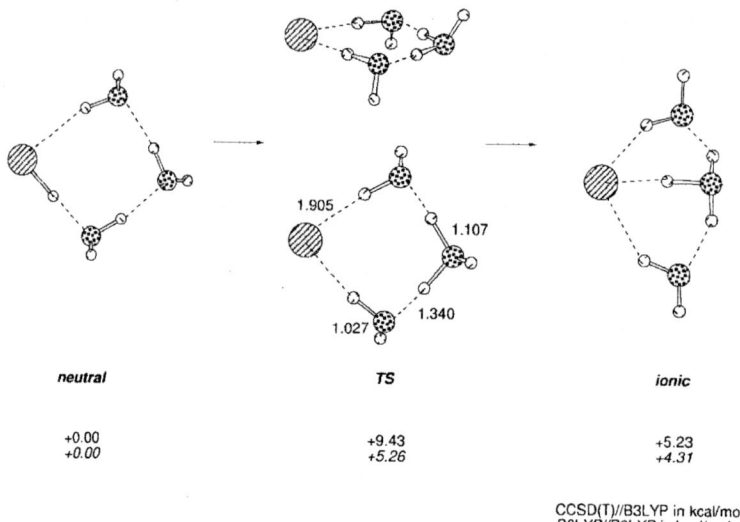

neutral   TS   ionic

+0.00   +9.43   +5.23
+0.00   +5.26   +4.31

CCSD(T)//B3LYP in kcal/mol
B3LYP//B3LYP in kcal/mol

Figure 1-20. Mechanism of the protolytic dissociation in the tetramer $(H_2O)_3HCl$

Since the ionic minimum does not correspond to the structure derived from X-ray experiments it was interesting to check[400] whether one can somehow mimic the crystal field effects, and calculate an effective structure of the tetramer in the presence of a continuum. In order to mimic the crystal field effects the ionic structure of the trihydrate was optimized[400] using the self-consistent reaction field method[248]. The SCRF geometry was found to be very different from the ionic geometry in the gas phase. In fact both the SCRF calculations and the X-ray diffraction measurements[397] predict globally the same, almost planar structure close to a rectangle, differing just by some details. Also the agreement between the computed and measured[398] infrared frequencies is rather good (with a root-mean-square deviation of $\approx 50\,\mathrm{cm}^{-1}$). This kind of agreement supports the conclusion[400] that the ionic dissociation in the trihydrate cannot be observed in the gas phase.

The barriers to the HCl dissociation in the trimer and tetramer are very high, making the dissociation process extremely slow. However, we all know that in solution HCl is fully dissociated. Therefore the authors of Ref. (386) tried to find how many water molecules are needed to dissociate HCl in the $(H_2O)_n HCl$ cluster. To this end the geometries corresponding to the neutral and ionic minima on the surface were optimized, and the lowest-energy pathways connecting these minima were determined. These pathways can be considered as mechanisms governing the chemical reaction of HCl dissociation in the presence of a few water molecules. Contrary to the tetramer, the global minimum of the pentamer $(H_2O)_4HCl$ is ionic. It corresponds to a zwitterion consisting of the Cl^- anion separated from the hydronium cation H_3O^+ by a 'crown' of three neutral water molecules. There is also a neutral minimum corresponding to a cyclic hydrogen-bonded structure with water molecules and HCl serving at the same time as proton donors and acceptors. This structure is roughly 2.4 kcal/mol less stable than the ionic structure. Note parenthetically that in Ref. (386) the optimizations were performed at the B3LYP level. However, the B3LYP energetic locations of the structures were confirmed by single-point CCSD(T) calculations[386].

In order to find the mechanism leading from the ion to the neutral pentamer, i.e. of the association, the reverse process of dissociation, the authors of Ref. (386) observed that the neutral system is cyclic while the ion has the cage-like structure of Figure 1-21. A likely ionic intermediate minimum is obtained by flipping a proton (marked by the hash '#' symbol in the global minimum structure shown on Figure 1-21), thus breaking the three-fold symmetry of the global minimum. This flip proceeds via the transition state 'TSflip0'. To get from the local minimum 'Ionic1' thus obtained to the next local minimum, denoted on the figure by 'Ionic2', we must cross a transition state referred to as TS1. This transition state is obtained by a proton transfer from the hydronium ion at the bottom to one of the water molecules in the crown, which now becomes a hydronium ion. It leads to a reorganization of the hydrogen bonds and to the formation of one hydrogen bond between the new hydronium ion and one of the waters in the crown. The structure 'Ionic2' is 2.67 kcal/mol higher in energy than the global minimum. Since two protons on neighboring water molecules are sticking down in 'Ionic2', it is energetically favorable to flip one proton, so that it points up. This gives the structure labeled by 'Ionic3'. This flipping motion proceeds via the transition state 'TSflip', where the flipping proton is again marked by a hash sign. The one but last step in the dissociation is the formation 'TS2', which leads to a rather high lying saddle point. Finally the Zundel cation donates a proton to Cl^-. During this proton transfer, the weak H bond breaks between the Cl^- anion and one of the water moieties with the distance increasing from 2.6 to 5.3 Å.

On the basis of the mechanism summarized above one can expect that the dissociation reaction in the pentamer will not be particularly fast. The barriers corresponding to the 'TS1' and 'TS2' structures are relatively high, but lower than the barriers in the dissociation pathways of the trimer and tetramer. One can expect that for the hexamer $(H_2O)_5HCl$ the barriers will be even lower, making the dissociation process very fast. The global minimum on the potential energy surface of the hexamer is again ionic[386]. The corresponding structure looks like an open book: one page is spanned by the ions and two water molecules, and the other again by the ions and the other two waters. Unlike in the pentamer, the hydronium and chloride ions are not separated by water molecules, but remain in a relatively close contact. There is also a neutral minimum corresponding to a cyclic hydrogen-bonded structure with water molecules and HCl serving at the same time as proton donors and acceptors. This structure is roughly 7.1 kcal/mol less stable than the ionic structure.

It is interesting to note that the analysis of the pair and three-body contributions to the total interaction energy quantitatively explains the relative stability of the ionic and neutral minima of the pentamer. This is illustrated in Table 1-12, where

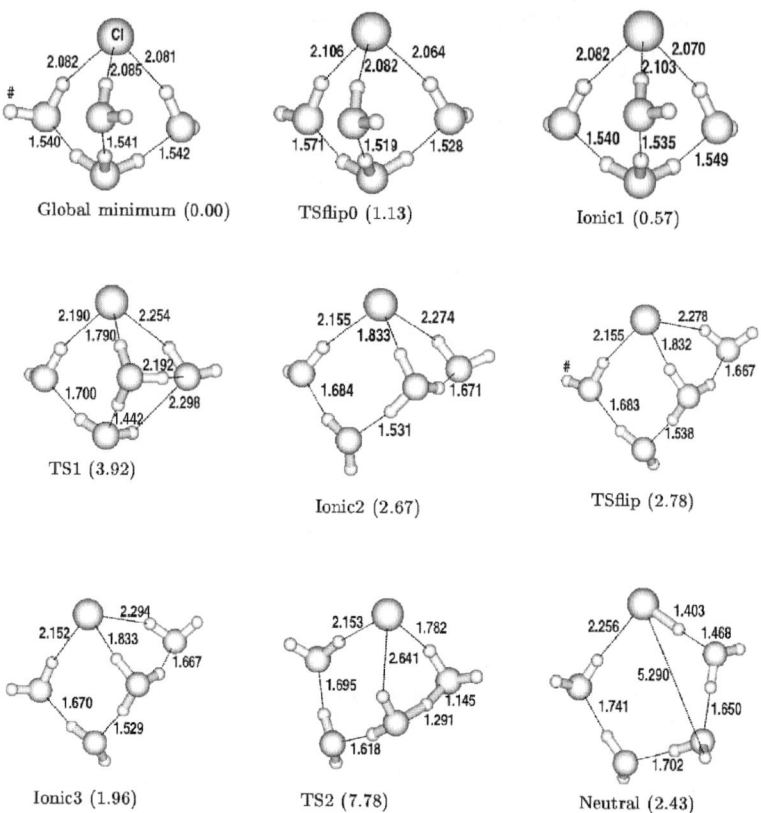

Figure 1-21. Mechanism of the protolytic dissociation in the pentamer $(H_2O)_4HCl$

two- and three-body contributions to the interaction energy for various subclusters of the ionic global minimum and neutral local minimum of the $(H_2O)_4HCl$ pentamer are reported. As expected, the main stabilizing term (-107.75 kcal/mol) is the interaction between the anion (monomer 5) and the cation (monomer 1). It gives about half of the total interaction energy. However, this contribution is not sufficient to stabilize the ionic cluster by itself, since the reaction

$$HCl + H_2O \rightarrow Cl^- + H_3O^+, \tag{1-288}$$

is endothermic by 163.05 kcal/mol at the CCSD(T) level. Thus, the ion–ion interaction is not sufficient to let the reaction proceed, but together with the pair interactions between the neutral water molecules in the crown (molecules 2, 3, and 4) and the hydronium cation (three times ≈ -25.4 kcal/mol) and Cl^- (three times ≈ -11.7 kcal/mol) the reaction becomes energetically possible. The nonadditive three-body effects are destabilizing the global minimum, and they contribute about $\approx 10\%$ of the pair interactions. Higher order terms in the many-body

Table 1-12. Two- and three-body contributions (in kcal/mol) to the total interaction energies for the ionic and neutral structures of the $(H_2O)_4HCl$ pentamer

Ionic				Neutral			
Subcluster	Pair	Subcluster	3-body	Subcluster	Pair	Subcluster	3-body
(1,5)	−107.75	(1,3,4)	5.21	(1,2)	−2.60	(3,4,5)	1.50
(1,2)	−25.37	(1,2,4)	5.20	(1,3)	−1.07	(2,4,5)	−2.44
(1,3)	−25.43	(1,2,3)	5.19	(1,4)	−0.99	(2,3,5)	−0.43
(1,4)	−25.40	(3,4,5)	1.46	(1,5)	−3.38	(2,3,4)	−1.72
(2,5)	−11.70	(2,4,5)	1.45	(2,3)	−3.21	(1,4,5)	−1.89
(3,5)	−11.69	(2,3,5)	1.45	(2,4)	−1.01	(1,3,5)	−0.84
(4,5)	−11.71	(1,4,5)	0.50	(2,5)	−1.43	(1,3,4)	−0.33
(2,3)	1.22	(1,3,5)	0.49	(3,4)	−3.54	(1,2,5)	−3.88
(2,4)	1.21	(1,2,5)	0.48	(3,5)	−0.68	(1,2,4)	−0.28
(3,4)	1.21	(2,3,4)	0.48	(4,5)	−0.30	(1,2,3)	−1.67
total	−215.41	total	21.91	total	−17.91	total	−11.98

expansion of the total interaction energy are negligible[386]. In the case of the neutral structure the situation is quite different. Although the largest contributions to the total pair interaction energy come from the closest water–water interactions, they contribute only 30%, and all other pair interactions including the distant ones have to be summed up to give the net interaction effect. The nonadditive effect is almost as large as the total pair interaction energy. Evidently, the three-body forces play an important stabilizing role in determining the structure of the neutral pentamer. Again higher-order many-body effects are small. They contribute $\approx 6.5\%$ of the total interaction energy at the CCSD(T) level.

The association process from the ionic to the neutral form proceeds in four steps. See Figure 1-22 for the details. In the first step we go to a transition state 'TS1', which contains the Zundel cation. A hydrogen bond on one page of the hexamer "book" is broken, and another is formed. The next structure occurring on the pathway is a local minimum 'Ionic1'. This structure closely resembles to the 'TS1' structure, except that there is no Zundel cation. The next step is the formation of the second transition state 'TS2'. This structure has a proton that is shared equally by a water molecule and the hydrogen chloride. In the last step a proton transfer towards the chloride ion occurs, leading to the neutral structure. It is worth noting that the dissociation process in the hexamer should be relatively fast. Indeed, starting from the neutral structure we have just to overcome two relatively low barriers.

In a very ingenious experiment the authors of Ref. (401) could get a direct probe of the HCl dissociation in water clusters pump-and-probe femtosecond spectroscopy. They observed that the dissociation process takes place spontaneously in the presence of five water molecules, while the dissociation of HCl in smaller clusters can only be photoinduced. Theoretical predictions reported in Ref. (386) fully confirm these experimental findings. In fact

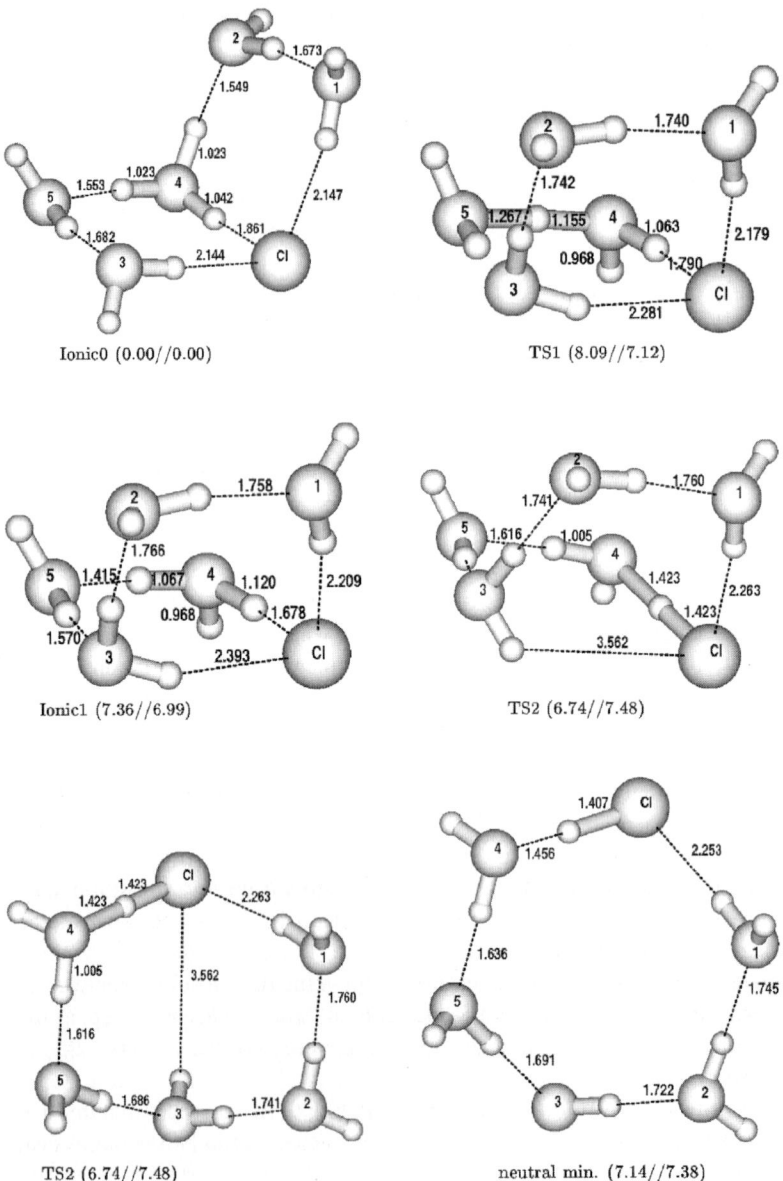

Figure 1-22. Mechanism of the protolytic dissociation in the hexamer $(H_2O)_5HCl$

theoretical results were published before the experimental paper, and were widely discussed in a commentary[402] on Ref. (401) in the Science Compass. The example of the HCl dissociation in small water clusters shows again that highly accurate methods of modern quantum chemistry together with the theory of intermolecular forces and theoretical spectroscopy are capable of correctly predicting such complicated phenomena like the mechanisms of acidic dissociation.

13.4. Collision-induced Properties and Modelling of Raman Spectra of Atomic Gases

The collision-induced light scattering in the helium gas has been the subject of many experimental studies[403-409] Most of these measurements were done at high densities[403-405], so the reported Raman intensities were affected by three-body contributions, and pure pair spectra had to be separated out[405] by applying simplified models. Only the polarized and depolarized Raman spectra reported by Proffitt, Keto, and Frommhold[409,410] were shown to be free from three-body contributions.

These experimental advances stimulated associated theoretical developments. In an extensive theoretical study Dacre and Frommhold[411] have checked the accuracy of the *ab initio* CISD (CISD stands for the Configuration Interaction method restricted to single and double substitutions) trace and anisotropy polarizabilities of He_2[412] by exposing them to the test of computing the observed Raman intensities. While the depolarized spectra computed from Dacre's polarizability[412] showed good agreement with the experiment, the theoretical polarized spectrum was much less intense than the spectrum derived from the experiment[411].

The reasons for the less satisfactory agreement between the theoretical and experimental polarized Raman spectra may be both on the theoretical and on the experimental sides. The experimental polarized spectrum is obtained as the difference of two nearly equal signals excited with different beam polarizations [409,410], and the accuracy of the polarized intensities deduced from the experiment is rather poor. On the other hand, the theoretical values of the interaction-induced trace may suffer from the size-inconsistency of the CISD method or from the basis-set superposition error.

In Ref. (316) the SAPT approach has been applied to compute the interaction-induced polarizability of the helium diatom. Before discussing the Raman spectra obtained from the SAPT collision-induced polarizabilities, let us discuss the importance of various physical contributions to the parallel and perpendicular components of the collision-induced polarizability tensor.

This is analyzed in Tables 1-13 and 1-14 where the computed values of $\Delta\alpha_{zz}$ and $\Delta\alpha_{xx}$ in terms of SAPT contributions at various interatomic distances are reported. The largest contributions to the components of the interaction-induced polarizability are given by the first-order terms. Except for the smallest interatomic distance ($R = 3$ bohr), the sum $\Delta\alpha_{ii,\mathrm{pol}}^{(1)} + \Delta\alpha_{ii,\mathrm{exch}}^{(1)}$, $i = z$ or x, reproduces more than 88% of the total

Table 1-13. SAPT contributions (in 10^{-3} a.u.) to the parallel component of the interaction-induced polarizability of He_2 as function of the interatomic distance R (in bohr)

R	3.0	5.6	7.0	10.0
$\Delta\alpha_{zz,pol}^{(1)}$	342.217	44.626	22.184	7.572
$\Delta\alpha_{zz,exch}^{(1)}$	−616.256	−9.498	−0.627	−0.001
$\Delta\alpha_{zz,ind}^{(2)}$	143.300	1.507	0.217	0.020
$\Delta\alpha_{zz,exch-ind}^{(2)}$	−129.820	−0.799	−0.034	−0.000
$\Delta\alpha_{zz,disp}^{(2)}$	57.033	2.976	0.603	0.045
$\Delta\alpha_{zz,exch-disp}^{(2)}$	−7.158	−0.250	−0.021	0.000
$\Delta\alpha_{zz}^{SAPT}$	−5.744	39.645	22.370	7.653

Table 1-14. SAPT contributions (in 10^{-3} a.u.) to the perpendicular component of the interaction-induced polarizability of He_2 as function of the interatomic distance R (in bohr)

R	3.0	5.6	7.0	10.0
$\Delta\alpha_{xx,pol}^{(1)}$	−114.194	−21.407	−11.037	−3.785
$\Delta\alpha_{xx,exch}^{(1)}$	−112.193	−1.140	−0.057	−0.000
$\Delta\alpha_{xx,ind}^{(2)}$	10.974	0.203	0.047	0.005
$\Delta\alpha_{xx,exch-ind}^{(2)}$	1.300	0.000	0.000	0.000
$\Delta\alpha_{xx,disp}^{(2)}$	18.550	0.923	0.217	0.021
$\Delta\alpha_{xx,exch-disp}^{(2)}$	−2.888	−0.047	−0.003	0.000
$\Delta\alpha_{xx}^{SAPT}$	−185.932	−21.442	−10.851	−3.752

interaction-induced polarizability. The interatomic correlation contributions are of relatively modest importance. For example, in the region of the potential minimum ($R = 5.6$ bohr) the sum of the dispersion and exchange-dispersion terms contributes to $\Delta\alpha_{zz}$ and $\Delta\alpha_{xx}$ only 7% and 4%, respectively.

The computed polarizability invariants have been analytically fitted and used in quantum-dynamical calculations of the binary collision-induced Raman spectra. The results of the dynamical calculations are summarized in Figure 1-23. An inspection of this figure shows that the agreement of the theoretical and measured[410] depolarized Raman intensities is satisfactory. Most of the intensities agree within 3% or better. Only at very low and high frequency shifts this good agreement deteriorates somewhat. Still, the predicted intensities at high frequencies are within the experimental error bars. At very low frequencies the theoretical results are outside the experimental error bars, but these discrepancies are consistent with the estimated (combined) error of the SAPT and dynamical calculations.

The theoretical polarized Raman intensities agree with the experiment within the large (± 50–60%) experimental error bars over a wide range of the frequency shifts. Except for the low frequency region, the predicted polarized spectrum is

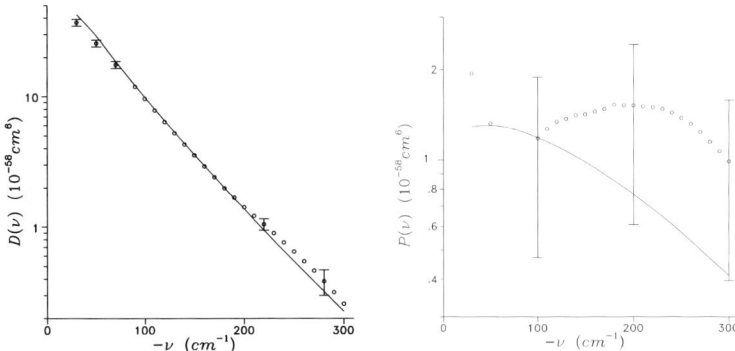

Figure 1-23. Theoretical (full lines) and experimental (open circles) depolarized (left panel) and polarized (right panel) Raman spectra of the He gas at room temperature

much less intense. It is worth noting that our results for the depolarized and polarized Raman intensities are in good agreement with those generated from the CISD polarizability invariants[411]. Both these observations suggest that the theoretical results are rather well converged, and that the error in the *ab initio* polarized intensities is considerably smaller than the experimental error of 50 to 60%. It was concluded in Ref. (316) that the improvement of the agreement between theory and experiment for these intensities should come mainly from the experimental side.

In 2000 Chrysos and collaborators[413−416] from the University of Angers in France reported new measurements of the depolarized and polarized Raman spectra of the helium gas at room temperature[413−415]. The agreement between the new experimental data and the calculations of Ref. (316) was excellent. In particular, the computed and measured polarized spectra showed a remarkable agreement over the whole frequency range scanned in the experiment, cf. Figure 1-24. In fact the experimental and theoretical curves on Figure 1-24 are almost indistinguishable. Moreover, low temperature measurements[416] were also in a perfect agreement with the theoretical results generated from the polarizability invariants of Ref. (316) . This level of agreement between theory and experiment suggests that invariants of the collision-induced polarizability tensor for the helium diatomic collisional complex, as computed by SAPT[316], are indeed very accurate. Similarly good agreement was obtained for the Raman spectra of the neon diatom[417,418].

13.5. Modelling of Dielectric and Refractive Properties of Atomic Gases

The dielectric properties of the helium gas are of great experimental interest, and it is not surprising that since the early 1960's increasingly accurate measurements[419−428] are reported in the literature. In Ref. (317) a detailed study of the importance of the quantum effects and of the applicability of the semiclassical expansion has been

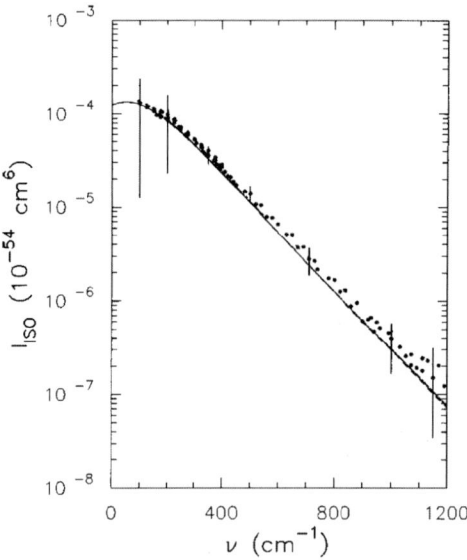

Figure 1-24. Theoretical (full line) and new experimental (circles) polarized spectrum of the He gas at room temperature

reported. This study started from the *ab initio* SAPT trace polarizability[316] and it involved semiclassical and full quantum calculations of the second dielectric virial coefficient for the ^{4}He gas at various temperatures.

Before comparing theory and experiment let us discuss the convergence of the semiclassical expansion of the dielectric second virial coefficient. In Table 1-15 the classical dielectric virial coefficient the first and second quantum corrections, and the full quantum result are reported. An inspection of this table shows that the quantum effects are small for temperatures larger than 100 K, and $B_\epsilon(T)$ can be approximated by the classical expression with an error smaller than 2.5%. At lower temperatures the dielectric virial coefficient of ^{4}He starts to deviate from the classical value. Still, for $T \geq 50$ K the quantum effects can be efficiently accounted for by the sum of the first and second quantum corrections. Indeed, for $T = 50, 75,$ and 100 K the series $B_\epsilon^{(0)}(T) + B_\epsilon^{(1)}(T) + B_\epsilon^{(2)}(T)$ reproduces the exact results with errors smaller than 2%. At temperatures below 50 K the semiclassical expansion of the second dielectric virial coefficient in powers of $\hbar^2$ starts to diverge. Given the overall pattern of convergence of the semiclassical expansion, it was interesting to check whether any rational approximations involving the low-order quantum corrections can reproduce the converged quantum result. Since only the three terms in the expansion of $B_\epsilon(T)$ as a power series in $\hbar^2$ were computed, only the simplest [1/1] approximant could be used. The values of this approximant at various temperatures are also reported in Table 1-15. Except for the lowest temperatures, the simple [1/1] Padé approximant works surprisingly well. For $T = 15$ and 20 K the sum of the

Table 1-15. Second dielectric virial coefficient of ^{4}He (in cm^6mol^{-2}) as function of the temperature (in K)

T	$B_\epsilon^{(0)}(T)$	$B_\epsilon^{(1)}(T)$	$B_\epsilon^{(2)}(T)$	$\sum_{n=0}^{2} B_\epsilon^{(n)}(T)$	[1/1]	$B_\epsilon(T)$
4	−0.0842	0.7178	−6.5461	−5.9125	−0.0133	−0.0081
5	−0.0512	0.3040	−2.0226	−1.7699	−0.0115	−0.0081
7	−0.0298	0.1028	−0.4354	−0.3624	−0.0102	−0.0085
10	−0.0209	0.0403	−0.1088	−0.0894	−0.0100	−0.0093
15	−0.0174	0.0169	−0.0282	−0.0287	−0.0111	−0.0108
20	−0.0170	0.0100	−0.0121	−0.0191	−0.0125	−0.0121
30	−0.0183	0.0054	−0.0041	−0.0171	−0.0153	−0.0152
40	−0.0203	0.0036	−0.0021	−0.0187	−0.0179	−0.0179
50	−0.0223	0.0028	−0.0012	−0.0208	−0.0204	−0.0204
75	−0.0274	0.0018	−0.0005	−0.0261	−0.0260	−0.0260
100	−0.0320	0.0013	−0.0003	−0.0309	−0.0309	−0.0309
125	−0.0362	0.0011	−0.0002	−0.0353	−0.0353	−0.0353
150	−0.0401	0.0009	−0.0001	−0.0393	−0.0393	−0.0393
175	−0.0438	0.0008	−0.0001	−0.0431	−0.0431	−0.0431
200	−0.0472	0.0007	−0.0001	−0.0466	−0.0466	−0.0466
250	−0.0536	0.0006	−0.0001	−0.0530	−0.0530	−0.0530
300	−0.0593	0.0005	−0.0000	−0.0588	−0.0588	−0.0588

classical term and first and second quantum corrections overestimates the exact result by 265% and 58%, quantum corrections overestimates the exact result by 265% and 58%, respectively, while the [1/1] approximant reproduces the quantum results with errors of the order of 3%. This result is gratifying since the calculation of the quantum corrections is much simpler than full quantum-statistical calculations. It remains to be seen, however, if this optimistic result holds for other systems as well.

The comparison of the theoretical and experimental values of the second dielectric virial coefficient can serve as a further check of the accuracy of the *ab initio* trace polarizability. An example of such a comparison is shown on Figure 1-25, where the theoretical and experimental second dielectric virial coefficients for the ^{4}He gas at various temperatures are reported.

At high temperatures the *ab initio* results agree well with the data from indirect measurements[422,426–428]. The only exception is the value at $T = 242.95$ K. Here the theoretical result is slightly outside the experimental error bars. The agreement with the results of direct measurements[420,421–425] is less satisfactory. The *ab initio* results agree very well with the old experimental data of Orcutt and Cole[420], and disagree with the data of Vidal and Lallemand[423,424]. Since the theoretical values agree with the majority of the high-temperature experimental data, and since the second dielectric virial coefficient changes very slowly with temperature, it is very likely that the results of direct measurements reported by Vidal and Lallemand[424] are contaminated by nonadditive three-body effects.

At low temperatures the situation is more complex. The *ab initio* result at 77.4 K agrees very well with the value from indirect measurements reported by Huot and

Bose[427]. Other low temperature data were obtained from direct measurements[425] and show much scatter. At $T = 13.804$ K the *ab initio* value agrees with the measurement, while at $T = 7.198$ K the theoretical result is almost within the experimental error bars. At other temperatures the disagreement is quite substantial, and it is unlikely that the present value of the theoretical polarizability trace differs from the exact one to the extent sufficient to explain the observed differences. Therefore, these data should probably be remeasured. In fact, given the level of agreement between the theoretical and experimental polarized Raman spectra[414,415] the theoretical values of the dielectric virial coefficient are probably more accurate than some of the experimental data.

The low temperature refractive properties of the He gas have not been studied extensively. However, the second virial Kerr coefficient can be related to the zeroth moment of the polarized Raman spectrum, and thus deduced from the Raman experiment. For the helium gas at the liquid nitrogen temperature the experiment gives 1.46 a.u.[416], the full quantum calculation 1.45[328], while the classical result computed according to Eq. (1-260) gives 1.63[328]. This shows that also for the Kerr effect the quantum corrections are important. A systematic study of these corrections and of the convergence of the semiclassical expansion has not been reported thus far, even though all necessary expressions are derived[328].

Recently, Rizzo and collaborators reported *ab initio* calculations of the polarizability invariants for the helium[318,319], neon[320], and argon[318,319] dimers, followed by quantum-statistical and classical calculations of the dielectric second

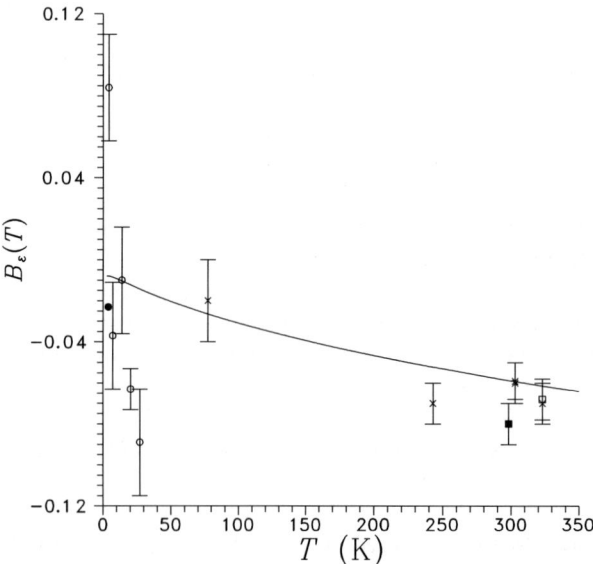

Figure 1-25. Theoretical (full line) and experimental second dielectric virial coefficients of the He gas at various temperatures

virial coefficients. For the helium dimer the polarizability invariants reported in Refs. (318,319) are in a fair agreement with the older data from SAPT calculations[316], but the agreement between the computed ad experimental dielectric second virial coefficients is about the same as in Ref. (317) . For heavier gases the quantum effects were shown to be very small[318-320]. Rizzo and collaborators also reported calculations of some other density dependent properties of atomic gases[322]. In particular, they found the second virial coefficient related to the electric field induced second harmonic generation should be large in the binary mixtures of He/Ar and Ne/Ar over a wide range of temperatures. Howeve, these findings have to confirmed by experiments.

14. CONCLUSIONS AND OUTLOOK FOR THE FUTURE

The aim of this chapter was to show the reader that much can be learned about intermolecular forces from *ab initio* quantum-mechanical studies, and from sound comparisons between the theory and high precision experiments. We paid special attention to the theory and computational methods required to calculate accurate interaction potentials, interaction-induced properties, bound states, spectra, and collisional characteristics of Van der Waals and hydrogen-bonded complexes, as well as dielectric, refractive, and thermodynamic properties of bulk phases. The examples given in the latter part of this chapter were meant to illustrate how the experimentalist and theoretician working in a tandem can gather a large amount of useful quantitative informations about the interactions between molecules. We hope that we succeeded in convincing the reader that the field of intermolecular forces and of the dynamics of bound and collisional Van der Waals and hydrogen-bonded complexes is a branch of science well worth pursuing, both from the theoretical and experimental sides.

Let us stress that several topics related to intermolecular forces were not covered by the present review. First of all, the problem of intramolecular degrees of freedom was not discussed. As far as the calculations of the interaction potentials are concerned, the dependence of the potential on the internal degrees of freedom must be considered. This does not introduce any significant complications in the calculations if the geometries of the monomers are not strongly distorted from their equilibrium values. *Ab initio* calculations of full potential energy surfaces (including the intramonomer coordinates) are tedious, and this explains why fully dimensional potentials are restricted to simple systems[346,354,429,430]. See, however, Ref. (431) for an *ab initio* calculation of the full flexible potential for the water dimer. Dynamics of Van der Waals complexes including the intramonomer excitations is much more complicated. Again, for simple cases fully dimensional calculations can easily be done[347,354,429,432-434], but for larger systems new ideas are needed. Some work in this direction has been done by Leforestier and collaborators[435]. Still, the advances in the computer power are so fast that nowadays it is possible to perform nearly exact dynamical calculations for systems like He–HC$_3$N (HC$_3$N is the cyanoacetylene molecule) including all bending and stretching motions of HC_3N[436]. The problem of averaged geometries of the monomers that would somehow take into account, in an effective way, the intramonomer vibrations is only important for monomers with large rotational constants like H_2 or HF. For these monomers the equilibrium geometry is quite different from the vibrationally averaged geometry, and a judicious choice of the average geometry must be done[429]. For systems with small rotational constants the problem of the intramonomer degrees of freedom and of the average geometry is important only in the case of "soft degrees of freedom", such as coordinates describing the low frequency bending modes in cyanoacetylene. Again, new ideas as far as the description of the coupling between these soft intramolecular modes and the intermolecular modes are needed.

The subject of interactions between open-shell monomers[437–441] was not discussed even though these interactions play an important role in the theory of chemical reactions[442]. For instance, the kinetics of a chemical reaction strongly depends on the Van der Waals interactions in the entrance channel, far from the region where the true reaction occurs[443]. Open-shell interactions and dynamics are very important for modeling of processes occurring in the atmospheric chemistry. Important progress has been done in this field, especially for high-spin open-shell complexes[444] like the oxygen dimer. Open-shell complexes are characterized by surfaces that show conical intersections[43]. The nonadiabatic couplings are due to these intersections, and strongly modify the dynamics of open-shell Van der Waals complexes. As rightly stated by Yarkony[43], "nonadiabatic chemistry is an intellectually demanding area of research, one of limited number of areas of the electronic structure theory that has not been trivialized by standardized electronic structure codes". Also interactions of atoms or molecules in their excited states are revisited now. This is especially interesting since some new types of interactions, e.g. resonant interactions, appear[445,446]. All these topics are subject of very intense theoretical and experimental studies, and much progress in this field is being done and has to be done. An important problem in this context is the dynamics of open-shell complexes on multiple potential energy surfaces. This problem was considered by Dubernet and Hutson[447–450] 10 years ago for the bound states within the diabatic approximation, and by Alexander and collaborators for the collisional processes[451–455].

Actually, rigorous diabatization procedures are highly needed. The diabatic representation of the Schrödinger equation for the nuclear motions is thought to take into account as much as possible of the nonadiabatic couplings. Unfortunately, the diabatization procedures employed at present are to some extent arbitrary, and minimize just the angular or radial couplings. As a matter of fact, the diabatization procedures employed in high-accuracy *ab initio* calculations[437–441] are rather dictated by the availability of some procedures in the quantum-chemical programs, and not the analysis of the physical problem.

Trimers and larger clusters involving open-shell monomers with spatially degenerate ground states are experimentally investigated by high-resolution optical spectroscopies and by photoelectron spectroscopy. On the theory side Chalasinski and collaborators[456–459] proposed an approximate treatment of nonadditive three-body effects in open-shell trimers. These authors suggested that the nonadditive effect in spatially degenerate trimers can partly be assigned to the so-called orientational nonadditivity related to the orientation of partially occupied degenerate orbitals with respect to the closed-shell monomer in the trimer, and partly to the so-called genuine nonadditive contribution similar in nature to closed-shell interactions. The definitions of the orientational and genuine nonadditivities are to some extent arbitrary. Visibly, much work in this direction is needed.

Very little is known about the relativistic and QED interactions[445,460,461]. These interactions, although very small, modify the long-range behavior of the potential energy surfaces, and have a strong influence on some quantities determining very low-energy scattering cross sections, e.g. the scattering length[462,463]. Recently, the forty years old theory developed by Meath has been revisited[464,465], and important advances in this direction are bewing done.

Among the relativistic interactions the spin-orbit coupling plays a special role. This term couples various electronic states, and thus modify the spectra, already for diatomic molecules[466,42]. The spin-orbit coupling also affects the long-range behavior of the potential energy surfaces. In general, the spin-orbit coupled potentials have a different asymptotics from the clamped-nuclei (Born-Oppenheimer) potentials[467]. Theory of intermolecular forces in the (a), (b), or (c) Hund's coupling cases is being developed. The long-range behaviour in the multipole approximation is well understood[468–471] if only the spin-orbit coupling of the isolated monomers is concerned. The effects of the intermolecular spin-orbit coupling, thus far neglected in *ab initio* calculations of the long-range interactions, are being investigated now[42,467].

All the subjects quoted above are very important for the studies of ultracold molecules, a new emerging field at the border of chemistry and physics. In recent years experimental developments in cooling and trapping of atoms and molecules have opened the possibility of studying collisional dynamics at ultralow temperatures. Experimental techniques based on the buffer gas cooling[472] or Stark deceleration[473] produce cold molecules with a temperature well below 1 K. Optical techniques, based on the laser cooling of atoms to ultralow temperatures and photoassociation spectroscopy to create molecules[474,475], lead to temperatures of the order of a few μK or lower. Experimental investigations of the collisions between ultracold atoms lead to precision measurements of atomic properties and interactions. Such collisions also produce ultracold molecules that can be used in high-resolution spectroscopic experiments to study inelastic and reactive processes at very low temperatures. In particular, dynamical studies of chemical species and reactions in the ultracold regime will open a new era in our understanding of basic chemical processes. Moreover, spectroscopy at ultralow temperatures will show an unprecedented accuracy since the Doppler broadening effects on the spectral lines will almost be absent. Such precision measurements will be a big challenge for theory, and will certainly lead to new theoretical and computational developments. It seems that this new emerging field of cold molecules in 2000's will be as exciting as the field of Van der Waals molecules in 1990's.

We wish to end our story of intermolecular forces by saying that it is exciting that challenging subjects remain to be solved. However, it is good to note that the present level theory can be applied with trust to more and more complicated systems, such as biological systems, and can give qualitative, or some time even semi-quantitative, explanations of the complicated processes encountered in the living matter[271,272 476–479].

ACKNOWLEDGMENTS

I would like to thank Bogumił Jeziorski for reading and commenting on the manuscript, Edyta Małolepsza and Konrad Piszczatowski for their invaluable help at all stages of this work, and Marek Orzechowski for useful discussions on the empirical force fields and computer simulations techniques. I am also indebted to Drs. Robert Bukowski, A. Robert W. McKellar, and Konrad Patkowski for providing me with the figures, and with their results prior to publication. This work was supported by the European Research Training Network "Molecular Universe" (contract no. MRTN-CT-2004-512302).

REFERENCES

1. Jeziorski B, Moszynski R, Szalewicz K (1994) Perturbation theory approach to intermolecular potential energy surfaces of Van der Waals complexes. Chem Rev 94:1887–1930
2. Jeziorski B, Szalewicz K (1998) Intermolecular interactions by perturbation theory. In: von Ragué Schleyer P, Allinger NL, Clark T, Gasteiger J, Kollman PA, Schaefer III HF, Schreiner PR (eds) Encyclopedia of computational chemistry, vol 2. Wiley, New York, pp1376–1398
3. Moszynski R, Wormer PES, Van derAvoird A (2000) Symmetry adapted perturbation theory applied to the computation of intermolecular forces. In: Bunker PR, Jensen P (eds) Computational molecular spectroscopy, Wiley, New York, pp69–109
4. Jeziorski B, Szalewicz K (2003) Symmetry-adapted perturbation theory. In: Wilson S (ed) Handbook of molecular physics and quantum chemistry, vol 3. Wiley, New York, pp 232–279
5. Chalasinski G, Szczesniak MM (1994) Origins of structure and energetics of Van der Waals clusters from ab initio calculations. Chem Rev 94:1723–1765
6. Chalasinski G, Szczesniak MM (2000) State of the art and challenges of the ab initio theory of intermolecular interactions. Chem Rev 100:4227–4252
7. Van der Avoird A, Wormer PES, Moszynski R (1994) From intermolecular potentials to the spectra of Van derWaals molecules, and vice versa. Chem Rev 94:1931–1974
8. Van der Avoird A, Wormer PES, Moszynski R (1997) Theory and computation of vibration, rotation and tunneling motions of van der waals complexes and their spectra. In: Scheiner S (ed) Molecular interactions: FromVan derWaals to strongly bound complexes, Wiley, New York, pp105–153
9. Moszynski R, Heijmen TGA, Wormer PES, Van der Avoird A (1997) Theoretical modeling of spectra and collisional processes of weakly interacting complexes. Adv Quantum Chem 28:119–140
10. Wormer PES, van der Avoird A (2000) Intermolecular potentials, internal motions, and spectra of van der waals and hydrogen-bonded complexes. Chem Rev 100:4109–4144

11. Le Roy RJ, Van Kranendonk J (1974) Anisotropic intermolecular potentials from an analysis of spectra of H_2-and D_2-inert gas complexes. J Chem Phys 61:4750

12. Kreek H, Le Roy RJ (1975) Intermolecular potentials and isotope effects for molecular hydrogen-inert gas complexes. J Chem Phys 63:338–344

13. Hutson JM, Howard BJ (1981) The intermolecular potential energy surface of Ar-HCl. Mol Phys 43:493–516

14. Le Roy RJ, Hutson JM (1987) Improved potential energy surfaces for the interaction of H_2 with Ar, Kr, and Xe. J Chem Phys 86:837–853

15. Hutson JM (1988) The intermolecular potential of Ar-HCl: Determination from high-resolution spectroscopy. J Chem Phys 89:4550–4557

16. Hutson JM (1992) Vibrational dependence of the anisotropic intermolecular potential of Ar-HF. J Chem Phys 96:6752–6767

17. Chuaqui CE, Le Roy RJ, McKellar ARW (1994) Infrared spectrum and potential energy surface of He-CO. J Chem Phys 101:39–61

18. Bissonnette C, Chuaqui CE, Crowell KG, Le Roy RJ, Wheatley RJ, Meath WJ (1996) A reliable new potential energy surface for H_2-Ar. J Chem Phys 105:2639–2653

19. Born M, Heisenberg W (1924) The quantum theory of molecules. Ann Phys (Leipzig) 74:1–31

20. Born M, Oppenheimer JR (1927) Quantum theory of molecules. Ann Phys (Leipzig) 84:457–484

21. Born M (1951) Kopplung der Elektronen-und Kernbewegung in Molek uln und Kristallen. Nachr. Akad. Wiss. Göttingen 6:1–3

22. Born M, Huang K (1954) Dynamical theory of crystal lattices. Oxford University Press, New York

23. Bunker PR, Jensen P (2000) The Born-Oppenheimer approximation. In: Jensen P, Bunker PR (eds) Computational molecular spectroscopy, Wiley, New York, pp 3–13

24. Kolos W (1970) Adiabatic approximation and its accuracy. Adv Quantum Chem 5:99–133

25. Kolos W, Wolniewicz L (1964) Accurate adiabatic treatment of the ground state of the hydrogen molecule. J Chem Phys 41:3663–3673

26. Kolos W, Wolniewicz L (1965) Potential-energy curves for the $X^1 \sum_g^+$, $b^3 \sigma_u^+$, and $C^1 \prod_u$ states of the hydrogen molecule. J Chem Phys 43:2429–2441

27. Kolos W, Wolniewicz L (1966) Potential-energy curve for the $B^1 \sum_u^+$ state of the hydrogen molecule. J Chem Phys 45:509–514

28. Kolos W, Wolniewicz L (1968) Confirmation of the discrepancy between the theoretical and experimental ground-state energies of H_2. Phys Rev Lett 20:243–244

29. Kolos W, Wolniewicz L (1968) Improved theoretical ground-state energy of the hydrogen molecule. J Chem Phys 49:404–410

30. Kolos W (1978) Recent theoretical developments in the spectroscopy of the hydrogen molecule. J Mol Struct 46:73–92

31. Kolos W (1993) Hydrogen molecule. Test of quantum chemistry. Polish J Chem 67:553–566

32. Csaszar AG, Allen WD, Yamaguchi Y, Schaefer III HF (2000) Ab initio determination of accurate ground electronic state potential energy hypersurfaces for small molecules. In: Jensen P, Bunker PR (eds) Computational molecular spectroscopy, Wiley, United Kingdom, pp 15–69

33. Wolniewicz L (1966) Vibrational-rotational study of the electronic ground state of the hydrogen molecule. J Chem Phys 45:515–523

34. Epstein ST (1966) Ground-state energy of a molecule in the adiabatic approximation. J Chem Phys 44:836–837, 4062

35. Eckart C (1935) Some studies concerning rotating axes and polyatomic molecules. Phys Rev 47:552–558

36. Watson JKG (1968) Simplification of themolecular vibration-rotation Hamiltonian. Mol Phys 15:479–490
37. VanVleck JH (1936) On the isotope corrections inmolecular spectra. J Chem Phys 4:327–338
38. Pack RT (1966) PhD thesis, University of Wisconsin, Wisconsin
39. Karl G, Poll JD (1967) On the quadrupole moment of the hydrogen molecule. J Chem Phys 46:2944–2950
40. Pack RT, Hirschfelder JO (1970) Energy corrections to the Born-Oppenheimer approximation. The best adiabatic approximation. J Chem Phys 52:521–534
41. Lefebvre-Brion H, Field RF (1986) Perturbations in the spectra of diatomic molecules. Academic, London
42. B. Bussery-Honvault, Moszynski R (2006) Theoretical spectroscopy of the calcium dimer in the $A^1 \sum_u$, $c^3 \prod_u$, and $a^3 \sum_u$ manifolds: An ab initio nonadiabatic treatment. J Chem Phys 125:114315-1–15
43. Yarkony DR (1995) Electronic structure aspects of nonadiabatic processes in polyatomic systems. In: Yarkony DR (ed) Modern electronic structure theory. Part I, World Scientific, Singapore, pp 642–721
44. Bartlett RJ (1981) Many-body perturbation theory and coupled cluster theory for electron correlation in molecules. Ann Rev Phys Chem 32:359–401
45. Roos BO, Siegbahn PEM (1980) A direct CI method with a multiconfigurational reference state. Int J Quantum Chem 17:485–500
46. Olsen J, Roos BO, Jørgensen P, Jensen HJA (1988) Determinant based configuration interaction algorithms for complete and restricted configuration interaction spaces. J Chem Phys 89:2185–2192
47. Shavitt I (1977) The method of configuration interation. In Schaefer III HF (ed) Methods of electronic structure theory, Plenum, NewYork, p 189
48. Andersson PA, Malmqvist K, Roos BO, Sadlej AJ, Wolinski K (1990) Second-order perturbation theory with a CASSCF reference function. J Phys Chem 94:5483–5488
49. Pople JA, Seeger R, Krishan R (1979) Variational configuration interaction methods and comparison with perturbation theory. Int J Quantum Chem Symp 11:149–163
50. Langhoff SR, Davidson ER (1974) Configuration interaction calculations on the nitrogen molecule. Int J Quantum Chem 8:61–72
51. Meissner L (1988) Size-consistency corrections for configuration interaction calculations. Chem Phys Lett 146:204–210
52. Jeziorski B, Monkhorst HJ (1981) Coupled-cluster method for multideterminantal reference states. Phys Rev A 24:1668–1681
53. Jeziorski B, Paldus J (1990) Valence universal exponential ansatz and the cluster structure of multireference configuration interaction wave function. J Chem Phys 90:2714–2731
54. Saxon RP, Kirby K, Liu B (1980) Excited states of CH^+: Potential curves and transition moments. J Chem Phys 73:1873–1879
55. Freed KF, Sheppard MG (1982) Ab initio treatments of quasidegenerate many-body perturbation theory within the effective valence shell Hamiltonian formalism. J Phys Chem 86:2130–2133
56. Pople JA, Krushnan R, Schlegel JS, ans Binkley HB (1978) Electron correlation theories and their application to the study of simple reaction potential surfaces. Int J Quantum Chem 14:545–560
57. Bartlett RJ and Purvis III GD (1978) Many-body perturbation theory, coupled-pair many-electron theory, and the importance of quadruple excitations for the correlation problem. Int J Quantum Chem 14:561–581
58. Coester F (1958) Bound states of a many-particle system. Nucl Phys 7:421–424

59. Coester F, Kümmel H (1960) Short-range correlations in nuclear wave functions. Nucl Phys 17:477–485

60. Cizek J (1960) On the correlation problem in atomic and molecular systems. Calculation of wavefunction components in Ursell-type expansion using quantum-field theoretical methods. J Chem Phys 45:4256–4266

61. Cizek J (1969) On the use of the cluster expansion and the technique of diagrams in calculations of correlation effects in atoms and molecules. Adv Chem Phys 14:35–89

62. Bartlett RJ (1989) Coupled-cluster approach to molecular structure and spectra: a step toward predictive quantum chemistry. J Phys Chem 93:1697–1708

63. Purvis III GD, Bartlett RJ (1982) A full coupled-cluster singles and doubles model: The inclusion of disconnected triples. J Chem Phys 76:1910–1918

64. Urban M, Noga J, Cole SJ, Bartlett RJ (1985) Towards a full CCSDT model for electron correlation. J Chem Phys 83:4041–4046

65. Raghavachari K, Trucks GW, Pople JA, Head-Gordon M (1989) A fifthorder perturbation comparison of electron correlation theories. Chem Phys Lett 157:479–483

66. Noga J, Bartlett RJ (1987) The full CCSDTmodel formolecular electronic structure. J Chem Phys 86:7041–7050

67. Korona T, Williams HL, Bukowski R, Jeziorski B, Szalewicz K (1997) Helium dimer potential from symmetry-adapted perturbation theory calculations using large Gaussian geminal and orbital basis sets. J Chem Phys 106:5109–5122

68. Rode M, Sadlej J, Moszynski R, Wormer PES, Van der Avoird A (1999) The importance of high-order correlation effects for the CO–CO interaction potential. Chem Phys Lett 314: 326–332

69. Kucharski SA, Bartlett RJ (1986) First-order many-body perturbation theory and its relationship to various coupled-cluster approaches. Adv Quantum Chem 18:281–344

70. Jeziorski B, Moszynski R (1993) Explicitly connected expansion for the average value of an observable in the coupled-cluster theory. Int J Quantum Chem 48:161–183

71. Becke AD (1988) Density-functional exchange-energy approximation with correct asymptotic behavior. Phys Rev A 38:3098–3100

72. Lee C, Yang W, Parr RG (1988) Development of the Colle-Salvetti correlation-energy formula into a functional of the electron density. Phys Rev B 37:785–789

73. Becke AD (1993) Density-functional thermochemistry. III. The role of exact exchange. J Chem Phys 98:5648–5652

74. Milet A, Korona T, Moszynski R, Kochanski E (1999) Anisotropic intermolecular interactions in Van der Waals and hydrogen-bonded complexes: What can we get from density functional calculations? J Chem Phys 111:7727–7735

75. Hirschfelder JO (1967) Perturbation theory for exchange forces, I. Chem Phys Lett 1: 325–329

76. Claverie P (1971) Theory of intermolecular forces. I. On the inadequacy of the usual Rayleigh-Schrödinger perturbation method for the treatment of intermolecular forces. Int J Quantum Chem 5:273–296

77. Jeziorski B (1978) quoted in Ref. (78)

78. Kutzelnigg W (1980) The "primitive" wave function in the theory of intermolecular interactions. J Chem Phys 73:343–359

79. Chalasinski G, Jeziorski B, Szalewicz K (1977) On the convergence properties of the Rayleigh-Schrödinger and the Hirschfelder-Silbey perturbation expansions for molecular interaction energies. Int J Quantum Chem 11:247–257

80. Cwiok T, Jeziorski B, Kolos W, Moszynski R, Rychlewski J, Szalewicz K (1992) Convergence properties and large-order behavior of the polarization expansion for the interaction energy of hydrogen atoms. Chem Phys Lett 195:67–76

81. Korona T, Moszynski R, Jeziorski B (1997) Convergence of symmetry adapted perturbation theory for the interaction between helium atoms and between a hydrogen molecule and a helium atom. Adv Quantum Chem 28:171–188

82. Adams WH (1990) Perturbation theory of intermolecular interactions: what is the problem, are there solutions? Int J Quantum Chem Sym 24:531–547

83. Patkowski K, Korona T, Jeziorski B (2001) Convergence behavior of the symmetry-adapted perturbation theory for states submerged in Pauli forbidden continuum. J Chem Phys 115: 1137–1152

84. Jeziorski B, Kolos W (1982) Perturbation approach to the study of weak intermolecular interactions. In: Ratajczak H, Orville-Thomas WJ (eds) Molecular Interactions, vol 3. Wiley, New York, pp1–46

85. Morgan III JD, Simon B (1980) Behavior of molecular potential energy curves for large nuclear separations. Int J Quantum Chem 17:1143–1166

86. Jeziorski B, Schwalm WA, Szalewicz K (1980) Analytic continuation in exchange perturbation theory. J Chem Phys 73:6215–6224

87. Jeziorski B, Chalasinski G, Szalewicz K (1978) Symmetry forcing and convergence properties of perturbation expansions for molecular interaction energies. Int J Quantum Chem 14:271–287

88. Jeziorski B, Kolos W (1977) On the symmetry forcing in the perturbation theory of weak intermolecular interactions. Int J Quantum Chem (Suppl. 1), 12:91–117

89. Cwiok T, Jeziorski B, Kolos W, Moszynski R, Szalewicz K (1992) On the convergence of the symmetrized Rayleigh–Schrödinger perturbation theory formolecular interaction energies. J Chem Phys 97:7555–7559

90. Cwiok T, Jeziorski B, Kolos W, Moszynski R, Szalewicz K (1994) Symmetry-adapted perturbation theory of potential-energy surfaces for weakly bound molecular complexes. J Mol Struct (Theochem) 113:135–151

91. Murrell JN, Shaw G (1967) Intermolecular forces in the region of small orbital overlap. J Chem Phys 46:1768–1772

92. Musher JI, Amos AT (1967) Theory of weak atomic and molecular interactions. Phys Rev 164:31–43

93. Jeziorski B, Szalewicz K, Jaszunski M (1979) Padé approximants and the convergence problem in the perturbation theory of intermolecular interactions. Chem Phys Lett 61:391–395

94. Eisenschitz R, London F (1930) Über das Verhältnis der van der Waalsschen Kräfte zu den homopolaren Bindungskräften. Z Phys 60:491–527

95. Hirschfelder JO (1967) Perturbation theory for exchange forces, II. Chem Phys Lett 1:363–368

96. Van der Avoird A (1967) Perturbation theory for intermolecular interactions in the wave-operator formalism. J Chem Phys 47:3649–3653

97. Peierls R (1973) Perturbation theory for projected states. Proc Royal Soc (London) A, 333: 157–170

98. Hirschfelder JO, Silbey R (1966) New type of molecular perturbation treatment. J Chem Phys 45:2188–2192

99. Polymeropoulos EE, Adams WH (1978) Exchange perturbation theory. I. General definitions and relations. Phys Rev A 17:11–17

100. Polymeropoulos EE, Adams WH (1978) Exchange perturbation theory. II. Eisenschitz-London type. Phys Rev A 17:18–23

101. Polymeropoulos EE, Adams WH (1978) Exchange perturbation theory, III. Hirschfelder-Silbey type. Phys Rev A 17:24–29
102. Chipman DM (1977) Localization in exchange perturbation theory. J Chem Phys 66:1830–1834
103. Klein DJ (1987) Exchange perturbation theories. Int J Quantum Chem 32:377–396
104. Korona T, Jeziorski B, Moszynski R, Diercksen GHF (1999) Degenerate symmetry-adapted perturbation theory of weak interactions between closed- and open-shell monomers: application to Rydberg states of helium hydride. Theor Chem Acc 101:282–291
105. Jeziorska M, Jeziorski B, Cizek J (1987) Direct calculation of the Hartree-Fock interaction energy via exchange-perturbation expansion. The He... He interaction. Int J Quantum Chem 32:149–164
106. Moszynski R, Heijmen TGA, Jeziorski B (1996) Symmetry-adapted perturbation theory for the calculation of Hartree-Fock interaction energies. Mol Phys 88:741–758
107. Patkowski K, Jeziorski B, Korona T, Szalewicz K (2002) Symmetry forcing procedure and convergence behavior of perturbation expansions for molecular interaction energies. J Chem Phys 117:5124–5134
108. Adams WH (2002) Two new symmetry-adapted perturbation theories for the calculation of intermolecular interaction energies. Theor Chem Acc 108:225–231
109. Patkowski K, Jeziorski B, Szalewicz K (2001) Symmetry-adapted perturbation theory with regularized Coulomb potential. J Mol Struct (Theochem) 547:293–307
110. Patkowski K, Jeziorski B, Szalewicz K (2005) Intermolecular interactions via perturbation theory: from diatoms to biomolecules. Struct Bond 116:43–117
111. Claverie P (1978) Elaboration of approximate formulas for the interactions between large molecules. Applications in organic chemistry. In: Pullman B (ed) Intermolecular interactions:From diatomics to biopolymers, Wiley, New York, pp 69–305
112. Van Duijneveldt-Van de Rijdt JGCM, Van Duijneveldt FB (1972) Double-exchange contributions to the first-order interaction energy between closed-shell molecules. Chem Phys Lett 17:425–427
113. Ahlrichs R (1973) Comments on the convergence of the ordinary Rayleigh- Schrödinger perturbation expansion. Chem Phys Lett 18:67–68
114. Williams DR, Schaad LJ, Murrell JN (1967) Deviations from pairwise additivity in intermolecular potentials. J Chem Phys 47:4916–4922
115. Jeziorski B, Bulski M, Piela L (1976) First-order perturbation treatment of the short-range repulsion in a system of many closed-shell atoms or molecules. Int J Quantum Chem 10:281–297
116. Moszynski R, Jeziorski B, Rybak S, Szalewicz K, Williams HL (1994) Many-body theory of exchange effects in intermolecular interactions. Density matrix approach and applications to He–F-, He–HF, H_2–HF, and Ar–H_2 dimers. J Chem Phys 100:5080–5092
117. Jørgensen P, Simons J (1981) Second quantization-based methods in quantum chemistry. Academic, New York
118. Oddershede J (1978) Polarization propagator calculations. Adv Quantum Chem 11:275–352
119. Moszynski R, Cybulski SM, Chalasinski G (1994) Many-body theory of intermolecular induction interactions. J Chem Phys 100:4998–5010
120. Salter EA, Trucks GW, Fitzgerald G, Bartlett RJ (1987) Theory and application of MBPT(3) gradients: The density approach. Chem Phys Lett 141:61–70
121. Trucks GW, Salter EA, Sosa C, Bartlett RJ (1988) Theory and implementation of the MBPT density matrix. An application to one-electron properties. Chem Phys Lett 147:359–366
122. Trucks GW, Salter EA, Noga J, Bartlett RJ (1988) Analytic many body perturbation theory MBPT(4) response properties. Chem Phys Lett 150:37–44
123. Salter EA, Trucks GW, Bartlett RJ (1989) Analytic energy derivatives in many-body methods. I. First derivatives. J Chem Phys 90:1752–1766

124. Salter EA, Bartlett RJ (1989) Analytic energy derivatives in many-body methods. II. Second derivatives. J Chem Phys 90:1767–1773

125. Koch H, Jørgen H, Jensen A, Jørgensen P, Helgaker T, Scuseria GE, Schaefer III HF (1990) Coupled cluster energy derivatives. Analytic Hessian for the closed-shell coupled cluster singles and doubles wave function: Theory and applications. J Chem Phys 92:4924–4940

126. Handy NC, Amos RD, Gaw JF, Rice JE, Simandiras ES (1985) The elimination of singularities in derivative calculations. Chem Phys Lett 120:151–158

127. Harrison RJ, Fitzgerald G, Laidig WD, Bartlett RJ (1985) Analytic MBPT(2) second derivatives. Chem Phys Lett 124:291–294

128. Helgaker T, Jørgensen P, Handy NC(1989) A numerically stable procedure for calculating Møller-Plesset energy derivatives, derived using the theory of Lagrangians. Theor Chim Acta 76:227–245

129. Chalasinski G, Jeziorski B (1976) On the exchange polarization effects in the interaction of two helium atoms. Mol Phys 32:81–91

130. Chalasinski G, Jeziorski B (1977) Exchange polarization effects in the interaction of closed-shell systems. The beryllium-beryllium interaction. Theor Chim Acta 46:277–290

131. Casimir HBG, Polder D (1948) The influence of retardation on the londonvan der waals forces. Phys Rev 73:360–372

132. McLachlan AD (1963) Retarded dispersion forces between molecules. Proc Roy Soc (London) Ser A 271:387–401

133. Longuet-Higgins HC (1965) Spiers memorial lecture. Intermolecular forces. Discuss Faraday Soc 40:7–18

134. Zaremba E, Kohn W (1976) Van der Waals interaction between an atom and a solid surface. Phys Rev B 13:2270–2285

135. Dmitriev Yu, Peinel G (1981) Coupled perturbation theory within the antisymmetrized product of separated geminals (APSG) framework. Int J Quantum Chem 19:763–769

136. McWeeny R (1984) Weak interactions between molecules. Croat Chem Acta 57:865–878

137. Moszynski R, Jeziorski B, Szalewicz K (1993) Møller-Plesset expansion of the dispersion energy in the ring approximation. Int J Quantum Chem 45:409–431

138. Olsen J, Jørgensen P (1985) Linear and nonlinear response functions for an exact state and for an MCSCF state. J Chem Phys 82:3235–3264

139. Moszynski R (1994) unpublished results

140. Dalgarno A, Lewis JT (1956) Representation of long-range forces by series. Proc Phys Soc A 69:57–64

141. Rybak S, Jeziorski B, Szalewicz K (1991) Many-body symmetry-adapted perturbation theory of intermolecular interactions. H_2O and HF dimers. J Chem Phys 95:6576–6601

142. Patkowski K et al (2006) to be published.

143. London F (1930) On some properties and applications of molecular forces. Z Phys Chem (B) 11:222

144. London F (1930) Zur Theorie und Systematik der Molekularkräfte. Z Phys 63:245–279

145. Ahlrichs R (1976) Convergence properties of the intermolecular force series (1/R-expansion). Theor Chim Acta 41:7–15

146. Erdelyi A (1956) Asymptotic expansions. Dover, New York

147. Hirschfelder JO, Curtiss CF, Bird RB (1954) Molecular theory of gases and liquids. Wiley, New York

148. Brink DM, Satchler GR (1975) Angular Momentum. Clarendon, Oxford

149. Wormer PES (1975) Intermolecular forces and the group theory ofmany-body systems. PhD thesis, University of Nijmegen, Nijmegen

150. Mulder F, Huiszoon C (1977) The dimer interaction and lattice energy of ethylene and pyrazine in the multipole expansion; a comparison with atom-atom potentials. Mol Phys 34:1215–1235

151. Isnard P, Robert D, Galatry L (1976) On the determination of the intermolecular potential between a tetrahedral molecule and an atom or a linear or a tetrahedral molecule-application to CH_4 molecule. Mol Phys 31:1789–1811

152. Coope JAR, Sinder RF, McCourt FR (1965) Irreducible Cartesian tensors. J Chem Phys 43:2269–2275

153. Coope JAR, Sinder RF (1970) Irreducible Cartesian tensors. II. General formulation. J Math Phys 11:1003–1017

154. Coope JAR (1970) Irreducible Cartesian tensors. III. Clebsch-Gordan reduction. J Math Phys 11:1591–1612

155. Stone AJ (1975) Transformation between Cartesian and spherical tensors. Mol Phys 29:1461–1471

156. Stone AJ (1976) Properties of Cartesian-spherical transformation coefficients. J Phys A 9:485–497

157. Mulder F (1978) Ab initio calculations of molecular multipoles, polarizabilties and Van der Waals interactions. PhD thesis, University of Nijmegen, Nijmegen

158. Mulder F, Van Dijk G, Huiszoon C (1979) Ab initio calculations of multipole moments, polarizabilities and long-range interaction coefficients for the azabenzene molecules. Mol Phys 38:577–603

159. Huiszoon C, Mulder F (1979) Long range C, N and H atom-atom potential parameters from ab initio dispersion energies for different azabenzene dimers. Mol Phys 38:1497–1506

160. Huiszoon C (1986) Ab initio calculation of multipole moments, polarizabilities and isotropic long range interaction coefficients for dimethylether, methanol, methane, and water. Mol Phys 58:865–885

161. Wormer PES, Mulder F, Van der Avoird A (1977) Quantum theoretical calculations of Van der Waals interactions between molecules. Anisotropic long range interactions. Int J Quantum Chem 11:959–970

162. Van der Avoird A, Wormer PES, Mulder F, Berns RM (1980) Ab initio studies of the interactions in Van derWaals molecules. Top Curr Chem 93:1–51

163. Heijmen TGA, Moszynski R, Wormer PES, Van der Avoird A (1996) Symmetry-adapted perturbation theory applied to interaction-induced properties of collisional complexes. Mol Phys 89:81–110

164. Leavitt RP (1980) Erratum: An irreducible tensor method of deriving the longrange anisotropic interactions between molecules of arbitrary symmetry [J. Chem. Phys. **72**, 3472 (1980)]. J Chem Phys 73:2017–2017

165. Leavitt RP (1980) An irreducible tensor method of deriving the long-range anisotropic interactions between molecules of arbitrary symmetry. J Chem Phys 72:3472–3482

166. Stone AJ, Tough RJA (1984) Spherical tensor theory of long-range intermolecular forces. Chem Phys Lett 110:123–129

167. Mulder F, Van der Avoird A, Wormer PES (1979) Anisotropy of long range interactions between linear molecules H_2-H_2 and H_2-He. Mol Phys 37:159–180

168. Mulder F, Van Dijk G, Van der Avoird A (1980) Multipole moments, polarizabilities and anisotropic long range interaction coefficient for N_2. Mol Phys 39:407–425

169. Jahn HA (1949) Note on the Bhagavantam-Suranarayana method of enumerating the physical constants of crystals. Acta Crystallogr 2:30–33

170. Berns RM, Wormer PES (1981) Finite field configuration interaction calculations on the distance dependence of the hyperpolarizabilities of H_2. Mol. Phys 44:1215–1227

171. Buckingham AD (1959) Molecular quadrupole moments. Quart Rev Chem Soc (London) 13:183–214

172. Buckingham AD (1967) Permanent and induced molecular moments and long range intermolecular forces. Adv Chem Phys 12:107–166

173. Buckingham AD, Orr BJ (1967) Molecular hyperpolarizabilities. Quart Rev Chem Soc (London) 21:195–212

174. Buckingham AD, Utting B (1970) Intermolecular forces. Ann Rev Phys Chem 21:287–316

175. Gray CG, Lo BWN (1976) Spherical tensor theory of molecular multipole moments and polarizabilities. Chem Phys 14:73–87

176. Kielich S, Zawodny R (1971) Tensor elements of the molecular electric multipole moments for all point group symmetries. Chem Phys Lett 12:20–24

177. Wormer PES, Hettema H (1992) Many-body perturbation theory of frequency-dependent polarizabilities andVan derWaals coefficients:Application to H_2O–H_2O and Ar–NH_3. J Chem Phys 97:5592–5606

178. Wormer PES, Hettema H (1992) POLCOR Package

179. Thakkar AJ, Hettema H, Wormer PES (1992) Ab initio dispersion coefficients for interactions involving rare-gas atoms. J Chem Phys 97:3252–3257

180. Wormer PES, Hettema H, Thakkar AJ (1993) Intramolecular bond length dependence of the anisotropic dispersion coefficients for H_2–rare gas interactions. J Chem Phys 98:7140–7144

181. Hettema H, Wormer PES, Thakkar AJ (1993) Intramolecular bond length dependence of the anisotropic dispersion coefficients for interactions of rare gas atoms with N_2, CO, Cl_2, HCl and HBr. Mol Phys 80:533–548

182. Korona T, Przybytek M, Jeziorski B. Time-independent coupled cluster theory of the polarization propagator. Implementation and application of the singles and doubles model to dynamic polarizabilities and Van der Waals constants, 2006. Submitted to Mol. Phys 104:2302–2316

183. Stolarczyk LZ, Piela L (1979) Invariance properties of the multipole expansion with respect to the choice of the coordinate system. Int J Quantum Chem 15:701–711

184. Langlet J, Claverie P, Caillet J, Pullman A (1988) Improvements of the continuum model. 1. Application to the calculation of the vaporization thermodynamic quantities of nonassociated liquids. J Phys Chem 92:1617–1631

185. Dalgarno A, Lewis JT (1955) The exact calculation of long-range forces between atoms by perturbation theory. Proc Roy Soc (London) A 233:70–74

186. Young RH (1975) Divergence of the R-1 expansion for the second-order H-H interaction. Int J Quantum Chem 9:47–50

187. Stone AJ (1981) Distributed multipole analysis, or how to describe amolecular charge distribution. Chem Phys Lett 83:233–239

188. Stone AJ, Alderton M (1985) Distributed multipole analysis: methods and applications. Mol Phys 56:1047–1064

189. Stone AJ (1985) Distributed polarizabilities. Mol Phys 56:1065–1082

190. Stone AJ (1991) Classical electrostatics in molecular interactions. In: Maksic ZB (ed) Theoretical models of chemical bonding, vol 4. Springer, New York, pp 103–131

191. Sokalski WA, Poirier RA (1983) Cumulative atomic multipole representation of the molecular charge distribution and its basis set dependence. Chem Phys Lett 98:86–92

192. Sokalski WA, Sawaryn A (1987) Correlated molecular and cumulative atomic multipole moments. J Chem Phys 87:526–534

193. Sokalski WA, Sneddon SF (1991) Efficient method for the generation and display of electrostatic potential surfaces from ab-inito wave functions. J Mol Graphics 9:74–77

194. Sokalski WA, Sawaryn A (1992) Cumulative multicenter multipole moment databases and their applications. J Mol Struct (Theochem) 256:91–112

195. Le Sueur CR, Stone AJ (1993) Practical schemes for distributed polarizabilities. Mol Phys 78:1267–1291

196. Bader RFW (1994) Atoms in molecules. A quantum theory. Clarendon Press, Oxford

197. Ángyán JG, Jansen G, Loos M, H attig C, Hess BA (1994) Distributed polarizabilities using the topological theory of atoms in molecules. Chem Phys Lett 219:267–273

198. Stone AJ (1997) The theory of intermolecular forces. Clarendon Press, Oxford

199. Buehler RJ, Hirschfelder JO (1951) Bipolar expansion of Coulombic potentials. Phys Rev 83:628–633

200. Buehler RJ, Hirschfelder JO (1952) Bipolar expansion of Coulombic potentials. Addenda. Phys Rev 85:149–149

201. Koide A (1976) A new expansion for dispersion forces and its application. J Phys B 9:3173–3184

202. Linder B, Lee KF, Malinowski P, Tanner AC (1980) On the relation between charge-density susceptibility, scattering functions, and Van der Waals forces. Chem Phys 52:353–361

203. Malinowski P, Tanner AC, Lee KF, Linder B (1981) Van der Waals forces, scattering functions and charge density susceptibility. II. Application to the He-He interaction potential. Chem Phys 62:423–438

204. Koide A, Proctor TR, Allnatt AR, Meath WJ (1986) Charge overlap effects for first-order molecule-molecule interactions, through high partial wave order, using the N_2-N_2 interaction as a model. Mol Phys 59:491–507

205. Koide A, Meath WJ, Allnatt AR (1981) Second order charge effects and damping functions for isotropic atomic and molecular interactions. Chem Phys 58:105–119

206. Krauss M, Neumann DB (1979) Charge overlap effects in dispersion energies. J Chem Phys 71:107–112

207. Rosenkrantz ME, Krauss M (1985) Damped dispersion interaction energies for He-H_2, Ne-H_2, and Ar-H_2. Phys Rev A 32:1402–1411

208. Knowles PJ, Meath WJ (1987) A separable method for the calculation of dispersion and induction energy damping functions with applications to the dimers arising from He, Ne and HF. Mol Phys 60:1143–1158

209. Moszynski R, Jeziorski B, Diercksen GHF, Viehland LA (1994) Symmetry-adapted perturbation theory potential for the HeK+ molecular ion and transport coefficients of potassium ions in helium. J Chem Phys 101:4697–4707

210. Tang KT, Toennies JP (1977) A simple theoretical model for the Van der Waals potential at intermediate distances. I. Spherically symmetric potentials. J Chem Phys 66:1496–1506

211. Tang KT, Toennies JP (1978) A simple theoretical model for the Van der Waals potential at intermediate distances. II. Anisotropic potentials of He-H_2 and Ne-H_2. J Chem Phys 68:5501–5517

212. Tang KT, Toennies JP (1984) An improved simple model for the Van der Waals potential based on universal damping functions for the dispersion coefficients. J Chem Phys 80:3726–3741

213. Szalewicz K, Jeziorski B (1979) Symmetry-adapted double-perturbation analysis of intramolecular correlation effects in weak intermolecular interactions. The He-He interaction. Mol Phys 38:191–208

214. Jeziorski B, Moszynski R, Rybak S, Szalewicz K (1989) Many-body theory of van der waals interactions. In: Kaldor U (ed) Many-body methods in quantum chemistry, volume 52 of Lecture Notes in Chemistry, Springer, New York, pp 65–95

215. Moszynski R, Jeziorski B, Szalewicz K (1994) Many-body theory of exchange effects in intermolecular interactions. Second-quantization approach and comparison with full configuration interaction results. J Chem Phys 100:1312–1325

216. Moszynski R, Jeziorski B, Ratkiewicz A, Rybak S (1993) Manybody perturbation theory of electrostatic interactions betweenmolecules: Comparison with full configuration interaction for four-electron dimers. J Chem Phys 99:8856–8869

217. Williams HL, Szalewicz K, Moszynski R, Jeziorski B (1995) Dispersion energy in the coupled pair approximation with noniterative inclusion of single and triple excitations. J Chem Phys 103:4586–4599

218. Korona T, Moszynski R, Jeziorski B (2002) Electrostatic interactions between molecules from relaxed one-electron density matrices of the coupled cluster singles and doubles model. Mol Phys 100:1723–1734

219. Zuchowski PS (2006) to be published.

220. Bukowski R, Jankowski P, Jeziorski B, Jeziorska M, Kucharski SA, Moszynski R, Rybak S, Szalewicz K, Williams HL, Wormer PES (1996) SAPT96: An ab initio program for many-body symmetry-adapted perturbation theory calculations of intermolecular interaction energies. University of Delaware and University of Warsaw

221. Jeziorski B, Moszynski R, Ratkiewicz A, Rybak S, Szalewicz K, Williams HL (1993) SAPT: A program for many-body symmetry-adapted perturbation theory calculations of intermolecular interaction energies. In: Clementi E (ed) Methods and techniques in computational chemistry: METECC-94, vol B. STEF, Cagliari, pp 79–129

222. Hesselmann A, Jansen G (2002) First-order intermolecular interaction energies from Kohn–Sham orbitals. Chem Phys Lett 357:464–470

223. Hesselmann A, Jansen G (2002) Intermolecular induction and exchangeinduction energies from coupled-perturbed Kohn–Sham density functional theory. Chem Phys Lett 362:319–325

224. Hesselmann A, Jansen G (2003) Intermolecular dispersion energies from time-dependent density functional theory. Chem Phys Lett 367:778–784

225. Hesselmann A, Jansen G, Schütz M (2005) Density-functional theorysymmetry- adapted intermolecular perturbation theory with density fitting: A new efficient method to study intermolecular interaction energies. J Chem Phys 122:014103

226. Williams HL, Chabalowski CF (2001) Using Kohn-Sham orbitals in symmetry-adapted perturbation theory to investigate intermolecular interactions. J Phys Chem A 105:646–659

227. Misquitta AJ, Jeziorski B, Szalewicz K (2003) Dispersion energy from density-functional theory description of monomers. Phys Rev Lett 91:033201,1–4

228. Podeszwa R, Szalewicz K (2005) Accurate interaction energies for argon, krypton, and benzene dimers from perturbation theory based on the Kohn-Sham model. Chem Phys Lett 412:488–493

229. Chalasinski G, Szczesniak MM (1988) On the connection between the supermolecular Møller-Plesset treatment of the interaction energy and the perturbation theory of intermolecular forces. Mol Phys 63:205–224

230. Cybulski SM, Chalasinski G, Moszynski R (1990) On decomposition of second-order møller–plesset supermolecular interaction energy and basis set effects. J Chem Phys 92:4357–4363

231. Moszynski R, Rybak S, Cybulski SM, Chalasinski G (1990) Correlation correction to the hartree-fock electrostatic energy including orbital relaxation. Chem Phys Lett 166:609–614

232. Cybulski SM, Chalasinski G (1992) Perturbation analysis of the supermolecule interaction energy and the basis set superposition error. Chem Phys Lett 197:591–598

233. Moszynski R (1992) unpublished results

234. Visentin T, Moszynski R, Dedieu A, Kochanski E (2001) Interaction of dichloromethane with palladium complexes: A comparative symmetry-adapted perturbation theory, supermolecule, and self-consistent reaction field study. J Phys Chem A 105:2031–2038

235. Boys SF, Bernardi F (1970) The calculation of small molecular interactions by the differences of separate total energies. Some procedures with reduced errors. Mol Phys 19:553–566

236. Bulski M, Chalasinski G (1977) On basis set effects in SCF calculations of the interaction energy between closed-shell atoms. Theor Chim Acta 44:399–404

237. Gutowski M, Van Lenthe JH, Verbeek J, Van Duijneveldt FB, Chalasinski G (1986) The basis set superposition error in correlated electronic structure calculations. Chem Phys Lett 124:370–375

238. Gutowski M, Van Duijneveldt FB, Chalasinski G, Piela L (1986) Does the boys and bernardi function counterpoise method actually overcorrect the basis set superposition error? Chem Phys Lett 129:325–328

239. Gutowski M, Van Duijneveldt FB, Chalasinski G, Piela L (1987) Proper correction for the basis set superposition error in SCF calculations of intermolecular interactions. Mol Phys 61:233–247

240. Gutowski M, Chalasinski G (1993) Critical evaluation of some computational approaches to the problem of basis set superposition error. J Chem Phys 98:5540–5554

241. Gutowski M, Van Duijneveldt-Van der Rijdt JGCM, Van Lenthe JH, Van Duijneveldt FB (1993) Accuracy of the Boys and Bernardi function counterpoise method. J Chem Phys 98:4728–4737

242. Van Duijneveldt FB, Van Duijneveldt-Van der Rijdt JGCM, Van Lenthe JH (1994) State of the art in counterpoise theory. Chem Rev 94:1873–1885

243. Certain PN, Hirschfelder JO (1970) New partitioning perturbation theory. III. Applications to electron exchange. J Chem Phys 52:5992–5999

244. Chalasinski G, Jeziorski B (1973) Exact calculation of exchange polarization energy for H_2^+ ion. Int J Quantum Chem 7:63–73

245. Bukowski R, Jeziorski B, Szalewicz K (1996) Basis set superposition problem in interaction energy calculations with explicitly correlated bases: saturated second- and third-order energies for He_2. J Chem Phys 104:3306–3319

246. Scuseria GE, Miller MD, Jensen F, Geertsen J (1991) The dipole moment of carbon monoxide. J Chem Phys 94:6660–6663

247. Raghavachari K, Pople JA, Repogle ES, Head-Gordon M (1990) Fifth order Moller-Plesset perturbation theory: comparison of existing correlation methods and implementation of newmethods correct to fifth order. J Phys Chem 94:5579–5586

248. Barone V, Cossi M, Tomassi J (1998) Geometry optimization of molecular structures in solution by the polarizable continuum model. J Comput Chem 19:404–417

249. Rinaldi D, Rivail JL (1973) Polarisabilités moléculaires et effet diélectrique de milieu á l'état liquide. Étude théorique de la molécule d'eau et de ses diméres. Theor Chim Acta 32:57–70

250. Rivail JL, Rinaldi D (1976) A quantum chemical approach to dielectric solvent effects in molecular liquids. Chem Phys 18:233–242

251. Rivail JL, Terryn B (1982) Energie libre d'une distribution de charges électriques séparée d'un milieu diélectrique infini par une cavité elipso quelconque. Application à l'étude de solvatation des molécules. J Chim Phys 79:1–15

252. Rinaldi D, Ruiz-Lopez M, Rivail JL (1983) Ab initio SCF calculations on electrostatically solvated molecules using a deformable three axes ellipsoidal cavity. J Chem Phys 78:834–838

253. Costa Cabral J, Rinaldi D, Rivail JL (1984) Sur le calcul du terme de dispersion de l'énergie libre de solvatation au moyen des modéles à cavité. CR Acad Sci (Paris) 298:495–498

254. Rinaldi D, Costa Cabral BJ, Rivail JL (1986) Influence of dispersion forces on the electronic structure of a solvated molecule. Chem Phys Lett 125:495–499

255. Rivail JL (1990) Calcul des effets de corrélation électronique dans une molécule solvaté par un milieu continu. CR Acad Sci (Paris) 311:307–311

256. Chipot C, Rinaldi D, Rivail JL (1992) Intramolecular electron correlation in the self-consistent reaction field model of solvation. A MP2/6-31G** ab initio study of the NH_3–HCl complex. Chem Phys Lett 191:287–292

257. Ángyan J (1992) Common theoretical framework for quantum chemical solvent effect theories. J Math Chem 10:93–137

258. Morokuma K (1971) Molecular orbital studies of hydrogen bonds. III. C=O...H—O hydrogen bond in H_2CO... H_2O and H_2CO... $2H_2O$. J Chem Phys 55:1236–1244

259. Sadlej AJ (1980) Long range induction and dispersion interactions between Hartree-Fock subsystems. Mol Phys 39:1249–1264

260. Gutowski M, Piela L (1988) Interpretation of the Hartree-Fock interaction energy between closed-shell systems. Mol Phys 64:337–355

261. Suhai S, Bagus PS, Ladik J (1982) An error analysis for Hartree-Fock crystal orbital calculations. Chem Phys 68:467–471

262. Sokalski WA, Harlharan PC, Kaufman JJ (1983) A self-consistent field interaction energy decomposition study of 12 hydrogen-bonded dimers. J Chem Phys 87:2803–2810

263. Sokalski S, Roszak WA, Harlharan PC, Kaufman JJ (1983) Improved scf interaction energy decomposition scheme corrected for basis set superposition effect. Int J Quantum Chem 23:847–854

264. Kitaura K, Morokuma K (1976) A new energy decomposition scheme for molecular interactions within the Hartree-Fock approximation. Int J Quantum Chem 10:325–340

265. Cybulski SM, Scheiner S (1990) Comparison of Morokuma and perturbation theory approaches to decomposition of interaction energy. (NH_4^+) ... NH_3. Chem. Phys Lett 166:57–64

266. Stevens WJ, Fink WH (1987) Frozen fragment reduced variational space analysis of hydrogen bonding interactions. Application to the water dimer. Chem Phys Lett 139:15–22

267. Jeziorski B, Bulski M, Piela L (1976) First-order perturbation treatment of the short-range repulsion in a system of many closed-shell atoms or molecules. Int J Quantum Chem 10:281–297

268. Jaszunski M (1980) Coupled Hartree-Fock calculation of the induction energy. Mol Phys 39:777–780

269. Sokalski WA, Roszak S, Pecul K (1988) An efficient procedure for decomposition of the SCF interaction energy into components with reduced basis set dependence. Chem Phys Lett 153:153–159

270. Sokalski WA, Roszak S (1991) Efficient techniques for the decomposition of intermolecular interaction energy at scf level and beyond. J Mol Struct (Theochem) 80:387–400

271. Dyguda E, Grembecka J, Sokalski WA, Leszczynski J (2004) Origins of the activity of PAL and LAP enzyme inhibitors: Toward ab initio binding affinity prediction. J Am Chem Soc 127:1658–1659

272. Szefczyk B, Mulholland AJ, Ranaghan KE, Sokalski WA (2004) Differential transition-state stabilization in enzyme catalysis: Quantum chemical analysis of interactions in the chorismate mutase reaction and prediction of the optimal catalytic field. J Am Chem Soc 126:16148–16159

273. Wheatley RJ, Meath WJ (1993) On the relationship between first-order exchange and Coulomb interaction energies for closed shell atoms and molecules. Mol Phys 79:253–275

274. Born M, Mayer JE (1932) Zur Gittertheorie der Lanenkristalle. Z Phys 75:1–6

275. Schmuttenmaer CA, Cohen RC, Saykally RJ (1994) Spectroscopic determination of the inter-molecular potential energy surface for Ar-NH_3. J Chem Phys 101:146–173

276. Zeiss GD, Meath WJ (1975) The H_2O-H_2O dispersion energy constant and the dispersion of the specific refractivity of dilute water vapour. Mol Phys 30:161–169

277. Zeiss GD, Meath WJ (1977) Dispersion energy constants $C_6(A,B)$, dipole oscillator strength sums and refractivities for Li, N, O, H_2, N_2, O_2, NH_3, H_2O, NO and N_2O. Mol Phys 33:1155–1176

278. Thomas GF, Meath WJ (1977) Dipole spectrum, sums and properties of ground-state methane and their relation to the molar refractivity and dispersion energy constant. Mol Phys 34:113–125

279. Zeiss GD, Meath WJ, MacDonald JCF, Dawson DJ (1977) Dipole oscillator strength distributions, sums, and some related properties for Li, N, O, H_2, N_2, O_2, NH_3, H_2O, NO, and N_2O. Can J Phys 55:2080–2200

280. Margoliash DJ, Meath WJ (1978) Pseudospectral dipole oscillator strength distributions and some related two body interaction coefficients for H, He, Li, N, O, H_2, N_2, O_2, NO, N_2O, H_2O, NH_3, and CH_4. J Chem Phys 68:1426–1431

281. Margoliash DJ, Proctor TR, Zeiss GD, Meath WJ (1978) Triple-dipole energies for H, He, Li, N, O, H_2, N_2, O_2, NO, N_2O, H_2O, NH_3 and CH_4 evaluated using pseudo-spectral dipole oscillator strength distributions. Mol Phys 35:747–757

282. Mulder F, Thomas GF, Meath WJ (1980) A critical study of some methods for evaluating the C_6, C_8 and C_{10} isotropic dispersion energy coefficients using the first row hydrides, CO, CO_2 and N_2O as models. Mol Phys 41:249–269

283. Zeiss GD, Meath WJ, MacDonald JCF, Dawson DJ (1980) On the additivity of atomic and molecular dipole properties and dispersion energies using H, N, O, H_2, N_2O_2, NO, N_2O, NH_3 and H_2O as models. Mol Phys 39:1055–1072

284. Jhanwar BL, Meath WJ, MacDonald JCF (1981) Dipole oscillator strength distributions and sums for C_2H_6, C_3H_8, n-C_4H_{10}, n-C_5H_{12}, n-C_6H_{14}, n-C_7H_{16}, and n-C_8H_{18}. Can J Phys 59:185–197

285. Mulder F, Meath WJ (1981) Multiple sumrules-ab initio SCF calculations for H_2, BH_3, CH_4, NH_3, H_2O, HF, N_2, CO, CO_2 and N_2O. Mol Phys 42:629–653

286. Jhanwar BL, Meath WJ (1982) Dipole oscillator strength distributions, sums, and dispersion energy coefficients for CO and CO_2. Chem Phys 67:185–199

287. Jhanwar BL, Meath WJ, MacDonald JCF (1983) Dipole oscillator strength distributions and related properties for ethylene, propene and 1-butene. Can J Phys 61:1027–1034

288. Jhanwar BL, Meath WJ (1984) Dipole oscillator strength distributions and properties for methanol, ethanol, and n-propanol. Can J Chem 62:373–381

289. Kumar A, Meath WJ (1985) Pseudo-spectral dipole oscillator strengths and dipole-dipole and triple-dipole dispersion energy coefficients for HF, HCl, HBr, He, Ne, Ar, Kr and Xe. Mol Phys 54:823–833

290. Kumar A, Fairley GRG, Meath WJ (1985) Dipole properties, dispersion energy coefficients, and integrated oscillator strengths for SF6. J Chem Phys 83:70–77

291. Kumar A, Meath WJ (1985) Dipole oscillator strength distributions and properties for SO_2, CS_2 and OCS. Can J Phys 63:417–427

292. Kumar A, Meath WJ (1985) Integrated dipole oscillator strengths and dipole properties for Ne, Ar, Kr, Xe, HF, HCl, and HBr. Can J Phys 63:1616–1630

293. Pazur RJ, Kumar A, Thuraisingham RA, Meath WJ (1988) Dipole oscillator strength properties and dispersion energy coefficients for H_2S. Can J Chem 66:615–619

294. Meath WJ, Kumar A (1990) Reliable isotropic and anisotropic dipolar dispersion energies, evaluated using constrained dipole oscillator strength techniques, with application to interactions involving H_2, N_2, and the rare gases. Int J Quantum Chem 24:501–520

295. Le Roy RJ, Bissonnette C, Wu TH, Dham AK, Meath WJ (1994) Improved modelling of atom-molecule potential-energy surfaces: illustrative application to He-CO. Faraday Discuss 97:81–94

296. Korona T, Moszynski R, Thibault F, Launay JM, Bussery-Honvault B, Boissoles J, Wormer PES (2001) Spectroscopic, collisional, and thermodynamic properties of the He-CO_2 complex from an ab initio potential: Theoretical predictions and confrontation with the experimental data. J Chem Phys 115:3074–3084

297. Price SL, Stone AJ (1980) Evaluation of anisotropicmodel intermolecular pair potentials using an ab initio SCF-CI surface. Mol Phys 40:805– 822

298. Millot C, Stone AJ (1992) Towards an accurate intermolecular potential for water. Mol Phys 77:439–462

299. Pearlman DA, Case DA, Caldwell JW, Ross WR, Cheatham III TE, DeBolt S, Ferguson D, Seibel G, Kollman P (1995) AMBER, a computer program for applying molecular mechanics, normal mode analysis, molecular dynamics and free energy calculations to elucidate the structures and energies of molecules. Comp Phys Commun 91:1–41

300. Cornell WD, Cieplak P, Bayly CI, Gould IR, Merz KM Jr, Ferguson DM, Spellmayer DC, Fox T, Caldwell JW, Kollman PA (1995) A second generation force field for the simulation of proteins, nucleic acids, and organic molecules. J Am Chem Soc 117:5179–5197

301. Cieplak P, Caldwell J, Kollman P (2001) Molecular mechanical models for organic and biological systems going beyond the atom centered two body additive approximation: aqueous solution free energies ofmethanol and n-methyl acetamide, nucleic acid base, and amide hydrogen bonding and chloroform/water partition coeficients of the nucleic acid bases. J Comput Chem 22:1048–1057

302. Moszynski R, Wormer PES, Jeziorski B, Van der Avoird A (1995) Symmetry-adapted perturbation theory of nonadditive three-body interactions in Van der Waals molecules. I. General theory. J Chem Phys 103:8058–8074

303. Kaplan IG, Santamaria R, Novaro O (1995) Nonadditive interactions and the relative stability of neutral and anionic silver clusters. Int J Quantum Chem 55:237–243

304. Kaplan IG, Santamaria R, Novaro O (1995) Non-additive forces in atomic clusters. The case of Ag_n. Mol Phys 84:105–114

305. Kaplan IG, Hernandez-Cobos J, Ortega-Blake I, Novaro O (1996) Many body forces and electron correlation in small metal clusters. Phys Rev A 53:2493–2500

306. Kaplan IG (1999) Nature of binding in small metal clusters. Int J Quantum Chem 74:241–247

307. Kaplan IG, Roszak S, Leszczynski J (2000) Nature of binding in the alkaline-earth clusters: Be_3, Mg_3, and Ca_3. J Chem Phys 113:6245–6252

308. Jakowski J (2001) Nonadditive forces in open-shell Van der Waals complexes. PhD thesis, University of Warsaw, Warsaw

309. Jeziorski B (1974) Perturbation theory of many-body effects in the interaction of atoms or molecules. PhD thesis, University of Warsaw, Warsaw (in Polish).

310. Korona T, Moszynski R, Jeziorski B (1996) Convergence of symmetryadapted perturbation theory expansions for pairwise nonadditive interatomic interactions. J Chem Phys 105: 8178–8186

311. Axilrod BM, Teller E (1943) Interaction of theVan derWaals type between three atoms. J Chem Phys 11:299–300

312. Muto Y (1943) Force between nonpolar molecules. Proc Phys Soc Jpn 17:629

313. Moszynski R, Heijmen TGA, Wormer PES, Van der Avoird A (1998) Symmetry-adapted perturbation theory of nonadditive three-body interactions in Van derWaals molecules. II. Application to the Ar_2–HF interaction. J Chem Phys 108:579–589

314. Wormer PES, Moszynski R, Van der Avoird A (2000) Intramonomer correlation contributions to first-order exchange nonadditivity in trimer. J Chem Phys 112:3159–3169

315. Jeziorski B, Van Hemert M (1976) Variation-perturbation treatment of the hydrogen bond between water molecules. Mol Phys 31:713–729

316. Moszynski R, Heijmen TGA, Wormer PES, Van der Avoird A (1996) Ab initio collision-induced polarizability, polarized and depolarized Raman spectra, and second dielectric virial coefficient of the helium diatom. J Chem Phys 104:6997–7007

317. Moszynski R, Heijmen TGA, van der Avoird A (1995) Second dielectric virial coefficient of helium gas: quantum-statistical calculations from an ab initio interaction-induced polarizability. Chem Phys Lett 247:440–446

318. Hättig C, Larsen H, Olsen J, Jørgensen P, Koch H, Fernández B, Rizzo A (1999) The effect of intermolecular interactions on the electric properties of helium and argon. I. Ab initio calculation of the interaction induced polarizability and hyperpolarizability in He_2 and Ar_2. J Chem Phys 111:10099–10107

319. Koch H, Hättig C, Larsen H, Olsen J, Jørgensen P, Fernández B, Rizzo A (1999) The effect of intermolecular interactions on the electric properties of helium and argon. II. The dielectric, refractivity, Kerr, and hyperpolarizability second virial coefficients. J Chem Phys 111:10108–10118

320. Hättig C, L'opez Cacheiro J, Fernández B, Rizzo A (2003) Ab initio calculation of the refractivity and hyperpolarizability second virial coefficients of neon gas. Mol Phys 101:1983–1995

321. L'opez Cacheiro J, Fernández B, Marchesan D, Coriani S, Hättig C, Rizzo A (2004) Coupled cluster calculations of the ground state potential and interaction induced electric properties of the mixed dimers of helium, neon and argon. Mol Phys 102:101–110

322. Rizzo A, Coriani S, Marchesan D, L'opezCacheiro J, Fern'andez B, Hättig C (2006) Density dependence of electric properties of binary mixtures of inert gases. Mol Phys 104:305–318

323. Prengel AT, Gornall WS (1976) Raman scattering from colliding molecules and Van der Waals dimers in gaseous methane. Phys Rev A 13:253–262

324. Frommhold L, Hong-Hong K, Proffitt MH (1978) Absolute cross sections for collision-induced depolarized scattering of light in krypton and xenon. Mol Phys 35:665–700

325. Buckingham AD, Pople JA (1955) The dielectric constant of an imperfect non-polar gas. Trans Faraday Soc 51:1029–1035

326. de Boer J (1949) Molecular distribution and equation of state of gases. Rept Prog Phys 12:305–374

327. Buckingham AD, Dunmur DA (1968) Kerr effect in inert gases and sulphur hexafluoride. Trans Faraday Soc 64:1776–1783

328. Hättig C, Moszynski R, Rizzo A (2002) unpublished results

329. Herzberg G (1945) Molecular spectra and molecular structure. II. Infrared and Raman spectra of polyatomic molecules. Van Nostrand, New York

330. Bratoz S, Martin ML (1965) Infrared spectra of highly compressed gas mixtures of the type HCl+X. A theoretical study. J Chem Phys 42:1051–1062

331. Moszynski R, Jeziorski B, Wormer PES, Van der Avoird A (1994) Rovibrational spectra of $Ar-H_2$ and $Ar-D_2$ Van der Waals complexes from an ab initio SAPT potential. Chem Phys Lett 221:161–166

332. Tennyson J, Sutcliffe BT (1982) The ab initio calculation of the vibrational-rotational spectrum of triatomic systems in the close-coupling approach, with KCN and H_2Ne as examples. J Chem Phys 77:4061–4072

333. Sutcliffe BT, Tennyson J (1986) Ageneralized approach to the calculation of ro-vibrational spectra of triatomic molecules. Mol Phys 58:1053–1066

334. Brocks G, Van der Avoird A, Sutcliffe BT, Tennyson J (1983) Quantum dynamics of non-rigid systems comprising two polyatomic fragments. Mol Phys 50:1025–1043

335. Percival C, Seaton MJ (1957) The partial wave theory of electronhydrogen atom collisions. Proc Camb Phil Soc 53:654–662

336. Arthurs AM, Dalgarno A (1960) The theory of scattering by a rigid rotator. Proc Royal Soc (London) A 256:540–551

337. McCourt FRW, Kohler WE, Beenakker JJM, Kuscer I (1990) Nonequilibrium phenomena in gases: Dilute gases. Clarendon Press, Oxford

338. McCourt FRW, Kohler WE, Beenakker JJM, Kuscer I (1991) Nonequilibrium phenomena in gases: Cross sections, scattering, and rarefied gases. Clarendon Press, Oxford

339. Pack RT (1983) First quantum corrections to second virial coefficients for anisotropic interactions: Simple, corrected formula. J Chem Phys 78:7217–7222

340. Wormer PES (2005) Second virial coefficients of asymmetric top molecules. J Chem Phys 122:184301

341. Brooks CL, Karplus M, Pettit BM (1988) Proteins: A theoretical perspective of dynamics, structure, and thermodynamics. Wiley, New York

342. Wood WW (1968) Monte Carlo studies of simple liquid models. In: Temperley HNV, Rowlinson JS, Rushbrooke GS (eds) The physics of simple liquids, Amsterdam, North Holland, pp115–230

343. Alder BJ, Hoover WJ (1968) Numerical statistical mechanics. In: Temperley HNV, Rowlinson JS, Rushbrooke GS (eds) The physics of simple liquids, Amsterdam, North Holland, pp79–114

344. Car R, Parrinello M (1985) Unified approach for molecular dynamics and density functional theory. Phys Rev Lett 55:2471–2474

345. Laasonen K, Sprik M, Parrinello M, Car R (1993) "ab initio" liquid water. J Chem Phys 99:9080–9089

346. Williams HL, Szalewicz K, Jeziorski B, Moszynski R, Rybak S (1993) Symmetry-adapted perturbation theory calculation of the Ar–H_2 intermolecular potential energy surface. J Chem Phys 98:1279–1292

347. Mrugala F, Moszynski R (1998) Near-infrared absorption spectrum of the Ar-HD complex: Confrontation of theory with experiment. J Chem Phys 109:10823–10837

348. Moszynski R, Wormer PES, Jeziorski B, Van der Avoird A (1994) Symmetry-adapted perturbation theory calculation of the He–HF intermolecular potential energy surface. J Chem Phys 101:2811–2824

349. Moszynski R, Jeziorski B, Van der Avoird A, Wormer PES (1994) Nearinfrared spectrum and rotational predissociation dynamics of the He–HF complex from an ab initio symmetry-adapted perturbation theory potential. J Chem Phys 101:2825–2835

350. Moszynski R, deWeerd F, Groenenboom GC, Van der Avoird A (1996) He–HF scattering cross sections from an ab initio SAPT potential: Confrontation with experiment. Chem Phys Lett 263:107–112

351. Mas EM, Szalewicz K, Bukowski R, Jeziorski B (1997) Pair potential for water from symmetry-adapted perturbation theory. J Chem Phys 107:4207–4218

352. Groenenboom G, Van der Avoird A, Wormer PES, Mas EM, Bukowski R, Szalewicz K (2000) Water pair potential of near spectroscopic accuracy. II. Vibration-rotation-tunneling levels of the water dimer. J Chem Phys 113:6702–6715

353. Moszynski R, Korona T, Wormer PES, Van der Avoird A (1995) Ab initio potential energy surface, infrared spectrum, and second virial coefficient of the He–CO complex. J Chem Phys 103:321–332

354. Heijmen TGA, Moszynski R, Wormer PES, Van der Avoird A (1997) A new He–CO interaction energy surface with vibrational coordinate dependence. I. Ab initio potential and infrared spectrum. J Chem Phys 107:9921–9928

355. Moszynski R, Korona T, Wormer PES, Van der Avoird A (1997) Ab Initio Potential Energy Surface and Infrared Spectrum of the Ne-CO Complex. J Phys Chem A 101:4690–4698

356. Moszynski R, Wormer PES, Van der Avoird A (1995) Ab initio potential energy surface and near-infrared spectrum of the He–C_2H_2 complex. J Chem Phys 102:8385–8397

357. Heijmen TGA, Moszynski R, Wormer PES, Van der Avoird A, Buck U, Ettischer I, Krohne R (1997) Total differential cross sections and differential energy loss spectra for He–C_2H_2 from an ab initio potential. J Chem Phys 107:7260–7265

358. Heijmen TGA, Moszynski R, Wormer PES, Van der Avoird A, Rudert AD, Halpern JB, Martin J, Gao WB, Zacharias H (1999) Rotational state-to-state rate constants and pressure broadening coefficients for He– C_2H_2 collisions: Theory and experiment. J Chem Phys 111:2519–2531

359. Bemish RJ, Oudejans L, Miller RE, Moszynski R, Heijmen TGA, Korona T, Wormer PES, Van der Avoird A (1998) Infrared spectroscopy and ab initio potential energy surface for $Ne-C_2H_2$ and $Ne-C_2HD$ complexes. J Chem Phys 109:8968–8979

360. Heijmen TGA, Korona T, Moszynski R, Wormer PES, Van der Avoird A (1997) Ab initio potential-energy surface and rotationally inelastic integral cross sections of the $Ar-CH_4$ complex. J Chem Phys 107:902–913

361. Heijmen TGA, Moszynski R, Wormer PES, Buck U, Steinbach C, Hutson JM (1998) Total differential cross sections for $Ar-CH_4$ from an ab initio potential. J Chem Phys 108:4849–4853

362. Heijmen TGA, Wormer PES, Van der Avoird A, Miller RE, Moszynski R (1999) The rotational and vibrational dynamics of argon– methane. I. A theoretical study. J Chem Phys 110:5639–5650

363. Miller RE, Heijmen TGA, Wormer PES, Van der Avoird A, Moszynski R (1999) The rotational and vibrational dynamics of argon– methane. II. Experiment and comparison with theory. J Chem Phys 110:5651–5657

364. Moszynski R, Korona T, Heijmen TGA, Wormer PES, Van der Avoird A, Schramm B (1998) Second virial coefficients for atom-molecule complexes from ab initio SAPT potentials. Polish J Chem 72:1479–1496

365. Weida MJ, Sperhac JM, Nesbitt DJ, Hutson JM (1994) Signatures of large amplitude motion in a weakly bound complex: High-resolution IR spectroscopy and quantum calculations for $HeCO_2$. J Chem Phys 101:8351–8363

366. Tang J, McKellar ARW, Mezzacapo F, Moroni S (2004) Bridging the gap between small clusters and nanodroplets: Spectroscopic study and computer simulation of carbon dioxide slovatedwith heliumatoms. Phys Rev Lett 92:145503

367. Nauta K, Miller RE (2001) Rotational and vibrational dynamics of CO_2 and N_2O in helium nanodroplets. J Chem Phys 115:10254–10260

368. Buck U, Schleusener J, Malik DJ, Secrest D (1981) On the argon-methane interaction from scattering data. J Chem Phys 74:1707–1717

369. Chapman WB, Schiffman A, Hutson JM, Nesbitt DJ (1996) Rotationally inelastic scattering in CH_4 + He, Ne, and Ar: State-to-state cross sections via direct infrared laser absorption in crossed supersonic jets. J Chem Phys 105:3497–3516

370. Miller RE (1994) In General disscusion. Faraday Discuss. 97:177–178

371. Wormer PES, Moszynski R (1996) SAPT3 package University of Nijmegen and University of Warsaw

372. Milet A, Moszynski R, Wormer PES, Van der Avoird A (1999) Hydrogen bonding in water clusters; pair and many-body interactions from symmetry-adapted perturbation theory. J Phys Chem A 103:6811–6819

373. Gregory JK, Clary DC (1995) Three-body effects onmolecular properties in the water trimer. J Chem Phys 103:8924–8930

374. Kollman P (1985) Theory of complex molecular interactions: computer graphics, distance geometry, molecularmechanics, and quantummechanics. Acc Chem Res 18:105–111

375. Bukowski R et al (2006) to be published

376. Mas EM, Bukowski R, Szalewicz K (2003) Ab initio three-body interactions for water. II. Effects on structure and energetics of liquid. J Chem Phys 118:4404–4413

377. Soper AK (2000) The radial distribution functions of water and ice from 220 to 673 K and at pressures up to 400 MBa. Chem Phys 258:121–137

378. Niesar U, Corongiu G, Huang M-J, Dupuis M, Clementi E (1989) Preliminary observations on a new water-water potential. Int J Quantum Chem Sym 23:421–443

379. Corongiu G, Clementi E (1992) Liquid water with an ab initio potential: X-ray and neutron scattering from 238 to 368 K. J Chem Phys 97:2030– 2038

380. Handon DR, Ravishankara AR (1992) Investigation of the reactive and nonreactive processes involving nitryl hypochlorite and hydrogen chloride on water and nitric acid doped ice. J Phys Chem 96:2682–2691

381. Chu LT, Leu MT, Keyser LF (1993) Heterogeneous reactions of hypochlorous acid + hydrogen chloride $\rightarrow$ Cl_2 + H_2O and chlorosyl nitrite + HCl $\rightarrow$ Cl_2 + HNO_3 on ice surfaces at polar stratospheric conditions. J Phys Chem 97:12798–12804

382. McCoustra MRS, Horn AB (1994) Towards a laboratory strategy for the study of heterogeneous catalysis in stratospheric ozone depletion. Chem Soc Rev 23:195–204

383. Kroes G-J, Clary DC (1992) Adsorption of HCl on ice under stratospheric conditions: A computational study. Geophys Res Lett 19:1355–1358

384. Kroes G-J, Clary DC (1992) Sticking of hydrogen chloride and chlorine hydroxide to ice: A computational study. J Phys Chem 96:7079–7088

385. Wang L, Clary DC (1996) Time-dependent wave-packet studies on the sticking of HCl to an ice surface. J Chem Phys 104:5663

386. Milet A, Struniewicz C, Moszynski R, Wormer PES (2001) Theoretical study of the protolytic dissociation of HCl in water clusters. J Chem Phys 115:349–356

387. Legon AC, Willoughby LC (1983) Identification andmolecular geometry of a weakly bound dimer (H_2O, HCl) in the gas phase by rotational spectroscopy. Chem Phys Lett 95:449–452

388. Kisiel Z, Pietrewicz BA, Fowler PW, Legon AC, Steiner E (2000) Rotational spectra of the less common isotopomers, electric dipole moment and the double minimum inversion potential of H_2O...HCl. J Phys Chem A 104:6970–6978

389. Kisiel Z, Bialkowska-Jaworska E, Pszczolkowski L, Milet A, Struniewicz C, Moszynski R, Sadlej J (2000) Structure and properties of the weakly bound trimer $(H_2O)_2HCl$ observed by rotational spectroscopy. J Chem Phys 112:5767–5776

390. Kisiel Z, Kosarzewski J, Pietrewicz BA, Pszczolkowski L (2000) Electric dipole moments of the cyclic trimers $(H_2O)_2HCl$ and $(H_2O)_2HBr$ from Stark effects in their rotational spectra. Chem Phys Lett 325:523–530

391. Milet A, Struniewicz C, Moszynski R, Sadlej J, Kisiel Z, Bialkowska-Jaworska E, Pszczolkowski L (2001) Structure and properties of the weakly bound trimer $(H_2O)_2HCl$. Theoretical predictions and comparison with high-resolution rotational spectroscopy. Chem Phys 271:267–282

392. Milet A, Struniewicz C, Wormer PES, Moszynski R (2000) Nature and importance of three-body interactions in the $(H_2O)_2HCl$ trimer. Theor Chem Acc 104:195–198

393. Bunker PR (1979) Molecular symmetry and spectroscopy. Academic Press, New York

394. Struniewicz C, Korona T, Moszynski R, Milet A (2001) Theoretical predictions of vibration-rotation-tunneling dynamics of the weakly bound trimer $(H_2O)_2HCl$. Chem Phys Lett 343:588–596

395. Milet A, Struniewicz C, Moszynski R (2002) unpublished results

396. Siegbahn PEM (1996) Models for the description of the H_3O^+ and OH^- ions in water. J Comput Chem 17:1099–1107

397. Lundgren J-O, Olovsson I (1967) Hydrogen bond studies.XVI. The crystal structure of chloride trihydrate. Acta Crys 23:971–976

398. Gilbert N, Sheppard AS (1973) Infra-red spectra of the hydrates of hydrogen chloride and hydrogen bromide. Absorption bands of the $H_5O_2^+$ species. J Chem Soc Faraday Trans II 69:1628–1642

399. Amirand C, Maillard D (1988) Spectrumand structure ofwater-rich waterhydracid complexes from matrix isolation spectroscopy: evidence for proton transfer. J Mol Struct 176: 181–201

400. Struniewicz C, Milet A, Sadlej J, Moszynski R (2002) Theoretical study of the hydrogen chloride trihydrate. Int J Quantum Chem 90:1151–1162

401. Hurley SM, Dermonta TE, Hydutsky DP, Castelman AW Jr (2002) Dynamics of hydrogen bromide dissolution in the ground and excited states. Science 298:202–204

402. Robertson WH, Johnson MA (2002) Caught in the act of dissolution. Science 298:69–69

403. Le Duff Y (1979) Collision-induced scattering in helium. Phys Rev A 20:48–53

404. Barocchi F, Mazzinghi P, Zoppi M (1978) Collision-induced light scattering in gaseous helium. Phys Rev Lett 41:1785–1788

405. Barocchi F, Zoppi M (1981) Experimental determination of two-body collision-induced light scattering spectrum of helium. In: Van Kranendonk J (ed) Intermolecular spectroscopy and dynamical properties of dense systems, Proceedings of the International School 'Enrico Fermi', Course LXXV, Amsterdam, North-Holland, pp263–274

406. Proffitt M, Frommhold L (1979) New measurement of the trace of the helium diatom polarizability from the collision-induced, polarized, Raman spectrum. Phys Rev Lett 42:1473–1475

407. Frommhold L, Proffitt M (1979) Concerning the anisotropy of the helium diatom polarizability. J Chem Phys 70:4803–4804

408. Proffitt M, Frommhold L (1980) The collision-induced polarized and depolarized raman spectra of heliumand the diatompolarizability. J Chem Phys 72:1377–1384

409. Proffitt M, Keto JW, Frommhold L (1980) Collision-induced spectra of the helium isotopes. Phys Rev Lett 45:1843–1846

410. Proffitt M, Keto JW, Frommhold L (1981) Collision-induced Raman spectra and diatom polarizabilities of the rare gases-an update. Can J Phys 59:1459–1474

411. Dacre PD, Frommhold L (1982) Rare gas diatompolarizabilities. J Chem Phys 76:3447–3460

412. Dacre PD (1982) On the pair polarizability of helium. Mol Phys 45:17–32

413. Rachet F, Chrysos M, Guillot NC, Le Duff Y (2000) Unique case of highly polarized collision-induced light scattering: The very far spectral wing by the helium pair. Phys Rev Lett 84:2120–2123

414. Rachet F, Le Duff Y., Guillot NC, Chrysos M (2000) Absolute isotropic spectral intensities in collision-induced light scattering by helium pairs over a large frequency domain. Phys Rev A 61:062501

415. Guillot NC, Chrysos M, Le Duff Y, Rachet F (2000) Depolarized collision-induced light scattering by gaseous helium. J Phys B 33:569–580

416. Guillot NC, Le Duff Y, Rachet F, Chrysos M (2002) Anisotropic and isotropic light scattering in gaseous low-temperature helium. Phys Rev A 66:012505

417. Rachet F, Chrysos M, Lothon G, Moszynski R, Milet A (2002) Collisioninduced light scattering by gaseous neon. In: AIP Conference Proceedings, vol 645. AIP Press New York, pp 174–180

418. Rachet F, Chrysos M, Lothon G, Moszynski R, Milet A (2003) Extended wavenumber domain collision-induced Raman scattering by gaseous neon. J Raman Spectr 34:972–976

419. Johnston DR, Oudemans GJ, Cole RH (1960) Dielectric constants of imperfect gases. I.Helium, argon, nitrogen, andmethane. J Chem Phys 33:1310–1317

420. Orcutt RH, Cole RH (1967) Dielectric constants of imperfect gases. III. Atomic gases, hydrogen, and nitrogen. J Chem Phys 46:697–702

421. Kerr EC, Sherman RH (1970) The molar polarizability of 3He at low temperatures and its density dependence. J Low Temp Phys 3:451–461

422. Kirouac S, Bose TK (1976) Polarizability and dielectric properties of helium. J Chem Phys 64:1580–1582

423. Vidal D, Lallemand M (1976) Evolution of the Clausius–Mossotti function of noble gases and nitrogen, at moderate and high density, near room temperature. J Chem Phys 64:4293–4302

424. Lallemand M, Vidal D (1977) Variation of the polarizability of noble gases with density. J Chem Phys 66:4776–4780

425. Gugan D, Michel GW (1980) Dielectric constant gas thermometry from 4.2 to 27.1K. Metrologia 16:149–167

426. Achtermann HJ, Magnus G, Bose TK (1991) Refractivity virial coefficients of gaseous CH_4, C_2H_4, C_2H_6, CO_2, SF_6, H_2, N_2, He, and Ar. J Chem Phys 94:5669–5684

427. Huot J, Bose TK (1991) Experimental determination of the dielectric virial coefficients of atomic gases as a function of temperature. J Chem Phys 95:2683–2687

428. Achtermann HJ, Hong JG, Magnus G, Aziz RA, Slaman J (1993) Experimental determination of the refractivity virial coefficients of atomic gase. J Chem Phys 98:2308–2318

429. Jeziorska M, Jankowski P, Jeziorski B, Szalewicz K (2000) On the optimal choice of monomer geometry in calculations of intermolecular interaction energies. rovibrational spectrum of ar–hf from two- and three-dimensional sapt potentials. J Chem Phys 113:2957–2968

430. Klopper W, Quack M, Suhm MA (1998) HF dimer: Empirically refined analytical potential energy and dipole hypersurfaces from ab initio calculations. J Chem Phys 108:10096–10115

431. Mok DKW, Handy NC, Amos RD (1997) A density functional water dimer potential surface. Mol Phys 92:667–675

432. Krause PJ, Clary DC (1998) Vibrational predissociation of D_2HF and H_2HF with a newpotential energy surface. Mol Phys 93:619–25

433. Vissers GWM, Groenenboom GC, van der Avoird A (2003) Spectrum and vibrational predissociation of the HF dimer. I. Bound and quasibound states. J Chem Phys 119:277–285

434. Vissers GWM, Groenenboom GC, van der Avoird A (2003) Spectrum and vibrational predissociation of the HF dimer. II. Photodissociation cross sections and product state distributions. J Chem Phys 119:286–292

435. Leforestier C, Gatti F, Fellers RS, Saykally RJ (1997) Determination of a flexible (12D) water dimer potential via direct inversion of spectroscopic data. J Chem Phys 117:8710–8722

436. Shirkov L, Dubernet ML, Moszynski R (2006) to be published

437. Klos J, Chalasinski G, Szczesniak MM, Werner H-J (2001) Ab initio calculations of adiabatic and diabatic potential energy surfaces of cl... HCl Van der Waals complex. J Chem Phys 115:3085–3098

438. Klos J, Chalasinski G, Szczesniak MM (2002) Ab initio calculations and modeling of 3-dimensional adiabatic and diabatic potential energy surfaces of F... H_2 Van der Waals complex. Int J Quantum Chem 90:1038

439. Klos J, Chalasinski G, Szczesniak MM (2002) Ab initio calculations and modeling of 3-dimensional adiabatic and diabatic... h_2 van der waals complex. J Phys Chem A 106: 7362–7368

440. Klos J, Chalasinski G, Szczesniak MM (2002) Modeling of adiabatic and diabatic potential energy surfaces of cl... h_2 van der waals complex from ab initio calculations. J Chem Phys 117:4709–4719

441. Rode J, Klos J, Rajchel L, Szczesniak MM, Chalasinski G, Bouchachenko A (2005) Interactions of open-shell clusters: ab initio study of pre-reactive complex $O(^3P)+HCl$. J Phys Chem A 109:11484

442. Klos J, Szczesniak MM, Chalasinski G (2004) Paradigm pre-reactive van der Waals complexes: X–HX and X–H_2 (X=F, Cl, Br). Int Rev Phys Chem 23:541–571

443. Skouteris D, Manolopoulos DE, Bian W, Werner HJ, Lai L-H, Liu K (1999) Van der Waals interactions in the Cl + HD reaction. Science 286:1713–1716

444. Zuchowski PS, Bussery-Honvault B, Moszynski R, Jeziorski B (2003) Dispersion interaction of high-spin open-shell complexes in the random phase approximation. J Chem Phys 119:10497–10511

445. Hirschfelder JO, Meath WJ (1967) The nature of intermolecular forces. Adv Chem Phys 12:3–106

446. Bussery-Honvault B, Moszynski R (2006) Ab initio potential energy curves, transition dipole moments, and spin-orbit coupling matrix elements for the first twenty states of the calcium diatom. Mol Phys 104:2387–2402

447. Dubernet ML, Flower D, Hutson JM (1991) The dynamics of open-shell Van der Waals complexes. J Chem Phys 94:7602–7618

448. Dubernet ML, Hutson JM (1993) Potential energy surfaces for Ar-OH ($X^2 \prod$) obtained by fitting to high-resolution spectroscopy. J Chem Phys 99:7477–7486

449. Dubernet ML, Hutson JM (1994) Atom-molecule Van der Waals complexes containing open-shell atoms. I. General theory and bending levels. J Chem Phys 101:1939–1958

450. Dubernet ML, Hutson JM (1994) Atom-molecule Van der Waals complexes containing open-shell atoms. II. The bound btates of Cl-HCl. J Phys Chem 98:5844–5854

451. Alexander MH (1993) Differential and integral cross sections for the inelastic scattering of NO ($X^2 \prod$) by Ar based on a new ab initio potential energy surface. J. Chem Phys 99:7725–7738

452. Dagdigian PJ, Patel-Misra D, Berning HJ, nad Werner A, Alexander MH(1993) A joint experimental and theoretical study of $A^2 \prod$ to $X^2 \prod$ electronic energy transfer in the cn molecule induced by collisions with helium. J Chem Phys 98:8580–8592

453. Alexander MH, Gregurick S, Dagdigian PJ, Lemire GW, Mc-Quaid MJ, Sausa RC (1994) Potential energy surfaces for the interaction of CH($X^2 \prod, B^2 \sum^-$)withAr and an assignment of the stretch-bend levels of the ArCH(B) van der Waals molecule. J. Chem Phys 101:4547–4560

454. Moonbong-Yang, Alexander MH (1995) Ab initio potential energy surfaces and quantum scattering studies of NO($X^2 \prod$) with He: Lambda-doublet resolved rotational and electronic fine-structure transitions. J Chem Phys 103:6973–6983

455. Yang X, Hwang E, Dagdigian PJ, Yang M, Alexander MH (1995) Experimental and theoretical study of the B-Ne nonbonding interaction: The free-bound $B^2\Sigma^+$- $X^2 \prod$ electronic transition. J Chem Phys 103:2779–2786

456. Jakowski J, Klos J, Chalasinski G, Severson MW, Szczesniak MM, Cybulski SM (2000) Structure and energetics of ArnNO$^-$ clusters from ab initio calculations. J Chem Phys 112:10895–10904

457. Jakowski J, Chalasinski G, Szczesniak MM, Cybulski SM (2003) Modeling of the three-body effects in the neutral trimers in the quartet state by ab initio calculations. H$_3$, Na$_3$, and Na$_2$B. Collect. Czech. Chem Commun 68:587–626

458. Jakowski J, Chalasinski G, Cybulski SM, Szczesniak MM (2003) Modeling of the three-body effects in the Ar$_2$O$^-$ trimer from ab initio calculations. J Chem Phys 118:2731–2747

459. Jakowski J, Chalasinski G, Gallegos J, Severson MW, Szczesniak MM (2003) Characterization of ArnO$^-$ clusters from ab initio and diffusion Monte Carlo calculations. J Chem Phys 118:2748–2759

460. Meath WJ, Hirschfelder JO (1966) Relativistic intermolecular forces, moderately long range. J Chem Phys 44:3197–3209

461. Meath WJ, Hirschfelder JO (1966) Long-range (retarded) intermolecular forces. J Chem Phys 44:3210–3215

462. Bussery-Honvault B, Launay J-M, Moszynski R (2003) Cold collisions of ground state calcium atoms in a laser field: A theoretical study. Phys Rev A 68:032718

463. Przybytek M, Jeziorski B (2005) Bounds for the scattering length of spinpolarized helium from high-accuracy electronic structure calculations. J Chem Phys 123:134315–1–9

464. Moszynski R, Lach G, Jaszunski M, Bussery-Honvault B (2003) Longrange relativistic interactions in the Cowan-Griffin approximation and their QED retardation: Application to helium, calcium, and cadmium dimers. Phys Rev A 68:052706

465. Pachucki K (2005) Relativistic corrections to the long-range interaction between closed-shell atoms. Phys Rev A 72:062706
466. Bussery-Honvault B, Launay J-M, Moszynski R (2005) Photoassociation of cold calcium atoms through the $A^1 \sum_u^+ (1^1S+1^1D)$, $c^3 \prod_u (1^3P+1^1S)$, and $a^3 \sum_u^+ (1^3P+1^1S)$ states: An ab initio nonadiabatic treatment. Phys Rev A 71:012702
467. Bussery-Honvault B, Moszynski R (2006) to be published.
468. Saute M, Bussery B, Aubert-Frecon M (1984) Coefficients d'interaction à grande distance pour les 23 états moléculaires de Cl_2 et de Br_2. Mol Phys 51:1459–1474
469. Bussery B, Aubert-Frecon M (1985) Multipolar long-range electrostatic, dispersion and induction energy terms for the interactions between two identical alkali atoms, Li, Na, K, Rb and Cs in various electronic states. J Chem Phys 82:3224–3234
470. Bussery B, Aubert-Frecon M (1985) Calculated long-range coefficients Cn (n = 6, 8, 10) for the interactions Na(3s) + K(4s) and Na(3s) + K(4p). J Phys B 18:L379–L381
471. Bussery B, Aubert-Frecon M (1985) Potential energy curves and vibration-rotation energies for the two purely long-range bound states 1u and O-g of the alkali dimers M_2 dissociating to $M(ns^2S_{1/2}) + M(ns^2P_{3/2})$ with M=Na, K, Rb and Cs. J Mol Spectrosc 111:21–27
472. Weinstein JD, de Carvalho R, Guillet T, Friedrich B, Doyle JM (1998) Magnetic trapping of calcium monohydride molecules at milliKelvin temperatures. Nature 395:148–150
473. Bethlem HL, Berden G, Meijer G (1999) Decelerating neutral dipolar molecules. Phys Rev Lett 83:1558–1561
474. Thorsheim HR, Weiner J, Julienne PS (1987) Laser-induced photoassociation of ultracold sodium atoms. Phys Rev Lett 58:2420–2423
475. Weiner J, Bagnato VS, Zilio S, Julienne PS (1999) Experiments and theory in cold and ultracold collisions. Rev Mod Phys 71:1–85
476. Sokalski WA, Kedzierski P, Grembecka J (2001) Ab initio study of the physical nature of interactions between enzyme active site fragments in vacuo. Phys Chem Chem Phys 3:657–663
477. Grembecka J, Kedzierski P, Sokalski WA (1999) Non-empirical anlysis of the nature of the inhibitor-active-site interactions in leucine aminopeptidase. Chem Phys Lett 313:385–392
478. Grembecka J, Sokalski WA, Kedzierski P (2001) Quantum chemical analysis of the interactions of transition state analogs with leucine aminopeptidase. Int J Quantum Chem 84:302–310
479. Grembecka J, Sokalski WA, Kedzierski P (2000) Computer-aided design and activity prediction of leucitine aminopeptidase inhibitors. J Comput Aided Mol Des 14:531–544

CHAPTER 2

HOHENBERG-KOHN-SHAM DENSITY FUNCTIONAL THEORY

The formal basis for a family of succesful and still evolving computational methods for modelling interactions in complex chemical systems.

TOMASZ A. WESOŁOWSKI

University of Geneva, Switzerland

Abstract: The emergence of a family of computational methods, known under the label 'density functional theory' or 'DFT', revolutionized the field of computer modelling of complex molecular systems. Many computational schemes belonging to the DFT family are currently in use. Some of them are designed to be universal (nonempirical) whereas other to treat specific systems and/or properties (empirical). This review starts with the introduction of the formal elements underlying all these methods: Hohenberg-Kohn theorems, reference system of noninteracting electrons, exchange-correlation energy functional, and the Kohn-Sham equations. The main roads to approximate the exchange-correlation-energy functional based on: local density approximation (LDA), generalized gradient approximation (GGA), meta-GGA, and adiabatic connection formula (hybrid functionals), are outlined. The performance of these approximations in describing molecular properties of relevance to intermolecular interactions and their interactions with environment in condensed phase (ionization potentials, electron affinities, electric moments, polarizabilities) is reviewed. Developments concerning new methods situated within the general Hohenberg-Kohn-Sham framework or closely related to it are overviewed in the last section

Keywords: computer modelling, density functional theory, dipole moment, dipole polarizability, electron affinity, empirical methods, exchange-correlation energy functional, hydrogen bonding, intermolecular interactions, ionization potential, Kohn-Sham equations, nonempirical methods, van der Waals complex

W. A. Sokalski (ed.), Molecular Materials with Specific Interactions, 153–201.

1.   INTRODUCTION

Works of Hohenberg and Kohn[1] and Kohn and Sham[2] provided the formal framework underlying a group of very successful computational methods. These methods are used to obtain ground-state properties of molecules and materials in Born-Oppenheimer approximation. In the literature, the Hohenberg-Kohn-Sham framework is referred frequently as density functional theory (DFT). One of its key elements is the use of explicit functionals* of electron density to express certain components of the total electronic energy. Here, we refer to it as Hohenberg-Kohn-Sham DFT. Compared to wave-function based methods, the Hohenberg-Kohn-Sham formal framework involves several new ideas: (*i*) the interpretation of the ground-state energy of a given system as a minimum of a functional which depends explicitly on electron density ($E[\rho]$), (*ii*) the reference system of noninteracting electrons, (*iii*) the one-electron equations to minimize the total energy functional, and (*iv*) the exchange-correlation-energy functional referred to in this work as $E_{xc}[\rho]$. Initially, these new concepts met no or only lukewarm interest in the theoretical chemistry community for two principal reasons:

(a) Opposite to the hierarchical structure of wavefunction-based methods making it possible to approach the exact results with arbitrary accuracy (at least for small systems), approximations to $E_{xc}[\rho]$ needed in practical calculations cannot be ordered in such a hierarchy. $E_{xc}[\rho]$ is defined implicitly but its analytic form is not known. Various theoretical considerations or empirical strategies lead to different approximations to this functional. Therefore, the Kohn-Sham formalism cannot be considered as a finished computational scheme but rather as a general theoretical framework encompassing various possible computational schemes. Among them, there are such which use only fundamental physical constants and exact mathematical conditions but also such which use extensively empirical data. This makes it very difficult to order all these methods in a series approaching systematically the exact solution.

(b) Replacing wavefunction by electron density as the fundamental variable makes a clear break with the main-stream tradition in quantum chemistry and the accumulated computational experience. In some cases, where wavefunction-based methods do not encounter any fundamental difficulties, practical implementations of the Kohn-Sham framework lead to qualitatively erroneous results. For instance: (*i*) in one electron systems, for which the Coulomb- and exchange energy match perfectly, whereas the use of an approximated density-dependent functional results in a spurious energy component known as the self-interaction error[4]; (*ii*) anionic systems are frequently not stable[5]; (*iii*) artificial splitting of degenerate energy levels differing in their electron densities, which is the results of different errors of the exchange-correlation-energy functional for different states[6]; (*iv*) the process of dissociation into radical fragments[7]; and (*v*)

---

* "By a functional, we mean a correspondence which assigns a definite (real) number to each function (or curve) belonging to some class."[3]

accounting for long-range intermolecular attraction (London dispersion forces). The fact that removal of these errors cannot be built-in into the Kohn-Sham framework in a straightforward manner is, therefore, rather discouraging.

The ideas of Hohenberg, Kohn, and Sham were, however, promptly adopted in the solid-state physics community. Even the simplest approximations to the exchange-correlation functional brought computational schemes of great potential in describing properties of solids. Nevertheless, a gradual increase of interest in the Hohenberg-Kohn-Sham DFT in the theoretical chemistry community resulted in the emergence of various practical computational methods, which dominate nowadays the field of computer modelling of molecular systems at quantum mechanical level (see Figure 2-1).

It is worthwhile to indicate here that the label "density functional theory" might be used in two[*] different contexts: pragmatic and methodological.

Pragmatic DFT: DFT methods make it possible to calculate properties of molecular systems with *acceptable accuracy at lower computational cost* than

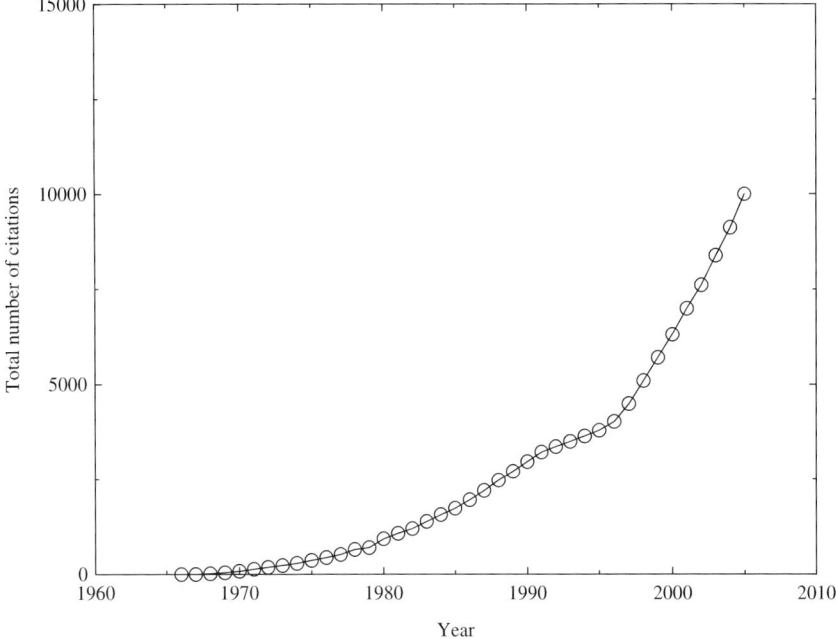

Figure 2-1. Total number of citations of the paper by Kohn and Sham.[2] Source: WSI Web of Science, Thomson Corp. 2005

---

[*] Although not covered in this review, it is worthwhile to mention the third face of DFT, the so-called *conceptual density functional*, which aims at linking mathematical objects of DFT with intuitive concepts of chemistry used to rationalize reactivity of chemical molecules (see Chermette,[8] for review).

the conventional wavefunction-based methods without, however, providing a practical strategy to approach the *exact* solution even for the simplest chemical systems. Moreover, the accuracy of the results can be improved further for a particular class of systems and/or properties owing to the possibility of using empirically fitted parameters. Each method obtained within the empirical strategy involves, however, such possible dangers as: lack of correlation among the quality of different observables calculated for the same system and/or rapid deterioration of the quality of the results once the investigated system falls out its original domain of applicability.

Methodological DFT: Hohenberg-Kohn-Sham density functional theory repre- sents an alternative strategy to solve Schrödinger equation in Born-Oppenheimer approximation. Converting it into an efficient computational tool involves *new challenges for many-body* such as identification of properties of certain mathematical objects of key importance in DFT or formal justifi- cation for some intuitive or *ad hoc* assumptions made in practical calculations. Works on these challenges may lead to practical benefits such as: (*a*) identifi- cation of the physical origin of failures of some approximations to $E_{xc}[\rho]$, (*b*) improved confidence of the results concerning new types systems and problems for which empirical methods have not been previously testes, and (*c*) more rational balance between the accuracy and computational costs in designing a computer modelling experiment.

The two faces of DFT, methodological and pragmatic, are obviously linked by a common ultimate objective – a reliable computational method for modelling polyatomic systems at quantum mechanical level. Progress in each of these domains proves to be beneficial. The exact mathematical properties of the exchange-correlation functional (and others closely related objects in DFT) provide guidelines for designing better approximations. Alternatively, successful approximations based on intuition or *ad hoc* assumptions prompt frequently the research aimed at finding their formal origin.

In view of the rapid growth of applications of DFT methods, a general overview of the pragmatic face of DFT becomes impossible. Such task should be left for specialized reviews dealing with particular systems or properties. This work concen- trates of the methodological issues and covers its pragmatic aspects in a selective way. The assumptions/approximations underlying the overviewed methods are given and discussed in the context of their strengths and weaknesses in describing inter- actions between a molecule and its environment in condensed phase. Properties of molecular systems such as *electric moments* and *polarizabilities* are considered here because they are the simplest observables, the quality of which relates directly to the accuracy of the exchange-correlation effective potential (one of the key approx- imate quantity in the Hohenberg-Kohn-Sham framework). *Ionization potentials, electron*, and *bonding parameters in weak intermolecular complexes* are discussed here as the simplest energy-related quantities which depend on the accuracy of both the exchange-correlation energy functional and the exchange-correlation effective potential.

The present review is organized in the following way. It starts with the key elements of the Hohenberg-Kohn-Sham density functional theory. In the following parts, the mainstream strategies to construct approximate methods based on the Kohn-Sham equations are overviewed. The overall performance of each group of approximate methods in reproducing molecular properties, which are of key importance in describing the interactions of a molecule with its environment, is reviewed in the subsequent section. The final part concerns current developments.

Except for occasional discussions of the basis set dependence of the results, the numerical implementation issues such as: grid integration techniques, electron-density fitting, frozen-cores, pseudopotentials, and linear-scaling techniques, are omitted.

Atomic units are used in all equations and all considerations concern non-relativistic quantum mechanics in Born-Oppenheimer approximation. Square brackets, as in E[ρ] for instance, are used to indicate that the relevant quantity is a functional i.e. the correspondence between a function in real space $\rho = \rho(\mathbf{r})$ and a real number (energy in this example). Abbreviations or acronyms denoting different approximate exchange-correlation functionals reflect their common usage in the literature. They are collected in Appendix. Unless specified, the equations are given for the spin-compensated case.

2. THE KOHN-SHAM EQUATIONS

The formal framework of all the computational methods considered in this review and branded commonly in the literature as "DFT methods" consists of the following elements: *(i)* the first Hohenberg-Kohn theorem defining the *density functional* E[ρ], *(ii)* the second Hohenberg-Kohn theorem introducing *variational principle* according to which the ground state electron density and energy can be obtained via minimization of E[ρ], subject only to the requirement that E[ρ] integrates to a given number of electrons, *(iii)* the *reference system* of noninteracting electrons of the electron density which is the same as that of the *real system*. The last element proved to be of inestimable value. Opposite to the real system of interacting electrons, for which the singledeterminantal wavefunction is only an approximation, the exact wavefunction has such a simple form in the fictitious system of noninteracting electrons. The exact properties of such a system can be thus used as guidelines in designing approximations applicable for real - interacting – system.

Kohn and Sham introduced yet another functional of the total energy $(E^{KS}[\phi_1, \phi_2, .., \phi_N])$, which unlike that of Hohenberg-Kohn (E[ρ]) does not depend explicitly on the electron density (see Eq. 2). $E^{KS}[\phi_1, \phi_2, .., \phi_N]$ is expressed analytically by means of orthogonal one-electron functions $\{\phi_i^{KS}\}$ (i = 1, N) yielding the electron density ρ.

$$\rho = 2 \sum_{i=1}^{N} \phi_i^{KS^*}(\vec{r}) \phi_i^{KS}(\vec{r}) \qquad (2\text{-}1)$$

It is possible that different sets of one-electron functions yield the given electron density ρ. Kohn-Sham orbitals are defined as this set which minimizes the kinetic energy of noninteracting electrons. The procedure leading to these orbitals for an arbitrary ρ is known in as *Levy constrained search*[9,10]. These orbitals can be used to construct the *exact* wavefunction for the reference system of *noninteracting electrons*, which has the form of a single determinant. However, the resulting wavefunction is not the optimal singledeterminantal function for *the real system* of interacting electrons*. The singledeterminantal wavefunction derived from Hartree-Fock calculations leads obviously to a lower energy for the same system. This energy lowering originates from the fact that the Hartree-Fock electron density (ρ_{HF}) differs from the exact one ($\rho_{HF} \neq \rho$).

Following the second Hohenberg-Kohn theorem[1], the ground-state electron density can be obtained by means of the minimization of the total energy functional. The introduction of orbitals corresponding to the reference-system of noninteracting electrons, makes it possible to perform the search for the energy minimum, not directly among all admissible electron densities, but among of the Kohn-Sham orbitals. This search procedure corresponds to minimizing the $E[\phi_1, \phi_2, .., \phi_N]$ functional which reads:

$$E^{KS}[\varphi_1, \varphi_2, \ldots, \varphi_N] = -2\sum_{i=1}^{N} \int \varphi_i \frac{1}{2}\nabla^2 \varphi_i d\vec{r}$$

$$+ \int d\vec{r}\, v_{ext}(\vec{r})\rho(\vec{r})$$

$$+ \frac{1}{2}\int d\vec{r}\int d\vec{r}' \frac{\rho(\vec{r})\rho(\vec{r}')}{|\vec{r}-\vec{r}'|} \tag{2-2}$$

$$+ E_{xc}[\rho]$$

The orthogonal orbitals, which minimize the Kohn-Sham energy functional are obtained from the following set of one-electron equations[2†] for $i = 1$, N:

$$\left[-\frac{1}{2}\nabla^2 + V_{eff}^{KS}[\rho](\vec{r})\right]\phi_i^{KS}(\vec{r}) = \varepsilon_i^{KS}\phi_i^{KS}(\vec{r}) \tag{2-3}$$

where the multiplicative potential $V_{eff}^{KS}[\rho](\mathbf{r})$ reads:

$$V_{eff}^{KS}[\rho](\vec{r}) = v_{ext}(\vec{r}) + \int d\vec{r}' \frac{\rho(\vec{r}')}{|\vec{r}-\vec{r}'|} + v_{xc}[\rho](\vec{r}) \tag{2-4}$$

and is the same for all orbitals.

---

* The determinant constructed from Kohn-Sham orbitals is referred to sometimes as "Kohn-Sham wavefunction". This terminology might be misleading because such a wavefunction does not correspond to the real system but to the fictitious system of noninteracting electrons.
† It is assumed here that the exact electron density of the *real (interacting) system* is also the exact electron density of the *fictitious system of noninteracting electrons*. Such electron densities are *pure-state noninteracting v-representable*. In practice, this assumption cannot be easily verified.

The analytic form of the first two terms in the Kohn-Sham effective potential ($V_{eff}^{KS}[\rho](\mathbf{r})$) is known. They represent the external potential (v_{ext} which is the nuclear attraction potential in most cases) and Coulomb repulsion between electrons. The second term is an explicit functional of electron density. The last term, however, represents the quantum many-body effects and has a traditional name of exchange-correlation potential. v_{xc} is the functional derivative* of the component of the total energy functional called conventionally *exchange-correlation energy* ($E_{xc}[\rho]$):

$$v_{xc}[\rho](\vec{r}) = \frac{\delta E_{xc}[\rho]}{\delta \rho} \tag{2-5}$$

$E_{xc}[\rho]$ is defined as the component of the total energy functional, which remains after subtracting all terms in the Hohenberg-Kohn total energy functional, which can be evaluated exactly:

$$E_{xc}[\rho] = E[\rho] - T_s[\rho] - J[\rho] - V[\rho] \tag{2-6}$$

The quantity $T_s[\rho]$ denotes the kinetic energy in the reference system of noninteracting electrons defined in the Levy constrained search procedure. The numerical value of $T_s[\rho]$ can be obtained exactly provided the orbitals minimizing the kinetic energy in the reference system of noninteracting electrons are known. This is the case of the orbitals derived from the Kohn-Sham equations. Therefore, the analytic form of $T_s[\rho]$ is not needed in Kohn-Sham calculations. The above definition of $E_{xc}[\rho]$ shows clearly that the name exchange-correlation functional might be confusing – it contains also the contribution from the kinetic energy because the numerical values of $T_s[\rho]$ and $T[\rho]$ (the kinetic energy of the real i.e. interacting system) are not equal.

It is worthwhile to underline that the exact v_{xc} is the functional of ρ ($v_{xc} = v_{xc}[\rho]$) not just a function of ρ ($v_{xc} = v_{xc}(\rho)$) as assumed in most of the practical methods. Therefore, its non-local dependence on ρ is not excluded.

The physical meaning of the left-hand-side of Eq. (6) is given in its adiabatic connection definition,[11–13] which links smoothly the artificial system of noninteracting electrons with the real one by means of a coupling strength parameter $0 \le \lambda \le 1$ (λ is the parameter multiplying electron-electron repulsion energy in the Schrödinger equation):

$$E_{xc}[\rho] = \int_0^1 U_{xc}^\lambda[\rho] d\lambda \tag{2-7}$$

where $U_{xc}^\lambda[\rho] = <\Psi^\lambda[\rho]|v_{ee}|\Psi^\lambda[\rho]> - U[\rho]$ ($\Psi^\lambda[\rho]$ is the ground-state wavefunction at coupling strength λ, v_{ee} is the Coulomb repulsion, and $U[\rho]$ is the Hartree energy).

---

* Note that the exchange-correlation functional is not linear, therefore, $E_{xc}[\rho] \ne \int \rho v_{xc} d\mathbf{r}$.

Each of the two equivalent definitions of $E_{xc}[\rho]$ (Eqs. 6 and 7) as well as possible alternative ones, some of them will be discussed in the last section of this review, can be used as the basis for construction of approximations. The Kohn-Sham equations and the Kohn-Sham energy expression involve two approximate quantities: $E_{xc}[\rho]$ and v_{xc}. At a given external potential V_{ext}, defined by nuclear charges and positions of nuclei in most cases, any deviation from the exact ground-state electron density can be attributed to errors in v_{xc} applied in practical calculations. Therefore, the quality of the calculated one-electron properties at a given geometry of the nuclei depends exclusively on the quality of v_{xc}. Moreover, since the energy gradients with respect to nuclear coordinates (R_N) can be expressed as:

$$\frac{\partial E}{\partial \vec{R}_N} = \int \frac{\delta E}{\delta \rho}(\vec{r}) \frac{\partial \rho}{\partial \vec{R}_N}(\vec{r}) d\vec{r} = \int v_{xc}(\vec{r}) \frac{\delta \rho}{\delta \vec{R}_N}(\vec{r}) d\vec{r} \qquad (2\text{-}8)$$

The quality of forces acting on nuclei depends directly on the accuracy of the used approximation for v_{xc}. The situation is more involved in the case of energy differences, where the quality of the applied approximation to $E_{xc}[\rho]$ influences the accuracy of both the electron density (through v_{xc} which is its functional derivative) and the total energy calculated for this density. Currently, various approximations for these two quantities are in use. In the following part, the most common approximations to $E_{xc}[\rho]$ are reviewed.

3. COMMONLY USED APPROXIMATIONS TO THE EXCHANGE-CORRELATION-ENERGY FUNCTIONAL

The groups of approximations discussed below are ordered according to the classification introduced by Perdew known as *"Jacob's ladder"* of approximate functionals. At the lowest rung, the exchange-correlation energy depends explicitly on electron density only. Moving to higher rungs introduces also other quantities, which are used to approximate the exchange-correlation energy. Usually, no distinction is made between the approximation for v_{xc} and $E_{xc}[\rho]^*$ in the literature because these two quantities are linked by Eq. (5).

3.1. The Starting Point: Local Density Approximation

For the uniform electron gas, all the functionals defined in the previous section can be obtained exactly or calculated with arbitrary accuracy. The expression for the exchange energy of noninteracting electrons given by Dirac[14] and introduced

---

* Such a distinction is made here for two reasons: a) In the cases where the theoretical considerations lead to a given v_{xc} directly, the "parent" functional is not known. b) In the process of developing approximations to the exchange-correlation functional, it is frequently the case that the functional is tested on electron densities obtained with a potential corresponding to another exchange-correlation potential i.e. not self-consistently.

into quantum chemistry by Slater[15], represents the largest contribution to E_{xc}. The remaining component defines the correlation energy functional in this case ($E_c[\rho] = E_{xc}[\rho] - E_x[\rho]$). Ceperley and Alder[16] applied Quantum Monte Carlo simulations to obtain numerical values of $E_c[\rho]$ for the uniform electron gas and densities spanning a wide range ($1 \leq r_s \leq 200$, where $r_s = (3/4\pi\rho)^{1/3}$). We recall here that the analytic dependence of an approximate $E_{xc}[\rho]$ on the electron density is needed for deriving the associated exchange-correlation effective potential. For this purpose, an analytic fit for the dependence of the density of the correlation energy ($\varepsilon_c(\rho)$) on the electron density is needed[17,18]. Currently, the most commonly used fit to the Ceperley-Alder data is the one made by Vosko et al.[17] Unless specified, the label LDA will be used for this functional dependence throughout this review*.

The essence of the local density approximation is the assumption that the exchange-correlation energy of a non-uniform electron density can be approximated as a sum of contributions from small volume elements each characterized by uniform electron density ($\varepsilon_{xc}(\rho)(\mathbf{r})$). Since the exchange-correlation energy of uniform electron gas is available with arbitrary accuracy, the LDA exchange-correlation functional takes the following simple form:

$$E_{xc}^{LDA}[\rho] = \int d\vec{r}\,\rho(\vec{r})\varepsilon_{xc}^{LDA}(\rho) \tag{2-9}$$

Deviation of the homogeneity can be considered small if the gradient of electron density satisfies the following condition:

$$\frac{|\nabla\rho|}{\rho} << (3\pi\rho)^{1/3} \tag{2-10}$$

Unfortunately, the above condition does not hold for atomic and molecular electron densities. Surprisingly however, LDA appears to be a very reasonable approximation as far as many properties are concerned (geometries of molecules, vibrational properties, electric moments, for instance). As a rule, however, the energy differences derived from LDA are not acceptable for chemical applications. Its practical advantage over other approximations originates from the fact that it is computationally the cheapest one. It has to be born in mind that it is a functional obtained without any empirical data. The Kohn-Sham equations applying LDA can be seen as the entry-level density-functional-theory formalism, which plays a similar role in Hohenberg-Kohn-Sham DFT, as does the Hartree-Fock theory among wavefunction-based methods. However, there is no simple relation between the domains of applicability of Hartree-Fock and LDA Kohn-Sham methods.

---

* Different fits were obtained by Vosko, Wilk, and Nusair for the spin-compensated and spin-polarized cases.

3.2.   The First Breakthrough: Generalized Gradient Approximation

A straightforward step beyond Local Density Approximation consists of introducing electron-density gradient ($\nabla\rho$) in the explicit dependence of the exchange-
correlation energy on electron density.[1] For slowly varying electron densities,
the gradient expansion approximation (GEA) leads to the following form of the
exchange-correlation energy functional:

$$E_{xc}^{GEA}[\rho] = \int d\vec{r}\, A_{xc}(\rho)\rho^{4/3} + \int d\vec{r}\, C_{xc}(\rho)\frac{|\nabla\rho|^2}{\rho^{4/3}} + \ldots \qquad (2\text{-}11)$$

where the coefficients A_{xc} and C_{xc} are functions of electron density and can be in
principle obtained from formal considerations (see Perdew[19] for review). They can
also be replaced by fitted constants in Eq. (11).*

Unfortunately, neither way leads to a noticeable improvement over LDA.[21-23] In
the eighties, a new route to approximate the exchange-correlation energy has been
proposed which also uses electron-density gradients but the gradient-dependence is
not build in by means of Eq. (11) but in a different – more general – way called
generalized gradient approximation (GGA):

$$E_{xc}^{GGA}[\rho] = \int d\vec{r}\, \rho(\vec{r})\varepsilon_{xc}^{GGA}(\rho, \nabla\rho) \qquad (2\text{-}12)$$

Opposite to local density approximation, for which the density of exchange-
correlation energy can be obtained following a unique strategy (i.e. using reference
data for the uniform electron gas), many strategies can lead to different analytic
forms of $\varepsilon_{xc}^{GGA}(\rho, \nabla\rho)$. They range from the ones, which use exact physical or mathematical properties of the exchange-correlation functional leaving none or only a few
free parameters (opening the way for a possible empirical data fit) in the analytical
expression for $\varepsilon_{xc}^{GGA}(\rho, \nabla\rho)$ to the ones relaying on experimental data more extensively. The B88 exchange functional, which is now one of the most common among
the GGA functionals, was obtained by imposing the correct asymptotic distance
dependence of the density of the exchange energy for atomic systems.[24] Perdew and
co-workers used the properties of the exchange-correlation hole in the construction
of several approximate GGA functionals: PW86[25] for exchange, P86 for correlation,[26] PW91[18,27] and PBE[28] for exchange and correlation. The popular LYP
correlation functional was derived based on the properties of the exact Hartree-
Fock orbital of the helium atom.[29,30] Handy, Tozer, and collaborators explored the
empirical strategy to construct exchange-correlation functional of the GGA form
in which a large number of empirical parameters are used to express ε_{xc}^{GGA} in
Eq. (12) as an explicit function of electron density and its gradient[31]. The parameters were obtained by means of the least-squares procedure using the training

---

* The purely empirical constant (β) value of the $C_{xc}(\rho)$ in the case of exchange energy has been in use
in the X$\alpha\beta$ extension[20] of the Slater's Xα method.[15]

data comprising reference numerical values of: total energies, ionization potentials, energy gradients, and numerical values of the exchange-correlation potential. The HCTH and HCTH-A functionals, obtained using this strategy, depend on 15 and 12 parameters, respectively.[31]

Exchange-correlation functionals of the GGA form have been proposed in the literature, which take the empirical information into account in different extent. Without claiming completeness, the list includes such functionals as: B86,[32] G96,[33] FT97,[34] mPW91,[35] revPBE,[36] RPBE,[37] and OPTX,[38] and mPBE.[39]

3.3. Meta-GGA

In both LDA and GGA, $\varepsilon_{xc}(\mathbf{r})$ at a given position in space $(\mathbf{r})$ is determined by local properties of electron density at the same position. In LDA, only electron density at $\mathbf{r}$ is needed $(\varepsilon_{xc}^{LDA}(\mathbf{r}) = \varepsilon_{xc}^{LDA}(\rho(\mathbf{r})))$. In the GGA case, both electron density and its gradient are required $(\varepsilon_{xc}^{GGA}(\mathbf{r}) = \varepsilon_{xc}^{GGA}(\rho(\mathbf{r}), \nabla\rho(\mathbf{r})))$. Including the dependence on higher derivatives leads to another class of semi-local functionals taking the following general form:

$$E_{xc}^{MGGA}[\rho] = \int d\vec{r}\ \rho(\vec{r})\varepsilon_{xc}^{MGGA}(\rho, \nabla\rho, \nabla^2\rho, \tau) \tag{2-13}$$

where

$$\tau = 2\sum_{i=1}^{N} -\frac{1}{2}\varphi_i^*(\vec{r})\nabla^2\varphi_i(\vec{r}) \tag{2-14}$$

is the density of the kinetic energy and $\nabla^2\rho$ is Laplacian of the electron density.

Today, meta-GGA functionals are less commonly used than the ones of the GGA type discussed in the previous sections. However, the interest in this type of approximation to the exchange-correlation energy is increasing. These developments parallel the ones concerning GGA, which happened more than a decade ago. The near future will bring probably a more systematic knowledge of strengths and weaknesses of the meta-GGA route. Therefore, these functionals are also discussed here. Similarly as in the case of the generalized gradient approximation, the density of the meta-GGA exchange-correlation energy can take different analytic forms. The exchange-energy component of $\varepsilon_{xc}^{MGGA}\left(\rho, \nabla\rho, \nabla^2\rho, \tau\right)$ can be obtained by means of Taylor expansion of the spherically averaged exchange hole.[40,41] Perdew, Scuseria, and collaborators[42] designed a nonempirical meta-GGA (TPSS), which in fact does not depend on Laplacian at all $\varepsilon_{xc}^{TPSS} = \varepsilon_{xc}^{MGGA}\left(\rho, \nabla\rho, \tau\right)$. The elimination of the dependence on Laplacian is justified by the fact that, although both Laplacian and the kinetic energy density appear in the Taylor expansion of the exchange-correlation hole, they carry the same information in the limit of slowly varying electron densities. Usually, however, the empirical functionals of the meta-GGA type depend on both quantities. Proynov et al.[43] proposed a Laplacian-dependent correlation functional. This functional was recommended to be used in combination with two

GGA exchange-energy functionals (B88 and PW86). Filatov and Thiel[44] introduced the exchange-correlation functional depending explicitly on the Laplacian of the electron density in its exchange- and correlation parts. By construction, the exchange part of the proposed functional reproduces the gradient expansion for the exchange energy at small gradients up to the fourth order and leads to the exchange effective potential with the correct asymptotic behavior $(-1/r)$. The free parameters in their model were fitted using exact exchange- and correlation energies in atoms.

3.4.   Hybrid Functionals

Introduction of the orbital dependence into the exchange-correlation energy is an apparent step away from the original Hohenberg-Kohn-Sham formulation of density functional theory[45] where only the kinetic energy part of the total energy involves explicit orbital dependence. Moreover, the evaluation of the exchange-correlation potential is not possible in a straightforward manner (i.e. as a functional derivative of $E_{xc}[\rho]$ with respect to ρ) because of the explicit dependence of the exchange-correlation energy on orbitals. The general form of the hybrid exchange-correlation functional reads:*

$$E_{xc}^{hyb}[\varphi_1, \varphi_2, \ldots, \varphi_N] = a \left(E_x[\varphi_1, \varphi_2, \ldots, \varphi_N] - E_x^{GGA}[\rho]\right) + E_{xc}^{GGA}[\rho]$$

$$(2\text{-}15)$$

where $E_x[\varphi_1, \varphi_2, \ldots, \varphi_N]$ is the Hartree-Fock expression for the exchange energy.

The formal basis for expressing the exchange-correlation energy as an explicit functional of Kohn-Sham orbitals is given in its adiabatic-connection definition (Eq. 7). Moreover, explicit orbital-dependence appears naturally in the Görling-Levy[46] perturbational formulation of density functional theory. For an overview of the formal aspects, see Görling.[47] The first approximate exchange-correlation functional constructed based on the adiabatic connection formula was the "Half-and-Half" functional by Becke.[48] The constant $a = 0.5$, which was chosen arbitrarily, corresponds to replacement of the integral in Eq. (7) by the numerical values of U_{xc} at only two points at $\lambda = 0$ (where $E_x[\varphi_1, \varphi_2, \ldots, \varphi_N]$ is exactly equal to U_{xc}) and at $\lambda = 1$ (where $E_{xc}[\rho]$ is approximately equal to U_{xc}). Another functional based

---

* In the Kohn-Sham equations, the use of such a functional is not straightforward because of the explicite orbital dependency (see the section dealing with optimized effective potential). Strictly speaking, therefore, most of common computer implementations do not solve the associate Kohn-Sham equations.

on the adiabatic connection formula (B3PW91) was introduced by Becke[49] shortly afterwards. It reads:*

$$E_{xc}^{B3PW91}[\varphi_1, \varphi_2, \ldots, \varphi_N] = a_0 \left(E_x[\varphi_1, \varphi_2, \ldots, \varphi_N] - E_x^{LDA}[\rho] \right)$$
$$+ E_{xc}^{LDA}[\rho] + a_x E_x^{B88}[\rho] + a_c E_c^{PW91}[\rho] \qquad (2\text{-}16)$$

The parameters $a_0 = 0.2$, $a_x = 0.72$, and $a_c = 0.81$ were obtained by a linear least-square fit to the 56 atomization energies, 42 ionization potentials, 8 proton affinities, and 10 first-row total atomic energies. The approximate GGA functionals in the above expression (B88 for exchange and PW91 for correlation) were introduced in the previous sections. Stephens et al.[50] applied the LYP correlation functional instead of PW91 in the above expression retaining the same numerical values of the parameters a_0, a_x, and a_c. The resulting functional, known under the label B3LYP, is probably the most commonly used approximation to the exchange-correlation energy nowadays.

Other commonly used hybrid functionals include: B97,[51] mPW1PW[35] and PBE0,[52,53] O3LYP,[38] and the one constructed by Kafafi.[54]

3.5. Beyond Meta-GGA

The hybrid exchange-correlation functionals discussed so far apply one of the possible GGA functionals as their orbital-free component. Using meta-GGA for this purpose leads to the functionals branded as hyper-GGA by Perdew – the convention adopted also in this review. Such functionals, take the following general form

$$E_{xc}^{HGGA}[\rho] = \int d\vec{r} \; \rho(\vec{r}) \; \varepsilon_{xc}^{HGGA}(\rho, \nabla\rho, \nabla^2\rho, \tau, \varepsilon_x) \qquad (2\text{-}17)$$

where ε_x is the density of the 'exact exchange' energy evaluated using the Hartree-Fock expression and Kohn-Sham orbitals.

The TPPSh functional[42] is a one-parameter combination of the parameter-free meta-GGA (TPSS) with 'exact exchange'. Other functional of this type were also proposed.[55–57]

In the previous sections, the following nonempirical functionals were mentioned: LDA, PBE, and TPSS. These functionals represent the lowest three rungs on the Jacob's ladder: LDA, GGA, and meta-GGA. At the GGA-and meta-GGA levels, the approximate exchange-correlation functionals were obtained, not directly, but through construction of approximations to the exchange-correlation hole ($n_{xc}(\mathbf{r}, \mathbf{r}')$)

---

* The term depending explicitly on the Kohn-Sham orbitals is called sometimes "exact-exchange". This nomenclature might be misleading because it uses the exact expression for the exchange energy from the Hartree-Fock theory. However, the numerical value of the exchange energy as defined in density functional theory can be obtained from this expression only for a particular set of orbitals.[9,10]

in a way enforcing selected known properties of the exact hole.[58] So far, no hyper-GGA functional was constructed following the same strategy. The construction of a completely nonempirical hyper-GGA functional is expected in the next future.[58] The performance of the Perdew's nonempirical series of approximations to the exchange-correlation energy (LDA, PBE, TPSS, and PBE0) representing the lowest four rungs of the Jacob's ladder is analyzed together with that of various popular empirical functionals.

4.   PERFORMANCE OF COMMON APPROXIMATIONS TO THE EXCHANGE-CORRELATION ENERGY

Every year brings more than one thousand of applications of density-functional-theory based methods in different fields (see Figure 2-1). A comprehensive review of their performance in all domains of applicability is not practical. This overview focuses on nonbonded intermolecular interactions. Separate sections deal with: (*i*) ionization potentials, (*ii*) electron affinities, (*iii*) electric moments, (*iv*) polarizabilities – key quantities on the theory of intermolecular interactions, and the bonding parameters for two groups of weak intermolecular complexes: (*v*) hydrogen-bonded complexes, and (*vi*) van der Waals complexes. A representative group of studies in which the density-functional-theory results were tested against either high-level wavefunction-based calculations or experiment (or both) was selected for each section.

4.1.   Electric Properties: Electric Moments

Local density approximation* was used by Dickson and Becke[59] to derive dipole moments for CO, HF, H_2O, and NH_3 and quadrupole moments for H_2, N_2, O_2, CO, HF, H_2O, NH_3, and CH_4. These studies have a benchmark character because the analyzed properties were calculated at the basis-set limit using the numerical density-functional code NUMOL.[60] The results are in a very satisfactory agreement with experimental data (see Tables 2-1 and 2-2). For HF, H_2O, and NH_3, the agreement with experiment is excellent. The largest absolute and relative errors of the dipole moment occur for the CO molecule. Keeping in mind the fact that the sign of the small dipole moment of this molecule corresponding to C^-O^+ polarization has been attributed traditionally to the correlation effects, the Kohn-Sham LDA result predicting the correct polarization was noted as a success of density functional theory although this effect is significantly overestimated.

Studies using the GGA- (BLYP) and hybrid- (HCTH, B3LYP, B97, B97-1) functionals combined with the Sadlej's basis set demonstrate that all considered functionals also lead to very reasonable dipole moments for a representative a set

---

* The LDA correlation functional used in these calculations was not the VWN parameterization of the Ceperly and Alder reference data for the uniform electron gas but that of Perdew and Wang.[18]

Table 2-1. Calculated (LDA) and experimental dipole moments (in atomic units) of small molecules[59],

Molecule	μ^{calc}	μ^{exp}
CO	0.0907	0.048
HF	0.7064	0.707
H_2O	0.7300	0.727
NH_3	0.6010	0.579

of small organic molecules (CO, H_2O, H_2S, HCl, HF, LiH, LiF, NH_3, PH_3, and SO_2).[61] Among the considered functionals, B3LYP, B97, and B97-1, lead to similar absolute mean errors (0.0177, 0.0158, and 0.0145 atomic units, respectively), which are about two times smaller then the ones of the BLYP and HCTH results (0.0361 and 0.0355 atomic units, respectively). For comparison, the errors of Hartree-Fock and second-order Møller-Plesset dipole moments equal 0.0675 and 0.0181 atomic units, respectively. For quadrupole moments, all considered methods (DFT and wavefunction based) lead to similar errors in the order of 0.10 atomic units.

Jasien and Fitzgerald[62] demonstrated that the LDA* dipole moments of such molecules as HF, H_2O, NH_3, formamide, imidazole, pyridine, cytosine, match very closely the experimental ones (the relative errors between 1 and 7%). For uracil and thymine, and adenine, the differences between LDA and experimental dipole moments are slightly larger (relative errors up to 12%) and compared better to the ones derived from second-order Møller-Plesset calculations. The authors underlined the noticeable effect the inclusion of the hydrogen $2p$ polarization

Table 2-2. Calculated (LDA) and experimental zz components of the trace-less quadrupole moment calculated with respect of the center of mass (in atomic units) of small molecules.[59]

Molecule	Θ_{zz}^{calc}	Θ_{zz}^{exp}
H_2	0.437	-
N_2	-1.137	-1.09
O_2	-0.356	0.25
CO	-1.478	-1.44
HF	1.707	1.75
NH_3	-2.251	-2.45
H_2O	-0.111	-0.10

---

* The LDA exchange-correlation potential used in this study was not the common Vosko, Wilk, Nusair parameterization of the Ceperley-Adler data for the uniform electron gas but the potential of Heidin and Lundqvist.[68]

functions into the basis set. The good overall performance of the LDA- and GGA functionals in reproducing experimental dipole moments of small organic molecules has been demonstrated.[63−66] Almost invariably, LDA leads to significantly better dipole moments than the ones derived from Hartree-Fock calculations. Typically, the choice of the basis set rather than the type approximation for $E_{xc}[\rho]$ determines the quality of the dipole moments as demonstrated in a comprehensive study of a representative set of biologically relevant molecules by Rashin et al.[64] and St.-Amant et al.[63] and for a set of over 100 first-row and second-row molecules by Scheiner et al.[67]

Filatov and Thiel[44] compared the performance of their meta-GGA exchange-correlation functional with that of B3LYP in reproducing experimental dipole moments for 26 small organic molecules. In most cases, the dipole moments are slightly overestimated. The largest magnitude (about 0.2–0.3 atomic units) of this overestimation takes place for NH_3, N_2H_4, H_2O_2, and CH_3Cl. For CN and SiO, the dipole moments are underestimated by about 0.2 atomic units. The average accuracy of the B3LYP results for the same set of molecules is similar.

4.2.   Electric Properties: Polarizabilities

McDowell et al.[69] analyzed the applicability of the LDA- and GGA (BLYP) functionals for calculating polarizabilities of such molecules as: HF, HCl, F_2, Cl_2, NH_3, PH_3, H_2O, H_2S, SO_2, CO_2, and C_2H_4. The Kohn-Sham results were compared with experiment and the polarizabilities obtained from conventional wavefunction-based methods (Hartree-Fock, second- and fourth-order Møller-Plesset perturbation theory, and coupled-cluster method using Brueckner orbitals). The studied properties were calculated either analytically or using finite-field method. Sadlej's basis set was used. Both LDA- and GGA average dipole polarizabilities are systematically overestimated (typically by about 10%) whereas the dipole polarizability anisotropies are slightly less underestimated. The magnitude of this error is similar that of the corresponding Hartree-Fock polarizabilities. The errors are, however, opposite in sign. This indicates that the way correlation effects are taken into account in the Kohn-Sham equations is rather unsatisfactory at the LDA- and GGA levels.

Dickson and Becke[59] performed finite-field LDA calculations of the dipole polarizabilities and hyperpolarizabilities of the following compounds: H_2, N_2, O_2, CO, HF, H_2O, NH_3, and CH_4. These studies have a benchmark character (for dipole polarizabilites and first hyperpolarizabilities). The calculated dipole polarizabilities are systematically overestimated (see Table 2-3). Other studies reveal the similar trend* that LDA overestimates the dipole polarizabilities of small organic molecules.[69−72]

---

* There are exceptions from this trend. A noted example is the polarizability of sodium clusters significantly underestimated by semi-local functionals.[73]

Table 2-3. Calculated (LDA) and experimental average dipole polarizabilites ($\alpha = 1/3(a_{xx} + \alpha_{yy} + \alpha_{zz})$), in atomic units) of small molecules.[59]

Molecule	α_{calc}	α_{exp}
H_2	5.91	5.43
N_2	12.3	11.74
O_2	10.93	10.78
CO	13.70	13.08
HF	6.23	5.60
H_2O	10.60	9.64
NH_3	15.54	14.56
CH_4	17.69	17.27

The calculated first hyperpolarizabilities (see Table 2-4) are surprisingly close to the experimental data, which is probably fortuitous because they were calculated without taking into account vibrational effect. These studies demonstrated also that the double-zeta basis set augmented by field-induced polarization functions, although sufficient for calculations of dipole and quadrupole moments of the studied molecules at the Kohn-Sham LDA level, is not sufficient in the case of hyperpolarizabilities.

It should be underlined that the comparisons with experiment are not straightforward. It has been demonstrated that zero-point vibrations result in the increase of the average dipole polarizabilities.[74,75] The effect of neither zero-point vibrations nor anharmonicity was taking into account in Dickson and Becke calculations. Therefore, a good agreement between the calculated and experimental results cannot be considered as a proof for the adequacy of these approximations. It indicates rather the flaws of the LDA- and GGA functionals. Studies applying the B3LYP hybrid functional[71,72,76] indicate that this overestimation can be only partially reduced by inclusion of the 'exact exchange'. B97-1, which is a reparameterized variant of the hybrid functional of Becke (B97) was shown to perform better.

Table 2-4. Calculated (LDA) and experimental average first dipole hyperpolarizabilites ($\beta = 3/5(\beta_{xxx} + \beta_{yyy} + \beta_{zzz})$), in atomic units) of small molecules.[59]

Molecule	β_{calc}	β_{exp}
CO	30.5	30(3)
HF	−9.2	−11
H_2O	−24.8	−22
NH_3	−55.(6)	−48

In the discussed before studies on a set of ten small organic molecules, Cohen and Tantirungrotechai[61] demonstrated that the average molecular polarizabilities derived using GGA- and hybrid functionals are quite reasonable although overestimated in line with previously discussed trends. The hybrid functionals (HCTH, B3LYP, B97, B97-1) lead to similar errors (about 1.5 atomic units) whereas the errors derived from GGA(BLYP) are larger (2.25 atomic units). As far as the anisotropy of dipole polarizability is concerned, the B3LYP, B97, and B97-1 functionals lead to results which are even better that the ones derived from the Hartree-Fock and second-order Møller-Plesset calculations. The BLYP and HCTH functionals performed noticeably worse.

The GGA functionals applied in the coupled perturbed Kohn-Sham calculations to obtain electric response properties of the HF molecule perform quite reasonably.[77] The reference value of the dipole moment (0.707 atomic units) derived from coupled cluster calculations is slightly underestimated by most GGA functionals. It varies between 0.673 and 0.685 atomic units. Dipole polarizabilities tend to be overestimated by up to 10–20%, whereas the hyperpolarizabilities show a strong dependence on the chosen GGA functional (β_{zzz} varying between -6.09 and -12.23 atomic units as compared to the coupled cluster value of -9.869 atomic units). The authors investigated also the basis set effect on the calculated properties showing that some of the response properties require larger than standard basis sets. For instance, the dipole moment decreases within less than 15% with the increase of the basis set from double zeta of Dunning (DZ) to that of Sadlej (pVTZ). The static dipole polarizabilities, however, is affected by more than factor five. The effect of the basis set on the hyperpolarizabilities is also significant (factor two when going from DZ to pVTZ.

The overestimation of the average dipole polarizabilities by the GGA and LDA functionals has been attributed to their incorrect asymptotic behavior at molecular tails.[69,78] The exact ν_{xc} is long-ranged. In the asymptotic region, the exact exchange-correlation potential is proportional to $-1/r$, where r is the distance from the atom or the center of the molecule. Compared to the $-1/r$ behavior, the GGA- and LDA exchange-correlation potentials are noticeably more short-ranged. As a result, the electrons in the outmost shell are bound too weakly, which leads to the increase in their responses if calculated at the GGA and LDA levels. The numerical values of exact ν_{xc} are available for some atoms and small molecules obtained by means of special techniques to construct it from electron density derived from high-level wavefunction-based methods (see, for instance, Zhao et al.[79]). Instead of constructing the exact ν_{xc}, the correct asymptotic behavior at long range can be built in into an approximated functional by construction as it is the case of the LB94 potential.*,[80] Indeed, calculations using LB94 potential lead to responses, which are much closer to the ones obtained by means of the exact ν_{xc}. Moreover,

---

* Since the LB94 exchange-correlation potential was constructed directly and not as the functional derivative of some approximate expression for $E_{xc}[\rho]$, there exist no corresponding exchange-correlation energy functional.

Table 2-5. Average dipole polarizabilites (in atomic units) of noble-gas atoms and small molecules calculated using different exchange-correlation potentials and the reference data (benchmark wavefunction calculations or experimental).[83]

atom/molecule	α_{LDA}	α_{LB94}	$\alpha_{Vxc(exact)}$	$\alpha_{reference}$
He	1.6576	1.3896	1.3824	1.3832
Be	43.79	42.87	39.57	37.73
Ne	3.049	2.590		2.670
H_2	5.9	5.61	5.16	5.1816
N_2	12.27	11.46	11.68	11.74
HF	6.20	5.31	5.49	5.52
HCl	18.63	17.86	17.25	17.39
H_2O	10.53	9.20	9.45	9.64
CO	13.87	12.62	12.86	13.08

application of the exact and the LB94 exchange-correlation potentials leads to much better response properties than does the LDA potential (see Table 2-5).

The crucial role of the asymptotic behavior of v_{xc} for obtaining the correct response properties has been recently demonstrated for various cases (see for instance Mori-Sanchez et al.[81]; Hirata et al.[82]).

The practical difficulties in describing all response properties by means of a given approximation to the exchange-correlation potential have been illustrated in the recent report by Jacob et al.[84] concerning complexation-induced dipole moments in weak complexes of the $CO_2 \ldots Rg$ (Rg=He, Ne, Ar, Kr, Xe, and Hg) type. The dominant effect responsible for emergence of the dipole moment of such a complex is the polarization of the rare gas atom by the electric field generated by the quadrupole moment of CO_2. As discussed previously, the LDA- and GGA functionals lead to overestimated dipole polarizabilities. This overestimation is even larger than 10% for some rare-gas atoms. The same approximations, however, lead to a very reasonable description of the quadrupole moment of CO_2 (1.55 atomic units in the PW91 case compared to the experimental value of 1.595 atomic units). The opposite tendencies occur for the SAOP exchange-correlation potential,[85] which leads to significantly better dipole polarizabilities of rare gas atoms (errors reduced by a factor of two) but which leads to much worse quadrupole moment of CO_2 (1.85 atomic units). As a result, neither PW91 nor SAOP exchange-correlation potentials lead to the complexation-induced dipole moment in the whole $CO_2 \ldots Rg$ series with uniform accuracy.

4.3. Ionization Potentials and Electron Affinities

In Kohn-Sham density functional theory, the ionization potential is the negative of the eigenvalue of the highest occupied Kohn-Sham orbital.[86-88] The IP $= -\varepsilon_{HOMO}$ relation holds, however, only for the exact exchange-correlation potential. Numerical confirmations for this relation exist for model systems such as the helium atom[89] This relation does not hold, however, for approximate exchange-correlation potentials. Due to the too rapid decay of the LDA- and GGA exchange-correlation potentials in the asymptotic region, the ionization potentials calculated as $IP = -\varepsilon_{HOMO}$ by means of these approximations can be expected to be systematically underestimated. Indeed, numerical calculations confirm this[80,85 90−92]). For similar reasons, the situation is even worse for electron affinities. Frequently, LDA or GGA do not lead to stable states of negative ions.[5,90,93]

Instead of using the $IP = -\varepsilon_{HOMO}$ relation, the ionization potential can be also calculated as the energy difference between neutral and charged species. This strategy leads typically to better numerical results. For atoms, the numerical values of $-\varepsilon_{HOMO}$ are about two times smaller than the first ionization potential if derived using LDA- and GGA potentials (see Table 2-6).

Similarly, as the case with response properties discussed in the previous section, imposing the correct asymptotic behavior of v_{xc} improves the agreement between the numerical values of $-\varepsilon_{HOMO}$ and the experimental ionization potentials.[85,90,91,94] For these reasons, the ionization potentials and electron affinities are usually obtained as energy differences (ΔSCF) in calculations using common approximations to the exchange-correlation functional. The discussed hereafter numerical values were obtained in this way.

Table 2-6. LDA and GGA(BLYP or PW91) atomic ionization potentials (in Hartree) calculated as energy differences (ΔSCF) or as $-\varepsilon_{HOMO}$ ([a]LDA results from Vydrov and Scuseria[98]; [c]LDA results from Perdew et al.[97], [b]LDA results from van Leeuwen and Baerends[80]; GGA results from Grabo and Gross[90]).

Atom	ΔSCF[a] LDA	$-\varepsilon_{HOMO}$ LDA	ΔSCF BLYP	$-\varepsilon_{HOMO}$ BLYP	ΔSCF PW91	$-\varepsilon_{HOMO}$ PW91	Exp.
He	0.892	0.571[b]	0.912	0.585		0.583	0.903
Li	0.200	0.116[c]	0.203	0.111	0.207	0.119	0.198
Be	0.331	0.206[b]	0.330	0.201	0.333	0.207	0.343
B	0.315	0.225[c]	0.309	0.143	0.314	0.149	0.305
C	0.429		0.425	0.218	0.432	0.226	0.414
N	0.548		0.542	0.297	0.551	0.308	0.534
O	0.508	0.273[c]	0.508	0.266	0.505	0.267	0.500
F	0.659	0.381[c]	0.656	0.376	0.660	0.379	0.640
Ne	0.812	0.490[b]	0.808	0.491	0.812	0.494	0.792
Na	0.195	0.113[c]	0.191	0.106	0.198	0.113	0.189
Mg	0.283		0.280	0.168	0.281	0.174	0.281
Al	0.220		0.212	0.102	0.221	0.112	0.220
Si	0.302	0.168[c]	0.294	0.160	0.305	0.171	0.300
P	0.386		0.376	0.219	0.389	0.233	0.385
S	0.386	0.224[c]	0.379	0.219	0.379	0.222	0.381
Cl	0.484	0.300[c]	0.476	0.295	0.482	0.301	0.477
Ar	0.585	0.381[b]	0.576	0.372	0.583	0.380	0.579

The relation between other than ε_{HOMO} eigenvalues of the *exact* Kohn-Sham orbitals and higher ionization potentials is currently an object of studies by Baerends and collaborators.[95,96]

Benchmark atomic ionization energies and electron affinities (from H to Zn) calculated as energy differences (ΔSCF) for LDA- and three nonempirical GGA functionals developed by Perdew and collaborators[97] indicate that errors of these quantities depend strongly on the choice of the GGA functional.

Ernzerhof and Scuseria[53] analyzed the electron affinities and ionization potentials obtained using the LDA-, PBE-, and PBE1PBE functionals with the reference date taken from the G2-1 data set.[99] Whereas the PBE functional leads to significantly more accurate results than does LDA, no noticeable improvement occurs at the meta-GGA level.

The asymptotic behavior of the exchange-correlation potential far from the molecule has been identified as the key factor determining the accuracy of the ionization potentials of anions and electron affinities of neutral molecules.[5] Recently, Wu et al.[91] proposed a variational method, which enforces the correct long-range behavior of v_{xc}. Indeed, a noticeable improvement compared to the Kohn-Sham results derived using conventional approximations (LDA-, GGA-, and hybrid functionals) was reported for atoms (H, He, Li, Be, B, C, N, O, and F) and diatomics (BeH, CH, NH, OH, CN, BO, NO, OO, FO, and FF). The still significant discrepancies between the experimental and calculated ionization potentials (or electron affinities) were attributed to errors of the exchange-correlation potential in the molecular interior.

The discrepancies between the calculated values of $-\varepsilon_{HOMO}$ and experimental ionization potentials are frequently attributed to self-interaction error. Indeed, applying the Perdew-Zunger technique[4] to correct this error of the LDA-, GGA- (PBE), and meta-GGA (TPSS) functionals improves the numerical values of $-\varepsilon_{HOMO}$.[98] Interestingly, these studies showed that the Perdew-Zunger correction does not improve ionization potentials and electron affinities if calculated as energy differences (ΔSCF).

Staroverov et al.[100] tested the performance of nonempirical and empirical exchange-correlation functionals of the LDA- GGA-, hybrid-, and meta-GGA types on 58 electron affinities and 86 ionization potentials. In all calculations, the 6-311++G(3df,3pd) basis set was applied and the reported results were obtained at the B3LYP optimized geometries. Experimental data was used as a reference. Among the nonempirical functionals, the quality of the obtained results improves upon introduction of gradients but no improvement follows the introduction of τ-dependence into $E_{xc}[\rho]$ (meta-GGA functionals). The mean absolute errors of the electron affinities amount to: 0.244 eV for LDA, 0.118 eV for GGA (PBE), and 0.137 eV for meta-GGA (TPSS). For comparison, the corresponding values for empirical functionals amount to: 0.187 eV (HCTH) and 0.124 eV (B3LYP). As far as ionization potentials are concerned, the mean absolute errors are rather constant for nonempirical functionals: 0.232 eV (LDA), 0.235 eV (PBE), and 0.242 eV (TPSS), 0.232 eV (HCTH). They are slightly smaller (0.184 eV) in the case of B3LYP.

Joantéguy et al.[101] performed a dedicated study of the performance of LDA-, GGA-, and hybrid functionals in determining the ionization potentials of unsaturated molecules. The authors concluded that the accuracy better than 0.1 eV cannot be reached using the considered functionals: LDA, BP86, B3P86, B3LYP, and B3PW91.

The benchmark calculations of ionization potentials and electron affinities of the atoms and molecules in the G2 data set[99] calculated using the hybrid functional (B97) show that this functional is adequate. The average absolute deviation from experimental data amounts to 0.055 eV and 0.056 eV for ionization potential and electron affinity, respectively.[51]

Hoe et al.[102] analyzed the performance of the OPTX exchange functional, which has the GGA form, used in combination with LYP for correlation (OLYP) and in the hybrid scheme (O3LYP), in reproducing ionization energies (first and second) of atoms from H to Ar. In both schemes, replacing the original Becke (B88) approximation by OPTX reduces the errors in the calculate ionization potential significantly. The performance of the OPTX exchange functional in obtaining various properties including ionization potentials and electron affinities was also analyzed by Xu and Goddard.[103] The mean absolute deviation from experimental ionization potential for the G2 set of molecules amount to: 0.187 eV (BLYP), 0.185 eV (B3LYP), 0.168 eV (OLYP), and 0.139 eV (O3LYP). For electron affinities, the corresponding deviations are equal to: 0.106 eV (BLYP), 0.133 eV (B3LYP), 0.103 eV (OLYP), and 0.107 eV (O3LYP).

LDA- and GGA functionals do not lead to a satisfactory description of negative anions (some even do not exist at the LDA- and GGA levels), which was attributed to the wrong asymptotic behaviour of the corresponding v_{xc}[5] and/or self-interaction error.[92,104,105] Nevertheless, quite reasonable numerical values of electron affinities calculated using hybrid functionals were reported for several anionic systems.[106,107] The Half-and-HalfLYP functional slightly underestimates the experimental values for this quantity, whereas B3LYP slightly overestimates it. The admixture of a small amount of the 'exact exchange' appears to be crucial.

4.4.   Intermolecular Interactions

4.4.1.   Hydrogen bonding

The interest in hydrogen bonding arises from its key effect on the structure of biomolecules, molecular crystals, and clusters of various sizes. Kohn-Sham calculations applying local density approximation are rather useless in practical computer simulation studies as shown by dedicated studies on model hydrogen-bonded complexes. LDA equilibrium donor-acceptor distances (r_e) are systematically slightly underestimated whereas the LDA well depths (D_e) are significantly overestimated. This tendency was demonstrated for: water dimer[108,109] ammonia dimer,[110] formamide dimer,[111] N-methylacetamide-water complex[112], and for a number of other hydrogen-bonded complexes.[109,113]

For various hydrogen-bonded complexes, it was shown that common GGA functionals such as BLYP, BPW86, PW86P86 perform better than LDA.[108-111], [113-117] The hybrid functional B3LYP perform as good or better then the GGA functionals.[109],[112-115],[118,119]

The GGA functionals were also shown to perform rather well even in cases of weaker hydrogen bonds. For hydrogen bonds involving aromatic π-acceptors (H_2O-C_6H_6, NH_3-C_6H_6, HCl-C_6H_6, H_2O-indole, and H_2O-methylindole), Zhao et al.[120] demonstrated a reasonably good performance of PW91, which was inferior, however, to that of the empirical hyper-GGA functional PWB6K introduced by these authors.

The bonding parameters of the dispersion-dominated complex of square-planar platinum(II) and water derived from second-order Møller-Plesset calculations are reasonably reproduced in the GGA calculations. For several orientations of the water ligand, both considered functionals (PW91 and BLYP) led to qualitatively correct potential-energy curves deviating slightly from the Møller-Plesset reference data. Systematically, the PW91 functional overestimates, whereas BLYP underestimates, the strength of the interaction.[121]

Tsuzuki and Lüthi (2001)[221], analyzed the performance of two GGA functionals (BLYP and PW91) and the B3LYP hybrid functional on a representative set of 12 hydrogen-bonded complexes. The DFT results were compared with the ones derived from second-order Møller-Plesset- and coupled-cluster (CCSD(T)) calculations. The effect of the basis set up was carefully examined. Basis sets as large as cc-pV5Z were used. Using the agreement with the CCSD(T) results as the accuracy criterion indicates that the PW91 functional leads to the best interaction energies (relative errors amount to 5% for $(NH_3)_2$, HCOOH-HCOOH, $HCONH_2$-$HCONH_2$, H_2O-CH_3OH, HCN-HF, and H_2O-$HCONH_2$ complexes. The B3LYP and BLYP functionals lead to noticeably underestimated interaction energies. The largest relative errors of the interaction energy occur in the CH_3OCH_3-H_2O case for all considered approximations to the exchange-correlation energy which amount to 17%, 34%, and 45% for PW91, B3LYP, and BLYP, respectively.

Milet et al.[122] investigated the performance of common GGA functionals (BP86, BLYP, and BPW91) and the corresponding hybrid approximations to the exchange-correlation energy in reproducing angular dependence of the potential energy surface in the OH-H_2O and $(H_2O)_2$ hydrogen-bonded intermolecular complexes. The comparison with benchmark results obtained using wavefunction-based methods show that the considered functionals are not capable to reproduce this feature of the potential energy surface.

Xu and Goddard[123] studied the performance of the hybrid functional (X3LYP) using an empirical combination of B88, PW91, and 'exact exchange' for various systems and properties including bonding properties of the water dimer. The results obtained using other GGA functionals (BP86, BLYP, BPW91, PW91, mPW91, PBE, XLYP) as well as hybrid approximations to the exchange-correlation energy (Half-and-HalfLYP, B3P86, B3LYP, B3PW91, PW1PW, mPW1PW, O3LYP) were also reported. Due to large contributions arising from anharmonicity and zero-point vibration effects, extracting parameters of the potential energy surface from experimental data is not straightforward. For the water dimer, however, accurate bonding parameters derived from high-level wavefunction-based calculations are available.[124] It was found that GGA functionals lead to the OO equilibrium distance within 0.04 Å of the reference value ($r_e = 2.912 \pm 0.005$ Å) and the value of the well depth D_e within 0.04 eV of the reference value ($D_e = 0.218 \pm 0.004$ eV). The best parameters were obtained by means of the PBE1PBE hybrid functional. The agreement with the reference data was within 0.016 Å and 0.002 eV for r_e and D_e, respectively.

The mPBE functional[39] is a modification of the nonempirical PBE functional into which one empirical parameter was reintroduced. Compared to PBE, mPBE leads to significantly worse interaction energy for the water dimer, whereas the interaction energies in the cases of hydrogen fluoride and hydrogen chloride dimers are only slightly improved.

The recent report by Zhao and Truhlar[113] provides a wealth of numerical data for analysis of the performance of commonly used approximations to $E_{xc}[\rho]$ in describing nonbonded interactions and hydrogen-bonding in particular. LDA-, GGA-, meta-GGA-, hybrid-, and hybrid meta-GGA (hyper-GGA in Perdew's nomenclature) functionals were considered. Literature benchmark well depths (D_e) derived from high-level wavefunction-based calculations were used as reference in testing the performance of the considered DFT methods. Among the GGA functionals, PBE was found to perform the best. This functional does not depend on empirical parameters and its better performance than that of empirical GGA functionals is noteworthy. However, the TPSS functional, which is the nonempirical meta-GGA, does not perform better than PBE – its nonempirical GGA cousin. Except for LDA, the best functional in each group of approximate functionals leads to better average structural and energetic performance than second-order Møller-Plesset calculations.

Staroverov et al.[100] tested the performance of nonempirical and empirical exchange-correlation functionals of the LDA-, GGA-, hybrid-, meta-GGA type on ten hydrogen-bonded systems $(HF)_2$, $(HCl)_2$, $(H_2O)_2$, HF-HCN, HF-H_2O, CN^--H_2O, OH^--H_2O, HCC^--H_2O, H_3O^+-H_2O, and NH_4^+-H_2O. In all calculations, the 6-311++G(3df,3pd) basis set was applied. Second-order Møller-Plesset dissociation energies were used as the reference. Among nonempirical functionals, the quality of the obtained results parallels their position on the "Jacob's ladder" of approximate functionals. The mean absolute errors of the dissociation energy amount to: 5.78 kcal/mol for LDA, 1.00 kcal/mol for GGA (PBE), and 0.59 kcal/mol for meta-GGA (TPSS). For comparison, the corresponding values for empirical functionals equal 0.91 kcal/mol and 0.43 kcal/mole for HCTH and B3LYP, respectively.

4.4.2.   *Van der Waals complexes*

Even qualitatively, the correct description of the whole potential energy surface of a weakly bound intermolecular complex is not possible by means of the commonly used approximations to the exchange-correlation energy. The semi-local approxima-tions (LDA, GGA, meta-GGA) are fundamentally flawed. At large intermolecular separations, only electrostatic- and induction components of the interaction energy can be accounted for by means of these approximations. The accuracy of these two components of the interaction energy depends obviously on the accuracy of such quantities as electric moments and polarizabilities derived using approx-imate functionals – the subject covered in one of the proceeding sections. The dispersion energy-component of the asymptotic behavior proportional to $-r_{AB}^{-6}$ (where r_{AB} is intermolecular distance) cannot be obtained from Kohn-Sham calcu-lations applying semi-local approximations to the exchange-correlation energy because these functionals are additive in the case of non-overlapping electron densities (ρ_A and ρ_B):

$$E_{xc}^{semi-local}[\rho_A + \rho_B] = E_{xc}^{semi-local}[\rho_A] + E_{xc}^{semi-local}[\rho_B] \qquad (2\text{-}18)$$

Current hybrid functionals do not improve this situation. Their non-local component (Hartree-Fock exchange) cannot give rise to any attraction. To describe quantita-tively the long-range interactions, either a non-local* approximation to $E_{xc}[\rho]$ must be applied within the Kohn-Sham framework or methods using other-than-Kohn-Sham formalism should be used. Some of such approaches will be discussed in the last section of this review. Here, we mention an especially promising combination of symmetry adapted perturbation theory with of the Kohn-Sham orbitals.[125]

Close to the equilibrium geometry, however, there is no principal reason why semi-local approximations should be bound to fail. In practice, however, description of the equilibrium geometry part of the potential energy surface of van der Waals complexes lays at the border area of applicability of the most common approximate exchange-correlation functionals. Empirical GGA exchange-energy functionals such as the one of Lacks and Gordon[126] or mPW91[35] were parameterized using the potential-energy-surface data for such complexes.

For obtaining interaction energies and equilibrium geometries, local density approximation is even less adequate than it is in the case of hydrogen-bonded complexes. The intermolecular distances are too short and the interaction energies are overestimated.[113,123, 127−129, 130,131] The overestimation of the interaction energy in the case of noble-gas dimers by factor three as it is the case for Ar_2 or even ten for He_2[127] makes LDA rather useless for this type of systems.

For van der Waals complexes, the interaction energy depends very strongly on the choice of the GGA functional.[126,129] This trend is in a sharp contrast to a rather uniform – and usually acceptable – performance of various GGA functionals in

---

* The term "non-local" is used sometimes in the literature in association with gradient-dependent (GGA) functionals. This nomenclature is not applied in this work. The LDA-, GGA-, and meta-GGA functionals are referred to as semi-local as they do not account for any long-range non-locality of the exchange-correlation energy density: $\varepsilon_{xc}^{semi-local}(\mathbf{r}) = \varepsilon_{xc}(\rho(\mathbf{r}), \nabla\rho(\mathbf{r}), \nabla^2\rho(\mathbf{r}), \tau(\mathbf{r}))$ whereas $\varepsilon_{xc}^{non-local}(\mathbf{r}) = \varepsilon_{xc}[\rho](\mathbf{r})$.

describing hydrogen-bonded complexes. In particular, it is the exchange component of the $E_{xc}^{GGA}[\rho]$, which is responsible for this erratic behavior.[126,129,132,133] Using the B88 exchange functional leads frequently to lack of any bonding in cases where bonded structures exist in reality: noble-gas dimers or π-stacking for instance.[54,103,115,127,128 134−136] As far as hybrid functionals are concerned, the situation is similar. Typically, B3LYP does not lead to satisfactory results for noble-gas dimers and π-stacked complexes.[103,137,138]

Numerical experience shows that the Lieb-Oxford bound,[139] which is the condition satisfied by the exact exchange-correlation-energy functional, should be used as the selection criterion among GGA functionals. Such functionals (PBE, PW91, RBPE, mPW91, for instance) lead usually to better bonding parameters than the ones obtained by means of the functionals that violate the Lieb-Oxford bound.[35,126,129,132,133,136,140] Tsuzuki and Lüthi (2001) confirmed numerically this regularity for a representative sample of nonbonded complexes.

The weak nonbonding interactions play also a key role in the conformational equilibria of flexible molecules. In the case of 2-phenylethanol and *n*-butylbenzene in neutral- and cationic forms, interactions between the side-chain and the aromatic ring are sterically possible. Using the second-order Møller-Plesset relative energies of different conformers, Patey and Dessent[140] demonstrated that the performance of PW91 functional is not uniform. It leads to very reasonable relative energies of different conformers of *n*-butylbenzene whereas it fails in the case of the 2-phenylethanol$^+$.

The important type of systems, where common LDA- and GGA functionals were shown to fail, includes the charge-transfer complexes formed by ethylene or ammonia interacting with a halogen molecule (C_2H_4-X_2, NH_3-X_2, X=F, Cl, Br, and I). These approximations lead to an unacceptable overestimation of the binding energy[141,142] regardless the exchange functional satisfies the Lieb-Oxford bound or not.

Milet et al.[122] demonstrated also that the common approximations to the exchange-correlation energy of the GGA type: BP86, BLYP, and BPW91, as well as the corresponding hybrids cannot account satisfactorily for angular dependence of the interaction energy in the $CO-H_2O$, $He-CO_2$ complexes. All the considered functionals lead to potential-energy curves lying significantly above the exact ones (the SAPT and CCSD(T) potential-energy curves, which are indistinguishable, were used as the reference). The relative energies at the maxima of the potential-energy curves were especially unsatisfactory (overestimation by at least factor two) for the $He-CO_2$ case and all considered approximations to the exchange-correlation-energy functionals except for B3LYP.

Tao and Perdew[130] reported recently a systematic study of the performance of nonempirical functionals (LDA, PBE, and TPSS) and a one-parameter hyper-GGA (TPSSh) in reproducing bonding parameters of noble-gas dimers (He_2, Ne_2, Ar_2, Kr_2, HeNe, HeAr, HeKr, NeAr, NeKr, and ArKr). The mean absolute error in equilibrium distance amounts to 0.86, 0.28, 0.60, and 0.61 bohr for LDA, PBE, TPSS, and TPPSh, respectively. The corresponding mean absolute errors in binding energy are equal to 0.35, 0.08, 0.10, 0.11 kcal/mol. The improvement upon LDA following the introduction of gradients is evident. However, introduction of the kinetic-energy-density dependence does not improve the calculated bonding parameters but worsens them slightly. Moving to the highest rung of the "Jacob's ladder" of approximate functionals (TPSSh) causes even a smaller effect on calculated bonding parameters. For each van der Waals complex considered, the TPSS and TPSSh results are very similar. The results of Tao and Perdew[130] supplement the similar analyses by Ruzsinszky et al.[131] concerning noble-gas and alkaline-earth dimers.

The empirical mPBE functional[39] discussed in the previous section leads to slightly better equilibrium interatomic distances for the He_2, Ne_2, and Ar_2 dimers without, however, affecting the interaction energies compared to the parent functional (PBE).

Kamiya et al.[143] applied the mPW91PW91 and mPW1PW91 functionals to rare gas dimers. The GGA exchange component of these funcitnals was parameterized originally using the exchange energy in He_2 and Ne_2 dimers. Although the distance-dependence of the interaction energy in the helium and neon dimers was shown to be reasonably good, the results for the argon dimer are disappointing.[143] The minimum energy is underestimated by about a factor of four and the equilibrium geometry is shifter by almost one Ångstrom towards longer interatomic distances compared to the reference data.

Zhao and Truhlar[138] analyzed the performance of several recently developed approximate exchange-correlation functionals in describing stacking interactions in systems of biological interest: nucleic acid bases complexes (adenine-thymine, guanine-cytosine, cytosine-cytosine, uracil-uracil) and stacked amino acids pairs (phenyloalanine-phenyloalanine, phenyloalanine-lysine, phenyloalanine-leucine, phenyloalanine-tyrosine). It is important to underline that the applicability of empirical approximations to the exchange-correlation functional for this type of systems has been object of studies for more than a decade. It is rather clearly established that the B3LYP and most of the GGA functionals lead to quantitative wrong potential-energy curves in this case.[115,135,138,144] In the case of nucleic acid bases complexes, Zhao and Truhlar used benchmark results derived from wavefunction-based calculations as a reference to test the performance of the functionals they proposed. For complexes formed by amino acids, the reference geometry was extracted from X-ray crystallographic structure of relevant proteins. The empirical functional dubbed PWB6K[57] leads to the best results.

The discussed in the previous section report by Zhao and Truhlar[113] provides a comprehensive overview of the performance of commonly used approximations to $E_{xc}[\rho]$ in describing also van der Waals complexes. The performance of empirical and nonempirical approximations to the exchange-correlation energy situated on all five rungs of the Jacob's ladder was analyzed, using the benchmark interaction energies derived from high-level wavefunction-based calculations as a reference, for the following type of complexes: HeNe, HeAr, Ne_2, NeAr, CH_4-Ne, C_6H_6-Ne, $(CH_4)_2$, $(C_2H_2)_2$, and $(C_2H_4)_2$. As expected, the errors are the largest for LDA. Among the GGA functionals, the mean absolute error of D_e was the smallest for PBE (0.27 kcal/mol). Even highly empirical HCTH functional leads to larger deviations from the reference data. None among the seven considered meta-GGA functionals leads to better accuracy than PBE. B97-1 leads to the smallest errors (0.19 kcal/mol for B97-1) among the hybrid functionals and the best hyper-GGA functional (MPWB1K) leads to very similar mean absolute error in D_e (0.20 kcal/mol).

The fact that semi-local approximations to the exchange-correlation energy cannot account for non-electrostatic components of the interaction energy in intermolecular complexes suggests a practical solution to obtain better interaction energies than the ones derived from the Kohn-Sham calculations applying such functionals. The missing term proportional to the $-R_{AB}^{-6}$, damped at short interatomic separations to reduce double-counting, has been used by several authors for this purpose.[135], [145–148] Wu and Yang[149] demonstrated that without taking into account the long-range dispersion forces, conventional approximations to the exchange-correlation functional lead to unrealistic relative energies of different conformers of flexible biopolymers. The energy difference between the 3_{10}-hellical and α-helical conformations of polypeptides (n = 2, 3, 5, 8), calculated by means of GGA (BLYP, BPW91, or PW91) and hybrid (B3LYP) functionals, was shown in qualitative disagreement with second-order Møller-Plesset results. Therefore, adding terms proportional to $-R_{AB}^{-6}$, on top of the calculated Kohn-Sham interaction energies was also recommended as a practical solution.

Another interesting empirical solution was proposed recently by Lilienfeld et al.[150] A special term taking the form of an atom-centered pseudopotential was added to the effective Kohn-Sham potential to correct the energies derived from the Kohn-Sham GGA calculations. The method was tested on noble-gas and benzene dimers.

The Hartree-Fock-Kohn-Sham scheme, in which the total energy is represented as the sum of the Hartree-Fock energy (orbital-dependent) and a density-dependent correlation contribution, was applied by Pérez-Jordá et al.[151] to derive the bonding parameters in noble-gas dimers: He_2, Ne_2, Ar_2, HeNe, HeAr, and NeAr. The authors tested several approximate expressions for the term representing the correlation functional and found that the functional of Wilson and Levy[152] performs the best: equilibrium geometries agreed within less than 0.2 Å with the reference values whereas the relative errors of the interaction energies are smaller than 25% for all cases except for the helium dimer for which the calculations overestimate the interaction energy by about 50%.

5.   ONGOING DEVELOPMENTS

Density functional theory of atoms and molecules is a lively area as evidenced by reports dealing with methodological developments appearing regularly in the literature as well as a constantly growing body of numerical results, which provide useful guidelines concerning applicability of a given approximation/parameterization. Ongoing methodological developments are mainly motivated by still unresolved issues which lay at the origin of spectacular failures of methods based on the Kohn-Sham equations: lack of dispersion attraction at long range, wrong description

radical dissociation processes, for instance. Reducing the errors in calculated properties also in cases where current methods lead to acceptable accuracy is also desired. The overview of the performance various approximations to the exchange-correlation functional shows that the semi-local series (LDA, GGA, meta-GGA) cannot handle such problems in a satisfactory manner. Both formal considerations (adiabatic connection and Levy-Görling perturbation formulation of DFT) as well as numerical examples indicate that explicit dependence of the exchange-correlation energy on orbitals is unavoidable. Recently, Baerends[153] demonstrated that going even further, namely expressing the exchange-correlation energy by means of unoccupied Kohn-Sham orbitals, lead to a correct description of the dissociation of H_2 molecule – a problem for which a practical solution within the Kohn-Sham framework remains to be found.

The principal objective in development of new DFT methods is the elimination or reduction of practical consequences of these flaws of current approximations. Current methodological works can be expected to lead to new methods/approximations, which will soon be refined to such extent that they will find the way to standard program packages and will enrich the still growing family of DFT methods available for a computational chemist. The developments reviewed in sections: *optimized effective potential, weighted density approximation, exchange-correlation-energy functional van der Waals density functional of Langreth and Lundqvist*, and *current-dependent exchange-correlation functional* aim at better approximations to the exchange-correlation potential and/or energy in the Kohn-Sham equations. Other sections deal with methods not involving these equations but using closely relevant ideas and concepts.

5.1.   Optimized Effective Potential

In wavefunction-based methods of quantum chemistry such as Hartree-Fock, many-body perturbation theory (Møller-Plesset), and coupled-cluster formalisms for instance,* the total energy can be expressed analytically as an explicit function of occupied and/or unoccupied orbitals. In such a case, the functional derivative of energy with respect to each orbital can be obtained analytically. Such derivatives have no direct use in the Kohn-Sham equations because v_{xc} is the functional derivative of one of the components of the total energy ($E_{xc}[\rho]$) with respect to electron density. *Optimized effective potential* (OEP) is a concept providing the link between the two types of derivatives.[154] In principle, one can construct OEP corresponding to any conventional quantum chemistry framework: Hartree-Fock-, Møller-Plesset-, coupled-cluster-, or even configuration interaction methods (see Hirata et al.[155], for a general overview). The basic equation linking OEP to the quantities occurring in the parent wavefunction-based method was given by Sharp and Horton[154] and by Talman and Shadwick.[156]

---

* Exchange-correlation functionals, which depend explicitly on Kohn-Sham orbitals such as meta-GGAs, hybrid- and hyper functionals discussed before fall also in this cathegory.

For many years, OEPs have been constructed using the orbital dependence of the energy taken from the Hartree-Fock theory (*exchange-only OEP*). The interest in exchange-only OEP increased owing to an additional approximation introduced by Krieger, Lee, and Iafrate [157] allowing one to construct the exchange-only OEP very accurately at low computational cost. As a result, OEP is associated frequently with *exchange-only* case. Studies on exchange-only OEP brought many benefits. It has been demonstrated that one can obtain the Hartree-Fock energy almost perfectly using a multiplicative potential. [158,159] The fact, that exchange-only OEP and Hartree-Fock methods do not led to *exactly* the same solution, results from the reasons which will not be discussed here further. We underline, however, that the numerical differences are of no chemical importance. Exact-exchange OEP studies provide also important data for interpretation of eigenvalues in the Kohn-Sham equations which are considered auxiliary quantities in the orthodox interpretation of the Kohn-Sham theory except for the ones corresponding to the highest occupied one. [86]

Current post-Hartee-Fock OEP methods were shown to handle successfully such cases for which other strategies to approximate the exchange-correlation potential fail. [82] They are very promising although their efficient numerical implementations making it competitive with the corresponding parent correlated wavefunction-based methods have not been developed yet.

5.2.   Weighted Density Approximation

Weighted density approximation [160–162] provides a strategy to construct non-local approximation to the exchange-correlation-energy functional using the relation between $E_{xc}[\rho]$ and the exchange-correlation hole $n_{xc}(\mathbf{r}, \mathbf{r}')$:

$$E_{xc}[\rho(\vec{r})] = \frac{1}{2} \int \rho(\vec{r})d\vec{r} \int \frac{n_{xc}(\vec{r}, \vec{r}')}{|\vec{r} - \vec{r}'|} d\vec{r}' \qquad (2\text{-}19)$$

In order to satisfy the exact condition:

$$\int n_{xc}(\vec{r}, \vec{r}')d\vec{r}' = -1 \qquad (2\text{-}20)$$

n_{xc} is approximated as:

$$n_{xc}^{WDA}(\vec{r}, \vec{r}') = \bar{\rho}(\vec{r}')G^{WDA}[\vec{r}, \vec{r}'; \rho^0(\vec{r})] \qquad (2\text{-}21)$$

where the function $G^{WDA}(\mathbf{r}, \mathbf{r}'; \rho^0)$ has a postulated analytic form, which depends parametrically and on yet another function $\rho^0(\mathbf{r})$. The exact condition given in Eq. (20) leads to the following equation for $\rho^0(\mathbf{r})$:

$$\int \bar{\rho}(\vec{r}')G^{WDA}[\vec{r}, \vec{r}'; \rho^0(\vec{r})]d\vec{r}' = 1 \qquad (2\text{-}22)$$

For a given function $G^{WDA}(\mathbf{r}, \mathbf{r}'; \rho^0)$, obtaining $\rho^0(\mathbf{r})$ involves a significant numerical effort because, at each point in space $\mathbf{r}$, $\rho^0(\mathbf{r})$ depends on electron density ρ in the whole space. Once $\rho^0(\mathbf{r})$ is obtained, it is used in Eqs. (19) and (21) to derive the corresponding $E_{xc}[\rho]$.

Weighted density approximation has been shown to improve upon LDA- and GGA results as far as ground-state properties of such materials as SrO, CaO, KNbO$_3$, KTaO$_3$, BaTiO$_3$,[163,164] silicon[165] and the dielectric response of oxides[166] are concerned. So far, only a few applications of the WDA strategy to molecules were reported.[167]

5.3. Exchange-correlation Energy-functional from Adiabatic Connection Fluctuation-dissipation Theorem

Adiabatic-connection fluctuation-dissipation theorem allows one to express the exchange-correlation energy-functional by means of imaginary-frequency density response function (χ_λ) of the system with the scaled Coulomb potential $(\lambda/|\mathbf{r} - \mathbf{r}'|)$[11,13]:

$$E_{xc}[\rho] = -\frac{1}{2}\int\limits_0^1 d\lambda \int d\vec{r} \int d\vec{r}' \frac{1}{|\vec{r} - \vec{r}'|}\left[\frac{1}{\pi}\int\limits_0^\infty du\chi_\lambda(\vec{r}, \vec{r}' : iu) + \rho(\vec{r})\delta(\vec{r} - \vec{r}')\right]$$

(2-23)

Recent years brought the whole series of works aimed at constructing approximations to the exchange-correlation-energy functional based on Eq. (23) in which different levels of approximations are used for χ_λ. Especially useful is the random phase approximation applied to derive: the whole exchange-correlation functional, its long-range, or short-range parts.[168−176] Furche, 2005[223]; The functionals of this type can, in principle, provide a practical solution to the outstanding problems faced by commonly used approximations: the correct dissociation of open shell fragments[177,222]; or description of van der Waals interactions,[168,170,171] for instance. However, further approximations concerning (χ_λ) are indispensable in order to use this formalism in practical simulations.[169, 174−176]

5.4. Van der Waals Density Functional of Langreth and Lundqvist

As indicated in the discussion of the performance of the common approximate exchange-correlation functionals in describing bonding of van der Waals complexes, the long-range behavior of the potential-energy curves cannot be properly described by means of semi-local functionals. Langreth, Lundquist, and collaborators proposed a solution of this deficiency by building in non-locality into the exchange-correlation functional.[174,175] The correlation-functional is split into two components: local and non-local. Whereas the local part is approximated using LDA, the non-local component is constructed in such a way that the correct long-range behavior of the correlation energy is built in minimizing also the possible double-counting. The exchange energy is approximated by means of the revPBE functional of the GGA type. The method has been applied originally for systems of layered geometries,[174] Ar_2, Kr_2, and $(C_6H_6)_2$ dimers,[175] dimers of polycyclic aromatic hydrocarbons,[178] parallel polymers,[179] for instance.

Hirao and collaborators,[143,180] proposed a closely linked approach to treat weakly bound complexes in which van der Waals density functional of Langreth, Lundqvist, and collaborators provides a key ingredient. The results of the tests on model systems (He_2, Ne_2, and Ar_2) are very encouraging.

5.5. Current-dependent Exchange-Correlation Functional

The aforementioned trend that semi-local functionals leads to overestimated dipole polarizabilities suggests a common origin of this flaw. In the case of small molecules, imposing correct asymptotic behavior of v_{xc} solves the problem to certain degree. In the case of extended systems, however, the calculated responses are even less reliable. Common approximations to the exchange-correlation potential fail dramatically for extended polymers.[18,181−184] Unfortunately, imposing the correct asymptotic behavior of v_{xc} does not help. The wavefunction-based methods predict that the dependence of the polarizability on the polymer length is sub-linear. Semi-local and hybrid approximations for the first- and second functional derivatives of the exchange-correlation energy, which are needed in linear-response time-dependent DFT calculations (for review, see Casida[185]), lead to a significant overestimation of the dipole polarizability especially for long chains. For the chain of ten monomers in polyethylene, the polarizability calculated using such approximations is two times too high (van Faassen et al., 2003[184]). Similar overestimation has been reported for polysilane, polysilene, polymethinemine, polybutatriene, polydiacetylene,[184] polymethineimine,[186] and a model system comprising a chain of hydrogen atoms[81,183] for which accurate CCSD(T) or fourth-order Møller-Plesset polarizabilities are available. For hyperpolarizabilities, a similar failure has been reported.[182] A highly non-local and collective nature of electronic polarization has been identified as the cause of this qualitative failure.[187] Both the current-dependent exchange-correlation functional[188] and the optimized effective potential based on 'exact exchange' account for the non-local effects more appropriately than semi-local functionals (LDA, GGA, or meta-GGA) and, indeed, lead to significantly better description of the polarizability of long polymer chains.[81,183] These methods, however, are computationally more expensive than the conventional ones.

The introduction of current-dependence into the exchange-correlation energy functional proved to be beneficial in yet another area. The fact, that the exchange-correlation energy is approximated, leads to a strong dependence of the degenerate open-shell atomic energies on the occupancy of the atomic orbitals.[6] This leads to uncertainties of the order of 3 to 5 kcal/mol in the atomic ground state energy of second and third period main group elements and the first transition series.[189] In this respect, GGA is a worse approximation than LDA because it leads to larger uncertainties.[190] Becke[190] showed that, expressing the exchange-correlation energy as a functional depending also on current reduces the spread between DFT calculated energies of different degenerate states to below 1.0 kcal/mole for such atoms as B, C, O, F, Al, Si, S, and Cl.

5.6. Density Functional Theory without the System of Noninteracting Electrons

The Kohn-Sham reference system corresponds to a singledeterminantal wavefunction. It is the *exact* wavefunction for an *artificial* system of noninteracting electrons. Such flaws of current approximations to $E_{xc}[\rho]$, which lead to qualitative wrong description of states described by multideterminantal wavefunction (diradicals for instance) are commonly attributed to this fact. The idea of combining the second Hohenberg-Kohn theorem with other reference state than a system of noninteracting electrons is, therefore, very appealing. In principle, one can construct various *beyond-Kohn-Sham* formulations of density functional theory depending on the choice of the reference system (see the pioneering work by Savin[191] for an overview). However, the decomposition of the total energy into density functionals of known and unknown analytic form depends on the choice of the reference state. Using functionals, which were developed as approximations to the exchange-correlation energy defined in the Kohn-Sham framework, in other frameworks involves the risk of double-counting. Currently, different practical realizations of multideterminantal reference state based strategy are under development in several groups.[192−199]

5.7. Dispersion Interactions from the Analysis of the Dipole Moment of the Exchange Hole

Becke and Johnson[200,201] proposed recently a new method to treat van der Waals complexes, to be used in association with either Hartree-Fock or Kohn-Sham orbitals. In this model, the total energy is expressed as:

$$E_{tot} = E_{HF} + E_C^{BR} + E_{disp} \tag{2-24}$$

where E_{HF} is the Hartree-Fock energy, E_C^{BR} is the correlation functional of the meta-GGA type,[202] and E_{disp} is a special parameter-free term obtained from the analysis of the position-dependent dipole associate with the exchange hole. This dipole moment depends on electron density, its gradients and Laplacian, and the density of the kinetic energy. The model predicts the C_6 dispersion coefficients with a very good accuracy.[200] The mean percentage error of these coefficients calculated for 178 cases amounts to only 11.1%. The same model leads also to very good interaction energies in weakly bound systems.[201] The mean percentage errors of the interaction energy based on representative complexes amount to: 10% for 24 dispersion bound complexes ranging from the helium- to benzene dimers, 8% for seven complexes comprising one polar and one nonpolar molecule (CH_4-NH_4, C_2H_4-HF, complexes for instance), and 6% for the six complexes formed by two polar molecules.

5.8.   Subsystem Formulation of DFT

Cortona[203] proposed an alternative strategy to approximate the Hohenberg-Kohn energy functional compared to that of Kohn and Sham. In this formulation, construction of the Kohn-Sham orbitals for the whole system under investigation is not needed because the total electron density of the system under consideration is constructed using several sets of one-electron functions (*embedded orbitals*). Each set corresponds to a subsystem comprising an integer number of electrons. The orbitals of each subsystem (denoted as A below) are derived from Kohn-Sham-type one-electron equations.*

$$\left[-\frac{1}{2}\nabla^2 + V_{eff}^{KSCED}[\rho_A, \rho_{Total} - \rho_A](\vec{r}) \right] \phi_i^A(\vec{r}) = \varepsilon_i^A \phi_i^A(\vec{r}) \qquad (2\text{-}25)$$

The effective potential for each set of embedded orbitals reads:

$$V_{eff}^{KSCED}[\rho_A, \rho_{Total} - \rho_A](\vec{r}) = V_{eff}^{KS}[\rho_A](\vec{r}) + V_{eff}^{emb}[\rho_A, \rho_{Total} - \rho_A](\vec{r}) \qquad (2\text{-}26)$$

where,

$$V_{eff}^{emb}[\rho_A, \rho_{Total} - \rho_A](\vec{r}) = -\sum_\alpha^{N_B} \frac{Z_\alpha^B}{|\vec{R}_\alpha^B - \vec{r}|} + \int \frac{\rho_{Total}(\vec{r}') - \rho_A(\vec{r}')}{|\vec{r} - \vec{r}'|} d\vec{r}'$$

$$+ \left. \frac{\delta E_{xc}[\rho]}{\delta\rho} \right|_{\rho=\rho_{Total}} - \left. \frac{\delta E_{xc}[\rho]}{\delta\rho} \right|_{\rho=\rho_A} + \left. - \frac{\delta T_s^{nad}[\rho_1, \rho_{Total} - \rho_1]}{\delta\rho_1} \right|_{\rho_1 = \rho_A} \qquad (2\text{-}27)$$

In the above formula, $T_s^{nad}[\rho_1, \rho_2] = T_s[\rho_1 + \rho_2] - T_s[\rho_1] - T_s[\rho_2]$ denotes the bifunctional, i.e. the functional of two electron densities, representing the nonadditivity of $T_s[\rho]$.

In the original work of Cortona,[203] the subsystems correspond to spherically symmetric atoms in solids. This formalism can be also applied for polyatomic subsystems – interacting molecules in particular.[204]

It is worthwhile to underline the key difference between the Cortona and Kohn-Sham frameworks to obtain ground-state energy and density in practical calculations

---

* The label KSCED standing for Kohn-Sham Equations with Constrained Electron Density is used here to indicate that, despite the similarity to Kohn-Sham equations, the effective potential and the one-electron functions differ from the corresponding quantities in these two frameworks.

based on Hohenberg-Kohn theorems. They differ only in the treatment of the kinetic energy in the reference system of noninteracting electrons. In practical Kohn-Sham calculations, this quantity is calculated exactly whereas it is calculated in a combined way in the subsystem-based framework (exact for $T_s[\rho_1]$ and $T_s[\rho_2]$ and approximate for $T_s^{nad}[\rho_1, \rho_2]$). Since the exchange-correlation energy is treated in the same way in both formalisms, the results of the calculations are prone to the same flaws of applied approximations to $E_{xc}[\rho]$. In particular, the treatment of dispersion forces is not possible using semi-local approximations to this functional. The approximations to the nonadditive kinetic energy is not relevant in this context because the exact $T_s^{nad}[\rho_1, \rho_2]$ disappears for non-overlapping electron densities. For overlapping electron densities, however, even simple approximations such as LDA and GGA applied simultaneously to both approximate components of the total energy functional (i.e. $E_{xc}[\rho]$ and $T_s^{nad}[\rho_1, \rho_2]$) lead to very reasonable interaction energies.[205,206]

Equations (25–26) provide also the formal foundations of the orbital-free and first-principles based multi-level type of computer simulations in which only one subsystem is described at the orbital level (denoted with A in Eqs. (25)–(27)), whereas the other subsystem(s) is (are) described using simpler methods for obtaining ρ_{Total}-ρ_A.[207] For a more complete review, see Dulak et al.[208] and Wesolowski.[209]

5.9. Density-matrix Functional Theory

Density-matrix functional theory is a natural extension of density functional theory of Hohenberg, Kohn, and Sham. Zumbach and Maschke[210] introduced the analog of the Kohn-Sham equations in which the exchange-correlation energy is expressed as a functional of one-particle density matrix (Γ). Similarly, as in the case of the Kohn-Sham equations, practical applications of such a scheme hinge on approximation to the exchange-correlation functional ($E_{xc}[\Gamma]$). Several theoretical groups are working on such approximations.[211–217] For recent tests of the performance of various approximations to $E_{xc}[\Gamma]$, see also Staroverov and Scuseria[218] (2002) and Cohen and Baerends.[219] Pernal and Cioslowski[220] investigated formally the applicability of some approximations to $E_{xc}[\Gamma]$ to describe the interactions in the case of two weakly interacting systems. Their analysis showed that the simple approximations based on natural spin orbitals are not capable to recover the dispersion-energy component of the interaction energy.

6. CONCLUDING REMARKS

The Hohenberg-Kohn-Sham density functional theory provides the common formal framework for various computational methods. Since each of the methods in use involves approximations, the calculated properties are not exact. Nevertheless, these methods proved to be very useful in chemistry and materials science. The huge and ever growing number of applications (see Figure 2-1) speaks for itself. Frequently, the errors of calculated quantities are acceptable for practical purposes. Obviously, reducing further these errors is very desirable. There are cases, however, where the results of the Kohn-Sham calculations are qualitatively wrong. For these reasons, further developments are needed. Using nonbonding interactions as an illustration, we have shown that development of methods situated within the Hohenberg-Kohn-Sham formulation of density functional theory is a dynamic field in which various strategies are being applied to address these unresolved issues. Noncovalently bonded complexes define one of the frontiers of applicability of density-functional-theory methods. At the local density approximation level, the Kohn-Sham equations lead to a very reasonable description of strong chemical bonds. Introduction of gradient-dependent functionals made it possible to extend the domain of applicability of the Kohn-Sham equations to hydrogen-bonded complexes more than one decade ago. Van der Waals complexes close to the equilibrium geometry belong to the domain of applicability of Hohenberg-Kohn-Sham DFT only if especially tailored or selected methods are used. Description of interactions between non-overlapping electron-densities (London dispersion forces) is not possible by means of the methods, which are in common use today. New formalisms, which might lead to efficient computational methods to describe such interactions, are in view.

This review, attempted also to provide a roadmap in the jungle of currently used approximations to the exchange-correlation energy functional. Eq. (6) provides its formal definition but does not give direct hints concerning approximating it. Each of equivalent definitions of the exchange-correlation functional (Eqs. (7), (19), and (23), for instance), opens the door for a possible strategy to approximate the exchange-correlation energy functional. Relying only on first-principles, leads to *nonempirical functionals* within each strategy. However, following a given strategy leaves usually space for parameters which can be determined using reference data for molecular systems leading thus to *empirical functionals*. In this review, the functionals were grouped also according the Perdew's hierarchy of methods known as *Jacob's ladder* of functionals. Each group: LDA, GGA, meta-GGA, and hyper-GGA corresponding to the ascending rungs of the ladder, collects functionals of the exchange-correlation energy which depends explicitly on a particular set of quantities such as electron density, electron-density gradients, density of the kinetic energy, and finally the Kohn-Sham orbitals (occupied or all). In principle, moving upwards on the ladder should increase the accuracy of the calculated properties. Frequently, it is the case but not always because each rung collects usually several functionals (empirical and nonempirical) of different accuracy. Except for the lowest rung (LDA), where the exact results for the uniform electron gas lead to a unique functional, each other rung on the ladder collects various functionals which use empirical parameters to various extent. The family of functionals developed by Perdew, Scuseria, and collaborators represent a nonempirical series covering all but the highest rung occupied currently by the TPSSh functional, which depends on one empirical parameter. Moving from the lowest rung (LDA) to the next one (PBE) leads invariably to the improvement of accuracy of the calculated observables. Moving to even higher level (TPSS) also reduces errors but exceptions occur (bonding distance in noble-gas dimers for instance). This might indicate that an alternative nonempirical construction is possible at the meta-GGA level or that the series starting from LDA, GGA, meta-GGA, and ending with hyper-GGA approaches the exact exchange-correlation energy in the oscillatory manner.

Owing to their flexibility, empirical functionals can outperform the nonempirical ones for some systems and properties. The growing number of systematic studies of the accuracy of various observables in different systems obtained by means of empirical approximations to the exchange-correlation energy might bring not only practical rules concerning their applicability but also more insight into the meaning of some empirical parameters. Such interaction between both practical and methodological aspects of present-day density functional theory can be only beneficial for the whole field of computer modelling of complex chemical systems and materials.

APPENDIX

Table 2-7. Acronyms and abbreviations used for approximate exchange, correlation, and exchange-correlation functionals discussed in this work and corresponding to their common use in the literature. In some cases, the same label is applied for exchange, correlation, or exchange-correlation functionals (PW91 for instance). In some cases, the labels used in the text are obtained as combinations of the ones included here: B3LYP denotes the combination of the B3 exchange- and LYP correlation functionals, for instance

Acronym used in text	Explicit dependence	Energy component	Reference
S or Dirac	ρ	exchange	Dirac,[14] Slater[15]
VWN	ρ	correlation	Vosko et al.[17]
WL	ρ	correlation	Wilson and Levy,[152]
PW92	ρ	correlation	Perdew and Wang,[18]
B86	$\rho, \nabla\rho$	exchange	Becke,[32]
PW86	$\rho, \nabla\rho$	exchange	Perdew and Yue,[25]
P86	$\rho, \nabla\rho$	correlation	Perdew,[26]
B or B88	$\rho, \nabla\rho$	exchange	Becke,[24]
LYP	$\rho, \nabla\rho$	correlation	Lee et al.[30]
PW91 or PW	$\rho, \nabla\rho$	exchange-correlation	Perdew,[27]
RPBE	$\rho, \nabla\rho$	exchange	Hammer et al.[37]
G96	$\rho, \nabla\rho$	exchange	Gill,[33]
PBE	$\rho, \nabla\rho$	exchange-correlation	Perdew et al.[28]
HCTH	$\rho, \nabla\rho$	exchange-correlation	Hamprecht et al.[31]
revPBE	$\rho, \nabla\rho$	exchange	Zhang and Yang,[36]
mPW91 or mPW	$\rho, \nabla\rho$	exchange	Adamo and Barone,[35]
HCTH-A	$\rho, \nabla\rho$	exchange-correlation	Hamprecht et al.[31]
OPTX or O	$\rho, \nabla\rho$	exchange	Handy and Cohen,[38]
mPBE	$\rho, \nabla\rho$	exchange-correlation	Adamo and Barone,[39]
LAP	$\rho, \nabla^2\rho$	correlation	Proynov et al.[43]
TPSS	$\rho, \nabla\rho, \tau$	exchange-correlation	Tao et al.[42]
Half-and-Half	$\rho, \nabla\rho, \phi_i$	exchange	Becke,[48]
B3	$\rho, \nabla\rho, \phi_i$	exchange	Becke,[49]
PBE1PBE	$\rho, \nabla\rho, \phi_i$	exchange-correlation	Perdew et al.[28]
B97	$\rho, \nabla\rho, \phi_i$	exchange-correlation	Becke,[51]
mPW1	$\rho, \nabla\rho, \phi_i$	exchange	Adamo and Barone,[35]
PBE0	$\rho, \nabla\rho, \phi_i$	exchange-correlation	Adamo and Barone,[52] Ernzerhof and Scuseria,[53]
O3	$\rho, \nabla\rho, \phi_i$	exchange	Handy and Cohen,[38]
X	$\rho, \nabla\rho, \phi_i$	exchange	Xu and Goddard[123]
TPSSh	$\rho, \nabla\rho, \tau, \phi_i$	exchange-correlation	Tao et al.[42]
PWB6K	$\rho, \nabla\rho, \nabla^2\rho, \phi_i, \tau$	exchange-correlation	Zhao and Truhlar,[57]

REFERENCES

1. Hohenberg P, Kohn W (1964) Inhomogeneous electron gas, Phys Rev B136: 864–871
2. Kohn W, Sham LJ (1965) Self-consistent equations including exchange and correlation Effects, Phys Rev, 140A: 1133–1138
3. Gelfand IM, Fomin SV (2000) Calculus of Variations, Dover Publications Inc., Mineola, New York, p 1
4. Perdew JP, Zunger A (1981) Self-interaction correction to density-functional approximations for many-electron systems, Phys Rev B, 23: 5048–5079
5. Shore HB, Rose JH, Zaremba E (1977) Failure of the local exchange approximation in the evaluation of the H^- ground state, Phys Rev B, 15: 2858–2861
6. Ziegler T, Rauk A, Baerends EJ (1977) Calculation of multiplet energies by Hartree-Fock-Slater method, Theor Chim Acta, 43: 261–271
7. Bally T, Sastry GN (1997) Incorrect dissociation behavior of radical ions in density functional calculations, J. Phys. Chem. A, 101: 7923–7925
8. Chermette H (1999) Chemical reactivity indexes in density functional theory, J. Comput. Chem, 20: 129–154
9. Levy M (1979) Universal variational functionals of electron densities, first-order density matrices, and natural spin-orbitals and solution of the v-representability problem, Proc Natl Acad Sci USA, 76: 6062–6065
10. Levy M (1982) Electron densities in search of Hamiltonians, Phys Rev A, 26: 1200–1208
11. Langreth DC, Perdew JP (1975) Exchange-correlation energy of metallic surface, Solid State Commun, 17: 1425–1429
12. Gunnarsson O, Lundqvist BI (1976) Exchange and correlation in atoms, molecules, and solids by spin-density functional formalism, Phys Rev B, 13: 4274–4298
13. Langreth DC, Perdew JP (1977) Exchange-correlation energy of a metallic surface – wave-vector analysis, Phys Rev B, 15: 2884–2901
14. Dirac PAM (1930) Note on exchange phenomena in the Thomas atom, Proc Cambridge Philos Soc, 26: 376–385
15. Slater JC (1951) A simplification of the Hartree-Fock method for the exchange energy ($E_x[\rho]$) in the uniform electron gas, Phys Rev, 81: 385–390
16. Ceperley DM, Alder BJ (1980) Ground State of the Electron Gas by a Stochastic Method, Phys. Rev. Lett, 45: 566–569
17. Vosko SH, Wilk, Nusair M (1980) Accurate spin-dependent electron liquid correlation energies for local spin density calculations: a critical analysis, Can J Phys, 58: 1200–1211
18. Perdew JP, Wang Y (1992) Accurate and simple analytic representation of the electron-gas correlation energy, Phys Rev B, 45: 13244–13249
19. Perdew JP (1991a) Generalized gradient approximations for exchange and correlation – A look backward and forward, Physica B, 172: 1–6
20. Herman F, Van Dyke JP, Ortenburger IB (1969) Improved statistical exchange approximation for inhomogeneous many-electron systems, Phys Rev Lett, 22: 807–811
21. Ma SK, Brueckner KA (1968) Correlation energy of an electron gas with a slowly varying high density, Phys Rev, 165: 18–31
22. Perdew JP, Langreth DC, V Sahni (1977) Corrections to the local density approximation: Gradient expansion versus wave-vector analysis for the metallic surface problem, Phys Rev Lett, 38: 1030–1033
23. Sahni V, Gruenebaum J, Perdew JP (1982) Study of the density-gradient expansion for the exchange energy, Phys Rev B, 26: 4371–4377

24. Becke AD (1988) Density-functional exchange-energy approximation with correct asymptotic behavior, Phys. Rev. A, 38: 3098–3100

25. Perdew JP, Yue W (1986) Accurate and simple density functional for the electronic exchange energy: Generalized gradient approximation, Phys Rev B, 33: 8800–8802

26. Perdew JP (1986) Density-functional approximation for the correlation energy of the inhomogeneous electron gas, Phys Rev B, 33: 8822–8824

27. Perdew JP (1991b) Unified theory of exchange and correlation beyond local density approximation, in: P.Ziesche and H.Eschrig (eds)., Electronic Structure of Solids'91, Akademie Verlag, Berlin, p11

28. Perdew JP, Burke K, Ernzerhof M (1996) Generalized gradient approximation made simple, Phys Rev Lett, 77: 3865–3868

29. Colle R, Salvetti O (1975) Approximate calculation of correlation energy for closed shells, Theoretica Chimica Acta, 37: 329–334

30. Lee C, Yang W, Parr RG (1988) Development of the Colle-Salvetti correlation-energy formula into a functional of the electron density, Phys Rev B, 37: 785–789

31. Hamprecht FA, Cohen AJ, Tozer DJ, Handy NC (1998) Development and assessment of new exchange-correlation functionals, J Chem Phys, 109: 6264–6271

32. Becke AD (1986) Density functional calculations of molecular-bond energies, J. Chem. Phys, 84: 4524–4529

33. Gill PWM (1996) A new gradient-corrected exchange functional, Mol Phys, 89: 433–445

34. Filatov M, Thiel W (1997) A new gradient-corrected exchange-correlation density functional, Mol Phys, 91: 847–859

35. Adamo C, Barone V (1998) Exchange functionals with improved long-range behavior and adiabatic connection methods without adjustable parameters: The mPW and mPW1PW models, J. Chem. Phys, 108: 664–675

36. Zhang Y, Yang W (1998) Comment on "Generalized Gradient Approximation Made Simple", Phys Rev Lett, 80: 890

37. Hammer B, Hansen LB, Nørskov JK (1999) Improved adsorption energetics within density-functional theory using revised Perdew-Burke-Ernzerhof functionals, Phys Rev B, 59: 7413–7421

38. Handy NC, Cohen A (2001) Left-right correlation energy, Mol Phys, 99: 403–412

39. Adamo C, Barone V (2002) Physically motivated density functionals with improved performances: The modified Perdew-Burke-Ernzerhof model, J. Chem. Phys, 116: 5933–5940

40. Becke AD (1983) Hartree-Fock exchange energy of an inhomogeneous electron-gas, Int. J. Quant. Chem, 23: 1915–1922

41. Becke A, Roussel ME (1989) Exchange holes in inhomogeneous systems: A coordinate-space model, Phys. Rev. A, 39: 3761–3767

42. Tao J, Perdew JP, Staroverov VN, Scuseria GE (2003) Climbing the density functional ladder: Nonempirical meta-generalized gradient approximation designed for molecules and solids, Phys Rev Lett, 91: 146401

43. Proynov EI, Vela A, Salahub DR (1994) Nonlocal correlation functional involving the Laplacian of the density, Chem Phys Lett, 230: 419–428

44. Filatov M, Thiel W (1998) Exchange-correlation density functional beyond the gradient approximation, Phys Rev A, 57: 189–199

45. Gill PW (2001) Obituary: density functional theory (1927–1993), Austr J Chem, 54: 661–662

46. Görling A, Levy M (1994) Exact Kohn-Sham scheme based on perturbation theory, Phys Rev A, 50: 196–204

47. Görling A (2005) Orbital- and state-dependent functionals in density-functional theory, J Chem Phys, 123: 062203

48. Becke AD (1993a) A new mixing of Hartree-Fock and local density-functional theories., J. Chem. Phys., 107: 8554–8560

49. Becke AD (1993b) Density-functional thermochemistry 3. The role of exact exchange., J. Chem. Phys, 98: 5648–5652

50. Stephens PJ, Devlin FJ, Chabalowski CF, Frisch MJ (1994) Ab Initio calculation of vibrational absorption and circular dichroism spectra using density functional force fields, J Phys Chem, 98: 11623–11627

51. Becke AD (1997) Density-functional thermochemistry 5. Systematic optimization of exchange-correlation functionals, J. Chem. Phys, 98: 1372–1377

52. Adamo C, Barone V (1999) Toward reliable density functional methods without adjustable parameters: The PBE0 model, J. Chem. Phys, 110: 6158–6170

53. Ernzerhof M, Scuseria GE (1999) Assessment of the Perdew-Burke-Ernzerhof exchange-correlation functional, J Chem Phys, 110, 5029–5036

54. Kafafi SA (1998) Novel density functional methodology for the computation of accurate electronic and thermodynamic properties of molecular systems and improved long-range behavior, J Phys Chem A, 102: 10404–10413

55. Tao J (2002) An accurate MGGA-based hybrid exchange-correlation functional, J Chem Phys, 116: 2335–2337

56. Toulouse J, Adamo C (2002) A new hybrid functional including a meta-GGA approach, Chem Phys Lett, 362: 72–78

57. Zhao Y, Truhlar DG (2004) Hybrid meta density functional theory methods for thermochemistry, thermochemical kinetics, and noncovalent interactions: The MPW1B95 and MPWB1K models and comparative assessments for hydrogen bonding and van der Waals interactions, J Phys Chem A, 108: 6908–6918

58. Perdew JP, Ruzsinsky A, Tao J, Starovyerov V, Scuseria GE, Csonka GI (2005) Prescription for the design and selection of density functional approximations: More constraint satisfaction with fewer fits, J Chem Phys, 123: 062201

59. Dickson RM, Becke AD (1996) Local density-functional polarizabilities and hyperpolarizabilities at the basis-set limit, J Phys Chem, 100: 16105–16108

60. Becke AD, Dickson RM (1990) Numerical-solution of Schrödinger equation in polyatomic molecules, J. Chem. Phys, 92: 3610–3612

61. Cohen AJ, Tantirungrotechai Y (1999) Molecular electric properties: an assessment of recently developed functionals, Chem. Phys. Lett, 299: 465–472

62. Jasien PG, Fitzgerald G (1990) Molecular dipole moments and polarizabilities from local density functional calculations: Applications to DNA base pairs, J Chem Phys, 93: 2554–2560

63. St.-Amant A, Cornell WD, Kollman PA, Halgren TA (1995) Calculation of molecular geometries, relative conformational energies, dipole-moments, and molecular electrostatic potential fitted charges of small organic-molecules of biochemical interest by density-functional theory, J Comput Chem, 16: 1483–1506

64. Rashin AA, Young L, Topol IA, Burt SK (1994) Molecular dipole moments calculated with density functional theory, Chem Phys Lett, 230: 182–188

65. Bakalarski G, Grochowski P, Kwiatkowski JS, Lesyng B, Leszczynski J (1996) Molecular and electrostatic properties of the N-methylated nucleic acid bases by density functional theory, Chem. Phys, 204: 301–311

66. de Luca G, Russo N, Sicilia E, Toscano M (1996) Molecular quadrupole moments, second moments, and diamagnetic susceptibilities evaluated using the generalized gradient approximation in the framework of Gaussian density functional method, J Chem Phys, 105: 3206–3210

67. Scheiner AC, Baker J, Andzelm J (1997) Molecular energies and properties from density functional theory: Exploring basis set dependence of Kohn-Sham equation using several density functionals, J Comput Chem, 18, 775–795

68. Heidin L, Lundqvist BI (1971) Explicit local exchange-correlation potentials, J Phys C, 4: 2064–2083

69. McDowell SAC, Amos RD, Handy NC (1995) Molecular polarisabilities - a comparison of density functional theory with standard *ab initio* methods, Chem Phys Lett, 235: 1–4

70. Guan J, Duffy P, Carter JT, Chong DP, Casida KC, Casida ME, Wrinn M (1993) Comparison of local-density and Hartree–Fock calculations of molecular polarizabilities and hyperpolarizabilities, J Chem Phys, 98: 4753–4765

71. van Caillie C, Amos RD (1998) Static and dynamic polarisabilities, Cauchy coefficients and their anisotropies: a comparison of standard methods, Chem. Phys. Lett, 291: 71–77

72. Calaminici P, Jug K, Koster AM (1998) Density functional calculations of molecular polarizabilities and hyperpolarizabilities, J. Chem. Phys, 109: 7756–7763

73. Guan J, Casida ME, Salahub DR (1995) ll-Electron Local and Gradient-Corrected Density-Functional Calculations of Na_n Dipole Polarizabilities for n=1–6, Phys Rev B, 52: 2185–2200

74. Russell AJ, Spackman MA (1995) Vibrational averaging of electrical-properties development of a routine theoretical method for polyatomic molecules, Mol Phys, 84: 1239–1255

75. Russell AJ, Spackman MA (1997) An ab initio study of vibrational corrections to the electrical properties of the second-row hydrides, Mol Phys, 90: 251–264

76. Wilson PJ, Bradley TJ, Tozer DJ (2001) Hybrid exchange-correlation functional determined from thermochemical data and *ab initio* potentials, J Chem Phys, 115: 9233–9242

77. Sophy KB, Pal S (2003) Density functional response approach for the linear and nonlinear electric properties of molecules, J Chem Phys, 118: 10861–10866

78. van Gisbergen SJA, Osinga VP, Gritsenko OV, van Leeuven R, Snijders JG, Baerends EJ (1996) Improved density functional theory for frequency-dependent polarizabilities, by the use of an exchange-correlation potential with correct asymptotic behavior, J Chem Phys, 105: 3142–3151

79. Zhao Q, Morrison RC, Parr RG (1994) From electron densities to Kohn-Sham kinetic energies, orbital energies, exchange-correlation potentials, and exchange-correlation energies, Phys Rev A, 50: 2138–2142

80. van Leeuwen, Baerends EJ (1994) Exchange-correlation potential with correct asymptotic behavior, Phys Rev A, 49: 2421–2431

81. Mori-Sanchez P, Wu Q, Yang WT (2003) Accurate polymer polarizabilities with exact exchange density-functional theory, J Chem Phys, 119: 11001–11004

82. Hirata S, Ivanov S, Bartlett RJ, Grabowski I (2005) Exact-exchange time-dependent density-functional theory for static and dynamic polarizabilities, Phys. Rev. A, 71: 032507

83. van Gisbergen SJA, Kootstra F, Schipper PRT, Gritsenko OV, Snijders JG, Baerends (1998) Density-functional-theory response-property calculations with accurate exchange-correlation potentials, Phys Rev A, 57: 2556–2570

84. Jacob CR, Wesolowski TA, Visscher L (2005) Orbital-free embedding applied to the calculation of induced dipole moments in CO_2.. X (X=He, Ne, Ar, Kr, Xe, Hg) van der Waals complexes, J Chem Phys, 123: 174104

85. Gritsenko O, Schipper PRT, Baerends EJ (1999) Approximation of the exchange-correlation Kohn-Sham potential with a statistical average of different orbital model potentials, Chem Phys Lett, 302: 199–207

86. Janak JF (1978) Proof that $\partial E / \partial n_i = \varepsilon_i$ in density functional theory, Phys Rev B, 18: 7165–7168

87. Perdew JP, Parr RG, Levy M, Balduz Jr JL (1982) Density-functional theory for fractional particle number: Derivative discontinuities of the energy, Phys Rev Lett, 49: 1691–1694

88. Perdew JP, Levy M (1995) Comment on "Significance of the highest occupied Kohn-Sham eigenvalue", Phys Rev B, 56: 16021–16028

89. Lindgren I, Salomonson S, Möller F (2005) Construction of accurate Kohn-Sham potentials for the lowest states of the helium atom: Accurate test of the ionization-potential theorem, Int J Quant Chem, 102: 1010–1017

90. Grabo T, Gross EKU (1995) Density-functional-theory using an optimized exchange-correlation potential, Chem Phys Lett, 240: 141–150

91. Wu Q, Ayers PW, Yang WT (2003) Density-functional theory calculations with correct long-range potentials, J Chem Phys, 119: 2978–2990

92. Weimer M, Delle Sala F, Görling A (2004) The Kohn-Sham treatment of anions via the localized Hartree-Fock method, Chem Phys Lett, 372, 538–547

93. Schwarz K (1978) Instability of stable negative ions in Xα method or other local density functional schemes, Chem Phys Lett, 57: 605–607

94. Hamel S, Casida ME, Salahub DR (2002b) Exchange-only optimized effective potential for molecules from resolution-of-the-identity techniques: Comparison with the local density approximation, with and without asymptotic correction, J Chem Phys, 116: 8276–8291

95. Gritsenko O, Baerends EJ (2002) The analog of Koopmans' theorem in spin-density functional theory, J Chem Phys, 117: 9154–9159

96. Chong DP, Gritsenko OV, Baerends EJ (2002) Interpretation of the Kohn-Sham orbital energies as approximate vertical ionization potentials, J. Chem. Phys, 116: 1760–1772

97. Perdew JP, Chevary JA, Vosko SH, Jackson KA, Pederson MR, Singh DJ, Fiolhais C (1992) Atoms molecules, solids, and surfaces: applications of the generalized gradient approximation for exchange and correlation, Phys Rev B, 46: 6671–6687

98. Vydrov OA, Scuseria GE (2005) Ionization potentials and electron affinities in the Perdew-Zunger self-interaction corrected density-functional theory, J Chem Phys, 122: 184107

99. Curtiss LA, Raghavachari K, Redfern PC, Pople JA (1997) Assessment of Gaussian-2 and density functional theories for the computation of enthalpies of formation, J Chem Phys, 106: 1063–1079

100. Staroverov VN, Scuseria GE, Tao JM, Perdew JP (2003) Comparative assessment of a new nonempirical density functional: Molecules and hydrogen-bonded complexes, J Chem Phys, 119: 12129–12137

101. Joantéguy S, Pfister-Guillouzo G, Chermette H (1999) Asessment of density functional methods for the calculation of ionization potentials of unsaturated molecules, J Phys Chem A, 103, 3505–3511

102. Hoe WM, Cohen A, Handy NC (2001) Assessment of a new local exchange functional OPTX, Chem Phys Lett, 341: 319–328

103. Xu X, Goddard III WA (2004a) Assessment of Handy-Cohen optimized exchange density functional (OPTX), J Phys Chem A, 108: 8495–8506

104. Cole LA, Perdew JP (1982) Calculated electron affinities of the elements, Phys. Rev. A, 25, 1265–1271

105. Rösch N, Trickey SB (1997) Comment on "Concerning the applicability of density functional methods to atomic and molecular negative ions", [J Chem Phys, 105, 862 (1996)], J Chem Phys, 106, 8940–8941

106. Galbraith JM, Schaefer HF (1996) Concerning the applicability of density functional methods to atomic and molecular negative ions, J Chem Phys, 105, 862–864

107. Tschumper GS, Schaefer HF (1997) Predicting electron affinities with density functional theory: Some positive results for negative ion, J Chem Phys, 107: 2529–2541

108. Sim F, St-Amant A, Papai I, Salahub DR (1992) Gaussian density functional calculations on hydrogen-bonded systems, J Am Chem Soc, 114: 4391–4400

109. Novoa JJ, Sosa C (1995) Evaluation of the density functional approximation on the computation of hydrogen bond interactions, J Phys Chem, 99: 15837–15845

110. Zhu T, Yang W (1994) Structure of the ammonia dimer studied by density functional theory, Int J Quant Chem, 49: 613–623

111. Florián J, Johnson B (1995) Structure, energetics, and force fields of the cyclic formamide dimer: MP2, Hartree-Fock, and density functional study, J Phys Chem, 99: 5899–5908

112. Han WG, Suhai S (1996) Density Functional Studies on N-Methylacatemide-Water Complex, J Phys Chem, 100: 3942–3949

113. Zhao Y, Truhlar DG (2005a) Benchmark databases for nonbonded interactions and their use to test density functional theory, J Chem Theory Comput, 1: 415–432

114. Pudzianowski AT, (1995) A systematic appraisal of density functional methodologies for hydrogen bonding in binary ionic complexes, J Phys Chem, 100: 4781–4789

115. Hobza P, Sponer J, Reschel T (1995) Density-functional theory and molecular clusters, J Comp Chem, 16: 1315–1325

116. Mele F, Mineva T, Russo N, Toscano M (1995) Hydrogen-bonded and van-der-Waals complexes studied by a gaussian density-functional method – the case of (HF)2, ArHCl, Ar2HCl systems, Theor Chim Acta, 91: 169–177

117. Süle P, Nagy A (1996) Density Functional study of strong hydrogen-bonded systems: The hydrogen diformate complex, J Chem Phys, 104: 8524–8534

118. Kim K, Jordan KD (1994) Comparison of density functional and MP2 calculations on the water monomer and dimer, J Phys Chem, 98: 10089–10094

119. Del Bene JE, Person WB, Szczepaniak K (1995) Properties of hydrogen-bonded Complexes Obtained from B3LYP Functional with 6–31G(d,p) and 6–31+G(d,p) Basis Sets: Comparison with MP2/631+G(d,p) Results and Experimental Data, J Phys Chem, 99: 10705–10707

120. Zhao Y, Tishchenko O, Truhlar DG (2005) How well can density functional methods describe hydrogen bonds to π acceptors?, J Phys Chem B, 109: 19046–19051

121. Bergès J, Caillet J, Langlet J, Kozelka J (2001) Hydration and 'inverse hydration' of platinum(II) complexes: an analysis using the density functionals PW91 and BLYP, Chem. Phys. Lett, 344: 573–577

122. Milet A, Korona T, Moszynski R, Kochaniski E (1999) Anisotropic intermolecular interactions in van der Waals and hydrogen-bonded complexes: What can we get from density functional calculations? J Chem Phys, 111: 7727–7735

123. Xu X, Goddard III WA (2004b) The X3LYP extended functional for accurate description of nonbond interactions, spin states, and thermochemical properties, Proc Natl Acad Sci USA, 101: 2673–2677

124. Klopper W, van Duijneveldt-van de Rijdt JGCM, van Duijneveldt FB (2000) Computational determination of equilibrium geometry and dissociation energy of the water dimer, Phys Chem Chem Phys, 2: 2227–2234

125. Misquitta AJ, Jeziorski B, Szalewicz K (2003) Dispersion energy from density-functional theory description of monomers, Phys Rev Lett, 91: 033201

126. Lacks DJ, Gordon RG (1993) Pair interactions of rare-gas atoms as a test of exchange-energy-density functionals in regions of large density gradients, Phys Rev A, 47: 4681–4690

127. Pérez-Jordá JM, Becke AD (1995) A density-functional study of van-der-Waals forces – rare-gas diatomics, Chem Phys Lett, 233: 134–137

128. Meijer EJ, Sprik M (1996) A density-functional study of the intermolecular interactions of benzene, J Chem Phys, 105: 8684–8689

129. Wesolowski TA (2000) Comment on "Anisotropic intermolecular interactions in van der Waals and hydrogen-bonded complexes: What can we get from density-functional calculations?" [J Chem Phys 111: 7727 (1999)], J Chem Phys, 116: 515–524

130. Tao J, Perdew JP (2005) Test of a nonempirical density functional: Short-range part of the van der Waals interaction in rare-gas dimers, J Chem Phys, 122: 11402

131. Ruzsinszky A, Perdew JP, Csonka GI (2005) Binding energy curves from nonempirical density functionals II. van der Waals bonds in rare-gas and alkaline-earth diatomics, J Phys Chem A, 109: 11015–11021

132. Zhang Y, Pan W, Yang W (1997) Describing van der Waals Interaction in diatomic molecules with generalized gradient approximations: The role of the exchange functional, J Chem Phys, 107: 7921–7925

133. Patton DC, Porezag DV, Pederson MR (1997) Simplified generalized-gradient approximation and anharmonicity: Benchmark calculations on molecules, Phys Rev B, 55: 7454–7459

134. Kristyan S, Pulay P (1994) Can (semi)local density-functional theory account for London dispersion forces, Chem Phys Lett, 229: 175–180

135. Šponer J, Leszczynski J, Hobza P (1998) Base staking in cytosine dimer. A comparison of correlated ab initio calculations with three empirical models and density functional theory calculations, J Comp Chem, 17: 841–850

136. Ye XY, Li ZH, Wang WN, Fan KN, Xu W, Hua ZY (2004) The parallel pi-pi stacking: a model study with MP2 and DFT methods, Chem Phys Lett, 397: 56–61

137. Small D, Zaitsev V, Jung Y, Rosokha SV, Head-Gordon M, Kochi JK (2004) Intermolecular π-to-π bonding between stacked aromatic dyads. Experimental and theoretical binding energies and near-IR optical transitions for phenalenyl radical/radical versus radical/cation dimerizations, J Am Chem Soc, 126: 138–13858. (B3LYP results are included as Supporting Materials to this article)

138. Zhao Y, Truhlar DG (2005b) How well can new-generation density functional methods describe stacking interactions in biological systems, Phys Chem Chem Phys, 7: 2701–2705

139. Lieb EH, Oxford S (1981) Improved lower bound on the indirect Coulomb energy, Int J Quant Chem, 19: 427–439

140. Patey MD, Dessent CEH (2002) A PW91 density functional study of conformational choice in 2-pentylethanol, n-butylbenzene, and their cations: Problems for density functional theory?, J Phys Chem A, 106: 4623–4631

141. Ruiz E, Salahub DR, Vela A (1995) Defining the domain of density functionals – charge-transfer complexes, J Am Chem Soc, 117: 1141–1142

142. Ruiz E, Salahub DR, Vela A (1996) Charge-transfer complexes: Stringent tests for widely used density functionals, J Phys Chem, 100: 12265–12276

143. Kamiya M, Tsuneda T, Hirao K (2002) A density functional study of van der Waals interactions, J Chem Phys, 117: 36010–6015

144. Šponer J, Jurecka P, Hobza P (2004) Accurate interaction energies of hydrogen-bonded nucleic acid base pairs, J Am Chem Soc, 126: 10142–10151

145. Wu X, Vargas MC, Nayak S, Lotrich V, Scoles G (2001) Towards extending the applicability of density functional theory to weakly bound systems, J Chem Phys, 115: 8748–8757

146. Elsner M, Hobza P, Frauenheim T, Suhai S, Kaxiras E (2001) Hydrogen bonding and stacking interactions of nucleic acid base pairs: A density-functional-theory based treatment, J Chem Phys, 114: 5149–5155

147. Grimme S (2004) Accurate description of van der Waals complexes by density functional theory including empirical corrections, J Comp Chem, 25: 1463–1473

148. Zimmerli U, Parrinello M, Koumoutsakos P (2004) Dispersion corrections to density functionals for water aromatic interactions, J Chem Phys, 120: 2693–2699

149. Wu Q, Yang W (2002) Empirical correction to density functional theory for van der Waals interactions, J Chem Phys, 116: 515–524

150. von Lilienfeld OA, Tavernelli I, Rothlisberger U, Sebastiani D (2004) Optimization of effective atom centered potentials for London dispersion forces in density functional theory, Phys Rev Lett, 93: 153004

151. Pérez-Jordá JM, San-Fabián E, Pérez-Jiménez AJ (1999) Density-functional study of van der Waals forces on rare-gas diatomics: Hartree–Fock exchange, J Chem Phys, 110: 1916–1920

152. Wilson LC, Levy M (1990) Nonlocal Wigner-like correlation-energy density functional through coordinate scaling, Phys Rev B, 41: 12930–12932

153. Baerends EJ (2001) Exact exchange-correlation treatment of dissociated H_2 in density functional theory, Phys. Rev. Lett, 87: 133004

154. Sharp RT, Horton GG (1953) A variational approach to the unipotential many-electron problem, Phys Rev, 30: 317–317

155. Hirata S, Ivanov S, Grabowski I, Bartlet RJ, Burke K, Talman JD (2001) Can optimized effective potentials be determined uniquely?, J Chem Phys, 115: 1635–1649

156. Talman JD, Shadwick WF (1972) Optimized effective atomic central potential, Phys Rev A, 14: 36–40

157. Krieger JB, Li Y, Iafrate GJ (1992a) Systematic approximations to the optimized effective potential: Application to orbital-density-functional theory, Phys Rev A, 46: 5453–5458

158. Garza J, Nichols JA, Dixon DA (2000) The optimized effective potential and the self-interaction correction in density functional theory: Application to molecules, J Chem Phys, 112: 7880–7890

159. Hamel S, Duffy P, Casida ME, Salahub DR (2002a) Kohn-Sham orbitals and orbital energies: fictitious constructs but good approximations all the same, J Electr Spectr and Related Phenomena, 123: 345–363

160. Gunnarsson O, Jonson M, Lundqvist BI (1977) Exchange and correlation in inhomogeneous electron-systems, Solid State Commun, 24: 765–768

161. Alonso JA, Girifalco LA (1978) Nonlocal approximation to exchange potential and kinetic energy in an inhomogenous electron gas, Phys. Rev. B, 17: 3735–3743

162. Gunnarsson O, Jonson M, Lundqvist BI (1979) Descriptions of exchange and correlation effects in inhomogeneous electron systems, Phys Rev B, 20: 3136–3164

163. Singh DJ (1997) Density functional studies of $PbZrO_3$, $KTaO_3$ and $KNbO_3$, Ferroelectrics, 194: 299–322

164. Wu Z, Cohen RE, Singh DJ (2004) Comparing the weighted density approximation with the LDA and GGA for ground-state properties of ferroelectric perovskites, Phys Rev B, 70: 104112

165. Rushton PP, Tozer DJ, Clark SJ (2002) Nonlocal density-functional description of exchange and correlation in silicon, Phys Rev B, 65: 235203

166. Marzari N, Singh (2004) Dielectric response of oxides in the weighted density approximation, Phys Rev B, 62: 12724–12729

167. Sadd M, Teter MP (1996) Weighted density approximation applied to diatomic molecules, Phys Rev B, 54: 13643–13648

168. Kohn W, Meir Y, Makarov DE (1998) van der Waals energies in density functional theory, Phys Rev Lett, 80: 4153–4156

169. Hult E, Rydberg H, Lundqvist BI, Langreth DC (1999) Unified treatment of asymptotic van der Waals forces, Phys Rev B, 59: 4708–4713

170. Lein M, Dobson JF, Gross EKU (1999) Dobson JF, Toward the description of van der Waals interactions within density functional theory, J Comput Chem, 20: 12–22

171. Dobson JF, Wang J (1999) Successful test of a seamless van der Waals density functional, Phys Rev Lett, 82: 2123–2123

172. Yan ZD, Perdew JP, Kurth S (2000) Density functional for short-range correlation: Accuracy of the random-phase approximation for isoelectronic energy changes, Phys Rev B, 61: 16430–16439

173. Fuchs M, Gonze X (2002) Accurate density functionals: Approaches using adiabatic-connection fluctuation-dissipation theorem, Phys. Rev. B, 65: 235109

174. Rydberg H, Dion M, Jacobson N, Schroder E, Hyldgaard P, Simak SI, Langreth DC, Lundqvist BI (2003) Van der Waals density functional for layered structures, Phys Rev Lett, 91: 126402

175. Dion M, Rydberg H, Schroder E, Hyldgaard P, Langreth DC, Lundqvist BI (2003) Van der Waals density functional for general geometries, Phys Rev Lett, 92: 246401

176. Langreth DC, Dion M, Rydberg H, Schroder E, Hyldgaard P, Lundqvist BI (2005) Van der Waals density functional theory with applications, Int J Quant Chem, 101: 599–610

177. Fuchs M, Niquet YM, Gonze X, Burke K (2005) Describing static correlation in bond dissociation by Kohn-Sham density functional theory, J Chem Phys, 122: 094116

178. Chakarova SD, Schröder E (2005) van der Waals interactions of polycyclic aromatic hydrocarbon dimers, J. Chem. Phys, 122: 054102

179. Kleis J, Schröder E (2005) van der Waals interaction of simple, parallel polymers, J Chem Phys, 122: 164902

180. Likura H, Tsuneda T, Yanai T, Hirao K (2001) A long-range correction scheme for generalized-gradient-approximation exchange functionals, J Chem Phys, 115: 3540–3544

181. Champagne B, Perpete EA, van Gisbergen SJA, Baerends EJ, Snijders JG, Soubra-Ghaoui C, Robins KA, Kirtman B (1998) Assessment of conventional density functional schemes for computing the polarizabilities and hyperpolarizabilities of conjugated oligomers: An ab initio investigation of polyacetylene chains, J. Chem. Phys, 109: 10489–10498

182. van Gisbergen SJA, Schipper PRT, Gritsenko OV, Baerends EJ, Snijders JG, Champagne B, Kirtman B (1999) Electric field dependence of the exchange-correlation potential in molecular chains, Phys Rev Lett, 83: 694–697

183. van Faassen M, de Boeij PL, van Leeuwen R, Berger JA, Snijders JG (2002) Ultranonlocality in time-dependent current-density-functional theory: Application to conjugated polymers, Phys Rev Lett, 83: 694–697

184. van Faassen M, de Boeij PL, van Leeuwen R, Berger JA, Snijders JG (2003) Application of time-dependent current-density-functional theory to nonlocal exchange-correlation effects in polymers, J Chem Phys, 118: 1044–1053

185. Casida ME (1996) Time-Dependent Density Functional Response Theory of Molecular Systems: Computational methods, and Functionals. In: Recent Developments and Applications of Modern Density Functional Theory: Theoretical and Computational Chemistry, Vol. 4., J.M. Seminario ed., Elsevier Science, pp. 391–439

186. Jacquemin D, André J-M, and Perpéte EA (2004) Geometry, dipole moment, polarizability and first hyperpolarizability of polymethineimine: An assessment of electron correlation contributions, J Chem Phys, 121: 4389–4396

187. Berger JA, Snijders JG (2002) Ultranonlocality in time-dependent current-density-functional theory: Application to conjugated polymers, Phys. Rev. Lett, 83: 694–697

188. Vignale G, Kohn W (1996) Current-dependent exchange-correlation potential for dynamical linear response theory, Phys Rev Lett, 77: 2037–2040

189. Baerends EJ, Branchadell V, Sodupe M (1997) Atomic reference energies for density functional calculations, Chem. Phys. Lett, 265: 481–489

190. Becke AD (2002) Current density in exchange-correlation functionals: Application to atomic states, J. Chem. Phys, 117: 6935–6938

191. Savin A (1995) Beyond the Kohn-Sham Determinant. In: Recent Advances in Density Functional Methods, Part I., D. P. Chong (Ed.) Word Scientific, Singapore, pp 123–153

192. Malcolm NOJ, McDouall JJW (1996) Combining multiconfigurational wave functions with density functional estimates of dynamic electron correlation, J Phys Chem, 100: 10131–10134

193. Leininger T, Stoll H, Werner HJ, Savin A (1997) Combining long-range configuration interaction with short-range density functionals, Chem Phys Lett, 275: 151–160

194. Grimme S, Waletzke M (1999) A combination of Kohn-Sham density-functional theory and multi-reference configuration interaction methods, J Chem Phys, 111: 5645–5655

195. Filatov M, Shaik S (1999) A spin-restricted ensamble-referenced Kohn-Sham method and its application to diradical situations, Chem Phys Lett, 304: 429–437

196. Gräfenstein J, Cremer D (2000) The combination of density functional theory with multi-configurational methods – CAS-DFT, Chem Phys Lett, 316: 569–577

197. Pollet J, Savin A, Leininger T, Stoll H (2002a) Combining multideterminantal wave functions with density functionals to handle near-degeneracy in atoms and molecules, J Chem Phys, 116: 1250–1258

198. Pollet R, Colonna F, Leininger T, Stoll H, Werner HJ, Savin A (2003) Exchange-correlation energies and correlation holes for some two- and four-electron atoms along a nonlinear adiabatic connection in density functional theory, Int J Quant Chem, 91: 84–93

199. Gusarov S, Malmquist P-A, Lindh R, Roos BO (2004) Correlation potrentials for a multiconfigurational-based density functional theory with exact exchange, Theor Chem Acc, 112: 84–94

200. Becke AD, Johnson ER (2005a) A density-functional model of the dispersion interaction, J. Chem. Phys, 123: 154101

201. Becke AD, Johnson ER (2005b) Exchange-hole dipole moment and the dispersion interaction, J. Chem. Phys, 122: 154104

202. Becke AD (1994) Thermochemical tests of a kinetic-energy dependent exchange-correlation approximation, Int. J. Quant. Chem. Symp, 98: 625–632

203. Cortona P (1991) Self-consistently determined properties of solids without band-structure calculations, Phys. Rev. B, 44: 8454–8458

204. Wesolowski TA, Weber J (1996) Kohn-Sham equations with constrained electron density: An iterative evaluation of the ground-state electron density of interacting molecules, Chem Phys Lett, 248: 71–76

205. Wesolowski TA, Tran F (2003) Gradient-free and gradient-dependent approximations in the total energy bifunctional for weakly overlapping electron densities, J Chem Phys, 118: 2072–2080

206. Kevorkiants R, Dulak M, Wesolowski TA (2006) Interaction energies in hydrogen-bonded systems – a testing ground for subsystem formulation of density functional theory, J Chem Phys, 124: 024104

207. Wesolowski TA, Warshel A (1993) Frozen density-functional approach for ab-initio calculations of solvated molecules, J Phys Chem, 97: 8050–8053

208. Dulak M, Kevorkiants R, Tran F, Wesolowski TA (2005) Applications of the orbital-free embedding formalism to study electronic structure of atoms and molecules in condensed phase, CHIMIA, 59: 488–492

209. Wesolowski TA (2006) One-electron equations for embedded electron density: Challenge for theory and practical payoffs in multi-level modeling of complex polyatomic systems, in: Computational Chemistry: Reviews of Current Trends, vol. X, J. Leszczynski, (ed.) World Scientific, pp 1–82

210. Zumbach G, Maschke K (1985) Density-matrix functional theory for the N-particle ground state, J Chem Phys, 82: 5604–5607

211. Goedecker S, Umrigar CJ (1998) Natural orbital functional for the many-electron problem, Phys Rev Lett, 81: 866–869

212. Cioslowski J, Pernal K (1999) Constraints upon natural spin orbital functionals imposed by properties of a homogeneous electron gas, J. Chem. Phys, 111: 3396–3400

213. Mazziotti DA (2001) Energy functional of the one-particle reduced density matrix: a geminal approach, Chem Phys Lett, 338: 323–328

214. Buijse MA, Baerends EJ (2002) An approximate exchange-correlation hole density as a functional of the natural orbitals, Mol. Phys, 100: 401–421

215. Yasuda K (2002) Local approximation of the correlation energy functional in the density matrix functional theory, Phys Rev Lett, 88: 053001

216. Lathiotakis NN, Helbig N, Gross EKU (2005) Open shells in reduced-density-matrix-functional theory, Phys Rev A, 72: 030501

217. Gritsenko O, Pernal K, Baerends EJ (2005) An improved density matrix functional by physically motivated repulsive corrections, J Chem Phys, 122: 204102

218. Staroverov VN, Scuseria GE (2002) Assessment of simple exchange-correlation energy functionals of the one-particle density matrix, J Chem Phys, 117: 2489–2495

219. Cohen AJ, Baerends EJ (2002) Variational density matrix functional calculations for the corrected Hartree and corrected Hartree-Fock functionals, Chem. Phys. Lett, 364: 409–419

220. Pernal K, Cioslowski J (2002) Density matrix functional theory of weak intermolecular interactions, J Chem Phys, 116: 4802–4807

221. Tsuzuki S, Luthi HP (2001) Interaction energies of van der Waals and hydrogen bonded systems calculated using density functional theory: Assessing the PW91 model, J Chem Phys, 114(9): 3949–3957

222. Furche F (2001) Molecular tests of the random phase approximation to the exchange-correlation energy functional, Phys Rev B, 64(19): Art. No. 195120

223. Furche F, van Voorhis (2005) Flucluation-dissipateon theorem density – functional theory, J chem phys, 122(6): Art. No. 164106

CHAPTER 3

SELECTED MICROSCOPIC AND MEZOSCOPIC MODELLING TOOLS AND MODELS – AN OVERVIEW

MAGDALENA GRUZIEL[*1,2], PIOTR KMIEĆ[*1,2], JOANNA TRYLSKA[2] AND BOGDAN LESYNG[1,2]

[1] *Faculty of Physics, Warsaw University, Żwirki i Wigury 93, 02-089 Warsaw, Poland*
[2] *Interdisciplinary Centre For Mathematical And Computational Modelling, Pawińskiego 5a, 02-106 Warsaw, Poland*
[*] *The authors contributed equally to this work.*

Abstract: In order to model (bio)molecular systems and to simulate their dynamics one requires the potential energy functions at the *microscopic*, classical and/or quantum levels, as well as fast generators of the free-energy functions at the *mezoscopic* level. A brief overview of the methods which allow computations of the potential energy functions and the free energies is presented. The ongoing research is focused on designing molecular mezoscopic interaction potentials, applicable to nanoscale (bio)molecular systems, and on utilizing conformationally dependent atomic charges. In particular, the coupling of a fast quantum SCC-DFTB method with the Poisson-Boltzmann (PB) or Generalized Born (GB) models is discussed, and the role of the SCC-DFTB CM3 charges in computations of the mean-field electrostatic energies of molecular systems in real molecular environments is indicated. These charges reproduce very well molecular dipole moments, and are obtained from the Mülliken ones by applying a mapping procedure, using a quadratic function of the Mayer's bond orders. The PB and GB models give electrostatic reaction field energies of molecular environments, in particular, provide electrostatic contributions to the solvation energies. It is assumed that the solvation energy consists of the mean-field electrostatic and nonpolar (hydrophobic) energy contributions. Typically, the nonpolar term consists of the cavity formation free energy, and sometimes also of a mean van der Waals interaction energy of the molecular system with its environment. This allows to reproduce experimental solvation/hydration energies assuming different analytical forms of the nonpolar energy terms. Refined GB models, with new formulae for the Born radii are discussed. The nonpolar energies are quite well reproduced using the solvent accessible surface area (SASA), or a polynomial series depending on reciprocal values of the Born radii. Presence of the mean van der Waals energy on the quality of the fits is also discussed. Reliable mezoscopic models and theories play a key role in describing the functioning of nanoscale (bio)molecular systems

Keywords: microscopic models; density functional; SCC-DFTB; CM3 charges; mezoscopic models; Poisson-Boltzmann; Generalized Born; nonpolar interactions; solvent accessible surface area

W. A. Sokalski (ed.), Molecular Materials with Specific Interactions, 203–223.
© 2007 *Springer.*

1.   INTRODUCTION

Studies of complex (bio)molecular systems and materials span a broad range of temporal and spatial scales. Typically, quantum or quantum-classical dynamics describe proton and/or electron transfer processes in the time scales ranging from femtoseconds to picoseconds ($10^{-15} - 10^{-12}$ s). In turn, microscopic, atomic classical molecular dynamics simulations permit describing motions in the time scale of picoseconds to nanoseconds ($10^{-12} - 10^{-9}$ s). Motions of the entire molecular fragments, which can be described using effective, mezoscopic or coarse-grained dynamical models, are applicable in the time frame spanning a range from nanoseconds to microseconds ($10^{-12} - 10^{-9}$ s). Typically, longer time scales correlate with larger spatial scales of molecular objects. The diagram below presents temporal and spatial scales accessible by current simulation techniques.

Different scales presented in Figure 3-1 are related to different approximation levels. For an overview of conventional molecular modelling methods, (see e.g. [1−3]). Bridging the above mentioned disparate time scales for the description of biologically relevant collective motions requires hierarchical, multi-scale approaches. In practice, to describe real complex (bio)molecular or material systems and processes various models have to be coupled to each other. Selected coupling mechanisms will be briefly reviewed.

For the past several years we have been developing and applying quantum dynamics (QD) and quantum-classical molecular dynamics (QCMD) methods, which are based on the explicitly time-dependent Schroedinger equation. For an overview of the models and simulation results see e.g. [4] and the references cited

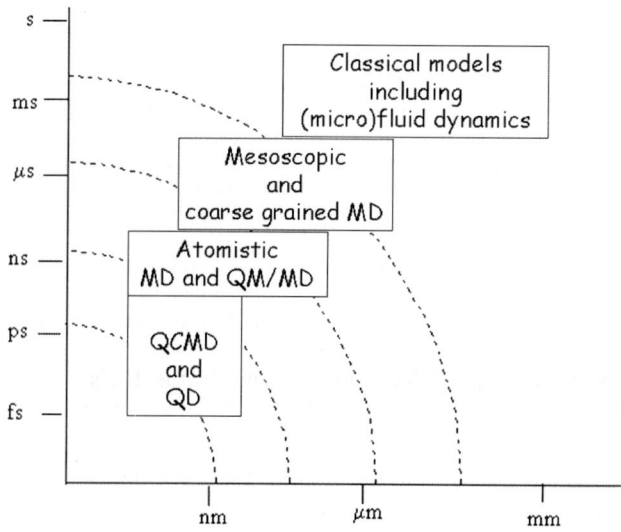

Figure 3-1. A diagram presenting spatial and temporal scales accessible by simulation methods

therein. Therefore, we will not be discussing these approaches. Also a very broad range of macroscopic models is outside the scope of this review. We will mostly concentrate on microscopic, atomistic and mezoscopic approaches, including the coarse-grained ones.

Both classical molecular mechanics (MM) and molecular dynamics (MD) with microscopic, atomistic representations are well established and routinely applied techniques. These will be briefly reported in Chapter 2.1. Mezoscopic coarse-grained models are presented in 2.2. An example of coupling of conventional MM or MD approaches with quantum mechanical (QM) ones will be discussed using the self-consistent charge density-functional tight-binding method (SCC-DFTB) as a well established test case. SCC-DFTB is widely and successfully applied in the studies of material and (bio)molecular systems. Quantum energy calculations allow determination of the potential energy function for molecular and macromolecular systems, which can be used simultaneously with MM, MD or QD simulations.

Influence of the molecular environment on the structure and dynamics of molecular subsystems will be outlined referring to the solvation free energy (Chapter 4). Implicit solvent models based on the Poisson-Boltzmann (PB) equation and the Generalized Born (GB) model is discussed in 5 and 6. The PB or GB models are used for studies of molecular electrostatic properties and allow proper assignments of positions of protons (hydrogen atoms) within the given (bio)molecular structure.

Mean-field hydrophobic interactions will be briefly presented as well. Finally, implementation of the mean-field electrostatic and hydrophobic potentials to an effective MD approach will be also presented.

The aim of challenging simulations is to study atomic/molecular objects and events which occur on spatial scales from nanometers to microns, and on a range of timescales from nanoseconds to milliseconds. Future perspectives will be briefly outlined.

2. MICROSCOPIC AND MEZOSCOPIC MOLECULAR MODELS

2.1. Atomic Resolution of Molecular Models and Microscopic Potential Energy Functions

A molecular system consists of electrons and nuclei. Their position vectors are denoted hereafter as r_{el} and q_α, respectively. The potential energy function of the whole system is $V(r_{el}, q_\alpha)$. For simplicity, we skip the dependence of the interactions on the spins of the particles. The nuclei, due to their larger mass, are usually treated as classical point-like objects. This is the basis for the so called Born-Oppenheimer approximation to the Schroedinger equation. From the mathematical point of view, the q_{nuc} variables of the Schroedinger equation for the electrons become the parameters. The quantum subsystem is described by the many-dimensional electron wave function $\Psi(r_{el}; q_\alpha)$.

$$H_{el}(r_{el}; q_{nuc}) \Psi(r_{el}; q_{nuc}) = U(q_{nuc}) \Psi(r_{el}; q_{nuc}) \tag{3-1}$$

Another possible description is given by the 3D electron density $\rho^{el}(r_{el}; q_{nuc})$ which is a scalar function of r_e and contains q_{nuc} as parameters. These two representations of the electron subsystem form the basis for the development of either conventional quantum chemistry methods or electron Density Functional Theory (DFT). The electron subsystem generates an effective potential, $U(q_{nuc})$, acting on the classical nuclei, which can be expressed as an average of the full potential V over the electron wave function Ψ, and written as:

$$U(q_{nuc}) = < \Psi|H_{el}|\Psi > . \qquad (3\text{-}2)$$

The possibility of assigning the effective potential to the molecular scaffold allows treatment of the molecular system as a mechanical object. This effective potential $U(q_{nuc})$ determines its atomic architecture (static properties) and atomic motions (dynamic properties).

One should note that the approximation presented above leads to a reduction of the dimensionality of the original potential and is a typical procedure applied in the studies of atomic and molecular systems. Integrating over the electronic degrees of freedom brings us from the "subatomic" to an "atomic" level. A further reduction of the dimensionality would lead to "mezoscopic" and "macroscopic" models, respectively. At the mezoscopic level, the objects under study are groups of atoms or groups of entire molecules. Macroscopic space dimensions are characteristic for cells and tissues and for all processes which involve those systems. One should note, however, that some molecular systems (or materials) and processes they involve cannot be assigned in a unique way to a given level mentioned above. For example, such objects like Bose-Einstein condensates are mezoscopic quantum systems existing in mezoscopic or macroscopic environments.

Popular microscopic theories and models are listed below:
- Molecular time-independent Quantum Mechanics (QM), including Density Functional Theory (DFT),
- Molecular time-dependent quantum mechanics, typically called Quantum Dynamics (QD),
- Classical Molecular Dynamics (MD),
- Classical Monte Carlo (MC),
- Quantum Monte Carlo (QMC),
- Classical Langevin Molecular Dynamics (LD),
- Classical (time-independent) Molecular Mechanics (MM), including multi-dimensional Energy Minimization (EM) techniques,
- Quantum-Classical Molecular Dynamics (QCMD).

Commonly applied mezoscopic models and theories are the following:
- Poisson-Boltzmann models (PB),
- Generalized Born models (GB),
- Brownian Dynamics (BD),
- Discrete or continuous coarse-grained models of (bio)polymers

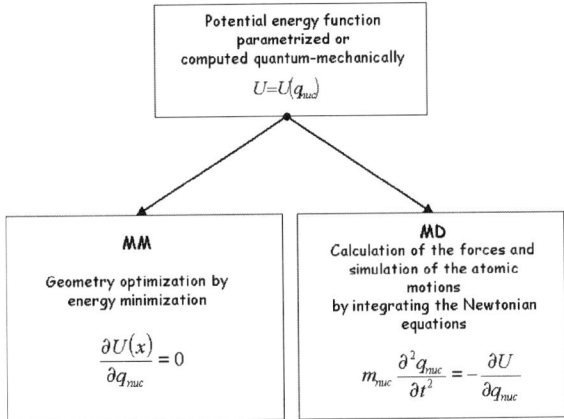

Figure 3-2. Molecular Mechanics (MM) and Molecular Dynamics (MD) methods. In coarse-grained models, groups of nuclei (atoms) are replaced by larger objects

Popular hybrid models:

- QM/MM models, with MM linked to a quantum method which generates the energy and forces for the molecular object or its fragment.

More detailed description of most of the above methods can be found in the book of Leach.[5]

In conventional MM the potential energy function (PEF) is parameterized and these optimized parameters are called "force-field" parameters. Such methods are widely applied in the studies of nucleic acids, proteins, their complexes and other biomolecular systems. A typical, simple force field of a molecular system is defined by the following equation:

$$U = V_{bond} + V_{ang} + V_{tors} + V_{impr} + V_{es} + V_{VdW}, \tag{3-3}$$

where the first four, so called bonded terms, are typically approximated with harmonic potentials for bond, bond-angle, dihedral and improper (out of plane) interactions. The V_{es} is the Coulombic pair-wise, charge-charge interaction, and V_{VdW} is the potential for the Van der Waals forces; both terms are classified as nonbonded ones.

Typically, MD equations of motion are integrated with a time-step of 1 femtosecond (10^{-15}s), which allows for overall simulation time of nanoseconds (10^{-9}s), and sometimes for smaller systems reaching microseconds (10^{-6}s). Longer simulations become numerically unstable. Coarse-grained models allow longer integration time-steps and, therefore, much longer simulation times than the microscopic ones, and can for example be applied to protein folding problems. Regardless of the molecular representation, in order to reach long simulation times one has to apply stable computational algorithms, in particular symplectic algorithms.[6] Another approach makes use of curvlinear degrees of freedom and applies the Lagrangian formalism.[7]

In order to obtain reliable simulation results of (bio)molecular systems at physiological conditions, protonation states of all ionizable groups have to be defined. This means that positions of all protons (hydrogen atoms) have to be properly assigned in the (bio)molecular structure. This requires application of the so called "computational titration" procedure. Such procedure is based on mezoscopic PB or GB models. Solving the PB differential equation allows for determination of local electrostatic fields inside and outside biomolecular structures, in particular in locations of mobile, dissociable protons. Solution of the PB equation requires the knowledge of the effective atomic charges, q_α, the dielectric scalar field, $\varepsilon = \varepsilon(\mathbf{x})$, resulting from molecular polarizabilities, and the concentration of ions present in a solvent surrounding the biomolecular structure. For an overview of this method and its applications to enzymes, see.[8,9] The GB model is a semi-analytical approximation to the PB model. It allows determination of the electrostatic energy of a set of atomic charges in a polarizable environment. The GB model is much faster than the PB one.

Biomolecular structures which are titrated using either the PB or GB methods are "electrostatically well-balanced". Therefore, they can further be studied using microscopic classical or quantum methods. However, since the titration procedure requires the optimal structure, the modelling procedure should be carried out in a self-consistent way. Such strategy is presented in Figure 3-3.

2.2. Coarse-grained Models for Biomolecules

Even though all-atom explicit solvent MD simulations are currently becoming feasible for larger systems and for time scales up to hundreds nanoseconds, a vast

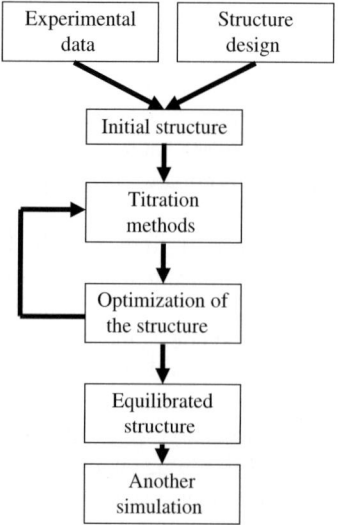

Figure 3-3. General scheme for modelling of biomolecular structures

number of interesting biological phenomena occur on the scales of micro- or milli-seconds. Additionally, such processes as translation, transcription or signal trans-duction involve complexes of macromolecules of nanoscale size. Due to the growing number of large structures available from X-ray crystallography, low resolution models of proteins and nucleic acids together with simplified force fields are needed to efficiently study their dynamical properties. These models involve reducing the degrees of freedom of the system by coarse-graining its representation; each residue is composed of one or few interacting centers/beads. However, lowering the level of atomic detail makes the formula for the effective potential energy (more accurately, a potential of mean force) very difficult to parameterize. Usually, such mezoscopic force fields are parameterized based on some reference configurations, therefore, their drawbacks may involve a bias towards a reference state and inability to transfer among different systems. But overall, the success of coarse-grained models in recent years caused growing interest in their further development. Rough classification and description of these models is presented in the following sections (for a more detailed reviews see e.g.[10,11]).

2.2.1.   Elastic Network Models

The simplest of these approaches includes Gaussian Network Models (GNM) or Elastic Network Models (ENM) which assume that the native state represents the minimum energy configuration. A structure is represented as a network of beads connected by harmonic springs.[12,13] One bead represents one residue and is usually centered on the position of the C_α carbon. Single parameter harmonic interactions are assigned to bead pairs which fall within a certain cutoff distance R_c. In case of proteins, R_c is usually around 8–10 Å. The representation of the molecule in the

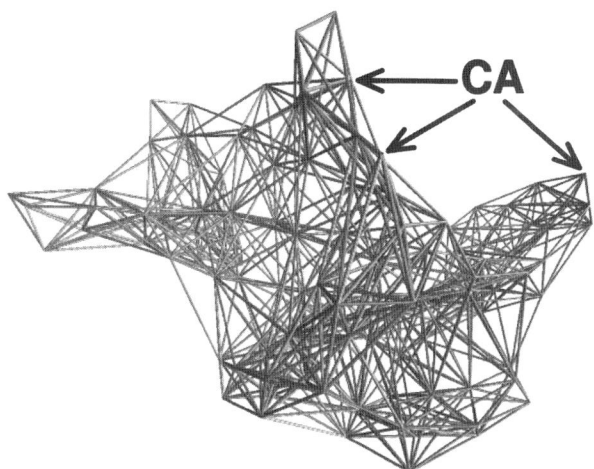

Figure 3-4. Representation of the molecule in the ENM. Each node represents the center of the bead which is the C_α carbon (CA) in case of one-bead protein models. The lines (pseudobonds) represent the harmonic interactions between beads that are within certain distance apart

ENM is presented in Figure 3-4. ENMs assume harmonicity of motions, and isotropic fluctuations with no directional preferences. Hence, the movements may be predicted only around a local minimum and dissociation of pseudobonds is not possible. The model is commonly used with Normal Mode Analysis techniques to predict principal modes, which are the ones that are believed to be functionally important. In spite of these approximations, the ENMs were able to predict global dynamics e.g., of the ribosome [14,15] or other macromolecules [16] and the swelling path of viruses [17] in accordance with experimental cryo-EM maps. They also succeeded in reproducing the X-ray crystallographic temperature factors (e.g. [18]). Many methods are recently being developed which are an extension of the ENM or GNM. An Anisotropic Network Model has been proposed which incorporates the anisotropy of motions. [19] Newer approaches involve a mixed level of coarse-graining in the ENM which enables to analyze different parts of the structure with different level of detail. [20]

2.2.2. Go-like models

Go model has been originally developed to study folding of proteins. [21,22] It is a one-bead model with no amino-acid specificity in which the native state is considered to be the energetic minimum. The bias toward a native configuration is maintained by setting attractive interactions between nonbonded beads that have a native contact i.e., whose mutual distance is below a predefined cutoff. These native contacts derived from the native conformation are stored in the contact matrix and are assumed stabilizing. Other nonbonded interactions, the so called non-native contacts, are set as repulsive. The force field may include bond, dihedral and van der Waals interactions depending on the type of the Go model. The model is used with molecular or Langevin dynamics, as well as with Monte Carlo simulations. [23–26] Go-type models have been successful in reproducing the thermodynamics and kinetics of folding because they represent a perfectly funneled toward the native state landscape. This indicates that many proteins have interactions that are minimally frustrated.

2.2.3. Knowledge-based potentials

In these models, the potential energy function is based on the molecular mechanics all-atom force field and includes the bond, angle, dihedral and non-bonded energy terms. The parameterization is based on the statistical analysis of sets of experimental structures. If a variable q describes a degree of freedom in the system (e.g., bond distances, angles, dihedrals) then, P(q), the probability distribution associated with this degree of freedom, is related to the potential of mean force, W(q), by the following equation

$$W(q) = -k_B T \ln(P(q)) + \text{constant} \tag{3-4}$$

where k_B is the Boltzmann constant and T temperature. For the derivation of the potential of mean force see Section 4.1. W(q) coincides with the potential energy associated with the variable q only in case of a single degree of freedom or if the degrees of freedom are uncorrelated. However, W(q) is often used as an approximation to the potential energy in the so-called knowledge-based force fields.[27,28] With such force fields the motions are often explored using molecular or Langevin dynamics.

A class of one-bead models which were based on classical molecular mechanics force fields including bonded and nonbonded energy terms were proposed originally for studies of proteins.[29,30,31] Other one-bead models are currently being developed allowing for simulations of global conformational changes where the bias towards the starting configuration is maintained only for local interactions. The one-bead model has been recently able to reproduce such large fluctuations as the flap opening in the HIV-1 protease[32] and to predict the decrease in flap opening time due to certain mutations.[33] The global motions of the ribosome were also explored by a one-bead model with half a microsecond molecular dynamics.[34] The use of anharmonic Morse potentials in the force field allowed for larger fluctuations from the starting conformation.

Two-bead models were originally proposed by Levitt.[35] A residue is represented by two centers, in case of proteins one is placed on C_α carbon and the other on the centroid of the side-chain. The parameters are amino-acid specific and based on all-atom force fields. The model of Bahar and Jernigan includes in the energy formula also the angle-dihedral correlations.[28] A similar model with parameterization based on the all-atom simulations for oligopeptides was developed by Scheraga group (see[36] and references therein). A two-bead model for the DNA with two beads representing a nucleotide was also proposed.[37] The base-base breakable interactions represent the hydrogen bonds of pairing nucleotides therefore the model was able to correctly reproduce DNA denaturation process. A two-bead model for the RNA, with one bead placed on the position of the phosphorus atom and additional bead placed in RNA helical regions, to fill the volume in the center of the helix, was applied to the small ribosomal subunit assembly.[38,39]

The multiple-bead models represent coarse-graining to a lesser extent. There are a few developed multiple-bead models for proteins.[40,41] The advantage of these models is that they treat explicitly the atoms of the protein backbone allowing for the description of hydrogen bonds. The multiple-bead models for DNA or RNA are also a subject of studies.[42–44]

In case of the dynamics, the simplification of the models and force fields allows to reach the spatial and temporal scales which are close to biological ones. However, one must be careful to choose an appropriate coarse-grained model in order to get rid of only those degrees of freedom that are not relevant to the problem under study. Future directions for the reduced biological models will include focusing on making the force field the most transferable with the least set of parameters involved.

3. SCC-DFTB METHOD AND CM3 CHARGES

The SCC-DFTB is a very fast density functional method applicable to large (bio)molecular systems.[45] Current implementations of this approach allow to carry out quantum-mechanical calculations for molecular systems containing several hundred atoms. In order to precisely describe electrostatic interactions between the (bio)molecular object and its environment, the atomic charges of the object should precisely reproduce experimental molecular dipole moments. The conventional Mülliken atomic SCC-DFTB charges are defined as:

$$q_k^{Mulliken} = \sum_{\mu \in k} (PS)_{\mu\mu} \tag{3-5}$$

where P is the density matrix, and S is the overlap matrix. Although dipole moments obtained from the Mülliken charges are typically of poor quality, the error contributed from each type of bond is systematic. It turns out that systematic corrections for the bond dipole moments for each type of bond can be designed. Such a simple idea works very well.[46] The procedure of mapping the SCC-DFTB Mülliken charges to the charge model 3 (CM3) is described in detail in,[47] and is given by the formulae:

$$q_k^{CM3} = q_k^{Mulliken} + \sum_{k' \neq k} T_{kk'} (B_{kk'}) \tag{3-6}$$

where $T_{kk'}$ is a quadratic function of the Mayer's bond orders $B_{kk'}$:

$$T_{kk'} = D_{Z_k Z_{k'}} B_{kk'} + C_{Z_k Z_{k'}} (B_{kk'})^2 \tag{3-7}$$

The C and D coefficients are parameters fitted to reproduce a set of experimental dipole moments. The atomic CM3 charges are conformationally dependent. They can either be averaged over a number of representative conformations and fixed, or computed "on the fly" for example, in the course of MD simulations.

Since the CM3 charges reproduce the dipole moments very well, they can reproduce also other electrostatic properties. These charges can, in particular, be used in the PB and GB models (see next sections). These models provide electrostatic reaction field energies of the molecular environment, in particular, giving electrostatic contributions to solvation energies.

4.   SOLVATION FREE ENERGY

Once the computational model of the molecule is created, it is of most interest to study its properties in the natural environment, in particular, water solvent. Surrounding the molecule with water, allows us to study the solvation process. Like molecules, the solvent may be also described with different levels of accuracy. Beginning with all-atom models of water,[48,49] which allow for the studies of solvent structure around solutes but are time consuming and the results are model dependent, to continuous dielectric models,[50–52] which are faster but less accurate and give no knowledge about the solvent itself. Thus, the difference in the level of description for both models is either an advantage or a drawback. These models are commonly known as explicit or implicit solvent models, respectively.

One of the crucial parameters describing the solvation phenomena is the free energy change. The main idea in most implicit solvation models is the decomposition of the solvation free energy, ΔG_{solv} into the electrostatic and nonpolar part,

$$\Delta G_{solv} = \Delta G_{el} + \Delta G_{np}, \tag{3-8}$$

The electrostatic term (ΔG_{el}) results from the decomposition of the solvation process into the work performed to discharge the solute in vacuum and the work performed to recharge it again in the solvent. The nonpolar term (ΔG_{np}) arises from the energetic cost of the insertion of the solute shaped cavity into continuum solvent. The most accurate solution for the ΔG_{el} brings us to the solution of the Poisson-Boltzmann equation.[53,54] However, for large systems and for studies of the dynamics, more useful is the Generalized Born[55,56] method because of its relatively small computer power and time requirements, and its reliability despite the simplicity of the model. Both GB and PB methods will be described in more detail later in this section.

The nonpolar (or hydrophobic) component, ΔG_{np} may be decomposed into a surface component, which is proportional to the solvent accessible surface area (SASA) of the solute and the component representing the solute-solvent non-electrostatic interactions,[57]

$$\Delta G_{np} = \Delta G_{SASA} + \Delta G_{VdW}. \tag{3-9}$$

The methods to calculate the surface component of the nonpolar part of the solvation free energy have been greatly evolving during the recent years. From the most simple models, assuming that the ΔG_{SASA} is proportional to SASA,[58] to empirical models based on weighted solvent accessible surface area[59] (WSAS) or atomic surface tensions[50] (SMx models of Cramer and Truhlar). The relatively new analytic implicit solvent model (AGBNP),[57] uses simple weighted SASA but also accounts for the van der Waals solute-solvent interactions. A more detailed description of the latter model will be given later in this section. The above mentioned work of Cramer and Truhlar[50] gives a very extensive description of implicit solvent models with an enormous list of references.

Even though fast, implicit solvent approach does not give any knowledge about the solvent structure around the solute. This kind of information can be obtained only from explicit solvent simulations. Such explicit system has many degrees of freedom and thus fluctuates. It is therefore described on a microscopic level with the tools of statistical physics. The Gibbs free energy is then given by the following formula:

$$G = -k_B T \ln Z, \tag{3-10}$$

$$Z = \int \exp(-\beta \mathcal{P} V) dV \int dp^{3N} \, dq^{3N} \exp(-\beta H(q^{3N}, p^{3N})) \tag{3-11}$$

where Z is a partition function in the isobaric-isothermic ensemble, p and q stand for the positions and momenta of elements of the system, H is the hamiltonian of the system, $\mathcal{P}$ and $\mathcal{V}$ are the pressure and volume of the system. The integrations are taken over the whole phase space $\Gamma(q^{3N}, p^{3N})$ and over the volume from 0 to infinity. For real systems such as liquid solutions, integral in the equations (3–11) cannot be analytically solved. From computer simulations of the system with explicit solvent, one may though estimate the free energy change, ΔG (rather than absolute value of G) using one of the common approaches: thermodynamic integration, free energy perturbation, potential of mean force[60] (PMF), Landau free energy[61–63] (LFE) or Widom particle insertion method.[64,65] The PMF and LFE methods will be described in the following section. It is worth noting, that the results obtained from the simulations are generally dependent on the choice of the water model.[48,49,66]

In between the implicit and explicit solvent models, there are mixed models, such as the solvation shell approximation.[67–69] This model describes explicitly only the first solvation shell molecules and treats as implicit the solvent region beyond the first solvation shell. Such treatment both provides the information about the solvent structure near the solute and allows for faster computation.

4.1. Potential of Mean Force and Landau Free Energy

Both the PMF and LFE express the free energy as a function of a certain degree of freedom. In case of PMF, as it has already been mentioned, it is usually one of the configurational degrees of freedom. The LFE is a function of arbitrary order parameters which form the additional degrees of freedom.

If the free energy function argument is ξ, then the probability density of ξ is described by the following equation:

$$\rho(\xi) = Z_\xi^{-1} \times \left[\int \exp(-\beta \mathcal{P} \mathcal{V}) d\mathcal{V} \int \exp(-\beta V(q^{3(N-m)}, \xi)) dq^{3(N-m)} \right]$$

(3-12)

where Z_ξ is a configurational partition function of the full $(q^{3(N-m)}, \xi)$ system (the momenta are skipped in the above and following equations because they are either not considered as in the Monte Carlo simulations or do not contribute, in general, to the free energy change as in MD simulations):

$$Z_\xi = \int \exp(-\beta \mathcal{P} \mathcal{V}) d\mathcal{V} \iint \exp(-\beta V(q^{3(N-m)}, \xi)) dq^{3(N-m)} d\xi \quad (3-13)$$

Further analysis will be divided into two topics, PMF and LFE.

For the PMF ($m > 0$, $\xi = q^{3m}$), Z_ξ becomes simply Z, a configurational partition function of a system of N particles. In a special case, if $m = 2$, then q^{3m} for the pairwise potential V may be denoted as R – the distance between two chosen particles. Using this for calculating the intergral of Eq. (3-12) denoted in parentheses, leads to the following form:

$$\rho(R) = \exp(-\beta W(R)) \times Z^{-1} \quad (3-14)$$

where $W(R)$ is the PMF, $\rho(R)$, after multiplying it by the factor of $\rho^2/(N(N-1))$ (where ρ is a bulk density of a system considered), is a radial distribution function, usually denoted as $g(R)$. Thus the PMF may be written as:

$$W(R) = -k_B T \ln g(R) + \text{constant,} \qquad (3\text{-}15)$$

where $g(R)$ corresponds to $P(q)$ of equation (3-4) in case of R being the distance between particles.

For Landau free energy ($m = 0$, ξ – the order parameter), Z_ξ is a partition function of an extended system with additional variable ξ. The integral denoted in square brackets in Eq. (3-12) is simply the configurational partition function of the system with a fixed value of ξ-Z. The statistical definition of the Gibbs free energy function combined with Eq (12) results in the following expression:

$$G(\xi) = -k_B T \ln Z = -k_B T \ln \rho(\xi) - k_B T \ln Z_\xi = -k_B T \ln \rho(\xi) + \text{const.} \qquad (3\text{-}16)$$

In practice, it is rarely possible to sample reasonable range of ξ within single simulation. To enhance sampling one usually adds a biasing potential, which directs the sampling to the desired range but also reduces the variance allowing for a shorter simulation time (smaller sample but the same error). Because one changes the Hamiltonian by adding the biasing potential, one has to subtract it from the free energy $G(\xi)$,

$$G(\xi) = -k_B T \ln \rho_{bias}(\xi) - V_{bias}(\xi) + \text{const} \qquad (3\text{-}17)$$

where ρ_{bias} is the probability density function in the biased ensemble and V_{bias} is some biasing potential.

Besides "unbiasing", one also has to reweigh any averaged values:

$$\langle Y \rangle = \langle Y \exp(-\beta V_{bias}(\xi)) \rangle_{bias} \times [\langle \exp(-\beta V_{bias}(\xi)) \rangle_{bias}]^{-1} \qquad (3\text{-}18)$$

where $\langle Y \rangle$ is the ensemble average of Y and $\langle Y \rangle_{bias}$ is the average of Y in the biased ensemble.

When the range of ξ to be sampled is too large, one divides this range into smaller overlapping windows and performs consecutive simulations in each window. This approach forms the basis for the widely used umbrella sampling[70] method.

In Figure 3-5 we present an example[71] of LFE as a function of solute parameters: charge – q, Lennard-Jones diameter – σ, or Lennard-Jones potential well-depth – ε. Figure 3-5 shows examples of $G(\sigma)$ and $G(\varepsilon)$ derived from the results of Monte Carlo simulations of single solutes in explicit solvent together with fitted functions. Each curve was created *via* umbrella sampling. Such parameterization of the free energy, $G(\sigma)$, (and $G(q)$ – not presented here), was also widely reported in the literature.[61,72–75]

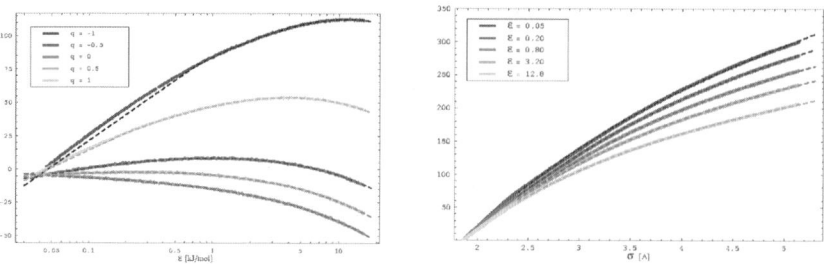

Figure 3-5. Free energy as a function of the solute parameters ε and σ; a) $G(\varepsilon)$ – left plot, fixed $\sigma = 0.35$ nm, the five curves correspond to different fixed charge values denoted in the legend, b) $G(\sigma)$ – right plot, fixed q $= +1e$, the five curves correspond to different fixed ε values given in the legend. All plots show the data acquired in the Monte Carlo simulations together with fitted functions

5.   OUTLINE OF THE POISSON-BOLTZMANN (PB) MODEL

The electrostatic field in the stationary state is described by the Poisson-Boltzmann equation. The PB model constitutes the fundamental equation of electrostatics and is based on the differential Poisson equation which describes the electrostatic potential $\Phi(\mathbf{r})$ in a medium with a charge density $\rho(\mathbf{r})$ and a dielectric scalar field $\varepsilon(\mathbf{r})$:

$$\nabla(\varepsilon(\vec{r})\ \nabla\phi(\vec{r})) = -4\pi\rho(\vec{r}) \qquad (3\text{-}19)$$

In the mezoscopic model the system is composed of two regions described by an interior and exterior charge density (see Figure 3-6). The interior region, represents a (bio)molecule as a collection of fixed point charges placed at positions of atoms $\vec{r}_i$. A low dielectric constant, typically between 2 and 20 is assigned to this region. The exterior is modeled in a continuum manner by surrounding the molecule with an implicit solvent representation characterized by a high dielectric value, typically around 80. To account for the distribution of the ions around the molecule, the mobile ion density is approximated by a Boltzmann distribution at temperature T.

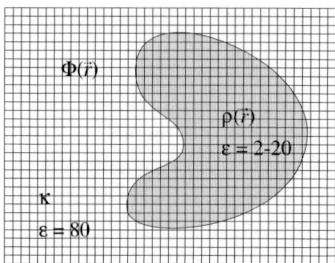

Figure 3-6. A diagram of mapping of the molecular system onto a grid in the PB model. The molecule represented with a fixed point charges is depicted in gray. The exterior is described by parameter κ dependent on the ionic strength. The electrostatic potential is solved in the nodes of the 3D grid

Assuming the above, the Poisson equation takes the following form:

$$\nabla(\varepsilon(\vec{r})\nabla\phi(\vec{r})) - 4\pi \sum_i q_i\delta(\vec{r} - \vec{r}_i) + \lambda(\vec{r})\overline{\kappa}^2 \sinh(\phi(\vec{r})/k_B T) \qquad (3\text{-}20)$$

where λ is 0 inside the molecule and 1 elsewhere, Δ is the Dirac function. The $\overline{\kappa}^2$ is proportional to the ionic strength of the exterior I, and is related to the Debye-Hückel parameter $(\overline{\kappa}^2 = \kappa^2/\varepsilon)$ defined as

$$\overline{\kappa}^2 = \frac{8\pi e^2 N_A I}{1000 k_B T} \qquad (3\text{-}21)$$

where N_A is the Avogadro number and e is the charge of the electron.

Finite difference method (FD) and finite element method (FE) are both well known algorithms for solving the PB equation. More detailed description of this equation is available in.[53,54] The methods mentioned above are commonly used in various software packages, for example in the University of Houston Brownian Dynamics[76] (UHBD) and Delphi[77] programs which implements FD algorithm, and in the Adaptive Poisson-Boltzmann Solver (APBS)[78] which implements as well the multilevel finite element method. Solving the PB equation enables one for example to visualize the electrostatic potential around the molecule and to obtain the electrostatic energy of the system.

6. OUTLINE OF THE GENERALIZED BORN (GB) MODEL

Unfortunately, due to large computational costs, calculations based on numerical solutions of the PB equation are of limited applicability. Therefore, several approximations to the original Born model[79] have been developed. The GB model is one of the semi-analytical approaches. In the GB model[55,56] each atom is represented as a sphere with the radius ρ and with a charge q placed at its center. In this model, the energetic cost of transferring the charge distribution into the solvent (the electrostatic part of the solvation energy) is defined as:

$$\Delta G_{elec} = -\frac{1}{2}\left(\frac{1}{\varepsilon_p} - \frac{1}{\varepsilon_w}\right) \sum_{i,j} \frac{q_i q_j}{\sqrt{r_{ij}^2 + R_i R_j \exp\left(\frac{-r_{ij}^2}{4R_i R_j}\right)}} \qquad (3\text{-}22)$$

where r_{ij} is the distance between atom i and atom j, ε_p and ε_w are the dielectric constant of the interior and exterior of the molecule, respectively. Denominator is a smooth, analytical function, which fulfills two limiting conditions: when the two atoms are infinitely separated $r_{ij} \to \infty$, and when the atoms are totally overlapped $r_{ij} = 0$. R_i are the so called effective Born radii, and reproduce the electrostatic free energy of a molecule when the charge of the atom i is "turned on".

$$\Delta G^i_{single} = -\frac{1}{2}\left(\frac{1}{\varepsilon_p} - \frac{1}{\varepsilon_w}\right)\frac{q_i^2}{R_i} \qquad (3\text{-}23)$$

6.1.   Models for Computing the Generalized Born Radii

A number of popular methods for computing the GB radii were proposed. The most popular one is based on the Coulomb field approximation, in which the generalized Born radii are defined by the equation:

$$\frac{1}{R_i} = \frac{1}{4\pi} \int\limits_{solvent} \frac{1}{r^4} d^3\vec{r} \qquad (3-24)$$

Exchanging the integral areas from the solvent area to the solute one, it is possible to get a more convenient form for calculations[55]:

$$\frac{1}{R_i} = \frac{1}{\rho_i} - \frac{1}{4\pi} \int\limits_{solute} \theta(|\vec{r}| - \rho_i)\frac{1}{r^4} d^3\vec{r} \qquad (3-25)$$

where Θ is a step function, and ρ_i is the VdW radius of atom i. Another method to compute the GB radius is based on the Kirkwood model and was recently proposed by Grycuk.[80] In this case, the GB radii are defined as:

$$\frac{1}{R_i} = \left(\frac{3}{4\pi} \int\limits_{solvet} \frac{1}{r^6} d^3\vec{r} \right)^{\frac{1}{3}} \qquad (3-26)$$

A few analytical formulae to compute the GB radii in the above two models were proposed. Recently Wojciechowski and Lesyng[81] proposed a generalization of this model mapping the exponents of 6 and 1/3 into n and 1/(n–3), respectively. In their model, the parameter n depends on the ratio of the solvent and solute dielectric constant. At present, this model reproduces the PB energy in the best way.

One should note also that Feig and others proposed another method to compute the GB radii, depending on the solvent and solute dielectric constants[82,56]:

$$R_i = -\frac{1}{C_oA_4 + C_1\left(\frac{3\varepsilon_{ex}}{3\varepsilon_{ex}+2\varepsilon_{in}}\right)A_7} + D + \frac{E}{\varepsilon_{ex}+1} \qquad (3-27)$$

7.   NONPOLAR CONTRIBUTION TO THE FREE ENERGY OF SOLVATION

Models for which the parameterizations are based on the solvent accessible surface area (SASA) are widely known in literature.[58] The nonpolar component of the free energy of solvation is described in these models as a polynomial of A_i, where A_i is the SASA of the atom i. A very good example of such approach is the Surface Generalized Born/Nonpolar Model (SGB/NP), proposed by Gallicchio and Levy.[83] The nonpolar contribution is expressed as:

$$\Delta G_{np} = \sum_{i=1}^{N}[\gamma(t_i)A_i + \alpha(t_i)] \qquad (3-28)$$

where $\gamma(k_i)$ and $\alpha(k_i)$ are adjustable parameters, which are dependent on the atom type (k_i). The first term $\gamma(k_i)A_i$ in the equation (3-8) is proportional to the SASA of the solute, and the γ parameters are interpreted as surface tension coefficients. The AGBNP[57] is the second and more flexible model of fitting the nonpolar contribution G_{np} to the free energy of solvation, proposed by the same authors. The nonpolar part of the hydration free energy G_{np} is divided into two terms:

$$\Delta G_{np} = \Delta G_{cav} + \Delta G_{vdW} = \sum_{i=1}^{N}\left[\gamma(t_i)A_i + \alpha(t_i)\frac{a_i}{(B_i + R_W)^3}\right] \qquad (3\text{-}29)$$

The first term (ΔG_{cav}) is connected with the surface area model, and the γ parameter has the same physical meaning as the one in the SGB/NP model. The second term (ΔG_{vdW}) is the solute-solvent van der Waals free energy term, which introduces another set of parameters. The first one, B_i corresponds to the Born radius of the atom i, the second, $R_w = 1.4\,\text{Å}$ is the radius of the water molecule, the next one $a_i = -\frac{16}{3}\pi\rho_w\varepsilon_{iw}\sigma_{iw}^6$, where $\sigma_{iw} = \sqrt{\sigma_i\sigma_w}$, $\varepsilon_{iw} = \sqrt{\varepsilon_i\varepsilon_w}$ and $\rho_w = 0.033428\,\text{Å}$, is the number density of water at standard conditions. ρ_w and σ_w are the OPLS Lennard-Jones parameters for the atom i, $\varepsilon_w = 0.155\text{kcal/mol}$, and $\sigma_w = 3.15365\,\text{Å}$ are the Lennard-Jones parameters of the TIP4P water oxygen. The last, dimensionless parameter $\alpha(k_i)$ is fitted in this model.

8. CONCLUSIONS

We briefly overviewed some promising modelling methods and simulation techniques, being developed and applied in the studies of large (bio)molecular systems. The coherent use of the interdisciplinary microscopic and mezoscopic ("coarse-grained") approaches allows for a more precise computation of the free energy of (bio)molecular systems, and for simulations of their dynamics at the mezoscopic level. This in turn allows for a better description of specific (bio)molecular recognition processes and spontaneous formation of functional (bio)molecular and nano-structures, amongst them proteins, nucleic acids, protein-RNA or protein-DNA complexes, and nanotubes. Merging the techniques for various time and spatial scales enables better description and understanding of functioning of enzymes or biomolecular complexes such as e.g. ribosomes. These methods may also help to design functional nanomachines, based on the knowledge of the existing "biomolecular solutions", which were selected in the evolutionary process. The newly developed methodologies should also allow for better description of energy transfer processes and molecular memory mechanisms.

ACKNOWLEDGEMENTS

These studies were supported by the BST funds 115/E-343/BST-1076/ICM/2005 of ICM and BST-975 of Department of Biophysics, Warsaw University, as well as by the European CoE for Multiscale Modelling, Bioinformatics and Applications (MAMBA, QLRI-CT-2002-90383).

REFERENCES

1. Lesyng B, McCammon JA (1993) Molecular Modeling Methods, Basic Techniques and Challenging Problems. Pharmac. Ther 60(149):167

2. Lesyng B, Rudnicki W (2003) Molecular Modelling in Drug Design, in Optimization of Aerosol Drug Delivery. Kluwer, Dordrecht, 23–48.

3. McCammon, JA, Harvey SC (1987) Dynamics of Proteins and Nucleic Acids. Cambridge University Press, New York.

4. Bala P, Grochowski P, Lesyng B, McCammon JA (1995) Quantum Classical Molecular Dynamics. Models and Applications. In: Field M (ed) Quantum Mechanical Simulation Methods for Studying Biological Systems. Les Houches Physics School Series, Springer Verlag and Les Houches Editions de Physique, 119–156

5. Leach AR (2001) Molecular Modelling. Principles and Applications, Addison Wesley Longman Limited 1996 2nd edn. Prentice Hall

6. Grochowski P, Lesyng B (2003) Extended Hellmann-Feynman Forces, Canonical Representations, and Exponential Propagators in the Mixed Quantum-Classical Molecular Dynamics. J. Chem. Phys. 119:11541–11555

7. Rudnicki W, Bakalarski G, Lesyng BA (2000) Mezoscopic Model of Nucleic Acids. Part 1. Lagrangian and Quaternion Molecular Dynamics. J. Biomol. Struct. and Dynamics 17:1097–1108

8. Briggs JM, McCammon JA (1992) Computation Unravels Mysteries of Molecular Biophysics, Comp. Phys. 6:238–243

9. Antosiewicz J, Błachut-Okrasińska E, Grycuk T, Lesyng BA (2000) Correlation Between Protonation Equilibria in Biomolecular Systems and their Shapes: Studies Using a Poisson-Boltzmann Model, Math. Sciences & Applications, GAKUTO International Series, 14, pp 11–17

10. Lattanzi G (2004) Application of coarse-grained models to the analysis of macromolecular structures. Comput. Materials Science 30:163–171

11. Tozzini V (2005) Coarse-grained models for proteins. Curr. Opin. Struct. Biol. 15:144–150

12. Tirion M (1996) Large amplitude elastic motions in proteins from a single-parameter, atomic analysis. Phys. Rev. Lett. 77:1905–1908

13. Haliloglu T, Bahar I, Erman B (1997) Gaussian dynamics of folded proteins. Phys. Rev. Lett. 79:3090–3093

14. Tama F, Valle M, Frank J, Brooks CL III (2003) Dynamic reorganization of the functionally active ribosome explored by normal mode analysis and cryo-electron microscopy. Proc. Natl. Acad. Sci. USA 100:9319–9323

15. Wang Y, Rader AJ, Bahar I, Jernigan RL (2004) Global ribosome motions revealed with elastic network model. J. Struct. Biol. 147:302–314

16. Chacon P, Tama F, Wriggers W (2003) Mega-dalton biomolecular motion captured from electron microscopy reconstructions. J. Mol. Biol. 326:485–492

17. Tama F, Brooks CL III (2002) The mechanism and pathway of ph induced swelling in cowpea chlorotic mottle virus. J. Mol. Biol. 318:733–747

18. Bahar I, Atilgan AR, Erman B (1997) Direct evaluation of thermal fluctuations in proteins using a single parameter harmonic potential. Fold. Des. 2:173–181

19. Atilgan AR, Durell SR, Jernigan RL, Demirel MC, Keskin O, Bahar I (2001) Anisotropy of fluctuations dynamics of proteins with an elastic network model. Biophys. J. 80:505–515

20. Kurkcuoglu O, Jernigan RL, Doruker P (2004) Mixed levels of coarse-graining of large proteins using elastic network model succeeds in extracting the slowest motions. Polymer 45:649–657

21. Ueda Y, Taketomi H, Go N (1978) Studies on protein folding, unfolding and fluctuations by computer simulation. A three-dimensional lattice model of lysozyme. Biopolymers 17:1531–1548

22. Go N, Abe H (1981) Noninteracting local-structure model of folding and unfolding transition in globular proteins. I. formulation, Biopolymers 20:991–1011

23. Hoang TX, Cieplak M (2000) Sequencing of folding events in go-type proteins, J. Chem. Phys. 113:8319–8328

24. Cecconi F, Micheletti C, Carloni P, Maritan A (2001) Molecular dynamics studies on hiv-1 protease drug resistance and folding pathways. Proteins, Struct., Funct., Genet. 43:365–372

25. Cieplak M, Hoang TX, Robbins MO (2002) Folding and stretching in a go-like model of titin. Proteins, Struct., Funct., Genet. 49:114–124

26. Faisca PFN, da Gama MMT (2005) Native geometry and the dynamics of protein folding. Biophys. Chem. 115:169–175

27. Sippl MJ (1995) Knowledge-based potentials for proteins. Curr. Op. Struct. Biol. 5:229–235

28. Bahar I, Jernigan RL (1997) Inter-residue potentials in globular proteins and the dominance of highly specific hydrophilic interactions at close separation. J. Mol. Biol. 266:195–214

29. Levitt M, Warshel A (1975) Computer simulation of protein folding. Nature 253:694–698

30. McCammon JA, Northrup SH, Karplus M, Levy RM (1980) Helix-coil transitions in a simple polypeptide model. Biopolymers 19:2033–2045

31. Bahar I, Kaplan M, Jernigan RL (1997) Short-range conformational energies, secondary structure propensities, and recognition of correct sequence-structure matches. Proteins 29:292–308

32. Tozzini V, McCammon JA (2005) A coarse-grained model for the dynamics of the flap opening in hiv-1 protease. Chem. Phys. Lett. 413:123–128

33. Chang C, Shen T, Trylska J, Tozzini V, McCammon JA (2006) Gated binding of ligands to HIV-1 protease, Brownian dynamics simulations in a coarse-grained model. Biophys. J. Vol. 90, pages 3880–3885

34. Trylska J, Tozzini V, McCammon JA (2005) Exploring global motions and correlations in the ribosome, Biophys. J. 89:1455–1463

35. Levitt M (1976) A simplified representation of protein conformations for rapid simulation of protein folding. J. Mol. Biol. 104:59–107

36. Liwo A, Khalili M, Scheraga HA (2005) Ab initio simulations of protein-folding pathways by molecular dynamics with the united-residue model of polypeptide chains. Proc. Natl. Acad. Sci. USA 102:2362–2367

37. Drukker K, Wu G, Schatz GC (2001) Model simulations of DNA denaturation dynamics. J. Chem. Phys. 114:579–590

38. Stagg SM, Mears JA, Harvey SC (2003) A structural model for the assembly of the 30S subunit of the ribosome, J. Mol. Biol. 328:49–61

39. Cui Q, Case DA (2005) Low-resolution modeling of the ribosome assembly of the 30s subunit by molecular dynamics simulations, Abstracts of Papers of the American Chemical Society, 229:U779

40. Nguyen HD, Hall CK (2004) Molecular dynamics simulations of spontaneous fibril formation by random-coil peptides. Proc. Natl. Acad. Sci. USA 101:16180–16185

41. Fujitsuka Y, Takada S, Luthey-Schulten ZA, Wolynes PG (2004) Optimizing physical energy functions for protein folding. Proteins, Struct., Funct., Genet. 54:88–103

42. Zhang F, Collins MA (1995) Model simulations of DNA dynamics, Phys. Rev. E 52:4217–4224

43. Maciejczyk M, Rudnicki WR, Lesyng BA (2000) mezoscopic model of nucleic acids. part 2. an effective potential energy function for DNA, J. Mol. Struct. and Dynamics 17:1109–1115

44. Tepper HL, Voth GA (2005) A coarse-grained model for double-helix molecules in solution, Spontaneous helix formation and equilibrium properties. J. Chem. Phys. 122:124909

45. Frauenheim T, Seifert G, Elstner M, Hajnal Z, Jungnickel G, Porezag D, Suhai S, Scholz R (2000) A Self-Consistent Charge Density-Functional Based Tight-Binding Method for Predictive Materials Simulations in Physics, Chemistry and Biology Phys. Stat. Sol. (b) 217:41

46. Li J, Zhu T, Cramer CJ, Truhlar DG (1998) J. Phys. Chem. A, New Class IV Charge Model for Extracting Accurate Partial charges from Wave Functions 102:1820–1831

47. Kalinowski JA, Lesyng B, Thompson JD, Cramer CJ, Truhlar DG (2004) Class IV Charge Model for the Self-Consistent Charge Density-Functional Tight-Binding Method. J. Phys. Chem A, 108:2545–2549

48. Chaplin M http//www.lsbu.ac.uk/water/

49. Guillot B (2002) A reappraisal of what we have learnt during three decades of computer simulations on water. J. Mol. Liq. 101:219–260

50. Cramer CJ, Truhlar DG (1999) Implicit Solvation Models, Equilibria, Structure, Spectra and Dynamics. Chem. Rev. 99:2161–2200

51. Zhang L, Gallicchio E, Levy RM (1999) Implicit Solvent Models for Protein-Ligand Binding, Insights Based on Explicit Solvent Simulations, (AIP Conference Proceedings, Simulation and Theory of Electrostatic Interactions in Solutions), 492:451–472

52. Roux B (2001) Implicit solvent models. In: Becker O, MacKerrel AD, Roux B, Watanabe M (eds) Computational Biophysics. Marcel Dekker Inc, New York

53. Davis ME, McCammon JA (1990) Electrostatics in biomolecular structure and dynamics. Chem. Rev. 90:509–521

54. Sharp KA, Honig B, (1990) Calculating Total Electrostatics energies with the Nonlinear Poisson-Boltzmann Equation, J. Phys. Chem. 94:7684–7692

55. Onufriew A, Bashford D, Case DA (2000) Modification of Generalized Born Model Suitable for Macromolecules J. Phys. Chem. B 104:3712

56. Bashford D, Case DA (2000) Generalized Born Models of Macromolecular Solvation Effects Annu. Rev. Phys. Chem. 51:129–152

57. Gallicchio E, Levy RM (2004) AGBNP, An Analytic Implicit Solvent Model Suitable for Molecular Dynamics Simulations and High-Resolution Modeling. J. Comput. Chem. 25:479–499

58. Lee B, Richards FM (1971) The interpretation of protein structures. Estimation of static accessibility. J. Mol. Biol. 55:379–400

59. Junmei Wang, Wei Wang, Shuanghong Huo, Lee M, Kollman PA (2001) Solvation Model Based on Weighted Solvent Accessible Surface Area, J. Phys. Chem. B 105:5055–5067

60. Beveridge DL, DiCapua FM (1989) Free energy via molecular simulation. In: van Gunsteren WF, Weiner PK (eds) A primer, Computer Simulation of Biomolecular Systems, Vol 1, Theoretical and Experimental Applications, ESCOM, Leiden, 1–26

61. Lynden-Bell RM, Rasaiah JC (1997) From hydrophobic to hydrophylic behaviour, A simulation study of solvation entropy and free energy of simple solutes. J. Chem. Phys. 107:1981–1991

62. Landau LD, Lifshitz EM (1980) Statistical Physics. Pergamon Press, Oxford.

63. Lynden-Bell RM (1995) Landau free energy, Landau entropy, phase transitions and limits of metastability in an analytical model with a variable number of degrees of freedom. Mol. Phys. 86:1353–1374

64. Widom B (1963) Some Topics in the Theory of Fluids. J. Chem. Phys. 39:2808–2812

65. Widom B (1982) Potential-Distribution Theory and the Statisctical Mechanics of Fluids. J. Phys. Chem. 86:869–872

66. van der Spoel D, van Maaren PJ, Berendsen HJC (1998) A systematic study of water models for molecular simulation, Derivation of water models optimized for use with a reaction field. J. Chem. Phys. 108:10220–10230

67. Matubayashi N, Reed LH, Levy RM (1994) Thermodynamics of the Hydration Shell, 1. Excess Energy of a Hydrophobic Solute. J. Phys. Chem. 98:10640–10649

68. Matubayashi N, Levy RM (1996) Thermodynamics of the Hydration Shell, 2. Excess Volume and Compressibility of a Hydrophobic Solute. J. Phys. Chem. 100:2681–2688

69. Matubayashi N, Gallicchio E, Levy RM (1998) On the local and nonlocal components of solvation shell models. J. Chem. Phys. 109:4864–4872
70. Torrie GM, Valeau JP (1977) Nonphysical sampling distributions in Monte Carlo free energy estimation – Umbrella sampling. J. Comput. Chem. 23:183–199
71. Gruziel M, (unpublished results)
72. Jayaram B, Fine R, Sharp K, Honig B (1989) Free energy calculations of ion hydration, an analysis of the Born model in terms of microscopic simulations. J. Phys. Chem. 93:4320–4327
73. Jin-Kee Hyun, Toshiko Ichiye (1997) Understanding the Born Radius via Computer Simulations and Theory. J. Phys. Chem. B 101:3596–3604
74. Rajamani S, Ghosh T, Garde S (2004) Size dependent ion hydration, its asymmetry, and convergence to macroscopic behavior. J. Chem. Phys. 120:4457–4466
75. Roux B, Hsiang-Ai Yu, Karplus M (1990) Molecular Basis for the Born Model of Ion Solvation. J .Phys. Chem. 94:4683–4688
76. Madura JD, Briggs JM, Wade RC, Davis ME, Luty BA, Ilin A, Antosiewicz J, Gilson MK, Bagheri B, Scott LR, McCammon JA (1995) Electrostatics and diffusion of molecules in solution: Simulations with the University of Houston Brownian Dynamics program Comp. Phys. Comm. 91:57–95
77. Delphi, Rocchia W, Alexov E, Honig B (2001) Extending the Applicability of the Nonlinear Poisson-Boltzmann Equation: Multiple Dielectric Constants and Multivalent Ions. J. Phys. Chem. B 105(28):6507–6514
78. Baker NA, Sept D, Joseph S, Holst MJ, McCammon JA (2001) Electrostatics of nanosystems, application to microtubules and the Ribosome. Proc. Natl. Acad. Sci. USA 98:10037–10041
79. Born M, (1920) Z. Phys. 1:45–48
80. Grycuk TJ (2003) Deficiency of the Coulomb-field approximation in the generalized Born model, An improved formula for Born radii evaluation. Chem. Phys. 119:4817
81. Wojciechowski M, Lesyng B (2004) Generalized Born Model, Analysis, Refinement and Applications to Proteins, J. Phys. Chem. B 108:18368–18376
82. Feig M, Im W, Brooks CL (2004) Implicit solvation based on Generalized Born theory in different dielectric environments. J. Chem. Phys. 120:903–911
83. Gallicchio E, Linda Yu Zhang, Levy RM (2002) The SGB/NP Hydration Free Energy Model Based on the Surface Generalized Born Solvent Reaction Field and Novel Nonpolar Hydration Free Energy Estimators. J. Comput. Chem. 23:517–529

CHAPTER 4

MODELING CHEMICAL REACTIONS
WITH FIRST-PRINCIPLE MOLECULAR DYNAMICS

ARTUR MICHALAK[1] AND TOM ZIEGLER[2]

[1]*Department of Theoretical Chemistry, Faculty of Chemistry, Jagiellonian University, R. Ingardena 3, 30-060 Cracow, Poland*
[2]*Department of Chemistry, University of Calgary, University Drive 2500, Calgary, Alberta, Canada T2N 1N4*

Abstract: Density functional theory (DFT)-based molecular dynamics (MD) has established itself as a valuable and powerful tool in studies of chemical reactions. Thanks to the rapid increase in power of modern computers, *ab initio* MD has nowadays become practical. Within the Car-Parinello approach, first-principle MD is already quite popular methodology in molecular modeling. MD reveals the dynamical effects at finite temperatures and is particularly useful in probing the potential energy surfaces. Also, it can be utilized to directly determine the reaction free-energy barriers, as it explicitly includes temperature and thus the entropic effects. The first part of the chapter provides a brief introduction to *ab initio* MD, within the Born-Oppenheimer and Car-Parinello approaches. Here, we introduce basic concepts of Car-Parinello MD, with focus on the practical aspects of the simulation. The next part of the chapter summarizes the approaches used to overcome high-energy barriers in a simulation, and thus to probe the part of the potential energy surface relevant for chemical reactions (from the reactants to products through transition states). A special emphasis is placed on the MD simulation along the intrinsic reaction path. The last part of the chapter presents examples from CP-MD simulations from the studies on a complex catalytic process: copolymerization of ethylene with polar monomers catalyzed by late transition-metal-complexes

Keywords: First-Principle Molecular Dynamics, Car-Parinello Molecular Dynamics, Density Functional Theory, Reaction Paths, Olefin Polymerization

1.   INTRODUCTION

Design of molecular materials with specific properties often requires interdisciplinary research involving various experimental and theoretical techniques. Molecular modeling by *ab initio* methods based on quantum-mechanics is now commonly used in such studies. However, theoretical investigations are still dominated by traditional, static approaches in which the stationary points on the respective potential

225

W. A. Sokalski (ed.), Molecular Materials with Specific Interactions, 225–274.

energy surfaces are determined from calculations modeling the molecular systems at 0K. In the studies on complex chemical processes, however, the dynamical effects at finite temperature are often important and should not be neglected.

Molecular dynamics (MD) has been traditionally linked to purely classical modeling, e.g. based on empirical force fields, or with classical trajectory calculations, based on predetermined potential energy surfaces. Unfortunately, such potentials are extremely expensive to evaluate for chemically interesting systems. Nevertheless, attempts to conduct *ab initio* molecular dynamics, in which the classical description of nuclear motion is combined with quantum-mechanical determination of the forces, dates back several decades.

Ab initio MD simulations were not practical until the last decades of the 20th century due to their computational cost. In addition to rapid increase in computational power of modern computers, the factor of great importance for expansion of the *ab initio* molecular dynamics was a novel, practical scheme proposed by Car and Parinello.[1] This methodology gave rise to a growing interest in *ab initio* MD and thus, quickly dominated the field.

The original Car-Parinello article[1] was published 20 years ago. Since that time, first-principle molecular dynamics (MD) has already established itself as a valuable and powerful tool in molecular modeling. A wide range of applications covers various areas of chemistry and physics. Examples of applications of the *ab initio* MD methodology can be found in all the fields in which other computational techniques of molecular modeling are used. The number of publications reporting *ab initio* MD studies is quickly growing; the recent review articles[2-9] summarize some of the most important studies. It is worth mentioning that first principle MD can already be successfully applied to relatively large molecular systems, such as biologically active compounds, or the real intermediates in catalytic processes. Implementations of the hybrid quantum-mechanics-molecular-mechanics (QM/MM) approaches and the solvation models open ways toward exploring even larger and more realistic models of reactive systems.

One of the main advantages of the MD over the static quantum chemical approaches is that it can be utilized to directly determine the reaction free energy barriers, as it explicitly includes entropic effects. An estimation of the free energy via a normal (static) DFT approach requires frequency calculations that are relatively expensive for large molecular systems. Such an approach assumes in addition the harmonic (normal mode) approximation, which breaks down for processes where weak intermolecular forces dominate.[10]

The main purpose of this chapter is to present the basics of *ab initio* molecular dynamics, focusing on the practical aspects of the simulations, and in particular, on modeling chemical reactions. Although CP-MD is a general molecular dynamics scheme which potentially can be applied in combination with any electronic structure method, the Car-Parinello MD is usually implemented within the framework of density functional theory with plane-waves as the basis set. Such an approach is conceptually quite distant from the commonly applied static approaches of quantum-chemistry with atom-centered basis sets. Therefore, a main emphasis of the first part of the chapter will be put on the basic 'magic' parameters of *ab initio* MD methodologies. The second part of the chapter will briefly summarize the approaches used to overcome high-energy barriers in a simulation, and thus to probe the part of the potential energy surface relevant for chemical reactions (from the reactants to the products through the transition states). A special emphasis will be put on the MD simulations along the intrinsic reaction paths, illustrated by a few examples. The last part of the chapter will present an example of MD application taken from studies on a complex chemical process: the copolymerization of olefins catalyzed by late transition-metal-complexes.

2. BASIC CONCEPTS AND PRACTICAL ASPECTS OF CAR-PARINELLO MD

In this section we will briefly present the basic concepts of *ab initio* molecular dynamics within the Born-Oppenheimer and Car-Parinello approach. It is not our intention to cover the theoretical background of the Car-Parinello MD scheme in details. Instead we would like to concentrate on the practical aspects of the simulation and only briefly comment on the physical meaning of the basic parameters that must be specified in the input for a simulation. A more detailed discussion of the theoretical basis for the CP MD can be found in an excellent review article by Marx and Hutter.[2]

2.1. Born-Oppenheimer MD and Car-Parinello MD

The idea of a classical treatment of the nuclear motion within the *molecular dynamics* (MD) scheme with *ab initio* determined, quantum-mechanical forces acing on nuclei is as old as quantum mechanics.[11,12] The commonly used Born-Oppenheimer approximation[12] introduces the concept of *potential energy surface* (PES). Different time-scales for nuclear and electronic motion allows for the *adiabatic separation* of the nuclear and electronic wave-function. In the *Born-Oppenheimer molecular dynamics* (BO-MD) the nuclei move according to Newton laws, while the quantum mechanics is required to determine the potential for this motion:

$$M_\alpha \ddot{R}_\alpha = -\nabla_\alpha E_0(\Psi; \mathbf{R})$$
$$\hat{H}_e(\mathbf{R})\Psi_e(\mathbf{R}) = E_0(\Psi; \mathbf{R})\Psi_e(\mathbf{R}) \tag{4-1}$$

In the equations above $\hat{H}_e$ is the electronic Hamiltonian, Ψ_e denotes the electronic wave-function, and $\mathbf{R}$ is a vector of nuclear coordinates, whereas $E_0(\mathbf{R})$ defines the PES.

Thus, in BO MD the fully converged electronic wave–function and the forces acting on the nuclei must be dermined at each timestep by solving the time-independent Schrödinger equation. The time consuming evaluation of the wave-function and the gradients is the main drawback of the BO MD approach (see Figure 4-1).

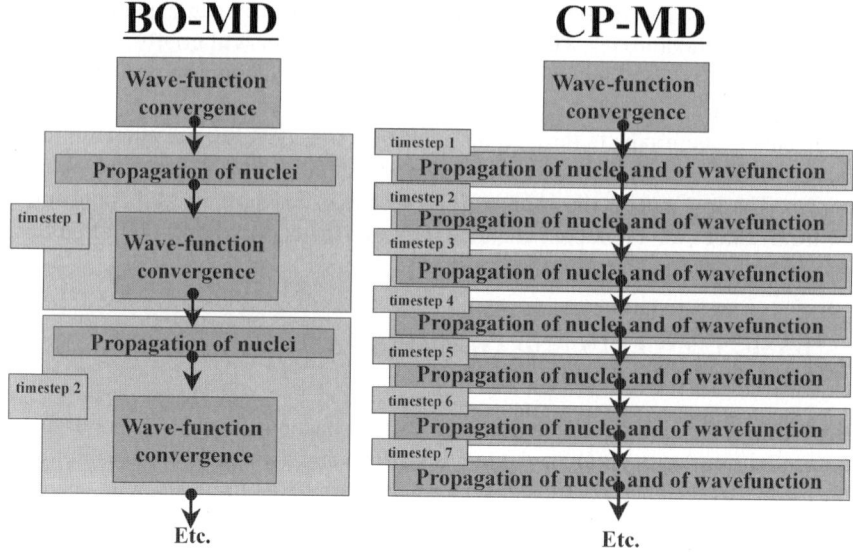

Figure 4-1. Basic ideas of the Born-Oppenheimer MD and Car-Parinello MD

The scheme proposed by Car and Parinello[1] in 1985 offers an attractive solution to this problem, by propagating the wave-function together with the nuclei. The ingenious idea of Car and Parinello was to include the fictitious kinetic energy term describing the 'wave-function motion' into the classical Lagrangian:

$$L = L(R, \dot{R}) = \frac{1}{2}\mu \sum_i \left\langle \dot{\psi}_i | \dot{\psi}_i \right\rangle + \frac{1}{2} \sum_\alpha M \dot{R}_\alpha^2 + E_0(\Psi; \mathbf{R}) + \text{constraints}$$

(4-2)

The first term in this Lagrangian contains the fictitious mass of the wave-function, μ. This fictitious kinetic energy term should not be confused with the real kinetic energy of electrons included in the electronic Hamiltonian.

The Euler-Lagrange equation leads to the Car-Parinello equations of motion of the form:

$$M_\alpha \ddot{R}_\alpha = -\nabla_\alpha E_0(\Psi; \mathbf{R}) + \frac{\partial}{\partial R_a}\{\text{constraints}\}$$

$$\mu \ddot{\psi}_i = -\frac{\delta}{\delta \psi_i} E_0(\Psi; \mathbf{R}) + \frac{\partial}{\partial \psi i}\{\text{constraints}\}$$

(4-3)

Thus, in the Car-Parinello MD scheme the initially converged wave function is propagated according to Eq. (2), and does not need to be re-optimized at every timestep (see Figure 4-1).

The consequence of the presence of the fictitious kinetic energy term in the CP Lagrangian is that the physical energy, i.e. the sum of kinetic energy of the nuclei and the potential energy is not conserved in CP MD. The conserved energy E_{cons} contains in addition the fictitious kinetic energy of the wave-function. This is illustrated in Figure 4-2.

The practical implication is the fact that in the CP MD simulation the molecular system does not evolve right on the Born-Oppenheimer PES, but stays close to it. A measure of deviations from the BO PES is the fictitious kinetic energy (wave-function temperature). Figure 4-2 demonstrates that this deviation is minor, as the electronic (fictitious) temperature is relatively low. The wave function stays 'cold' (compared to the 'hot' nuclei); in the MD terminology the term 'cold electrons' is often used in this context.

The evolution of the wave-function 'slows down' the nuclei, as the kinetic energy of the nuclei is lower in CP MD than in BO MD. This is illustrated in Figure 4-3, that compares the kinetic and potential energy from BO and CP MD simulations started from the same nuclear configuration and the same wave function.

2.2.   Forces in *ab initio* MD; Plane-wave-based Electronic Structure Methods

The Car-Parinello MD approach is usually applied in combination with plane wave based electronic structure methods. Use of plane waves in Car-Parinello MD is in many ways easier than the atom-centered basis sets (Gaussian-type or Slater-type

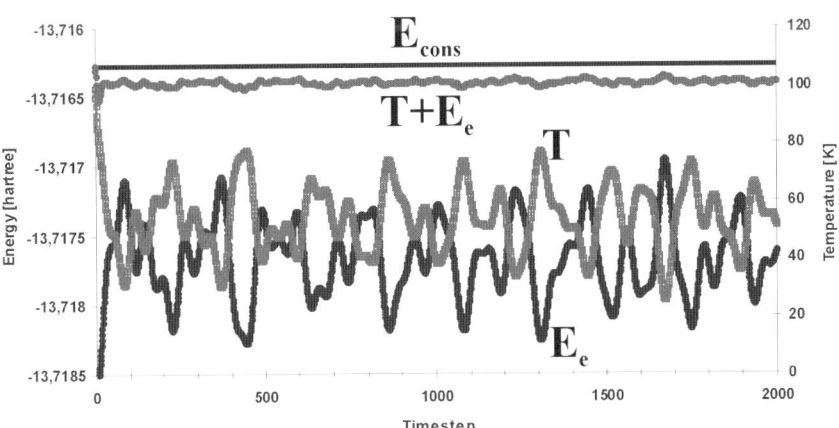

Figure 4-2. Energy conservation in CP-MD: the potential energy (E_c, main axis), temperature (kinetic energy, T, auxiliary, right-hand side axis), physical energy ($T + E_c$, auxiliary axis), and conserved energy (E_{cons}). The difference between E_{cons} and $T + E_c$ is the fictitious kinetic energy of the wavefunction. The data from the simulation for the ethylene molecule with the CPMD program [13] (Troullier-Martins pseudopotentials [14,15], time step of 4 a.u., fictitious mass 400 a.u., cut-off energy 70 Ry, unit cell 12 A x 12 A x12 A)

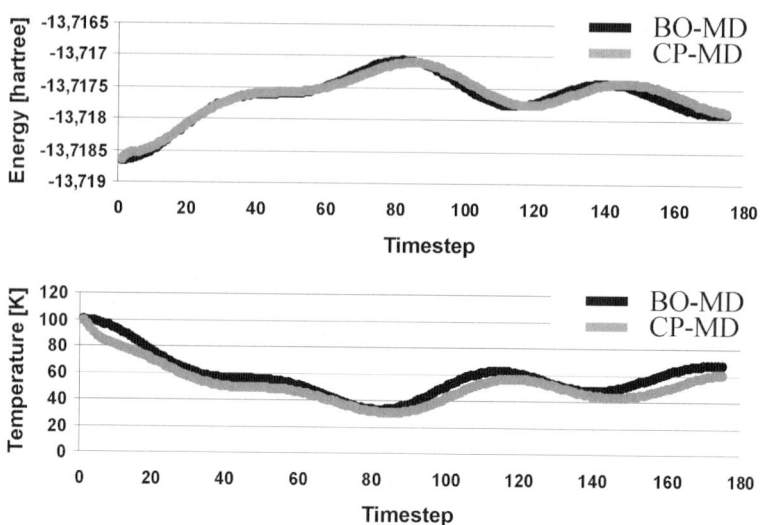

Figure 4-3. A comparison of the Car-Parinello and Born-Oppenheimer molecular dynamics: the potential energy (top) and temperature (kinetic energy, bottom) from the CP-MD and BO-MD simulations for ethylene, started from the same geometry and wave function. The results obtained form the simulations with the CPMD program [13] (Troullier-Martins pseudopotentials, [14,15] time step of 4 a.u., fictitious mass 400 a.u., cut-off energy 70 Ry, unit cell 12 A x 12 A x12 A)

orbitals). The difficulty with atom based basis sets is in the evaluation of the forces. Namely, the derivative of the expectation value of the Hamiltonian (electronic energy), $\langle \Psi | \hat{H} | \Psi \rangle$, with respect to nuclear positions, contains not only the terms originating from the Hamiltonian $\langle \Psi | (\partial \hat{H} / \partial R) | \Psi \rangle$, but also the contributions from the wave-function dependence on the nuclar coordinates, $\langle (\partial \Psi / \partial R) | \hat{H} | \Psi \rangle$ and $\langle \Psi | \hat{H} | (\partial \Psi / \partial R) \rangle$. These terms vanish if a complete basis set is used (Hellmann-Feynman theorem). [16,17] However, with an incomplete basis set these contributions (so called *Pulay forces*) [18,19] are non-negligible and must be evaluated.

The problem disappears in the complete basis set limit. Also, it does not exist for *origin-less basis set*, such as plane waves. For this reason the plane wave based methods are attractive for molecular dynamics simulations where the energy derivative must be evaluated many times.

The wave function expanded in plane waves takes on the form of periodic Bloch functions

$$\psi_{i,k} = \frac{1}{\Omega} e^{i\mathbf{k}\mathbf{r}} \sum_{\mathbf{G}} c_{\mathbf{G}}^{i\mathbf{k}} e^{i\mathbf{G}\mathbf{r}} \qquad (4\text{-}4)$$

where Ω is the system volume, $c_{\mathbf{G}}^{i\mathbf{k}}$ is a set of expansion coefficients, and $\mathbf{G}$ is a reciprocal lattice vector. In practical calculations the infinite sum over the $\mathbf{G}$-vectors must be truncated, and this determines the quality of the basis set. The *cutoff kinetic energy*, E_{cutoff} is the parameter controlling the truncation of the basis set: at each **k**-point in the basis set included are all the **G**-vectors for which

$$\frac{1}{2}|\mathbf{k}+\mathbf{G}|^2 \le E_{cutoff} \qquad (4\text{-}5)$$

In all-electron calculations with the plane wave basis set the cutoff energy must be very high, in order to correctly reproduce rapid oscillations of the wave-function in the region close to the nuclei. Certainly, this affects the computational time. Therefore, the plane wave based methods are combined with pseudopotentials or augmentation approaches, to reduce the computational cost connected with a description of the core-states, less important for chemistry.

2.3.   Finite Temperature Simulations: Thermostats

In standard molecular dynamic simulations the temperature is not constant. The MD simulation samples the microcanonical ensemble, or NVE ensemble, as the volume (unit-cell size) is assumed to be constant. The control of temperature is on the other hand especially important in the simulation of chemical reactions, when the excess of heat dissipated or adsorbed during the reaction strongly influences the kinetic energy (temperature) of the system.

Kinetic energy and temperature are related according to the following equation:

$$E_{kin} = \frac{3}{2} N k_B T \qquad (4\text{-}6)$$

where k_B is the Boltzman constant, and N is the total number of degrees of freedom. Thus, the most obvious way to control the temperature is to monitor the average kinetic energy and to scale up or down the velocities of the nuclei. Several approaches of this kind have been used in classical molecular dynamics.[20–22]

The most popular way to control the temperature in the CP MD simulation was introduced by Nose and Hoover.[23,24] This approach includes an extra friction term (velocity dependent) into the Car-Parinello equations of motions (cf. Eq. 3):

$$M_\alpha \ddot{R}_\alpha = -\nabla_\alpha E_0(\Psi; \mathbf{R}) + \frac{\partial}{\partial R_a}\{\text{constraints}\} - M_\alpha \dot{R}_\alpha \dot{\zeta}_1 \qquad (4\text{-}7)$$

and the thermostat friction variable ζ changes according to its own equation of motion:

$$Q_1 \ddot{\zeta}_1 = \sum_\alpha \left\lfloor M_\alpha \dot{R}^2 - N k_B T \right\rfloor \qquad (4\text{-}8)$$

thus altering periodically the system kinetic energy (velocities) with the frequency $\omega_1 = (N k_B T/Q_1)^{1/2}$.

The Nose-Hoover thermostat exhibits non-ergodicity problems for some systems, e.g. the classical harmonic oscillator. These problems can be solved by using *a chain of thermostats,*[25,26] i.e. introducing the second thermostat that is thermostatting the first thermostat, etc. This is done by introducing a friction term due to the second thermostat into the equation of motion of the first termostat, in the same way as the first thermostat modified the equations of motion of the nuclei (Eq. 7). Namely, Eq. 8 takes the form:

$$Q_1 \ddot{\zeta}_1 = \sum_\alpha \left\lfloor M_\alpha \dot{R}^2 - Nk_B T \right\rfloor - Q_2 \dot{\zeta}_1 \dot{\zeta}_2 \qquad (4\text{-}9)$$

and the equations of motion for the second, and all the other thermostats in the chain, are similar to Eq. 9 (or Eq. 8 for the last thermostat in the chain).

The Nose-Hoover thermostat, or chain of thermostats, can be used as well to control the wave function temperature, i.e. the fictitious kinetic energy. This prevents drifting of the wave function from the Born-Oppenheimer PES during long simulations. Wave function thermostats are introduced in a similar way to Eqs. 7–9.

It should be pointed out that the use of a thermostat affects the energy conservation in MD. Namely, in thermostatted dynamics the 'conserved energy' (kinetic and potential energy of nuclei plus the fictitious kinetic energy of the wave function) discussed in Section 2.1 is no longer conserved. Instead, the energy that includes additional terms due to the thermostats (nuclear and 'electronic') is constant. For example, for a system thermostatted by a chain of n nuclear thermostats, controlled by variables $\{\zeta_t\}$ and $\{Q_i\}$, the conserved energy takes the form:

$$E^{cons} = \frac{1}{2}\mu \sum_i \left\langle \dot{\psi}_i | \dot{\psi}_i \right\rangle + \frac{1}{2} \sum_\alpha M \dot{R}_\alpha^2 + E_0(\Psi; \mathbf{R})$$

$$+ \frac{1}{2} \sum_{m=1}^n Q_m \dot{\zeta}_m^{\,2} + \sum_{m=2}^n k_B T \zeta_m + Nk_B T \zeta_1 \qquad (4\text{-}10)$$

Another method of controlling the temperature that can be used in CP MD is the stochastic thermostat of Andersen.[27] In this approach the velocity of randomly selected nucleus is rescaled; this corresponds in a way to the stochastic collisions with other particles in the system. Therefore, this approach is often called *a stochastic collision method*. The Andersen thermostat has recently been shown[28] to perform very well in the Car-Parinello molecular dynamic simulations of bimolecular chemical reactions.

2.4.   Practical Aspects of Car-Parinello MD Simulation

2.4.1.   Elements of MD simulation

Typical molecular dynamics simulation requires a few initial steps; this is illustrated in Figure 4-4. Often each of these steps is realized as a separate run of the MD program. First of all, the wave function must be fully converged for the initial configuration of nuclei. If the simulation is supposed to sample the vicinity of

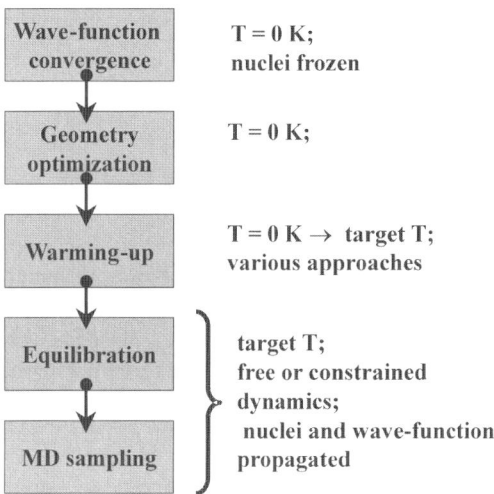

Figure 4-4. Steps in MD simulation

a minimum on the PES, the wave function convergence run is followed by the geometry optimization. Sometimes the geometry optimization is done by dynamics with a friction used for nuclei, to damp the oscillations; in such a case the user has to specify the friction coefficient.

Prior to the simulation at finite temperature, the system must be heated up to the target temperature and thermally equilibrated. The temperature should be distributed among all the normal modes in the system. Thermal equilibration usually requires running dynamics for a long period of time (of the order of picoseconds). This time may be shortened if the warm-up procedure does not displace the system far from equilibrium. Thus, the warm-up may be realized by a sequence of kinetic energy pulses, followed by a short relaxation (free dynamics). If these pulses are orthogonal, then different normal mode become excited. It should be emphasized also at this point, that prior to the constrained dynamics simulation, the warm-up and equilibration should be performed with the same constraints that will be used in the sampling simulation.

2.4.2. Parameters originating from equations of motion: time steps and the fictitious mass

The Car-Parinello equations of motion (Eq. 3) contain the basic parameter of the method, i.e the fictitious mass of the wave function, μ.

The equations of motion are integrated by means of the finite-difference methods, most often employing either the original *Verlet algorithm*, in which

$$\mathbf{R}(t+\Delta t) = 2\mathbf{R}(t) - \mathbf{R}(t-\Delta t) + \mathbf{A}(t)\Delta t^2 \qquad (4\text{-}11)$$

or the *velocity Verlet* method in which:

$$\mathbf{R}(t+\Delta t) = \mathbf{R}(t) + \mathbf{V}(t)\Delta t + \mathbf{A}(t)\Delta t^2$$

$$\mathbf{V}(t+\Delta t) = \mathbf{V}(t) + \frac{1}{2}\Delta t[\mathbf{A}(t) + \mathbf{A}(t+\Delta t)] \tag{4-12}$$

In the above equations $\mathbf{R}(t)$, $\mathbf{V}(t)$, and $\mathbf{A}(t)$ denote vectors of positions, velocities and accelerations at time t.

Thus, integration of the equations of motion introduces another basic parameter common for any MD simulation, i.e. the integration time-step, Δt.

The fictitious mass controls the magnitude of the drift from the BO PES; the smaller its value is, the smaller is deviation from the BO PES. However, lowering μ implies that one has to decrease the time step. The maximum time step is related to the fictitious mass and the cutoff energy according to the following rule:[2]

$$\Delta t_{max} \propto \left(\frac{\mu}{E_{cutoff}}\right)^{\frac{1}{2}} \tag{4-13}$$

The actual values used for the two parameters are $\mu = 500\text{--}1500$ a.u. and $\Delta t = 5\text{--}10$ a.u. It is worth mentioning that in BO MD significantly larger time steps can be used (up to 100 a.u.)

2.4.3.   *Parameters originating from the plane-wave methods: kinetic energy cutoff, unit cell*

Use of the plane wave based electronic structure methods introduces two basic parameters: the kinetic energy cutoff value, controlling the basis set quality, and the periodic unit-cell (*supercell*) size, present due to periodic nature of these approaches. Both of these parameters should be large enough to guarantee the convergence in the total energy and in all the physical quantities that are supposed to be determined from the simulation.

As we have already mentioned in Section 2.2, a huge number of plane waves is required for all electron calculations, in order to properly describe the core region. Therefore, different approaches are used to exclude the core states from the electronic problem to be solved. There is no universal rule concerning the cutoff energy; for each method different values are typically required for convergence.

In Figures 4-5 and 4-6 we present examples from calculations on the ethylene and butadiene molecules performed with the Trouillier-Martins pseudopotentials.[14,15] The total energy requires 70–80 Ry for convergence (Figure 4-5). Similarly, a 70 Ry cutoff is needed for the energy difference between the *cis* and *trans* isomers of butadiene; here, however, a 30 Ry value gives already a quite reasonable, qualitative estimate. The C-C bond distances (Figure 4-6) requite 70–80 Ry to achieve full convergence; again 30 Ry gives a rough qualitative estimate. In general, for the calculations with Trouillier-Martins[14,15] pseudopotentials the required cutoff value varies between 60 and 100 Ry. For the ultrasoft Vanderbilt[29] pseudopotentials much

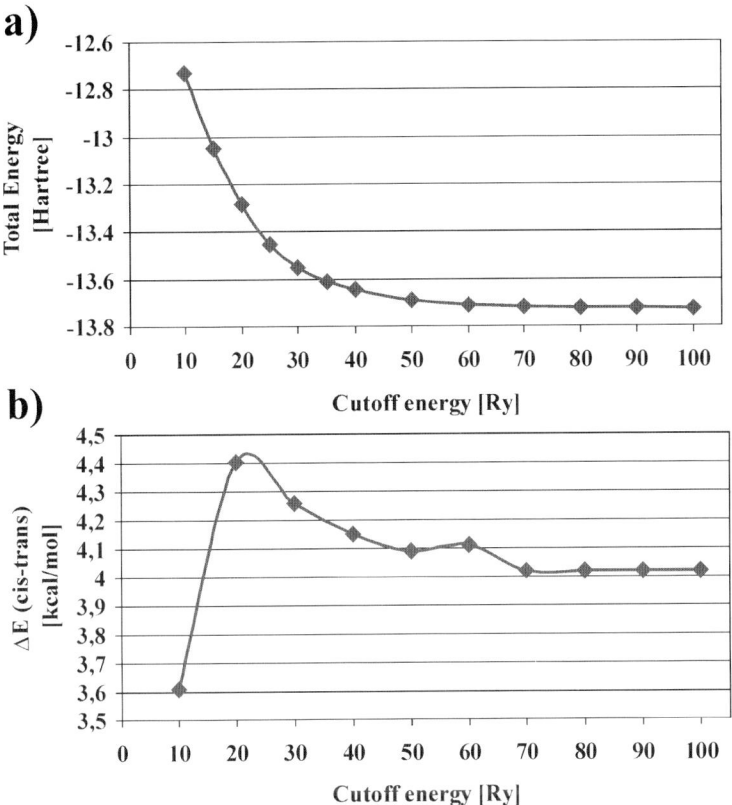

Figure 4-5. Plane-wave convergence of the ethylene total energy (**a**, in hartree) and the energy difference between *cis-* and *trans-* isomers of butadiene (**b**, in kcal/mol). Results of the simulations with the CPMD program [13] (Troullier-Martins pseudopotentials, [14,15] time step of 4 a.u., fictitious mass 400 a.u., unit cell 12 A x 12 A x12 A)

lower values are needed, typically in the range of 20–40 Ry. The relatively 'hard' Goedecker pseudopotentials [30] require usually 100–200 Ry. For the applications of the Projector-Augmented Wave (PAW) method by Blochl, [6,31,32] a low cutoff values can be used, comparable to those for the ultrasoft pseudopotentials.

Periodicity, natural for the plane wave based methods, implies interactions between the molecules in the neighboring cells. This is certainly a shortcoming for molecular calculations in the gas-phase. The unit-cell-size parameter must be large enough to prevent interactions between nearest neighbors. The correct size can be chosen by monitoring the convergence of the physical quantities. In Figure 4-7 an example for the ethylene molecule is presented. In this case the orthorhombic cell with an edge of at least 7–8 A is needed to achieve convergence in total energy. Typically, the cell-size parameter should be 4–5 A larger than the largest interatomic distance in the molecule.

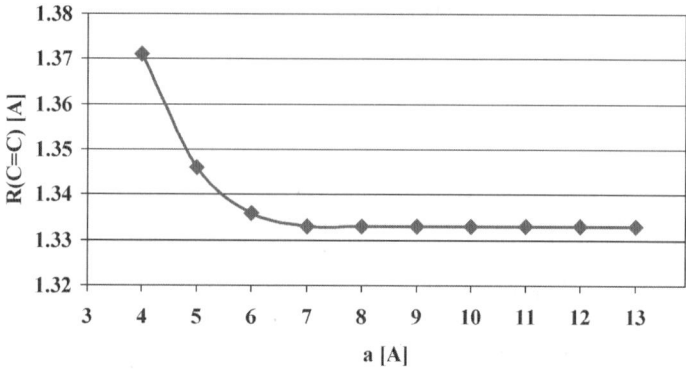

Figure 4-6. Plane wave convergence of the carbon-carbon bond length in the ethylene and butadiene molecules, from the simulations with the CPMD program[13] (Troullier-Martins pseudopotentials,[14,15] time step 4 a.u., fictitious mass 400 a.u., unit cell 12 A x 12 A x 12 A)

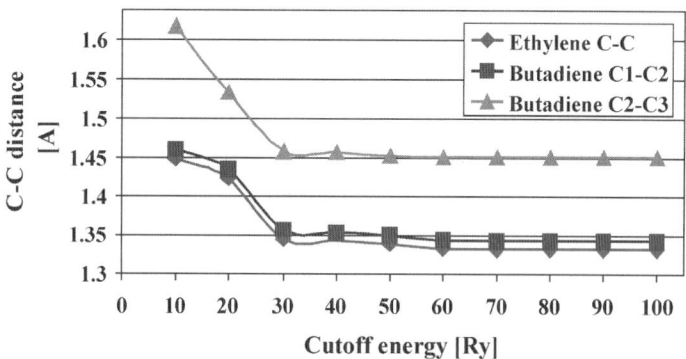

Figure 4-7. Unit-cell-size convergence for the total energy of ethylene. Results from simulations with the CPMD program[13] (Troullier-Martins pseudopotentials,[14,15] time step of 4 a.u., fictitious mass 400 a.u., cut-off energy 70 Ry)

2.4.4.   *Thermostat parameters*

Controlling the temperature implies specifying the parameters characteristic for the thermostat method used in the simulation. In the case of the most popular Nose-Hoover thermostat the basic quantities are the target temperature and the thermostat frequency.

The value of the frequency parameter controlling the temperature of nuclei should enable an efficient coupling of the thermostat to the molecular vibrations. For chain-thermostats the frequency for each item in the chain should be specified. In general, a larger number of thermostats in a chain makes it possible to couple to the normal modes of different vibrational frequencies. In the extreme case, each normal mode can be separately thermostated; in the MD terminology such a procedure is called 'massive' thermostating.[33]

The thermostats for the wavefunction should not be coupled to the nuclear thermostats. Therefore, the value of the frequency parameter should be a few times larger than that of maximum of the phonon spectrum.[2]

3. MODELING CHEMICAL REACTIONS; MD ALONG INTRINSIC REACTION PATHS

In this section we discuss first in general the techniques used to simulate the chemical reactions with relatively high activation barriers. After that we provide a more detailed discussion of the constrained dynamics approach. Special emphasis will be put on MD along the intrinsic reaction coordinate (IRC), illustrated by a few simple examples.[33]

3.1. Towards Overcoming High Energy Barriers

The typical time scale for the Car-Parinello MD simulation is presently of the order of picoseconds. This time scale is usually not sufficient to directly observe a chemical reaction in a single free dynamics simulation, due to relatively high activation-energy barriers. Thus, many approaches have been proposed to simulate such rare reactive events.

The two most popular simulation methods dealing with the reactive events are the constrained dynamics and the bias-potential (umbrella sampling) approaches.[34-38] The former method[34-37] introduces a geometrical constraints that freezes the movement along a selected pathway; the constraint value is changed in order to drag the system from the reactant to product; in the following section we will discuss this approach in details. In the latter method[38] the potential energy surface is modified by an additional bias potential, introduced in order to lower the activation barrier for the studied reaction.

In standard implementations of the umbrella sampling techniques the bias potential is defined in configurational space, as a function of selected geometrical variables. However, new attempts have recently been made to alternatively apply bias potential dependent on the electronic degrees of freedom.[39-42] Very promising initial applications have demonstrated that such approaches can be useful for exploring the regions on the potential energy surfaces separated by non-negligible barriers.

Another group of recently proposed methods is aimed at exploring directly the free-energy surface (FES)[43-45] rather than the corresponding potential energy surface. The approach by Fleurat-Lassard and Ziegler[43] is directed towards the determination of the minimum-free energy reaction path, while the metadynamics approach proposed by Parinello et al.[44,45] introduces additional fictitious terms into the Car-Parinello Lagrangian, to efficiently sample the regions on the free-energy surfaces corresponding to different species separated by high barriers, within a few picoseconds of simulation.

3.2.   Constrained Dynamics, Thermodynamic Integration, and Free-energy Barriers

Various methodologies have been developed to determine the free energy barriers from MD simulations. One of the common approaches applied in *ab initio* MD, known as the potential of mean force method has been derived from the thermo-dynamic integration technique.[34,35] In a canonical (NVT) ensemble the free energy difference, ΔA, between the two states, **0** and **1**, can be calculated as the integral

$$\Delta A_{(0\to 1)} = \int_0^1 \frac{\partial A(\lambda)}{\partial \lambda}d\lambda = \int_0^1 \left\langle \frac{\partial E(\mathbf{X}, \lambda)}{\partial \lambda} \right\rangle_\lambda d\lambda \qquad (4\text{-}14)$$

where the parameter λ is smoothly changing between the two states, E stands for the potential energy, and $\mathbf{X}$ denotes the coordinates of the atoms, $\mathbf{X} = \{X_i, i = 1, \ldots, 3N\}$; the subscript λ represents an ensemble average at fixed λ.

In a simulation performed to model a chemical reaction, the parameter λ corresponds to an arbitrarily chosen reaction coordinate, such as a bond length, an angle, a torsion, or a combination of these parameters. Such a reaction coordinate defines a constraint in configurational space. Constrained dynamics[36] at finite temperature can be performed for a series of fixed λ values, corresponding to a transition from the reactants to the product through the respective TS.

The idea of constrained dynamics performed for a set of points along such a 'reaction path', i.e. for a set of fixed values of the reaction coordinate, λ, is not specific to MD. Similar approaches have been commonly used in computational studies based on static quantum-chemical calculations. Such approaches are known as *linear transit* calculations, *reaction path* scans, etc. A set of constrained geometry optimizations with the constraint 'driving' the system from reactants to products is a popular way to 'bracket' a transition state, for instance.

From a constrained MD simulation the free energy difference can be calculated as

$$\Delta A = \sum_i^{n_{points}} \langle F_i \rangle_\lambda \Delta\lambda_i \qquad (4\text{-}15)$$

where F_j denotes the force required to satisfy the constraint corresponding to λ_j.

One of the requirements typical for this approach is the fact that the system must be properly equilibrated for each λ_j value. Thus, the simulation must be carried out for a long period of time, in order to obtain the converged value of the average force F_j. In practice, for each λ_j the simulation will consist of a following sequence of the MD runs: initial calculation of the converged wave function, constrained geometry optimization, warming-up the system, long-time equilibration (constrained as well), and finally a long sampling simulation, performed to eventually calculate the average restraint force.

Alternatively in the *slow-growth* approach[37] the constraint value is changed in a continuous manner from the initial to the final state. Thus, all the initial steps, including a time-consuming thermal equilibration needs to be performed only once, for the initial geometry. If the system is initially equilibrated, and the rate of the constraint change is small, the system stays (approximately) equilibrated during the simulation. It should be emphasized that the a *small* rate for the change of the constraint is crucial. Preferably the difference between the constraint value at two subsequent points should be negligible compared to the thermal-motion displacement in one time-step. i.e. it should be a few orders of magnitude smaller. In practice, the rate of the constraint change is often determined by the number of time steps chosen to model the whole reactive event. This number is chosen to obtain a compromise between the computational cost and the accuracy. The typical number of time-steps used in a simulation modeling a reaction that involves a bond formation/bond breaking varies between 20 000 and 200 000. Thus, if the inter-atomic distance is used as a reaction coordinate, changing by 4–10 a.u., the rate of the constraint change will vary between 1^*10^{-5} and 5^*10^{-4} a.u./timestep.

However, even with a small rate for the constraint change, the reaction barriers obtained from thermodynamic integration by the slow-growth simulations are dependent on the choice of the reaction coordinate. First of all, an unfortunate choice of the reaction coordinate may correspond to an unfavorable reaction path, which does not pass the transition state region, and thus leads to a substantial overestimation of the barrier.

Further, even if the reaction path goes trough the transition state, a large hysteresis in the free energy profile can be observed. This can happen if the RP does not run along, but crosses the bottom of the valley connecting the transition state with reactant/product. We schematically illustrate this in Figure 4-8. Note that in a constrained simulation the system is allowed to spontaneously evolve in all the directions perpendicular to the corresponding reaction path, defined by the fixed value of the reaction coordinate. If, as in panel **a** of the figure, the reaction path goes along the bottom of the valley and the reaction coordinate is changing slowly, the cut of the potential energy surface in the perpendicular direction practically does not change for the two consecutive points on the RP. Thus, the system regularly oscillates around the minimum and stays equilibrated during the whole simulation. However, if the reaction path crosses the bottom of the valley (Figure 4-8 **b**), then the projection of the gradient on the direction perpendicular to the RP can be large for some points. For such points the fixed constraint value in fact does not freeze the motion along the valley, and thus it does not prevent the system from 'sliding' downhill the potential energy surface. In other words, the cut of the potential energy surface in the direction perpendicular to the RP may change dramatically for the two consecutive points on the RP. For example, with the constraint defining the path shown in Figure 4-8b, from one point the system will spontaneously evolve towards the TS, while from another point towards the minimum. Eventually, the system can easily loose its equillibration as a result of 'chaotic' movements following the change in the constraint value. This leads to a hysteresis in the free energy profile for a forward and backward scan. It should be pointed out, however, that for an infinite sampling, i.e. infinitesimal change in the constraint value, the hysteresis problem would theoretically disappear.

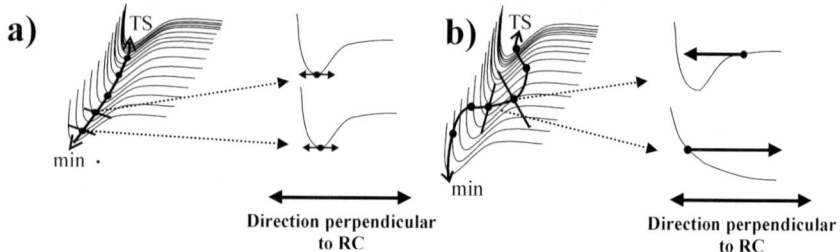

Figure 4-8. Implications of the choice of the reaction coordinate on the constrained MD simulation. In panel **a** the assumed reaction path goes along the bottom of a valley connecting TS and the reactants/products. In panel **b** RP crosses the bottom of the valley

3.3.   MD along Intrinsic Reaction Paths

As we mentioned previously, in practical applications the reaction coordinate is usually chosen as a single geometrical variable (bond length, bond, angle, torsion angle) or at most as a linear combination of few variables. Except from the simplest reactions, such an *a proiri* choice of the reaction coordinate cannot guarantee neither that the RP reaches the true TS region, nor that it goes along the bottom of the valley connecting TS with the reactants/products.

Therefore, an *a posteriori* approach seems to be an attractive alternative, in which the finite temperature MD simulation is performed along the *pre-determined reaction paths.*[33] Such an approach seems to cost more computational time, as it requires determination of the reaction path prior to MD simulation. However, it can be often less expensive than repeating a simulation due to unexpected problems, e.g. a pronounced hysteresis.

Such an *a posteriori* approach can be implemented quite easily and naturally using the machinery of constrained dynamics. The point is in using a proper constraint that freezes the motion along the predetermined, reference reaction path. Such a constraint was defined,[33] based on the fact that in order to freeze the motion in a direction given by a vector $\mathbf{r}^{ref}$, the projection of the displacement vector $\mathbf{r}$ on $\mathbf{r}^{ref}$ must be zero, $\mathbf{r} \bullet \mathbf{r}^{ref} = \mathbf{0}$.

The choice of reaction path definition used as the reference for such a constrained dynamics is arbitrary; any path may be used in practice. However, a natural choice in order to ensure that the simulation moves along the bottom of the potential energy valley connecting reactants/products with TS is the *intrinsic reaction path* (IRP) of Fukui.[46,47] IRP by definition goes along the bottom of such a valley. IRP simply corresponds to a steepest descent path in a mass-weighted coordinates:

$$dx_i = -\frac{\partial E}{\partial x_i} dt \qquad\qquad (4\text{-}16)$$

where x_i stands for the mass-weighted cartesian coordinate $x_i = m_i^{1/2} X_i, i = 1, .., 3N$ and m_i corresponds to the atomic mass of the atom with a position described by the i-th coordinate.

To summarize, the steps involved in the slow-growth MD simulation along the IRP or any other predetermined reaction path are as follows: (i) determination of the geometries of the reactant/product and the TS; (ii) determination of IRP (or other path used as reference for the simulation), recording the path; (iii) constrained, slow-growth MD simulation with a constraint that freezes the motion along the reference path. It should be pointed out, that before starting the slow-growth simulation, the system corresponding to the first point on the path must be thermally equilibrated (with the same constraint).

3.4. Illustrative Examples

In this section we will present results[33] from the MD simulations along the IRC performed for five model reactions: the HCN $\rightarrow$ CNH isomerization reaction, the conrotatory ring opening of cyclobutene, ethylene-butadiene cycloaddition, the prototype SN2 reaction: $Cl^- + CH_3Cl \rightarrow ClCH_3 + Cl^-$, and the chloropropene isomerization: $Cl - CH_2 - CH{=}CH_2 \rightarrow CH_2{=}CH - CH_2Cl$.

All these results were obtained with the Car-Parrinello projector augmented wave (PAW) code developed by Blöchl.[6,31,32] The wave function was expanded in plane waves up to an energy cutoff of 30 Ry. The frozen core approximation has been employed; a Ne core has been used for Cl, and a He core for the first-row elements. Periodic boundary conditions were used, with a unit cell spanned by the lattice vectors ([0 R R] [R 0 R] [R R 0]; R=7–10 Å, depending on the size of the system). All simulations were performed using the local density approximation in the parametrization of Perdew and Zunger[48] with gradient corrections due to Becke and Perdew.[49–51] To prevent electrostatic interactions between neighboring unit cells, the charge isolation scheme of Blöchl was used.[32]

To achieve an evenly distributed thermal excitation, the nuclei were brought to a temperature of 300 K by applying a sequence of 30 sinusoidal pulses, each of which was chosen to raise the temperature by 10 K. Each of the excitation vectors was chosen to be orthogonal to the already excited modes. The warmed-up systems were equilibrated for the 10 000 timesteps. The time step of 7 au. was used. Constraints were maintained by SHAKE algorithm.[36] A temperature of 300 K was controlled by a Nose' thermostat.[23,24] The fictitious kinetic energy of the electrons was controlled in a similar fashion by a Nose' thermostat.[52]

Prior to the MD simulations, the transition state (TS) structures were determined by the method described by Blöchl.[8] The intrinsic reaction paths were determined by the steepest descent in mass-weighted coordinates going from the TS to the reactants and products; this corresponds to a 0 K simulations with a friction coefficient of 1.0. To increase the efficiency of the IRP determination, the nuclear displacements were scaled up by a factor of 2.0, until the average displacement in the geometry reached a value of 0.001 A or an increase in the potential energy was noticed. The slow-growth MD simulations were performed with a constraints freezing the motion along IRP, for the equidistant points on the IRP, with an increment of $0.0002 \, amu^{-1/2} \times bohr$ at every timestep. This corresponds to a total number of 39 300, 66 450, 22 600, and 55 300 timesteps for simulations of the HCN→ CNH isomerization, ring opening of cyclobutene, SN2 reaction, and the chloropropene isomerization, respectively. In case of the last two reactions, only the path from leading from the TS to the products was studied.

3.4.1.   *HCN → CNH isomerization*

As the simplest example of the MD simulation along the IRC, the isomerization of hydrogen cyanide to hydrogen isocyanide has been chosen, This reaction has been extensively studied by both experimental and theoretical methods[53–59] and widely used as a test case for the IRP determination methods.[60–65]

The structures of HCN, CNH, and the triangular TS are shown in Figure 4-9. The IRC energy profile is shown in Figure 4-10a; the free energy profile together with the average potential energy profile obtained from the MD simulations at 300K are presented in Figure 4-10b. The average potential energy in Figure 4-10b has been calculated as running average with the window of 200 timesteps. The portion of the IRP from HCN to TS is shorter than the path from TS to CNH, with lengths of IRP s = 3.58 and s = 4.28 amu$^{-1/2}$ × bohr, respectively. From the simulations the reaction is endothermic with $\Delta E = +14.0$ kcal/mol, with the activation barrier of 44.7 kcal/mol. These values agree well with the results of static DFT studies. The average potential energy profile of Figure 4-10b is practically indistinguishable from the IRC profile, within the accuracy of 0.1 kcal/mol.

In Figure 4-11a the changes in the interatomic distances along the IRP are presented, while Figure 4-11b shows the corresponding changes during the 300K MD simulation along the IRP. The geometries of the reactant, product and the TS are in a very good agreement (within 0.005A) with both, the previous static DFT calculations[64,65] and the experimental data. For HCN the CN, CH, and NH distances are 1.16, 1.08 and 2.25 A, respectively; the corresponding values for the CNH molecule and the TS are 1.17, 2.20, 1.01 A, and 1.20, 1.21, 1.38 A, respectively.

Figure 4-11b demonstrates that during the simulation all geometrical parameters regularly oscillate around the IRP values. It is illustrative to compare the results of the MD along the IRP with the similar simulations performed with the *a priori* assumed reaction coordinate. Here the difference between NH and CH distances was chosen as a reaction coordinate, i.e. the substitution constraint was used, $R_{NH} - R_{CH} = $ const. The MD simulations from both approaches were

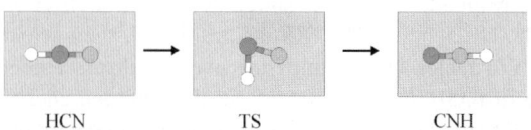

HCN   TS   CNH

Figure 4-9. Isomerisation of hydrogen cyanide to hydrogen isocyanide: the structures of the reactant, TS, and the product

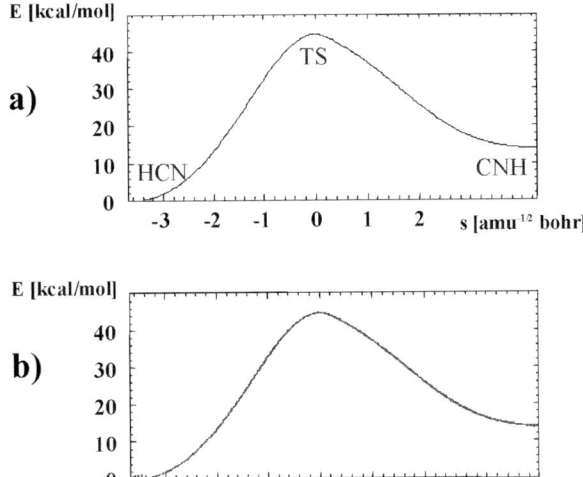

Figure 4-10. Isomerisation of hydrogen cyanide to hydrogen isocyanide: the IRC energy profile (panel **a**), and the free-energy profile together with the average potential energy profiles (panel **b**), calculated from the 300K MD simulations along the IRP

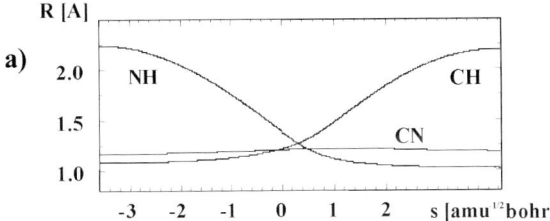

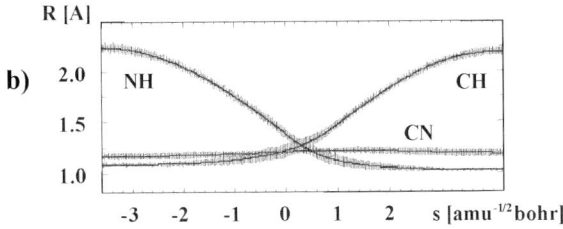

Figure 4-11. Isomerisation of hydrogen cyanide to hydrogen isocyanide: the changes in the CH, NH, and CN distances along the IRP, during the 300K MD simulation

performed with similar settings, including the same number of timesteps. In the simulations with *a priori* RC, the constraint value was smoothly changed between the values characteristic for the HCN and CNH molecules.

The computed reaction barrier practically does not differ from the value obtained from the simulation along IRP, the reaction free energy is higher by c.a. 1.2 kcal/mol than the change in the average potential energy between HCN and CNH. However, in the region after passing the TS ($R_{NH} - R_{CH}$ = from 0.6 to 1.2 bohr) the reaction path defined by the constraint visibly deviate from the IRC. Since the applied constraint does not represent the minimum energy path, the molecular system spontaneously evolve toward lower energies, partially loosing equilibration. After passing TS, the substitution constraint does not prevent going downhill the potential energy surface; this is reflected by damped oscillations between the constraint values of 0.2 and 0.6 bohr. Afterwards, the unequilibrated systems exhibits increased variation in all the interatoomic distances between the constraint values of 0.6 and 1.2 bohr.

This effect is clearly illustrated in Figure 4-12, in which we compare the hydrogen paths from the two simulations, plotted together with the hydrogen path

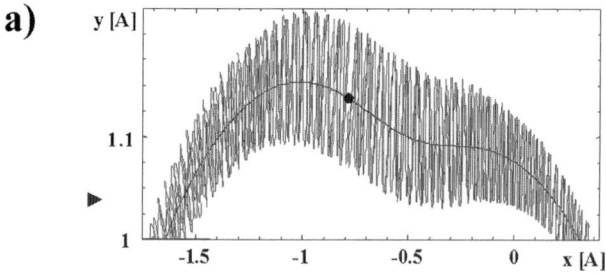

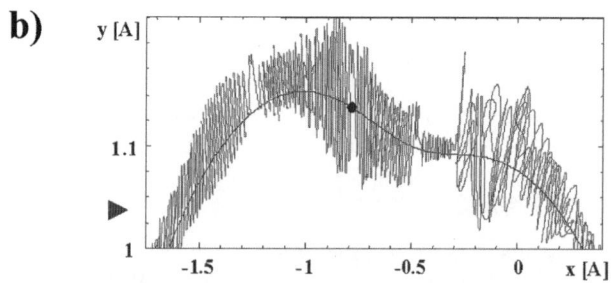

Figure 4-12. Isomerisation of hydrogen cyanide to hydrogen isocyanide: relative motion of the atoms during the isomerisation of hydrogen cyanide to hydrogen isocyanide calculated from the 300K MD simulations along the IRP (oscillating line in panel **a**) and from the 300 K MD simulation with the constraint R_{NH}-R_{CH} = const. (oscillating line in panel **b**). The corresponding data along the IRP are given as a solid middle-line. The nitrogen atom is fixed at the origin (0,0); the carbon atom is located on the x-axis; the lines correspond to the trajectories of the hydrogen atom relative to CN. Only the vicinity of the TS is included in the plots. For the whole pathway see Ref. 33

corresponding to IRP. For clarity of the figure only the vicinity of TS is shown here. While the hydrogen path from the simulation along the IRP (Figure 4-12a) oscillates very regularly around IRP, in the case of the simulations with the substitution constraint (Figure 4-12b) strong irregularities can be observed. Here, after passing the TS the oscillations are first damped – the hydrogen atom runs away from the IRP, and then increased in a quite chaotic fashion – a sign that the system has lost its equilibration. In this simple reaction involving a small, triatomic molecule, the problem discussed above does not affect the reaction free energy very much, since the molecule regains equilibration relatively quickly. In the large molecular systems with many degrees of freedom, however, the equilibration lost at some point can lead to inaccurate estimation of the reaction free energies as well as the reaction barriers.

3.4.2.   Conrotatory ring opening of cyclobutene

The reaction of cyclobutene ring opening leading to 1,3-cyclobutadiene has been extensively studied by experimental and theoretical methods.[64,66−71]

The ring opening of cyclobutadiene is predicted by symmetry rules to proceed as a conrotatory process. The kinetic product of the reaction is *gauche*-1,3-butadiene with deviation from planarity of ca. 30°. The structures of the cyclobutene, TS, and *gauche*-1,3-butadiene are shown in Figure 4-13.

The kinetic product can be further transformed into the most stable *trans*- isomer. Here, we study only the pathway leading to a *gauche*- rotamer. The mechanism of the reaction involves four main processes: (i) breaking of the $C_1 - C_4$ σ-bond; (ii) rehybridiration of the carbon skeleton, i.e. formation of the conjugated π-electron system by partial breaking of the $C_2 = C_3$ π-bond and partial formation of the $C_1 = C_2$ and $C_3 = C_4$ π-bonds; (iii) skewing of the carbon skeleton by rotation around $C_2 - C_3$ bond; (iv) conrotatory movement of the two methylene groups. Recent static DFT IRC studies[64] demonstrated that the reaction proceeds in a concerted fashion rather than according to a stepwise mechanism. Since the reaction path involves concerned changes in all geometrical parameters, it represents an interesting test case for the MD along the IRP.

The results of our calculations are summarized in Figure 4-14 and 4-15. The results are in a good agreement with the recent, static DFT studies[64] and the

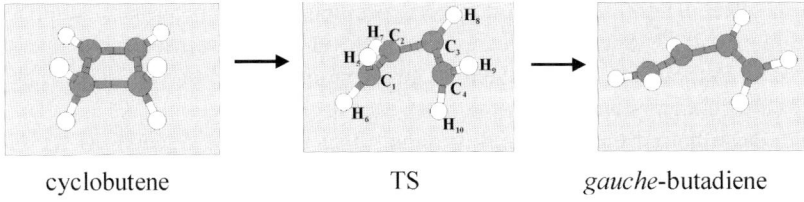

cyclobutene   TS   *gauche*-butadiene

Figure 4-13. Conrotatory ring opening of cyclobutene: the reactant, TS, and the product geometries

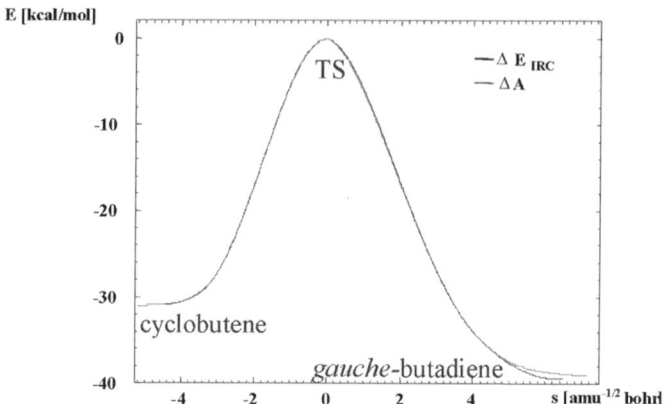

Figure 4-14. Conrotatory ring opening of cyclobutene: IRC energy profile and free energy profile from the MD simulation at 300K

experimental data.[66–68] We will not repeat here the detailed comparison and the discussion from the original article.[33] We would only like to emphasize that the MD along the IRP leads to smooth free energy profiles (Figure 4-14) even in cases where the reaction involves concerted changes in many geometrical parameters. All the geometrical variables regularly oscillate around their IRP values (Figure 4-15) and the system practically stays equilibrated during the whole slow-growth MD simulation.

3.4.3.   *Ethylene-butadiene cycloaddition*

The basic Diels-Alder cycloaddition reaction involving ethylene and butadiene is probably one of the best known chemical reactions. It has been studied extensively by a variety of both, experimental and theoretical techniques.[72] We have performed MD along the IRC for the key reaction pathway leading from reactants to the boat-like product of C_s symmetry (Figure 4-16); the isomerization of the product was not studied here.

Similarly to the previous reaction, the ethylene-butadiene cycloaddition is a nice example of a reaction in which the finite temperature pathway (300K here) follows the IRC. The changes in the crucial geometrical parameters are summarized in Figure 4-17. It can be clearly seen that all the geometrical parameters oscillate around their IRC values. Quite large deviations are observed only for the 'reactive' C-C-C-C torsion angle in the initial part of the simulation. This is understandable, since before the C-C bond forming, there is a lot of freedom as far as the mutual orientations of the reactants are concerned. For the same reason, there exists deviation from the IRC for the C(Et)-C(Btd) distances in the very first part of the reaction pathway.

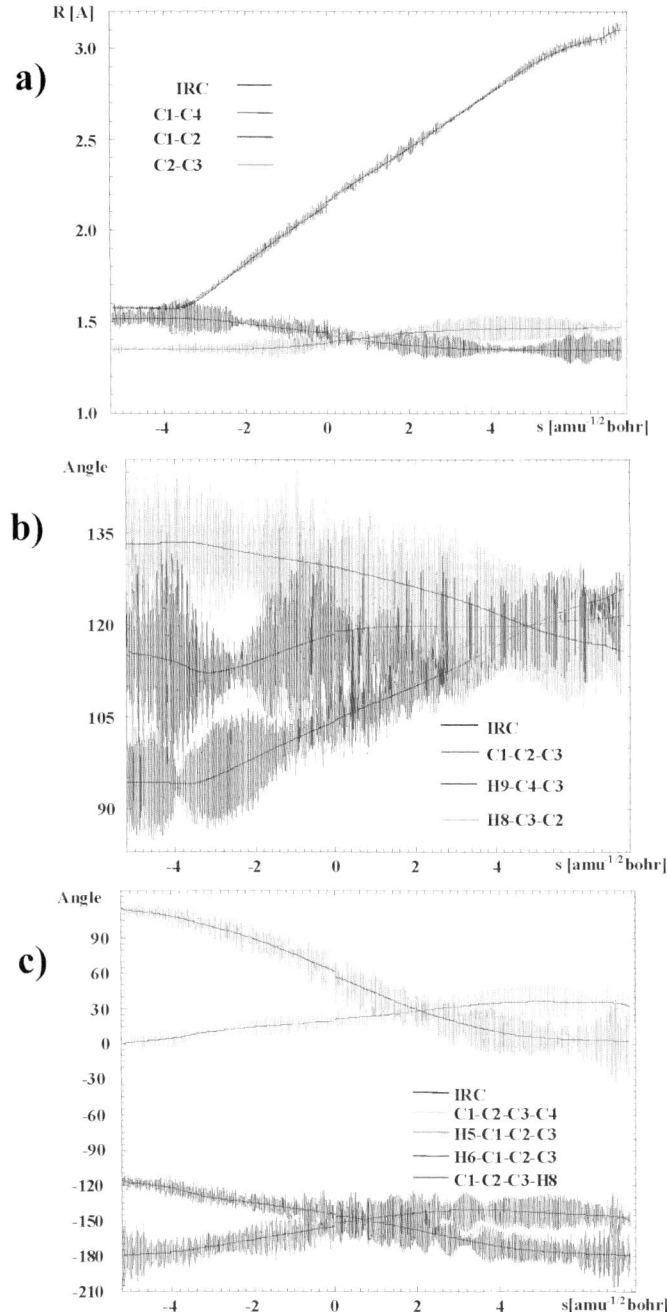

Figure 4-15. Conrotatory ring opening of cyclobutene: the changes in the interatomic distances (a), angles (b), and torsional angles (c) along the IRC and from the 300K MD simulation

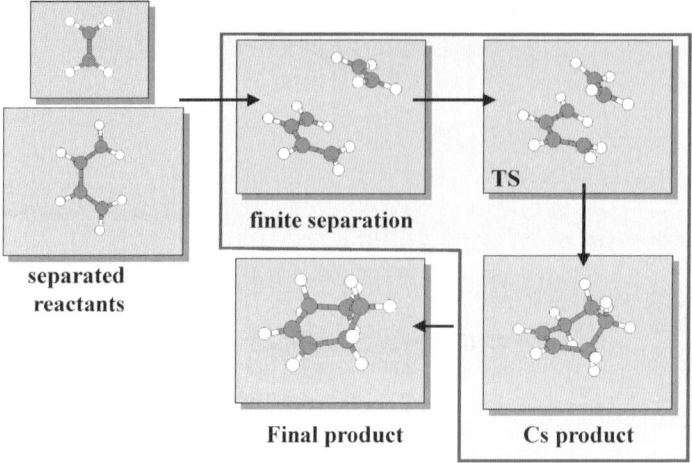

Figure 4-16. Ethylene-butadiene cycloaddition: reactants, products and TS structures

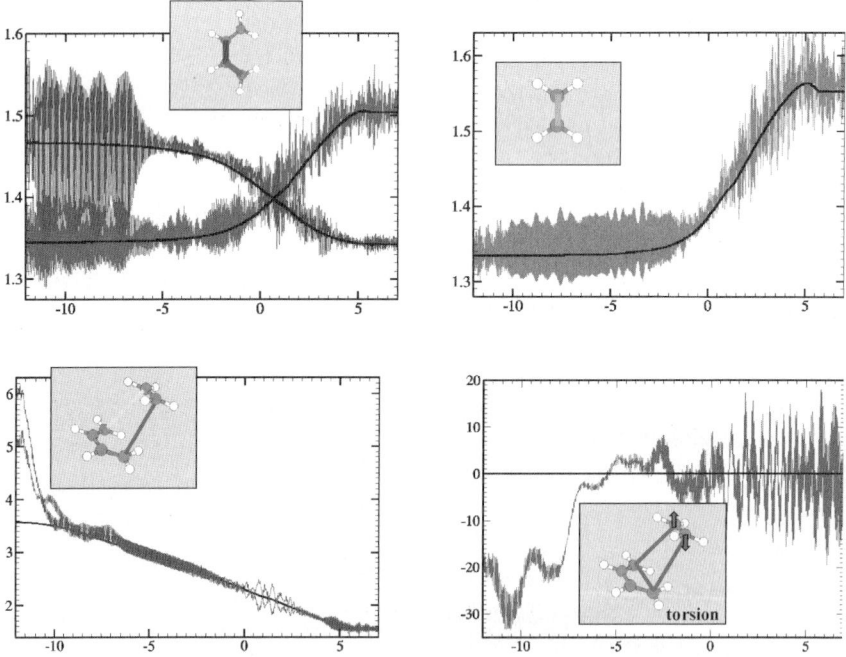

Figure 4-17. Ethylene-butadiene cycloaddition: the changes in the interatomic distances (a), angles (b), and torsional angles (c) along the IRC and from a 300K MD simulation

3.4.4. *Prototype SN2 reaction* $Cl^- + CH_3Cl \rightarrow ClCH_3 + Cl^-$

The halogen exchange reactions between methyl halides and a halogen anion, $X^- + CH_3X \rightarrow XCH_3 + X^-$ (X=F, Cl, Br, I), have been studied by many experimental and theoretical techniques as a prototype SN2 reaction.[64,73-76] We have performed the MD simulations along IRC for the chlorine exchange reaction (Figure 4-18).

The reaction profile is characterized by a double-well shape with two minima, corresponding to Van der Waals complexes (VdW), X^-----CH_3X, located symmetrically around the barrier. Using MD we studied the major part of the reaction, i.e. the path between VdW complex and the TS; the reaction path corresponding to the formation of the VdW complex from the isolated chlorine anion and chloromethane has not been studied. The structures of the two VdW complexes and the TS are shown in Figure 4-18. Due to the reactants-products symmetry, in the following we present the results of the MD simulation along the part of the IRP leading from the transition state to a VdW complex.

The results of our simulation are summarized in Figure 4-19. The detailed discussion was presented in the original article.[33] Here we will only discuss some major points. The reaction mechanism involves Walden inversion of chloromethane, with the planar CH_3 group in the symmetrical transition state geometry. The IRP involves linear movement of the carbon and two chlorine atoms, accompanied by bending the C-H bonds toward tetrahedrical orientation.

At 300K, in the TS region the atomic movements are dominated by two soft modes: the bending of the Cl-C-Cl angle, i.e. the movement of the CH_3 fragment perpendicular to the Cl-Cl axis, and the symmetrical stretching of the two C-Cl bonds. The plots of Figure 4-19 show that during the MD simulation all bonded interatomic distances oscillate around their IRP values with relatively small amplitudes. In the case of non-bonded distances, however, a large deviations from their IRP values may be observed for the RC values corresponding to the region after the C-Cl bond-formation, i.e. $s > 2.8 \, amu^{-1/2} \times bohr$. This is because the dynamic VdW complex at 300K does not resemble its static image at 0K. After the formation of the C-Cl bond, the chloromethane molecule starts rotating. Thus, the Cl-C-Cl angle adapts values from the whole range between 0° and 180°. In other words, at 300K in the VdW stage all mutual orientations of the chlorometane and chlorine anion are possible, incuding the linear H_3C-Cl----Cl^- complex. This is reflected by large changes in the non-bonded C-Cl and Cl-Cl distances. In Figure 4-19c are

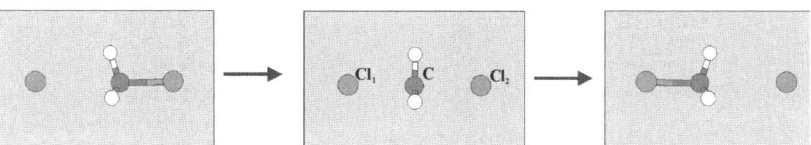

Figure 4-18. SN2 reaction: the key structures

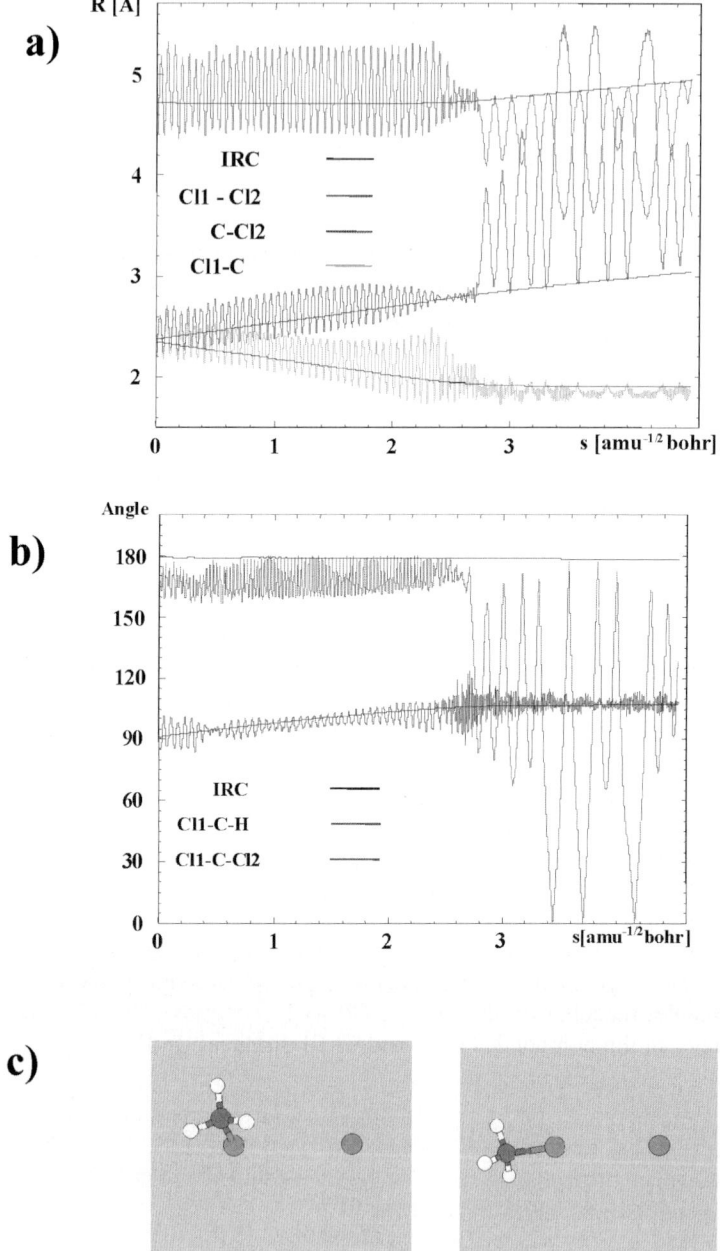

Figure 4-19. SN2 reaction: the changes in the selected interatomic distances (a) and angles along the IRC and the 300K MD simulation. Example geometries from the trajectory corresponding to the van der Waals complex are shown in panel **c**

presented two representative geometries of the complex at the final stage of the reaction, illustrating the above discussion.

The example of the prototype SN2 reaction demonstrates that the algorithm of MD along the IRP applied here works correctly also in the case of a reaction with relatively large deviations of the molecular geometries at finite temperatures from the corresponding structures at the zero-temperature IRP.

3.4.5. $Cl\text{-}CH_2\text{-}CH=CH_2 \rightarrow CH_2=CH\text{-}CH_2Cl$ isomerization

The isomerization of 3-chloropropene (allyl chloride) involves a transfer of the chlorine atom between two terminal carbon atoms, through an allylic TS state. The structures of the reactant, TS and the kinetic product are shown in Figure 4-20. Again, due to the reactant-product symmetry, only the part of the reaction path leading from the TS to a kinetic product will be discussed here.

The most stable conformation of the allyl chloride is a *gauche*-rotamer. The energy difference between *gauche*- and *cis*-species is 0.9 kcal/mol according to our calculations. Recent experimental data[93] suggest an energy difference of 0.3–0.4 kcal/mol, while recent RHF and MP2 calculations give the value of 0.9–1.3 kcal/mol. The *gauche*- and *cis*-rotamers are separated by a barrier of 2.9 kcal/mol. Here, our result reproduces the experimental value.[93] However, the IRP for the chlorine transfer reaction leads from the allylic TS to a minimum corresponding to a *cis*-conformer, which can later rotate to form a *gauche*-species. Therefore, in the following we present the results of the MD simulation along the IRP leading to a *cis*-rotamer.

The results are summarized in Figure 4-21. During the MD simulation, all the bonded inter-atomic distances and bond angles oscillate around their IRP values. The plots of Figure 4-21, however, exhibit quite interesting features. Namely, at $s = 7.5$ amu$^{-1/2}$ × bohr one can observe large deviations of the non-bonded C-Cl distance and the Cl-C_1-C_2-C_3 torsion from the corresponding IRP values (see Figure 4-21). Despite the fact that the IRP leads from TS to the *cis*-rotamer, at 300K the system spontaneously evolves towards the *gauche*-geometry. This leads to a decrease in the free energy, and is accompanied by a deviation of the non-bonded inter-atomic distances from their IRP values. After reaching the region with the geometry closer to the *gauche*-rotamer, the molecule is brought back to the IRP and eventually reaches the geometry of the *cis*-product, since the constraint used in the MD simulations forces the system to move along the IRP.

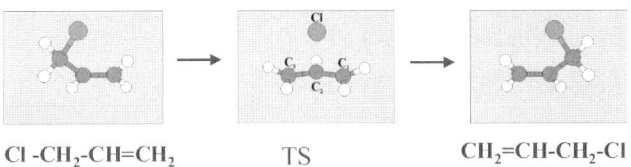

Cl -CH$_2$-CH=CH$_2$ TS   CH$_2$=CH-CH$_2$-Cl

Figure 4-20. Chloropropene isomerization: structures of the reactant, TS and the product

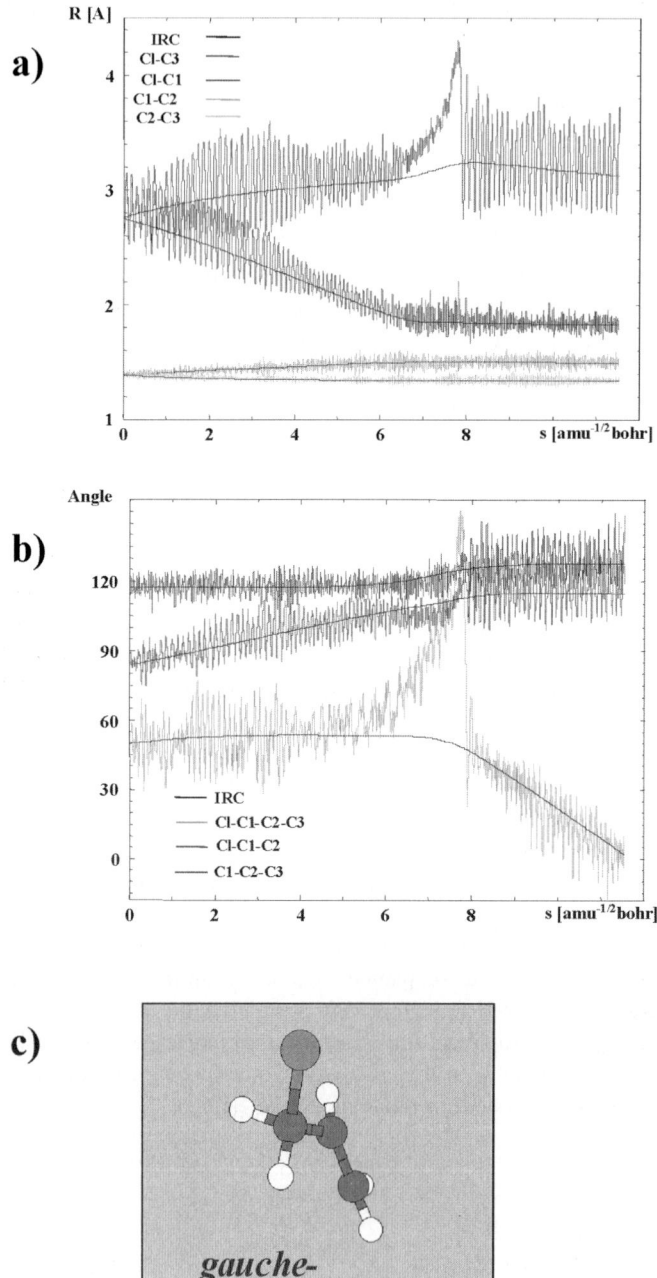

Figure 4-21. Chloropropene isomerization: the changes in the selected interatomic distances (a) and the torsions (b) along the IRC and the 300K MD simulation. The *gauche*-rotamer structure, formation of which is responsible for deviation from IRP is shown in panel **c**

Thus, the chloropropene isomerization reaction represents an interesting example of a system for which MD reveals a 'thermal shortcut', i.e. the 300 K path does not follow the zero-temperature IRP, leading directly to a different, more stable conformer. This example emphasizes the strength of the finite-temperature MD simulations that can often reveal the reaction paths not noticeable by the static (zero-temperature) investigations. At the same time, this example demonstrates that the algorithm used here works well even in the cases when the system geometry deviates far from the IRP during the finite-temperature simulation.

4. MOLECULAR DYNAMICS IN THE STUDIES OF THE ETHYLENE – METHYL ACRYLATE COPOLYMERIZATION

In this section we will present results of *ab initio* molecular dynamics simulations performed for more complex chemical reactions. Catalytic copolymerization of α-olefins with polar group containing monomers, chosen here as an example, is a complex process involving many elementary reactions. While for many aspects of such a process the standard approach by static quantum chemical calculations performed for the crucial reaction intermediates provides often sufficient information, for some aspects it is necessary to go beyond static computations. In the case of the process presented here, MD was priceless in exploring the potential energy surfaces for a few elementary reactions that were especially difficult for a static approach, due to a large number of alternative transition states and thus, alternative reaction pathways.[77]

This section is organized as follows: first the mechanism of the polar copolymerization process will be presented, then the results from the static calculations will be briefly summarized, and finally the MD results will be discussed.

4.1. The Polar Copolymerization Process and its Mechanism

Polyolefins are nowadays the materials of great importance, commonly known not only from many industrial applications, but as well from their use in everyday life. The properties of the polymeric material depend on numerous factors. The choice of monomer is obviously of major importance, although it does not uniquely determine the polymer. Among the primary factors is the molecular weight of macro-molecules in the polymer, as well as the molecular weight distribution. Further, the architecture of the macromolecules, i.e. the overall degree of branching as well as the topology of branches strongly influences the polymer properties. These factors are only partly determined by the general mechanism of a process (radical, anionic, cationic, coordination-type polymerization), as they can be often strongly influenced by a specific mechanism, e.g. following the choice of catalyst/initiator, the thermodynamic parameters of the process (temperature, pressure), etc. This implies that a great variety of new polymeric materials can be potentially designed. Polyethylene homopolymers or copolymers with other olefins are good example, as various classes of polyethylenes (HDPE, LDPE, LLDPE) are produced in different processes, often controlled by structure of ligands on the catalyst, temperature, pressure, etc.[78]

The development of a single-site Ziegler-Natta-type catalysts capable of copolymerizing α-olefins with monomers containing polar groups is one of the major goals in a design of new polymerization catalysts.[79] Of special interest are the oxygen and nitrogen containing monomers, such as vinyl alcohols, ethers, esters, nitriles, etc. Even a small incorporation of polar comonomer into the polyolefin chain can strongly modify the properties of the copolymer, compared to homopolymer. The traditional heterogeneous Ziegler-Natta catalyst[80−83] and the early-transition metal complexes[84−88] are poisoned by polar monomers. Introducing the new generation of non-metalocene catalysts based on the late-transition-metals gives potentially a chance to develop systems that are active as catalysts for polar copolymetrizations. The late-transition-metal-based complexes[78−90] are more functional group tolerant. The first successful example was the Brookhart Pd-diimine catalyst,[78] capable of copolymerizing ethylene with acrylates.[91,92] Interestingly, the analogous Ni-based catalyst is not active in polar copolymerization under the same reaction conditions, although it is an effective catalyst for the α-olefin homopolymerization. Thus, the first step toward designing new catalysts, is to understand the differences between the palladium- and nickel based systems.

The mechanistic aspects of the homopolymerization processes catalyzed by the diimine catalysts have been studied extensively by experimental[78,91−98] and theoretical methods[99−106]. Molecular modeling by a combined approach that links quantum-chemical calculations (static DFT and MD) with the stochastic (Monte-Carlo like) simulations of the polymer growth was helpful in understanding the relationship between the catalyst, temperature, and pressure of the process and the topology of the resulting macromolecules. This topic was reported in a series of articles,[101,102,104,107,108] and summarized in monographic reviews,[109−111] so it will not be discussed here. This section will cover the polar copolymerization of ethylene with methyl acrylate, focused mainly on investigations by the molecular dynamics approach; the main goal of these studies was to understand the aforementioned differences between Pd- and Ni-diimine catalysts. The computational study summarized here was followed by the applications of molecular modeling techniques to other copolymerization processes.[112−116]

The mechanism of polar copolymerization (Figures 4-22 and 4-23) involves in the first step binding of the non-polar or polar monomer to the catalyst. In the standard Coose-Arlman mechanism[117,118] for α-olefin polymerization the monomer insertion follows the complexation of the olefin by its double C=C bond (π-complex, **2** and **4** in Figure 4-22). Alternatively, the polar monomer can be coordinated by its functional group, e.g. carbonyl oxygen atom in the case of acrylates (σ-complex, **3**). The competition between the two binding modes of acrylate is one of the important factors for the catalyst activity. This is because the insertion of the polar monomer in a random copolymerization mechanism may start from the π-complex only. A formation of a stable σ-complex results in poisoning of the catalyst.

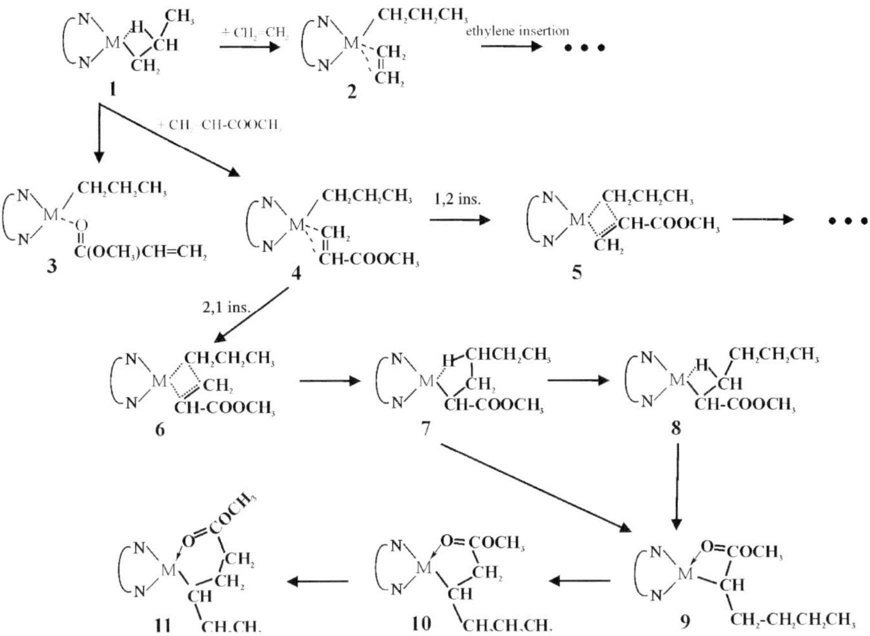

Figure 4-22. Mechanism of ethylene-methyl acrylate copolymerization: monomer insertion

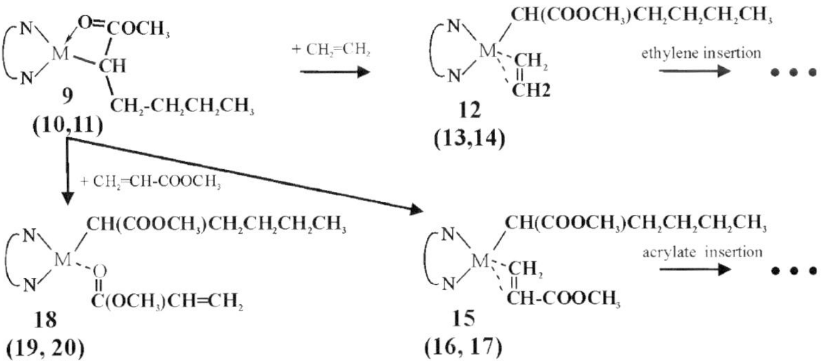

Figure 4-23. The steps in the copolymerization mechanism following the acrylate insertion

The ethylene insertion leads to alkyl agostic complexes (γ- and β-agostic), a starting points for further insertions. The polar monomer insertion results in the formation of chelates (**9, 10, 11**), more stable than the corresponding agostic complexes (γ- and β-agostic, **7** and **8**). It has been found experimentally[8] that the 6-member chelate (**11**) is a resting state for the catalyst in acrylate polymerization catalyzed by Pd-diimine complexes. Formation of the chelates is followed by the insertion of the next monomer, again starting from the respective π-complexes

(**12–14** and **15–17** in Figure 4-23). Here as well, the polar monomer may be bound alternatively by its functional group (**18–20**) and potentially poison the catalyst. Thus, the potential factors of importance for polar copolymerization that should be considered are the competition between the two binding modes of the polar olefin, the competition between polar and non-polar olefin, the difference in their insertion barriers, and difficulty of the opening chelates by ethylene coordination, following the acrylate insertions. This factors were studied by a combined static DFT and MD studies.

4.2.   DFT and MD Studies on the Monomer Binding and Insertion

DFT calculations with Becke-Perdew exchange-correlation functional were carried out for all the reaction intermediates present in the catalytic cycle of polar copolymerization, based on the simplified diimine catalyst in which the bulky diimine substituents were replaced by hydrogen atoms (model catalyst: $N^\wedge N-M^+$, $N^\wedge N =$ $-N(Ar)-CR-CR-N(Ar)-$ R=H, Ar=H, M=Ni, Pd). Some of the calculations, for the most important structures, were repeated using the real catalyst, containing bulky diimine substituents [real catalyst: $Ar=C_6H_3(o\text{-}i\text{-}Pr)_2$, $R=CH_3$].

Figure 4-24 presents the energy profile for the acrylate insertion, resulting from the calculations performed for the structures based on the model catalyst. The lowest energy reaction intermediates for the 2,1-insertion are present in the figure, as the 2,1-insertion was found to have higher barriers for both, the Ni- and Pd-based systems.

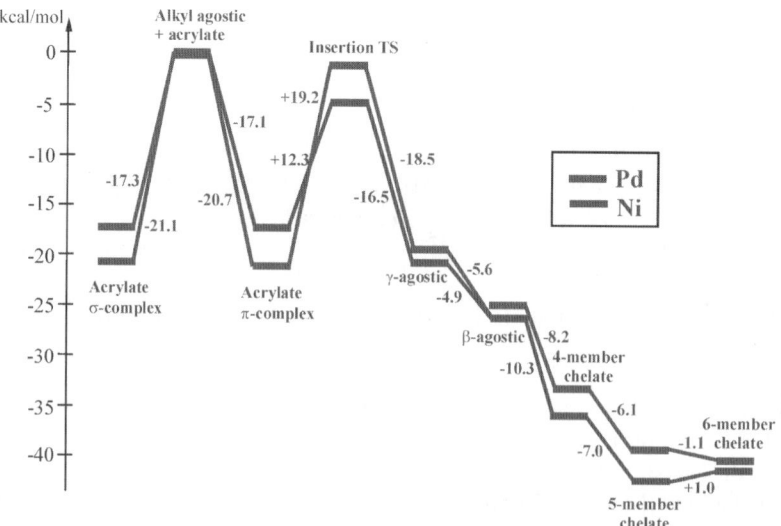

Figure 4-24. An energy profile for the 2,1-patwhway of methyl acrylate insertion into the M-alkyl bond (M = Ni, Pd) calculated for the model catalyst

The first step in the copolymerization mechanism involves a competition between the complexation of the polar and non-polar olefin, and in the former case a competition between the two binding modes. Formation of the π-complex, **4**, is necessary for the monomer insertion. The σ- complex **3** can poison the catalyst, if it is strongly preferred energetically. The relative stability of the two binding modes has been found to be different for the Ni- and Pd- complexes.[103,105] For the Pd system, active in the polar copolymerization, the π-complex is indeed preferred by 3.6 (model catalyst) and 3.0 kcal/mol (real catalyst), while for the Ni catalyst the σ-complexes have lower energy by 4.0 (generic catalyst) or 3 kcal/mol (real catalyst). Thus, the Ni-catalyst is initially poisoned by formation of the 'inactive' σ-complex, while in the case of the Pd-complexes a formation of the acrylate π-complex leads to its insertion. The structures of the alternative complexes are shown in Figure 4-25.

The difference in the preferred binding mode observed for the Pd- and Ni-based catalysts can be the crucial factor determining activity/inactivity of these two systems in polar copolymerization. However, the question arises about the stability of the alternative binding modes at finite temperature. If the minima were separated by relatively low barriers and fast interconversion between the two isomer complexes could occur, then this difference would be of minor importance. In order to check the stability of the two modes and get the insight into the mechanism of possible interconversions, a series of molecular dynamics simulations was performed.

In the first stage of this study, the unconstrained MD simulations at T = 300 K were performed for all the complexes. In addition, for the higher energy complexes (Pd/σ-complex and Ni/π-complex) the unconstrained MD simulations were performed at T = 700K, in order to check whether they would spontaneously evolve towards the corresponding global minima. Figure 4-26 presents the changes in the crucial geometric parameters (metal-carbon and metal-oxygen distances) along the respective MD trajectories. The results clearly demonstrate that both, the π- and σ-complexes of methyl acrylate are stable at 300K and 700K for both the metals. That is, all the simulations represent thermal vibrations around the starting equilibrium structure. Thus, the thermal motion is not able to carry out an interconversion between equilibrium structures of different isomers. For all the systems the Me-C and Me-O distances oscillate around their equilibrium values with relatively small amplitudes. In the π-complexes the Me-C distances change by ±0.001Å,

a)   b)

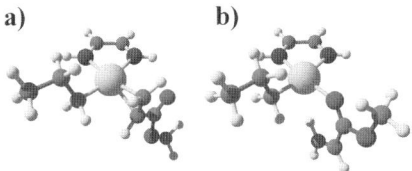

Figure 4-25. The π- and σ- complexes of methyl acrylate with the model catalyst

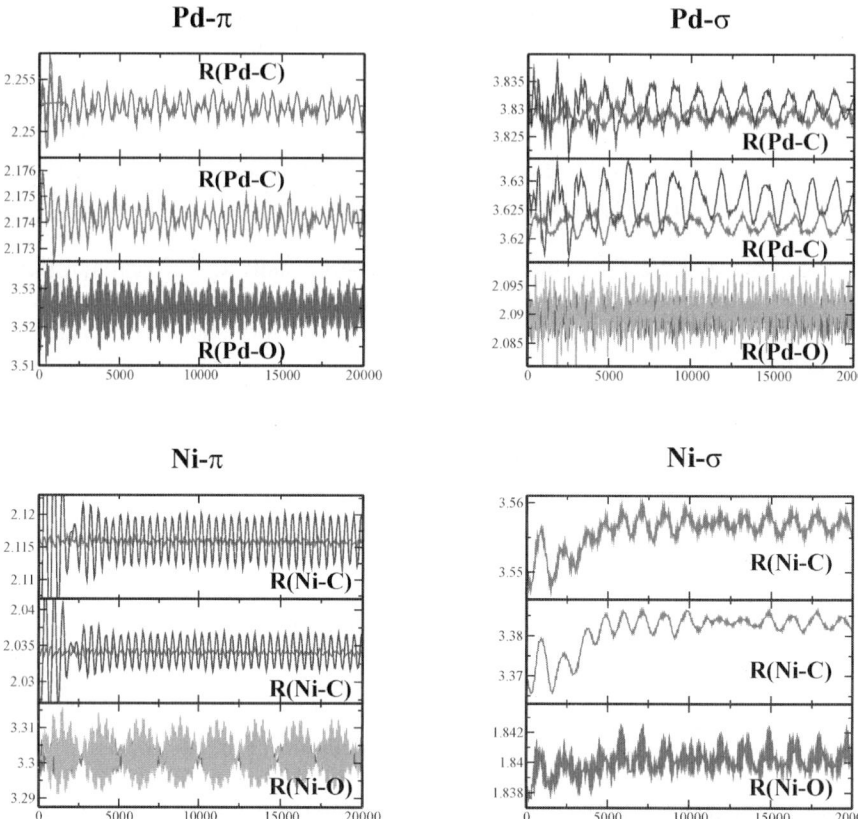

Figure 4-26. Unconstrained MD simulations for methyl acrylate bound to the palladium/nickel diimine complex through the oxygen (top-right/bottom-right) and the C=C (top-left/bottom-left) functionalities. The three panels in each of the four graphs represent variations in the metal-carbon (top two panels) and metal-oxygen (bottom panel) distances. The simulations were carried out at 300 K for all the systems, and 700 K for the local minima, as indicated

while the Me-O distances change by ± 0.008Å at 300K. In both the Ni- and Pd-π-complexes the amplitudes for the Me-C distances are much smaller than for the Me-O distances, since the latter represent the separation between non-bonded atoms, and are affected by the phases of all the vibrations in the system. Obviously, in the O-complexes this trend is inverted: here, the amplitudes for the Me-O bond lengths are substantially smaller (± 0.001Å) than for the Me-C non-bonding distances (up to ± 0.005Å).

The trajectories obtained from the simulations at 700 K performed for the Pd-O- and Ni-π-complex are different from those at 300K only by the amplitude of the oscillations. From the plots of Figure 4-26, there is no doubt that these local minima are stable on the 700K free energy surface and do not tend to transform into the global minima geometries.

In the next stage, the MD simulations were performed, in order to qualitatively model the possible inter-conversion reactions: $\pi \rightarrow \sigma$ and $\sigma \rightarrow \pi$. Our MD studies of the four interconversion processes consist for each reaction of two parts. In the first (constrained) part the system is dragged along a reaction coordinate (the constraint $RC = R_{Pd-C}-R_{Pd-O}$) from one form to the other while the thermal motion (trajectory) of all the other normal modes are recorded along with the change in energy. An unconstrained (relaxation) MD simulation is carried out in the second part after the system has reached the final value of the reaction coordinate RC. In this part thermal motion is allowed along all the normal modes including RC.

An example of such a simulation starting from the acrylate π-complex is summarized in Figure 4-27, in which the crucial interatomic distances and selected geometries are displayed. The data in the figure show that at the end of the simulation the σ-complex is not formed. Instead, the methyl acrylate molecule dissociate. Similar results were obtained for the system with Pd-catalyst. Thus, MD simulations suggest a dissociative pathway for the interconversion reactions between the two binding modes.

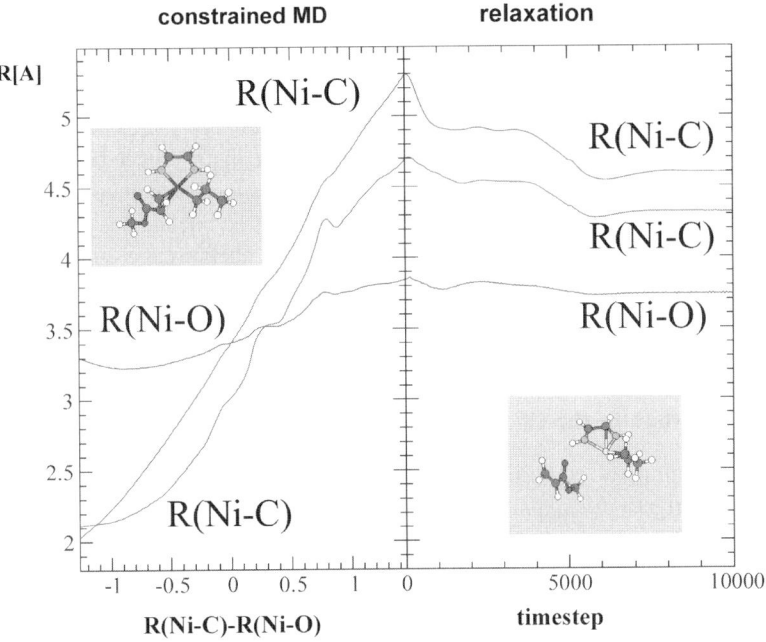

Figure 4-27. Changes in the two distances between the olefin carbons and the metal ($M-C(\pi)$ and $M-C'(\pi)$) as well as the metal-oxygen distance M-O from constrained MD and the following relaxation (unconstrained) simulations for the methyl acrylate π-complex $->$ σ-complex interconversion with the Ni-based diimine catalyst. The initial and final geometries are shown in the figure

The origin of the $\pi - /\sigma-$ binding mode preference was investigated by applying the Ziegler-Rauk fragment analysis of the interaction energy as well as the analysis of two-electron Fukui functions and the molecular electrostatic potential of the catalyst and monomer based on the static DFT results.[103,119] All these analyses led to similar conclusions: the preference of the acrylate σ-complex has an electrostatic origin. Thus, use of neutral or anionic catalyst instead of the cationic systems should result in stabilization of the π-complex compared to the σ-complex.

Concerning the polar monomer insertion, it was found out that the activation barriers are lower for the Ni- than for the Pd-diimine catalysts (see Figure 4-24). The 2,1-acrylate insertion barrier for the model Ni-catalyst (12.2 kcal/mol) is substantially lower than for the Pd-analogue (19.2 kcal/mol). For the real catalyst the barriers are relatively close for both systems: 12.4 kcal/mol for Pd-, and 13.5 kcal/mol for Ni-complex. For comparison, the geometry of the ethylene insertion TS was determined with the real Ni-catalyst. The ethylene insertion barrier is 14.2 kcal/mol for the Ni-, while for the analogue Pd-catalyst it is 16.7 kcal/mol. Thus, it is not the acrylate insertion that makes the copolymerization difficult in the Ni-case, as for both catalysts the acrylate inserion barriers are lower than the ethylene insertion barriers.

The stability of the insertion products is also quite similar for Ni- and Pd-catalysts. For both systems, insertion leads to a complex with a γ-agostic M-H interaction, **7**, that can easily isomerize to form more stable β-agostic complex, **8**, or can directly lead to the structure involving the chelating M-O bond, **9**. The 4-member chelate, **9**, is more stable than the γ-agostic complex by 15.2 kcal/mol, and is less stable than the structures with a 5- and 6-member ring, **10** and **11**. Comparing the thermodynamic stability of isomeric chelates, there is a slight difference between the two catalysts. Namely, for the Pd catalyst the complex with a 6-member ring is the most stable isomer, whereas in the Ni case, the 5-membered chelate has the lowest energy. It should be emphasized however, that all the chelates are substantially lower in energy compared to the agostic complexes for the Ni-catalyst than for Pd (see Figure 4-24). This again reflects higher oxophilicity of the Ni-system.

4.3.   MD Studies on the Chelate Opening by Ethylene

The chelate opening by coordination of the next monomer was expected to be a factor of importance for explaining the activity/inactivity of the Pd-/Ni-system. Therefore, the chelate opening reactions were investigated in details by both, static DFT and the molecular dynamics approach.

It should be pointed out that in the lowest energy structures of the π-complexes of the monomer formed from the chelates, the chelating bond M-O is still present, with the oxygen atom moved to the axial position. This is true for all sorts of chelates (4-, 5-, 6-membered), for both the monomers, ethylene and acrylate, as well as for both metals, Ni and Pd. The structure of the lowest energy ethylene complex formed from the 6-member chelate is shown in the first part of Figure 4-28.

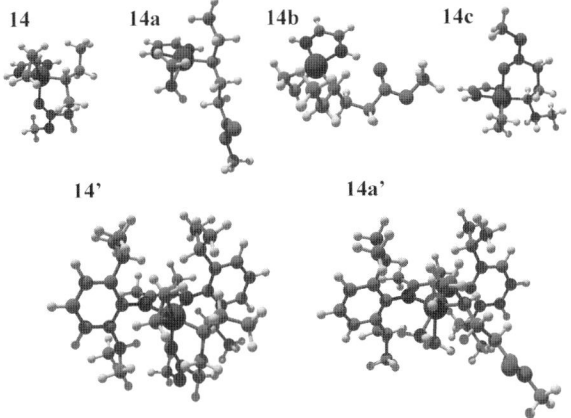

Figure 4-28. The π- complexes of ethylene formed from the 6-member chelate (after methyl acrylate insertion) with the model (**14**, **14a**, **14b**, **14c**) and real catalyst (**14′**, **14a′**)

On the other hand, it can be expected that on the respective potential energy surfaces, there should exist a variety of the isomeric structures without the M-O bond. Two examples (**14a**, **14b**) are shown in Figure 4-28. Thus, in order to understand the details of the mechanism, one has to consider the one-step chelate opening reaction by the monomer insertion, as well as the two-step process in which the M-O bond is broken (chelating ring is opened) prior to the monomer insertion, and the insertion starts from the higher energy complex without a M-O bond.

The static DFT calculations were not conclusive on the chelate opening,[105] due to a huge variety of the structures with and without M-O bond, and thus a huge variety of possible reaction pathways. Dozens of isomeric structures were optimized by DFT calculations for each type of chelate. Static calculations suggested two-step chelate opening, however the optimized insertion transition states were doubtful, as they were not necessarily connected to the most favorable of many possible pathways. Therefore, applying the MD approach was expected to be very helpful at this point. The additional advantage of the MD approach was that the simulations allow one to determine directly the free energy differences, not only the energy differences.

Among the main goals of these MD simulations were: (i) determine the stability of the local minima – higher energy structures without the M-O bond at finite temperatures; (ii) determine the monomer insertion pathways and their energetics for the insertion starting from the global minimum structure of the π-complex; here the TS determination by the static DFT approach was especially difficult; (iii) determine the energetics for the monomer insertion pathways starting from the alternative, higher energy complexes with a M-O bond; (iv) determine the energetics for the opening of the chelates prior to the ethylene insertion, i.e. at the monomer π-complex stage; (v) determine the energetics for the monomer insertion pathways starting from the alternative, higher energy complexes without M-O bond. The set of MD simulations performed for the Pd- and Ni-complexes was expected to reveal differences in their abilities as copolymerization catalysts due to a difference in their aptitude towards the chelate opening.

All the structures without the M-O chelating bond are higher in energy compared to the most stable, chelated structures. In Figure 4-28 two examples of such isomers for the Ni-catalyst are presented; the structures for the Pd-systems are qualitatively similar. The complexes **14a** and **14b** are higher in energy then **14** by 16.9 kcal/mol and 6.3 kcal/mol, respectively. The corresponding Pd-complexes were found to be higher in energy then the most stable isomer by 11.8 and 1.7 kcal/mol, respectively. Thus, the energy differences between the analogous chelated and non-chelated structures are larger for Ni then for Pd by ca. 5 kcal/mol. This difference can be attributed to an increased stability of the Ni-chelates compared to Pd- (see Figure 4-24), reflecting the higher oxophilicity of the Ni-systems.

In order to determine the stability of the non-chelated complexes **14a** and **14b** at finite temperatures, we have performed unconstrained MD simulations (at 300K and 1300K) for the respective Pd- and Ni-based systems. The results showed that for both metals the geometries oscillate in the vicinity of the corresponding local minima. No spontaneous isomerization towards chelated structures was observed, nor towards any other isomers. Thus, it can be concluded that the non-chelated isomers are stable on their free-energy surfaces, and separated from other isomers by a non-negligible barriers.

The opening of the six-member chelate prior to ethylene insertion was studied by a slow-growth MD simulation for Ni and Pd ethylene-chelate complexes **14**. The M-O distance was used as a reaction coordinate, increasing from the values in **14** (2.29 Å for Pd and 2.08 Å for Ni) up to 4.23 Å (8 bohr). The TS regions were located at a M-O distance of 3.12 Å and 3.71 Å for Ni and Pd, respectively. The activation energies and free energies are: $\Delta E^{\#} = 6.8$ kcal/mol, $\Delta G^{\#} = 11.3$ kcal/mol for Pd, and $\Delta E^{\#} = 10.9$ kcal/mol, $\Delta G^{\#} = 14.4$ kcal/mol for the Ni-catalyst. Thus, for the Pd catalyst the activation barrier was found to be slightly lower than for the Ni complex.

The MD simulations were performed for the model catalyst, in which the presence of the bulky substituents was neglected. It might be expected that the steric bulk would facilitate the chelate opening, since the chelating oxygen occupies an axial position and must interact strongly with the catalyst substituents. In order to investigate the steric effect, the static DFT calculations were carried out for the complexes **14′** and **14a′** with the real Ni- and Pd- diimine catalysts (analogous to **14** and **14a**). The optimized example geometries of **14′** and **14a′** with the Ni-catalyst are presented in the bottom part of Figure 4-24. The results show that indeed, the energy difference between **14′** and **14a′** is strongly decreased compared to the model systems (**14**, **14a**) for both, the Ni- and Pd- complexes. The opened-chelate complex **14a′** is higher in energy than **14′** by 4.2 kcal/mol and 7.5 kcal/mol, for the Ni- and Pd-catalyst, respectively. Thus, for the real systems the chelate opening prior to insertion, **14′** → **14a′** becomes more facile compared to the model systems; the corresponding reaction energies are decreased by 12.7 kcal/mol and 4.3 kcal/mol for the Ni- and Pd-system, respectively. This effect appears as a result of a strong destabilization of the chelated systems in the congested geometries of **14′**. The systems without addional chelating bonds, **14a′**, are only slightly influenced by the bulky substituents on the catalyst. It is not surprising that the effect of the steric bulk is substantially larger for Ni than for Pd, as all the bonds involving Ni are typically shorter than those containing Pd by 0.1–0.2 Å, and thus, the nickel systems are more congested. These results clearly demonstrate that the steric bulk in real systems facilitate the chelate opening.

In order to determine the ethylene insertion starting from the chelated complexes, the slow growth MD simulations were performed, with the distance between the α-carbon of the chain and an olefin carbon chosen as a reaction coordinate.. The activation barriers obtained from the simulations are presented in Table 4-1. The results clearly show that in each case the barriers are substantially lower for the Ni- than for the Pd-catalyst. For all the systems, the ethylene insertion reactions starting from the most stable chelate structures **12**, **13**, and **14** have very high barriers (38–53 kcal/mol and 32–41 kcal/mol, for Pd and Ni, respectively). These values are much higher than the 'standard' ethylene insertion barriers into the metal-alkyl bond ($\Delta E^{\#} = 16.8$ kcal/mol and 14.2 kcal/mol for Ni and Pd, respectively). These high barriers demonstrate that the ethylene insertion definitely cannot proceed from the most stable ethylene-chelate structure.

The example trajectories from the reaction with 6-membered chelate are summarized in Figure 4-29. The Figure summarizes the structures, together with crucial inter-atomic distances (metal-oxygen and the metal-γ-hydrogen) along the MD trajectories. The results demonstrate that for both metals, the chelating bond is practically present along the whole insertion pathway. The Ni-O and Pd-O bonds are weakened after passing the TS region; they are extended by 0.07 Å and 0.35 Å compared to the initial structures, respectively. In both cases, the insertion leads directly to the 8-member chelate, as reflected by the Ni-O and Pd-O bond shortening after passing the TS. For both, Ni and Pd, catalyst there is practically no

Table 4-1. Ethylene Insertion Barriers after the Acrylate Insertion Obtained from the MD simulations

Initial complex	activation barriers[a]			
	Ni – catalyst		Pd-catalyst	
	$\Delta E^{\#}$	$\Delta G^{\#}$	$\Delta E^{\#}$	$\Delta G^{\#}$
4-chelate:				
12	28.1	32.9	35.8	38.2
5-chelate:				
13	33.9	36.5	41.2	44.7
6-chelate:				
14	38.8	40.8	49.9	53.4
14a	18.8	20.4	24.0	30.4

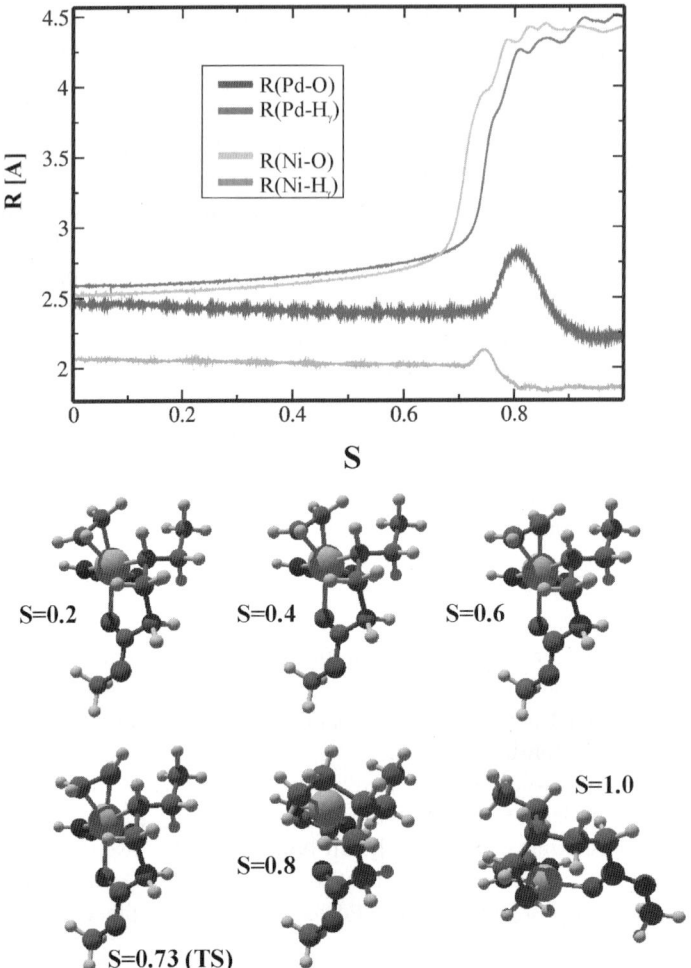

Figure 4-29. Changes in the interatomic distances for M-O (carbonyl) and M-Hg (M = Ni, Pd) along the MD trajectory obtained from the ethylene insertion starting from the complex **14** [top] and examples of the structures observed along the trajectory [bottom]; S denotes a reaction progress variable

interaction between the metal and the γ-hydrogen (i.e. α-hydrogen in the initial structure); the distance between the metal atom and the γ-hydrogen increase as the insertion proceed. This is different from the usual olefin insertions observed in the homopolymerization reactions, where the TS is usually stabilized by a γ-hydrogen interacting with the metal. It seems very likely that the lack of this γ-hydrogen stabilization is responsible for the very high insertion barriers observed here. In the most stable structures of **14**, **15** and **16** the α-hydrogen of –CH(R^1)(R^2)COOCH$_3$ group (i.e. the γ-hydrogen of the product) points towards ethylene, and during the

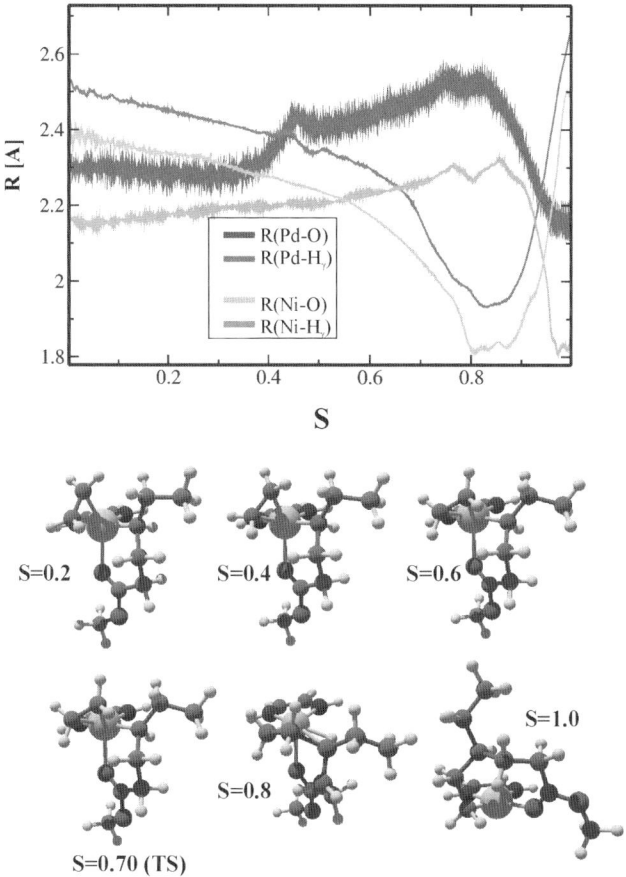

Figure 4-30. Changes in the interatomic distances for M-O (carbonyl) and M-H$_g$ (M = Ni, Pd) along the MD trajectory obtained from the ethylene insertion starting from the complex **14c [top]** and examples of the structures observed along the trajectory [bottom]; *S* denotes a reaction progress variable

insertion pathways the –CHR^{1}R^2 group rotates in a direction that increases the metal-hydrogen distance.

Therefore, we also performed simulations for the alternative isomers of **14**, in which the hydrogen is located 'on the opposite side' of ethylene, so that there exist a possiblility of the γ-agostic interactions (structure **14c** in Figure 4-28). Figure 4-30 reports the trajectories from the MD simulation modeling the insertions starting from the Ni- and Pd-complexes **14c**: the metal-oxygen and metal-γ-hydrogen distances are presented, together with the example structures along the corresponding trajectories. Figure 4-30 shows that indeed, during the reaction the γ-hydrogen approaches the metal, and in the TS region the Pd-H$_γ$ and Ni-H$_γ$ distances are typical for γ-agostic interactions (1.93 and 1.81 A for Pd and Ni, respectively). The system spontaneously evolves towards an 8-member chelate for both metals after passing the TS as reflected in a decrease in the metal-oxygen distances. It should be pointed out here, that again the chelating metal-oxygen bond is present along the whole pathway. In the TS region the metal-oxygen distance is extended only by $0.16\,\text{Å}$ and $0.22\,\text{Å}$ for Ni and Pd, respectively. The conclusion drawn on the basis of our finite temperature MD calculations differ from those obtained previously[105] from static DFT calculations where the chelating bond breaks before reaching the insertion TS. This demonstrates that the MD approach is especially valuable for studying reactions, for which there exist a large number of isomeric structures, giving rise to many alternative pathways with different isomeric TS geometries.

The activation barriers for the insertion starting from **14c** are visibly lowered compared to those obtained for **14**. (Table 4-1). Decrease in the barriers demonstrates a role of the γ-agostic interactions in stabilization of the TS for the olefin insertion reactions. However, for both metals the barriers are still substantially higher than the activation barriers for olefin insertion observed in the homopolymerization processes.

Finally, for the Pd-catalyst the MD simulations were performed for yet another ethylene insertion pathway, starting form the higher energy π-complex **14a**, in which the chelating Pd-O bond does not exist. For such an insertion pathway the earlier[105] static DFT calculations demonstrated that the barriers for insertion starting from π-complexes without chelating M-O bonds are comparable to the barriers of ethylene insertion in the homo-polymerization processes. Also, the DFT calculations suggested that in such a case the oxygen atom practically does not play a role in the insertion.

Figure 4-31 shows the Pd-O and $Pd-H_\gamma$ distances along the MD trajectory, the activation barriers are listed in Table 4-1. The results demonstrate that at $T = 300K$ the insertion proceeds according to the mechanism typical for ethylene homopolymerization. The distance between the metal and the γ-hydrogen decreases along the pathway, and the insertion product is a γ-agostic complex. The presence of carbonyl oxygen does not influence the reaction: this oxygen atom stays remote from the metal ($5.3–6\,\text{Å}$). The calculated activation barrier [$\Delta E^{\#} = 17.1\,\text{kcal/mol}$, $\Delta G^{\#} = 19.1\,\text{kcal/mol}$] for the insertion is comparable with the homopolymerization activation barriers. Very similar results were obtained for the insertion starting from **14b**: $\Delta E^{\#} = 17.02\,\text{kcal/mol}$, $\Delta G^{\#} = 22.3\,\text{kcal/mol}$.

The results of the MD simulations clearly demonstrate that the insertion starting from the higher energy isomers of the ethylene-chelate complexes in which the chelating bond has been broken have much smaller activation barriers, that are comparable to those observed in ethylene homopolymerization. This, however, does not explain the differences in the copolymerization activity of Pd and Ni-diimine complexes, as the barriers for the ethylene insertion into Ni-alkyl bond are smaller ($14.2\,\text{kcal/mol}$) than those for Pd-alkyl bond ($16.8\,\text{kcal/mol}$). Thus, it may be concluded that the ethylene insertion following the insertion of the polar monomer is not a crucial factor for the diimine catalyst copolymerization activity. It is the initial poisoning of the catalyst by formation of the

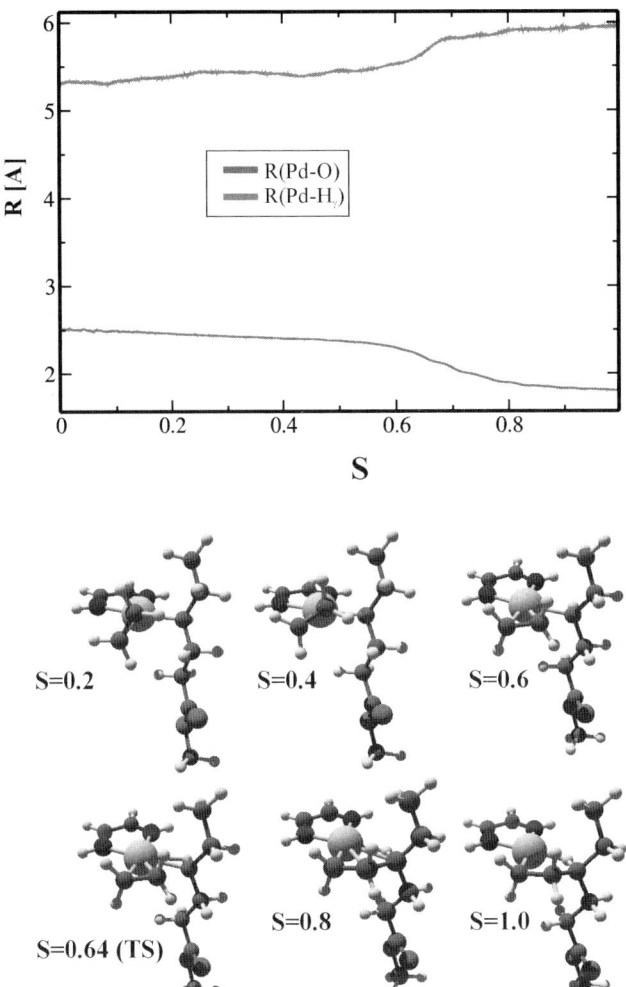

Figure 4-31. Changes in the interatomic distances for M-O (carbonyl) and M-H$_g$ (M = Ni, Pd) along the MD trajectory obtained from the ethylene insertion starting from the complex **14a [top]** and examples of the structures observed along the trajectory **[bottom]**; S denotes a reaction progress variable

O-complexes and the chelate opening prior to the ethylene insertion that seems to be responsible for differences in catalytic activity of the Ni- and Pd-diimine complexes.

The example of the complex catalytic process studied here shows that *ab initio* MD can be applied as a very useful supplementary methodology combined with the 'standard' static DFT approach. Many details of the molecular mechanisms of such processes can be disclosed and rationalized on the basis of the static computations and do not require additional computational effort introduced by

MD. However, MD is a very helpful approach when, for instance, there exist many alternative pathways linking the same reactants with the products and it is difficult to draw conclusions without exploring such pathways at finite temperature.

5.   CONCLUDING REMARKS

The first-principle molecular dynamics has reached the stage in which it can be applied in modeling of relatively large molecular systems. Although the time-scale of typical *ab initio* MD simulations is still not sufficient to observe the chemical reactions occurring spontaneously, a development of various approaches towards overcoming the high activation barriers makes it possible to investigate molecular mechanisms of complex chemical reactions. The number of successful applications of *ab initio* MD in wide areas of chemistry and physics is quickly growing. It seems very likely that soon further developments of 'routine pathways' for certain types of applications and further improvements of existing MD software will lead to a revolution similar to the one that has been taking place in the area of applications of static quantum-chemical programs. DFT-based molecular dynamics already has become one of standard molecular modeling tools used to support experimental investigations and to stimulate new experiments. Already now, the role of *ab initio* MD methodologies in interdisciplinary studies directed towards a rational design of new molecular materials cannot be underestimated.

REFERENCES

1. Car R, Parrinello M, Unified Approach for Molecular Dynamics and Density-Functional Theory, Phys Rev Lett, 55, 2471 (1985)
2. Marx D, Hutter J, Ab Initio Molecular Dynamics Theory and Implementation, In Modern Methods and Algorithms of Quantum Chemistry, edited by J Grotrndorst, NIC Series, Vol 1 (John von Neumann Institute of Computing: Jülich, 2000), pp 301–449; and refs therein
3. Rothlisberger U, 15 Years of Car-Parrinello Simulations in Physics, Chemistry and Biology In: Computational Chemistry Reviews of Current Trends, Vol 6, edited by J Leszczynski (World Scientific, Singapore, 2001), pp 33–68
4. Carloni P, Rothlisberger U, Parrinello M, The Role and Perspective of Ab-initio Molecular Dynamics in the Study of Biological Systems, AccChem Res 35, 455–464 (2002)
5. Andreoni W, Curioni A, New advances in chemistry and materials science with CPMD and parallel computing, Parallel Comput 26, 819–842 (2000)
6. Blöchl PE, Först CJ, Schimpl J, The Projector Augmented Wave Method: Ab-initio Molecular Dynamics Simulations with Full Wave Functions, Bull Mater Sci 26, 33 (2003)
7. Magistrato A, Togni A, Rothlisberger U, Woo TK, Molecular Modeling of Enantioselective Hydrosilylation by Chiral Pd Based Homogeneous Catalysts with First-Principles and Combined QM/MM Molecular Dynamics Simulations In: Computational Modelling of Homogeneous Catalysis, edited by: F Maseras, A Lledos (Kluwer Academic, Dordrecht, 2002), pp 213–252
8. Blöchl P E, Senn H M, Togni A, Molecular Reaction Modeling from Ab-Initio Molecular Dynamics, In: Transition State Modeling for Catalysis, edited by D G Truhlar, K Morokuma, ACS Symposium Series 721 (American Chemical Society, Washington DC, 1998), pp 88–99

9. Woo TK, Margl PM, Deng L, Cavallo L, Ziegler T, Transition state Modeling for Catalysis, edited by D G Truhlar, K Morokuma, ACS Symposium Series 721 (American Chemical Society, Washington DC, 1998), pp 173–186

10. Baveridge DL, DiCapua FM, Free Energy Via Molecular Simulation: Applications to Chemical and Biomolecular Systems, Ann Rev Biophys Chem 18, 431–492 (1989)

11. Ehrenfest P, Bemerkung über die angenäherte Gültigkeit der klassischen Mechanik innerhalb der Quantenmechanik, Z Phys, 45, 455–457 (1927)

12. Born M, Oppenheimer JR, On the Quantum Theory of Molecules, Ann Phys, 84, 457 (1927)

13. CPMD, Copyright IBM Corp 1990–2001, Copyright MPI fur Festkorperforschung Stuttgart 1997–2005

14. Trouiller N, Martins JL, Efficient pseudopotentials for plane-wave calculations, Phys Rev B, 43, 1993 (1991)

15. Trouiller N, Martins JL, Efficient pseudopotentials for plane-wave calculations II Operators for fast iterative diagonalization, Phys Rev B, 43, 8861 (1991)

16. Hellmann H, Zur Rolle der kinetischen Elektronenenergie für die zwischenatomaren Kräfte, Z Phys, 85, 180–190 (1933)

17. Feynman R P, Forces in Molecules, Phys Rev, 56, 340 (1939)

18. Pulay P, Ab initio Calculation of Force Constants and Equilibrium Geometries I Theory, Mol Phys, 17, 197 (1969)

19. Pulay P, Derivative Methods in Quantum Chemistry, Adv Chem Phys, 69, 241 (1987)

20. Leach AR, Molecular Modelling Principles and Applications (Pearson Education, 2001); and refs therein

21. Woodcock LV, Isothermal molecular dynamics calculations for liquid salts, Chem Phys Lett, 10, 257–261 (1971)

22. Berendsen HJC, Postma JPM, van Gunsteren WF, Di Nola A, Haak JR, Molecular dynamics with coupling to an external bath, J Chem Phys, 81, 3684 (1984)

23. Nose S , Constant-temperature molecular dynamics, Mol Phys, 57, 187 (1986)

24. Hoover WG, Canonical dynamics: Equilibrium phase-space distributions, Phys Rev A, 31, 1695 (1985)

25. Tuckerman ME, Parrinello M, Integrating the Car–Parrinello equations I Basic integration techniques, J Chem Phys, 101, 1302 (1994)

26. Hutter J, Tuckerman ME, Parrinello M, Integrating the Car–Parrinello equations III Techniques for ultrasoft pseudopotentials, J Chem Phys, 102, 859 (1995)

27. Andersen HC, Molecular dynamics simulations at constant pressure and/or temperature, J Chem Phys, 72, 2384–2393 (1980)

28. Kelly E, M Seth, T Ziegler, Calculation of free energy profiles for elementary bimolecular reactions by a initio molecular dynamics: Sampling methods and thermostat considerations, J Phys Chem A, 108, 2167–2180 (2004)

29. Vanderbilt D, Phys Rev B, Soft self-consistent pseudopotentials in a generalized eigenvalue formalism, 41: 7892 (1990)

30. Goedecker S , K Maschke, Transferability of pseudopotentials, Phys Rev A, 45: 88 (1992)

31. Blöchl PE, Projector augmented-wave method, Phys Rev B, 50: 17953 (1994)

32. Blöchl PE, Electrostatic Decoupling of Periodic Images of Plane-Wave-Expanded Densities and Derived Point Charges, J Chem Phys, 103, 7422–7428 (1995)

33. Michalak A, Ziegler T, First-principle molecular dynamic simulations along the intrinsic reaction paths, J Phys Chem A, 105, 4333–4343 (2001)

34. Carter EA, Ciccotti G, Hynes JT, Kapral R, Constrained reaction coordinate dynamics for the simulation of rare events, Chem Phys Lett, 156, 472–477 (1989)

35. Paci E, Ciccotti G, Ferrario M, Kapral R, Activation energies by molecular dynamics with constraints, Chem Phys Lett, 176, 581–587 (1991)

36. Ryckaert JP, Ciccotti G, Berendsen HJC, Numerical integration of the cartesian equations of motion of a system with constraints: molecular dynamics of n-alkanes , J Comput Phys, 23, 327–341 (1977)

37. Straatsma TP, Berendsen HJC, Postma JPM, Free energy of hydrophobic hydration: A molecular dynamics study of noble gases in water, J Chem Phys 85, 6720–6727 (1986)

38. Sprik M, Ciccotti G, Free energy from constrained molecular dynamics, J Chem Phys 109, 7737–7744 (1998)

39. Vuilleumier R, Sprik M, Electronic control of reactivity using density functional perturbation methods, Chem Phys Lett 365, 305–312 (2002)

40. VandeVondele J, Rothlisberger U, Accelerating Rare Reactive Events by Means of a Finite Electronic Temperature, J Am Chem Soc, 124, 8163–8171 (2002)

41. Mosey NJ, Hu A, Woo TK, Ab initio molecular dynamics simulations with a HOMO–LUMO gap biasing potential to accelerate rare reaction events, Chem Phys Lett, 373, 498–505 (2003)

42. Guidoni L, Rothlisberger U, Scanning reactive pathways with orbital biased Molecular Dynamics, J Chem Theory Comput, 1, 554–560 (2005)

43. Fleurat-Lessard P, Ziegler T, Tracing the minimum-energy path on the free-energy surface, J Chem Phys, 123, 084101 (2005)

44. Laio A, Parrinello M, Escaping free-energy minima, ProcNatl Acad Sci, 99, 12562 (2002)

45. Ianuzzi M, Laio A, Parrinello M, Efficient Exploration of Reactive Potential Energy Surfaces Using Car-Parrinello Molecular Dynamics, Phys Rev Lett, 90, 238302 (2003)

46. Fukui K, The path of chemical reactions – the IRC approach, Acc Chem Res, 14, 363–368 (1981)

47. Tachibana A, Fukui K, Novel variational principles of chemical reaction, Theor Chem Acta (Berl), 57, 81–94 (1980)

48. Perdew JP, Zunger A, Self-interaction correction to density-functional approximations for many-electron systems, Phys Rev B, 23, 5048 (1981)

49. Becke A, Density-functional exchange-energy approximation with correct asymptotic behavior, Phys RevA 38, 3098 (1988)

50. Perdew JP, Density-functional approximation for the correlation energy of the inhomogeneous electron gas, Phys Rev B, 33, 8822 (1986)

51. Perdew JP, Erratum: Density-functional approximation for the correlation energy of the inhomogeneous electron gas, Phys Rev B 34, 7406 (1986)

52. Blöchl PE, Parinello M, Adiabaticity in first-principles molecular dynamics, Phys Rev B 45, 9413 (1992)

53. Pearson PK, Schaefer HF III, Potential energy surface for the model unimolecular reaction HNC → HCN, J Chem Phys, 62, 350–354 (1975)

54. Redmon LT, Purvis GD III, RJ Bartlett, Correlation effects in the isomeric cyanides: HNC HCN, LiNC LiCN, and BNC BCN, J Chem Phys 72, 986–991 (1980)

55. Nobels RH, Random L, HOC+: An observable interstellar species? A comparison with the isomeric and isoelectronic HCO+, HCN and HNC, Chem Phys , 60, 1–10 (1981)

56. Peric, Maledenivic M, Ab initio study of the isomerization HNC → HCN I Ab initio calculation of the HNC HCN potential surface and the corresponding energy levels, Chem Phys, 82, 317–336 (1983)

57. Murrell JN, Carter S, Halonen LO, Frequency optimized potential energy functions for the ground-state surfaces of HCN and HCP, J Mol Spectrosc 93,307–316 (1982)

58. Holme TA, Hutchinson JS, Vibrational energy flow into a reactive coordinate: A theoretical prototype for a chemical system J Chem Phys, 83, 2860–2869 (1985)

59. Pan CF, Hehre WJ, Heat of formation of hydrogen isocyanide by ion cyclotron double resonance spectroscopy, J Phys Chem, 86, 321 (1982)

60. Ishida K, Morokuma K, Komornicki AJ, The intrinsic reaction coordinate An ab initio calculation for HNCHCN and H–+CH4CH4+H–, J Chem Phys 66, 2153–2156 (1977)

61. Muller K, Brown LD, Location of saddle points and minimum energy paths by a constrained simplex optimization procedure, Theoret Chim Acta (Berl) 53, 75–93 (1979)

62. Garret BC et al, Algorithms and accuracy requirements for computing reaction paths by the method of steepest descent, J Phys Chem, 92, 1476–1478 (1988)

63. Gonzalez C, Schegel HBJ, Reaction path following in mass-weighted internal coordinates, J Phys Chem , 94, 5523–5527 (1990)

64. Deng L, Exploring Potential Energy Surfaces using Density Functional Theory and Intrinsic Reaction Coordinate Methods, *PhD Thesis* (University of Calgary, Calgary, 1996)

65. Deng L, Ziegler T, Combining Density Functional Theory and Intrinsic Reaction Coordinates, Int JQuant Chem, 52, 731–765 (1994)

66. Car RW Jr, Walters WDJ, The Thermal Isomerization of Cyclobutene, J Phys Chem, 69, 1073 (1965)

67. Wiberg KB, Fenoglio RA, Heats of formation of C4H6 hydrocarbons J Am Chem Soc, 90, 3395–3397 (1968)

68. Lipnick RL, Garbisch EW Jr, Conformational analysis of 1,3-butadiene, J Am Chem Soc, 95, 6370–6375 (1973)

69. Hsu K, Buenker RJ, Peyerimhoff SD, Theoretical determination of the reaction path in the prototype electrocyclic transformation between cyclobutene and cis-butadiene Thermochemical process J Am Chem Soc 93, 2117–2127 (1971)

70. Hsu K, Buenker RJ, Peyerimhoff SD, Role of ring torsion in the electrocyclic transformation between cyclobutene and butadiene Theoretical study J Am Chem Soc 94, 5639–5644 (1972)

71. Brulet J, Schaefer HF III, J Am Chem Soc, Conrotatory and disrotatory stationary points for the electrocyclic isomerization of cyclobutene to cis-butadiene 106, 1221–1226 (1984)

72. Wiest O, Houk KN, Density Functional Theory Calculations of Pericyclic Reaction Transition Structures. In: Density Functional Theory IV, edited by RF Nalewajski, Topics in Curr Chem 183(Springer, Berlin, Heilderberg, 1996), pp 2–24, and refs therein

73. Vetter R, Zulicke L, Theoretical study of potential wells and barriers for SN2 rearrangement in the systems (XCH3X)- with X = F, Cl, and Br, J Am Chem Soc,112, 5136–5142 (1990), and refs therein

74. Tucker SC, Truhlar DG, Ab initio calculations of the transition-state geometry and vibrational frequencies of the SN2 reaction of chloride with chloromethane, J Phys Chem 93, 8138–8142 (1989), and refs therein

75. Shi Z, Boyd RJ, Intrinsic barriers of some model SN2 reactions: second-order Moeller-Plesset perturbation calculations, J Am Chem Soc, 113, 2434–2439 (1991) and refs therein

76. Hu W, Truhlar DG, Structural Distortion of CH3I in an Ion-Dipole Precursor Complex, J Phys Chem, 98, 1049– 1052 (1994), and refs therein

77. Michalak A, Ziegler T, Organometallics, 22, 2660 (2003)

78. Ittel SD, Johnson LK, Brookhart M, Late-Metal Catalysts for Ethylene Homo- and Copolymerization, Chem Rev, 100, 1169–1204 (2000), and refs therein

79. Boffa LS, Novak BM, Copolymerization of Polar Monomers with Olefins Using Transition-Metal Complexes, Chem Rev, 100, 1479–1494 (2000), and refs therein

80. Ziegler K, Holtzkamp E, Martin H, Breil, H, Das Mülheimer Normaldruck-Polyäthylen-Verfahren, Angew Chem, 67, 541 (1955)

81. Ziegler K, Holtzkamp E, Breil H, Martin H, Polymerisation Äthylen und Anderen Olefinen, Angew Chem, 67, 426 (1955)

82. Natta GJ, Une Nouvelle Classe de Polymeres d'a-Olefines ayant une Regularite de Structure Exceptionelle, Polym Sci, 16, 143 (1955);

83. Natta GJ, Stereospezifische Katalysen und isotaktische Polymere, Angew Chem 68, 393 (1956)

84. Kaminsky W, Kulper K, Brintzinger HH, FRWP Wild, Polymerization of Propene and Butene with a Chiral Zirconocene and Methyl Aluminoxane as Cocatalyst, Angew Chem Int Ed Engl, 24, 507 (1985)

85. Brintzinger HH, Fischer D, Mülhaupt R, Rieger B, Waymouth RM, Stereospecific Olefin Polymerization with Chiral Metallocene Catalysts, Angew Chem Int Ed Engl 34, 1143 (1995), and refs Therein

86. McKnight AL, Waymouth RM, Group 4 ansa-Cyclopentadienyl-Amido Catalysts for Olefin Polymerization, Chem Rev 98, 2587–2598 (1998), and refs therein

87. Alt HG, Koppl A, Effect of the Nature of Metallocene Complexes of Group IV Metals on Their Performance in Catalytic Ethylene and Propylene Polymerization, ChemRev, 100, 1205–1222 (2000), and refs Therein

88. Coates GW, Precise Control of Polyolefin Stereochemistry Using Single-Site Metal Catalysts, ChemRev, 100, 1223–1252 (2000)

89. Britovsek GJP, Gibson VC, Wass DF, The Search for New-Generation Olefin Polymerization Catalyst: Life Beyond Metallocenes, Angew Chem Int Ed, 38, 428 (1999), and refs therein

90. Beyond Metallocenes Next-Generation Polymerization Catalysts, edited by AO Patil, GG Hlatky, ACS Symp Ser 857 (ACS, Washington DC, 2003)

91. Johnson LK, Mecking S, Brookhart M, Copolymerization of Ethylene and Propylene with Functionalized Vinyl Monomers by Palladium(II) Catalysts, J Am Chem Soc, 118, 267–268 (1996)

92. Mecking S, Johnson LK, Wang L, Brookhart M, Mechanistic Studies of the Palladium-Catalyzed Copolymerization of Ethylene and -Olefins with Methyl Acrylate, J Am Chem Soc, 120, 888–899 (1998)

93. Tempel DJ, Johnson LK, Huff RL, White PS, Brookhart M, Mechanistic Studies of Pd(II)–Diimine-Catalyzed Olefin Polymerizations, J Am Chem Soc, 122, 6686–6700 (2000)

94. Shultz LH, Brookhart M, Measurement of the Barrier to -Hydride Elimination in a -Agostic Palladium-Ethyl Complex: A Model for the Energetics of Chain-Walking in (-Diimine)PdR+ Olefin Polymerization Catalysts, Organometallics, 20, 3975–3982 (2001)

95. Shultz LH, Tempel DJ, Brookhart M, Palladium(II) -Agostic Alkyl Cations and Alkyl Ethylene Complexes: Investigation of Polymer Chain Isomerization Mechanisms, J Am Chem Soc, 123, 11539–11555 (2001)

96. Musaev DG, Froese RDJ, Morokuma K, Molecular Orbital and IMOMM Studies of the Chain Transfer Mechanisms of the Diimine-M(II)-Catalyzed (M = Ni, Pd) Ethylene Polymerization Reaction, Organometallics, 17, 1850–1860 (1998)

97. Froese RDJ, Musaev DG, Morokuma K, Theoretical Study of Substituent Effects in the Diimine-M(II) Catalyzed Ethylene Polymerization Reaction Using the IMOMM Method, J Am Chem Soc, 120, 1581–1587 (1998)

98. Schenck H, Stromberg S, Zetterberg K, Ludwig M, Akermark B, Svensson M, Insertion Aptitudes and Insertion Regiochemistry of Various Alkenes Coordinated to Cationic (-R)(diimine) palladium(II) (R = -CH3, -C6H5) A Theoretical Study, Organometallics, 20, 2813–2819 (2001)

99. Deng L, Margl P, Ziegler T, J Am Chem Soc, A Density Functional Study of Nickel(II) Diimide Catalyzed Polymerization of Ethylene, 119, 1094–1100 (1997)

100. Deng L, Woo TK, Cavallo L, Margl PM, Ziegler T, The Role of Bulky Substituents in Brookhart-Type Ni(II) Diimine Catalyzed Olefin Polymerization: A Combined Density Functional Theory and Molecular Mechanics Study, J Am Chem Soc, 119, 6177–6186 (1997)

101. Michalak A, Ziegler T, Pd-catalyzed Polymerization of Propene – DFT Model Studies, Organometallics, 18, 3998–4004 (1999)

102. Michalak A, Ziegler T, DFT studies on substituent effects in Pd-catalyzed olefin polymerization, Organometallics, 19, 1850–1858 (2000)

103. Michalak A, Ziegler T, DFT Studies on the Copolymerization of a-Olefins with Polar Monomers: Comonomer Binding by Nickel- and Palladium-Based Catalysts with Brookhart and Grubbs Ligands, Organometallics, 20, 1521–1532 (2001)

104. Michalak A, Ziegler T, DFT Studies on the Copolymerization of a-Olefins with Polar Monomers: Ethylene-Methyl Acrylate Copolymerization Catalyzed by a Pd-based Diimine Catalyst, J Am Chem Soc, 123, 12266–12278 (2001)

105. Michalak A, Ziegler T, Stochastic simulations of polymer growth and isomerization in the polymerization of propylene catalyzed by Pd-based diimine catalysts, J Am Chem Soc, 124, 7519–7528 (2002)

106. Deubel D, Ziegler T, DFT Study of Olefin versus Nitrogen Bonding in the Coordination of Nitrogen-Containing Polar Monomers to Diimine and Salicylaldiminato Nickel(II) and Palladium(II) Complexes Implications for Copolymerization of Olefins with Nitrogen-Containing Polar Monomers, Organometallics, 21, 1603–1611 (2002)

107. Michalak A, Ziegler T, Exploring the scope of possible microstructures accessible from polymerization of ethylene by late transition metal single-site catalysts A theoretical study, Macromolecules, 36, 928–933 (2003)

108. Michalak A, Ziegler T, Polymerization of ethylene catalyzed by a Nickel(+2) anilinotropone-based catalyst: DFT and stochastic studies on the elementary reactions and the mechanism of polyethylene branching, Organometallics, 22, 2069–2079 (2003)

109. Margl P, Michalak A, Ziegler T, Theoretical Studies on Copolymerization of Polar Monomers. In: Catalytic Synthesis of Alkene – Carbon Monoxide Copolymers and Cooligomers; edited by A Sen (Kluwer Academic Publishers, 2002), pp 265–307

110. Michalak A, Ziegler T, The Key Steps in Olefin Polymerization Catalyzed by Late Transition Metals. In: Computational Modeling of Homogeneous Catalysis, edited by F Maseras, A Lledos (Kluwer Academic Publishers, 2002)

111. Michalak A, Ziegler T, Theoretical Studies on the Polymerization and Copolymerization Processes Catalyzed by the Late-Transition Metal Complexes, in: Beyond Metallocenes Next-Generation Polymerization Catalysts; ACS Symp Series 857, edited by AO Patil, GG Hladky (American Chemical Society, Washington 2003), pp 154–172

112. Szabo MJ, Jordan RF, Michalak A, Piers WE, Weiss T, Sheng-Yong Yang, Ziegler T, Organometallics, 23, 5565 (2004)

113. Sheng-Yong Yang, Szabo MJ, Michalak A, Weiss T, Piers WE, Jordan RF, Ziegler T, The Exploration of Neutral Azoligand-Based Grubbs Type Palladium(II) Complex as Potential Catalyst for the Copolymerization of Ethylene with Acrylonitrile: A Theoretical Study Based on Density Functional Theory, Organometallics, 24, 1242–1251 (2005)

114. Szabo M J, Galea NM, Michalak A, Sheng-Yong Yang, Groux LF, Piers W E, Ziegler T, Copolymerization of Ethylene with Polar Monomers by Anionic Substitution A Theoretical Study Based on Acrylonitrile and the Brookhart Diimine Catalyst Organometallics, 24, 2147–2156 (2005)

115. Haras A, Michalak A, Rieger B, Ziegler T, Theoretical analysis of factors controlling the non-alternating CO/C2H4 copolymerization, J Am Chem Soc, 127, 8765–8774 (2005)

116. Szabo MJ, Galea NM, Michalak A, Sheng-Yong Yang, Groux LF, Piers W E, Ziegler T, Copolymerization of Ethylene with Polar Monomers:Chain Propagation and Side Reactions A DFT Theoretical Study Using Zwitterionic Ni(II) and Pd(II) Catalysts, J Am Chem Soc, 127, 14692–14702 (2005)

117. Coose P, Ziegler-Natta Catalysis I Mechanism of Polymerization of a-Olefins with Ziegler-Natta Catalyst, J Catal, 3, 80 (1964)

118. Arlman EJ, Coose P, Ziegler-Natta Catalysis III Stereospecific Polymerization of Propene with the Catalyst System $TiCl_3-AlEt_3$, J Catal, 3, 99 (1964)

119. Michalak A, Two-reactant Fukui function and molecular electrostatic potential analysis of the methyl acrylate binding mode in the complexes with the Ni- and Pd-diimine catalysts, Chem Phys Lett, 386, 346–350 (2004)

CHAPTER 5

COMPUTATIONAL ENZYMOLOGY: INSIGHTS INTO ENZYME MECHANISM AND CATALYSIS FROM MODELLING

ADRIAN J. MULHOLLAND AND IAN M. GRANT

Centre for Computational Chemistry, School of Chemistry, University of Bristol, Bristol BS8 1TS, UK
E-mail: Adrian.Mulholland@bris.ac.uk

Abstract: Modern modelling methods can now give a uniquely detailed understanding of enzyme-catalysed reactions, including analysing mechanisms and identifying determinants of specificity and catalytic efficiency. A new field of computational enzymology has emerged, which has the potential to contribute significantly to structure-based design, and in developing predictive models of drug metabolism; for example, in predicting the effects of genetic polymorphisms. This review outlines important techniques in this area, including quantum chemical model studies, and combined quantum mechanics/molecular mechanics (QM/MM) methods. Some recent applications to enzymes of pharmacological interest are also covered, showing the types of problems that can be tackled, and the insight they can give

Keywords: QM/MM, molecular dynamics, transition state, drug metabolism, polymorphism, cytochrome P450, pharmacogenomics

1. INTRODUCTION

Computational modelling of the mechanisms of enzyme-catalysed reactions has advanced significantly over recent years, and has matured to the point where a new field of computational enzymology has emerged.[1-3] Computer simulation and modelling methods can investigate fundamental questions about enzyme mechanism and catalysis that cannot be easily analysed by experiment. A number of different types of modelling methods have provided insight into enzyme reactions. In particular, combined quantum mechanics/molecular mechanics (QM/MM) methods are increasingly important in this growing field of computational enzymology.[4] Recent years have seen a large and continuing increase in the computational analysis of enzyme mechanisms. In the early– to mid-1990s, the number of computational

275

W. A. Sokalski (ed.), Molecular Materials with Specific Interactions, 275–304.
© 2007 *Springer.*

studies of enzymic reactions was relatively small.[5-7] Today, the number of computational studies of enzyme mechanisms is so large that it is practically impossible to cover the majority—even of just recently published studies—in a single review. There has also been a significant increase in the sophistication and reliability of these studies.

In addition to identifying probable chemical mechanisms, modelling can also analyse questions of enzyme specificity, predict the effects of mutations or genetic variation, and assist with the derivation of structure–activity relationships. Altogether, computer modelling is making an increasingly important practical contribution to enzymology, and biochemistry more widely.[8]

The possible range of applications is huge. The large amount of data provided by biological research in genomics, proteomics, glycomics, and structural biology poses a great challenge. Modelling should aid in using this information to develop new drugs, medical therapies, and biologically based technologies (e.g. in nanotechnology). One exciting, developing area is computational protein design,[9,10] which promises a route to new catalysts, and to components for biologically inspired nanotechnology and molecular medicine. Computer modelling is increasingly important for the interpretation of experimental data from the range of sophisticated physical techniques that are being applied to the study of biological systems.

2.   AIMS IN MODELLING ENZYME REACTIONS

Experimental data, particularly protein structures, usually form the essential starting point for modelling an enzyme-catalysed reaction. Using structural data (usually a protein structure from X-ray crystallography, although in some cases a homology model could be sufficiently reliable), modelling can investigate mechanistic and other questions that are difficult to answer by experiment alone.[11]

A first, vital step in studying an enzymic reaction is to establish its chemical mechanism. This requires identifying the roles of catalytic residues, which are often not obvious (indeed even exactly which residues are involved may not be certain). One very important advantage of modelling is that it can analyse transition states directly. Transition states are obviously central to questions of chemical reactivity and catalysis in enzymes: they cannot be studied directly experimentally because of their vanishingly short lifetimes.

Next, to understand a catalytic mechanism, any specific interactions that stabilize transition states or intermediates in the reaction must also be identified and analysed. Interactions of this type are often not apparent from experimental structures. In addition to providing detailed, atomic-level analysis of the reactive processes in an enzyme, identifying interactions of this type may help in the design of ligands. Many enzymes show very high apparent binding affinities for transition states and intermediates, and therefore, such interactions may offer potentially enhanced affinity if they can be exploited in designed ligands; for example, in the design of pharmaceutical lead compounds.

Modelling can pinpoint functional groups and analyse catalytic interactions. In several enzymes, catalytic interactions have been identified via calculation. For example, in the flavin-dependent monooxygenases, para-hydroxybenzoate hydroxylase and phenol hydroxylase, a conserved proline residue was found from QM/MM modelling, which specifically stabilizes the transition state for aromatic hydroxylation.[12,13]

Among the many challenges presented by enzymes to the modeller, is the fact that proteins have complex dynamics: as molecular dynamics simulations (with empirical 'molecular mechanics' potential functions) and spectroscopic experiments have shown, proteins and other biological macromolecules undergo a wide variety of complex internal motions, some of which are vital to their function.[14] Many enzymes also undergo large changes in conformation as part of their reaction cycles.[15] The function of these conformational changes, and their relationship to the chemical steps in an enzymic reaction, should be examined.[14]

There has been much recent debate about the possible contribution of protein dynamics to enzyme catalysis, but simulations indicate that the direct effect of protein dynamics in determining the chemical reaction rates of enzymes is generally relatively small.[16–18] It is nevertheless important, in general, to consider the effects of protein conformational fluctuations and variations on enzyme reactions; that is, where possible, to investigate a representative sample of conformations (which could, for example, be generated from molecular dynamics or Monte Carlo simulations). In many enzyme reactions involving hydrogen transfer, quantum effects such as nuclear tunnelling are also important.[19,20]

It is important to emphasize that, in order to understand why an enzyme is an effective *catalyst*, (i.e. to understand why the reaction in the enzyme is faster than the uncatalysed reaction), the enzymic reaction and its equivalent reaction in solutions (the 'reference reaction') should be compared. However, the appropriate reference reaction for comparison may not be obvious for all enzymes.

For practical applications, the motivation in modelling an enzymic reaction may be to predict the effects of a mutation—designed or natural—on activity, or perhaps in altering or broadening the specificity of an enzyme for alternative substrates. Overall, it is clear that modelling the mechanisms of enzyme-catalysed reactions, and understanding enzyme specificity and catalysis, presents many different levels of complexity. To investigate different types of questions in this area, different modelling or simulation approaches may be more appropriate, depending on the issue of interest.

3.   METHODS FOR MODELLING ENZYME-CATALYSED REACTION MECHANISMS

Perhaps the most obvious challenge in investigating enzyme reactions by computational modelling is posed by the very large size of enzymes, exacerbated by the need to include at least a representative part of their environment; that is, the surrounding solvent, perhaps membrane or other proteins, cofactors or DNA which may be bound to the enzyme.

Standard 'molecular mechanics' (MM) force fields have been developed that provide a good description of protein structure and dynamics,[21] but they cannot be used to model chemical reactions. Molecular dynamics simulations are very important in simulations of protein folding and unfolding,[22] an area in which they complement experiments and aid in interpretation of experimental data.[23] Molecular dynamics simulations are also important in drug design applications,[24] and particularly in studies of protein conformational changes,[25,26] simulations of the structure and function of ion channels and other membrane proteins,[27–29] and in studies of biological macromolecular assemblies such as F-1-ATPase.[30]

The first protein molecular dynamics simulation was carried out almost thirty years ago, on bovine pancreatic trypsin inhibitor (BPTI), in the gas phase.[31] Since that pioneering early work, a number of empirical molecular mechanics force fields have been developed for proteins, nucleic acids, lipids and other biological molecules. It is important to distinguish between programs used for biomolecular simulation, and the force fields that have been developed for them. Several good quality parameter sets have been developed, and may be applied with several different programs. Among the most widely-used computer *programs* for biological molecular dynamics simulations are AMBER,[32] CHARMM,[33] GROMOS,[34] NAMD,[35] and TINKER.[36] Several other molecular dynamics simulations packages are available, including commercial and academic programs.

A force field is made up of the parameters and the energy function used. Current protein force fields use similar, simple potential energy functions to allow large systems to be simulated over multinanosecond timescales. Bonds and valence angles are represented by harmonic terms. Electrostatic interactions are represented by invariant atomic point partial charges. Dispersion and exchange repulsion interactions are represented by a Lennard-Jones energy function (typically of the 12-6 variety). The simple molecular mechanics representation is obviously limited in many important respects. For example, the simple model including atomic point change model cannot capture the full electrostatic properties (e.g. multipole moments) of a molecule. Also, changes in electronic polarization are not included: the atomic charges do not change in response to changes in the molecular conformation or its environment. Kim *et al.* have analysed the effects of solvent and protein polarizability in simulations of bovine pancreatic trypsin inhibitor in explicit water.[38] The effects were found to be similar in relatively non-polar parts of the protein, but significant where relatively strong electrostatic fields occur in the protein (near charged amino acid residues).

The simple harmonic terms used to represent the energy of bond stretching in typical protein molecular mechanics force fields cannot model the making and breaking of chemical bonds. Also, molecular mechanics parameters are usually developed based on the properties of stable molecules, and so might not be applicable to transition states and intermediates. Molecular mechanics functions and parameters can be developed specifically for reactions, an approach that has been applied to many organic reactions in solution.[39] However, the parameters are generally applicable only to a specific reaction, meaning that laborious reparameterization is required in every case. The form of the potential function may be significantly limited; for example, electronic polarization may not be included.

Several reviews of protein force fields have been published recently.[40,41] All-atom force fields include all atoms, while united-atom force fields treat only heavy (non-hydrogen) atoms and polar hydrogen atoms explicitly, including non-polar hydrogen atoms only implicitly, as part of the carbon atom to which they are bonded. For proteins, OPLS/AA,[42,43] CHARMM22,[44] and AMBER (PARM99)[32,45] are among the most popular all-atom force fields. Modern protein MM force fields show comparable results in molecular dynamics simulations.[46] Consistent force fields have also been developed for other types of biological macromolecules (for example, lipids[47] and nucleic acids[48–52]). Polysaccharides present a particular challenge, because of their structural and conformational complexity, and the difficulty of balancing inter– and intramolecular interactions.[53] A QM/MM approach, treating the sugar by QM, may be preferable in some cases.[54] Standard semi-empirical molecular orbital quantum chemical methods such as AM1 and PM3 do not treat carbohydrate conformations well, but reparameterization can improve this situation (for example, the PM3CARB-1 model).[55]

Important united-atom protein force fields include GROMOS87 and 96,[34,56] CHARMM PARAM19,[57] OPLS/UA (united-atom),[58] and the original AMBER force field.[59] United-atom force fields were developed to reduce the computer time required for molecular dynamics simulations by reducing the number of atoms. They are still important today, in studies using either explicit or implicit solvation models. They are particularly widely used in studies of protein folding, often employing a continuum solvation description to reduce computation demands in these long timescale simulations, by avoiding the need to include explicit water molecules. Several implicit solvent models have been developed for use with the CHARMM PARAM19 force field, including EEF1 (effective energy function),[60] ACE (analytic continuum electrostatics),[61] models based on the Generalized Born approach,[62–64] and other fast implicit solvation models for molecular dynamics simulations.[65] Assessment of the performance (both accuracy and efficiency) of implicit solvent models (e.g. by comparison with explicit solvent simulations) is a highly active area of research.

Most biomolecular molecular mechanics force fields have been developed with simple point charge models of water, in particular the TIP3P water model,[66] and variants thereof. Electronic polarization is included only in an approximate, gross way in models such as TIP3P; for example, the dipole moment of such models is higher than that observed in the gas phase, thus including the effects of polarization in the condensed phase. Similarly, as mentioned above, protein MM force fields only include electronic polarization in an average, and invariant, way. Polarizable force fields for biological molecules are the subject of much current research and development effort.[67–74] The next generation of protein MM force fields will probably include electronic polarization explicitly. Other improvements to protein

MM force fields include the use of *ab initio* quantum chemical results to improve the potential energy surface for peptide backbone dihedral angle rotation.[75,76]

Quantum chemical methods aim to treat the fundamental quantum mechanics of electronic structure, and so can be used to model chemical reactions. Such quantum chemical methods are more flexible and more generally applicable than molecular mechanics methods, and so are often preferable and can be easier to apply. The major problem with electronic structure calculations on enzymes is presented by the very large computational resources required, which significantly limits the size of the system that can be treated. To overcome this problem, small models of enzyme active sites can be studied in isolation (and perhaps with an approximate model of solvation). Alternatively, a quantum chemical treatment of the enzyme active site can be combined with a molecular mechanics description of the protein and solvent environment: the QM/MM approach. Both will be described below.

3.1. Quantum Chemical Approaches to Modelling Enzyme Reactions: Cluster (or Supermolecule) Approaches

The active site of an enzyme is a relatively small region, containing the catalytic residues, substrate(s), and any cofactors. The substrate(s) are typically bound at the active site by multiple weak interactions, such as hydrogen bonds, electrostatic and van der Waals interactions. By focusing on the few key groups, quantum chemical modelling is possible. This is sometimes called the 'supermolecule' approach. For modelling reactions, quantum chemical techniques (e.g. *ab initio* molecular orbital or density functional theory calculations) can be used routinely to study non-periodic, molecular systems including tens of atoms. 'Cluster' models of around this size can represent essential features of enzyme reactions, and can identify likely mechanisms. Small molecules represent important functional groups (e.g. key amino acid side-chains involved in catalysis or binding the substrate, or cofactors, and so on), with their positions typically taken from a representative X-ray crystal structure of an enzyme complex. This approach has been particularly useful for metalloenzymes, where all the important chemical steps may take place at one metal centre (or a small number of metal ions bound at one site), and the metal also restrains its ligands, limiting the requirement for restraints to maintain the correct active site structure in calculations. Reliable, semi-quantitatively accurate calculations are practical using methods based on density functional theory—such as the widely used B3LYP hybrid functional—which give good results for many organometallic reactions. Some recent applications of calculations of this type are outlined below. The extensive work of Siegbahn *et al.*[77] on many metalloenzymes demonstrates the insight that such calculations can give for enzyme mechanisms. It is possible, for example, to discriminate between alternative proposed mechanisms: a mechanism can be excluded if the calculated barriers are significantly higher than the experimentally derived activation energy.

A small model might not include all the important functional groups. However, a larger model is not always a better model, as it may introduce greater conformational complexity: structural changes away from the active site might affect reaction energies in unrealistic ways. Including unshielded charged groups could also have unrealistically large effects on reaction energies. Environmental effects can be included approximately through continuum solvation models, although these cannot completely represent the heterogeneous electrostatic environment in a protein. [17]

Approximate electronic structure methods (e.g. the semi-empirical molecular orbital techniques AM1 and PM3) can be applied to larger molecular systems (of the order of hundreds of atoms). However, these approximate methods are inaccurate for many systems (giving very large errors in some calculated energies of reaction), and often cannot be applied straightforwardly to some types of system (e.g. many transition metals). Techniques such as 'linear-scaling' methods allow semi-empirical electronic structure calculations on whole proteins. [78-80] Similarly, the scaling properties of high-level quantum chemical methods are being improved to permit applications to larger molecular systems. Increasingly important in biomolecular simulations [81] is the *ab initio* molecular dynamics technique, first proposed by Car and Parrinello around 20 years ago, [82,83] which combines molecular dynamics simulation and density functional theory. The direct application of electronic structure allows the treatment of chemical reactions, and includes electronic polarization effects. The major practical limitation is that *ab initio* molecular dynamics simulations are extremely computationally demanding, so the sizes of the systems that can be simulated, and the timescale of feasible dynamics simulations, are relatively limited. For this reason, combined QM/MM techniques are attractive for *ab initio* molecular dynamics simulations. An example is a scheme for Car-Parrinello QM/MM molecular dynamics simulations with the CPMD and EGO programs. [84] With the QM/MM approach, Car-Parrinello *ab initio* molecular dynamics simulations of large systems can be performed, which explicitly include the steric and electrostatic effects of the protein and solvent.

3.2.   Empirical Valence Bond Methods

In the empirical valence bond method, resonance structures (e.g. ionic and covalent resonance forms) are chosen to represent the reaction, with the energy of each resonance form given by a simple empirical force field (with realistic treatment of stretching important bonds, for example). [85] The potential energy is given by solving the secular equation for the resonance forms. The Hamiltonian is calibrated to reproduce experimental—or alternatively *ab initio* quantum chemical—data for a given solution reaction. [86] The protein and solution are treated by an empirical force field. The free energy of activation for the reaction in solution, and in the enzyme, can be calculated using free energy perturbation simulations. [87] As in any valence bond representation, it is essential that the chosen valence bond forms should represent all the resonance forms that are important in the reaction. Recent applications include comparison of alternative nucleotide insertion mechanisms for T7 DNA polymerase,[88] and the reaction mechanism of human aldose reductase.[89]

3.3. Combined Quantum Mechanics/Molecular Mechanics (QM/MM) Methods

Enzyme–substrate complexes are very large, containing thousands of atoms, and are currently beyond even semi-empirical quantum chemical methods for modelling reactions. For modelling a reaction, it is necessary to optimize many important points (such as transition state structures), and preferably entire reaction pathways. To generate a representative collection of molecular structures, extensive conformational sampling may be needed: a single structure may not be truly representative.[90] These are significant challenges for large molecules. It is important also to treat the environment of the enzyme (e.g. aqueous solution, concentrated solutions, in membranes, or in complexes with other proteins or nucleic acid). To calculate free energy profiles (potentials of mean force),[91] a simulation method must be able to calculate molecular dynamics trajectories of at least many picoseconds. Combined quantum mechanics/molecular mechanics (QM/MM) methods are increasingly able to meet these challenges.[92,93] QM/MM techniques allow the study of large models, and, with low levels of QM theory, molecular simulations are feasible.[94] As well as activation free energies, quantum effects such as tunnelling and zero-point corrections can be calculated.[91] Transition state structures can also be optimized with QM/MM methods.[95,96]

The first QM/MM study of an enzyme-catalysed reaction was the seminal study of the reaction mechanism of hen egg-white lysozyme by Warshel and Levitt in 1976.[97] In recent years, the use—and sophistication—of QM/MM methods has grown significantly. In the QM/MM approach, a small part of the system is treated quantum mechanically; that is, by an electronic structure method, for example, at the *ab initio* or semi-empirical molecular orbital, or density functional theory QM level. This allows the bond-breaking and making, and electronic rearrangement, that is involved in a chemical reaction, to be modelled. For an enzymic reaction mechanism, the QM region would typically be the enzyme active site, including the reacting groups of the enzyme, substrate, and any co-factors. The large non-reactive part is described more simply by empirical molecular mechanics. The combination of the flexibility of a QM electronic structure method with the efficiency of a molecular mechanics force field permits the modelling of reactions in large systems. Molecular mechanics methods treat protein structure and interactions well, as described above. Different ways of coupling the QM and MM regions are possible. For application to enzymes, including interactions between the QM and MM regions is probably important.

QM/MM methods have proved their value for enzyme reactions in differentiating between alternative proposed mechanisms, and in analysing contributions to catalysis. A current example is the analysis of the contribution of conformational effects and transition state stabilization in the reaction catalysed by the enzyme chorismate mutase.[98,99] QM/MM calculations can be performed with *ab initio*[100,101] or semi-empirical[102] molecular orbital, density functional,[103] or approximate density functional (e.g. self-consistent charge density functional tight-binding (SCC-DFTB)) levels of QM electronic structure theory. QM/MM calculations at higher QM levels (e.g. *ab initio* or density functional level QM) are required for some systems, but can be very demanding of computational resources.

The most straightforward type of QM/MM coupling is a simple subtractive model (sometimes denoted as mechanical embedding), in which the electrostatic interaction between the QM and MM regions is treated by including MM atomic point charges for the QM atoms. If these charges do not change in response to the influence of their environment, then electronic polarization is not included. Enzymes are polar and heterogeneous, and so including polarization of the QM region by its MM environment is probably important in modelling an enzymic reaction. Most QM/MM studies of enzymic reactions have included polarization of the QM system by the MM system, by directly including the charges of the MM group in the QM calculation. In this way, the electronic structure calculation includes the effects of the MM atomic charge. The partial atomic charges of the MM atoms are included in the Hamiltonian for the QM region (through the one-electron integrals). No electrons are present on the MM atoms, and therefore it is necessary to include some representation of QM/MM dispersion and exchange repulsion interactions. Typically, MM (classical) van der Waals interactions (e.g. Lennard-Jones functions, as described above) between QM and MM atoms are included. MM van der Waals radii must therefore be chosen for the QM atoms. A limitation of this approach is that the same van der Waals parameters are used for the QM atoms throughout a simulation: in modelling a chemical reaction, the chemical nature of the groups involved (treated by QM) may change, altering their interactions, and so the use of unchanging MM parameters could be inappropriate and lead to inaccuracies. Riccardi *et al.* have recently tested the effects of van der Waals interactions in QM/MM simulations.[105] Condensed phase thermodynamic quantities (e.g. the calculated reduction potential and potential of mean force) were found not to be very sensitive to the van der Waals parameters used. This group recommended that work to improve the reliability of QM/MM methods for condensed phase properties should focus on other factors, such as the treatment of long-range electrostatic interactions.

Often, standard MM van der Waals (Lennard-Jones) parameters optimized for similar MM groups are used for QM atoms in QM/MM calculations. This is convenient, but it is important to test that these van der Waals parameters provide a reliable description of QM/MM interactions. Where necessary, the (MM) van der Waals parameters for the QM atoms can be optimized to reproduce experimental or high level *ab initio* results (e.g. interaction energies and geometries) for small molecular complexes. Van der Waals terms are important in differentiating MM atom types in their interactions with the QM system; that is, they are important in differentiating between MM atoms of the same charge (e.g. halide ions), which would otherwise be indistinguishable to the QM system. In general, the van der

Waals terms are important at close range, and play a determining role for QM/MM interaction energies and geometries.

The treatment of QM/MM electrostatic interactions is not as straightforward when semi-empirical QM methods are used, because semi-empirical molecular orbital methods such as AM1 and PM3 include only valence electrons directly. Core electrons are treated together with the nucleus as an atomic 'core'. In semi-empirical QM/MM methods (such as the AM1/CHARMM method developed by Field *et al.*[102]), QM/MM electrostatic interactions are calculated by treating the MM atoms exactly as if they where semi-empirical atomic cores.

Polarization of the MM environment by the QM region is usually not included, because MM protein force fields do not as yet allow for polarization, or indeed any changes in atomic charges. QM/MM methods which include polarization of the MM system have been developed for small systems.[106] QM/MM calculations should help in developing polarizable MM force fields; for example, in investigating polarization effects for small (QM) regions in large biomolecules.[107]

The total energy of a system in a QM/MM calculation can be written as the sum of four contributions:

$$E_{eff} = E_{QM} + E_{MM} + E_{QM/MM} + E_{Boundary}$$

The energy of the QM region, E_{QM}, is calculated as in an electronic structure calculation. The energy of the MM region, E_{MM}, is given by a molecular mechanics force field. The boundary energy, $E_{Boundary}$, arises (as in MM simulations) because the simulation system can only include a finite number of atoms, so terms to reproduce the effects of the bulk must be included. Typically, harmonic restraints are applied for the atoms towards the edge of the simulation system. Atoms still more distant from the centre of the simulation system under investigation may be held fixed. Molecular dynamics simulations can be carried out for truncated systems by the stochastic boundary molecular dynamics method.[108,109] To include the effects of dielectric screening, it may also be necessary to reduce charges at the outer boundary of the simulation system, to avoid overestimating the effects of charged groups on the reaction at the active site.[110] The QM/MM interaction energy, $E_{QM/MM}$, is made up of terms due to electrostatic interactions and van der Waals interactions, and any bonded interaction terms between the QM and MM regions. MM bonding terms (energies of bond stretching, angle bending, torsion angle rotation, and so on) are typically included for any QM/MM interactions that involve at least one MM atom. QM/MM electrostatic interactions are usually included in the electronic structure calculations; that is, in an *ab initio* QM/MM calculation. The MM atomic charges are typically included directly through the one-electron integrals as outlined above. The nuclei of the atoms in the QM system also interact with the MM atomic partial charges. QM/MM van der Waals interactions are usually calculated by a molecular mechanics (e.g. Lennard-Jones) energy function, as described above.

In a QM/MM calculation, the whole enzyme system is typically truncated, in order to reduce the computational effort required: for example, only a part of the whole protein might be included in the simulation. This might be an approximately spherical region around the active site. Ideally, the effects of longer range electrostatic interactions in the protein should be included explicitly. The generalized solvent boundary potential (GSBP) method[111] has been implemented for reliable treatment of electrostatics for spherical boundary conditions in QM/MM simulations of truncated macromolecules (at the self-consistent-charge density functional tight-binding QM/MM level).[112] QM/MM simulations using the generalized solvent boundary potential method have been found to be more consistent with available experimental data than standard stochastic boundary molecular dynamics simulations, which can produce artefacts. The use of truncation schemes for QM/MM electrostatic interactions can lead to problems, particularly where extensive conformational sampling is performed. For QM/MM simulations of periodic systems (periodic boundary conditions are now typical in MM molecular dynamics simulations), Nam *et al.* have developed an efficient linear-scaling Ewald method for long-range QM/MM electrostatic interactions in QM/MM calculations.[113]

3.3.1. QM/MM partitioning methods and schemes

For the majority of enzyme-catalysed reactions, covalently bonded parts of the system must be separated into QM and MM regions. There has been considerable research into methods for QM/MM partitioning of covalently bonded systems. Important methods include the local self-consistent field (LSCF) method,[114,115] and the generalized hybrid orbital (GHO) technique.[116] Alternatively a QM atom (or QM pseudo-atom) can be added to allow a bond at the QM/MM frontier; for example, the link atom method or the connection atom method.

The local self-consistent field (LSCF) method[117] has been used at the semi-empirical[118] and *ab initio*[117] levels. The LSCF approach avoids the need for dummy atoms and provides a reasonable description of the chemical properties of the frontier bond. The generalized hybrid orbital (GHO) method for QM/MM calculations uses hybrid orbitals as basis functions on the frontier atom of the MM fragment. This method removes the need for extensive specific parameterization, which is necessary with the LSCF method. A key aspect of the GHO method is that the parameters for the boundary atom are transferable. Garcia-Viloca and Gao have developed the GHO approach to combine the semi-empirical PM3 method with the CHARMM MM force field.[119] These workers have developed parameters—consistent with the PM3 method—for a carbon boundary atom. They found the combined GHO-PM3/CHARMM model to perform well on molecular structures and proton affinities for a number of organic molecules. More recently, the GHO approach has been developed at the *ab initio*,[120] self-consistent-charge density functional tight-binding (SCC-DFTB),[121] and density functional[122] QM levels.

Alternatively, to satisfy the valence of the frontier atom in the QM system, the approach of adding a (QM) 'dummy junction atom' or 'link atom' can be used.[123] Usually, the link atom is a hydrogen atom,[102] but other atom types have also been used. The link atom approach introduces additional, artificial, degrees of freedom associated with the link atom. Also, clearly a C–H bond is not chemically exactly equivalent to a C–C covalent bond. However, the simplicity of the link atom method has led to its widespread use in QM/MM modelling. The results of QM/MM calculations can be highly dependent on the positioning of the link atom, and can also depend on which MM atoms are excluded from the classical electrostatic field that interacts with the QM region. However, Reuter *et al.* found the LSCF and link atom approaches to give similar results for a variety of molecular properties in semi-empirical QM/MM calculations. [124] It has been recommended that a link atom should interact with all MM atoms except for those closest to the QM atom to which the link atom is bonded. Given a reasonable choice of the boundary between QM and MM regions (e.g. choosing the QM/MM boundary to lie across a carbon–carbon single bond, distant from chemical changes and also from highly charged MM atoms), the link atom method can give good results.

Another approach to treating the boundary between covalently bonded QM and MM systems is the connection atom method, [125,126] in which, rather than a link atom, a monovalent pseudo-atom is used. This 'connection atom' is parameterized to give the correct behaviour of the partitioned covalent bond, and has been implemented at semi-empirical molecular orbital (AM1 and PM3) [125] and density functional theory [126] levels of QM theory. It has been suggested that the connection atom approach is more accurate than the standard link atom approach. [125]

Brooks *et al.* have developed a Gaussian delocalization method for molecular mechanical charges in QM/MM calculations, [127] to include delocalization of electron density that should be present for atoms in the MM region. This approach could have the benefit that the MM host atom charge may not have to be excluded from the QM calculation, as would be necessary when treating it simply as a point charge. This group have also proposed a 'double link atom' method to overcome some of the problems of electrostatic interactions that can arise with the single link atom method.

Cui *et al.* have tested a number of different QM/MM partitioning methods based on the link atom approach, for enzymic reactions, with the SCC-DFTB QM method. [128] These workers have also developed a new QM/MM partitioning method, described as the divided frontier charge approach. In this method, the partial charge associated with the MM atom bonded to the QM atom is evenly distributed to the other MM atoms in the same molecular mechanical group. QM/MM-calculated proton affinities and deprotonation energies can be highly sensitive to the particular link atom scheme employed, which can lead to absolute errors of the order of 15–$20\,\text{kcal mol}^{-1}$ compared to pure QM calculations, though more sophisticated link atom schemes perform better. Activation barriers and reaction energies for proton transfer reactions (in the gas phase and in enzymes) were found to be relatively insensitive to the choice of link atom scheme (e.g. to within a range of 2–$4\,\text{kcal mol}^{-1}$) because of cancellation of errors. It is encouraging that the effect of using different link atom schemes in QM/MM simulations was found to be relatively small for chemical reactions in which the total charge does not change.

4. EXAMPLES OF RECENT MODELLING STUDIES OF ENZYMIC REACTIONS

To illustrate the practical contribution that modelling can make to understanding enzyme-catalysed reactions (and to demonstrate the capabilities of modelling methods), a few recent applications are outlined below. This is an evolving field, not yet capable of quantitative, exact predictions of reaction rates or the effects of mutation. For this reason, it is vital to try to connect modelling with experimental investigations of enzymes. This can help to validate predictions from modelling. For example, it can be useful to compare activation barriers for a series of alternative substrates with the activation energies derived from experimental rates. Demonstration of a correlation between calculated and experimental barriers can validate mechanistic calculations as being truly predictive.[129]

Enzyme structures from experiment are the usual starting point for modelling enzymic reactions. Most important are protein structures from X-ray crystallography, which has produced a large and ever-growing number of structures of biological macromolecules. The RCSB Protein Data Bank (PDB) is the standard source for three-dimensional structures of proteins and nucleic acids.[130] The resolution of a crystallographic protein structure is one indication of its precision, ranging from low resolution where perhaps just the overall shape of the protein may be revealed, to higher resolution (e.g. below 2Å resolution) where most heavy atom positions can be determined. However, even in high resolution structures, there can be considerable uncertainty due to the dynamic nature of proteins, which can give rise to conformational variability. The molecular model of a protein structure from crystallography represents an average over all the molecules in the crystal and over the time course of the experiment. This averaging is manifested in the presence of alternative conformations for amino acid side-chains in many protein crystal structures. Similarly, some parts of the structure (such as surface loops or terminal regions of the protein) may have no well-defined conformation or position, and so might not be resolved by crystallography. Such factors should be carefully considered when building a molecular model for computational studies. Another important factor is the selection of the correct protonation states for ionizable groups in the protein.

4.1. Chorismate Mutase: Analysing Fundamental Principles of Enzyme Catalysis

Recent investigations of the enzyme chorismate mutase show how modelling can contribute to fundamental debates in enzymology, such as analysing the importance of transition state stabilization in catalysis, and alternative proposals to explain enzyme catalytic proficiency.

Chorismate mutase catalyses the Claisen rearrangement of chorismate to form prephenate. It is an excellent system for analysing catalysis because the same reaction occurs in solution with the same reaction mechanism: no covalent catalysis by the enzyme is involved. The activation free energy ($\Delta^\ddagger G = 15.4\,\text{kcal mol}^{-1}$, $\Delta^\ddagger H = 12.7\,\text{kcal mol}^{-1}$) found experimentally for chorismate mutase from *Bacillus subtilis* enzyme is significantly lower than that for the uncatalysed reaction in aqueous solution ($\Delta^\ddagger G = 24.5\,\text{kcal mol}^{-1}$, $\Delta^\ddagger H = 20.7\,\text{kcal mol}^{-1}$).[131] This equates to a rate acceleration of 10^6 by the enzyme ($\Delta\Delta^\ddagger G = 9.1\,\text{kcal mol}^{-1}$). QM/MM calculations (at the approximate semi-empirical AM1/CHARMM or *ab initio* QM level) have shown stabilization of the transition state by the enzyme.[99,132–136] Similar studies have also shown that the conformation of chorismate bound to the enzyme is significantly different from that in solution, and is similar to the transition state.[98,137–140] It has recently been controversially argued that transition state stabilization is not involved in chorismate mutase catalysis: Bruice *et al.* have proposed that catalysis is almost entirely due to the selection of a reactive conformation, described as a near-attack conformation (NAC).[141,142] This proposal has changed over time, but has been forcefully promoted by these workers as a potentially generally significant effect in enzyme catalysis, creating considerable debate. The NAC proposal is similar to the hypothesis that enzymes function by distorting (or straining) their substrates into reactive conformations. Essentially, the proposal is that by binding a particular conformation, from which the reaction has a small barrier, very little additional stabilization of the transition state is needed. The reactive conformation is very improbable in solution, but is favoured in the enzyme. Estimates of the free energy cost of NAC formation (e.g. from MM molecular dynamics simulations—with or without restraints on the substrate) led to the proposal that catalysis in chorismate mutase is due entirely to its ability to maintain high populations of NACs.

The exact free energy cost of forming a NAC—and the catalytic benefit associated with forming such a conformation—will depend on the definition used. There is no unique or general definition of a NAC: several different proposals have been put forward. The lack of a general or rigorous definition makes this hypothesis weak, subjective, and unsatisfactory. There is a danger of the definition of a NAC being fitted to the catalytic effect it is designed to explain, making it a circular definition. Bruice *et al.* have calculated NAC populations from unrestrained molecular dynamics simulations (e.g. in solution and in the enzyme). Unfortunately, high energy conformations are sampled too infrequently (even in multinanosecond dynamics simulations), producing an unreliable, overestimated free energy cost for their formation, and so an overestimation of the catalytic benefit of NAC formation. Thermodynamic integration molecular dynamics simulations have also been performed to try to estimate the free energy cost of NAC formation in different environments,[143] but these used a different definition of a NAC. It is hard to judge these simulations, as the published technical details are limited. There is also a concern that the accuracy of the methods (particularly molecular mechanics) used by Bruice *et al.* have not been fully tested for their treatment of conformational energies and interactions.

It is nevertheless certainly useful and interesting to study the catalytic benefit of forming a potentially more reactive conformation of the substrate. Many modelling studies, dating back to the first QM/MM study of the enzyme,[132] have shown the structure of chorismate bound to the enzyme to be significantly altered from its conformation in solution or in the gas phase. The energetic cost of forcing the conformation of chorismate (in solution) into the more restricted conformation found in the enzyme has been calculated by free energy perturbation molecular dynamics simulations as 3.8–4.6 kcal mol^{-1}, or 5 kcal mol^{-1} by semi-empirical QM/MM (AM1/CHARMM) or empirical valence bond methods, respectively. There is good agreement[137,98] between these results, obtained with different theoretical methods. These findings indicate a catalytic contribution of the conformational effect of only around 40–55 per cent of the total $\Delta\Delta^{\ddagger}G$ between enzyme and solvent.

These results suggest that catalysis is not purely due to conformational effects. In agreement with earlier QM/MM findings for the enzyme-catalysed reaction, they imply that in the enzyme, the transition state is significantly stabilized relative to the bound substrate. These earlier studies generally employed relatively low level QM/MM methods, however, leading to questioning of their accuracy. The central issue of whether the transition state is stabilized relative to the bound substrate has recently been examined with high-level QM/MM methods (B3LYP/6-31G(d)/CHARMM).[144] Sixteen different adiabatic reaction pathways were calculated using a combination of the Jaguar[145] and Tinker[37] programs for QM/MM calculations.[146] These pathways were taken from structures derived from semi-empirical QM/MM (AM1/CHARMM and PM3/CHARMM) molecular dynamics simulations of the transition state.

The substrate was chosen as the QM region, and treated at the hybrid density functional B3LYP/6-31G(d) level of theory, which treats this reaction well.[134,99] Similar results are obtained when some active site amino acid side-chains (e.g. Glu78 and Arg90) are also included in the QM region.[147,148] Electrostatic interactions dominate at the active site of chorismate mutase, making a QM/MM treatment of the active site interactions appropriate and reliable.[136] The simulation system was an approximately 25Å radius sphere of protein and solvent, treated with the CHARMM force field.[44] The outer 5Å of the system was fixed, and all other atoms were free to move. The reaction coordinate used was the difference in length between the forming C–C and breaking C–O bonds, which has been shown to be appropriate for modelling the reaction.[99,134] Reaction pathways were calculated by a series of geometry optimizations along the reaction coordinate, with the reaction coordinate harmonically restrained. The average B3LYP/6-31G(d) barrier over the sixteen pathways was 12.0 kcal mol^{-1} (with a standard deviation of only 1.7 kcal mol^{-1}), in excellent agreement with the experimental activation enthalpy (12.7 kcal mol^{-1}). The variation in the calculated energy barriers between the different pathways (ranging from 9 to 15 kcal mol^{-1}) was due almost entirely to differences in the structure of protein environment. The average length of the breaking C–O bond at the transition state was 2.02Å, while the average length of the forming C–C bond was 2.63Å.

To analyse catalysis, ideally energy profiles in the enzyme and in solution should be compared. The barrier to reaction in solution, relative to the enzyme-bound conformation, has been found to be comparable to that in the gas phase.[99,136,137] Therefore, for chorismate mutase, the reaction in the gas phase can be used as a meaningful and convenient comparison. The difference between the QM/MM energy (i.e. the energy in the enzyme) and the gas-phase (QM-only) energy shows the stabilization of the reacting system by its (MM) protein environment. This stabilization is large and negative throughout the reaction, due to favourable Coulombic interactions between the di-anionic substrate and several positively-charged amino acid side-chains in the active site. Significantly, the stabilization energy was found to vary systematically along the reaction coordinate. For all the calculated paths, the transition state was stabilized significantly more than the reactant. The product was found to be destabilized (relative to the reactant) in most cases. The transition state stabilization energy was found to correlate very well with the calculated barrier height. This shows that the barrier to reaction in the enzyme is determined by the amount of transition state stabilization, not the conformation of the substrate. The calculated average stabilization of the transition state was $4.2 \, \text{kcal mol}^{-1}$ (relative to the reactant; that is, differential transition state stabilization). This stabilization is primarily due to specific electrostatic interactions at the active site.[132,136,137]

Through the use of high-level QM/MM calculations, and by studying multiple reaction pathways, this work provides a good estimate of transition state stabilization by chorismate mutase.[144] The average transition state stabilization[44] ($4.2 \, \text{kcal mol}^{-1}$) and the previously calculated cost of forming the bound conformation in the enzyme, compared to solution[137,98] ($3.8–5 \, \text{kcal mol}^{-1}$) add together to give a value very close to the catalytic rate acceleration by chorismate mutase as derived from experiment ($\Delta\Delta^{\ddagger}G = 9.1 \, \text{kcal mol}^{-1}$). This suggests that conformational effects and transition state stabilization (relative to the bound substrate) contribute similar amounts to catalysis in this crucial model enzyme. The same interactions are responsible for binding the substrate in the reactive conformation, and for transition state stabilization. Chorismate mutase is therefore a good example of an enzyme in which transition state stabilization is central to catalysis.

4.2. Cytochrome P450: Mechanism and Structure–Reactivity Relationships

Cytochrome P450 enzymes are important in pharmaceutical research and development, because of their roles in drug metabolism.[129] They are a ubiquitous class of haem enzymes, which act as monooxygenases in a wide variety of biological reactions.

Structure–reactivity relationships could help in predicting biotransformations of pharmaceuticals and other xenobiotics, and so help drug development.[3]

The catalytically active form of the enzyme for oxidation is believed to be a haem oxo-iron(IV) porphyrin radical cation, called Compound I. The key step in substrate oxidation involves hydrogen atom abstraction or C=C bond addition by the oxygen atom of the Compound I intermediate. Among the many reactions catalysed by P450 enzymes are the hydroxylation of alkanes and aromatic compounds, and the epoxidation of alkenes. Many of these reactions are potentially useful in synthetic and other practical applications. Shaik *et al.* have investigated the mechanism of alkane hydroxylation by cytochrome P450, using density functional theory-based quantum chemical and QM/MM modelling methods.[149–152] For example, the potential energy surface for the so-called 'rebound' mechanism (with methane as a substrate) was calculated for two spin states, the high spin (HS) quartet state and low spin (LS) doublet state. In this rebound mechanism, Compound I first abstracts a hydrogen atom from the alkane, followed by recombination of the hydroxo-radical on the iron with the alkyl radical, generating the ferric-alcohol complex. Calculations were carried out on a model of Compound I, with SH^- used to represent its cysteinate ($SCys^-$) ligand. Although it itself is probably not a P450 substrate, methane was used as a model alkane substrate. The B3LYP hybrid density functional theory method was used, as it has been found to accurately predict energetics and structures for many transition metal complexes, and particularly for bioinorganic systems such as P450 Compound I. Shaik and co-workers have also studied P450 reaction mechanisms of ethane,[153] such as epoxidation by Compound I.[154,155] The results showed the possibility of intermediates with significantly different lifetimes, and different electronic configurations.

Quantum chemical calculations on small models can help in developing structure–reactivity relationships, as recent research on aromatic hydroxylation by cytochrome P450 has shown. In drug metabolism, hydroxylation of C–H bonds is a particularly important class of reaction.[156] This type of reaction can activate pro-drugs, or affect the bioavailability of drugs. An important goal in pharmaceutical research is the development of structure–activity relationships to predict conversions of drugs in the body. Such relationships should allow the reliable prediction of the metabolism and toxicology (ADME/TOX) properties of drugs. Previous work showed that structure–activity relationships based only on substrate structures and properties are of limited use. More detailed models are required, which take into account the reaction mechanism and specificity of different cytochrome P450 isozymes.

An example of a development in this direction comes from modelling investigations of the hydroxylation of simple aromatic compounds by P450 Compound I.[157,158] The model used contained the porphyrin macrocycle ring (without side-chains), with the cysteinate iron ligand represented by a methyl mercaptide group (CH_3S^-). The calculations identified two different possible orientations of the substrate approach for the addition of Compound I to benzene: 'side on' and 'face on' (the second with a lower barrier). This had not previously been observed. Both orientations may be involved in the reactions of different drugs in different P450s. Analysis of spin and charge distributions showed that the transition state for aromatic hydroxylation has mixed radical and cationic character. The effects of substituents on the barrier for addition were also studied. This work produced a new structure–reactivity relationship for substituted aromatics, which can be used to predict barriers to aromatic hydroxylation for simple substituted aromatic compounds. This applied a two-parameter approach, combining simple empirical radical and cationic electronic descriptors, based on the calculated properties of the transition state.

The reactive properties of Compound I may be affected by the different protein environments in different cytochrome P450 enzymes. Different P450 isozymes exhibit very different hydroxylation preferences and substrate specificity. These differences could potentially be caused by orientation or binding effects,[159] and can also be affected by the intrinsic chemical reactivity of different positions in the substrates. Genetic polymorphism may also have a significant effect, which can affect drug metabolism.[160] It has also been suggested that the electronic properties of Compound I may be altered in specific and important ways by the protein environment, and that this could be a determining factor in the reactivity and specificity of cytochrome P450 enzymes. To investigate the effects of the protein on the electronic properties of Compound I, QM/MM methods can be used. The potential of QM/MM methods here has been highlighted by recent studies of the bacterial P450$_{cam}$ enzyme, which have led to considerable debate about the nature of factors behind the catalytic activity of the enzyme.[161,162]

The first QM/MM study of human cytochrome P450 enzymes (including modelling of complexes with the drugs diclofenac and ibuprofen) has recently been published.[163] Compound I has three unpaired electrons: two on the Fe–O centre, and one shared between the proximal cysteinyl sulfur atom and the porphyrin ring. The electronic and geometric structure of Compound I was studied with B3LYP/CHARMM QM/MM calculations. Three human P450s that are important in drug metabolism (P450 2C9, 2B4, and 3B4) were studied. The results showed that Compound I is remarkably similar in all the different P450 enzymes. The third unpaired electron was found mostly on the porphyrin ring. This result was found with little sensitivity to the density functional, the basis set, or the size of the QM region used in the QM/MM calculations. Substrate complexes were also studied, and it was found that the presence of drug molecules also has essentially no effect on this result. Some variability in the calculated spin density on the cysteinyl sulfur (from 30 to 47 per cent) was found, mostly due to details of the set up of the system; for example, the choice of protein starting structure in QM/MM minimization.[163] These conformational effects were found to be larger than the calculated differences between human P450s. These results indicate that the electronic properties of Compound I in the different human P450s are not distinguishable, which implies that observed differences in substrate selectivity are not caused by differences in their electronic properties.

4.3.   Other Recent Modelling Studies of Enzyme-Catalysed Reactions

To show the breadth and type of current modelling investigations of enzymic reactions, some interesting and representative studies are briefly mentioned here. For example, the mechanism of antibiotic breakdown by a β-lactamase enzyme has recently been investigated. β-lactamase enzymes are the most widespread cause of bacterial resistance against β-lactam antibiotics. They are therefore a serious and growing danger to the effectiveness of antibacterial chemotherapy, and pose a major threat to human health. The reaction mechanism of a Class A β-lactamase (with benzylpenicillin) has been investigated by QM/MM modelling. Glu166 was identified as the base in both acylation and deacylation reactions in the mechanism of breakdown of β-lactam antibiotics (such as penicillin) in the TEM1 β-lactamase enzyme, by QM/MM (AM1/CHARMM) modelling with high level (B3LYP hybrid density functional) QM energy corrections.[164–166] This QM/MM approach has also been used to investigate the reaction mechanism of the enzyme fatty acid amide hydrolase,[167] which is central to endocannabinoid metabolism, and a promising target in the treatment of disorders of the central and peripheral nervous systems.

QM/MM modelling using variational transition state theory with the small curvature approximation for tunnelling corrections has been used to investigate the proton transfer step in the reaction of methylamine to formaldehyde, catalysed by methylamine dehydrogenase. These variational transition state theory/small curvature tunnelling methods (VTST/SCT) allow kinetic isotope effects (KIEs) to be calculated for reactions in enzyme.[91] Alhambra *et al.*[168] have calculated kinetic isotope effects by these VTST/SCT techniques in QM/MM studies of methylamine dehydrogenase, using the PM3 method (with specific reaction parameters) combined with the CHARMM22 MM force field.[44] The classical activation free energy ($20.3\,\mathrm{kcal\,mol^{-1}}$) was reduced to $17.1\,\mathrm{kcal\,mol^{-1}}$ when the quantum mechanical vibrational energy was included. Including quantum tunnelling contributions produced an effective (phenomological) activation energy of $14.6\,\mathrm{kcal\,mol^{-1}}$, which agrees well with the experimental value of $14.2\,\mathrm{kcal\,mol^{-1}}$. The calculated hydrogen/deuterium primary KIE of 18.3 for the per-deuterated substrate also agrees well with the experimental result (17.2). The VTST/SCT method allows the separation of different contributions (e.g. of tunnelling). QM/MM methods (PM3 with specific reaction parameters, combined with the AMBER MM force field) and variational transition state theory combined with multi-dimensional tunnelling in the small curvature approximation corrections have also been applied separately to analyse proton transfer for two substrates (methylamine and ethanolamine) in methylamine dehydrogenase.[169] The calculated kinetic isotope effects covered a wide range, but were reasonably close to the experimental values. Two different structural configurations were found for the ethanolamine substrate, giving rise to quite different kinetic behaviours, with different amounts of tunnelling.

Bjelic and Åqvist have investigated the substrate binding mode and reaction mechanism of a malaria protease with a novel active site, which is a target for anti-malarial drug design, using a validated homology model.[170] The only amino acid residue found to be directly involved in the enzyme-catalysed reaction was an aspartate side-chain, with some stabilization by a histidine residue. The calculated reaction rate (for a hexapeptide substrate derived from human haemoglobin) agreed well with experimental kinetic data.

Other interesting recent studies include the application of density functional theory electronic structure methods to model the reaction mechanisms of class III ribonucleotide reductase,[171] naphthalene dioxygenase,[172] and 4-hydroxyphenylpyruvate dioxygenase.[173] Semi-empirical QM/MM methods (at the PM3/CHARMM level) have been applied to model 4-chlorobenzoyl-CoA dehalogenase; in particular, the formation of the Meisenheimer intermediate.[174] The isomerization of proline in cylcophilin (and mutants) has been studied using SCC-DFTB/CHARMM QM/MM methods.[175] QM/MM methods have also been used to investigate the contribution of the protein backbone in the mechanism in 4-oxalocrotonate tautomerase.[176] Kinetic isotope effects have been calculated with QM/MM methods for the enzymes chorismate mutase[177] and catechol O-methyltransferase,[178] among others. For macrophomate synthase, the Diels-Alder and the Michael-Aldol reaction mechanisms have been compared by QM/MM Monte Carlo free energy perturbation simulations.[179] Density functional theory calculations and QM/MM molecular dynamics simulations have been used to investigate a metallo β-lactamase.[180] QM/MM modelling demonstrated substrate autocatalysis in uracil DNA-glysosylase.[181] The free energy profile in chorismate mutase has been calculated from multiple steered molecular dynamics simulations with a density functional QM/MM technique.[182] The inhibition mechanisms of neutrophil elastase by peptidyl alpha-ketoheterocyclic inhibitors have been studied with QM/MM methods, highlighting the potential of QM/MM calculations in structure-based drug design.[183] *Ab initio* techniques have been used to investigate the catalytic site of galactose oxidase and a designed biomimetic catalyst.[184] The nature of the proton bottleneck in redox-coupled proton transfer in cytochrome c oxidase has been investigated.[185]

5.   CONCLUSIONS

Computer modelling and simulation are powerful and effective tools to investigate enzyme reaction mechanisms, specificity, and catalysis. A thriving field of computational enzymology has recently emerged, and continues to grow. Enzymes are complex and challenging systems to model, making care and detailed testing essential. Applied with care, modelling techniques can provide uniquely detailed, molecular-level analysis into the fundamental processes of enzyme catalysis, and practical insight into important biochemical and biological problems. The importance and impact of modelling and simulation methods in this area will undoubtedly continue to increase.

ACKNOWLEDGEMENTS

AJM thanks his co-workers and collaborators in some of the work described here (see individual references). He also thanks BBSRC, EPSRC, Vernalis plc, the Royal Society, and the IBM High Performance Computing Life Sciences Outreach Programme for support of his research.

REFERENCES

1. Cunningham MA, PA Bash (1997) Computational enzymology. Biochimie 79 (11): 687–689
2. Bruice TC, K Kahn (2000) Computational enzymology. Curr. Opin. Chem. Biol. 4 (5): 540–544
3. Mulholland AJ (2005) Modelling enzyme reaction mechanisms, specificity and catalysis. Drug Discovery Today 10 (20): 1393–1402
4. Field MJ (2002) Simulating enzyme reactions: Challenges and perspectives. J. Comput. Chem. 23 (1): 48–58
5. Mulholland AJ, GH Grant, WG Richards (1993) Computer Modeling of Enzyme Catalyzed Reaction-Mechanisms. Protein Eng. 6 (2): 133–147
6. Mulholland AJ, M Karplus (1996) Simulations of enzymic reactions. Biochem. Soc. Trans. 24 (1): 247–254
7. Åqvist J, A Warshel (1993) Simulation of Enzyme-Reactions Using Valence-Bond Force-Fields and Other Hybrid Quantum-Classical Approaches. Chem. Rev. 93 (7): 2523–2544
8. Perruccio F, L Ridder, AJ Mulholland (2003) Quantum-mechanical/molecular-mechanical methods in medicinal chemistry. In Quantum Medicinal Chemistry, edited by P Carloni and F Alber: Wiley-VCH.
9. Park S, JG Saven (2005) Computationally assisted protein design. Ann. Rep. Comp. Chem. 1: 245–253
10. Kuhlman B, G Dantas, GC Ireton, G Varani, BL Stoddard, D Baker (2003) Design of a novel globular protein fold with atomic-level accuracy. Science 302 (5649): 1364–1368
11. Martí S, M Roca, J Andrés, V Moliner, E Silla, I Tuñón, J Bertrán (2004) Theoretical insights in enzyme catalysis. Chem. Soc. Rev. 33 (2): 98–107
12. Ridder L, JN Harvey, IMCM Rietjens, J Vervoort, AJ Mulholland (2003) Ab initio QM/MM modeling of the hydroxylation step in p-hydroxybenzoate hydroxylase. J. Phys. Chem. B 107 (9): 2118–2126
13. Ridder L, AJ Mulholland, IMCM Rietjens, J Vervoort (2000) A quantum mechanical/molecular mechanical study of the hydroxylation of phenol and halogenated derivatives by phenol hydroxylase. J. Am. Chem. Soc. 122 (36): 8728–8738
14. Karplus M, YQ Gao, JP Ma, A van der Vaart, W Yang (2005) Protein structural transitions and their functional role. Phil. Trans. Roy. Soc. London Ser. A 363 (1827): 331–355
15. Fersht A (1999) Structure and Mechanism in Protein Science. A Guide to Enyzme Catalysis and Protein Folding. New York: Freeman.
16. Garcia-Viloca M, J Gao, M Karplus, DG Truhlar (2004) How enzymes work: Analysis by modern rate theory and computer simulations. Science 303 (5655): 186–195
17. Shurki A, A Warshel (2003) Structure/function correlations of proteins using MM, QM/MM, and related approaches: Methods, concepts, pitfalls, and current progress. Protein Sim.: Adv. Protein Chem. 66: 249–313
18. Olsson MHM, A Warshel (2004) Solute solvent dynamics and energetics in enzyme catalysis: The S(N)2 reaction of dehalogenase as a general benchmark. J. Am. Chem. Soc. 126 (46): 15167–15179
19. Kohen A, R Cannio, S Bartolucci, JP Klinman (1999) Enzyme dynamics and hydrogen tunnelling in a thermophilic alcohol dehydrogenase. Nature 399 (6735): 496–499
20. Masgrau L, A Roujeinikova, LO Johannissen, J Basran, KE Ranaghan, P Hothi, AJ Mulholland, MJ Sutcliffe, NS Scrutton, D Leys (2006) Atomic description of an enzyme reaction dominated by proton tunnelling. Science 312: 237–241
21. Karplus M, and J Kuriyan (2005) Molecular dynamics and protein function. Proc. Natl. Acad. Sci. U. S. A. 102 (19): 6679–6685
22. Daggett V, and A Fersht (2003) The present view of the mechanism of protein folding. Nature Reviews Mol. Cell. Biol. 4 (6): 497–502

23. Mayor U, NR Guydosh, CM Johnson, JG Grossmann, S Sato, GS Jas, SMV Freund, DOV Alonso, V Daggett, AR Fersht (2003) The complete folding pathway of a protein from nanoseconds to microseconds. Nature 421 (6925): 863–867

24. Wong CF, JA McCammon (2003) Protein simulation and drug design. Protein Sim.: Adv. Protein Chem. 66: 87–121

25. Woods CJ, MH Ng, S Johnston, SE Murdock, B Wu, K Tai, H Fangohr, P Jeffreys, S Cox, JG Frey, MSP Sansom, and JW Essex (2005) Grid computing and biomolecular simulation. Phil. Trans. Roy. Soc. London Ser. A 363 (1833): 2017–2035

26. Elber R (2005) Long-timescale simulation methods. Curr. Opin. Struct. Biol. 15 (2): 151–156

27. Sansom MSP, PJ Bond, SS Deol, A Grottesi, S Haider, ZA Sands (2005) Molecular simulations and lipid-protein interactions: potassium channels and other membrane proteins. Biochem. Soc. Trans. 33: 916–920

28. Roux B (2002) Computational studies of the gramicidin channel. Acc. Chem. Res. 35 (6): 366–375

29. Gumbart J, Y Wang, A Aksimentiev, E Tajkhorshid, K Schulten (2005) Molecular dynamics simulations of proteins in lipid bilayers. Curr. Opin. Struct. Biol. 15 (4): 423–431

30. Dittrich M, S Hayashi, K Schulten (2004) ATP hydrolysis in the beta(TP) and beta(DP) catalytic sites of F-1-ATPase. Biophys. J. 87 (5): 2954–2967

31. McCammon JA, BR Gelin, M Karplus (1977) Dynamics of Folded Proteins. Nature 267 (5612): 585–590

32. Case DA, TE Cheatham, III, T Darden, H Gohlke, R Luo, KM Merz, Jr, A Onufriev, C Simmerling, B Wang, and RJ Woods (2005) The Amber biomolecular simulation programs. J. Comput. Chem. 26 (16): 1668–1688. (See http://amber.scripps.edu/)

33. Brooks BR, RE Bruccoleri, BD Olafson, DJ States, S Swaminathan, M Karplus (1983) CHARMM - a Program for Macromolecular Energy, Minimization, and Dynamics Calculations. J. Comput. Chem. 4 (2): 187–217. (See http://www.charmm.org/)

34. Scott WRP, PH Hunenberger, IG Tironi, AE Mark, SR Billeter, J Fennen, AE Torda, T Huber, P Kruger, WF van Gunsteren (1999) The GROMOS biomolecular simulation program package. J. Phys. Chem. A 103 (19): 3596–3607. (See http://www.igc.ethz.ch/gromos/gromos.html)

35. Phillips JC, R Braun, W Wang, J Gumbart, E Tajkhorshid, E Villa, C Chipot, RD Skeel, L Kale, K Schulten (2005) Scalable molecular dynamics with NAMD. J. Comput. Chem. 26 (16): 1781–1802. (See http://www.ks.uiuc.edu/Research/namd/)

36. TINKER Software Tools for Molecular Design 4.0, Saint Louis, MO.

37. Ponder JW, FM Richards (1987) An Efficient Newton-Like Method for Molecular Mechanics Energy Minimization of Large Molecules. J. Comput. Chem. 8 (7): 1016–1024. (See http://dasher.wustl.edu/tinker/)

38. Kim BC, T Young, E Harder, RA Friesner, BJ Berne (2005) Structure and dynamics of the solvation of bovine pancreatic trypsin inhibitor in explicit water: A comparative study of the effects of solvent and protein polarizability. J. Phys. Chem. B 109 (34): 16529–16538

39. Lim D, C Jenson, MP Repasky, WL Jorgenson (1999) Solvent as Catalyst: Computational Studies of Organic Reactions in Solution. In Transition State Modeling for Catalysis, edited by D. G. Truhlar and K. Morokuma. Washington, DC: American Chemical Society.

40. MacKerell AD, Jr (2005) Empirical Force Fields for Proteins: Current Status and Future Directions. Ann. Rep. Comp. Chem. 1: 91

41. Ponder JW, DA Case (2003) Force fields for protein simulations. Protein Sim.: Adv. Protein Chem. 66: 27

42. Jorgensen WL, DS Maxwell, J Tirado-Rives (1996) Development and testing of the OPLS all-atom force field on conformational energetics and properties of organic liquids. J. Am. Chem. Soc. 118 (45): 11225–11236

43. Kaminski GA, RA Friesner, J Tirado-Rives, WL Jorgensen (2001) Evaluation and reparametrization of the OPLS-AA force field for proteins via comparison with accurate quantum chemical calculations on peptides. J. Phys. Chem. B 105 (28): 6474–6487

44. MacKerell AD, Jr, D Bashford, M Bellott, RL Dunbrack, JD Evanseck, MJ Field, S Fischer, J Gao, H Guo, S Ha, D Joseph-McCarthy, L Kuchnir, K Kuczera, FTK Lau, C Mattos, S Michnick, T Ngo, DT Nguyen, B Prodhom, WE Reiher, B Roux, M Schlenkrich, JC Smith, R Stote, J Straub, M Watanabe, J Wiórkiewicz-Kuczera, D Yin, and M Karplus (1998) All-atom empirical potential for molecular modeling and dynamics studies of proteins. J. Phys. Chem. B 102 (18): 3586–3616

45. Cornell WD, P Cieplak, CI Bayly, IR Gould, KM Merz, Jr, DM Ferguson, DC Spellmeyer, T Fox, JW Caldwell, PA Kollman (1995) A 2nd Generation Force-Field for the Simulation of Proteins, Nucleic-Acids, and Organic-Molecules. J. Am. Chem. Soc. 117 (19): 5179–5197

46. Price DJ, CL Brooks III (2002) Modern protein force fields behave comparably in molecular dynamics simulations. J. Comput. Chem. 23 (11): 1045–1057

47. Feller SE, DX Yin, RW Pastor, AD MacKerell, Jr (1997) Molecular dynamics simulation of unsaturated lipid bilayers at low hydration: Parameterization and comparison with diffraction studies. Biophys. J. 73 (5): 2269–2279

48. Cheatham TE, III (2005) Molecular Modeling and Atomistic Simulation of Nucleic Acids. Ann. Rep. Comp. Chem. 1: 75

49. Cheatham TE, III (2004) Simulation and modeling of nucleic acid structure, dynamics and interactions. Curr. Opin. Struct. Biol. 14 (3): 360–367

50. Foloppe N, AD MacKerell Jr (2000) All-atom empirical force field for nucleic acids: I. Parameter optimization based on small molecule and condensed phase macromolecular target data. J. Comput. Chem. 21 (2): 86–104

51. MacKerell AD, Jr, NK Banavali (2000) All-atom empirical force field for nucleic acids: II. Application to molecular dynamics simulations of DNA and RNA in solution. J. Comput. Chem. 21 (2): 105–120

52. Cheatham TE, III, P Cieplak, PA Kollman (1999) A modified version of the Cornell et al. force field with improved sugar pucker phases and helical repeat. J. Biomol. Struct. Dyn. 16 (4): 845–862

53. Imberty A, S Perez (2000) Structure, conformation, and dynamics of bioactive oligosaccharides: Theoretical approaches and experimental validations. Chem. Rev. 100 (12): 4567–4588

54. French AD, GP Johnson, AM Kelterer, MK Dowd, CJ Cramer (2001) QM/MM distortion energies in di- and oligosaccharides complexed with proteins. Int. J. Quantum Chem. 84 (4): 416–425

55. McNamara JP, AM Muslim, H Abdel-Aal, H Wang, M Mohr, IH Hillier, RA Bryce (2004) Towards a quantum mechanical force field for carbohydrates: a reparametrized semi-empirical MO approach. Chem. Phys. Lett. 394 (4–6): 429–436

56. Schuler LD, X Daura, WF van Gunsteren (2001) An improved GROMOS96 force field for aliphatic hydrocarbons in the condensed phase. J. Comput. Chem. 22 (11): 1205–1218

57. Neria E, S Fischer, M Karplus (1996) Simulation of activation free energies in molecular systems. J. Chem. Phys. 105 (5): 1902–1921

58. Jorgensen WL, J Tirado-Rives (1988) The OPLS Potential Functions for Proteins – Energy Minimizations for Crystals of Cyclic-Peptides and Crambin. J. Am. Chem. Soc. 110 (6): 1657–1666

59. Weiner SJ, PA Kollman, DA Case, UC Singh, C Ghio, G Alagona, S Profeta, P Weiner (1984) A New Force-Field for Molecular Mechanical Simulation of Nucleic-Acids and Proteins. J. Am. Chem. Soc. 106 (3): 765–784

60. Lazaridis T, M Karplus (1999) Effective energy function for proteins in solution. Proteins: Struct., Funct., Genet. 35 (2): 133–152

61. Schaefer M, C Bartels, F Leclerc, M Karplus (2001) Effective atom volumes for implicit solvent models: Comparison between Voronoi volumes and minimum fluctuation volumes. J. Comput. Chem. 22 (15): 1857–1879

62. Lee MS, FR Salsbury, CL Brooks III (2002) Novel generalized Born methods. J. Chem. Phys. 116 (24): 10606–10614

63. Lee MS, M Feig, FR Salsbury, CL Brooks III (2003) New analytic approximation to the standard molecular volume definition and its application to generalized born calculations (vol 24, pg 1348, 2003). J. Comput. Chem. 24 (14): 1821

64. Im WP, MS Lee, CL Brooks III (2003) Generalized born model with a simple smoothing function. J. Comput. Chem. 24 (14): 1691–1702

65. Ferrara P, J Apostolakis, A Caflisch (2002) Evaluation of a fast implicit solvent model for molecular dynamics simulations. Proteins: Struct., Funct., Genet. 46 (1): 24–33

66. Jorgensen WL, J Tirado-Rives (2005) Potential energy functions for atomic-level simulations of water and organic and biomolecular systems. Proc. Natl. Acad. Sci. U. S. A. 102 (19): 6665–6670

67. Gresh N, JP Piquemal, M Krauss (2005) Representation of Zn(II) complexes in polarizable molecular mechanics. Further refinements of the electrostatic and short-range contributions. Comparisons with parallel *ab initio* computations. J. Comput. Chem. 26 (11): 1113–1130

68. Vorobyov IV, VM Anisimov, AD MacKerell Jr (2005) Polarizable empirical force field for alkanes based on the classical drude oscillator model. J. Phys. Chem. B 109 (40): 18988–18999

69. Anisimov VM, G Lamoureux, IV Vorobyov, N Huang, B Roux, AD MacKerell, Jr (2005) Determination of electrostatic parameters for a polarizable force field based on the classical Drude oscillator. J. Chem. Theory Comput. 1 (1): 153–168

70. Patel S, AD MacKerell, Jr, CL Brooks III (2004) CHARMM fluctuating charge force field for proteins: II – Protein/solvent properties from molecular dynamics simulations using a nonadditive electrostatic model. J. Comput. Chem. 25 (12): 1504–1514

71. Harder E, BC Kim, RA Friesner, BJ Berne (2005) Efficient simulation method for polarizable protein force fields: Application to the simulation of BPTI in liquid. J. Chem. Theory Comput. 1 (1): 169–180

72. Kaminski GA, HA Stern, BJ Berne, RA Friesner (2004) Development of an accurate and robust polarizable molecular mechanics force field from *ab initio* quantum chemistry. J. Phys. Chem. A 108 (4): 621–627

73. Kaminski GA, HA Stern, BJ Berne, RA Friesner, YXX Cao, RB Murphy, RH Zhou, TA Halgren (2002) Development of a polarizable force field for proteins via *ab initio* quantum chemistry: First generation model and gas phase tests. J. Comput. Chem. 23 (16): 1515–1531

74. Ren PY, JW Ponder (2003) Polarizable atomic multipole water model for molecular mechanics simulation. J. Phys. Chem. B 107 (24): 5933–5947

75. MacKerell AD, Jr, M Feig, CL Brooks III (2004b) Improved treatment of the protein backbone in empirical force fields. J. Am. Chem. Soc. 126 (3): 698–699

76. MacKerell AD, Jr, M Feig, CL Brooks III (2004a) Extending the treatment of backbone energetics in protein force fields: Limitations of gas-phase quantum mechanics in reproducing protein conformational distributions in molecular dynamics simulations. J. Comput. Chem. 25 (11): 1400–1415

77. Himo F, PEM Siegbahn (2003) Quantum chemical studies of radical-containing enzymes. Chem. Rev. 103 (6): 2421–2456

78. Van der Vaart A, V Gogonea, SL Dixon, KM Merz, Jr (2000) Linear scaling molecular orbital calculations of biological systems using the semiempirical divide and conquer method. J. Comput. Chem. 21 (16): 1494–1504

79. Khandogin J, DM York (2004) Quantum descriptors for biological macromolecules from linear-scaling electronic structure methods. Proteins 56 (4): 724–737

80. Khandogin J, K Musier-Forsyth, DM York (2003) Insights into the regioselectivity and RNA-binding affinity of HIV-1 nucleocapsid protein from linear-scaling quantum methods. J. Mol. Biol. 330 (5): 993–1004

81. Carloni P, U Röthlisberger, M Parrinello (2002) The role and perspective of a initio molecular dynamics in the study of biological systems. Acc. Chem. Res. 35 (6): 455–464

82. Car R, and M Parrinello (1985) Unified Approach for Molecular-Dynamics and Density-Functional Theory. Phys. Rev. Lett. 55 (22): 2471–2474

83. Remler DK, PA Madden (1990) Molecular-Dynamics without Effective Potentials Via the Car-Parrinello Approach. Mol. Phys. 70 (6): 921–966

84. Eichinger M, P Tavan, J Hutter, M Parrinello (1999) A hybrid method for solutes in complex solvents: Density functional theory combined with empirical force fields. J. Chem. Phys. 110 (21): 10452–10467

85. Warshel A (2003) Computer simulations of enzyme catalysis: Methods, progress, and insights. Ann. Rev. Biophys. Biomol. Struct. 32: 425–443

86. Bentzien J, RP Muller, J Florian, A Warshel (1998) Hybrid *ab initio* quantum mechanics molecular mechanics calculations of free energy surfaces for enzymatic reactions: The nucleophilic attack in subtilisin. J. Phys. Chem. B 102 (12): 2293–2301

87. Warshel A (1997) Computer Modeling of Chemical Reactions in Enzymes and Solutions. New York: John Wiley & Sons

88. Florian J, MF Goodman, A Warshel (2003) Computer simulation of the chemical catalysis of DNA polymerases: Discriminating between alternative nucleotide insertion mechanisms for T7 DNA polymerase. J. Am. Chem. Soc. 125 (27): 8163–8177

89. Varnai P, A Warshel (2000) Computer simulation studies of the catalytic mechanism of human aldose reductase. J. Am. Chem. Soc. 122 (16): 3849–3860

90. Zhang YK, J Kua, JA McCammon (2003) Influence of structural fluctuation on enzyme reaction energy barriers in combined quantum mechanical/molecular mechanical studies. J. Phys. Chem. B 107 (18): 4459–4463

91. Gao JL, DG Truhlar (2002) Quantum mechanical methods for enzyme kinetics. Annu. Rev. Phys. Chem. 53: 467–505

92. Mulholland AJ (2001) The QM/MM Approach to Enzymatic Reactions. In Theoretical Biochemistry, edited by L. A. Erikkson. Amsterdam: Elsevier.

93. Friesner RA, V Guallar (2005) *Ab initio* quantum chemical and mixed quantum mechanics/molecular mechanics (QM/MM) methods for studying enzymatic catalysis. Annu. Rev. Phys. Chem. 56: 389–427

94. Ridder L, IMCM Rietjens, J Vervoort, AK Mulholland (2002) Quantum mechanical/molecular mechanical free energy Simulations of the glutathione S-transferase (M1-1) reaction with phenanthrene 9,10-oxide. J. Am. Chem. Soc. 124 (33): 9926–9936

95. Martí S, V Moliner (2005) Improving the QM/MM description of chemical processes: A dual level strategy to explore the potential energy surface in very large systems. J. Chem. Theory Comput. 1 (5): 1008–1016

96. Prat-Resina X, JM Bofill, A Gonzalez-Lafont, JM Lluch (2004) Geometry optimization and transition state search in enzymes: Different options in the microiterative method. Int. J. Quantum Chem. 98 (4): 367–377

97. Warshel A, M Levitt (1976) Theoretical Studies of Enzymic Reactions - Dielectric, Electrostatic and Steric Stabilization of Carbonium-Ion in Reaction of Lysozyme. J. Mol. Biol. 103 (2): 227–249

98. Ranaghan KE, AJ Mulholland (2004) Conformational effects in enzyme catalysis: QM/MM free energy calculation of the 'NAC' contribution in chorismate mutase. Chem. Comm. (10): 1238–1239

99. Ranaghan KE, L Ridder, B Szefczyk, WA Sokalski, JC Hermann, AJ Mulholland (2004) Transition state stabilization and substrate strain in enzyme catalysis: *ab initio* QM/MM modelling of the chorismate mutase reaction. Organic & Biomolecular Chemistry 2 (7): 968–980

100. Mulholland AJ, PD Lyne, M Karplus (2000) *Ab initio* QM/MM study of the citrate synthase mechanism. A low-barrier hydrogen bond is not involved. J. Am. Chem. Soc. 122 (3): 534–535

101. Woodcock HL, M Hodoscek, P Sherwood, YS Lee, HF Schaefer, BR Brooks (2003) Exploring the quantum mechanical/molecular mechanical replica path method: a pathway optimization of the chorismate to prephenate Claisen rearrangement catalyzed by chorismate mutase. Theor. Chem. Acc. 109 (3): 140–148

102. Field MJ, PA Bash, M Karplus (1990) A Combined Quantum-Mechanical and Molecular Mechanical Potential for Molecular-Dynamics Simulations. J. Comput. Chem. 11 (6): 700–733

103. Lyne PD, M Hodoscek, M Karplus (1999) A hybrid QM-MM potential employing Hartree-Fock or density functional methods in the quantum region. J. Phys. Chem. A 103 (18): 3462–3471

104. Cui Q, M Elstner, E Kaxiras, T Frauenheim, M Karplus (2001) A QM/MM implementation of the self-consistent charge density functional tight binding (SCC-DFTB) method. J. Phys. Chem. B 105 (2): 569–585

105. Riccardi D, GH Li, Q Cui (2004) Importance of van der Waals interactions in QM/MM Simulations. J. Phys. Chem. B 108 (20): 6467–6478

106. Jensen L, PT van Duijnen (2005) The first hyperpolarizability of p-nitroaniline in 1,4-dioxane: A quantum mechanical/molecular mechanics study. J. Chem. Phys. 123 (7):-

107. Greatbanks SP, JE Gready, AC Limaye, AP Rendell (1999) Enzyme polarization of substrates of dihydrofolate reductase by different theoretical methods. Proteins: Struct., Funct., Genet. 37 (2): 157–165

108. Poulsen TD, M Garcia-Viloca, JL Gao, DG Truhlar (2003) Free energy surface, reaction paths, and kinetic isotope effect of short-chain Acyl-CoA dehydrogenase. J. Phys. Chem. B 107 (35): 9567–9578

109. Brooks CL, III, M Karplus, BM Pettitt (1988) Proteins, A Theoretical Perspective of Dynamics, Structure and Thermodynamics. New York: Wiley.

110. Cui Q, M Karplus (2002) Quantum mechanics/molecular mechanics studies of triosephosphate isomerase-catalyzed reactions: Effect of geometry and tunneling on proton-transfer rate constants. J. Am. Chem. Soc. 124 (12): 3093–3124

111. Im W, S Berneche, B Roux (2001) Generalized solvent boundary potential for computer simulations. J. Chem. Phys. 114 (7): 2924–2937

112. Schaefer P, D Riccardi, Q Cui (2005) Reliable treatment of electrostatics in combined QM/MM simulation of macromolecules. J. Chem. Phys. 123 (1):-

113. Nam K, JL Gao, DM York (2005) An efficient linear-scaling Ewald method for long-range electrostatic interactions in combined QM/MM calculations. J. Chem. Theory Comput. 1 (1): 2–13

114. Monard G, M Loos, V Théry, K Baka, and JL Rivail (1996) Hybrid classical quantum force field for modeling very large molecules. Int. J. Quantum Chem. 58 (2): 153–159

115. Assfeld X, JL Rivail (1996) Quantum chemical computations on parts of large molecules: The *ab initio* local self consistent field method. Chem. Phys. Lett. 263 (1–2): 100–106

116. Gao JL, P Amara, C Alhambra, MJ Field (1998) A generalized hybrid orbital (GHO) method for the treatment of boundary atoms in combined QM/MM calculations. J. Phys. Chem. A 102 (24): 4714–4721

117. Ferre N, X Assfeld, JL Rivail (2002) Specific force field parameters determination for the hybrid *ab initio* QM/MM LSCF method. J. Comput. Chem. 23 (6): 610–624

118. Antonczak S, G Monard, MF Ruiz-Lopez, JL Rivail (1998) Modeling of peptide hydrolysis by thermolysin. A semiempirical and QM/MM study. J. Am. Chem. Soc. 120 (34): 8825–8833

119. Garcia-Viloca M, JL Gao (2004) Generalized hybrid orbital for the treatment of boundary atoms in combined quantum mechanical and molecular mechanical calculations using the semiempirical parameterized model 3 method. Theor. Chem. Acc. 111 (2–6): 280–286

120. Pu JZ, JL Gao, DG Truhlar (2004b) Generalized hybrid orbital (GHO) method for combining *ab initio* Hartree-Fock wave functions with molecular mechanics. J. Phys. Chem. A 108 (4): 632–650

121. Pu JZ, JL Gao, DG Truhlar (2004a) Combining self-consistent-charge density-functional tight-binding (SCC-DFTB) with molecular mechanics by the generalized hybrid orbital (GHO) method. J. Phys. Chem. A 108 (25): 5454–5463

122. Pu JZ, JL Gao, DG Truhlar (2005) Generalized hybrid-orbital method for combining density functional theory with molecular mechanicals. ChemPhysChem 6 (9): 1853–1865

123. Amara P, MJ Field (2003) Evaluation of an *ab initio* quantum mechanical/molecular mechanical hybrid-potential link-atom method. Theor. Chem. Acc. 109 (1): 43–52

124. Reuter N, A Dejaegere, B Maigret, M Karplus (2000) Frontier bonds in QM/MM methods: A comparison of different approaches. J. Phys. Chem. A 104 (8): 1720–1735

125. Antes I, W Thiel (1999) Adjusted connection atoms for combined quantum mechanical and molecular mechanical methods. J. Phys. Chem. A 103 (46): 9290–9295

126. Zhang YK, T-S Lee, WT Yang (1999) A pseudobond approach to combining quantum mechanical and molecular mechanical methods. J. Chem. Phys. 110 (1): 46–54

127. Das D, KP Eurenius, EM Billings, P Sherwood, DC Chatfield, M Hodoscek, BR Brooks (2002) Optimization of quantum mechanical molecular mechanical partitioning schemes: Gaussian delocalization of molecular mechanical charges and the double link atom method. J. Chem. Phys. 117 (23): 10534–10547

128. Konig PH, M Hoffmann, T Frauenheim, Q Cui (2005) A critical evaluation of different QM/MM frontier treatments with SCC-DFTB as the QM method. J. Phys. Chem. B 109 (18): 9082–9095

129. Ridder L, AJ Mulholland (2003) Modeling biotransformation reactions by combined quantum mechanical/molecular mechanical approaches: From structure to activity. Curr. Topics Med. Chem. 3 (11): 1241–1256

130. Berman HM, J Westbrook, Z Feng, G Gilliland, TN Bhat, H Weissig, IN Shindyalov, PE Bourne (2000) The Protein Data Bank. Nucleic Acids Res. 28 (1): 235–242. (See http://www.rcsb.org/pdb/)

131. Kast P, M Asif-Ullah, D Hilvert (1996) Is chorismate mutase a prototypic entropy trap? Activation parameters for the Bacillus subtilis enzyme. Tetrahedron Lett. 37 (16): 2691–2694

132. Lyne PD, AJ Mulholland, WG Richards (1995) Insights into Chorismate Mutase Catalysis from a Combined QM/MM Simulation of the Enzyme Reaction. J. Am. Chem. Soc. 117 (45): 11345–11350

133. Martí S, J Andrés, V Moliner, E Silla, I Tuñón, J Bertrán (2000) A QM/MM study of the conformational equilibria in the chorismate mutase active site. The role of the enzymatic deformation energy contribution. J. Phys. Chem. B 104 (47): 11308–11315

134. Ranaghan KE, L Ridder, B Szefczyk, WA Sokalski, JC Hermann, AJ Mulholland (2003) Insights into enzyme catalysis from QM/MM modelling: transition state stabilization in chorismate mutase. Mol. Phys. 101 (17): 2695–2714

135. Martí S, J Andrés, V Moliner, E Silla, I Tuñón, J Bertrán (2001) Transition structure selectivity in enzyme catalysis: a QM/MM study of chorismate mutase. Theor. Chem. Acc. 105 (3): 207–212

136. Szefczyk B, AJ Mulholland, KE Ranaghan, WA Sokalski (2004) Differential transition-state stabilization in enzyme catalysis: Quantum chemical analysis of interactions in the chorismate mutase reaction and prediction of the optimal catalytic field. J. Am. Chem. Soc. 126 (49): 16148–16159

137. Strajbl M, A Shurki, M Kato, A Warshel (2003) Apparent NAC effect in chorismate mutase reflects electrostatic transition state stabilization. J. Am. Chem. Soc. 125 (34): 10228–10237

138. Guo H, Q Cui, WN Lipscomb, M Karplus (2001) Substrate conformational transitions in the active site of chorismate mutase: Their role in the catalytic mechanism. Proc. Natl. Acad. Sci. U. S. A. 98 (16): 9032–9037

139. Martí S, J Andrés, V Moliner, E Silla, I Tuñón, J Bertrán (2003). Preorganization and reorganization as related factors in enzyme catalysis: The chorismate mutase case. Chem. Eur. J. 9 (4): 984–991

140. Guimarães CRW, MP Repasky, J Chandrasekhar, J Tirado-Rives, WL Jorgensen (2003) Contributions of conformational compression and preferential transition state stabilization to the rate enhancement by chorismate mutase. J. Am. Chem. Soc. 125 (23): 6892–6899

141. Hur S, TC Bruice (2003b) Enzymes do what is expected (chalcone isomerase versus chorismate mutase). J. Am. Chem. Soc. 125 (6): 1472–1473

142. Hur S, TC Bruice (2003a) Comparison of formation of reactive conformers (NACs) for the Claisen rearrangement of chorismate to prephenate in water and in the E-coli mutase: The efficiency of the enzyme catalysis. J. Am. Chem. Soc. 125 (19): 5964–5972

143. Hur S, TC Bruice (2003c) Just a near attack conformer for catalysis (chorismate to prephenate rearrangements in water, antibody, enzymes, and their mutants). J. Am. Chem. Soc. 125 (35): 10540–10542

144. Claeyssens F, KE Ranaghan, FR Manby, JN Harvey, AJ Mulholland (2005) Multiple high-level QM/MM reaction paths demonstrate transition-state stabilization in chorismate mutase: correlation of barrier height with transition-state stabilization. Chem. Comm. (40): 5068–5070

145. Jaguar 4.0. Schrödinger, Inc., Portland, Oregon.

146. Harvey JN (2004) Spin-forbidden CO ligand recombination in myoglobin. Faraday Discuss. 127: 165–177

147. Lee YS, SE Worthington, M Krauss, BR Brooks (2002) Reaction mechanism of chorismate mutase studied by the combined potentials of quantum mechanics and molecular mechanics. J. Phys. Chem. B 106 (46): 12059–12065

148. Crespo A, DA Scherlis, MA Marti, P Ordejon, AE Roitberg, DA Estrin (2003) A DFT-based QM-MM approach designed for the treatment of large molecular systems: Application to chorismate mutase. J. Phys. Chem. B 107 (49): 13728–13736

149. Meunier B, SP de Visser, S Shaik (2004) Mechanism of oxidation reactions catalyzed by cytochrome P450 enzymes. Chem. Rev. 104 (9): 3947–3980

150. Shaik S, D Kumar, SP de Visser, A Altun, W Thiel (2005) Theoretical perspective on the structure and mechanism of cytochrome P450 enzymes. Chem. Rev. 105 (6): 2279–2328

151. Ogliaro F, N Harris, S Cohen, M Filatov, SP de Visser, S Shaik (2000b) A model "rebound" mechanism of hydroxylation by cytochrome P450: Stepwise and effectively concerted pathways, and their reactivity patterns. J. Am. Chem. Soc. 122 (37): 8977–8989

152. Ogliaro F, S Cohen, M Filatov, N Harris, S Shaik (2000a) The high-valent compound of cytochrome P450: The nature of the Fe-S bond and the role of the thiolate ligand as an internal electron donor. Angew. Chem., Int. Ed. Engl. 39 (21): 3851-+

153. Yoshizawa K, T Kamachi, Y Shiota (2001) A theoretical study of the dynamic behavior of alkane hydroxylation by a compound I model of cytochrome P450. J. Am. Chem. Soc. 123 (40): 9806–9816

154. de Visser SP, F Ogliaro, N Harris, S Shaik (2001a) Multi-state epoxidation of ethene by cytochrome P450: A quantum chemical study. J. Am. Chem. Soc. 123 (13): 3037–3047

155. de Visser SP, F Ogliaro, S Sason (2001b) Stereospecific oxidation by Compound I of Cytochrome P450 does not proceed in a concerted synchronous manner. Chem. Comm. (22): 2322–2323

156. Guengerich FP (2001) Common and uncommon cytochrome P450 reactions related to metabolism and chemical toxicity. Chem. Res. Toxicol. 14 (6): 611–650

157. Bathelt CM, L Ridder, AJ Mulholland, JN Harvey (2003) Aromatic hydroxylation by cytochrome P450: Model calculations of mechanism and substituent effects. J. Am. Chem. Soc. 125 (49): 15004–15005

158. Bathelt CM, L Ridder, AJ Mulholland, JN Harvey (2004) Mechanism and structure-reactivity relationships for aromatic hydroxylation by cytochrome P450. Organic & Biomolecular Chemistry 2 (20): 2998–3005

159. de Groot MJ, SB Kirton, MJ Sutcliffe (2004) In silico methods for predicting ligand binding determinants of cytochromes P450. Curr. Topics Med. Chem. 4 (16): 1803–1824

160. Pirmohamed M, BK Park (2003) Cytochrome P450 enzyme polymorphisms and adverse drug reactions. Toxicology 192 (1): 23–32

161. Schoneboom JC, S Cohen, H Lin, S Shaik, W Thiel (2004) Quantum mechanical/molecular mechanical investigation of the mechanism of C–H hydroxylation of camphor by cytochrome P450(cam): Theory supports a two-state rebound mechanism. J. Am. Chem. Soc. 126 (12): 4017–4034

162. Guallar V, MH Baik, SJ Lippard, RA Friesner (2003) Peripheral heme substituents control the hydrogen-atom abstraction chemistry in cytochromes P450. Proc. Natl. Acad. Sci. U. S. A. 100 (12): 6998–7002

163. Bathelt CM, J urek, AJ Mulholland, JN Harvey (2005) Electronic structure of compound I in human isoforms of cytochrome P450 from QM/MM modeling. J. Am. Chem. Soc. 127 (37): 12900–12908

164. Hermann JC, L Ridder, AJ Mulholland, H-D Höltje (2003) Identification of Glu166 as the general base in the acylation reaction of class A beta-lactamases through QM/MM modeling. J. Am. Chem. Soc. 125 (32): 9590–9591

165. Hermann JC, C Hensen, L Ridder, AJ Mulholland, H-D Höltje (2005) Mechanisms of antibiotic resistance: QM/MM modeling of the acylation reaction of a class A beta-lactamase with benzylpenicillin. J. Am. Chem. Soc. 127 (12): 4454–4465

166. Hermann JC, L Ridder, H-D Höltje, AJ Mulholland (2006) Molecular mechanisms of antibiotic resistance: QM/MM modelling deacylation in a class A beta-lactamase. Organic & Biomolecular Chemistry 4 (2): 206–210

167. Lodola A, M Mor, JC Hermann, G Tarzia, D Piomelli, AJ Mulholland (2005) QM/MM modelling of oleamide hydrolysis in fatty acid amide hydrolase (FAAH) reveals a new mechanism of nucleophile activation. Chem. Comm. (35): 4399–4401

168. Alhambra C, ML Sanchez, JC Corchado, J Gao, DG Truhlar (2002) Quantum mechanical tunneling in methylamine dehydrogenase. Chem. Phys. Lett. 355 (3–4): 388–394

169. Tresadern G, H Wang, PF Faulder, NA Burton, IH Hillier (2003) Extreme tunnelling in methylamine dehydrogenase revealed by hybrid QM/MM calculations: potential energy surface profile for methylamine and ethanolamine substrates and kinetic isotope effect values. Mol. Phys. 101 (17): 2775–2784

170. Bjelic S, J Åqvist (2004) Computational prediction of structure, substrate binding mode, mechanism, and rate for a malaria protease with a novel type of active site. Biochemistry 43 (46): 14521–14528

171. Cho KB, V Pelmenschikov, A Gräslund, PEM Siegbahn (2004) Density functional calculations on class III ribonucleotide reductase: Substrate reaction mechanism with two formates. J. Phys. Chem. B 108 (6): 2056–2065

172. Bassan A, MRA Blomberg, PEM Siegbahn (2004) A theoretical study of the cis-dihydroxylation mechanism in naphthalene 1,2-dioxygenase. J. Biol. Inorg. Chem. 9 (4): 439–452

173. Borowski T, A Bassan, PEM Siegbahn (2004) 4-hydroxyphenylpyruvate dioxygenase: A hybrid density functional study of the catalytic reaction mechanism. Biochemistry 43 (38): 12331–12342

174. Xu DG, YS Wei, JB Wu, D Dunaway-Mariano, H Guo, Q Cui, JL Gao (2004) QM/MM studies of the enzyme-catalyzed dechlorination of 4-chlorobenzoyl-CoA provide insight into reaction energetics. J. Am. Chem. Soc. 126 (42): 13649–13658

175. Li GH, Q Cui (2003) What is so special about Arg 55 in the catalysis of cyclophilin A? Insights from hybrid QM/MM simulations. J. Am. Chem. Soc. 125 (49): 15028–15038

176. Cisneros GA, M Wang, P Silinski, MC Fitzgerald, WT Yang (2004) The protein backbone makes important contributions to 4-oxalocrotonate tautomerase enzyme catalysis: Understanding from theory and experiment. Biochemistry 43 (22): 6885–6892

177. Martí S, V Moliner, M Tuñón, IH Williams (2005) Computing kinetic isotope effects for chorismate mutase with high accuracy. A new DFT/MM strategy. J. Phys. Chem. B 109 (9): 3707–3710

178. Ruggiero GD, IH Williams, M Roca, V Moliner, I Tuñón (2004) QM/MM determination of kinetic isotope effects for COMT-catalyzed methyl transfer does not support compression hypothesis. J. Am. Chem. Soc. 126 (28): 8634–8635

179. Guimarães CRW, M Udier-Blagovic, WL Jorgensen (2005) Macrophomate synthase: QM/MM simulations address the Diels-Alder versus Michael-Aldol reaction mechanism. J. Am. Chem. Soc. 127 (10): 3577–3588

180. Park H, EN Brothers, KM Merz Jr (2005) Hybrid QM/MM and DIFT investigations of the catalytic mechanism and inhibition of the dinuclear zinc metallo-beta-lactamase CcrA from Bacteroides fragilis. J. Am. Chem. Soc. 127 (12): 4232–4241

181. Dinner AR, GM Blackburn, M Karplus (2001) Uracil-DNA glycosylase acts by substrate autocatalysis. Nature 413 (6857): 752–755

182. Crespo A, MA Marti, DA Estrin, AE Roitberg (2005) Multiple-steering QM-MM calculation of the free energy profile in chorismate mutase. J. Am. Chem. Soc. 127 (19): 6940–6941

183. Gleeson MP, IH Hillier, NA Burton (2004) Theoretical analysis of peptidyl alpha-ketoheterocyclic inhibitors of human neutrophil elastase: Insight into the mechanism of inhibition and the application of QM/MM calculations in structure-based drug design. Organic & Biomolecular Chemistry 2 (16): 2275–2280

184. Röthlisberger U, P Carloni, K Doclo, M Parrinello (2000) A comparative study of galactose oxidase and active site analogs based on QM/MM Car Parrinello simulations. J. Biol. Inorg. Chem. 5 (2): 236–250

185. Olsson MHM, PK Sharma, A Warshel (2005) Simulating redox coupled proton transfer in cytochrome c oxidase: Looking for the proton bottleneck. FEBS Lett. 579 (10): 2026–2034

CHAPTER 6

COMPUTATIONAL DETERMINATION
OF THE RELATIVE FREE ENERGY
OF BINDING – APPLICATION TO ALANINE
SCANNING MUTAGENESIS

IRINA S. MOREIRA, PEDRO A. FERNANDES AND MARIA J. RAMOS

REQUIMTE/Departamento de Química, Faculdade de Ciências da Universidade do Porto, Rua do Campo Alegre 687, 4169-007 Porto- Portugal

Abstract: Protein-protein recognition and complex formation are key issues in understanding cellular functions. Therefore, having in mind that it is of extreme importance to detect the functional sites in proteins interfaces, the present review focuses on computational approaches used to calculate the binding free energy contributions of each of the interface residues. Usually these methods do not allow the calculation of the contribution of each residue for binding in the wild type complex, but instead the difference in binding free energy between the wild type and a given residue. Although the first would be more meaningful from a phenomenological point of view, the second is the only one that is possible to measure experimentally. A number of quantitative models with different levels of rigor and speed are available for determination of the relative binding energy upon alanine mutation of residues in protein-protein interfaces. These algorithms can be divided essentially in two types: (a) empirical functions or simple physical methods and (b) fully atomistic methods

Computer simulations complement experimental analysis, and add molecular insight to the macroscopic properties, by allowing the decomposing of the binding free energy into contributions of the various energetic factors. The capacity of predicting protein-protein associations is essential in computational chemistry because it establishes the connecting bridge between structure and function of biomolecular systems, and it allows the characterization of the energetics of molecular complexes

Keywords: binding free energy; computational mutagenesis; empirical functions; fully atomistic methods; protein-protein association; MM-PBSA

W. A. Sokalski (ed.), Molecular Materials with Specific Interactions, 305–339.

1.   INTRODUCTION

Proteins tendency to bind to one another and to different ligands, forming specific stable complexes, is fundamental to all biological processes because protein-protein interactions networks are the basis of most cellular functions.[1-4] Hence, to understand and to target protein-protein interactions is an important challenge because it allows the design of new protein-protein interactions and the comprehensive knowledge of the physical basis of affinity, it permits to engineer new functions and adjust cellular behavior in a predictive manner, and it enables the understanding of molecular recognition.[5-16] Molecular recognition has many practical applications, which include the design of sensors[17] or separation techniques[18,19], and the rational design of new therapeutic agents.[1] Most known drugs bind specifically to a disease-causing bio-molecule and inhibit its function.[20-22] Therefore, the prediction and design of ligands that can reversibly bind to pharmaceutical targets is the key of structure-based drug design. Drug discovery is increasingly becoming more systematic and rational[23], and the ability to predict the binding affinity of a candidate molecule without having to synthesize it will save a lot of time and resources.[24-26]

The recent increase of information from crystallographic structures, alanine-scanning mutagenesis of protein-protein interfacial residues, and structural and thermodynamic studies[27-30] have enabled the understanding of the structure and chemistry involved in binding reactions.[31,32] It has been discovered that a larger value of the binding energy in a given complex is related only to several amino acids at the protein-protein interface: the hot spots.[33-41] These energetic determinants are compact, centralized regions of residues crucial for protein association.[42,43] Thus, hot spots[41,44,45] have been defined as those sites where alanine mutations cause a significant increase in the binding free energy of at least 4.0 kcal/mol[46], even though lower values are used for statistical analyses. The warm-spots are those with binding free energy differences ranging from 2.0 to 4.0 kcal/mol, and the null-spots are the residues with binding free energy differences lower than 2.0 kcal/mol. In Figure 6-1 it is represented a complex formed between an immunoglobulin and a hen egg lysozyme highlighting the computational mutated residues with a vdW representation.

As the experimental determination of hot spots is time consuming and costly, an effort has been made in achieving accurate, predictive computational methodologies for alanine scanning mutagenesis, capable of reproducing the experimental mutagenesis values. Therefore, it is important to accurately calculate the binding free energies of known three-dimensional structures and the effect of mutations in the corresponding affinities.[15] Binding free energy determination by computational alanine scanning mutagenesis methods allows a rational design of complexes of high affinity and specificity,[47] as well as of small molecules that can mimic the large interface that is typical of protein-protein complexes.[36]

Consequently, the capacity of predicting protein-protein associations is essential in computational chemistry. It is a very useful link between structure and function

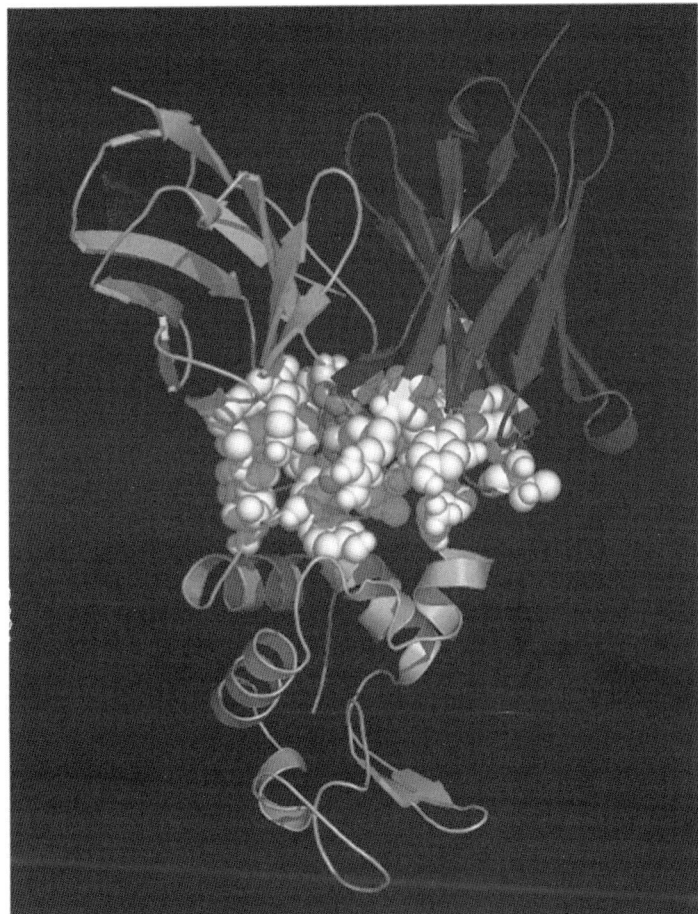

Figure 6-1. Complex formed between an immunoglobulin and a hen egg lysozyme highlighting the mutated residues with a vdW representation. The null-spots are represented in yellow (relative binding energy < 2.0 kcal/mol), the warm-spots in orange (relative binding energy between 2.0 and 4.0 kcal/mol), and the hot spots in red (residues with a relative binding energy higher than 4.0 kcal/mol)

of biomolecular systems, and it allows the characterization of the energetics of molecular complexes.[37,48−63]

2. COMPUTATIONAL CALCULATION OF THE RELATIVE BINDING ENERGY

The theoretical prediction of free binding energy differences, and the understanding of the physical foundations of affinity and specificity of complex interaction, prior to experimental design are crucial in computational biochemistry.[48,64] To apply a quantitative model of the binding affinity determination of a broad variety of protein-protein interfaces complements experimental analysis, and adds molecular insight to the macroscopic properties measured therein.[24,26,65−68]

In hot spot determination, it is sufficient to calculate a relative binding free energy (the difference in binding free energy between different ligands that bind to the same protein), and therefore the expensive calculation of the absolute binding free energy can be avoided.[69−71] Absolute binding free energies are more difficult to obtain because they contain significant entropic contributions, which are not easy nor simple to calculate.[25,66,72] These entropic contributions are usually cancelled when calculating relative free energies, as are a large number of inaccuracies resulting from the theoretical molecular models, hamiltonians, and calculation protocols, conferring to the relative free energies a much greater accuracy than to the absolute free energies.

The difference in binding free energies between the wild type and the alanine mutated complex is defined as:

$$\Delta\Delta G_{binding} = \Delta G_{binding-mutant} - \Delta G_{binding-wildtype} \qquad (6\text{-}1)$$

with the binding free energy described as:

$$\Delta G_{binding-molecule} = G_{complex} - (G_{receptor} + G_{ligand}) \qquad (6\text{-}2)$$

A huge number of methods with different levels of rigor and speed are accessible to measure the relative binding energy, and therefore to identify the hot spots. These algorithms of varied complexity, which have been employed to address the binding energy between biological molecules, can be divided essentially in two types. First, empirical functions or simple physical methods that use experimentally calibrated knowledge-based simplified models to evaluate the binding free energy. Second, versatile/universal fully atomistic methods that estimate the free energy of association or changes in the binding free energies, as a result of mutating the residues of the interacting molecules based only in the respective hamiltonians.[25]

All different types of methods have some degree of limitations and problems. An equilibrium must be achieved between the use of simple algorithms that allow fast calculations and the inclusion, conservation, and consideration of the important atomic detail of biomolecules.[73] Consequently, when deciding on the computational approach for predicting the binding free energies it is important to foresee the computational time required, without forgetting that sometimes it is affordable and advantageous to carry out more accurate time-consuming calculations because an atomic-detail description of biomolecules is often important in elucidating their structures and functions.[74]

The more rapid methods for estimation of binding free energies are the empirical or knowledge-based (statistical) scoring approaches, which are based on very simple energy functions[75−77] or on the frequency of occurrence of different atom-atom contact pairs in complexes of known structure.[78–83] These empirical methods in conjunction with simple physical models[65,84] are fast methodological approaches that have become increasingly useful for rapid protein-protein interface determinants analysis. However, in highly simplified approximations, the simplicity of the energy function along with the lack of conformational sampling and explicit water treatment makes these approaches very fast, but usually at the cost of accuracy by neglecting important components of the binding free energy.[74,85,86]

Methods that are more sophisticated like full atomistic simulations, rely on the adequate realistic sampling of the conformational space of the complex and the free molecules, which requires accurate force fields and simulation protocols.[25,87–90] These methods[48] include both the rigorous free energy perturbation (FEP)[91] and thermodynamic integration (TI)[92], and more approximate methods such as MM-PBSA.[69] λ-dynamic[93], chemical Monte Carlo/molecular mechanics[87,94] or ligand interaction scanning[95] are also methodological approaches, which have been proposed to identify the interfacial hot spots.

An important advantage of computer simulations over experiments is that they not only provide quantitative estimates but also, and mainly, they enhance our molecular understanding of the nature of complex formation in terms of the biophysical features of the process. They open the possibility of decomposing the binding free energy ($\Delta G_{binding}$) into contributions from the atoms of the residues that constitute the binding interface, as well as into the various energetic factors[25,96,97], such as the electrostatic energy $\Delta E_{electrostatic}$, the van der Waals energy ΔE_{vdw}, the free energy of polar solvation $\Delta G_{polar\ solvation}$, and the free energy of nonpolar solvation $\Delta G_{non-polar\ solvation}$.

There is another type of hot spot determination methods: structure-based qualitative methods. The PP_SITE method uses hydrogen bonds and hydrophobic characteristics to describe interactions between proteins and to decompose the contribution of atoms in hot spot residues.[4] Hu *et al* derived another qualitative method recently based in the sequence and structure alignments proteins. Residues are characterized as hot spots based in their conserved, polar characteristics.[98]

The purpose of this review is not an exhaustive description of all the methods in the literature to calculate relative binding differences, and therefore we are going to present only those, which have been applied to computational mutagenesis, especially alanine scanning mutagenesis.

3. EMPIRICAL APPROACHES AND SIMPLE PHYSICAL MODELS

The empirical approaches or knowledge based scoring functions[72,74,76–79,82–84,99,100] for calculating binding free energies for a given protein-ligand complex, include protein flexibility through conformational searches or sampling of rotamer libraries, as well as through empirical solvation models.[100] The simple functions on which they are based account for example for the hydrogen bonds, ionic interactions, the lipophilic protein-ligand contact surface and the number of rotatable bonds in the ligand.[85] Solvent screening and polarization effects are either not included or approximated by a distance dependent dielectric constant making these algorithms very fast.[47,100] The dataset for the calibration of the function consists of a small group of proteins, and therefore these scoring functions use generally adjustable parameters, which have been proven to be transferable to complexes other than those used for the parametrization.[101]

3.1.   Wallqvist Model

This model has been developed in 1994,[102] from a methodological analysis of 38 crystal complexes, to calculate the statistical probability for the occurrence of adjacent surface contacts between various atom types within a binding interface. Consequently, this model is used to calculate the individual atomic components of the interfacial interactions having as a basis an effective binding parameter, specific to each atomic interaction, weighted by an atomic pairwise surface burial. The strength of the atomic interactions is used to rank, by order of importance, each binding component with respect to its contribution to the total binding strength. Computational alanine mutations of these complexes are then used to compare the calculated changes in binding free energy with those obtained after the mutations take place.[103] The alanine mutation is constructed by simply deleting all the side-chain atoms, except the $C\alpha H_3$ methyl group of the mutated residue, without readjusting the surrounding protein environment. The model accounts for the change in binding free energy due to the direct effect of loosing interfacial residue atom contacts in the wild-type structure, and the indirect effect of releasing crystallographically ordered water molecules associated with the mutated residues, which are treated as ligands that stabilize the interaction of the bound complex.[103] The total binding free energy is calculated as a sum of the binding energy due to interactions between each member of the complex $\Delta G_{A:B}$ and the binding energy due to interactions between the crystallographic water molecules and the complex $\Delta G_{AB:water}$:

$$\Delta G_{calculated} = \Delta G_{A:B} + \Delta G_{AB:water} \tag{6-3}$$

The free energy is defined as a sum over all atom types i and j of molecules A and B, being α_{ij} an effective binding parameter and A_{ij} the mutual surface jointly buried by the atom types i and j in the complex. β is a constant for the entropy-penalty for the association of two molecules:

$$\Delta G_{pred} = \sum_{i \in A} \sum_{i \in B} \alpha_{ij} A_{ij} + \beta \tag{6-4}$$

This method was applied to the study of the interface of three complexes: barnase-barstar[104], D1.3-HEL[105] and D1.3-E5.2.[106] The method shows with some exceptions a theoretical result within ± 1.5 kcal/mol of the experimental ones.[103]

3.2. Molecular Statics (MS) Method

This computational method dissects the free energy of binding into two elements, nonpolar and polar. It consists in using the continuum solvation model and solving the Poisson-Boltzmann (PB) equation for the polar factor[107,108] and calculating the nonpolar contribution to the binding free energy with the surface area method. A limited rotamer search is done in order to derive the structures of alanine mutant proteins bound with substrate from the static X-ray crystallographic structures. When steric overlaps occur, caused by the structural perturbation, they are relieved by using the Ponder and Richards[109] side-chain rotamer library before energy evaluation.[47] It contrasts with other algorithms where the crystal structure is treated in a dynamic manner.[100,110,111]

The free energy is calculated as a sum of the polar and the nonpolar parts:

$$G_{molecule} = G_{polar\ solvation} + G_{nonpolar\ solvation} \qquad (6\text{-}5)$$

This methodology was applied to calculate the relative binding free energy of 63 pairs of nine different mutant proteins with seven substituted R-malate substances of *Escherichia Coli* isocitrate dehydrogenase. The average difference for the calculated and the observed relative binding difference was 0.5 kcal/mol.[47]

3.3. Partitioning Approach

This method avoids the convergence and accuracy problems of molecular dynamics or Monte Carlo simulations of systems containing explicit solvent molecules, by evaluating the electrostatic free energy of just one solute conformation surrounded by a dielectric continuum, and by adding the surface term and an estimate of the loss of the configurational entropy upon binding.[77]

The binding energy function is partitioned into three terms: the hydrophobic or cavity term ΔG_H, the electrostatic term ΔG_{el} and the entropic term ΔG_s. The hydrophobic term, which is generally the major driving force in biomolecular complexes, reflects the variation of water/non-water interface area. The electrostatic contribution that is very important for specificity is composed by the coulombic interactions and desolvation of partial charges transferred from an aqueous medium to a protein core environment. This term is calculated by solving the Poisson-Boltzmann (PE)-equation with an internal dielectric constant value of 8. The entropic term results from the decrease in the conformational freedom of functional groups buried upon complexation.

An additional constant term C accounts for the change of entropy of the system due to the decrease of free molecules concentration (critic factor) and the loss of rotational/translational degrees of freedom. Conformational strain and dynamics of the process are incorporated in the C parameter and can be adjusted from one set of complexes to another.[77] The free energy is defined as:

$$\Delta G = \Delta G_H + \Delta G_{el} + \Delta G_s + C \qquad (6\text{-}6)$$

This method predicts, within $\pm 2.5\,\text{kcal/mol}$ the binding energy of small molecule-protein, peptide-protein and protein-protein complexes.[77]

3.4.   Kortemme- Simple Physical Model

This model is based on an all-rotamer description of the side-chains. The energy function is dominated by the Lennard-Jones potential to describe atomic packing interactions, by an implicit solvation model[25], an orientation-dependent hydrogen-bonding potential derived from high-resolution protein structures[112], by statistical terms approximating the backbone-dependent amino acid-type and rotamer probabilities, and by an estimate of unfolded reference state energies:

$$\Delta G = W_{attr}E_{LJattr} + W_{rep}E_{LJrep} + W_{HB(sc\text{-}bb)}E_{HB(sc\text{-}bb)} + W_{HB(sc\text{-}sc)}E_{HB(sc\text{-}sc)}$$

$$+ W_{sol}G_{sol} + W_{\varphi\psi}E_{\varphi\psi}(aa) + \sum_{aa=1}^{20} n_{aa}E_{aa}^{ref} \qquad (6\text{-}7)$$

E_{LJattr} is the attractive part of a Lennard-Jones potential; E_{LJrep} is a linear distance-dependent repulsive term; $E_{HB(s\text{-}bb)}$ is the orientation-dependent side chain-backbone hydrogen bond potential; $E_{HB(sc\text{-}sc)}$ is the orientation-dependent side chain-side chain hydrogen bond potential; G_{sol} is the free energy of solvation calculated with the implicit solvation model, W is the relative weight of the different energy terms; $E_{\varphi\psi}(aa)$ is the amino acid type-dependent backbone torsion angle propensity; and E_{aa}^{ref} is the amino acid type-dependent reference energy (n_{aa} is the number of amino acids of a certain type).

Although this alanine scanning mutagenesis method has been applied to 19 complexes with a relative high success rate (69% hot spot correct detection), it is not a fully atomistic method and has some limitations. Some of the most important consist in the fact that, as the terms in the energy function are pairwise additive multiple mutations are always assumed to be additive which is not always the case.[113,114] Therefore, alanine shaving cannot be done, and cooperativity cannot be measured. It is also assumed that there are no conformational changes in the mutant upon binding, and cofactors, metal ions, hydrogen-bonding water molecules bridging side-chains in the protein interface, or other nonpeptide ligands or binding partners (such as nucleic acids) are not taken into account.[65,84]

4.   LINEAR INTERACTION ENERGY (LIE)

LIE is a semi empirical method first proposed by Åqvist,[115–118] which reduces the computational time involved in the binding free energy calculations by considering only the physical relevant states, instead of spending time on sampling intermediate states as Free energy perturbation (FEP) or Thermodynamic integration (TI) do. It is based on a linear response approximation in which the binding free energy is the combination of the weighted electrostatic and van der Waals interactions between the ligand and the receptor. It requires just two simulations, one of the solvated ligand-protein system and another of the ligand alone in solution.

Molecular dynamics or Monte Carlo simulations are used to generate the ensembles: one with the ligand free in solution and the other with the ligand bound to the macromolecule. The average differences between electrostatic interaction and vdW interaction energies of the ligand with other molecules are then calculated, in agreement with equation (6-8), using both trajectories [115]:

$$\Delta G = \alpha < \Delta E_{elec} > + \beta < \Delta E_{vdw} > \qquad (6\text{-}8)$$

α and β are semi empirical parameters: α is equal to 0.5 from linear response theory, adjusted for the presence of OH groups, and β varies from 0.15 to 1.0 depending on the hydrophobicity of the ligand binding site.[69] $< \Delta E_{elec} >$ and $< \Delta E_{vdw} >$ are the average differences between the electrostatic interactions and the van der Waals interactions in ligand bound and free trajectories. Several values have been proposed for the empirical parameter β. It has been found out that the more hydrophobic groups are buried after binding, the more favourable is the binding and the larger the value of β.[118] This can be justified by the fact that both the solute-solvent van der Waals energies and the hydrophobic solvation free energies depend on the same variables (such as the accessible surface area), and that average van der Waals energies scale almost linearly with solute size.[115]

LIE calculations have been performed in various systems using different programs, force fields and computational procedures, and the resulting optimizations can vary considerably.[117−126] Although different protocols are needed for different systems, a significant part of the large variety of parameter values obtained is dependent on the computational procedure used.[117] One of the critical technical issues in this type of calculations seems to be the treatment of electrostatic interactions, at least for charged ligands. The implementation of boundary conditions, cutoffs, sampling time and system neutrality is thus of considerable importance, as well as ensuring compatibility of the simulations of the bound and free states with respect to electrostatic solvation energies.[116] An interesting extension of the LIE method that employs the surface generalized Born model of Still *et al.* (1990)[127] for the solvent has also been reported.[128,129]

5.   MM-PBSA

Another methodological approach, which has become more attractive in the last few years for estimating binding free energies of protein-protein complexes, is the MM-PBSA method (Molecular Mechanics/Poisson-Boltzmann Surface Area). This method is a fully atomistic approach that combines molecular mechanics and continuum solvent, and has several appealing features as the possibility of being applied to a variety of systems not suitable for FEP such as very large protein-protein complexes.[69,129−137]

In this section, we shall focus our attention on some important points of this methodological approach to calculate relative binding free energies. It is important to have a more comprehensive knowledge about computational simulations, and therefore we first review the force-field methods existent for all type of complexes involving proteins, acid nucleic, carbohydrates and lipids, as well as the solvation models, before giving an overview of the MM-PBSA methodology.

5.1.   Force Fields for Bimolecular Simulations

Molecular dynamics (MD) simulations provide a detailed description of complex systems in a wide range of time and spatial scales.[138] Simulations involve a statistical uncertainty component as the result of the finite length of the simulation.[139–143] MD methods generate a series of time-correlated points in phase space by propagating a suitable starting set of coordinates and velocities according to Newton's second equation. This kind of computational simulations are useful in studies of time evolution of a variety of systems: biological molecules, polymers, or catalytic materials, and in a variety of states: crystal, aqueous solutions, or in the gas phase.

Molecular Mechanics (MM) force field-based methods represent a major tool for the theoretical understanding of biomolecular systems.[144,145] A force field is a mathematical expression that describes the dependence of the energy of a molecule on the coordinates of the atoms in the molecule. It is based in the observation that different molecules tend to be composed of units that are structurally similar. A force field is used in structure determination, conformational energies, rotational and pyramidal inversion barriers, vibrational frequencies, Monte Carlo and molecular dynamics. Examples of force fields are: Molecular Mechanic Force Field for Small Molecules (MM2/3/4)[146]; Chemistry at Harvard Macromolecular Mechanics (CHARMM)[147,148]; Assisted Model Building with Energy Refinement (AMBER)[149–151]; Optimized Parameters for Liquid Simulation (OPLS)[152]; Consistent Force Field (CFF)[153]; Valence Consistent Force Field[154]; Merck Molecular Force Field 94 (MMFF94);[155–157] Universal Force Field (UFF)[158]; Groningen Molecular simulation Program Package (GROMOS);[159,160] CVFF, the force field developed at BIOSYM (now ACCELRYS)[161]; the force field described by Daggett *et al.* within Levitt's simulation program ENCAD.[162] Such force fields have been developed for many types of molecules[163–171] but we will focus on those applied regularly for biomolecular studies.

The potential energy function applied in most force fields is an additive function of pairs of atoms such as:

$$U\left(\vec{R}\right) = \sum_{bonds} \frac{K_b}{2} (l - l_0)^2 + \sum_{angles} \frac{K_\theta}{2} (\theta - \theta_0)^2 + \sum_{dihedral} \frac{K_\chi}{2} [1 + \cos(n\chi - \phi)]$$

$$+ \sum_{improper} \frac{K_{imp}}{2} (\varphi - \varphi_0)^2 + \sum_{nonbonded} 4\varepsilon_{ij} \left[\left(\frac{\sigma_{ij}}{r_{ij}}\right)^{12} - \left(\frac{\sigma_{ij}}{r_{ij}}\right)^6 \right] + \frac{1}{4\pi\varepsilon} \frac{q_i q_j}{r_{ij}} \qquad (6\text{-}9)$$

Equation (9) is a collection of simple functions, which represent a minimal set of forces that can describe molecular structures. Bonds, angles, and out-of-plane distortions (improper dihedral angles) are treated harmonically and dihedral or torsional rotations are described by a sinusoidal term. In equation (9), the first four terms represent the intramolecular parameters.[172] l is the bond length, θ is the valence angle, χ is the dihedral or torsion angle, φ is the improper angle, r_{ij} is the distance between atoms i and j. K_b and l_0 are the bond force constant and reference distance respectively; K_θ and θ_0 are the valence angle force constant and reference angle respectively; K_χ, n, ϕ are the dihedral force constant, multiplicity and phase angle respectively; and K_φ and φ_0 are the improper force constant and reference improper respectively. The last two sets of parameters are the nonbonded parameters between atoms i and j, with q the partial atomic charges, ε_{ij} the Lennard-Jones well-depth, σ_{ij} depending on the atom pair involved and $2^{1/6}\sigma_{ij}$ the minimum interaction radius used to treat the vdW interaction. Parameterization plays a crucial role in reproducing the experimental data[173] leading to subtle differences between the different existent force fields.[24,25,174,175] Although not commonly used, equation (9) can have extra terms, namely the cross terms that describe any coupling between already existing ones.

Besides the amino acid force fields mentioned above, a number of force fields have been especially developed for carbohydrates, nucleic acids and lipids. They follow the same fundamental philosophy used in protein simulations. The nature of nucleic acids requires a more accurate treatment of the balance between the conformational energy and the interactions with the solvent.[129,172] Some of the most important force fields developed for nucleic acids are: parm94[149] and parm98/99[176] of the AMBER package and CHARMM27[177] of the CHARMM package. Other force fields are also available such as the one developed by Langley[178], GROMOS,[179,180] CVFF[161] and MMFF.[155−157] When treating nucleic acids it is important to use methods that smooth the truncation of the long-range electro-static interactions, either Ewald-like methods (PME[181] or PPPM[182]) or simpler methods like switching/shifting functions.[183] Lipids represent a major challenge, and therefore current lipid force fields include both all-atom and united extended-atom models.[184,185] There are course-grain lipid models designed for simulations of extended lipid bilayers.[186−187] In the area of carbohydrates, CHARMM,[188,189] AMBER,[190−191] and OPLS[192] have developed parameters for computational simulations.

Transferability, using the same set of parameters to model a series of related molecules, is an important feature of a force field. Concerning the specialized biomolecular force fields, their transferability varies. For example, AMBER is more transferable than CHARMM especially with recent efforts on more automated methods of parameter assignment.[129,193] Essentially the quality of a force field depends on how appropriate is the mathematical form of the energy expression and how accurate are the parameters.

5.2.   Solvation

Electrostatics and solvation play important roles in the structure and dynamics of biomolecules, and accurate descriptions of the effects of solvation are indispensable for computer simulations of protein-protein complexes.[194–199] Solvation can be described as the process of transferring one molecule from a fixed position in an ideal gas phase to a fixed position in the fluid phase, at constant temperature and pressure. The effect of a solvent may be divided in "short-range" effects, "mid-range" effects and "long-range" effects. The "short-range" effects are repulsive interactions, arising from Pauli's exclusion principle, $1/r^{12}$ dependent, and usually assumed to be proportional to the volume of the cavity. The "mid-range" effects, dispersive interactions also called van der Waals contributions, are $1/r^6$ dependent and assumed to be proportional to the contact surface area. The "macroscopic" or long-range electrostatics" effects, involving screening of charges, are $1/r$ dependent and responsible for the generation of a dielectric constant different from 1.[200]

The solvation free energy is the consequence of the transference of a molecule from vacuum to water and may be written as:

$$\Delta G_{solvation} = \Delta G_{cavity} + \Delta G_{dispersion} + \Delta G_{electrostatic} = \Delta G_{non\text{-}polar} + \Delta G_{electrostatic}$$

$$(6\text{-}10)$$

$\Delta G_{dispersion}$ is the energy of the van der Waals interactions between solvent and solute. It is usually favourable for solvation (negative), partially compensating for the entropic cost. Since the van der Waals interactions decrease quickly with distance ($1/r^6$), the first solvation shell contributes mostly to the $\Delta G_{dispersion}$, and it is roughly proportional to the solvent accessible surface area S. ΔG_{cavity} is a term which includes the entropy penalty for reorganizing the solvent molecules around a solute and the work against solvent pressure to create a cavity in the solvent to immerse a solute. The creation of the cavity costs energy, e. g. the loss of solvent-solvent van der Waals interactions, producing a positive ΔG_{cavity} value. The reorganizational entropy should be correlated with the number of water molecules in the first solvation shell, because the solvent molecules more affected by a solute are those in the first solvation shell, and therefore this term is proportional to the solvent accessible surface area of the solute. Thus, the $\Delta G_{non\text{-}polar}$ is the free energy of solvating a molecule from which all charges have been removed (partial charges of every atom are set to zero), and is equal to the sum of the dispersion and the cavity energies. $\Delta G_{electrostatic}$ is the free energy of first removing all charges in the vacuum, and then adding them back in the presence of the solvent environment.[201] Electrostatic interactions are long range and critically dependent on the boundary shape between a solute and solvent.

There are various methods for treating solvation, ranging from a detailed description at the molecular level to reaction field models where the solvent is modelled as a continuum method.[125]

The explicit water models most currently used are TIP3P, TIP4P[202], TIP5P[203], SPC and SPC/E.[204] The parameters in all models are empirically adjusted so that they reproduce the enthalpy of vaporization and the density of water.[129] TIP3P is probably the most commonly used model. However, when selecting a water model to use for a particular study the most important is that it is compatible with the biomolecular force field being used. This approach appears to give reasonable dynamic and thermodynamic properties with the presently available force fields. Even though explicit solvent representations offer a high degree of microscopic detail, they are the most computationally demanding of the solvation methods.[205−207] To include a large number of solvent molecules explicitly increases the number of atoms present in the simulation, and places severe limits in the type of problems that can be studied. However, in explicit solvent simulations, as long as sufficient conformational sampling is performed, many different aspects that contribute to relative binding free energies can be included automatically. These aspects involve changes in the rotational, translational, conformational and vibrational entropy of the partners; entropy changes associated with solvent ordering around hydrophobic or charged groups; solute conformational strain; changes in electrostatic and van der Waals interactions within and between the partners and the solvent; and counterions reorganization.[96]

However, to overcome the limitations of explicit solvent simulations especially the elevated computationally cost and to allow the calculation of the $\Delta G_{solvation}$, a high number of implicit solvation models for proteins have been developed that combine an empirical force field for the molecular solute interactions in vacuum with a solvation correction. Implicit solvent methods such as Poisson-Boltzmann (PB)[15,60,73,107,108,194,208−211] and Generalized Born model (GB)[208,212−214] offer significant computational savings by implicitly accounting for solvation effects via a simple dielectric model.[215] The GB model has been widely used and is incorporated in several molecular mechanics programs. The programs[216,217] most extensively used that solve the Poisson-Boltzmann equation[218,219] for a protein-solvent system are Delphi[220], Grasp[221] and UHBD.[222] The Poisson-Boltzmann equation is given by the expression:

$$\nabla(\varepsilon\,(\vec{r})\,\nabla\phi\,(\vec{r})) = -4\pi\rho\,(\vec{r}) \tag{6-11}$$

Where $\phi(\vec{r})$ describes the electrostatic potential in a medium with a charge density $\rho(\vec{r})$ and a dielectric scalar field $\varepsilon(\vec{r})$. The Poisson–Boltzmann equation is a second-order elliptic partial differential equation that can be solved analytically for regularly shaped solutes and numerically for most irregularly shaped slots in different ways. For example, in the Delphi program, it is used the Finite Difference Method in which the solute with its associated charges is mapped on to a grid and the electrostatic potential due to the presence of a dielectric continuum solvent is determined in each grid point. In this method, the protein is modelled as a dielectric continuum of low polarizability embedded in a dielectric medium of high polarizability.[223] It is assumed that the solvent is a homogeneous medium characterized by a single dielectric constant with a value usually near 80, which is taken to be equal to the bulk value for pure water. Separated by an abrupt interface it meets the solute that is represented as a dielectric body whose shape is defined by atomic coordinates and radii, and with an internal dielectric constant usually between 1 and 20. The protein dielectric constant, which accounts for responses to an electric field that are not treated explicitly, is an adjustable parameter.[224] It is not a universal constant but simply a parameter that depends on the model and the relevant region in the protein.[225]

Implicit solvation models are an emergent field, which is widely applied in many computational simulations with a reasonable time cost.[226-235]

5.3.   The MM-PBSA approach fundamental theory

The MM-PBSA method is based on partitioning the free energy into a sum of enthalpic and entropic contributions.[48,236] It describes the free energy of the complex and respective monomers as a sum of the internal energy (bond, angle and dihedral), the electrostatic and the van der Waals interactions, the free energy of polar solvation, the free energy of nonpolar solvation and the entropic contribution for the molecule free energy:

$$G_{molecule} = E_{internal} + E_{electrostatic} + E_{vdw} + G_{polar\ solvation} + G_{nonpolar\ solvation} - TS$$

$$(6\text{-}12)$$

A fundamental question with the MM-PBSA approach is how to best determine the contribution from the entropy change[237,238] upon binding. The entropy contribution of ligands to a common protein receptor is often assumed to cancel when the ligands are of similar size and when the interest is the relative binding free energy. However, if it is necessary to obtain the absolute binding free energy, the solute entropic contribution must be determined in a consistent fashion to yield meaningful results. There are several approaches to estimate the solute entropy, including the normal-mode analysis[239], the quasi-harmonic analysis[240-242] and the quasi-Gaussian approach.[243]

One of the most important applications of this method concerns computational fluorine or alanine-scanning mutagenesis.[129-131,135,244] As it was already pointed out, mutagenesis studies allow the determination of hot spots having a major impact in structure-based drug design. This method is based in a post-processing treatment of the complex by using its structure, and calculating the respective energies for the complex and all interacting monomers. To generate the structure of the mutant complex a simple truncation of the mutated side chain is made, replacing C_γ by a hydrogen atom, and setting the C_β-H bond direction to that of the former C_β-C_γ. The complexation free energy can be calculated using the thermodynamic cycle in Scheme 6-1, in which ΔG_{gas} is the interaction free energy between the ligand and the receptor in the gas phase and ΔG_{solv}^{lig}, ΔG_{solv}^{rec} and ΔG_{solv}^{cpx} are the solvation free energies of the ligand, the receptor and the complex respectively. The first three terms of equation 12 are calculated with no cutoff. The electrostatic solvation free energy is that of first removing all charges in vacuum, and then adding them back in the presence of the solvent environment.[92,201]

It is calculated by solving the Poisson-Boltzmann equation with the software Delphi v.4 or other equivalent software.[107,108,200,245,246] Normally, it is obtained by

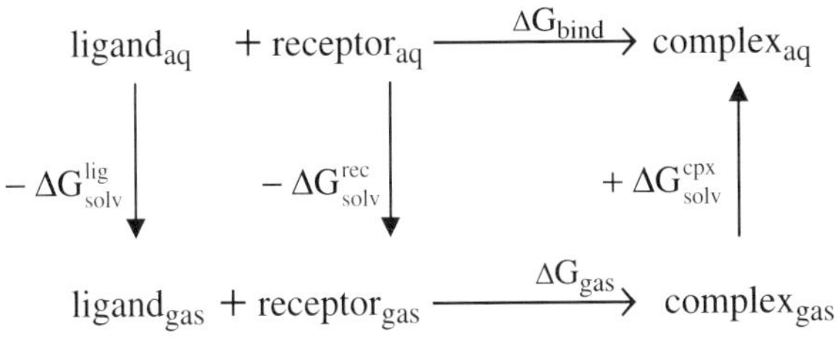

$$\text{ligand}_{aq} + \text{receptor}_{aq} \xrightarrow{\Delta G_{bind}} \text{complex}_{aq}$$

$$-\Delta G_{solv}^{lig} \qquad -\Delta G_{solv}^{rec} \qquad +\Delta G_{solv}^{cpx}$$

$$\text{ligand}_{gas} + \text{receptor}_{gas} \xrightarrow{\Delta G_{gas}} \text{complex}_{gas}$$

Scheme 6-1. Thermodynamic cycle used to calculate the complexation free energy

taking the difference between suitable states, in reaction field energies. This means that it has got to be solved twice, once in vacuum and a second time for the desired solution environment:

$$\Delta G_{electrostatic} = \frac{1}{2} \sum q_i (\phi_i^{80} - \phi_i^1) \qquad (6\text{-}13)$$

However, there is in the DelPhi package[245], an alternative way to calculate the electrostatic solvation free energy based on the concept of induced charges, which is obtained with a simple Coulomb calculation between the induced charges and the real charges as if they were in vacuum. The first method of calculation uses two runs, with the grid positions for both real and induced charges. The second method uses the actual position for real charges and the optimized position for induced charges, and utilizes the surface of the molecule where no fixed charges are present, to calculate the potential without requiring a second finite-difference run.

The nonpolar contribution to solvation free energy, due to van der Waals interactions between the solute and the solvent and cavity formation, is modelled as a term that is dependent in the solvent accessible surface area of the molecule. It is estimated using an empirical relation: $\Delta E_{nonpolar} = o \cdot A + b$, in which A is the solvent-accessible surface area and is calculated with programs based on an idea primarily developed by Mike Connolly.[247] o and b are empirical constants with values $0.00542 \, \text{kcal} \, \text{Å}^{-2} \, \text{mol}^{-1}$ and $0.92 \, \text{kcal} \, \text{mol}^{-1}$ respectively. The cavity term equals the work to create the cavity against the solvent pressure and the entropy penalty associated with the reorganization of solvent molecules around the solute. Therefore, the linear dependence of A can be explained by the fact that the solvent molecules more affected by this reorganization are in the first solvation shell, and this number is proportional to the accessible surface area of the solute. The solute-solvent van der Waals interaction energy falls off rapidly with distance and is consequently dependent upon the number of solvation molecules present in the first shell.

At this point it is necessary to emphasize that the success rate of this method is dependent on the protocol used. As it is resumed in Scheme 6-2, many different approaches have been tried.[71] They vary from a static to a dynamic analysis of the complexation process.[71,134,248]

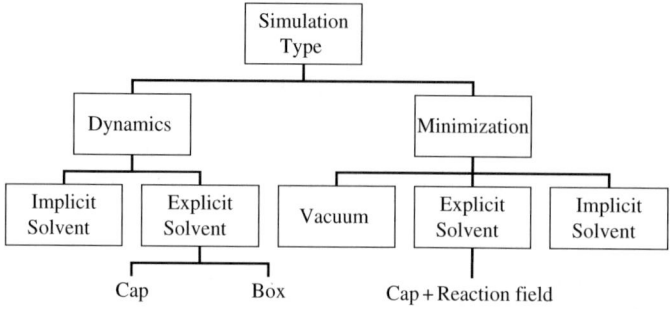

Scheme 6-2. Resume of the Methodological Approach for Computational Alanine Scanning Mutagenesis

    The simulation protocol can vary also from the use of implicit or explicit solvation models. The MM-PBSA method by Massova/Kollman[130,135] is a full atomistic method that, although not accurate enough in the original implementation, unlocked the possibility for the development of a newly improved methodology. In Scheme 2 we present a resume of our own computational alanine scanning study.[70,71] Initially we have tried different solvent representations (explicit or implicit), and different internal dielectric constants for proteins ($\varepsilon = 1$–15). Subsequently, we have tried protocols with a different number of dynamics simulations trajectories. Finally, we tried a less expensive method, a minimization approach. After careful analysis of the results we concluded that what gives a higher success rate are MD simulations using the Generalized Born Solvation model. A set of different internal dielectric constants used to calculate the first 3 terms of equation (12) as well as the free solvation energy is used, depending on the type of amino acid that is mutated. Therefore, for the charged amino acids (aspartic acid, glutamic acid, lysine, arginine) and histidine a constant of 4 should be used, for the remaining polar residues (aspargine, glutamine, cysteine, tyrosine, serine and threonine) not ionized at physiological pH the internal dielectric constant should be 3, and for the non-polar amino acids (valine, leucine, isoleucine, phenylalanine, methionine and tryptophan) the internal dielectric constant should be 2. The different internal dielectric constants mimic the different degrees of relaxation of the interface when different types of amino acids are mutated for alanine.[70,71]

    Wang *et al.* developed a Virtual mutagenesis (VM) method in 2000 which evaluates the relative free energy upon amino acid mutagenesis. It consists in taking one snapshot that has the closest binding free energy to the average binding free energy value obtained from the MD trajectory, and making mutations by a fast screening procedure. For each mutation, a systematic conformation search for a total of 100 conformations is performed. Only those conformations with no steric clash with other atoms are further investigated. Each surviving conformation is minimized with a distance dependent dielectric constant while all other residues in the molecule are kept fixed. The binding free energy is then calculated using

MM-PBSA. The final binding free energy for each mutation is the average value for all rotamers.[133]

A procedure, conceptually similar to the MM/PBSA[70,71,113,114,129−131,249−252] approach, was developed. A simpler energy function[87,253,254] was used in conjunction with Monte Carlo simulations[255−257] to sample the conformational space and adequately describe the multitude of the low-energy states available to the system. The resulting conformational states are evaluated with a detailed free energy model, which includes the molecular mechanics AMBER force field[258] and the solvation energy term based on the continuum generalized born model and the solvent accessible surface area (GB/SA) solvation model.[259−265] To represent alternate protein conformations it was applied a simulation approach to model ligand binding with the ensembles of multiple protein conformations. This method can account for protein flexibility by considering a finite number of protein states that have significant differences in both side-chain and main-chain conformations. Although the two methods give good qualitative results, the MM-PBSA approach can be used more accurately for a quantitative analysis.[4]

MM-PBSA, wherein only the initial and final states of the system are evaluated, is computationally less expensive than free energy perturbation methods, making it suitable for a greater variety of systems and problems, and representing a promising direction for evaluating binding affinities.[67,129]

6.   FREE ENERGY PERTURBATION (FEP) AND THERMODYNAMIC INTEGRATION (TI)

Free energy perturbation (FEP) and thermodynamic integration (TI) methods[266−273] have been successfully applied to predict the binding strength of proteic complexes.[91,266] These rigorous methods yield accurate free energy differences relying on equilibrium sampling of the entire transformation path, from an initial to a final state.[204] They are implemented numerically and utilize a thermodynamical cycle, and the fact that the free energy is a state function. Thus, sufficient statistical sampling must be carried out which turns them computationally very intensive and limits their use in screening effects of a large number of structural perturbations.[266,272,274−281] These methods follow the thermodynamic cycle in Scheme 6-3:

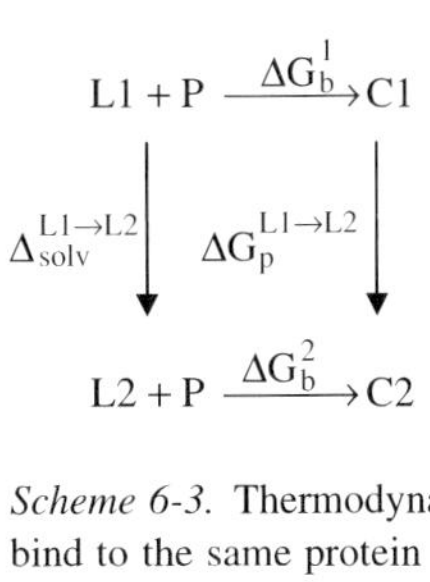

$$L1 + P \xrightarrow{\Delta G_b^1} C1$$

$$\Delta_{solv}^{L1 \to L2} \Big\downarrow \qquad \Delta G_p^{L1 \to L2} \Big\downarrow$$

$$L2 + P \xrightarrow{\Delta G_b^2} C2$$

Scheme 6-3. Thermodynamic cycle used to calculate the relative free energies between two ligands that bind to the same protein

In agreement with Scheme 3 the free energy of binding is defined by equation (14) as a difference between the binding free energy for ligand1 ΔG_b^1 and ligand2 ΔG_b^2, with the ΔG_{solv} and ΔG_p the non-physical transmutation free energies from ligand1 to ligand2 in free and bound states, respectively:

$$\Delta\Delta G = \Delta G_b^1 - \Delta G_b^2 = \Delta G_{solv}^{L1 \to L2} - \Delta G_p^{L1 \to L2} \qquad (6\text{-}14)$$

FEP calculations use the formulation present in equation (15):

$$\Delta G = -RT \sum_{i=1}^{N-1} \ln \left\langle \exp(-\frac{H(\lambda_{i+1}) - H(\lambda_i)}{RT}) \right\rangle_{\lambda_i} \qquad (6\text{-}15)$$

ΔG is the free energy difference between two states L1 and L2. The transformations are made by using a coupling parameter λ such that $\lambda = 0$ corresponds to one ligand and $\lambda = 1$ to the other. $H(\lambda_i)$ represents the Hamiltonian of system λ_i and $<>_{\lambda i}$ indicates an ensemble average. After deciding the simulation protocol (length of trajectories, equilibration, number of λ-points and their spacing) the difference between TI and FEP resides in the formula chosen to evaluate the relative free energy.[48] Therefore, within TI, the average of derivatives of the hamiltonian at each λ, $H(\lambda)$, are calculated and then numerical integration is used over λ to calculate the free energy difference between two states. The TI method follows equation (16):

$$\Delta G = \int_0^1 \left\langle \frac{\partial H(\lambda)}{\partial \lambda} \right\rangle d\lambda \qquad (6\text{-}16)$$

Molecular dynamics free energy simulations (MDFE)[282,283] use the FEP and TI methodologies. The MDFE results provide the opportunity to rank the importance of different amino acids to complex formation and to decompose the energy in their different types of terms of differential binding.[282] Molecular dynamics simulations together with Poisson-Boltzmann free energy calculations (PBFE)[284,285] represent a less expensive alternative to MDFE, suitable for examining multiple active site mutations.[130] They are useful to complement MDFE results if they are available.[286,287] MDFE, an alchemical transformation of the ligand or protein is present, and in contrast, in PBFE a chemical binding reaction is studied. Good agreement was found (1 kcal/mol) between MDFE and PBFE when an internal dielectric constant of four was used.[282]

The increasing number of applications of free energy calculations has shown that the use of these methods is not as simple as expected. One of its severe limitations is the sampling of the conformational space,[288] which must be long and in the correct region of space. Thus, large efforts for improving this methodology have been made, especially in improving the treatment of long-range electrostatic interactions and molecular polarization.[48,129] However, combined with other modelling tools, free energy calculation methods can be used in a broader range of research, such as to evaluate stability of folding structures, to determine hot spots in protein-protein interfaces and to design new therapeutic drugs.[67,129]

7. PROFEC AND OWFEG

PROFEC (pictorial representation of free energy changes)[289] is a methodological approach that provides a rapid estimate of docking free energy changes for modified ligands, based on a grid around the ligand determined by FEP analysis of a single MD trajectory of the complex, according to the traditional FEP equation:

$$\Delta G_{ins}(i, j, k) = -\beta^{-1} \ln < \exp(-\beta \Delta v(i, j, k)) >_0 \qquad (6\text{-}17)$$

in which $\Delta v(i, j, k)$ is the van der Waals interaction energy between the particle and the surrounding atoms. The electrostatic contribution can be examined by calculating the derivative of the binding free energy with respect to charge at each grid point, under the assumption that a particle has already been inserted. Two MD simulations, one of the protein-ligand complex and one of the ligand in solution are performed, and the free energies of all grid points are then displayed as contour surfaces around the protein. This PROFEC method could quantitatively suggest relatively more favourable regions for molecular change, and was shown to be promising.[289] Typically, PROFEC should be used in combination with one of the more detailed approaches such as traditional FEP, LIE or MM-PBSA to computationally validate the changes suggested prior to experimental design. However, PROFEC has limitations such as its inability to evaluate free energies when multiple sites are modified or when modifications induce large conformational changes.[48]

The one-window free energy grid (OWFEG)[129,290] methodology also utilizes a single-window ($\lambda = 0$) trajectory to create a grid from FEP to added probe atoms, but has introduced two modifications in the PROFEC approach. First, each grid point undergoes translation and rotation along with the atom of the ligand, taking into account the possible flexibility in the ligand by allowing the grid points to drift as if they were connected to the ligand. Second, three probes with neutral, positive, and negative charges were used instead of only a neutral probe to examine the desirability of introducing charged groups along the grid, providing suggestions as to what type of charges should be placed at that grid point.[48]

8. Λ - DYNAMICS AND CHEMICAL MONTE CARLO/MOLECULAR DYNAMICS (CMC/MD)

The most time-consuming and rigorous methods are based on molecular force fields and involve slow gradual transformations between the states of interest using either molecular dynamics (MD) or Monte Carlo (MC) simulations for generating ensemble averages.[291]

Molecular dynamics (MD) and Monte Carlo (MC) methods have provided dynamic and atomic insights to understand complex biological systems. Thus, many techniques such as the λ-dynamics and the chemical Monte Carlo/Molecular Dynamics (CMC/MD) method have been developed to improve their efficiencies.[129]

Inspired by the work of Tidor (1993)[292,293], Kong and Brooks (1996)[93] have proposed a new approach to multiple state free energy calculations. It is a variant of free energy simulations called λ-dynamics, that simulates a group of several molecules and directly yields estimates of their relative binding affinity in only a single run of simulation.[93,276,277,294–296] This method uses MD to propagate both the atomic coordinates and the chemical space (coupling parameter). It is, however, possible to use MC in order to sample the coupling parameter stochastically combined with MD for propagating the atomic coordinates.

The methodology is based on the idea that multiple ligands will compete for a common receptor based on their relative free energies, and that this can be explored using multiple copy simultaneous search approaches[297], providing a route to assess several free energies. Equation (18) in which the coupling parameter λ is treated as a dynamic variable, is used in this method:

$$V(\{\lambda\}) = V_{env} + \sum_{i=1}^{L} \lambda_i^2 (V_i - F_i) \qquad (6\text{-}18)$$

where L is the total number of ligands, V_{env} is the interaction involving the surrounding atoms, V_i is the interaction involving any of the atoms in ligand i, λ_i is the coupling parameter and F_i is a reference energy. In traditional free energy calculations a coupling parameter λ, provides the link between the initial and the final systems. In most free energy simulations, λ varies from 0 to 1. However, the λ-dynamics technique considers λ to be another particle in the simulation.

The chemical Monte Carlo/Molecular Dynamics (CMC/MD) methodology[95,279,280] was developed by Pitera and Kollman in order to rank binding affinities for several ligands in a single MD simulation, and is conceptually similar to the λ-dynamics. In a CMC/MD calculation, there are two parts of the simulated system: the surroundings and the Monte Carlo region. The MD is used to sample conformations of each ligand, and the MC is used to sample chemical space of all ligands.[292,298] The Metropolis algorithm[299–301] (generates a random walk of points distributed according to the Boltzmann probability distribution) is used to determine whether the substitution is accepted. At the end of the simulation, free energy differences between ligands are given by the ratio of the populations of each ligand.[95]

Both λ-dynamics and the CMC/MD have been successfully used to estimate relative binding free energies of similar compounds[129,280]. However, it is difficult to control the actual spatial coordinates for different ligands since in practice they cannot be allowed to drift away from each other. Restraining the various ligands to each other[294,295] can overcome this limitation, but impose other limitations concerning the amount of conformational space that can be explored.[48]

9.   CONCLUSION

Most protein-protein binding energies are related only to a group of a few amino acids at intermolecular protein interfaces: the hot spots. The characterization of the energetics of molecular complexes, especially the detection of these hot spots is essential to structure-based drug design.

The capacity of predicting protein-protein hot spots, and the characterization and understanding of the physical foundations of affinity and specificity of the complex interaction, prior to the experimental design, is crucial in computational biochemistry. To apply a quantitative model for the determination of the relative binding free energy upon mutation of residues in protein-protein interfaces complements experimental analysis, and adds molecular insight to the macroscopic properties measured therein. An important advantage of computer simulations over experiments is the possibility of decomposing the binding free energy into contributions of the various energetic factors.

A huge amount of methods with different levels of rigor and speed are accessible to measure the relative binding free energy upon computational mutagenesis. These algorithms can be divided essentially in two types: (a) empirical functions or simple physical methods and (b) fully atomistic methods. Empirical methods in conjunction with simple physical models are fast methodological approaches but usually at the cost of accuracy and versatility by neglecting important components of the binding free energy. Methods that are more sophisticated such as full atomistic simulations rely on the adequate realistic sampling of the conformational space of the complex and the free molecules, which requires accurate force fields and simulation protocols. We have highlighted the new possibilities of the improved MM-PBSA approach. [70,71] This improved all atom method can nowadays be applied to an alanine scanning mutagenesis study of a complete protein-protein interface presenting a high success rate. Being a fully atomistic method it also opens the possibility of a predictive behaviour (essential in computational biochemistry).

REFERENCES

1. Kortemme T, Baker D (2004) Computational design of protein-protein interactions, Curr Opin Chem Biol, 8: 91–97
2. Russel RB, Alber F, Aloy P, Davis FP, Korkin D, Pichaud M, Topf M, Sali A (2004) A structural perspective on protein-protein interactions, Curr Opin Struct Biol, 14: 313–324
3. Verkhivker GM (2004) Computational analysis of ligand binding dynamics at the intermolecular hot spots with the aid of simulated tempering and binding free energy calculations, Mol J Graph Model, 22(5): 335–348
4. Moreira IS, Fernandes PA, Ramos MJ (2006) Hot spots – a detailed review of protein-protein interface determinant amino acid residues (Submitted)
5. Baker D, Agard DA (1994) Kinetics versus thermodynamics in protein folding, Biochemistry, 33: 7505–7509
6. Braden BC, Poljak RJ(1995) Conservation of water molecules in an antibody-antigen interaction, Mol J Recogn, 9: 317–325
7. Braden BC, Fields BA, Ysern X, Goldbaum FA, Dall'Acqua W, Schwartz FP, Poljak RJ, Mariuzza RA (1996) Crystal structure of the complex of the variable domain of antibody D1.3 and the turkey egg with lysozyme: a novel conformational change in antibody CDF-L3 selects for antigen, J Mol Biol, 257: 889–894
8. Dall'Acqua W, Goldman E, Eisenstein E, Mariuzza RA (1996) A mutational analysis of the binding of two different proteins to the same antibody, Biochemistry, 35: 9667–9676

9. DeGrado WF, Summa CM, Pavone V, Nastri F, Lombardi A (1999) De novo design and structural characterization of proteins and metalloproteins, Annu Rev Biochem, 68: 779–819
10. Davis ME, Mccammon JA (1990) Electrostatics in biomolecular structure and dynamics, Chem Rev, 94: 7684
11. Frader X, de la Cruz X, CHTP Silva, Gelpi JL, Luque FJ and Orozco M(2002) Ligand-induced changes in the binding sites of proteins, Bioinformatics, 18(7): 939–948
12. Hill RB, Raleigh DP, DeGrado WF (2000) De novo design of helical bundles as models for understanding protein folding and function, Acc Chem Res, 33: 745–754
13. Lubienski MJ, Bycroft M, Freund SMV, Fersht AR (1994) 13C assignments and the three-dimensional solution structure of barstar using nuclear magnetic resonance spectroscopy, Biochemistry, 33: 8866–8877
14. Nakamura HQ (1996) Roles of electrostatic interactions in proteins, Rev Biophys, 29: 1–90
15. Sheinerman FB, Norel R, Honig B (2000) Electrostatics aspects of protein-protein interactions, Curr Opin Struct Biol, 10: 153–159
16. Schreiber G, Fersht AR (1995) Energetics of protein-protein interactions: analysis of the barnase-barstar interface by single mutations and double mutant cycles, J Mol Biol, 248: 478–486
17. de Silva AP, Gunaratne HQN , Gunnlaugsson T, Huxley AJM, McCoy CP, Rademacher JT, Rice TE (1997) Chem Rev, 97: 1515
18. Sellergren BJ (2001) Imprinted chiral stationary phases in high-performance liquid chromatography, Chromatogr A J, 906(1–2): 227–252, Review
19. Deng Q, German I, Buchanan D, Kennedy RT (2001) Retention and separation of adenosine and analogues by affinity chromatography with an aptamer stationary phase, Anal Chem, 73: 5415–5421
20. Anthonsen HW, Baptista A, Drablos F, Martel P, Petersen SB (1994) The blind watchmaker and rational protein engineering, J Biotechnol, 36: 185–220
21. Gordon DB, Marshall SA, Maya SL (1999) Curr Opin Struct Biol, 9: 509–513
22. Michielin O, Luescher I, Karplus M (2000) Modeling of the TCR-MHC-peptide complex, M J Mol Biol , 300: 1205–1235
23. Gane PJ, Dean PM (2000) Recent advances in structure-based rational drug design, Curr Opin Struct Biol, 10: 401
24. Ajay, Murcko MA (1995) Computational methods to predict binding free energy in ligand-receptor complexes, J Med Chem , 38: 4953–4967
25. Lazaridis T (2002) Binding affinity and specificity from computational studies, Current Organic Chemistry, 6: 1319–1332
26. McCammon JA (1998) Theory of Biomolecular Recognition, Curr Opin Struct Biol, 8: 245
27. Stites WE (1997) Protein–protein interactions: interface structure, binding thermodynamics and mutational analysis, Chem Rev, 97: 1233
28. Jones S, Thorton JM (1996) Principles of protein–protein interactions, Proc Natl Acad Sci USA, 93: 7
29. Davies DR, Cohen GH (1996) Interactions of protein antigens with antibodies, Proc Natl Acad Sci USA, 93: 7
30. Brooijmans N, Sharp KA, Kruntz ID (2002) Stability of macromolecular complexes, Proteins, 48: 645
31. Fersht AR (1987) The hydrogen bond in molecular recognition, Trends Biochem Sci, 12: 301–304
32. Perutz M (1992) Protein structure, new approaches to disease and therapy, New York: WH Freeman and Company
33. Wells JA (1991) Systematic mutational analyses of protein-protein interfaces, Methods Enzymol, 202: 390–411

34. Wells JA, De Vos AM (1993) Structure and function of human growth hormone: implications for the hematopoietins, Annu Rev Biophys Biomol Struct, 22: 329–351

35. Wells JA (1994) Structural and functional basis for hormone binding and receptor oligomerization, Curr Opin Cell Biol, 6: 163–173

36. Wells JA (1996) Binding in the growth hormone receptor complex, Proc Natl Acad Sci USA, 93:1–6

37. Verkhivker GM, Bouzida D, Gehlhaar DK, Rejto PA, Freer ST, Rose PM(2003) Computational detection of the binding-site hot spot at the remodelled human growth hormone-receptor interface, Proteins, 53: 201–219

38. Thornton JM (2001) The Hans Neurath Award lecture of The Protein Society: proteins – A testament to physics, chemistry, and evolution, Protein Sci, 10: 3–11

39. Cunnigham BC, Wells JA (1993) Comparison of a structural and functional epitope, J Mol Biol, 234: 554–563

40. Clackson T, Wells JA (1995) A hot spot of binding energy hormone-receptor interfaces, Science, 267: 383–386

41. Conte LL, Chothia C, Janin J (1999) The atomic structure of protein-protein recognition sites, J Mol Biol, 285: 2177–2198

42. Arkin MR, Wells AJ (2004) Small-molecules inhibitors of protein-protein interactions: progressing towards the dream, Drug Discovery, 3: 301–317

43. Arkin MR, Randal M, DeLano WL, Hyde J, Luong TN, Oslob JD, Raphael DR, Taylor L, Wang J, McDowell RS, Wells JA, Braisted AC (2003) Binding of small molecules to an adaptive protein–protein interface, Proc Natl Acad Sci USA, 100: 1603–1608

44. Bogan AA, Thorn KS (1998) Anatomy of hot spots in protein interfaces, J Mol Biol, 280: 1–9

45. Keskin O, Ma B, R Nussinov R (2005) Hot regions in protein-protein interactions: the organization and contribution of structurally conserved hot spot residues, J Mol Biol, 345: 1281–1294

46. Pons J, Rajpal A, Kirsch J (1999) Energetic analysis of an antigen/antibody interface: alanine scanning mutagenesis and double mutant cycles on the HyHEL-100lysozyme interaction, Protein Sci, 8: 958–968

47. Zhang T, Koshland DE (1996) Computational method for relative binding energies of enzyme-substrate complexes, Protein Sci, 5: 348–356

48. Åqvist J, Luzhkov B, Brandsda BO (2002) Ligand binding affinities from MD simulatio, Acct Chem Res, 35: 358–365

49. Vorobjev YN, Hermans J (2001) Free energies of protein decoys provide insight into determinants of protein stability, Protein Sci, 10: 2498–2506

50. Vorobjev YN, Almagro JC, Hermans J (1998) Discrimination between native and intentionally misfolded conformations of proteins: ES/IS, a new method for calculating conformational free energy that uses both dynamics simulations with an explicit solvent and an implicit solvent continuum model, Proteins, 32: 399–413

51. Wallqvist A, Smythers GW, Covel DG (1997) Identification of cooperative folding units in a set of native proteins, Protein Sci, 6: 1627–42

52. Thirumalai D, Klimov DK, Woodson SA (1997) Deciphering the timescales and mechanisms of protein folding using minimal off-lattice models, Theor Chem Acc, 96: 14–22

53. Sheinerman FB, Brooks III CL (1998) Molecular picture of folding of a small alpha/beta protein, Proc Natl Acad Sci USA, 95: 1562–7

54. Nolting B, Andert K (2000) Mechanism of protein folding, Proteins, 41: 288–298

55. Murphy KP, Bhakuni V, Xie D, Freire E (1992) Molecular basis of cooperativity in protein folding. III. Structural identification of cooperative folding units and folding intermediates, J Mol Biol, 227: 293–306

56. Moult J, Melamud E (2000) From fold to function, Curr Opin Struct Biol, 10: 384–9

57. Lazaridis T, Karplus M (1997) "New view" of protein folding reconciled with the old through multiple unfolding simulations, Science, 278: 1928–1931

58. Friedberg I, Margalit H (2002) Persistently conserved positions in structurally similar, sequence dissimilar proteins: roles in preserving protein fold and function, Protein Sci, 11: 350–360

59. Kuhlman B, Baker D (2000) Native protein sequences are close to optimal for their structures, Proc Natl Acad Sci USA, 97: 10383–10388

60. Honig B (1999) Protein folding: from the Levinthal paradox to structure prediction, J Mol Biol , 293: 283–293

61. Dill KA (1990) Dominant forces in protein folding, Biochemistry, 29: 7133–55

62. Dill KA and Chan HS (1997) From Levinthal to pathways to funnels, Nat Struct Biol, 4: 10–19

63. Dokholyan NV, Shakhnovich EI (2001) Understanding hierarchical protein evolution from first principles, J Mol Biol, 312: 289–307

64. Zhu H, Snyder M (2002) 'Omic' approaches for unraveling signaling networks, Curr Opin Cell Biol, 14: 173–179

65. Kortemme T, Baker D (2002) A simple physical model for binding energy hot spots in protein-proteins complexes, Proc Natl Acad Sci USA, 99: 14116–14121

66. Gilson MK, Given JA, Bush BL, McCammon JA (1997) The statistical-thermodynamic basis for computation of binding affinities: a critical review, Biophys J, 72: 1047

67. Swanson JMJ, Henchman RH, McCammon JA (2000) Revisiting free energy calculations: a theoretical connection to MM/PBSA and direct calculation of the association free energy, Biophys J, 86: 67–74

68. Simonson T, Archontis G, Karplus M (2002) Free energy simulations come of age: protein-ligand recognition, Acct Chem Res, 35: 430–437

69. Kollman PA, Massova I, Reyes C, Kuhn B, Huo S, Chong L, Lee M, Lee T, Duan Y, Wang W, Donini O, Cieplak P, Srinivasan J, Case DA, Cheatham III TE(2000) Calculating structures and free energies of complex molecules: combining molecular mechanics and continuum models, Acc Chem Res, 33: 889–897

70. Moreira IS, Fernandes PA, Ramos MJ (2006) Computational alanine scanning mutagenesis – an improved methodological approach, J Comput Chem, In Press

71. Moreira IS, Fernandes PA, Ramos MJ (2007) Unravelling Hot Spots – a comprehensive computational mutagenesis study, Theor Chem Acc, 117: 99–113

72. Janin J (1996) Quantifying biological specificity: the statistical mechanics of molecular recognition, Proteins, 25(4): 438–445

73. Luo R, David L, Gilson MK (2002) Accelerated Poisson-Boltzmann calculations for static and dynamic systems, J Comput Chem, 23: 1244–1253

74. Bjørn Brandsdal O, Österberg F, Almlöf M, Feierberg I, Luzhkov VB, and Åqvist J (2003) Free energy calculations and ligand binding advances in protein chemistry, 66: 123–158

75. Böhm HJ (1994) The development of a simple empirical scoring function to estimate the binding constant for a protein-ligand complex of known three-dimensional structure, J Comput Aided Mol Des, 8(3), 243–256

76. Eldridge MD, Murray CW, Auton TR, Paolini GV, Mee RP (1997) Empirical scoring functions: I. The development of a fast empirical scoring function to estimate the binding affinity of ligands in receptor complexes, J Comput Aided Mol Des, 11: 425–445

77. Schapira M, Totrov M, Abagyan R (1999) Prediction of the binding energy for small molecules, peptides and proteins J Mol Recog, 12: 177–190

78. Gillis D, Rooman M (1996) Stability changes upon mutation of solvent-acessible residues in protein evaluated by database-derived potentials J Mol Biol, 257: 1112–1126

79. Gohlke H, Hendlich M, Klebe G (2000) Knowledge-based scoring function to predict protein-ligand interactions, J Mol Biol, 295: 337–356
80. Jiang L, Gao Y, Mao F, Liu Z, Lai L (2002) Potential of mean force for protein-protein interaction studies, Proteins, 46(2): 190–196
81. Moont G, Gabb HA, Stenberg MJ (1999) Use of pair potentials across protein interfaces in screening predicted docked complexes, Proteins, 35: 364
82. Muegge I, Martin YC (1999) Atomtypes assigned by SMARTS rules. Problems with insufficient data in PDB for some atom types, J Med Chem, 42: 791–804
83. Sippl MJ, Ortner M, Jaritz M, Lackner P, Flockner H (1996) Helmholtz free energies of atom pair interactions in proteins, Fold Des, 1: 289–298
84. Kortemme T, Kim DE, Baker D (2004) Computational alanine scanning of protein-protein interfaces, SciSTKE, 219: 12–15
85. Böhm HJ, Stahl M (1999) Rapid empiring scoring functions in virtual screening applications, Med Chem Res, 9: 445–462
86. Oprea TI, Marshall GR (1998) Receptor based prediction of binding affinities, Perspect Drug Discov Des, 9: 35–41
87. Verkhivker GM, Bouzida D, Gehlhaar DK, Rejto PA, Freer ST, Rose PM (2002) Monte Carlo simulation of the peptide recognition at the consensus binding site of the constant fragment of the human immunoglobulin G: the energy landscape analysis of a hot spot at the intermolecular interface, Proteins, 48: 539–557
88. Kollman PA (1998) Recent advances in structure-based ligand design using molecular mechanics and monte carlo methods, Pharm Res, 15: 368–370
89. Lamb ML, Jorgensen WL (1997) Computational approaches to molecular recognition, Curr Opin Chem Biol, 1: 449–457
90. Åqvist J, Marelius J (2001) The linear interaction energy method for predicting ligand binding free energies, Comb Chem High T Scr, 4: 613–626
91. Kollman PA (1993) Free energy calculations—applications to chemical and biochemical phenomena, Chem Rev, 93: 2395–2317
92. Gouda H, Kuntz I, Case DA, Kollman PA (2002) Free energy calculations for theophylline binding to an RNA aptamer: comparison of MM-PBSA and thermodynamic integration methods, Biopolymers, 68: 16–34
93. Kong XJ, Brooks III CL (1996) Lambda dynamics—a new approach to free energy calculations, J Chem Phys, 105: 2414–23
94. Banba S, Guo Z, Brooks III CL (2001) New free energy based methods for ligand binding from detailed structure-function to multiple ligand screening, In Free Energy Calculations in Rational Drug Design, NY: Kluwer Academic/Plenum Publishers, 195–223
95. Eriksson MAL, Pitera J, Kollman PA (1999) Prediction of the binding free energies of new TIBO-like HIV-1 reverse transcriptase using a combination of PROFEC, PB/SA, CMC/MD and free energy calculations, J Med Chem, 42: 868–881
96. Archontis G, Simonson T, Karplus M (2001) Binding free energies and free energy components from molecular dynamics and Poisson-Boltzmann calculations, Application to amino acid recognition by aspartyl-tRNA synthetase, J Mol Biol, 306(2): 307–327
97. Moreira IS, Fernandes PA, Ramos MJ (2006) Hot spots computational identification – application to the complex formed between the hen egg-white lysozyme (HEL) and the antibody HyHEL-10, Int J Quantum Chem, 107: 299–310
98. Hu ZJ, Ma BY, Wolsfon H, Nussinov R (2000) Conservation of polar residues as hot spots at protein interfaces, Proteins, 39(4): 331–342

99. Novotny J, Bruccoleri R, Saul F (1989) On the attribution of binding energy in antigen-antibody complexes MCPC 603, D1.3, and HyHEL-5, Biochemistry, 28: 4735

100. Wilso C, Mace J, Agard D (1991) Computational method for the design of enzymes with altered substracte specificity, J Mol Biol, 220: 495

101. Vadja S, Sippl M, Novotny J (1997) Empirical potentials and functions for protein folding and binding, Curr Opin Struct Biol, 7: 222–228

102. Wallqvist A, Uller M (1994) A simplified amino acid potential for use in structure predictions of proteins, Proteins, 18: 267–280

103. Covell DG, Wallqvist A (1997) Analysis of protein-protein interactions and the effects of amino acid mutations on their energetics. The importance of water molecules in the binding epitope, J Mol Biol, 269: 281–297

104. Guillet V, Laphorn A, Hartley RW, Mauguen Y (1993) Recognition between a bacterial ribonuclease, barnase, and its natural inhibitor, barstar, Structure, 1: 165–177

105. Bhat TN, Bentley GA, Boulot G, Greene MI, Tello D, Dall'Acqua W, Souchon H, Schwarz FP, Mariuzza RA, Poljak RJ (1994) Bound water molecules and conformational stabilization help mediate an antigen-antibody association, Proc Natl Acad Sci USA, 9: 1089–1093

106. Braden BC, Fields BA, Ysern X, Goldbaum FA, Dall'Acqua W, Schwartz FP, Poljak RJ, Mariuzza RA (1996) Crystal structure of an Fv-Fv idiotope-antiidiope complex at 1.9 Å resolution, J Mol Biol, 264: 137–151

107. Sharp K, Honig B (1990) Calculating total electrostatic energies with the nonlinear Poisson-Boltzmann equation, J Phys Chem, 94: 7684–7692

108. Sharp K, Honig B (1991) Electrostatic interactions in macromolecules: theory and applications, Annu Rev Biophys Chem, 19: 301–332

109. Ponder J, Richards F (1987) Tertiary templates for proteins. Use of packing criteria in the enumeration of allowed sequences for different structural classes, J Mol Biol, 193: 775

110. Bash P, Singh U, Langridge R, Kollman PA (1987) Free energy calculations by computer simulation, Science, 236: 564

111. Warshel A, Sussman F, Hwang J (1998) Evaluation of the catalytic free energies in genetically modified proteins, J Mol Biol, 201: 139

112. Kortemme T, Morozov AV, Baker D (2003) An orientation-dependent hydrogen bonding potential improves prediction of specificity and structure for proteins and protein-protein complexes, J Mol Biol, 326: 1239–1259

113. Moreira IS, Fernandes PA, Ramos MJ (2006) Detailed microscopic study of the full ZipA:FtsZ interface, 63: 811–21

114. Moreira IS, Fernandes PA, Ramos MJ (2006) Unravelling the importance of protein-protein interaction: application of a computational alanine scanning mutagenesis to the study of the IgG1: streptococcal protein G (c2 Fragment) complex, J Phys Chem B, 110: 10962

115. Aqvist J, Medina C, Samuelsson JE (1994) A new method for predicting binding affinity in computer-aided drug design, Protein Eng, 7: 385–391

116. Aqvist J (1996) Calculations of absolute binding free energies for charged ligands and effects of ling-range electrostatic interactions, J Comput Chem, 17: 1587–1597 Proteins, 63: 811

117. Hansson T, Marelius J, Åqvist J (1998) Ligand binding affinity prediction by linear interaction energy methods, J Comput Aided Mol Des, 12: 27–35

118. Wang W, Wang J, Kollman PA (1999) What determines the van der Waals coefficient β in the LIE (linear interaction energy) method to estimate binding free energies using molecular dynamic simulation, Proteins, 34: 395–402

119. Gorse AD, Cready JE (1997) Molecular dynamics simulations of the docking of substituted N5-deazepterins to dihydrofolate reductase, Protein Eng, 10: 23–30

120. Jones-Hertzog DK, Jorgensen WL (1997) Binding affinities for sulfonamide inhibitors with human thrombin using Monte Carlo simulations with a linear response method, J Med Chem, 40:1539–1549

121. Ljungberg KB, Marelius J, Musil D, Svensson P, Norden B, Åqvist J (2001) Computaional modelling of inhibitor binding to human thrombin, Eur J Pharm Sci, 12: 441–446

122. Lamb ML, Tirado-Rives J, Jorgensen WL (1999) Estimation of the binding affinities of FKBP12 inhibitors using a linear response method, Bioorg Med Chem, 7: 851–860

123. Paulsen MD, Ornstein RL (1996) Binding free energy calculations for P450cam-substracte complexes, Protein Eng, 9: 567–571

124. Sham YY, Chu ZT, Tao H, Warshel A (1996) Examining methods for calculations of binding free energies: LRA, LIE, PDLD-LRA, and PDLD/S-LRA calculations of ligands binding to an HIV protease, Proteins, 39(4): 393–407

125. Smith PE, Pettit BM (1994) Modeling Solvent in Biomolecular Systems, J Phys Chem, 98: 9700

126. Wall ID, Leach AR, Salt DW, Ford MG, Essex JW (1999) Binding constants of neuraminidase inhibitors: an investigation of the linear interaction energy method, J Med Chem, 42: 5142–5152

127. Still WC, Tempczyk A, Hawley RC, Hendricksen TJ (1990) Semi-analytical treatment of solvation for molecular mechanics and dynamics, J Am Chem Soc, 112: 6189

128. Zhou R, Friesner RA, Ghosh A, Rizzo RC, Jorgensen WL, Levy J (2001) New linear inter-action method for binding affinity calculations using a continuum solvent mode, J Phys Chem B, 105: 10388–10397

129. Wang W, Donini O, Reyes CM, Kollman PA(2001) Biomolecular simulations: recent developments in force field, simulations of enzyme catalysis, protein-ligand, protein-protein, and protein-nucleic acid noncovalent interactions, A Rev Biophys Biomol Struct, 30: 211–243

130. Massova I, Kollman PA (1999) Computational alanine scanning to probe protein-protein interac-tions: a novel approach to evaluate binding free energies, J Am Chem Soc, 121: 8133–8143

131. Massova I, Kollman PA (2000) Combined molecular mechanical and continuum solvent approach (MM-PBSA/GBSA) to predict ligand binding, Perspectives in drug discovery and design, 18: 113–135

132. Wang J, Morin P, Wang W, Kollman PA (2001) Use of MM-PBSA in reproducing the binding free energies to HIV-1 RT of TIBO derivatives and predicting the binding mode to HIV-1 RT of efavirenz by docking and MM-PBSA, J Am Chem Soc, 123: 5221–5230

133. Wang W, Kollman PA (2000) Free energy calculations on dimer stability of the HIV protease using molecular dynamics and a continuum solvent model, J Mol Biol, 303: 567–582

134. Reyes CM, Kollman PA (2000) Investigating the binding specificity of U1A-RNA by computa-tional mutagenesis, J Mol Biol, 295: 1–6

135. Huo S, Massova I, Kollman PA (2002) Computational alanine scanning of the 1:1 human growth hormone-receptor complex, J Comput Chem, 23: 15–27

136. Srinivasan J, Cheatham T, Cieplak P, Kollman PA, Case DA (1998) Continuum solvent studies of the stability of the DNA, RNA, and phosphoramidate-DNA helices, J Am Chem Soc, 120: 9401–9409

137. Laitinen T, Kankare JA, Perakyla M (2004) Free energy simulations and MM-PBSA analyses on the affinity and specificity of steroid binding to antiestradiol antibody, Proteins, 55(1): 34–43

138. Elder R, Ghosh A, Cardenas A (2002) Long time dynamics of complex systems, Acc chem Res, 35: 396–403

139. Allan MP, Tildesley DJ (1987) Computer Simulations of Liquids, Clarendon Press Oxford Science Publications

140. Leach AR (1996) Molecular Modelling: Principles and Applications, Longman Addison Wesley Longman, Essex

141. Haile JM (1991) Molecular Dynamic Simulations, Wiley New York

142. Frenkel D, Smith B (1996) Undestanding Molecular Simulations, Academic Press Inc, San Diego
143. van Gunsteren WF, Berendsen HJC (1990) Computer simulation of molecular dynamics: methodology, applications and perspectives in chemistry, Angew Chem Int Ed, 29: 992
144. Karplus M, Petsko GA (1990) Molecular Dynamics Simulations in Biology, Nature, 347: 631
145. Becker OM, MacKerell Jr AD, Roux B, Watanabe M (2001) Computational Biochemistry and Biophysics, NY: Marcel-Dekker, Inc, 512
146. Allinger NL (1977) Conformational analysis 130. MM2. A hydrocarbon force field utilizing V1 and V2 torsional terms, J Am Chem Soc, 99: 8127–8134
147. Brooks BR, Bruccoleri RE, Olafson BD, States DJ, Swaminathan S, Karplus MJ (1983) CHARMM: a program for macromolecular energy, minimization, dynamics calculations, J Comput Chem, 4: 187–217
148. MacKerell AD, Bashford D, Bellott M, Dunbrack RL, Evanseck JD (1998) All-atom empirical potential for molecular modeling and dynamics studies of proteins, J Phys Chem B, 102: 3586–3616
149. Cornell WD, Cieplak P, Bayly CI, Gould IR, Merz Jr KM, Ferguson DM, Spellmeyer DC, Fox T, Caldwell JW, Kollman PA (1995) A second generation force field for the simulation of proteins, nucleic acids, organic molecules, J Am Chem Soc, 117: 5179–5197
150. Kollman PA (1997) Computer Simulations of Biomolecular Systems, ed. WF van Gunsteren, PK Weiner, A Wilkinson, Leiden: ESCOM
151. Wang W, Donin O, Reyes CM, Kollman PA (2001) Biomolecular simulations: recent developments in force fields, simulations of enzyme catalysis, protein-ligand, protein-protein, and protein-nucleic, acid noncovalent interactions, Annu Rev Biophys Biomol Struct, 30: 211–243
152. Kaminski G, Jorgensen WL (1996) Performance of the Amber94, Mmff94, and Opls-A a force fields for modeling organic liquids, J Phys Chem, 100: 18010–18013
153. Niketic SR and Rasmussen K (1997) The consistent force field: a documentation. Lectures Notes in Chemistry 3, Springer-Verlag, p. 212 Berlin
154. Jónsdóttir SO, Welsh WJ, Rasmunnen K, Klein RA (1999) The critical role of force field in property prediction, New J Chem, 153–163
155. Halgren TA (1996) Merck molecular force field. I basis, form, scope, parameterization, and performance of Mmff94, J Comp Chem, 17: 490–519
156. Halgren TA (1996) Merck molecular force field. II. Mmff94 van der Waals and electrostatic parameters for intermolecular interactions, J Comp Chem, 17: 520–552
157. Halgren TA (1996) Merck molecular force field. III. Molecular geometries and vibrational frequencies for Mmff94, J Comp Chem, 17: 553–586
158. Rappe AK, Casewit CJ, Colwell KS, Goddard III WA, Skiff WM (1992) UFF, a rule based full periodic table force field for molecular mechanics and molecular dynamics simulations, J Am Chem Soc, 114: 10024
159. van Gunsteren WF (1987) GROMOS Groningen molecular simulation program package, Groningen: University of Groningen
160. Stocker U, van Gunsteren WL (2000) Molecular dynamics simulation of hen egg white lysozyme: a test of the GROMOS96 force field against nuclear magnetic resonance data, Proteins: Struct Funct Genet, 40: 145–531
161. Ewig CS, Thacher TS, Hagler AT (1999) Derivation of class II force fields. VII. Nonbonded force field parameters for organic compounds, J Phys Chem B, 103: 6998–7014
162. Levitt M, Hirshberg M, Sharon R, Daggett V (1995) Potential energy function and parameters for simulations of the molecular dynamics of proteins and nucleic acids in solution, Comput Phys Commun, 91: 215–231

163. Engler EM, Andose JD and Schleyer PVR (1973) Critical evaluation of molecular mechanics, J Am Chem Soc, 95: 8005–8025

164. Gundertofte K, Palm J, Pettersson I Stamvik A (1991) A comparison of conformational energies calculated by molecular mechanics (MM2(85), Sybyl 5.1, Sybyl 5.21, ChemX) and semiempirical (AM1 and PM3) methods, J Comp Chem, 12: 200–208

165. Gundertofte K, Liljefors T, Norrby PO, Pattesson I (1996) Comparison of conformational energies calculated by several molecular mechanics methods, J Comp Chem, 17: 429–449

166. Hall D, Pavitt N (1984) An appraisal of molecular force fields for the representation of polypeptides, J Comp Chem, 5: 441–450

167. Hobza P, Kabelac M, Sponer J, Mejzlik P, Vondrasek J (1997) Performance of empirical potentials (AMBER, CFF95, CVFF, CHARMM, OPS, POLTEV), semiemprical quantum chemical methods (AM1, MNDO/M, PM3) and ab initio Hartree-Fock method for interaction of DNA bases: comparison of nonempirical beyond Hartree-Fock results, J Comp Chem, 18: 1136–1150

168. Kini RM, Evans HJ (1992) Comparison of protein models minimized by the all-atom and united atom models in the amber force field, J Biomol Structure and Dynamics, 10: 265–279

169. Roterman IK, Gibson KD, Scheraga HA (1989) A comparison of the CHARMM, AMBER, and ECEPP/2 potential for peptides I, J Biomol Struct Dynamics, 7: 391–419

170. Roterman IK, Lambert MH, Gibson KD, Scheraga HA (1989) A comparison of the CHARMM, AMBER, ECEPP/2 potential for peptides II, J Biomol Struct Dynamics, 7: 421–452

171. Whitlow M, Teeter MM (1986) A empirical examination of potential energy minimization using the well-determined structure of the protein crambin, J Am Chem Soc, 108: 7163–7172

172. Mackerell Jr AD (2004) Empirical force fields for biological macromolecules: overview and issues, J Comput Chem, 25: 1584–1604

173. Wang J, Wolf R, Caldwell JW, Kollman PA, Case DA (2004) Development and testing of a general amber force field, J Comput Chem, 25: 1157–1174

174. Agalarov SC, Prasad GS, Funke PM, Stout CD, Williamson Jr JR (2000) Structure of the S15,S18-rRNA complex: assembly of the 30S ribosome central domain, Science, 288: 107–112

175. Alhambra C, Wu L, Zhang ZY, Gao JL (1998) Walden-inversion-enforced transition-state stabilization in a protein tyrosine phosphatase, J Am Chem Soc, 120: 3858–3866

176. Cheatham TE, Cieplak P, Kollman PA (1999) A modified version of the Cornell et al force field with improved sugar pucker phases and helical repeat, J Biomol Struct, 16: 845–862

177. Foloppe N, Mackerell AD (2000) Allatom empirical force field for nucleic acids. I. Parameter optimization based on small molecule and condensed phase macromolecular target data, J Comp Chem, 21: 86–120

178. Langley DR (1998) Molecular dynamic simulations of environment and sequence dependent DNA conformations: the development of the BMS nucleic acid force field and comparison with experimental results, J Biomol Struct, 16: 487–509

179. Tapia O, Velasquez I (1997) Molecular Dynamics Simulations of DNA with Protein's Consistent GROMOS Force Field and the Role of Counterions' Symmetry, J Am Chem Soc, 119: 5934

180. Roxstrom G, Velazquez I, Paulino M, Tapia O (1998) DNA structure and fluctuations sensed from a 1.1ns molecular dynamics trajectory of a fully charged Zif268-DNAcomplex in water, J Biomol Struct, 16: 301–312

181. Essmann U, Perera L, Berkowitz ML, Darden T, Lee H, Pedersen LG (1995) A smooth particle mesh Ewald method, J Chem Phys, 103: 8577–8593

182. Beckers JVL, Lowe CP, DeLee SW (1998) An iterative PPPM method for simulating Coulombic systems on distributed memory parallel computers, Mol Simul, 20: 369–383

183. Cheatham TE, Brooks BR (1998) Recent advances in molecular dynamics simulation towards the realistic representation of biomolecules in solution, Theor Chem Acc, 99: 279–288

184. Berger O, Edholm O, Jahnig F (1997) Molecular dynamics simulations of a fluid bilayer of dipalmitoylphosphatidylcholine at full hydration, constant pressure, and constant temperature, Biophys J, 72: 2002–2013

185. Smondyrev AM, Berkowitz ML (2000) Molecular dynamics simulation of dipalmitoylphosphatidylcholine membrane with cholesterol sulphate, Biophys J, 78: 1672–1680

186. Shelley JC, Shelley MY, Reeder RC, Bandyopadhyay S, Moore PB, Klein ML (2001) Simulations of phospholipids using a coarse grain model, J Phys Chem B, 105: 9785–9792

187. Marrink SJ, Mark AE (2003) The mechanism of vesicle fusion as revealed by molecular dynamics simulations, J Am Chem Soc, 125(37): 11144–11145

188. Naidoo KJ, Brady JW(1999) Calculation of the Ramachandran potential of mean force for a disaccharide in aqueous solution, J Am Chem Soc, 121: 2244–2252

189. Reiling S, Schlenkrich M, Brickmann J (1996) Force field parameters for carbohydrates, J Comp Chem, 17: 450–468

190. Pathiaseril A, Woods RJ (2000) Relative energies of binding for antibody carbohydrate-antigen complexes computed from free-energy simulations, J Am Chem Soc, 122: 331–338

191. Simmerling C, Fox T, Kollman PA (1998) Use of locally enhanced sampling in free energy calculations: testing and application to the alpha beta anomerization of glucose, J Am Chem Soc, 120: 5771–5782

192. Dam W, Frontera A, Tirado-Rives J, Jorgensen WL (1997) OPLS all-atom force field for carbohydrates, J Comp Chem, 18: 1955–1970

193. JM Wang, P Cieplak, PA Kollman PA (2000) How well does a restrained electrostatic potential (RESP) model perform in calculating conformational energies of organic and biological molecules? J Comp Chem, 21: 1049–1074

194. Honig B, Nicholls A (1995) Classical electrostatics in biology and chemistry, Science, 268: 1144–1149

195. Davis ME, Mccammon JA (1990) Electrostatics in biomolecular structure and dynamics, Chem Rev, 90: 500–521

196. Davis ME, McCammon JA (1991) Solving the finite-difference linearized Poisson-Boltzmann equation: a comparison of relaxation and conjugate gradient methods, J Comp Chem, 10: 386

197. Bourne P, Weissig H, eds (2003) Structural Bioinformatics, Other Structure-Based Databases, NY: John Wiley & Sons, Inc, 427

198. Roux B, Simonson T (1999) Implicit solvent models, Biophys Chem, 78: 1–20

199. Tsui V, Case DA (2001) Theory and applications of the generalized Born solvation model in macromolecular simulations, Biopolymers, 56: 275–291

200. Moreira IS, Fernandes PA and Ramos MJ (2005) Accuracy of the numerical differentiation of the Poisson-Boltzmann equation, J Mol Struct (Theochem), 729: 11–18

201. Onufriev A, Bashford D, Case DA (2004) Exploring protein native states and large-scale conformational changes with a modified generalized Born model, Proteins, 55: 383–394

202. Jorgensen WL, Chandrasekhar J, Madura J , Impey RW, Klein ML (1983) Comparison of simple potential functions for simulating liquid water, J Chem Phys, 79: 926–935

203. Mahoney MW, Jorgensen WL (2000) A five-site model for liquid water and the reproduction of the density anomaly by rigid, nonpolarizable potential functions, J Chem Phys, 112: 8910–8922

204. Berendsen HJC, Grigera JR, Stratsma TP (1987) The missing term in effective pair potentials, J Phys Chem, 91: 6269–6274

205. Xia B, Tsui V, Case DA, Dyson J , Wright PE (2002) Comparison of protein solution structures refined by molecular dynamics simulations in vacuum, with a generalized Born model, and with explicit water, J Biomol NMR, 22: 317–331

206. Shen M, Freed KF (2002) Long time dynamics of met-enkephalin: comparison of explicit and implicit solvent models, Biophys J, 82: 1791–1808

207. Cheatham III TE, Kollman PA (1996) Observation of the A DNA to B DNA transition during unrestrained molecular dynamics in aqueous solution, J Mol Biol, 259: 434–444

208. Feig M, Onufriev A, Lee MS, Im W, Case DA, Brooks III CL (2004) Performance comparison of generalized Born and Poisson methods in the calculation of electrostatic solvation energies for protein structures, J Comput Chem, 25: 265–284

209. Baker NA, Sept D, Joseph S, Holst MJ, McCammon JA (2001) Electrostatics of nanosystems: application to microtubules and the ribosome, Proc Natl Acad Sci USA, 98: 10037–10041

210. Lipkowitz KB, Larter R, Cundari TR, eds (2003) Reviews in Computational Chemistry, The Poisson-Boltzmann Equation, Hoboken, NJ: John Wiley and Sons, Inc, 147

211. Wen EZ, Hsieh MJ, Kollman PA, Luo R (2004) Enhanced ab initio protein folding simulations in Poisson-Boltzmann molecular dynamics with self guiding forces, J Mol Graph Model, 22: 415–424

212. Onufriev A, Bashford D, Case DA (2000) Modification of the generalized born model suitable for macromolecules, J Phys Chem B, 104: 3712

213. Im W, Lee MS, Brooks III CL (2003) Generalized born model with a simple smoothing function, J Comput Chem, 24: 1691

214. Bashford D, Case DA (2000) Generalized born models of macromolecular solvation effects, Annu Rev Phys Chem, 51: 129–152

215. Wagoner J, Baker N (2004) Solvation forces on biomolecular structures: a comparison of explicit solvent and Poisson-Boltzmann models, J Comput Chem, 25: 1623–1629

216. Nielsen JE, Andersen KV, Honig B, Hooft RW, Klebe G, Vriend G, Wade RC (1999) Improving macromolecular electrostatics calculations, Protein Eng, 12(8): 657–662

217. Fogolari F, Brigo A, Molinari H (2002) The Poisson-Boltzmann equation for biomolecular electrostatics: a tool for structural biology, J Mol Recognit, 15: 377–392

218. Gouy M (1910) Sur la constitution de la charge électrique a la surface d'un électrolyte, J Phys, 9: 457–468

219. Chapman DL (1913) A contribution to the theory of electrocapillarity, Phill Mag, 25: 475–481

220. Nicholls A, Honig B (1991) Rapid finite difference alogrithm, utililizing successive over-relaxation to solve the Poisson-Boltzman equation, J Comput Chem, 12: 435–445

221. Nicholls A, Sharp K, Honig B (1991) Protein folding and association: insights from the interfacial and thermodynamic properties of hydrocarbons, Proteins, 11: 281–296

222. Madura JD (1995) Electrostatics and diffusion of molecules in solution: simulations with the University of Houston Brownian Dynamics Program, Comput Phys Commun, 91: 57–95

223. Höfinger S, Simonson T(2000) Dielectric Relaxation in Proteins: A Continuum Electrostatic Model Incorporating Dielectric Heterogeneity of the Protein and Time-Dependent Charges, J Comput Chem, 22: 290

224. Gilson MK, Honig B (1988) Calculation of the total electrostatic energy of a macromolecular system: solvation energies, binding energies, and conformational analysis, Proteins Struct Funct Genet, 4: 7–18

225. Gilson MK, Honig B (1988) Energetics of charge-charge interactions in proteins, Proteins Struct Funct Genet, 3: 32–52

226. Demchuk E, Wade RC (1996) Improving the continuum dielectric approach to calculating pKas of ionizable groups in proteins, J Phys Chem, 100: 17373–17387

227. Gallicchio E, Levy RM (2004) AGBNP, an analytic implicit solvent model suitable for molecular dynamics simulations and high-resolution modeling, J Comput Chem, 25: 479–499

228. Gallicchio E, Zhang LY, Levy RM (2002) The SGB/NP hydration free energy model based on the surface generalized born solvent reaction field and novel non-polar hydration free energy estimators, J Comput Chem, 23: 517–529

229. Glattli A, Daura X , van Gunsteren WF (2003) A novel approach for designing simple point charge models for liquid water with three interaction sites, J Comput Chem, 24: 1087–1096

230. Davies DR (1990) The structure and function of the aspartic proteinases, Annu Rev Biophys Chem, 19: 189–215

231. Florian J, Warshel A (1997) Langevin dipoles model for Ab initio calculations of chemical processes in solution: parametrization and application to hydration free energies of neutral and ionic solutes, and conformational analysis in aqueous solution, J Phys Chem B, 101: 5583–5595

232. Jayaram B, Sprous D, Beveridge DL (1998) Solvation free energy of biomacromolecules: parameters for a modified generalized born model consistent with the AMBER force field, J Phys Chem B, 102: 9571–9576

233. Banavalli NK, Roux B (2002) J Phys Chem B, 106: 11026

234. Nina M, Beglov D , Roux B (1997) Atomic Born radii for continuum electrostatic calculations based on molecular dynamics free energy simulations, J Phys Chem B, 101: 5239–5248

235. Wesson L, Eisenberg D (1992) Atomic solvation parameters applied to molecular dynamics of proteins in solution, Protein Sci, 1: 227–235

236. Vorobjec YN, Hermans J (1990) ES/IS: estimation of conformational free energy by combining dynamic simulations with explicit solvent with an implicit solvent continuum model, Biophys Chem, 78: 195–205

237. Dill KA (1997) Additivity principles in biochemistry, J Biol Chem, 272: 701–704

238. Mark AE, van Gunsteren WF (1994) Decomposition of the free energy of a system in terms of specific interactions. Implications for theoretical and experimental studies, J Mol Biol, 240: 167–176

239. Wilson EB, Decius JC, Cross PC (1955) Molecular Vibrations, New York: McGraw-Hill

240. Brooks BR, Janezic D, Karplus M (1995) Harmonic analysis of large sytems. I methodology, J Comput Chem, 16: 1522–1542

241. Janezic D, Brooks BR (1995) Harmonic analysis of large system. II. Comparison of different protein models, J Comput Chem, 16: 1543–1553

242. Janezic D, Venable RM, Brooks BR (1995) Harmonic analysis of large system. III. Comparison with molecular dynamics, J Comput Chem, 16: 1554–1566

243. Roccatano D, Amadei A, Apol MEF, Nola AD, Berendsen HJC (1998) Application of the quasi-Gaussian entropy theory to molecular dynamic simulation of Lennard-jones fluids, J Chem Phys, 109: 6358–6363

244. Kuhn B, Kollman PA (2000) Binding of a diverse set of ligands to avidin and streptavidin: an accurate quantitative prediction of their relative affinities by a combination of molecular mechanics and continuum solvent models, J Med Chem, 43(20): 3786–3791

245. Rocchia W, Sridharan S , Nicholls A, Alexov E, Chiabrera A, Honig B (2002) Rapid grid-based construction of the molecular surface for both molecules and geometric objects: applications to the finite difference Poisson-Boltzmann method, J Comp Chem, 23: 128–137

246. Rocchia W, Alexov E, Honig B (2001) Extending the applicability of the nonlinear Poisson-Boltzmann equation: multiple dielectric constants and multivalent ions, J Phys Chem B, 105: 6507–6514

247. Connolly ML (1983) Analytical molecular surface calculation, J Appl Cryst, 16: 548–558

248. Kuhn B, Gerber P, Schulz-Gasch T, Stahl M (2005) Validation and use of the MM-PBSA approach for drug discovery, J Med Chem, 48: 4040–4048

249. Chong LT, Duan Y, Wang L, Massova I, Massova PA (1999) Molecular dynamics and free-energy calculations applied to affinity maturation in antibody 48G7, Proc Natl Acad Sci USA, 96(25): 14330–14335

250. Lee TS, Kollman PA (2000) Theoretical studies suggest a new antifolate as a more potent inhibitor of thymidylate synthase, J Am Chem Soc, 122: 4385–93

251. Lee MR, Duan Y, Kollman PA (2000) Use of MM-PB/SA in estimating the free energies of proteins: application to native, intermediates, and unfolded villin headpiece, Proteins, 39(4): 309–316

252. Gohlke H, Case DA (2004) Converging free energy estimates: MM-PB(GB)SA studies on the protein-protein complex Ras-Raf, J Comput Chem, 25(2): 238–250

253. Verkhivker GM, Bouzida D, Gehlhaar DK, Rejto PA, Schaffer L, Arthurs S, Colson AB, Freer ST, Larson V, Luty BA, Marrone T, Rose PW (2001) Hierarchy of simulation models in predicting molecular recognition mechanisms from the binding energy landscapes: structural analysis of the peptide complexes with SH2 domains, Proteins, 45(4): 456–470

254. Verkhivker GM, Bouzida D, Gehlhaar DK , Rejto PA, Arthurs S, Colson AB, Freer ST, Larson V , BA Marrone T, Rose PW (2000) Deciphering common failures in molecular docking of ligand-protein complexes, J Comput Aided Mol Des, 14(8): 731–751

255. Bouzida D, Gehlhaar DK, Rejto PA, Schaffer L, Arthurs S, Colson AB, Freer ST, Larson V, Luty BA, Rose PW, Verkhivker GM (1999) Computer simulations of ligand-protein binding with ensembles of protein conformations: a Monte Carlo study of HIV-1 protease binding energy landscapes, Int J Quantum Chem, 73: 73–84

256. Bouzida D, Rejto PA, Verkhivker GM(1999) Monte Carlo simulations of ligand-protein binding energy landscapes with the weighted histogram analysis method, Int J Quantum Chem, 73: 113–121

257. Verkhivker GM, Rejto PA, Bouzida D, Arthurs S, Colson AB, Freer ST, Gehlhaar DK, Larson V, Luty BA, Marrone T, Rose PW (1999) Towards understanding the mechanisms of molecular recognition by computer simulations of ligand-protein interactions, J Mol Recognit, 12(6): 371–389

258. Weiner SJ, Kollman PA, Case DA, Singh UC, Chio C, Alagona G, Profeta S, Weiner PA (1984) A new force field for molecular mechanical simulation of nucleic acid and proteins, J Am Chem Soc, 106: 765–784

259. Weiser J, Weiser AA, Shenkin PS, Still WC (1998) Neighbour-list reduction: optimization for computation of molecular van der Waals and solvent accessible surface areas, J Comput Chem, 19: 797–808

260. Weiser J, Shenkin PS, Still WC (1999) Approximate atomic surfaces from linear combinations of pairwise overlaps (LCPO), J Comput Chem, 20: 217–230

261. Weiser J, Shenkin PS, Still WC (1999), Fast, approximate algorithm for detection of solvent-inacessible atoms, J Comput Chem, 20: 586–596

262. Tsui V, Case DA (2002) Molecular dynamic simulations of nucleic acids with a generalized born solvation model, J Am Chem Soc, 122: 2489–2498

263. Qiu D, Shenkin PS, Hollinger FP, Still WC (1997) The GB/SA continuum model for solvation: a fast analytical method for the calculation of approximate born radii, J Phys Chem A, 101: 3005–3014

264. Still WC, Tempczyk A, Hawley RC, Hendrickson T (1990) Semianalytical treatment of solvation for molecular mechanics and dynamics, J Am Chem Soc, 112: 6127–6129

265. Mohamadi F, Richards NGJ, Guida WC, Liskamp R, Lipton M, Caufield C, Chang G, Hendrickson T, Still WC (1990) MacroModel – an integrated software system for modelling organic and bioorganic molecules using molecular mechanics, J Comput Chem, 11: 440–467

266. Beveridge DL, Dicapua FM (1989) Free energy via molecular simulations: application to chemical and biochemical sytems, Annu Rev Biopys Chem, 18: 431–492

267. Kirkwood JG (1935) Statistical mechanics of fluid mixtures, J Chem Phys, 3: 300–313

268. Zwanzig RW (1954) High temperature equation of state by a perturbation method. I. nonpolar gases, J Chem Phys, 22: 1420–1426

269. Torrie GM, Valleau JP (1974) Monte Carlo free energy estimates using non-Boltzmann sampling: application to the subcritical Lennard-Jones fluid, Chem Phys Lett, 28: 578–581

270. Torrie GM, Valleau JP (1977) Nonphysical sampling distributions in Monte Carlo free energy estimation: umbrella sampling, J Comp Phys, 23: 187–199

271. Postma JPM, Berendsen HJC, Haak JR, Faraday (1982), Thermodynamics of cavity formation in water: a molecular dynamics study, Symp Chem Soc, 17: 55–67

272. Warshel A (1982) Dynamics of reactions in polar solvent. Semiclassical trajectory studies of electron-transfer and proton-transfer reactions, J Phys Chem, 86: 2218–2224

273. Mezei M, Swaminathan S, Beveridge DL (1978) Ab initio calculation of the free energy of liquid water, J Am Chem Soc, 100: 3255–3256

274. Lopez MA, Kollman PA (1993) Application of molecular dynamics and free energy perturbation methods to metalloporphyrin-ligand systems II: CO and dioxygen bynding to myoglobin, Protein Sci, 2: 1975–1986

275. Boresh S, Archontis G, Karplus M (1994) Free energy simulations: the meaning of the individual contributions from a component analysis, Proteins, 20: 25–33

276. Guo Z, Brooks CL, Kong X (1998) Efficient and flexible algorithm for free energy calculations using the lambda-dynamics approach, J Phys Chem B, 102: 2032–36

277. Guo ZY, Brooks CL (1998) Rapid screening of binding affinities: application of the lambda-dynamics method to a trypsin-inhibitor system, J Am Chem Soc, 120: 1920–21

278. Liu HY, Mark AE, Van Gunsteren WF (1996) Estimating the relative free energy of different molecular states with respect to a single reference state, J Phys Chem, 100: 9485–94

279. Pitera JW, Kollman PA (2000) Exhaustive mutagenesis in silico: multicoordinate free energy calculations on proteins and peptides, Proteins, 41: 385–397

280. Pitera J, Kollman PA (1998) Designing na optimum guest for a host using multimolecule free energy calculations: predicting the best ligand for Rebek's "tennis ball", J Am Chem Soc, 120: 7557–67

281. Hansson T, Aqvist J (1995) Estimation of binding free energies for HIV protease inhibitors by molecular dynamic simulations, Protein Eng, 8: 1137–1144

282. Archontis G, Simonson T, Moras D, Karplus M (1998) Specific amino acid recognition by aspartyl-tRNA synthetase studied by free energy simulation, J Mol Biol, 275: 823–846

283. Gao J, Kuczer K, Tidor B, Karplus M (1989) Hidden I thermodynamic of mutant proteins: a molecular dynamics analysis, Science, 244: 1069–1072

284. Warwicker J, Watson H (1982) Calculation of the electrostatic potential in the active site cleft due to α helix dipoles, J Mol Biol, 571: 671

285. Rogers N (1986) The modelling of the electrostatic interactions in the function of globular proteins, Prog Biophys Mol Biol, 48: 37–66

286. Gilson M, Honig B (1988) Calculating electrostatic interactions in bio-molecules: method an error assessment, J Comp Chem, 9: 327–335

287. Hendsch Z, Tidor B (1999) Electrostatic interactions in the GCN4 leucine zipper: substantial contributions arise from intramolecular interactions enhanced on binding, Protein Sci, 8: 1381–1392

288. Olson M, Reinke LT (2000) Modelling Implicit reorganization in continuum descriptions of protein-protein interactions, Proteins, 38: 115–119

289. Radmer RJ, Kollman PA (1998) The application of three approximate free energy calculations methods to structure based ligand design: trypsin and its complex with inhibitors, J Comput Aided Mol Des, 12: 215–227

290. Pearlman DA (1999) Free energy grids: a practical qualitative application of free energy perturbation to ligand design using the OWFEG method, J Med Chem, 42: 4313

291. McCammon JA (1991) Free energy from simulations, Curr Opin Struct Biol, 1: 196

292. Tidor B (1993) Simulated annealing on free energy surfaces by a combined molecular dynamics and monte carlo approach, J Phys Chem, 97: 1069–1073

293. Tidor B (1993) Simulated annealing on free energy surfaces by a combined molecular dynamics and Monte-Carlo approach, J Phys Chem, 97: 1069–73

294. Banba S, Guo Z, Brooks III CL (2000) Are many-body effects important in protein folding? J Phys Chem B, 104: 6903

295. Banba S, Brooks CL (2000) Energetic frustration and the nature of the transition state in protein folding, J Chem Phys, 113: 3423–3433

296. Jarque C, Tidor B (1997) Simulated annealing on coupled free energy surfaces: relative solvation energies of small molecules, J Phys Chem B, 101: 9362–9374

297. Miranker A, Karplus M (1991) Functionality maps of binding sites: a multiple copy simultaneous search method, Proteins: Struct Funct Genet, 11(1): 29–34

298. Bennett CH (1976) Efficient estimation of free energy differences from Monte Carlo data, J Comp Phys, 22: 245–268

299. Metropolis NA, Rosenbluth AW, Teller AH , Teller E (1953) Generalizing Swendsen-Wang to sampling – arbitrary posterior probabilities, J Chem Phys, 21: 1087

300. Jorgensen WL (1988) Energy profiles for organic reactions in solution, Adv Chem Phys, 70: 469

301. Jorgensen WL, Tirado-Rives J (1988) The OPLS potential functions for proteins, Energy minimizations for crystals of cyclic peptides and crambin, J Am Chem Soc, 110: 1657

CHAPTER 7

SUBSTRATE-ENZYME INTERACTIONS FROM MODELING AND ISOTOPE EFFECTS

RENATA A. KWIECIEŃ[1], ANDRZEJ LEWANDOWICZ[2], AND PIOTR PANETH[1*]

[1]*Institute of Applied Radiation Chemistry, Department of Chemistry, Technical University, Zeromskiego 116, 90-924 Lodz, Poland*
[2]*International Institute of Molecular and Cell Biology, 02-109 Warsaw, Trojdena 4 Street, Poland*

Abstract: Isotope effects provide a powerful tool for learning structures of transition states, species that are not amenable for direct observation. In the case of enzymatic processes, however, their application for the purpose of transition state structure elucidation is often obscured by reaction complexity. However, experimental measurements of isotope effects, enhanced by theoretical QM/MM modeling of the chemical step of enzymatic catalysis, allows study of the changes that occur upon conversion of substrates to transition states. Information obtained about the nature of specific interactions within the active site of an enzyme may be used for practical purposes. In this communication we will summarize studies of haloacid dehalogenases, ornithine decarboxylase, and methylmalonyl-CoA mutase to exemplify these studies. Studies of transition state structure will also be presented for purine nucleoside phosphorylases (PNP). Experimental measurements of kinetic IEs for this enzyme together with theoretical analysis of their values led to rational synthesis of new inhibitors of this enzyme. The application of transition state theory to PNP has led to the most potent and specific inhibitors known for this important enzyme

Keywords: QM/MM calculations, isotope effects, rational drug design, PNP, purine nucleoside phosphorylase, DADMe, Immucillin, nucleosidase, transition state analogue

1.   INTRODUCTION

The amazing rate enhancement observed in enzymatic catalysis results from stabilization of the transition states and/or destabilization of the substrates. These effects are achieved by interactions of the reactants with the protein residues. Depending on the particular reaction some interactions may play a dominant role, or many different types, such as electrostatics, hydrophobic interactions, geometric distortion, or hydrogen bonds, may concurrently contribute to catalysis. Understanding these interactions is the key factor in exploiting enzymatic reactions for the purpose

341

W. A. Sokalski (ed.), Molecular Materials with Specific Interactions, 341–363.
© 2007 *Springer.*

of medicine, bioremediation, industrial applications, and other uses. For stable molecules a plethora of physicochemical methods is available to extract information for atomic interactions. For short-lived intermediates and in particular for transition states the arsenal of chemical tools is much more limited. Two recently developed approaches used separately or together are most promising. One of these is molecular modeling in which catalytic mechanism is approached theoretically. The other is the use of isotope effects; especially kinetic isotope effects (KIEs) because their magnitudes can be directly related to properties of transition states. In this communication we present aspects of using these two tools. We first show how molecular modeling and isotope effects can be used in learning details of hydrogen bonds within the active sites of enzymes. Then we illustrate successful applications of these methods to rational synthesis of biologically active inhibitors.

Regardless of whether hydrogen bonds play a dominant role in an enzymatic reaction, they are always present in enzymatic systems. In extreme cases the presence or even direction of a single hydrogen bond may be responsible for a 100-million fold rate enhancement. For example, in penicillin-binding enzymes that are responsible for the bacterial cell growth the antibiotic binds covalently to the enzyme. Hydrolysis of this acyl-enzyme complex (deacylation) is retarded in these enzymes with the consequence of cell death. Active sites of β-lactamases, on the other hand, bind antibiotics in a similar way but facilitate hydrolysis to destroy antibiotics and lead to antibiotic resistance. B3LYP/6-31G(d)/OPLS-AA studies of the deacylation steps in both types of enzymes[1] revealed mechanistic differences in the ability of β-lactamases to preserve a hydrogen bond to an active site tyrosine (Tyr150) upon acylation, as opposite to penicillin-binding enzymes. Similarly, energies calculated at the MP2/6-31G(d) theory level on AM1/CHARMM optimized structures of transition state of chorismate mutase catalyzed isomerization shown[2] that rotation of a hydroxyl hydrogen from Cys75 to Glu78 may contribute as much as 9 kcal/mol to the stabilization of the transition state.

Both examples also illustrate the state-of-the-art methodology used in molecular modeling of enzymatic reactions. Due to the size of enzymes quantum-chemical theory levels cannot be currently applied to whole systems. As the remedy for this situation the system is usually divided into at least two zones. The smaller one includes reactants and catalytically important fragments of the enzyme and is treated at the quantum level. The remaining part, which usually consists of the remaining part of the enzyme and water molecules, is treated at the molecular mechanics level. This so called QM/MM approach, suffers from many conceptual pitfalls,[3-6] but still has proved to be highly successful in studying mechanisms of enzymatic reactions.

2.   BINDING CAN BE REFLECTED IN ISOTOPE EFFECTS

One of the first measures of how strongly hydrogen bonds can be reflected in isotope effects was the experimental determination of the oxygen isotope effect on binding of an inhibitor, oxamate, to lactate dehydrogenase.[7] The inverse isotope effect

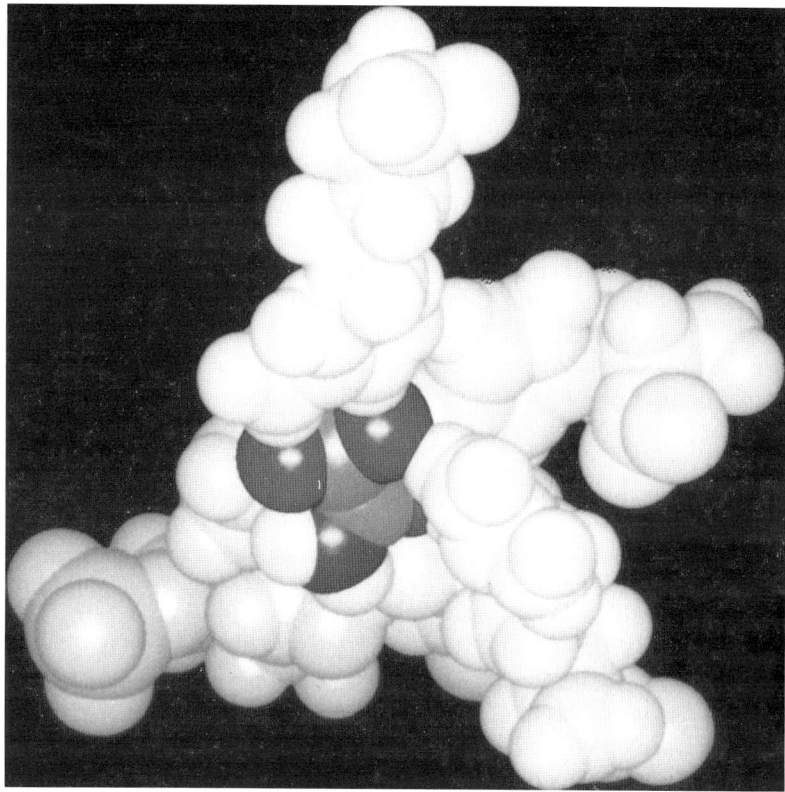

Figure 7-1. Model of the NAD-oxamate-LDH ternary complex

of 0.984 for both carboxylic oxygens indicated that these atoms are much more strongly hydrogen bonded in the enzyme than they are in aqueous solution. These experimental findings have been rationalized by semiempirical AM1 calculations.[8] The isotope effect has been matched theoretically using three active site residues, the inhibitor and truncated NAD^+ model shown in Figure 7-1 and ascribed to strong, bifurcated hydrogen bonds from the guanidinium moiety of Arg106 to oxamate, shown in the center of the figure.

3.   ISOTOPE EFFECTS AND HYDROGEN BONDING

Recently, we have modeled[9] intrinsic carbon kinetic isotope effects on the ornithine decarboxylase-catalyzed decarboxylations. Decarboxylations occur from the pyridoxal 5′-phosphate (PLP) - substrate complexes. These reactions provide a good model case since a number of ^{13}C kinetic isotope effects for the wild-type enzyme and its mutants, as well as for physiological and slow substrates, have been reported.[10] Using AM1/CHARMM/MD calculations on nearly 18000-atom models we have shown that different hydrogen bonding for the physiological substrate and the native enzyme compared to systems containing either slow substrate or mutated enzyme leads to different intrinsic isotope effects. In particular, intrinsic ^{13}C kinetic isotope effects of 1.041, 1.059, and 1.058 were calculated for wild-type enzyme and ornithine, wild-type enzyme and lysine (slow substrate), and Glu274Ala mutated enzyme and ornithine, respectively. The difference between the first value and the remaining two is very large and strongly influences interpretation of the observed isotope effects. All three transition states occupy similar positions of the reaction path and are characterized by similar activation barriers. The main difference between them lies in the different hydrogen bonding to the departing carbon dioxide. These are collected in Table 7-1. In case of slow substrate and mutated enzyme only one hydrogen bond has been found that originates from a water molecule or from the hydroxyl group of PLP. In case of the complex in wild-type enzyme involving physiological substrate, two such hydrogen bonds have been found (see Figure 7-2).

Hydrogen bonding networks may also be responsible for the difference in chlorine kinetic isotope effects on the DL-2-haloacid dehalogenase reaction. This enzyme catalyzes hydrolysis of both R- and S- stereoisomers of 2-chloropropionic acid. The experimental chlorine kinetic isotope effects differ by 20% indicating that reaction complexity masks the intrinsic isotope effect to different extent for the two stereoisomers or that the intrinsic isotope effects are different for these two molecules. SM5.4A/AM1 calculations that included three putative residues of the active site of the enzyme and substrates indicated that the latter explanation is plausible. The presence of a hydrogen donating group in contact with the carboxyl group of the R-isomer allows for hydrogen bonding contact to the departing chloride in the transition state, as shown in Figure 7-3, and lowers the chlorine kinetic isotope effect.

Observation of a variable number of hydrogen bonds to the active atoms for two substrates of the same enzyme and for a wild-type and mutated enzyme described above is new and striking. However, acceleration of enzymatic reactions by increased numbers and/or strengths of hydrogen bonds at the transition states have been documented previously.[11] In particular, using AM1/CHARMM calculations Gao and coworkers[12] showed increased hydrogen bonding at the transition state for the reaction catalyzed by a protein tyrosine phosphatase. Also transient

Table 7-1. Hydrogen bonds (in Å) to the departing carbon dioxide in PLP-substrate complexes

enzyme substrate	wild-type ornithine	wild-type lysine	Glu247Ala ornithine
H_2O952	2.90(2.69)a	-	-
H_2O88	2.81(3.11)	3.05(2.85)	-
PLP	-	-	2.69(2.64)

[a] values in parenthesis correspond to transition states

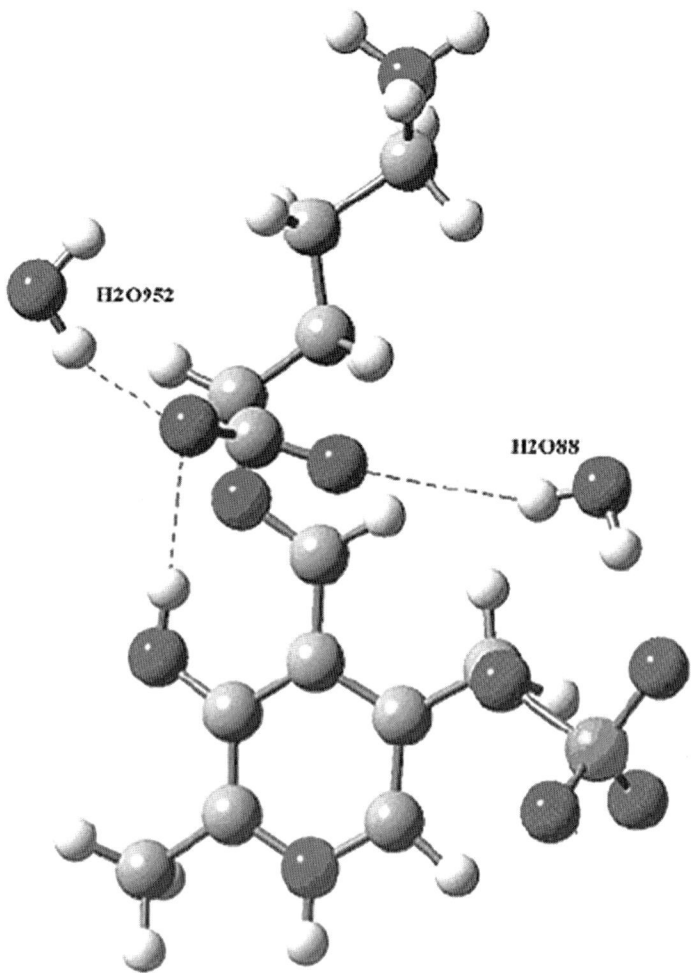

Figure 7-2. Hydrogen bonds to carboxylic oxygen atoms of ornithine

formation of hydrogen bonds to a transition state or highly reactive intermediates has been observed in molecular dynamics and AM1/CHARMM calculations. Thiel and coworkers suggested [13] that a hydrogen bond between the amide oxygen of Pro293 and the transferring OH forms temporarily near the transition state. This hydrogen bond seems to stabilize the transition state and shepherd the reaction. Similarly, recent studies on a class C β-lactamase [14] using Amber and HF/6-31G(d,p) theory levels showed the steering effect of hydrogen bonds. Increased strength of hydrogen bonds between a water molecule and Tyr150 in the acylation transition state and between the substrate, cofactor, and Ser64 in the deacylation transition state has been found. Both events are associated with a temporary weakening of hydrogen

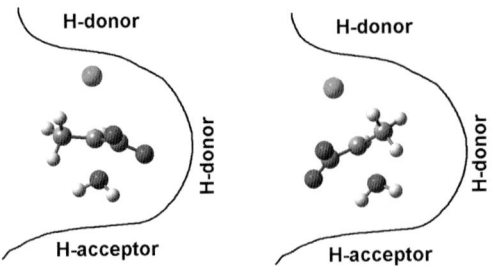

Figure 7-3. Orientations of 2-chloropropionate stereoisomers in a model of the DL-2-haloacid dehalo-genase active stite

bonds from water to the substrate and from the substrate to Ala318, respectively. This latter hydrogen bond reaches its maximum strength for a short-lived interme-diate between the two transition states.

Molecular modeling provides possibility of studying interactions within the active site even when corresponding intermediates are too short-lived to be detected experimentally. We have examined the hydrogen bonding network in the active site of methylmalonyl-CoA mutase (MCM). This enzyme catalyzes conversion of methylmalonyl-CoA to succinyl-CoA via radical intermediates that are initiated by carbon – cobalt bond homolysis of the B_{12} cofactor. Details of the substrate binding pattern are not amenable to experimental scrutiny since the process initiates homolysis and subsequent chemical conversion of the substrate. Computationally, however, it was possible to optimize the ternary complex and characterize hydrogen bonds for this structure, as well as for the transient intermediate, the transition state, and radical products of the homolysis step. We will describe these challenging calculations in more detail.

The active site of MCM is described by the model presented in Figure 7-4. The model was based on the 4REQ[15] crystal structure deposited in the Protein Data Bank and corresponds to the closed and reactive conformation of the enzyme with the reactant bound to the active site. The model (Figure 7-4) includes all amino acids (1372 atoms) within 15 Å from the cobalt atom of AdoCbl. Hydrogen atoms, not included in the 4REQ crystal structure, were added using GaussView program.[16] The N- and C-termini were capped with NHMe and C(O)Me moieties respectively, where protein chains were truncated. The quantum part includes the corrin ring without sidearm chains, ribose and imidazole as the upper and lower ligands respectively, giving 71 atoms. The remaining part of the cofactor, reactant, and the active site residues were included in the MM part of the model. The reactant (methylmalonyl-CoA, MCA) was truncated at the 15 Å boundary of the model. The 'link-atoms' formalism[17] was used to saturate the shells of QM-atoms covalently bonded to MM-atoms.

Geometry optimizations were carried out using Morokuma's ONIOM approach as implemented in Gaussian03.[18] QM-atoms (Figure 7-4 as balls) were treated using Turbomole's[19,20] spin unrestricted procedure with BP86 functional[21] and the def-SV(P) basis set.[22] For the DFT energy and gradients the resolution of the

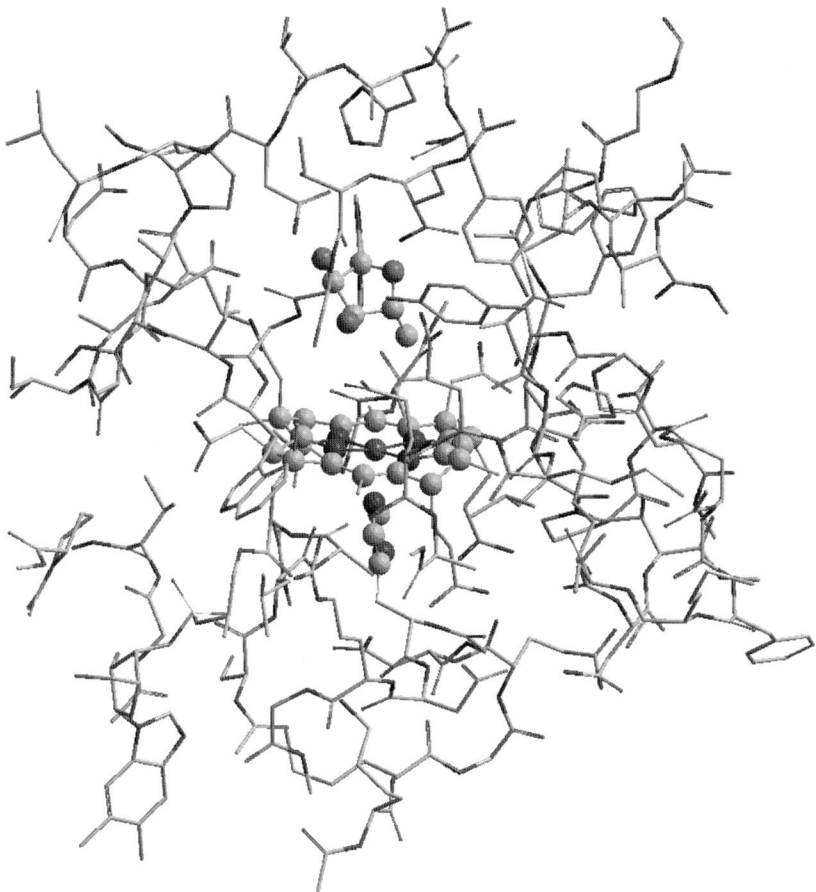

Figure 7-4. Model of MCM, balls indicate QM part and sticks indicate the MM part. Hydrogen atoms and water molecules are not shown

identity (RI) method[23–25] was employed. The MM-atoms were treated using the Amber force field.[26] Missing parameters for vitamin B_{12} were taken from literature,[27] while for the truncated substrate were generated on the basis of B3LYP/6-31G(d) optimization.[28–31]

The resulting wild-type model was also explored by a mutation Tyr89Phe. The mutant was fully optimized at the same theory level. On wild-type enzyme, relaxed potential energy scan (PES) calculations along the cobalt – carbon bond were carried out with steps 0.15 Å. We observed that the PES reached a maximum followed by a drastic drop indicating a discontinuity in the energy profile. We performed transition state searching, starting from the points on both sides of the discontinuity. The transition state (TS) for the homolysis step was characterized by only one normal mode with an imaginary frequency. This mode corresponds to

cobalt – carbon bond breaking. This assignment was confirmed by IRC calculations which located stationary points. It revealed that the product (P) is the radical
pair resulting from homolysis. The other stationary point of the IRC path was an
intermediate (I) that differs from the wild-type in dAdo conformation.

Mancia and Evans[15] suggested that binding of the substrate to MCM causes
conformational changes during which the most significant movement is associated
with Tyr89. According to investigations on the Tyr89Phe mutant,[32] the loss of
interactions involving the hydroxyl group of Tyr89 and substrate molecule increased
the rearrangement barrier between substrate- and product-derived radicals. Vlasie
and Banerjee[33] studied the role of Tyr89 in acceleration of cobalt – carbon bond
homolysis by creating two mutants Tyr89Ala and Tyr89Phe. They observed that
both mutants have significant influence on homolysis and subsequent substrate
radical generation. Moreover, Tyr89Phe mutation caused the lost of catalytic activity
of MCM and Tyr89 was found to be the major factor accelerating cobalt – carbon
homolysis rate.

McDonald and Thornton[34] studied hydrogen bond criteria in proteins and
proposed empirical rules of hydrogen bond identification: $D - A < 3.9 \text{Å}$, $H - A <
2.5 \text{Å}$, $D - H - A > 90.0°$, $AA - A - D > 90.0°$, $AA - A - H > 90.0°$. Based on
those rules, we identified atoms which may participate in stabilization of the active
site of MCM by forming hydrogen bonds with reactant (MCA) and dAdo molecules
in the four structures described above: wild-type, TS, P and Tyr89Phe. Distances
hydrogen – acceptor are collected in Table 7-2 for all models and Figure 7-5
illustrates the hydrogen bonds.

MCA is held in place by Arg207, His244 and Tyr89. The planar guanidinium
group of Arg207 participates in hydrogen bonds with the carboxyl group of MCA.
One of the two NH_2 groups and the NH group donates two hydrogens in the
'wild-type' model. The most significant changes in hydrogen bond network were
observed in Tyr89Phe. MCA looses one bond with Arg207 and all bonds with
His244. His244 is a hydrogen donor for carboxyl and carbonyl groups of MCA in
a 'wild-type' model. Arg207 forms a new hydrogen bond with the side arm amide

Table 7-2. Distances between hydrogen – acceptor (H–A) in Å

		wild-type	TS	P	Tyr89Phe
Arg207 –NH	MCA(COO⁻)	2.10	-	-	-
Arg207 –NH₂	MCA(COO⁻)	1.81	1.88	2.07	1.95
His244	MCA(COO⁻)	2.15	1.69	1.69	-
His244	MCA(=O)	2.00	-	-	-
Tyr89	MCA(COO⁻)	1.76	1.84	1.70	mutated
Tyr89	H₂O170	1.76	1.80	1.79	mutated
Phe117 amid	H₂O170	-	-	2.29	-
Arg207 –NH₂	amid-cor	-	2.40	2.10	2.15
Tyr243 –OH	Thr331 =O amid	1.70	1.68	1.68	1.69
Tyr243 –OH	H₂O576	2.09	-	-	1.98
dAdo –OH	H₂O576	-	1.89	1.87	-

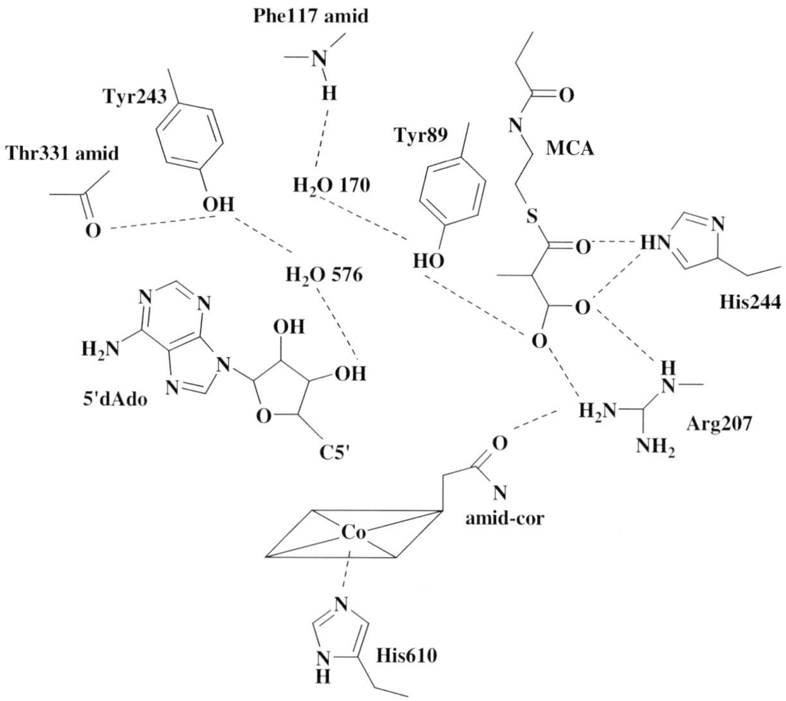

Figure 7-5. Hydrogen bonds identified in the active site of MCM models

group of the corrin ring. The loss of one hydrogen bond by MCA as a consequence of Tyr89Phe mutation causes its distortion and orientation changes (see Figure 7-6). Tyr243 participates in a weak hydrogen bond with the amide oxygen of Thr331 and with crystallographic water H_2O576 in the 'wild-type' and Tyr89Phe mutant.

Starting with the 'wild-type' enzyme we observed significant changes in the hydrogen bond network in the active site only between the starting model and intermediate I. Since there were no significant differences in hydrogen bonds between structures of intermediate I and the transition state TS, only the TS is shown in Table 7-2. The substrate molecule loses two H-bonds, one with Arg207 and one with His244. As the second Arg207 bond weakens, the bond with the corrin amide arm strengthens. This bond does not exist in the 'wild-type' model but is formed as a weak bond in intermediate I and the TS, and is the strongest in the product P. Also the bond between $MCA(COO^-)$ and His244 becomes stronger. His244 creates the strongest hydrogen bond with carboxyl group of MCA. We did not observe any significant changes in bonding of Tyr89 with carboxyl group of MCA, but it slowly loses the H-bond with crystallographic water H_2O170 in the favor of a weak bond with the Phe117 amide, which is present only in product P. The reactant molecule is strongly held in place by a hydrogen bond network in opposition to dAdo, which is bonded only with one crystallographic water H_2O576. It is worth noting, that

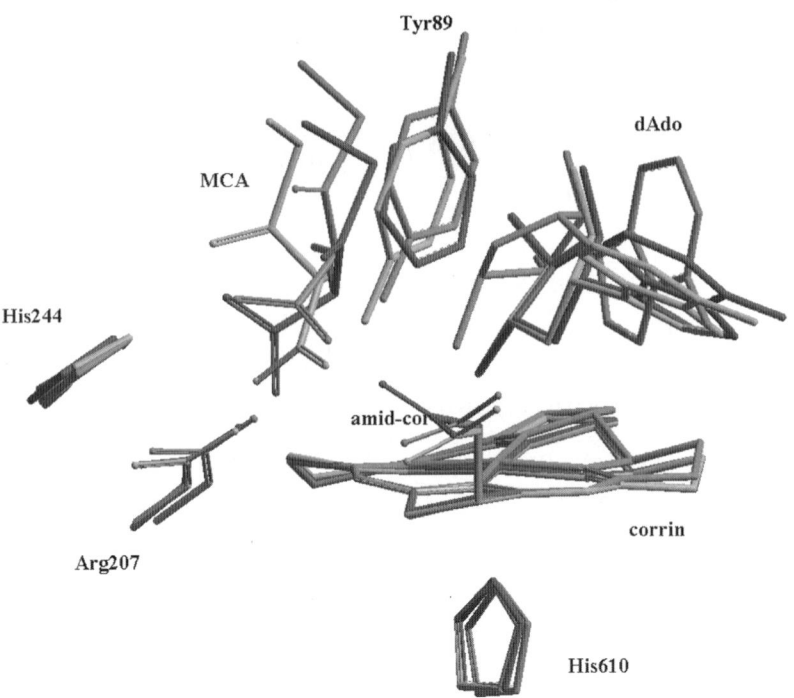

Figure 7-6. Overlay of the active site residues in 'wild-type' MCM (coloured in green), TS model (red) and Tyr89Phe (blue)

creation of the dAdo - H_2O576 hydrogen bond occurs simultaneously with disappearance of the Tyr243 - H_2O576 bond. Tyr243 creates a strong hydrogen bond with the Thr331 amide, which is stable during the full reaction.

Computational QM/MM studies presented thus far provide better understanding of enzymatic catalysis and description of interactions within the active sites. Comparison of experimentally determined isotope effects with corresponding values predicted theoretically serves to indicate that theoretical methods yield meaningful results. In the remaining part of this contribution we will show how information about properties of the transitions state gathered collectively from molecular modeling and measurements of kinetic isotope effects can be effectively used in devising new compounds with therapeutic applications.

4.   PURINE NUCLEOSIDE PHOSPHORYLASE – MULTIPLE KIEs STUDY AND TS ANALOGUES DESIGN

The purine nucleoside phosphorylases (PNPs) are N-ribosyltransferases (Figure 7-7) where transition state analogue design on the basis of kinetic isotope effects analysis has had success. The inhibition of phosphorylation catalyzed by human

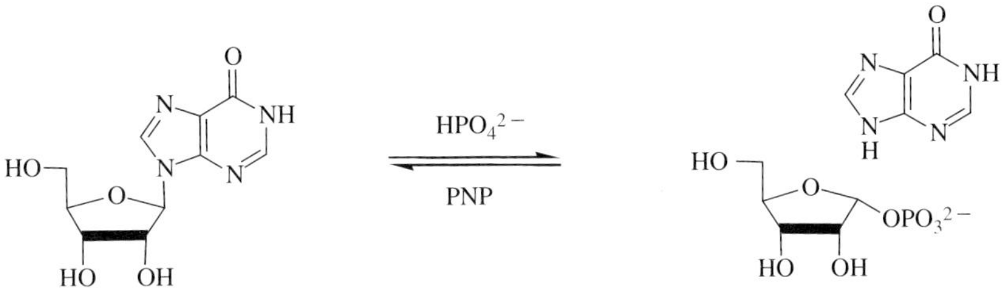

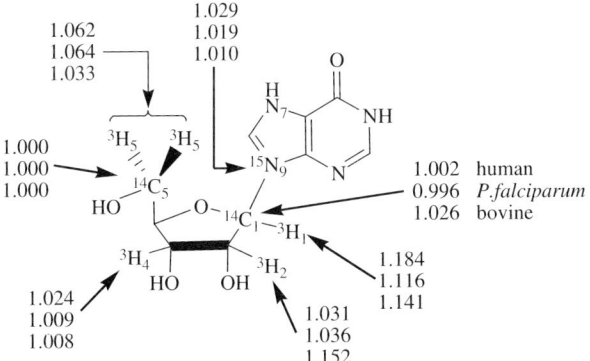

Figure 7-7. Purine nucleoside phosphorylase (PNP) catalyzed phosphorolysis is shown

PNP suppression of T-cell division and is considered to be crucial for T-cell derived cancers, autoimmune disorders or graft *versus* host disease treatment.[35−37] *Plasmodium falciparum* (*P. f.*) PNP has also been considered as a target for antibiotic design, especially where malaria drug resistance is increasing. The main source of purine for this purine auxotroph is hypoxanthine formed from inosine phosphorolysis catalyzed by PNP. Thus, parasites are sensitive to PNP inhibition. It was shown that blocking of PNP activity causes parasite death in erythrocyte cultures.[38−40] However it is unknown if the inhibition of the parasite isozyme alone is sufficient for therapy or if inhibition of the human host PNP is also required for parasite death. Specific inhibitors could help to answer this question, so analysis of transition state properties and design of specific analogues for those enzymes is important.

The transition state of bovine PNP was solved previously by multiple kinetic isotope effect analysis revealing partial bond order to the leaving group and very low bond order to the phosphate oxygen nucleophile.[41−45] Despite of the substantial sequence similarity (86 %) between human and bovine PNP their transition states look different as concluded from kinetic isotope effect analysis (Figure 7-8). The

Figure 7-8. Intrinsic isotope effects for human, *P. falciparum* and bovine PNPs. Numbering of atoms is shown as subscripts

P. falciparum PNP also proceeds through a TS different than bovine however similar to human isozyme. The TS properties and their interactions in the active site obtained from KIEs analysis provide useful information for inhibitor design.

4.1. What do KIEs Tell us about the TS Properties and its Interactions with Enzymes?

4.1.1. *Primary carbon KIE – insight into the strict reaction center*

The most important and highly informative KIE for transition state structure determination in nucleophilic substitution is the primary ^{14}C KIE at the electrophilic center. It is expected to be unity for fully dissociated S_N1 transition states but to be in the range of 1.02 to 1.04 for borderline S_N1 reactions or highly expanded S_N2 mechanisms and to approach a maximum of 1.13 for symmetric concerted nucleophilic transition states.[41] The measured $1'-^{14}C$ kinetic isotope effects for human and *P.f.* PNP isozymes is near unity (1.002 and 0.996) consistent with a $1'-^{14}C$ equilibrium isotope effect for formation of an isolated ribocation. Thus, the TS has weak interactions of Cl′ with the leaving and attacking groups[46] at distances greater than 3.0 Å, a distance corresponding to insignificant bond order.

This is a hint for TS analogue design suggesting that the leaving group mimic, should be separated from anomeric carbon by approximately 3.0 Å. The long distance of the Cl′ carbon to the leaving group, as well as to the attacking nucleophile was confirmed by the $1'-^{14}C$ kinetic isotope effect calculation at B1LYP/6-31G(d) theory. Residual N9-Cl′ bond order of 0.095 and nucleophile O-Cl′ bond order of 0.040 at the TS gave a large $1'-^{14}C$ KIE value of 1.11 and 1.10 for either unprotonated or protonated N7 nitrogen at the transition state (Figure 7-9). Thus, the measured $1'-^{14}C$ KIE of unity matches the calculated equilibrium isotope effect (EIE) of 1.0024 between free inosine and free oxacarbenium ion formed via a S_N1 mechanism, modeled at B1LYP/6-31G(d) theory.

4.1.2. *Hydrogen KIE as a more sensitive probe for TS structure*

Although primary carbon isotope effects tell us much about the nature of the electrophilic reaction center, the hydrogen isotope effects are more informative with regard to direct interactions with atoms of the active site. The $1'-^3H$ KIE depends on carbon hybridization, as well as being sensitive to noncovalent and remote *van der Waals* interactions. Large α-secondary hydrogen KIEs for human (1.184) as well as for *P. falciparum* PNPs (1.116) are in a similar range as other N-ribosyltransferases including bovine PNP (Figure 7-8) and are consistent with dissociative mechanisms where sp^3 to sp^2 rehybridization of Cl′ gives increased freedom to the out-of-plane bending mode, causing vibrational "looseness" around H1′.[47] In contrast, small α-secondary hydrogen KIE indicates S_N2 transition states[48] as recently obtained for thymidine phosphorylase catalyzed arsenolysis of thymidine.[49]

The α-secondary $1'-^3H$ KIE has the most striking difference of 7% between human and *P. f.* PNP isotope effects. We have concluded that the vibrational

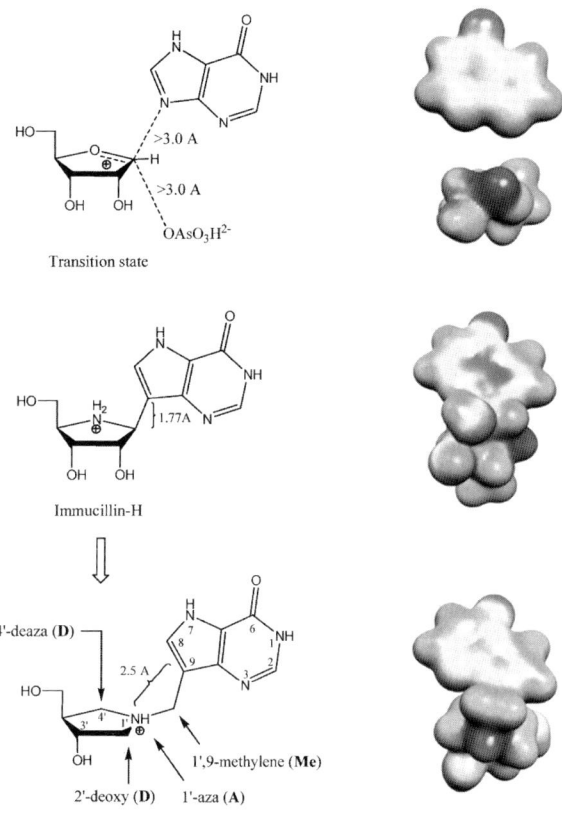

Figure 7-9. Oxacarbenium ion transition state, Immucillin-H and DADMe-Immucillin-H are shown. The nomenclature - "DADMe" is (4′-Deaza-1′-Aza-2′-Deoxy-1′,N-9-Methylene)-Immucillin-H. On the right, the molecular electrostatic potential surfaces (MEP) for the oxacarbenium/hypoxanthine pair at the transition state, Immucillin-H and DADMe-Immucillin-H are shown. MEP were calculated at HF/STO-3G theory level for optimized geometry at B1LYP/6-31G(d) theory level and visualized by Molekel 4.0 at electron density of 0.008 electron/bohr

environment at the transition state of *P. f.* PNP is stiffer (lower KIE) than for human PNP. It is difficult to determine what causes such a difference. However, it was proposed to originate from H1′ hydrogen bond interactions with peptide carbonyl oxygen of Ser91 in *P. f.* PNP (distance 3.33 Å) or Ala116 in human PNP (distance 4.02 Å), localized near the C1′ carbon of the transition state analogue crystallized in the catalytic site.[50] The unpolarized 1′-hydrogen of inosine is a poor candidate for hydrogen bond formation, however the strongly polarized 1′-C-H bond in the oxacarbenium ion transition state is a better proton donor for the carbonyl oxygen. This interaction could suppress the α-secondary 1′-^{3}H KIE through a stiffened vibrational environment. Computational modeling with a single carbonyl oxygen fixed at 3.3 Å from the H1′ hydrogen was unable to reproduce such an influence on 1′-^{3}H KIE value. However, that distance is taken from the stable transition state analogue complex rather than an actual transition state.

4.1.3.   The nitrogen KIE of the leaving group – can it predict protonation state required for tight binding?

The isotope effect of polar nitrogen atoms is a good probe for conjugated bond order patterns influenced by hydrogen bonding and protonation state in the active site even if protonation is remote. The appropriate hydrogen bond network is crucial to achieve tight binding of a TS inhibitor. Although the contribution of a single hydrogen bond may be only around 2–4 kcal/mol, the effect of hydrogen bond formation appears to be a cooperative process and loss of one of them may disrupt the whole hydrogen bond network resulting in large increases in the dissociation constant.[51]

The 9-^{15}N leaving group kinetic isotope effects (1.019) of *P. f.* PNP in comparison to human isozyme suggests a more vibrationally constrained *P. f.* PNP transition state around the purine. The 9-^{15}N KIE of 1.029 obtained for human PNP is in good agreement with the theoretical equilibrium isotope effect for hypoxanthine formation (1.025) computed at the B1LYP/6-31G(d) level of theory. Thus the interaction between leaving hypoxanthine and electrophilic C1′ carbon is negligible, as is also confirmed by the unity 1′-^{14}C KIE. Low 9-^{15}N KIE to the leaving group shows bond order preservation to the nitrogen and it can be accomplished either by bond order to the electrophilic C1′ carbon or via bond reconjugation in a dissociative S_N1 mechanism where N7 nitrogen protonation causes double bond formation to the N9 nitrogen. The small value of 9-^{15}N KIE for *P. f.* PNP and bovine PNP is attributed to the leaving group effect and leaving group activation. Proton transfer to the N7 nitrogen contributes to low KIE via conserved net bond order to N9. Neutral purine is a better leaving group than its ionized form. Restraints imposed on the leaving hypoxanthine by interactions with the active site together with protonation of N7 facilitate oxacarbenium ion stabilization at the transition state.[52] This is an important piece of information for transition state analogue design. If the leaving group is activated on the way to reach the TS, a proper mimic of the leaving group interaction may have greater influence on binding. Disruption of stabilizing interactions with the ribose ring in the active site should have a small destabilizing effect for hypoxanthine analogues than where binding relies upon ribose interactions. In fact, the 5′-methylthio modification in 5′-methylthio-Immucillin-H does not disable binding of this TS analogue to the *P. f.* isozyme although it prevents tight binding to human PNP.[50] Protonation of the N7 nitrogen before oxacarbenium transition state formation has been experimentally established for *P. falciparum* PNP by solvent deuterium isotope effects, where values of 1.6 and 1.3 were obtained for 2.5 mM and 50.0 mM arsenate, respectively.[46] Significantly higher values in the range of 3–6 would be expected where the proton is in transfer during transition state formation.[53] The higher values obtained for human PNP (1.8 and 2.5 for 2.5 mM and 50.0 mM arsenate, respectively) however, are still enough small to confirm full protonation at N7. The N7 nitrogen protonation state is crucial for tight binding of TS analogues. Protonation at N7 is an unequivocal requirement necessary for proper hydrogen bond network formation and low dissociation constants.[51]

4.1.4. *Unusual remote KIEs – interpretation more obscured*

While primary kinetic isotope effects, especially those originating from heavy atoms are little affected by conformational changes and substrate binding to the enzyme, the interpretation of secondary hydrogen isotope effects may not be so clear. It is difficult to unequivocally interpret the $5'-^3H$ IE values of around 6% for *P. f.*, human PNPs and many other N-ribosyltransferases.[54] Such a large value has been attributed to a contribution from 5'-hydroxyl in oxacarbenium ion stabilization at the stage of TS formation. The lone pair of the 5'-hydroxyl has been proposed to overlap with 4'-oxygen to polarize its electronic distribution and move electrons toward the electrophilic center.[55] This interaction causes the carbon-hydrogen bonds of the 5'-methylene to distort at the transition state and is reflected by the KIE value. This interpretation has been widely accepted for stepwise S_N1 mechanisms where the 5'-hydroxyl lone pair is involved in cationic center stabilization. However, interactions in the active site of thymidine phosphorylase require a totally different explanation for the large remote isotope effect of 1.061. Thymidine phosphorylase catalyzes inosine arsenolysis via an S_N2 mechanism based on the kinetic isotope effect analysis.[56] The large intrinsic remote secondary $5'-^3H$ IE originates from binding rather than from approaching the TS.[57,58] Thus to properly interpret TS structure from KIEs we also need to take into account the binding contribution to extract intrinsic values.

4.1.5. *From KIEs to enzymatic transition state analogues design*

Enzymatic transition state theory proposes that the enzyme binds the transition state tighter than substrate in the Michaelis complex by the factor equivalent to the enzymatic rate enhancement.[59–61] In the case of chemically stable transition state analogues, the corresponding energy is captured for binding instead of catalysis and TS analogue affinity corresponds to the free enthalpy of activation for the reaction proceeding through the transition state. Enzyme inhibitors representing transition state analogues are among the most tight binding inhibitors known,[62,63] and drug design based on TS structure is desirable.[64,65] However it should be emphasized that approaching an "ideal" mimic of the TS by stable analogues is physically impossible since atom distances in the TS are not in equilibrium, and thus not reproducible by substitution with covalent bonds. Although TS analogue binding is weaker than the theoretical limit for real transition states, they are potent enough to cause physiological effects in extremely small doses.[66]

Enzymatic transition state theory permits classification of enzymes for potential inhibition susceptibility. The simple rule suggests that the higher the catalytic rate enhancement, the tighter the binding the transition state analogue should be. Hence, slow enzymes demonstrating a small k_{cat}/k_{chem} is expected to bind weakly to TS analogues.

The basic rule of TS mimic design is to reproduce electrostatic and hydrogen bond interactions of the TS by stable molecules with geometry and electrostatic properties most compatible to the transition state. The similarity between TS structure and the analogue is by comparison of molecular electrostatic potential (MEP) surfaces for the TS and potential TS analogue candidates.[67–69] Although computational methods are being improved, the simple Bader's relationship between bond order n_{ij} and stretching force constants $F_{ij} = F_1 \cdot n_{ij}$ between i and j atoms (F_1 denotes the single bond force constant) are the elementary basis of TS structure modeling.[54]

4.1.6. The PNP case – most successful in inhibitors design

The PNP case demonstrates "pioneer" studies for translation of the knowledge of TS structure into potential pharmaceutical design and provides rationale for TS analogue modeling for other enzymes.[70,71] The first generation PNP TS analogue is Immucillin-H which was designed as a TS analogue on the basis of bovine PNP KIEs analysis (Figure 7-10). The C1'-N9 distance of 1.5 Å in Immucillin-H approximately mimics that of 1.77 Å determined for the bovine PNP early TS with a 0.32 Pauling bond order to the leaving group, while the interaction with the arsenate nucleophile is negligible with a O-C1' distance of 3.01 Å, corresponding to 0.02 bond order.[41] The

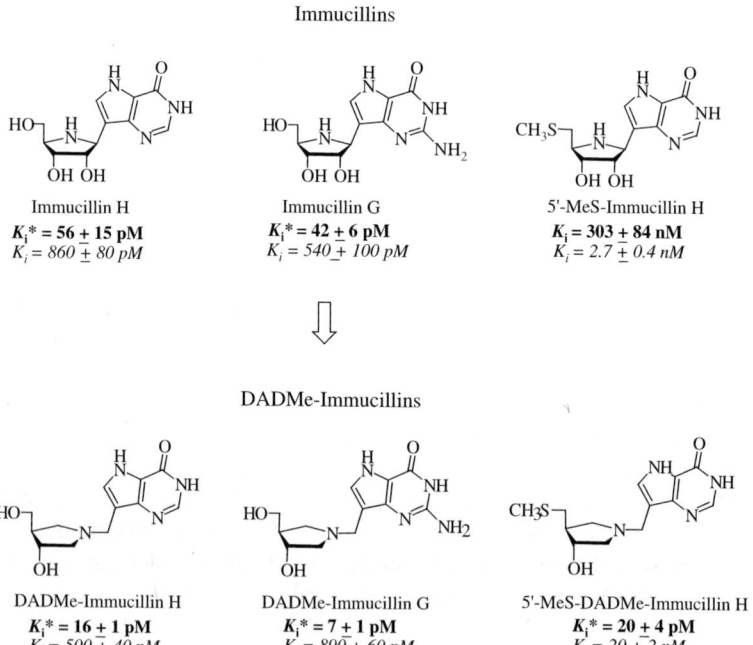

Figure 7-10. Immucillin transition state analogues and their corresponding DADMe-modifications. Initial (K_i) or slow-onset equilibrium dissociation constants (K_i^*), are shown for human (bold) and P. falciparum PNP (italic)

Immucillin-H binding has two phases with initial loose binding of the neutral form of the inhibitor followed by protonation at the N4' nitrogen.[72] A slow conformational changes is proposed to accompany protonation and formation of the tightly bound ternary complex (EI*) characterized by an equilibrium dissociation constant (K_i^*) of 23 pM.[51] The two-phase binding mechanism described as "slow-onset" takes place only in the case of high affinity inhibitors.[73]

4.1.7. DADMe – simple but logic modification leading to great improvement

Our analysis of KIEs for the human and *P. falciparum* PNP kinetic isotope effects points to fully developed carbocation at the transition state, e.g. an oxacarbenium ion (Figure 7-9). Therefore, the leaving group mimic was positioned far from the iminoribitol ring in the TS analogue. These modifications resulted in the recently synthesized DADMe derivatives[74] of Immucillin-H. The DADMe acronym is explained in Figure 7-9. This structural motif is based on incorporation of a methylene bridge between C1' and N9 and changing the iminoribitol nitrogen position from 4' as in Immucillin to the 1'-position and removing the 2'-hydroxyl for chemical stability. This structure mimics the more dissociative character of the transition state by extension of the linear distance between C1'-N9 from 1.5 Å to 2.5 Å. The DADMe modification also provides a favorable charge localization to place the cation at the 1'-position. The pK_a of 1'-pyrrolidine nitrogen in DADMe system is 8.5[75] and provides the cationic nitrogen to ion pair with phosphate in the active site of phosphorylases. The N4' nitrogen of Immucillin-H has a pK_a of 6.9[72] and is cationic in the active site, however the C1'-C9 distance is only 1.5 Å, not optimal for a fully-dissociated nucleoside at the transition state. The MEP analysis demonstrates the similarity of the analogues with positive charge at 1' to the transition state (Figure 7-9).[46] A crucial feature for tight binding of Immucillins as well as their DADMe analogues is the protonation state of N7 in the purine leaving group. Replacement of the N9 nitrogen with carbon in 9-deazahypoxanthine increases the pK_a of N7 to more than 10.[72]

The strong interaction between the cationic iminoribitol rings of the TS analogue with phosphate in the active site was established by IR/Raman studies.[76] Although there is negligible nucleophilic participation by phosphate at the transition state,[46] an ion pair is formed between phosphate and the ribosyl cation both in the transition state and the transition state analogues.

4.1.8. Specific P. f. PNP inhibitor – only one but found

Similar transition state structures from similar intrinsic kinetic isotope effects do not necessarily predict the same binding energy for transition state analogues. For example, human PNP has a k_{cat} of 31.0 s^{-1} while *Plasmodium falciparum* PNP has a k_{cat} of 1.7 s^{-1}[50] so the TS binding potency of *P. f.* PNP is intrinsicly lower. This explains the difference in Immucillin dissociation constants. With one exception, all of the TS analogues for PNP bind preferentially to the human isozyme. 5'-Methylthio-Immucilin-H prefers *P. falciparum* enzyme by a factor of 112 (Figure 7-11).[77] This specificity originates from spatial hindrance around the

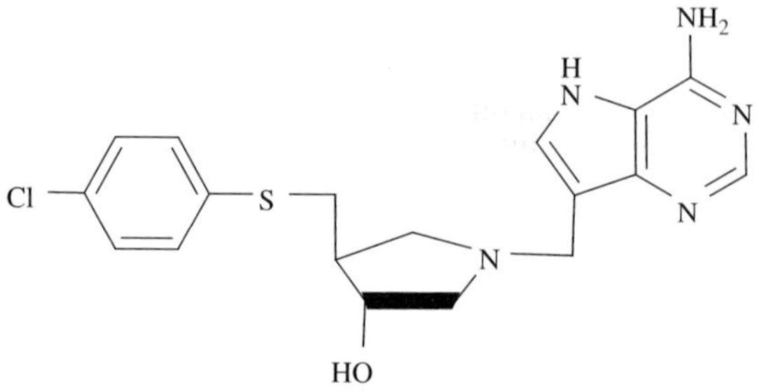

p-Cl-5′-phenylthio-DADMe-Immucillin A

$K_i = 2.6 \pm 0.1$ pM; $K_i^* = 47 \pm 3$ fM

Figure 7-11. 5′-p-Cl-phenylthio-DADMe-Immucillin-A is a femtomolar transition state analogue for E.coli 5′-methylthioadenosine/S-adenosylhomocysteine nucleosidase (MTAN). Initial (K_i) and equilibrium (K_i^*) dissociation constant are shown

5′-carbon, discriminating against the human PNP active site. However, it can be also proposed that breaking the favorable iminoribitol interactions due to replacement of the 5′-hydroxyl group causes a decrease of affinity. Human PNP has a stiffer vibrational environment around leaving group with a larger contribution to inhibitor anchoring as discussed above for the N9 nitrogen KIE values.

4.1.9.   Enzyme substrate interaction from isotope effects - farther achievements, perspectives and clinical translation

KIEs analysis and its practical application to the design of TS analogues were recently applied to other enzymes belonging to the N-ribosyltransferase family. Besides human PNP[78] and *M. tuberculosis* PNP[64] inhibitors in the picomolar dissociation constant range have been revealed for human 5′-methylthioadenosine phosphorylase (MTAP), an enzyme from the polyamine pathway considered as a target in proliferative diseases.[79,80]

The example of 5′-methylthioadenosine/S-adenosylhomocysteine nucleosidase from *E. coli* (MTAN) deserves special attention.[71] The analysis of kinetic isotope effects for this enzyme[81] points to fully dissociated oxacarbenium ion transition state with little involvement of the leaving group and attacking nucleophile water in a $D_N^*A_N$ (S_N1) mechanism. Thus the transition state is similar to those for human and *P. falciparum* PNPs.[46] Recently synthesized 5′-p-Cl-phenylthio-DADMe-Immucillin-A exhibits a femtomolar equilibrium dissociation constant of 47 fM, thus MTAN transition state analogues are among the tightest binding inhibitors ever known (Figure 7-11). MTAN hydrolyzes its substrates to form adenine and 5-methylthioribose (MTR) or S-ribosylhomocysteine (SRH), hence it is involved either in the polyamine pathway or in quorum sensing in Gram-negative bacteria.[70,71] MTAN inhibitors are thus potential candidates to disrupt crucial bacterial biochemical pathways.

It is important to emphasize that the insight into transition state – enzyme active site interactions by KIEs analysis and modeling has a real translation into medicine, especially in case of the PNPs. DADMe-Immucillin-H (BCX-4208) has recently entered into clinical trials for psoriasis.[82] Immucillin-H under the name of Fodosine, is presently in phase II clinical trials for patients with T-lymphocyte derived proliferative diseases.[36] Although the DADMe motif in 5′-methylthio-DADMe-Immucillin-H did not improve 5′-methylthio-Immucillin-H binding properties against *Plasmodium falciparum* PNP (Figure *7-10*), 5′-methylthio-Immucillin-H remains as the only inhibitor with specificity for *Plasmodium falciparum* PNP relative to human PNP. Research on the application of those compounds to malaria is ongoing.[77]

ACKNOWLEDGEMENTS

We thank Professor Vern L. Schramm for critical reading of the manuscript and helpful discussions. Access to supercomputer facilities at PCSS Poznan and Cyfronet Krakow are acknowledged. This work was supported by grants from the State Committee for Scientific Research, Poland (P.P.).

REFERENCES

1. Gherman BF, Goldberg SD, Cornish VW, Friesner RA (2004) Mixed quantum mechanical/molecular mechanical (QM/MM) study of the deacylation reaction in a penicillin binding protein (PBP) versus in a class C β-lactamase. J. Am. Chem. Soc. 126:7652–7664

2. Szefczyk B, Mulholland A, Sokalski WA (2004) Differential transition-state stabilization in enzyme catalysis: quantum chemical analysis of interactions in the chorismate mutase reaction and prediction of the optimal catalytic field. J. Am. Chem. Soc. 126:16148–16159

3. Klahn M, Braun-Sand S, Rosta E, Warshel A (2005) On possible pitfalls in ab initio quantum mechanics/molecular mechanics minimization approaches for studies of enzymatic reactions. J. Phys. Chem. B 109:15645–15650

4. König PH, Hoffmann M, Frauenheim Th, Cui Q (2005) A Critical evaluation of different QM/MM frontier treatments with SCC-DFTB as the QM method. J. Phys. Chem. B 109:9082–9095

5. Warshel A (2003) A Computer simulations of enzyme catalysis: Methods, progress, and insights. Annu. Rev. Biophys. Biomol. Struct. 32:425–443

6. Gao J, Truhlar DG (2004) Generalized Hybrid Orbital (GHO) Method for Combining Ab Initio Hartree-Fock Wave Functions with Molecular Mechanics. J. Phys. Chem. A 108:632–650

7. Gawlita E, Anderson VE (1994) Paneth P Semiempirical calculations of the oxygen equilibrium isotope effect on binding of oxamate to lactate dehydrogenase. Eur. Biophys. J. 23:353–360

8. Gawlita E, Anderson VE, Paneth P (1995) Equilibrium isotope effect on ternary complex formation of [1-^{18}O]oxamate with NADH and lactate dehydrogenase. Biochemistry 34:6050–6058

9. Sicińska D, Truhlar DG, Paneth P (2005) Dependence of transition state structure on substrate: the intrinsic C-13 kinetic isotope effect is different for physiological and slow substrates of the ornithine decarboxylase reaction because of different hydrogen bonding structures. J. Am. Chem. Soc. 127:5414–5422

10. Swanson T, Brooks HB, Osterman AL, O'Leary MH, Phillips MA (1998) Carbon-13 isotope effect studies of *Trypanosoma brucei* ornithine decarboxylase. Biochemistry 37:14943–14947

11. Garcia-Viloca M, Gao J, Karplus M, Truhlar DG (2004) How enzymes work: analysis by modern rate theory and computer simulations. Science 303:186–195

12. Alhambra C, Wu L, Zhang Y-Z, Gao J (1998) Walden-inversion-enforced transition-state stabilization in a protein tyrosine phosphatase. J. Am. Chem. Soc. 120:3858–3866

13. Senn HM, Thiel S, Thiel H (2005) Enzymatic hydroxylation in p-hydroxybenzoate hydroxylase: a case study for QM/MM molecular dynamics. J. Chem. Theory Comput. 1:494–505

14. Hata M, Tanaka Y, Fujii Y, Neya S, Hoshino T (2005) A theoretical study on the substrate deacylation mechanism of class C β-lactamase. J. Phys. Chem. B 109:16153–16160

15. Mancia F, Evans PR (1998) Conformational changes on substrate binding to methylmalonyl CoA mutase and new insights into the free radical mechanism. Structure 6:711–720

16. Pittsburgh PA (2000) GaussView 2. Gaussian, Inc.

17. Singh U, Kollman P (1986) A combined ab initio quantum mechanical and molecular mechanical method for carrying out simulations on complex molecular systems: applications to the $CH_3Cl + Cl^-$ exchange reaction and gas-phase protonation of polyethers. J. Comput. Chem. 7:718–730

18. Pittsburgh PA, Frisch MJ, et al. (2003) Gaussian 03, Revision A.1, Gaussian, Inc.

19. Ahlrichs R, Bär M, Häser M, Horn H, Kölmel C (1989) Electronic structure calculations on workstation computers: The program system turbomole. Chem. Phys. Lett. 162:165–169

20. Karlsruhe Quantum Chemistry Group, Turbomole homepage (2002) http://www.turbomole.com/

21. (a) Becke AD (1988) Density-functional exchange-energy approximation with correct asymptotic behaviour. Phys. Rev. A 38:3098–3100 (b) Perdew JP (1986) Density-functional approximation for the correlation energy of the inhomogeneous electron gas. Phys. Rev. B 33:8822–8824

22. Schaefer A, Horn H, Ahlrichs R (1992) Fully optimized contracted Gaussian basis sets for atoms Li to Kr. J. Chem. Phys. 97:2571–2577

23. Eichkorn K, Treutler O, Öhm H, Häser M, Ahlrichs R (1995) Auxiliary basis sets to approximate Coulomb potentials. Chem. Phys. Lett. 240:283–290; erratum: Chem. Phys. Lett. 242:652–660

24. Eichkorn K, Weigend F, Treutler O, Ahlrichs R (1997) Auxiliary basis sets for main row atoms and transition metals and their use to approximate Coulomb potentials. Theor. Chem. Acc. 97:119–124

25. Ahlrichs R, Elliot S, Huniar U (2000) In: J. Grotendorst J (ed) Ab Initio Treatment of Large Molecules in Modern Methods and Algorithms of Quantum Chemistry. Proceedings, 2nd edn. John von Neumann Institute for Computing, Jülich, NIC Series 3, pp 7–25

26. Cornell WD, Cieplak P, Bayly CI, Gould IR, Merz KM Jr, Ferguson DM, Spellmeyer DC, Fox T, Caldwell JW, Kollman PA (1995) A second generation force field for the simulation of proteins, nucleic acids, and organic molecules. J. Am. Chem. Soc. 117:5179–5197

27. (a) Marques HM, Ngoma B, Egan TJ, Brown KL (2001) Parameters for the AMBER Force Field for the Molecular Mechanics Modeling of the Cobalt Corrinoids. J. Mol. Struct. 561:71–91 (b) Marques HM, Brown KL (1995) A molecular mechanics force field for the cobalt corrinoids, J. Mol. Struct. (THEOCHEM) 340:97–124

28. Becke AD (1993) Density-functional thermochemistry. III. The role of exact exchange. J. Chem. Phys. 98:5648–5642

29. Stephens PJ, Devlin FJ, Chablowski CF, Frisch MJ (1994) Ab initio calculation of vibrational absorption and circular dichroism spectra using density functional force fields. J. Phys. Chem. 98:11623–11627

30. Ditchfield R, Hehre WJ, Pople JA (1971) Self-consistent molecular-orbital methods. IX. Extended Gaussian-type basis for molecular-orbital studies of organic molecules. J. Chem. Phys. 54:724–728

31. Hehre WJ, Ditchfield R, Pople JA (1972) Self-consistent molecular orbital methods. XII. Further extensions of Gaussian-type basis sets for use in molecular orbital studies of organic molecules. J. Chem. Phys. 56:2257–2261

32. Thomä NH, Meier TW, Evans PR, Leadlay PF (1998) Stabilization of radical intermediates by an active-site tyrosine residue in methylmalonyl-CoA mutase. Biochemistry 37:14386–14393

33. Vlasie M, Banerjee R (2003) Tyrosine 89 Accelerates Co-Carbon Bond Homolysis in Methylmalonyl-CoA Mutase. J. Am. Chem. Soc. 125:5431–5435

34. McDonald IK, Thornton IK (1994) Satisfying hydrogen-bonding potential in proteins. J. Mol. Biol. 238:777–793

35. Giblett ER, Ammann AJ, Wara DW, Sandman R, Diamond LK (1975) Nucleoside-phosphorylase deficiency in a child with severely defective T-cell immunity and normal B-cell immunity. Lancet 1:1010–1013

36. Bantia S, Ananth SL, Parker CD, Horn LL, Upshaw R (2003) Mechanism of inhibition of T-acute lymphoblastic leukemia cells by PNP inhibitor – BCX-1777. Int Immunopharmacol. 3:879–887

37. Schramm VL (2002) Development of transition state analogues of purine nucleoside phosphorylase as anti-T-cell agents. Biochim. Biophys. Acta 1587:107–117

38. Queen SA, Jagt DL, Reyes P (1990) In vitro susceptibilities of *Plasmodium falciparum* to compounds which inhibit nucleotide metabolism. Antimicrob Agents Chemother. 34:1393–1398

39. Kicska GA, Tyler PC, Evans GB, Furneaux RH, Schramm VL, Kim K (2003) Purine-less death in *Plasmodium falciparum* induced by Immucillin-H, a transition state analogue of purine nucleoside phosphorylase. J. Biol. Chem. 277:3226–3231

40. Kicska GA, Tyler PC, Evans GB, Furneaux RH, Kim K, Schramm VL (2002) Transition state analogue inhibitors of purine nucleoside phosphorylase from *Plasmodium falciparum*. J. Biol. Chem. 277:3219–3225

41. Kline PC, Schramm VL (1993) Purine nucleoside phosphorylase. Catalytic mechanism and transition-state analysis of the arsenolysis reaction. Biochemistry 32:13212–13219

42. Berti PJ, Tanaka KSE (2002) Transition state analysis using multiple kinetic isotope effects: mechanisms of enzymatic and non-enzymatic glycoside hydrolysis and transfer. Adv. Phys. Org. Chem. 37:239–314

43. Miles RW, Tyler PC, Furneaux RH, Bagdassarian CK, Schramm VL (1998) One-third-the-sites transition state inhibitors for purine nucleoside phosphorylase. Biochemistry 37:8615–8621

44. Furneaux RH, Limberg G, Tyler PC, Schramm VL (1997) Synthesis of transition state inhibitors for N-riboside hydrolases and transferases. Tetrahedron 53:2915–2930

45. Kline PC, Schramm VL (1995) Pre-steady-state transition-state analysis of the hydrolytic reaction catalyzed by purine nucleoside phosphorylase. Biochemistry 34:1153–1162

46. Lewandowicz A, Schramm VL (2004) Transition state analysis for human and Plasmodium falciparum purine nucleoside phosphorylases. Biochemistry 43:1458–1468

47. Pham TV, Fang Y-R, Westaway KC (1997) Using secondary deuterium kinetic isotope effects to determine the symmetry of SN2 transition states. J. Am. Chem. Soc. 119:3670–3676

48. Glad SS, Jensen F (1997) Transition state looseness and secondary kinetic isotope effects. J. Am. Chem. Soc. 119:227–232

49. Birck MR, Schramm VL (2004) Nucleophilic participation in the transition state for human thymidine phosphorylase. J. Am. Chem. Soc. 126:2447–2453

50. Shi W, Ting LM, Kicska GA, Lewandowicz A, Tyler PC, Evans GB, Furneaux RH, Kim K, Almo SC, Schramm VL (2004) Plasmodium falciparum purine nucleoside phosphorylase: crystal structures, immucillin inhibitors, and dual catalytic function. J. Biol. Chem. 279:18103–18106

51. Kicska GA, Tyler PC, Evans GB, Furneaux RH, Shi W, Fedorov A, Lewandowicz A, Cahill SM, Almo SC, Schramm VL (2002) Atomic dissection of the hydrogen bond network for transition-state analogue binding to purine nucleoside phosphorylase. Biochemistry 41:14489–14498

52. Fedorov A, Shi W, Kicska GA, Fedorov E, Tyler PC, Furneaux RH, Hanson JC, Gainsford GJ, Larese JZ, Schramm VL, Almo SC (2001) Transition state structure of purine nucleoside phosphorylase and principles of atomic motion in enzymatic catalysis. Biochemistry 40:853–860

53. Horenstein BA, Parkin DW, Estupinan B, Schramm VL (1991) Transition-state analysis of nucleoside hydrolase from *Crithidia fasciculate*. Biochemistry 30:10788–10795

54. Schramm VL, Purich DL (1999) In: Methods in Enzymol. Enzyme Kinetics and Mechanism Part E. Energetics of Enzyme Catalysis 308:301–355

55. Nunez S, Antoniou D, Schramm VL, Schwartz SD (2004) Promoting vibrations in human purine nucleoside phosphorylase. A molecular dynamics and hybrid quantum mechanical/molecular mechanical study. J. Am. Chem. Soc. 126:15720–15729

56. Birck MR, Schramm VL (2004) Nucleophilic participation in the transition state for human thymidine phosphorylase. J. Am. Chem. Soc. 126:2447–2453

57. Birck MR, Schramm VL (2004) Binding causes the remote [5'-³H]thymidine kinetic isotope effect in human thymidine phosphorylase. J. Am. Chem. Soc. 126:6882–6883

58. Lewis BE, Schramm VL (2003) Binding equilibrium isotope effects for glucose at the catalytic domain of human brain hexokinase. J. Am. Chem. Soc. 125:4785–4798

59. Miller BG, Wolfenden R (2002) Catalytic proficiency: the unusual case of OMP decarboxylase. Annu. Rev. Biochem. 71:847–885

60. Wolfenden R (1999) Conformational aspects of inhibitor design: enzyme-substrate interactions in the transition state. Bioorg. Med. Chem. 7:647–652

61. Schramm VL (1998) Enzymatic transition states and transition state analog design. Annu. Rev. Biochem. 67:693–720

62. Schramm VL (2001) Transition state variation in enzymatic reactions. Curr. Opin. Chem. Biol. 5:556–563

63. Schramm VL (2003) Enzymatic transition state poise and transition state analogues. Acc. Chem. Res. 36:588–596

64. Lewandowicz A, Shi W, Evans GB, Tyler PC, Furneaux RH, Basso LA, Santos DS, Almo SC, Schramm VL (2003) Over-the-barrier transition state analogues and crystal structure with mycobacterium tuberculosis purine nucleoside phosphorylase. Biochemistry 42:6057–6066

65. Evans GB, Furneaux RH, Schramm VL, Singh V, Tyler PC (2004) Targeting the polyamine pathway with transition-state analogue inhibitors of 5'-methylthioadenosine phosphorylase. J. Med. Chem. 47:3275–3281

66. Lewandowicz A, Tyler PC, Evans GB, Furneaux RH, Schramm VL (2003) Achieving the ultimate goal in transition state analogues for human purine nucleoside phosphorylase. J. Biol. Chem. 278:31465–31468

67. Politzer P, Truhlar G (1981) In: Chemical Applications of Atomic and Molecular Electrostatic Potentials. Plenum Press, New York, pp 1–6

68. Horenstein BA, Schramm VL (1993) Electronic nature of the transition state for nucleoside hydrolase. A blueprint for inhibitor design. Biochemistry 32:7089–7097

69. Bagdassarian CK, Schramm VL, Schwartz SD (1996) J. Am. Chem. Soc. 118:8825–8836

70. Singh V, Evans GB, Lenz DH, Mason JM, Clinch K, Mee S, Painter GF, Tyler PC, Furneaux RH, Lee JE, Howell PL, Schramm VL (2005) Femtomolar transition state analogue inhibitors of 5'-methylthioadenosine/S-adenosylhomocysteine nucleosidase from *Escherichia coli*. J. Biol. Chem. 280:18265–18273

71. Lee JE, Singh V, Evans GB, Tyler PC, Furneaux RH, Cornell KA, Riscoe MK, Schramm VL, Howell PL (2005) Structural rationale for the affinity of pico- and femtomolar transition state analogues of *Escherichia coli* 5'-methylthioadenosine/S-adenosylhomocysteine nucleosidase. J. Biol. Chem. 280:18274–18282

72. Sauve AA, Cahill SM, Zech SG, Basso LA, Lewandowicz A, Santos DS, Grubmeyer C, Evans GB, Furneaux RH, Tyler PC, McDermott A, Girvin ME, Schramm VL (2003) Ionic states of substrates and transition state analogues at the catalytic sites of N-ribosyltransferases. Biochemistry 42:5694–5705

73. Morrison JE, Walsh CT (1988) The behavior and significance of slow-binding enzyme inhibitors. Adv. Enzymol. Relat. Areas Mol. Biol. 61:201–301

74. Evans GB, Furneaux RH, Lewandowicz A, Schramm VL, Tyler PC (2003) Synthesis of second-generation transition state analogues of human purine nucleoside phosphorylase. J. Med. Chem. 46:5271–5276

75. Zhou G-C, Parikh SL, Tyler PC, Evans GB, Furneaux GB, Zubkova OV, Benjes PA, Schramm VL (2004) Inhibitors of ADP-ribosylating bacterial toxins based on oxacarbenium ion character at their transition states. J. Am. Chem. Soc. 126:5690–5698

76. Deng H, Lewandowicz A, Schramm VL, Callender R (2004) Activating the phosphate nucleophile at the catalytic site of purine nucleoside phosphorylase: a vibrational spectroscopic study. J. Am. Chem. Soc. 126:9516–9517

77. Ting LM, Shi W, Lewandowicz A, Singh V, Mwakingwe A, Birck MR, Taylor Ringia EA, Bench G, Madrid DC, Tyler PC, Evans GB, Furneaux RH, Schramm, VL, Kim K (2004) Targeting a novel Plasmodium falciparum purine recycling pathway with specific immucillins. J. Biol. Chem. 280:9547–9554

78. Lewandowicz A, Taylor Ringia EA, Ting LM, Kim K, Tyler PC, Evans GB, Zubkova OV, Mee S, Painter GF, Lenz DH, Furneaux RH, Schramm VL (2005) Energetic mapping of transition state analogue interactions with human and Plasmodium falciparum purine nucleoside phosphorylases. J. Biol. Chem. 280:30320–30328

79. Singh V, Shi W, Evans GB, Tyler PC, Furneaux RH, Almo SC, Schramm VL (2004) Picomolar transition state analogue inhibitors of human 5'-methylthioadenosine phosphorylase and X-ray structure with MT-immucillin-A. Biochemistry 43:9–18

80. Evans GB, Furneaux RH, Lenz DH, Painter GF, Schramm VL, Singh V, Tyler PC (2005) Second generation transition state analogue inhibitors of human 5'-methylthioadenosine phosphorylase. J. Med. Chem. 48:4679–4689

81. Singh V, Lee JE, Nunez S, Howell PL, Schramm VL (2005) Transition state structure of 5'-methylthioadenosine/S-adenosylhomocysteine nucleosidase from *Escherichia coli* and its similarity to transition state analogues. Biochemistry 44:11647–11659

82. http://www.biocryst.com/bcx_4208.htm.

CHAPTER 8

FROM INHIBITORS OF LAP TO INHIBITORS OF PAL
Lessons from Molecular Modeling and Experiment Interface

ŁUKASZ BERLICKI, JOLANTA GREMBECKA, EDYTA
DYGUDA-KAZIMIEROWICZ, PAWEŁ KAFARSKI,
W. ANDRZEJ SOKALSKI
*Department of Chemistry, Wrocław University of Technology, Wybrzeże Wyspiańskiego 27, 50-370
Wrocław, Poland*

Abstract: Computer-aided techniques of rational design of enzyme inhibitors were reviewed. *In silico* lead generation and optimization protocols were outlined and several methods of inhibitor potency estimation by both empirical scoring functions as well as *ab initio* based calculations were described. Two representative examples of successful computer-aided analysis and design of novel, highly potent inhibitors of leucine aminopeptidase and glutamine synthetase were demonstrated. In addition fully nonempirical and systematic analysis of the physical nature of enzyme active site interactions has been performed for series of leucine aminopeptidase (LAP) and phenylalanine ammonia lyase (PAL) inhibitors. Results derived from *ab initio* calculations indicate that inhibitory activity is controlled by interactions with limited number of active site residues. Examination of entire hierarchy of theoretical models indicates that the inhibitory activity could be well represented by electrostatic interactions, leading to so called "electrostatic key-lock" principle

Keywords: Drug design, molecular modeling, agrochemicals, enzyme inhibitors, *ab initio*, inter-molecular interactions, leucine aminopeptidase, glutamine synthetase, phenylalanine ammonia lyase

1. INTRODUCTION

Evidence of the use of medicines and drugs can be found as far back in time as the first Egyptian dynasty, 3100 B.C. For most of the time drugs had been used, discovering them was a trial-and-error process, which relied mostly on a wide variety of plants as sources of active substances. If such an approach to drug discovery continued, few diseases would be curable today. It is well illustrated by the fact that natural products make up only a small percentage of drugs in the current

W. A. Sokalski (ed.), Molecular Materials with Specific Interactions, 365–398.

market. Moreover, when a natural product is found to be active, it is frequently chemically modified in order to improve its properties.

It was not until the 1960s that some understanding began to develop about the quantitative relationship between chemical structure and biological activity. A fundamental assumption for rational drug design is that drug activity is obtained through the molecular binding of one molecule (the ligand) to the specific portion of another, usually larger, molecule (the receptor, commonly a protein). In their active conformations, the molecules exhibit geometric and chemical complementarity, both of which are essential for a successful drug action. By binding to the target macromolecules, drugs may modulate signal pathways (e.g. by altering a sensitivity to hormonal action) or alter metabolism (e.g. by interfering with the catalytic activity of an enzyme). Most commonly, this is achieved by binding in the specific cavity of the protein (receptor or active site) and thus preventing an access of the natural effector(s).

A typical research cycle from defining the target disease to drug marketing usually requires 10 years, at a cost which may range from 0.3 to 3 billion dollars. This represents an expenditure, which could be as low as one hundred thousand or as high as of one million dollars per day. Therefore, any technology that can shorten this process, make it more efficient, or increase the efficacy and novelty of the resultant drug is worth to be introduced in this cycle. Computer-aided drug design offers such a possibility.

Knowledge of biomolecular structures, which is increasing at an explosive rate, enables to apply computer-aided methods for *de novo* drug design (so called *in silico* drug design). Computer-aided drug design is a comprehensive subject based on the knowledge of chemistry, biology and computer science. With the help of computer and the structural information about drugs and their macromolecular targets, this methodology can guide and assist the design of new therapeutic agents by means of molecular modeling, theoretical calculation and prediction methods. In a course of nearly 30 years' development, computer-aided drug design has shown its potential advantages and plays a more and more important role in current pharmaceutical industry. However, modeling of a molecular structure is a complex task, in particular because most molecules are flexible and have an ability to adopt a number of different conformations that are of similar energy. Additionally, modeling of the binding process is also difficult as the characteristics of the receptor, the ligand, and the solvent have to be taken into account simultaneously. This is why being so close to realizing the pharmaceutical industry's dream to be able to design new drugs from the beginning, no single drug has been designed solely by computer techniques so far. On the other hand, the contribution of these methods to drug discovery is no longer a matter of dispute.

Some chemists believe that "organic solvents and computers don't mix". Likewise, medicinal and synthetic organic chemists working in drug discovery generally tend to toil at the bench, leaving the virtual work of computer-aided drug design to the theoreticians and synthesis of new molecules alongside with physiological studies to experimentalists. This could be partly due to the fact that most of the theoretical drug design techniques have an empirical character, limiting chances to systematically obtain a deeper insight into the factors determining drug activity at a molecular level. However, due to the recent progress in the theory of intermolecular interactions it is now possible to analyze drug activity from the first principles of quantum mechanics and to determine the key interactions crucial for inhibitory activity. In this chapter we present an experience gained by binding the gap between these two groups of researchers, which in turn enabled to built the system, in which design, synthesis of new molecules and testing their physiological activity is done in an "on-line" mode.

2. COMPUTER AIDED INHIBITOR DESIGN

Proteins might be regarded as modular structures assembled from a limited number of individual building blocks. Although the estimate of a number of proteins in humans range between 100 000 and 450 000 there is a common agreement that they fold into a limited number of structures. At present, around 600 fold types are recognized by classification of structurally characterized proteins available in protein data banks. In many examples it has been shown that protein families can have similar folds even though they at first seem to have completely different sequences. On the other hand, empirical observations confirm that related sequences have similar folds and enzymes catalyzing a similar reaction utilize similarly built active sites.

The basic principle of enzyme catalysis, formulated by Linus Pauling in 1946, is the selective stabilization of the transition state by strong preferential binding of the latter by an enzyme.[1,2] Therefore, stable compounds that resemble transition state of a reaction act as very effective inhibitors of the enzyme that catalyze the reaction. Thus, the three-dimensional structure of an active site is crucial for an understanding of the function of a given protein and certainly helps to design its new and effective inhibitors. Even more helpful is the knowledge of the structure of enzyme bound with a slowly reacting substrate or with a potent inhibitor.

Computational tools have become increasingly important in the drug discovery and design processes.[3,4] Computational chemistry methods are used routinely to study drug-receptor complexes in atomic detail[5] and to calculate properties of small molecule drug candidates.[6] Tools from information science and statistics are increasingly essential in organization and management of the huge chemical and biological activity databases. At the lowest level the computer models replace crude mechanical models visualizing complex structures, which much more accurately reflect molecular reality and are additionally capable of demonstrating motion and solvent effects. Beyond this, theoretical calculations permit to estimate binding free energies and to determine the modes of inhibitor binding to target proteins, using empirical force fields mimicking the electrostatic forces, hydrogen bonds and hydrophobic interactions of inhibitor fragments buried in enzyme surface areas. In the cases when a protein structure is not available, quantitative structure-activity relationships (QSARs), relating a molecule's biological activity to its structure and some more or less arbitrarily selected properties, are successfully used.

The process of discovery of a new clinical candidate for a novel drug consists of several steps, which have to be iteratively repeated in order to obtain the structure of high potency with drug-like properties. The first and crucial for the whole project step is a choice of the proper molecular target of the drug — an enzyme or a molecular receptor. Then lead generation process and its subsequent optimization results in some candidates for clinical trials.

2.1.   Lead Generation

Once the target is selected, a lead compound has to be found. Lead compound has to exhibit several features, which make design process easier. It should be a material with a considerable activity and relatively simple, synthetically accessible structure, where several potential modifications are possible. The most simple situation is when a promising lead compound is already known. It could be found by *in vitro* screening of a vast library of compounds using simple enzymatic test. On the other hand, *in silico* screening could be also used.[7] This computer-aided approach has several advantages. It significantly reduces costs and time of the process as the real compounds are required only for an experimental testing of candidates selected from the whole library of structures. However, due to the fact that process of inhibitor binding to the enzyme is rather complicated and not easily described by computational methods, *in silico* screening often generates false (not active) lead candidates or omits highly potent structures.

Process of computer-aided lead generation consists of several stages (Figure 8-1) and only a successful completing of each of them can result in a satisfactory lead compound. In order to perform computational screening, the structure of the receptor (most often protein) or its model has to be known. Three-dimensional structures of macromolecules can be obtained by few experimental methods: X-ray diffraction,[8] neutron diffraction or NMR.[9,10] Moreover, homology modeling based on the structures of similar proteins can be performed.[11] The second important step is library generation. It has to contain a large number of three-dimensional structures of compounds, which are already available or can be easily synthesized. This library of potential ligands has to be optimally positioned — docked into the active site of a receptor. In all recent algorithms the structure of a ligand is flexible, while the structure of a receptor – practically rigid. Theoretically, an application of flexible receptor model would lead to better results, but practical hit rate is lower due to the use of fast and thus relatively simple docking methods. Typical approach is rather the use of the structure of receptor taken from known ligand-receptor complex. There are several program packages[12] performing docking: DOCK,[13] GOLD,[14,15] FlexX,[16] Glide,[17–19] LUDI,[20] AutoDock.[21] In the next step ligand-receptor complexes should be evaluated in order to estimate the ligand affinity and thus obtain an ordered list of potential inhibitors.

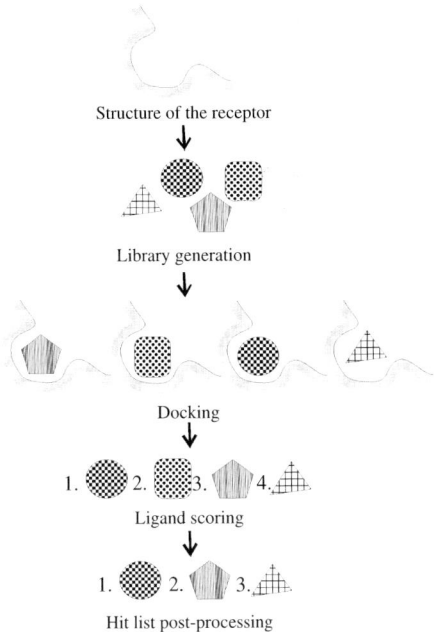

Structure of the receptor

Library generation

Docking

1. 2. 3. 4.

Ligand scoring

1. 2. 3.

Hit list post-processing

Figure 8-1. Computer-aided lead generation process

Scoring functions are used for computation of ligand affinity to the receptor based on the structure of the ligand-receptor complex.[22] Obtained values should give the possibility of prediction of experimental activity of the compound. However, although several attempts have been made to obtain the fast and accurate scoring function, the scoring problem still does not seem to be fully resolved. Three general approaches can be found in the literature: empirical scoring functions, force-field-based and knowledge-based ones.

Empirical scoring function is a weighted sum of relatively simple functions describing different interactions: hydrogen bonds, ionic interactions, hydrophobic interactions and binding entropy. All parameters are calibrated using a large set of ligand-receptor structures with experimentally measured binding affinities. The examples of empirical scoring functions are Böhm's LUDI scoring function,[23] ChemScore,[24] GOLD,[14,25] AutoDock,[21] and X-SCORE.[26]

Force-field-based scoring functions use arbitrary empirical estimates of inter-action energies obtained by molecular mechanics energy functions. This simple approximation, which takes into account only enthalpic contribution often correlates well with the experiment. Solvent effects are described by atom-based solvation parameters, which are computed for the surface of both ligand and receptor which is buried upon complexation. DOCK-chemical[27] and CHARMm scoring functions represent this class.

Another approach — knowledge-based scoring function uses a set of atom pairs potentials which describes favorable/non-favorable interactions of the atom from ligand and the atom from receptor. The functions describing these potentials are derived from interatomic distances distributions constructed using known ligand-receptor structures by the application of 'inverse' Boltzmann law. Examples of knowledge-based scoring functions are DrugScore[28] and PMF.[29]

Unfortunately, no scoring function is able to predict experimental results with a satisfactory quality, thus an approach combining results from several functions – consensus scoring – was developed.[30] It was shown that application of this method, particularly when employed along with statistical analysis, reduces considerably a number of false positives.[31,32]

Hit list of structures generated *in silico* is composed in such a manner that the compounds are listed in order of decreasing potential activities. The last step before chemical synthesis relies on the selection of the structures, which are simple enough to be synthesized readily in organic laboratories. Compounds of complex structures (e.g. those containing numerous chiral centers, or those possessing many functional groups located in a close proximity) are removed from the list or left for the next step of research. The rest of them are considered as potential "lead substances" and are being synthesized and evaluated for a target physiologic activity.

2.2.   Lead Optimization

Structure-based inhibitor design relies on the known inhibitor-receptor 3D structure. This could be obtained either by previously mentioned experimental methods or by docking the lead into the active site of the free receptor. On the basis of this structure, the lead compound is modified in a way that adds groups to enhance binding to the receptor.

There are several computer programs performing *de novo* generation of ligands.[33] Although this method does not require a lead compound, practically, in order to obtain highly potent inhibitor, it is desirable to know some lead structures. The algorithm of a ligand discovery is different in each program, however, some general scheme can be found (Figure 8-2).

In the first step (Figure 8-2, process A) the active site — search region is defined and should be represented in the computer memory in a way which makes the search faster and easier. The common method is a construction of the interaction sites, namely the set of virtual N-H, C=O or C moieties which would interact favorably with the receptor.

In the second step, the lead is modified. This could be performed in some variants. The lead compound can be enlarged by individual atoms or by molecular fragments. Both approaches have several advantages and disadvantages.[34] Building up a new molecule atom by atom can produce any possible structure, and thus a complete space of possible molecules is searched but, unfortunately, most of the obtained structures are not easily accessible by means of chemical synthesis. On the other hand, fragment based approach generates mainly structures which are

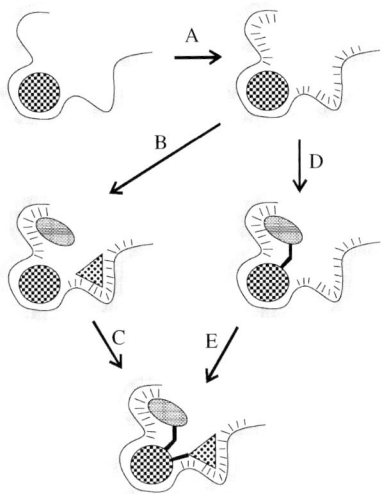

Figure 8-2. Schematic representation of design process

synthetically available but several interesting ligands could be omitted. Atom based approach was implemented in programs: LEGEND,[35] GROWMOL,[36] GenStar,[37] MCDNLG,[38] SPROUT,[39] CONCEPTS.[40] As more advantageous, most of the available programs are fragment-based ones: Grow,[41] NEWLEAD,[42] GroupBuild,[43] BUILDER,[44] HOOK,[45] CONCERTS,[46] SMoG,[47,48] and LUDI.[49,50]

There are two strategies applied in fragment-based design. First, molecular fragments are positioned independently in the search region in order to produce the best possible interactions (Figure 8-2, process B). Then program is trying to find linkers between these fragments and gathers the molecule (Figure 8-2, process C). The advantage of this approach is the possibility of positioning of groups without any restraints, thus, in a correct place, where it could interact most favorably. In most cases, the obtained structures are rigid ones, which is an additional benefit. Unfortunately, the linker generation is not an easy task and production of reasonable structures often fails. The second general strategy is building up the molecule step-by-step, by sequential addition of new fragments (Figure 8-2, processes D and E). The most important advantage of this approach is generation of synthetically feasible molecules, as some elements of chemical knowledge can be build into the growing procedure.[51] For example, interesting results were obtained by using amino acids fragments, that generated peptide-like structures. This method tends to produce undesired flexible structures. Additionally, some problems can be found when two possible interaction sites are located in some distance and none of the fragments have an incorporated spacer which could bridge this gap.[43]

Finally, generated ligand-receptor structures are ranked using scoring function and obtained hit-list undergoes further processing in the same way as during the lead generation (see Chapter 2.1).

2.3.   The Physical Nature of Ligand Binding

Given the fundamental importance of molecular recognition resulting from the strength and specificity of ligand binding, the process of complex association has remained an active area of theoretical research. Since the experimental measurements provide Gibbs free energy of binding, quantification of interactions requires both the enthalpic and entropic contributions to be calculated. On the other hand, when similar set of ligands is considered, they are assumed to exhibit comparable characteristics in terms of desolvation, protein (ligand) reorganization energy and the conformational entropy loss. In fact, we do not regard these effects as negligible, only that they are relatively constant across the series. Thus, computationally-demanding study of association free enthalpy may be replaced by the more affordable analysis of a binding energy that constitutes the most characteristic contribution to the observed binding affinity.

There are two general approaches to the evaluation of binding energy. Probably the most intuitive one is to subtract the energies of isolated monomers from the total energy of a complex (the supermolecular method). The apparent advantage of presented approach is its independence on the choice of a particular energy calculation method or a size of model (especially when considering the extent of the intermolecular distances that are still handled properly). However, all the system configurations need to be treated in the same way and this rule is not followed if different basis sets are applied for each of the reacting molecules – which results in so called Basis Set Superposition Error (BSSE). The common way to overcome this problem is to calculate monomers energies in the basis set of a dimer (counterpoise correction of Boys and Bernardi,[52] CP).

In revealing the nature of interactions in a non-empirical manner, the accurate contributions of particular components to a binding energy are most important, whereas the total energy itself obtained within supermolecular method is insufficient. The interaction energy partitioning is enabled with an use of perturbational approach, where the intermolecular interaction is introduced as a perturbation into the $A + B$ supermolecule Hamiltonian and subsequent corrections to the initial energy of an isolated molecules unperturbed state can be attributed a particular physical meaning. Accordingly, the first-order energy being the expectation value of a perturbation operator for an unperturbed wavefunction gives directly electrostatic energy and hence describes Coulomb's interaction of the two unperturbed (fixed) charge distributions. Likewise, the second order correction consists of two terms, namely induction and dispersion originating from mutually induced charge distribution fluctuations.

The key assumption in a foregoing approach is the polarization approximation neglecting the possible intermolecular electron exchange. This simplification is reasonable in case of long intermolecular distances only. However, for shorter distances the exchange repulsion cannot be ignored: neither interaction energy term in polarization perturbation theory accounts for this effect. To consider electron exchange, the wavefunction for a combined system should be antisymmetrized product of the two molecules wavefunctions, that is its sign should revert when two electrons are exchanged. Symmetry-Adapted Perturbation Theories (SAPT) have been developed to handle this issue and deal with intermediate and short range of intermolecular distances. Herein, the corrections of each order consist of more terms than in a precedent treatment: the above-listed interaction energy components are supplemented by corresponding exchange repulsion terms. Making allowance for an additional exchange term prevents the non-physical effect of infinite interaction energy lowering when the two neutral molecules approach each other. SAPT methods enable probably the most accurate yet very expensive interaction energy calculations, their application is therefore limited by a size of investigated systems.

While no unique way of energy partitioning exists and a multitude of formulations can be encountered, the variation-perturbation scheme[53] utilized in our research constitutes a reasonable compromise between accuracy and computational cost. The SCF interaction energy, determined as in a supermolecule method and calculated in a dimer basis set to correct BSSE, constitutes a starting point of energy partitioning. After subtracting the first-order Heitler-London term, $E^{(1)}$, representing an interaction of monomers with frozen electronic density distributions, the remaining delocalization component expresses an effect of interaction on the electronic distribution relaxation of the monomer and comprises higher-order SAPT corrections including induction:

$$E_{SCF} = E^{(1)} + E_{DEL}^{(R)}$$

According to SAPT formulation of the first-order interaction energy, the Heitler-London term consists of electrostatic and exchange contributions (the former obtained from the perturbation theory formula):

$$E^{(1)} = E_{EL}^{(1)} + E_{EX}^{(1)}$$

SCF entirely excludes dispersion and an accurate description of all types of weak interactions (i.e., determination of the correlation term) requires the correlated *ab initio* methods to be used.

In summary, the MP2 interaction energy decomposition reveals the following terms:

$$E_{MP2} = E_{EL}^{(1)} + E_{EX}^{(1)} + E_{DEL}^{(R)} + E_{CORR}^{(R)}$$

As one can easily note, a well-defined hierarchy of successive interaction energy approximations, varying from the most expensive MP2 method to the various electrostatic energy representation (the more simplified the theory, the less computationally demanding calculation), demonstrates the utility of this decomposition scheme (Figure 8-3):

$$E_{EL}^{(1)} < E^{(1)} < E_{SCF} < E_{MP2}$$

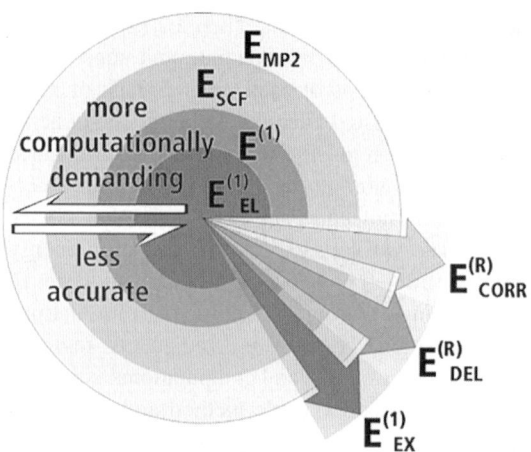

Figure 8-3. Variational-perturbational scheme of interaction energy decomposition. The circles denote the consecutive levels of theory while the arrows designate the corresponding correction terms

It performs exceptionally well when derivation and testing of simplified yet reasonable models by a subsequent elimination of less important and more time-consuming contributions is required. Moreover, owing to the full counterpoise correction in each of the interaction energy component (all of them are evaluated consistently in a dimer basis set) the basis set dependency has been significantly reduced. Finally, additional advantage of a presented treatment comes from the implementation of a direct SCF technique, so that the storage of integrals during calculations can be avoided and the efficient study of relatively large models is possible. Currently molecular systems containing up to circa 100 atoms could be investigated, opening the possibility to analyze the physical nature of inhibitor interactions in enzyme active sites.

Variational-perturbational energy decomposition scheme was successfully employed in a number of diverse phenomena investigations including the influence of mutations within enzyme active site[54], the contribution of particular active site residues to catalytic effects[55] and the connection between inhibitory activity and the interaction energy[56–58]. The latter will be presented in what follows.

3.   LEUCINE AMINOPEPTIDASE INHIBITORS

Leucine aminopeptidase (LAP, E.C.3.4.11.1) is one of the first discovered and the most widely studied aminopeptidase with respect to sequence, structure and mechanism of action.[59–63] LAP is a zinc containing exopeptidase that catalyzes the removal of amino acids from the N-terminus of peptides or proteins. Similar to other aminopeptidases, this enzyme is of significant biological and medical importance because of its key role in protein modification, activation, and degradation as well as in the metabolism of biologically active peptides and activity regulation of hormonal and non-hormonal peptides.[64] Altered activity of human leucine aminopeptidase has been associated with several pathological disorders, such as cancer[65,66] eye lens aging and cataracts,[67-69] myeloid leukemia and blood cell counts.[70] This enzyme may also be important in the processing of antigenic peptides and in the determination of the immunodominance of various peptides.[71] Moreover, LAP seems to play an important role in the early stages of HIV infection, and thus serum activity of this enzyme may be useful as a surrogate marker for HIV infection and as a response to chemotherapy.[72]

Leucine aminopeptidase is an enzyme for which the most systematic and detailed computational studies regarding enzyme-inhibitor interactions have been performed.[54,56-73,74] This results both from the availability of the crystal structure of LAP with bound inhibitors, including phosphonic acid analogue of leucine - LeuP (structure encoded as 1lcp in PDB, Figure 8-4), and from the existence of binding data for many LAP inhibitors. Such a set of experimental results enabled to evaluate the effectiveness of theoretical methods: both empirical and quantum chemical in designing and activity prediction of LAP inhibitors.

The Böhm method implemented in the computer program LUDI[20,23,75] was applied to design new LAP inhibitors, predict their binding affinities and analyze the interactions with the enzyme.[73] A very important feature of the LUDI program was the successful reproduction of the most active known aminophosphonic LAP inhibitors, including LeuP (**1**), therefore validating the method applied. In addition, numerous potential inhibitors of leucine aminopeptidase with higher theoretical activity than those known from the literature were designed in this way (Figure 8-5).[73] New designed LAP inhibitors represented aminophosphonic acids obtained by the replacement of the isobutyl moiety of LeuP with the fragments from Ludi_link library.

The designed ligands have a shape and charge distribution complementary to the S1 pocket of LAP, which engages the side chains of Met270, Ala451, Thr359, Gly362 and Met454 near the active site, and some polar groups relatively far away from the active center. Consequently, the majority of newly designed LAP inhibitors

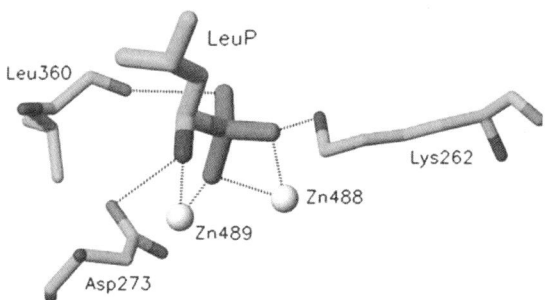

Figure 8-4. The binding mode of phosphonic acid analogue of leucine (LeuP) by LAP (1lcp in PDB); the interactions with two zinc ions and hydrogen bonds are indicated as dashed lines

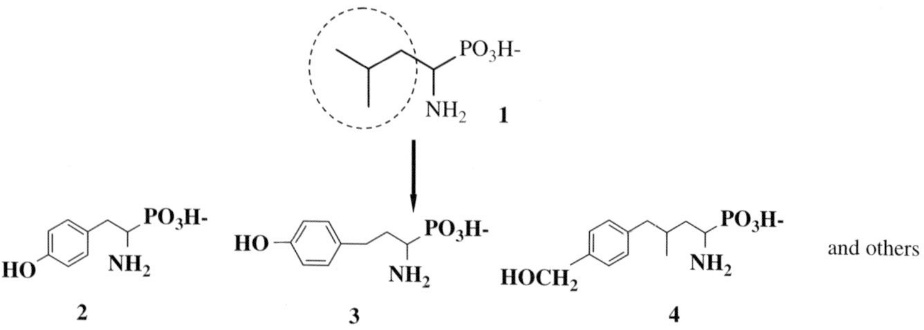

Figure 8-5. Examples of LAP inhibitors, LeuP analogues, designed using LUDI approach

contained long hydrophobic chain and terminal polar group (Figure 8-5), which might be involved in hydrogen bonds with Asp365 or Ala451. Based on LUDI's empirical scoring function the designed ligands were expected to be more potent enzyme inhibitors than other known aminophosphonates. Selected compounds were synthesized and tested for their inhibition of LAP. Out of them, compounds **2** (phosphonic analogue of tyrosine) and **3** (phosphonic analogue of homotyrosine) were found to be the most potent, with $K_i = 330$ nM and 120 nM for RS mixtures of **2** and **3**, respectively.[76] Therefore, **3** indeed represents the most potent aminophosphonic inhibitor of LAP discovered thus far, with K_i at least two times better comparing to the best till now L-LeuP ($K_i = 230$ nM). Due to strong substrate specificity of LAP toward L isomers of such analogues, L enantiomer of **3** is expected to be up to twice more potent than the racemic mixture of this compound. These results clearly confirm that LUDI represents a powerful approach to the design of new potent protein inhibitors.

The experimentally measured activities of known LAP inhibitors, which structure were reproduced during LUDI searches, were used to verify the usefulness of LUDI scoring function for binding affinity prediction. LUDI_Score calculated for these compounds was correlated with their binding affinities (Figure 8-6), resulting in a very good correlation (correlation coefficient, $R = 0.93$). The resulting approach, which includes the specific effects for LAP-inhibitor interactions, can be further applied to predict binding affinities of new LAP inhibitors. In addition, the K_i values derived from Score_all were calculated for these compounds (**1, 5–12** in Figure 8-6). This resulted in a very good agreement between experimental and theoretical binding affinities (**1**, LeuP: $K_{i,calc} = 0.21 \mu M$, $K_{i,exp} = 0.23 \mu M$; **5**, PheP: $K_{i,calc} = 0.07 \mu M$, $K_{i,exp} = 0.42 \mu M$; **6**: $K_{i,calc} = 0.41 \mu M$, $K_{i,exp} = 0.47 \mu M$; **7**: $K_{i,calc} = 0.31 \mu M$, $K_{i,exp} = 1.0 \mu M$; **8**: (ValP): $K_{i,calc} = 0.34 \mu M$, $K_{i,exp} = 1.2 \mu M$; **9**: $K_{i,calc} = 0.50 \mu M$, $K_{i,exp} = 3.6 \mu M$). Larger differences between the predicted and measured inhibition constants were obtained for weak LAP inhibitors (**10–12**, $K_{i, exp} > 100 \mu M$), which structures were not found by the LUDI program.[73]

Another example of a very successful application of LUDI for designing new LAP inhibitors was discovery of phosphinate dipeptide analogues, representing a new class

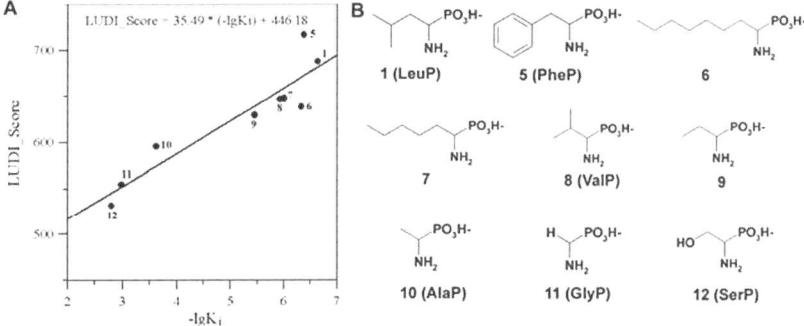

Figure 8-6. a) LUDI_Score as a function of experimentally measured activity of LAP inhibitors (R = 0.93). Numbers of particular points correspond to inhibitors designation introduced in the right panel; b) structures of LAP inhibitors included in correlation

of very potent inhibitors of the enzyme.[74] The replacement of the hydroxyl group in LeuP (**1**) by new structural fragments from LUDI_link library interacting with the S1′ pocket of LAP led to identification of these compounds (**13–17**, Table 8-1). They appeared to be the most active LAP inhibitors among the compounds containing phosphorus atom in the structure, with the binding affinities in nanomolar range. Simultaneous modification at the P1 position of such analogues, specifically the replacement of Leu by homophenylalanine (hPheP) side chain which fits better to the hydrophobic S1 pocket of LAP, resulted in an additional enhancement of the activity ($K_i = 66\,nM$ and 67 nM for **16** and **17**; values correspond to the mixture of four isomers). The increased activities of **16** and **17** comparing to LeuP arise most likely from the additional hydrogen bond between carboxyl group of these ligands and Gly362 and also from the additional hydrophobic interactions with the S1′ pocket of LAP. Based on LAP substrate specificity the binding affinities of the appropriate diastereoisomers of the phosphinate dipeptide analogues discussed here are expected to be even higher. This new class of tight-binding, competitive inhibitors of leucine aminopeptidase may be considered as new lead compounds and offer the possibility of further modifications at P1 and P1′ positions, which should result in more active and selective compounds. Furthermore, the binding affinities for these compounds calculated using LUDI empirical scoring function remain in excellent agreement with the experimental values (Table 8-1), again confirming the usefulness of this approach in the drug design process.

Summing up, all these results confirm that LUDI is a very powerful tool for designing protein inhibitors, predicting their affinity and determining the nature of the interactions in the ligand-receptor system. However, some of the systems, particularly with dominant electrostatic interactions, may require a more precise description of the molecular charge distribution and engagement of more accurate methods might improve binding affinity estimates.

More rigorous, non-empirical approach has been successfully applied to analyze the interactions of LAP active site with another class of LAP inhibitors (**1, 18–25**,

Table 8-1. Structures of LAP inhibitors, phosphinate dipeptide analogues, designed using LUDI. Experimental and predicted activities are presented for each compounds. [a] Value corresponding to the mixture of two diastereomers (1:1). [b] Binding affinity for the mixture of four diastereomers. [c] Value corresponding to the racemic mixture. [d] Predicted binding affinity for one isomer which preferentially interacts with the protein

LAP inhibitor	$K_{i,\ exp}$ [nM]	$K_{i,\ LUDI}$ [nM]
13	65[a] 110[b]	48[d]
14	330[c]	250[d]
15	74[b]	19[d]
16	66[b]	20[d]
17	67[b]	15[d]

Figure 8-7) using the variation-perturbation decomposition of the interaction energy.[53] This approach provided the opportunity to study the physical nature of LAP-inhibitor interactions and to derive and validate simplified models of inhibitory activity by stepwise neglecting the LAP active site residues and stabilization energy components of minor importance.[54–58] These studies allowed, for the first time, the detailed and accurate analysis of interactions between inhibitors and active site residues of the metaloenzyme. All inhibitors included in the analysis contained the same hydrophobic side chain interacting with the S1 pocket of LAP and they differed in the fragment interacting with the enzyme active site. LAP active site (Figure 8-4) was modeled by the enzyme residues and zinc ions, which interact with

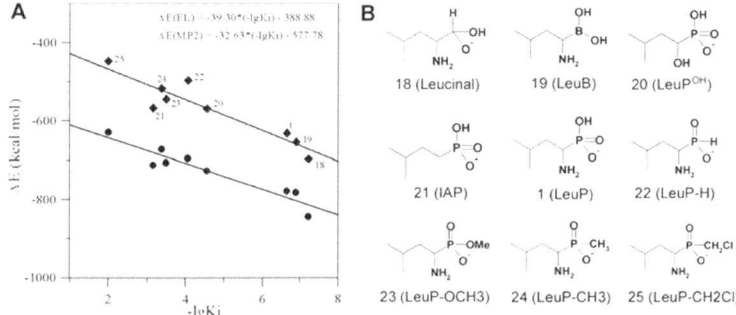

Figure 8-7. A. Binding energy (ΔE_{MP2}, circles, and ΔE_{EL}, squares) as a function of inhibitory activity for the model comprising LAP active site residues. Numbers of particular points correspond to inhibitors designation introduced in right panel. B. Structures of LAP inhibitors included into binding energy calculations presented in left panel

the phosphonic group (or corresponding groups in other ligands) and the amine (or hydroxyl) group of the inhibitors: $Zn^{2+}488$, $Zn^{2+}489$, Lys^+262 (represented by NH_4^+), Asp^-273 (represented by $HCOO^-$) and Leu360 (represented by HCHO). Total interaction energies constituted the sum of stabilization energy obtained in a pairwise manner. The total interaction energy calculated at the second-order Moller-Plesset level of theory was partitioned according to the variation-perturbation scheme[53] discussed in the Section 2.3.

The stabilization energies obtained at various theory levels correlated very well with the experimentally determined inhibition constants ($-lgK_i$) for nine LAP inhibitors taken into consideration.[56,57] The best correlation was observed for E_{MP2} (R = 0.95), while for E_{SCF} and E_{EL} [1] the results were only slightly worse (R = 0.94 and 0.92, respectively) (Figure 8-7). The equation resulting from the correlation of the interaction energy calculated at the MP2 level (Figure 8-7) was applied for activity prediction of the studied inhibitors.

The theoretical activities obtained in this way were compared with experimentally measured ones, resulting in a very good agreement for most of the inhibitors (Table 8-2). The agreement of calculated activities with experimental data for leucinal, LeuB and Leu^{OH} is at least one order of magnitude better comparing to the results obtained by applying empirical scoring functions. Slightly bigger differences in the experimental and predicted activities for leucinal and IAP suggest that other effects (entropy or solvation effects) besides the interactions with the enzyme might influence the binding affinities of these two compounds. Furthermore, similar approach applied for PheP (phosphonic analogue of phanylalanine) and its analogues with modified phosphonic group resulted in an equally good correlation of the interaction energies calculated at various levels with their binding affinities toward LAP.[54,56]

Interestingly, simplified models of inhibitory activity by stepwise neglecting the LAP active site residues demonstrated that out of five active site residues

Table 8-2. Measured and calculated activities of LAP inhibitors. [a] values for L enantiomers

LAP inhibitor	$K_{i,exp}$ (μM)	$K_{i,E(MP2)}$ [a] (μM)
Leucinal	0.06 (L)	0.006
LeuB	0.13 (DL)	0.54
LeuP^{OH}	28.5 (DL)	27.5
IAP	700	72
LeuP	0.23 (L)	0.70
LeuP–H	87 (DL)	245
LeuP–OCH$_3$	320 (DL)	109
LeuP–CH$_3$	425 (DL)	1288
LeuP–CH$_2$Cl	10000 (DL)	26303

only $Zn^{2+}488$ and Lys^+262 were recognized as essential for relative stabilization energy.[56,58] It is remarkable that, among the five mostly charged LAP residues, each of the two is sufficient to determine the relative binding affinity for LeuP analogues (compounds **1, 22–25**, for E_{MP2} R = 0.95 both for the interactions with $Zn^{2+}488$ and Lys^+262; Figure 8-8).

The above presented studies revealed that these two LAP active site residues are responsible for the observed differences in the activity of the considered inhibitors, although the interactions with $Zn^{2+}489$ are the strongest.

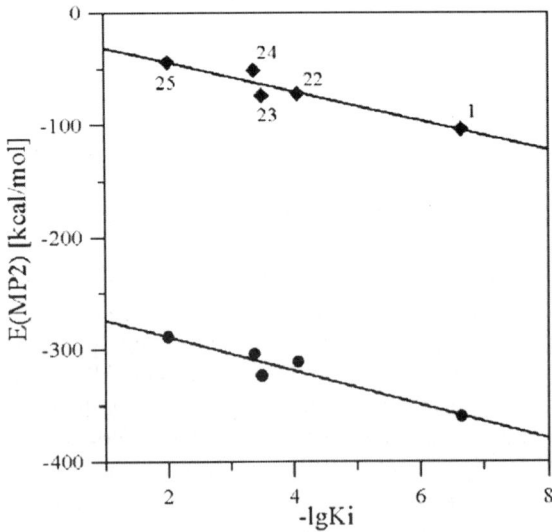

Figure 8-8. Correlation of inhibitory activities with ΔE_{MP2} interaction energies calculated for Zn488 (circles; R = 0.95) and Lys262 (squares; R = 0.95). Numbers of particular points correspond to inhibitors designation introduced in Figure 8-7

Summing up, we presented one of the first successful attempts of the application of nonempirical interaction energy calculations for binding affinity prediction of enzyme inhibitors bound in a similar manner by the enzyme. We confirm that the higher level of theory is applied for description of intermolecular interactions the better degree of correlation with experiment is observed. The application of variational-perturbational decomposition of the interaction energy allows to analyze the physical nature of the inhibitor binding forces with the enzyme active site and identify the receptor residues that are crucial for specificity. As a consequence, models of binding affinity prediction can be constructed at the *ab initio* level, providing an approach which can be utilized in the drug design process for binding affinity predictions of newly developed ligands.

4. GLUTAMINE SYNTHETASE INHIBITORS

Glutamine synthetase (GS, EC 6.3.1.2) is an enzyme which catalyzes the formation of glutamine (**27**) from glutamate (**26**) and ammonium ion in a presence of ATP (Scheme 8-1).[77]

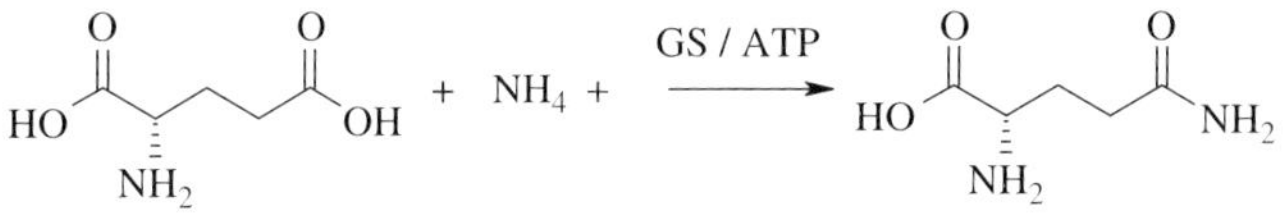

26　　　　　　　　　　　　**27**

Scheme 8-1. Reaction catalyzed by glutamine synthetase

Due to an essential role of glutamine in nitrogen metabolism, GS is of high importance for living cell, and its regulation is crucial for the organism.[78] This stimulates an interest in novel inhibitors of GS, which could be both practically applicable and useful for experimental analysis of biochemical role of the enzyme. The most widely studied biological activity of GS inhibitors is the herbicidal one.[79] The most potent inhibitor of GS — phosphinothricin (**28**) is the active component of commercially available total herbicide.[80] Irreversible inhibition of plant GS produces severe elevation of ammonia concentration which, in turn, causes plant death.[81–85] Moreover, disruption of nitrogen metabolism is the origin of photo-synthesis and photorespiration breakdown.[86] The second highly important possible application of GS inhibitors is as anti-tuberculosis drugs.[87,88] It was shown that L-methionine sulfoximine (**29**) — a potent GS inactivator causes death of *Mycobac-terium tuberculosis*, the pathogen giving rise to this disease. It blocks biosynthesis of polyglutamate/glutamine cell wall structure. Interestingly, this phenomenon is observed only in the case of pathogenic bacteria, thus this class of compounds could be considered as very promising and selective drug candidates.

Bacterial glutamine synthetase is a huge enzyme which consists of 12 subunits forming two hexameric rings positioned face-to-face.[89,90] There are 12 active

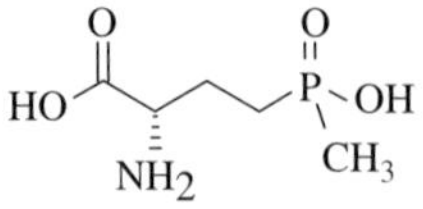

28

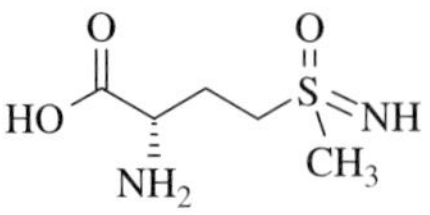

29

sites present in this dodecamer, each one placed between two subunits. On a basis of several crystal structures of glutamine synthetase with substrate, products, and inhibitors a mechanism of enzymatic reaction was postulated (Scheme 8-2).[91]

First, glutamate (**26**) and ATP are bound to protein and phosphorylation of glutamate proceeds. Resulting γ-glutamyl phosphate (**30**) is a reactive intermediate, which is able to undergo a nucleophilic substitution by ammonia via tetrahedral transition state and thus glutamine (**27**) is formed.

Similarly to biosynthetic reaction, some of highly active inhibitors also can be phosphorylated by ATP in the active site of the enzyme (Scheme 8-3).[92] In the case of phosphinothricin (**28**) and methionine sulfoximine (**29**), the formed phosphates are actual inhibitors of GS and are irreversibly bound to the protein.[93]

Most probably, phosphorylated forms of these inhibitors are transition state analogues, in which methyl group represents $-NH_3^+$ substituent. Crystal structures of both MetSox-GS[91] and PPT-GS[94] complexes were measured and solved, however, due to the isoelectronic structure of oxygen atom and methyl group at

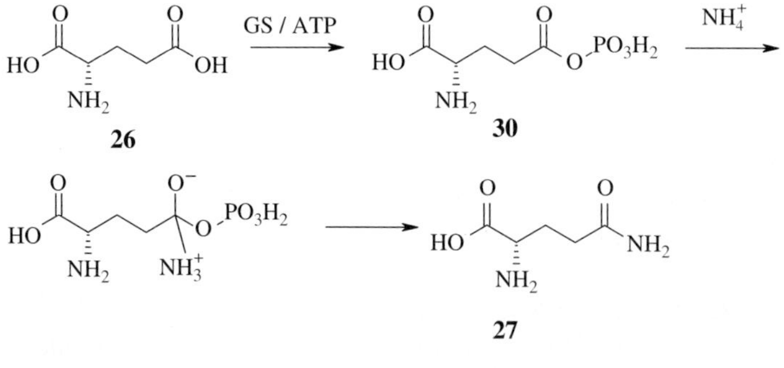

26   GS / ATP   **30** NH_4^+

27

Scheme 8-2. Mechanism of reaction catalyzed by GS

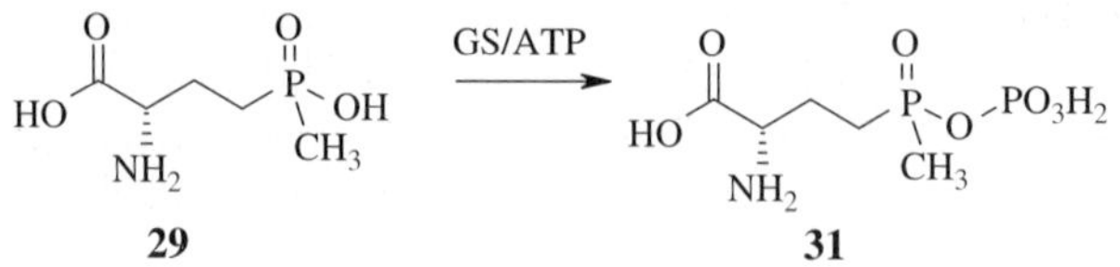

29   GS/ATP   **31**

Scheme 8-3. Mechanism of GS inhibition by phosphinothricin (**29**)

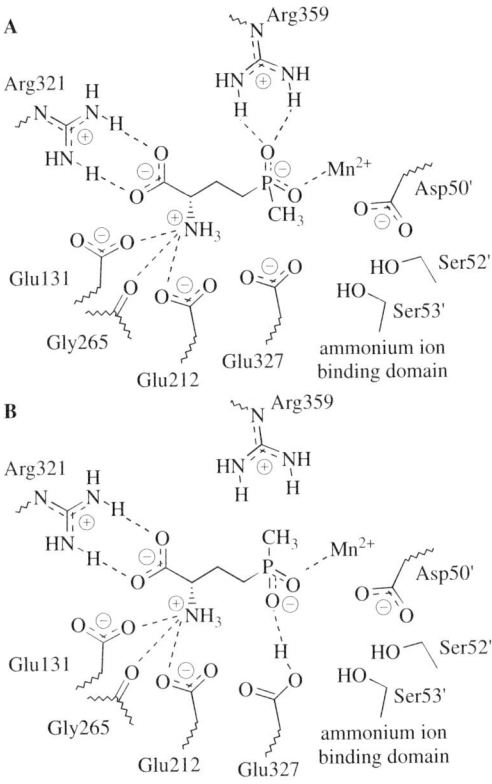

Figure 8-9. Possible conformations of PPT in GS active site

least two different inhibitor conformations in the active site can be postulated (Figure 8-9). In conformation A phosphinothricin is a transition state analogue, where oxygen atoms attached to phosphorus atom interact with Arg359 and metal ion while methyl group is placed in an ammonium ion biding site. In conformation B methyl group is placed near Arg359 and one of the oxygen atoms forms a hydrogen bond with Glu327.

Scoring of both conformations resulted in conclusion that the first one (A) is correct. Higher than that of conformation B affinity of the inhibitor results from additional strong interaction with Arg359 — similar to that found in glutamate-GS complex.[95] In order to support that conclusion several other known inhibitors of GS were docked into the active site of the enzyme and scored with eight different functions. In all cases good correlation was found, which confirmed the correctness of the obtained structures and particularly the correctness of the conformation **A** of PPT in GS active site.

As the majority of potent GS inhibitors are phosphorylated upon enzymatic process, these forms of inhibitors were docked as well. It was found out that correlation between experiment and scores obtained for these complexes were better then in case of nonphosphorylated inhibitors (mean R^2 value 0.72 versus 0.64).

Finally, a structural and electronic features required for highly potent GS inhibitor were determined. Comparison of electrostatic potential on a molecular surface of enzymatic reaction transition state and phosphorylated forms of phosphinothricin and methionine sulfoximine revealed high similarity, while less potent inhibitors showed some differences. On the basis of these results, it was concluded that these inhibitors are typical transition state analogues. Moreover, charge distribution near tetrahedral atom of inhibitor (phosphorus or sulfur) should consist of two negatively charged centers and additional positively charged or neutral one.

In the next step, phosphinothricin was chosen as a lead compound for designing novel inhibitors.[96] This structure has some advantages as a lead. It is the compound of a high potency combined with a relatively simple structure, which could be easily modified in several ways. Moreover, crystal structure of PPT-GS complex was solved,[94] and thus it could be used as a template for rational, computer-aided design of novel inhibitors.

Using LUDI program[20,49,50] in Link_Mode several substituents of phosphinothricin methyl group were designed. Among the structures proposed by the program, four compounds (**32–35**) combining molecular economy with potentially high affinity, were chosen. This set of inhibitors was synthesized in enantiomerically pure form and tested against *E. coli* glutamine synthetase. As predicted by LUDI, the highest potency was found for compound **32** ($K_i = 0.59\mu M$), where methyl group was replaced with methylamino moiety. Modeled structure of **32**-glutamine synthetase complex (Figure 8-10) showed that additional amino group interacts favorably with negatively charged moieties forming an ammonium ion binding site. Unfortunately, affinities of novel compounds were not considerably higher than that of lead compound, which resulted most probably from steric reasons as the space available in the active site is highly restricted. On

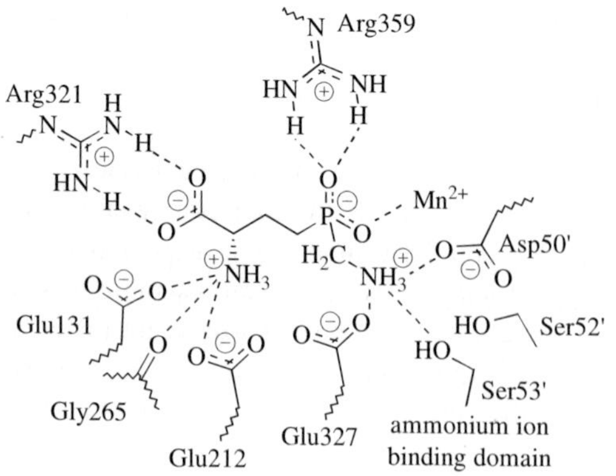

Figure 8-10. Modeled mode of binding of compound **32** to glutamine synthetase

the other hand, compound **32** ranks as one of the best inhibitors of glutamine synthetase found so far and this confirms usefulness of applied computational methodology.

5. L-PHENYLALANINE AMMONIA LYASE INHIBITORS

Phenylalanine ammonia-lyase (PAL, E.C. 4.3.1.5) is a key enzyme of plant and fungi phenylpropanoid metabolism. It catalyzes a nonoxidative, stereospecific elimination of ammonia from L-phenylalanine to give trans-cinnamic acid, therefore providing plants with a mechanism for the diversion of large amounts of carbon from aromatic amino acid metabolism into the biosynthesis of products based on phenylpropane skeleton. The products of phenylpropanoid metabolism provide a highly efficient UV screening (flavonoids and small phenolic compounds), control of water loss and structural rigidity (lignins) as well as defense against pathogens (nearly all classes of phenylpropanoid compounds), all traits that are essential for the success of land plants.[97,98] PAL, being the connection between natural product biosynthesis and primary metabolism, gains an interest as a possible control site of desirable compound accumulation. Moreover, phenylalanine ammonia-lyase is a potential target for herbicides and, surprisingly, its application in a treatment of genetic disease phenylketonuria was recently suggested.[99]

Presently, some information is available on the structure of the active site of PAL and its mechanism of action. This knowledge was advanced with an aim of studies on PAL mutants obtained by site-directed mutagenesis,[100–102] comparison of its structure with the recently determined three-dimensional structure of histidine ammonia-lyase [103,104] as well as the detailed analysis of structure-activity relationship found for its inhibitors.[105–108] Despite a recent knowledge of PAL three-dimensional structure,[109,110] the detailed mechanism of this particular enzyme action remains unresolved.

Taking an advantage of the existence of kinetic measurements data for several inhibitors of PAL from potato,[111,112] *ab initio* investigation of the physical nature of enzyme-inhibitor interactions and the existence of correlation between particular energy components and inhibitors activity was conducted.[58] Similarly to the above presented LAP study,[54–58] the validity of successive approximations was tested to qualify a permissible level of simplification and – first of all – to reveal which constituents of interaction energy contribute the most to the specific inhibitors binding within an enzyme active site. In contrast to the LAP active site, no metal ion is present in the PAL binding pocket while the inhibitors considered in both LAP and PAL case are essentially similar.

Since the tertiary structure of PAL from potato is unavailable, it was modelled on the basis of homology to *Pseudomonas putida* histidine ammonia-lyase, HAL[58,103] (by the time this research was being conducted, none experimental PAL structure existed). Remarkably, our model of PAL active site is in agreement with recently revealed X-ray structure of PAL from *Rhodospiridium toruloides*.[110] Six highly conserved active site residues in the vicinity of a variable part of inhibitors were selected to represent an enzyme environment (Figure 8-11). In particular, two asparagine (Asn187[A], Asn311[A]; superscripts indicate the corresponding monomer of PAL homotetramer) and glutamine (Gln275[B]) residues were represented by an acetamide molecule, arginine (Arg⁺281[B]) was modelled by a methylguanidinium cation, whereas tyrosine residues (Tyr35[A], Tyr278[B]) were truncated at Cβ atoms and mimicked by p-cresol. All the inhibitors considered herein are the phosphonic analogues of phenylalanine (with the exception of compound **1**, that is phenylalanine with an α-amino group replaced by an aminooxy moiety – see Figure 8-12). In Figure 8-11 given is the model of PAL active site with the docked inhibitor **2** molecule.

The computational protocol utilized here is similar to the one employed in case of LAP.[54–58] The total binding energy was taken from a pairwise analysis of the interactions between the PAL active site residues and particular inhibitor molecules (model **A**). With the goal of deriving some justified approximations in mind, the decrese of correlation with experimental data was monitored that resulted in a

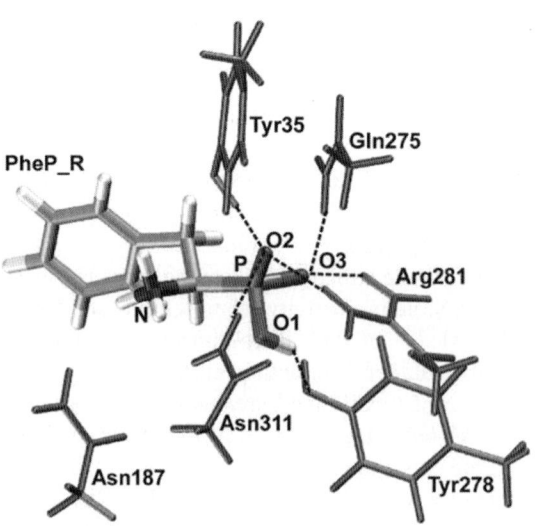

Figure 8-11. PAL active site model (residues shown in black) and the mode of PheP_R, 2 inhibitor binding. The closest atomic contacts with PAL residues are marked as dashed lines

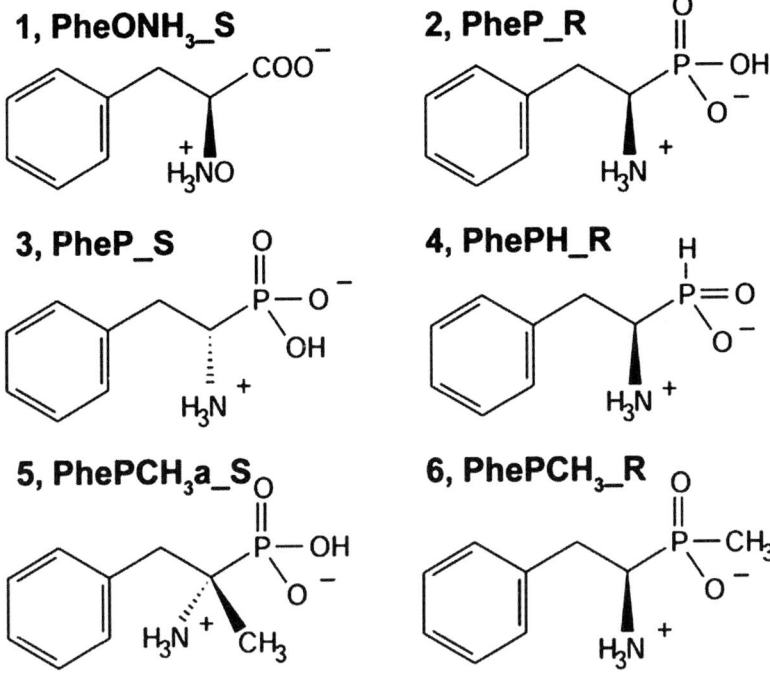

Figure 8-12. The structures of PAL inhibitors in the order of decreasing inhibitory activity

selection of the four active site residues serving as a limited size model of binding pocket, namely Tyr35, Asn187, Tyr278, and Arg281 (model **B**).

To verify the quality of our models as well as their usefulness in binding affinity prediction, we took advantage of the availability of experimental inhibition constants, K_i.[111,112] The remarkable correlation with experiment (*i.e.*, with the value of $-\log K_i$) was revealed for subsequent interaction energy components including the first-order electrostatic term (Table 8-3 and Figure 8-13). This finding justifies our assumption regarding insignificant contribution of entropic and solvation effects to the free enthalpy of binding. Noticeably, the higher the level of theory applied to the description of interactions, the more pronounced the relationship of resulting binding energies with experimental data.[111,112] The most complete interaction energy representation at MP2 level is very closely reflected in both models by its SCF counterpart, indicating a minor influence of correlation effects.

The corresponding correlation coefficients (Table 8-3) indicate negligible change at MP2 and SCF levels when passing from model **A** to **B**. Although the analogous values for $E^{(1)}$ and $E^{(1)}_{EL}$ are affected, neglecting delocalization and exchange terms still produces reasonable agreement. Overall, model **B** provides a sufficiently accurate description of the experimentally observed characteristics of PAL inhibitors binding.

Table 8-3. Correlation coefficients of the relationship between experimental inhibitory activity[111,112] ($-\log K_i$) and the interaction energies at various levels of theory. In case of LAP, model **A** corresponds to the whole LAP active site model given in Figure 8-4, while models **B** and **C** denote the single residue's models: Zn488 and Lys262, respectively

Method	PAL active site model		LAP active site model		
	Model A	Model B	Model A	Model B	Model C
E_{MP2}	0.99	0.99	0.97	0.95	0.95
E_{SCF}	0.99	0.98	0.95	0.92	0.95
$E^{(1)}$	0.92	0.69	0.94	0.91	0.93
$E^{(1)}_{EL}$	0.88	0.78	0.93	0.91	0.86

The above conclusions are further confirmed by an excellent accuracy of the predicted inhibition constants based on the equations derived from correlation analysis of MP2 interaction energy (Table 8-4).

As noted above, our analysis suggests that a satisfactorily precise description of this particular set of PAL inhibitors binding is provided by the PAL active site model encompassing four out of the initial number of six amino acid residues, especially

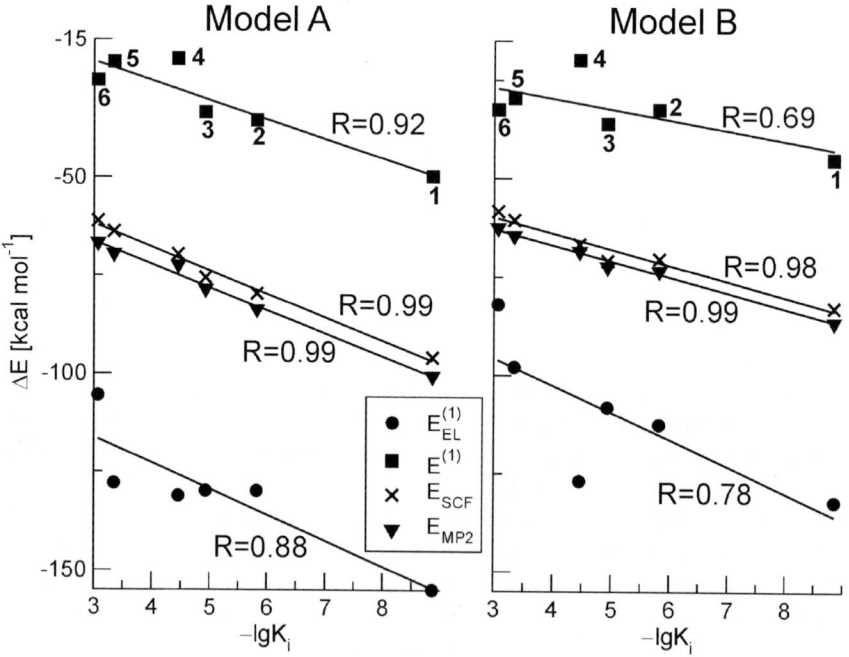

Figure 8-13. Interaction energy at consecutive levels of theory as a function of inhibitory activity.[111,112] Numbers of particular points correspond to the inhibitors designation introduced in Figure 8-12. Correlation coefficients, R, correspond to the consecutive regression curves

Table 8-4. Comparison of the experimental[111,112] and predicted inhibition constants

PAL inhibitor	$K_{i,exp}$ (μM)	$K_{i,E(MP2)}$ (μM)	
		Model **A** $\Delta E_{MP2} = -5.82 \cdot (-\log K_i) - 48.98$	Model **B** $\Delta E_{MP2} = -4.00 \cdot (-\log K_i) - 50.90$
1, PheONH3_S	0.0014	0.0014	0.0014
2, PheP_R	1.5	1.1	2.4
3, PheP_S	11.6	8.4	4.6
4, PhePH_R	35	95	42
5, PhePCH3a_S	450	390	290
6, PhePCH3_R	850	1230	870

when the SCF (or higher) level of theory is employed. Another important question might arise, namely which PAL active site residues are crucial for the observed specificity of binding? Surprisingly, despite Arg281 exhibiting the strongest binding with inhibitors, Tyr278 interactions are more representative for the observed differences in consecutive ligands stabilization (Table 8-5). This conclusion applies not only to interaction energy at SCF and MP2 theory levels, but even more significant loss of correlation is found for electrostatic term. On the other hand, elimination of Arg281 does not alter the latter. Noticeably, omitting the two residues assumed to be of a minor importance while constructing model **B** (i.e., Gln275 and Asn311) does not imply the loss of correlation.

Since phenylalanine ammonia-lyase constitutes the second example of an application of non-empirical interaction energy decomposition to the inhibitors binding, it is interesting to compare the above results for PAL with our previous LAP study.[54–58] In both cases, phosphonic analogues of phenylalanine (for LAP and PAL) as well as leucine (for LAP) were considered. However, active site composition is significantly different in that LAP is a metalloenzyme containing the two zinc ions interacting with inhibitors and mostly charged the remaining active site constituents (except of Leu360; see Figure 8-4), while amino acid residues forming PAL binding pocket (Figure 8-11) are relatively neutral (except of positively charged Arg281).

Table 8-5. Correlation coefficients obtained in the absence of individual active site constituents – particular PAL residues are ranked according to the decreasing contribution to stabilization energy

Residue	$E^{(1)}_{EL}$	$E^{(1)}$	E_{SCF}	E_{MP2}
Tyr278	0.61	0.94	0.84	0.81
Arg281	0.87	0.73	0.86	0.86
Asn187	0.44	0.92	0.93	0.96
Tyr35	0.84	0.94	0.95	0.96
Asn311	0.84	0.81	0.97	0.98
Gln275	0.81	0.82	0.99	0.99

In Table 8-3 given is the comparison of the PAL and LAP correlation coefficients. It is remarkable, that the general trends in LAP inhibitors binding can be predicted based on an analysis of the interaction between the latter and the single LAP active site component : either Zn488 ion or Lys262 residue. Analogous predictions for PAL require at least four residues to be taken into account. Furthermore, the electrostatic term gives sufficiently accurate results in case of LAP, while exchange-repulsion effects are also necessary for reasonable reproduction of experimentally observed binding affinities of PAL inhibitors. Seemingly, metalloenzyme inhibitors' binding is governed by interactions with one of a few charged residues, whereas in enzymes with a generally neutral binding site it is the coincident interaction with several residues that controls the inhibitory activity (Figure 8-14). In addition to the electrostatics effects, steric complementarity also has to be considered in the latter case.

As emphasized above, what is probably more important than the analysis of physical nature of interactions itself is the first principles-based derivation of approximate yet reasonable models for binding affinity prediction. The use of empirical scoring functions only, although often successful, does not necessarily prove the validity of a computational model, that is the agreement with experiment might result from, for example, an accidental cancellation of errors. On the other hand, nonempirical approach can also lead to simple models relating the binding energy to quantities that may be rapidly evaluated (e.g. electrostatic potential in some selected positions). In such a case, however, the limit of accuracy has already been established by the highest level of theory and the following systematic neglect of the least significant contributions is well founded. To address this idea we focused on the model comprising electrostatic potential generated by inhibitors' molecules and atomic point charges of binding site residues. Strikingly, significant correlation was found described by

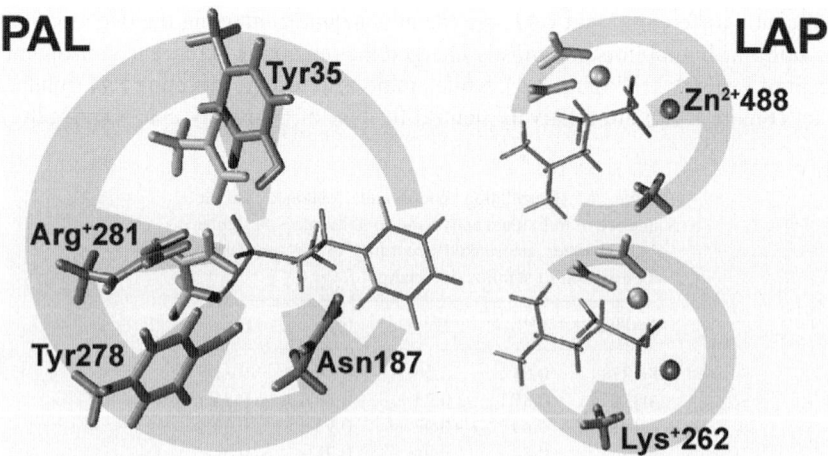

Figure 8-14. PAL and LAP active site constituents essential for the inhibitors binding. The minimal size binding site models are composed of the labelled residues

correlation coefficients of 0.96 and 0.95 for Mulliken and potential-derived CHELP[113] charges. By consecutive elimination of components contributing the least to total electrostatic energy, six PAL active site atoms were selected to mimic the electrostatic properties of the entire PAL binding site (see the equation in Figure 8-15).

Correlation coefficients evaluated for an interaction between these six atoms point charges and molecular electrostatic potential of particular inhibitors were only slightly lowered – 0.83 and 0.91 for Mulliken and CHELP charges. Finally, atomic point charges were arbitrarily assigned for the above presented set of six atoms. In particular, charges of +0.4 and −1.0 described hydrogen and nitrogen atoms, respectively (total charge of a system, +1, reflects the summary charge of a full model). Electrostatic energy arising from these six atomic point charges interaction with inhibitors' electrostatic potential (Figure 8-15) was in remarkable agreement with experimental inhibitory activity (R = 0.88). Such an analysis demonstrates a stepwise generation of simplified models, upon which knowledge of the factors neglected plays as an important role as the final results themselves.

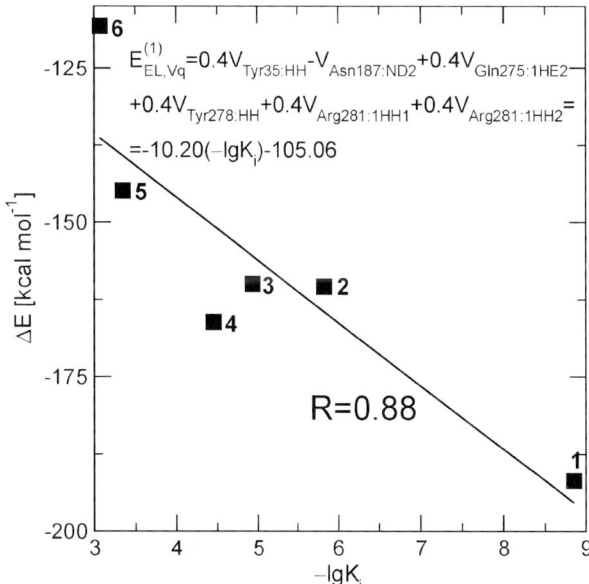

Figure 8-15. Electrostatic binding energy (evaluated according to the equation inserted in the top part of plot) as a function of inhibitory activity.[111,112] V denotes molecular electrostatic potential generated by inhibitors' molecules in the position of any PAL atom given as a subscript; R is the correlation coefficient corresponding to the linear regression curve. Points enumeration corresponds to the inhibitors designation introduced in Figure 8-12

6.   CONCLUSIONS

Application of modern computer-aided design techniques enhances significantly the process of discovery of novel, highly active enzyme inhibitors. The analysis of known inhibitor-enzyme complexes allowed to find important intermolecular interactions and to define steric and electronic features of a potent inhibitor. Subsequent ligand design with the use of LUDI program resulted in novel inhibitors of leucine aminopeptidase and glutamine synthetase. Their enhanced potency was achieved by incorporation in the chosen lead compounds structure of the additional groups interacting favorably with the active cleft. This resulted in the most potent low-molecular weight inhibitors of both enzymes.

Lead generation and optimization constitute effective empirical drug design techniques, which could be further refined by more precise representation of specific molecular charge distribution derived from quantum chemical wavefunctions. Drug-receptor interactions could be systematically analyzed within variation-perturbation partitioning of stabilization energy leading to the entire hierarchy of gradually more approximate theoretical models of inhibitory activity. Correlation of the observed inhibition constants for LAP and PAL leads to the minimal model consisting of one (LAP) or four (PAL) residues. Apparently, in the case of the neutral active site (PAL), interactions with more residues are essential, in contrast to charged metalloenzyme active center (LAP). The simplest theoretical model introduced by Naray-Szabo[114] has been derived here systematically from *ab initio* results. It may be represented by inhibitor molecular electrostatic potentials at the shortest contacts with the most essential active cleft residues, constituting an alternative to arbitrary QSAR (Quantitative Structure-Activity Relationships).

ACKNOWLEDGEMENTS

This work was supported by Wrocław University of Technology and the Polish Committee of Scientific Research (grant no. 2P04B 00729). Use of Accelrys software in WCSS under nationwide licence is also acknowledged as well as use of resources of Wrocław (WCSS), Poznań (PCSS) and Warsaw (ICM) supercomputer centers.

REFERENCES

1. L Pauling (1946) Molecular architecture and biological reactions, Chem Eng News 24, 1375–1377
2. L Pauling (1948) Chemical achievement and hope for the future, Am Sci 36: 51–58
3. WL Jorgensen (2004) The many roles of computation in drug discovery, Science 303 (5665), 1813–1818
4. Ł Berlicki, P Kafarski (2005) Computer-Aided Analysis and Design of Phosphonic and Phosphinic Enzyme Inhibitors as Potential Drugs and Agrochemicals, Curr Org Chem 9 (18):1829–1850
5. F Ooms (2000) Molecular Modeling and Computer Aided Drug Design Examples of their Applications in Medicinal Chemistry, Curr Med Chem 7:141–158
6. I Muegge (2003) Selection criteria for drug-like compounds, Med Res Rev 23 (3):302–321

7. R Abagyan, M Totrov (2001) High-throughput docking for lead generation, Curr Opin Chem Biol 5 (4):375–382

8. KE Goodwill, MG Tennant, RC Stevens (2001) High-throughput x-ray crystallography for structure-based drug design, Drug Discovery Today 6 (2):113–118

9. KH Gardner, LE Kay (1998) The use of H-2, C-13, N-15 multidimensional NMR to study the structure and dynamics of proteins, Ann Rev Biophys Biomol Struct 27:357–406

10. V Kanelis, JD Forman-Kay, LE Kay (2001) Multidimensional NMR methods for protein structure determination, IUBMB Life 52 (6):291–302

11. A Hillisch, LF Pineda, R Hilgenfeld (2004) Utility of homology models in the drug discovery process, Drug Discovery Today 9 (15):659–669

12. RD Taylor, PJ Jewsbury, JW Essex (2002) A review of protein-small molecule docking methods, J Comput Aided Mol Des 16 (3):151–166

13. TJA Ewing, S Makino, AG Skillman, ID Kuntz (2001) DOCK 40: Search strategies for automated molecular docking of flexible molecule databases, J Comput Aided Mol Des 15 (5):411–428

14. G Jones, P Willett, RC Glen (1995) Molecular recognition of receptor sites using a genetic algorithm with a description of desolvation, J Mol Biol 245 (1):43–53

15. ML Verdonk, JC Cole, MJ Hartshorn, CW Murray, RD Taylor (2003) Improved Protein-Ligand Docking Using GOLD, Proteins 52(4):609–623

16. M Rarey, B Kramer, T Lengauer, G Klebe (1996) A Fast Flexible Docking Method using an Incremental Construction Algorithm, J Mol Biol 261(3):470–489

17. P Kirkpatrick (2004) Virtual screening: Gliding to success, Nature Reviews Drug Discovery 3:299–299

18. T A Halgren, R B Murphy, R A Friesner, H S Beard, L L Frye, W T Pollard, J L Banks (2004) Glide: A New Approach for Rapid, Accurate Docking and Scoring 2 Enrichment Factors in Database Screening, J Med Chem 47(7):1750–1759

19. R A Friesner, J L Banks, R B Murphy, T A Halgren, J J Klicic, D T Mainz, M P Repasky, E H Knoll, M Shelley, J K Perry, D E Shaw, P Francis, P S Shenkin (2004) Glide: A New Approach for Rapid, Accurate Docking and Scoring 1 Method and Assessment of Docking Accuracy, J Med Chem 47(7):1739–1749

20. H-J Böhm (1993) A novel computational tool for automated structure-based drug design, J Mol Rec 6(3):131–137

21. G M Morris, DS Goodsell, RS Halliday, R Huey, WE Hart, RK Belew, AJ Olson (1998) Automated Docking Using a Lamarckian Genetic Algorithm and Empirical Binding Free Energy Function, J Comput Chem 19:1639–1662

22. H Gohlke, G Klebe, Approaches to the Description and Prediction of the Binding Affinity of Small-Molecule Ligands to Macromolecular Receptors, Angew Chem Int Ed 41(15):2644–2676

23. HJ Böhm (1994) The development of a simple empirical scoring function to estimate the binding constant for a protein-ligand complex of known three-dimensional structure, J Comput Aided Mol Design 8(3):243–256

24. MD Eldridge, CW Murray, TR Auton, GV Paolini, RP Mee (1997) Empirical scoring functions: I The development of a fast empirical scoring function to estimate the binding affinity of ligands in receptor complexes, J Comput Aided Mol Des 11(5):425–445

25. G Jones, P Willett, RC Glen, AR Leach, R Taylor (1997) Development and Validation of a Genetic Algorithm for Flexible Docking, J Mol Biol 267(3):727–748

26. R Wang, L Lai, S Wang (2002) Further development and validation of empirical scoring functions for structure-based binding affinity prediction, J Comput Aided Mol Des 16(1):11–26

27. TJA Ewing, ID Kuntz (1997) Critical evaluation of search algorithms for automated molecular docking and database screening, J Comput Chem 18(9):1175–1189

28. H Gohlke, M Hendlich, G Klebe (2000) Knowledge-based scoring function to predict protein-ligand interactions, J Mol Biol 295(2):337–356

29. I Muegge, Y C Martin (1999) A General and Fast Scoring Function for Protein-Ligand Interactions: A Simplified Potential Approach, J Med Chem 42(5):791–804

30. P S Charifson, J J Corkery, M A Murcko, W P Walters (1999) Consensus Scoring: A Method for Obtaining Improved Hit Rates from Docking Databases of Three-Dimensional Structures into Proteins, J Med Chem 42(25):5100–5109

31. R Wang, Y Lu, S Wang (2003) Comparative Evaluation of 11 Scoring Functions for Molecular Docking, J Med Chem 46(12):2287–2303

32. M Jacobsson, P Liden, E Stjernschantz, H Bostrom, U Norinder (2003) Improving Structure-Based Virtual Screening by Multivariate Analysis of Scoring Data, J Med Chem 46(26):5781–5789

33. AC Anderson (2003) The Process of Structure-Based Drug Design, Chem Biol 10(9):787–797

34. RS Bohacek, C McMartin (1997) Modern computational chemistry and drug discovery: structure generating programs, Curr Opin Chem Biol 1(2):157–161

35. Y Nishibata, A Itai (1993) Confirmation of usefulness of a structure construction program based on three-dimensional receptor structure for rational lead generation, J Med Chem 36(20):2921–2928

36. RS Bohacek, C McMartin (1994) Multiple Highly Diverse Structures Complementary to Enzyme Binding Sites: Results of Extensive Application of a de Novo Design Method Incorporating Combinatorial Growth, J Am Chem Soc 116(13):5560–5571

37. SH Rotstein, MA Murcko (1993) GenStar: a method for de novo drug design, J Comput Aided Mol Des 7(1):23–43

38. DK Gehlhaar, KE Moerder, D Zichi, CJ Sherman, RC Ogden, ST Freer (1995) De novo design of enzyme inhibitors by Monte Carlo ligand generation, J Med Chem 38(3):466–472

39. V Gillet, AP Johnson, P Mata, S Sike, P Williams (1993) SPROUT: a program for structure generation, J Comput Aided Mol Des 7(2):127–153

40. DA Pearlman, MA Murcko (1993) CONCEPTS: New dynamic algorithm for de novo drug suggestion, J Comput Chem 14(10):1184–1193

41. JB Moon, WJ Howe (1991) Computer design of bioactive molecules: A method for receptor-based de novo ligand design, Proteins: Struct, Funct, Genet 11(4):314–328

42. V Tschinke, NC Cohen (1993) The NEWLEAD program: a new method for the design of candidate structures from pharmacophoric hypotheses, J Med Chem 36(24):3863–3870

43. SH Rotstein, MA Murcko (1993) GroupBuild: a fragment-based method for de novo drug design, J Med Chem 36(12):1700–1710

44. DC Roe, ID Kuntz (1993) BUILDER v2: improving the chemistry of a de novo design strategy, J Comput Aided Mol Des 9(3):269–282

45. MB Eisen, DC Wiley, M Karplus, RE Hubbard (1994) HOOK: A program for finding novel molecular architectures that satisfy the chemical and steric requirements of a macromolecule binding site, Proteins: Struct, Funct, Genet 19(3):199–221

46. DA Pearlman, MA Murcko (1996) CONCERTS: Dynamic Connection of Fragments as an Approach to de Novo Ligand Design, J Med Chem 39(8):1651–1663

47. R S DeWitte, E I Shakhnovich (1996) SMoG: de Novo Design Method Based on Simple, Fast, and Accurate Free Energy Estimates 1 Methodology and Supporting Evidence, J Am Chem Soc 118(47):11733–11744

48. R S DeWitte, A V Ishchenko, E I Shakhnovich (1997) SMoG: de Novo Design Method Based on Simple, Fast, and Accurate Free Energy Estimates 2 Case Studies in Molecular Design, J Am Chem Soc 119(20):4608–4617

49. H-J Böhm (1992a) The computer program LUDI: a new method for the de novo design of enzyme inhibitors, J Comput Aided Mol Design 6(1):61–78

50. H-J Böhm (1992b) LUDI: rule-based automatic design of new substituents for enzyme inhibitor leads, J Comput Aided Mol Design 6(6):593–606

51. HJ Böhm (1996) Towards the automatic design of synthetically accessible protein ligands: peptides, amides and peptidomimetics, J Comput Aided Mol Design 10(4):265–272

52. S F Boys, F Bernardi (1970) Calculation of small molecular interactions by differences of separate total energies – some procedures with reduced errors, Mol Phys 19(4): 553

53. WA Sokalski, S Roszak, K Pecul (1988) An efficient procedure for decomposition of the SCF interaction energy into components with reduced basis set dependence: Chem Phys Lett 153(2,3): 153–159

54. WA Sokalski, P Kedzierski, J Grembecka (2001) Ab initio study of the physical nature of interactions between enzyme active site fragments in vacuo, Phys Chem Chem Phys 3(5): 657–663

55. B Szefczyk, A J Mulholland, K E Ranaghan, W A Sokalski (2004) Differential transition-state stabilization in enzyme catalysis: quantum chemical analysis of interactions in the chorismate mutase reaction and prediction of the optimal catalytic field, J Am Chem Soc 126 (49): 16148–16159

56. J Grembecka, P Kedzierski, WA Sokalski (1999) Non-empirical analysis of the nature of the inhibitor-active-site interactions in leucine aminopeptidase, Chem Phys Lett 313(1): 385–392

57. J Grembecka, WA Sokalski, P Kafarski (2001) Quantum chemical analysis of the interactions of transition state analogs with leucine aminopeptidase, Int J Quantum Chem 84(2), 302–310

58. E Dyguda, J Grembecka, WA Sokalski, J Leszczynski (2005) Origins of the activity of PAL and LAP enzyme inhibitors: Toward ab initio binding affinity prediction, J Am Chem Soc 127(6):1658–1659

59. A Taylor (1993) Aminopeptidases: towards a mechanism of action, Trends Biochem Sci 18:167–171

60. EL Smith, RL Hill, (1960) in: *The enzymes, Academic Press, New York*, pp 37–62

61. N Strater, WN Lipscomb (1995) Two-metal ion mechanism of bovine lens leucine aminopeptidase: active site solvent structure and binding mode of L-leucinal, a gem-diolate transition state analogue, by X-ray crystallography, Biochemistry 34:14792–14800

62. H Kim, WN Lipscomb (1994) Structure and mechanism of bovine lens leucine aminopeptidase, Adv Enzymol Relat Areas Mol Biol 68:153–213

63. N Strater, L Sun, ER Kantrowitz, WN Lipscomb (1999) A bicarbonate ion as a general base in the mechanism of peptide hydrolysis by dizinc leucine aminopeptidase, Proc Natl Acad Sci USA 96:11151–11155

64. A Taylor (1993) Aminopeptidases: structure and function, FASEB J 7:290–298

65. H Umezawa (1980) Screening of small molecular microbial products modulating immune responses and bestatin, Recent Results Cancer Res 75:115–125

66. SK Gupta, M Aziz, AA Khan (1989) Serum leucine aminopeptidase estimation: a sensitive prognostic indicator of invasiveness in breast carcinoma, Indian J Pathol Microbiol 32:301–305

67. A Taylor, M Daims, J Lee, T Surgenor (1982) Identification and quantification of leucine aminopeptidase in aged normal and cataractous human lenses and ability of bovine lens LAP to cleave bovine crystallins, Curr Eye Res 2:47–56

68. A Taylor, MJ Brown, MA Daims, J Cohen (1983) Localization of leucine aminopeptidase in normal hog lens by immunofluorescence and activity assays, Invest Ophthalmol Vis Sci 24: 1172–1180

69. A Taylor, T Surgenor, DK Thomson, RJ Graham, H Oettgen (1984) Comparison of leucine aminopeptidase from human lens, beef lens and kidney, and hog lens and kidney, Exp Eye Res 38:217–229

70. CS Scott, M Davey, A Hamilton, DR Norfolk (1986) Serum enzyme concentrations in untreated acute myeloid leukaemia, Blut 52:297–303

71. J Beninga, KL Rock, AL Goldberg (1998) Interferon-gamma can stimulate post-proteasomal trimming of the N terminus of an antigenic peptide by inducing leucine aminopeptidase, J Biol Chem 273:18734–18742

72. G Pulido-Cejudo, B Conway, P Proulx, R Brown, CA Izaguirre (1997) Bestatin-mediated inhibition of leucine aminopeptidase may hinder HIV infection, Antiviral Res 36:167–177

73. J Grembecka, WA Sokalski, P Kafarski (2000) Computer-aided design and activity prediction of leucine aminopeptidase inhibitors, J Comput Aid Mol Des 14(6) 531–544

74. J Grembecka, A Mucha, T Cierpicki, P Kafarski (2003) The most potent organophosphorus inhibitors of leucine aminopeptidase Structure-based design, chemistry, and activity, J Med Chem 46(13):2641–2655

75. H-J Böhm (1998) Prediction of binding constants of protein ligands: A fast method for the prioritization of hits obtained from de novo design or 3D database search programs, J Comput Aided Mol Design 12(4):309–323

76. M Drag, J Grembecka, M Pawelczak, P Kafarski (2005) alpha-aminoalkylphosphonates as a tool in experimental optimisation of P1 side chain shape of potential inhibitors in S1 pocket of leucine - and neutral aminopeptidases, Eur J Med Chem 40(8):764–771

77. D Eisenberg, HS Gill, GM Pfluegl, SH Rotstein (2000) Structure-function relationships of glutamine synthetases, Biochim Biophys Acta 1477(1):122–145

78. ER Stadtman (2001) The Story of Glutamine Synthetase Regulation, J Biol Chem 276(48): 44357–44364

79. GM Kishore, DM Shah (1988) Amino acid biosynthesis inhibitors as herbicides, Ann Rev Biochem 57:627–663

80. E Bayer, K H Gugel, K Hägele, H Hagenmaier, S Jessipow, W A Köonig, H Zähner (1972) Metabolic products of microorganisms 98 Phosphinothricin and phosphinothricyl-alanyl-analine, Helv Chim Acta 55:224–239

81. K Tachibana (1987) Herbicidal characteristics of bialphos, in Pesticide Science and Biotechnology, R Greenhalgh T.R. Roberts (Eds) pp 145–148

82. A Wild, H Sauer, W Rühle (1987) The effect of phosphinothricin (glufosinate) on photosynthesis I: Inhibition of photosynthesis and accumulation of ammonia, Z Naturforsch 42:263–269

83. K Tachibana, T Watanabe, Y Sekizawa, T Takematsu (1986) Accumulation of ammonia in plants treated with bialphos, J Pest Sci 11:33–37

84. PJ Lea, KW Joy, JL Ramos, MG Guerrero (1984) The action of 2-amino-4-(methylphosphinyl)-butanoic acid (phosphinothricin) and its 2-oxo-derivative on the metabolism of cyanobacteria and higher plants, Phytochemistry 23:1–6

85. A Wild, R Manderscheid (1984) The effect of phosphinothricin in the assimilation of ammonia in plants, Z Naturforsch 39:500–504

86. H Sauer, A Wild, W Rühle (1987) The effect of phosphinothricin (glufosinate) on photosynthesis II: The causes of inhibition of photosynthesis, Z Naturforsch 42:270–278

87. G Harth, MA Horwitz (1999) An inhibitor of exported Mycobacterium tuberculosis glutamine synthetase selectively blocks the growth of pathogenic mycobacteria in axenic culture and in human monocytes: extracellular proteins as potential novel drug targets, J Exp Med 189(9):1425–1436

88. G Harth, MA Horwitz (2003) Inhibition of Mycobacterium tuberculosis glutamine synthetase as a novel antibiotic strategy against tuberculosis: demonstration of efficacy in vivo, Infect Immun 71(1):456–464

89. RJ Almassy, CA Janson, R Hamlin, NH Xuong, D Eisenberg (1986) Novel subunit-subunit interactions in the structure of glutamine synthetase, Nature 323(6086):304–309

90. MM Yamashita, RJ Almassy, CA Janson, D Cascio, D Eisenberg (1989) Refined atomic model of glutamine synthetase at 35 A resolution, J Biol Chem 264:17681–17690

91. S H Liaw, D Eisenberg (1994) Structural model for the reaction mechanism of glutamine synthetase, based on five crystal structures of enzyme-substrate complexes, Biochemistry 33(3):675–681

92. RA Ronzio, A Meister (1968) Phosphorylation of Methionine Sulfoximine by Glutamine Synthetase, Proc Natl Acad Sci USA 59(1):164–170

93. J A Colanduoni, J J Villafranca (1986) Inhibition of E coli glutamine synthetase by phosphino-thricin, Bioorg Chem 14:163–169

94. H S Gill, D Eisenberg (2001) The Crystal Structure of Phosphinothricin in the Active Site of Glutamine Synthetase Illuminates the Mechanism of Enzymatic Inhibition, Biochemistry 40(7):1903–1912

95. Ł Berlicki, P Kafarski (2006), Computer-aided analysis of the interactions of glutamine synthetase with its inhibitors, Bioorg Med Chem 14(13):4578–4585

96. Ł Berlicki, A Obojska, G Forlani, P Kafarski (2005) Design, Synthesis, and Activity of Analogues of Phosphinothricin as Inhibitors of Glutamine Synthetase, J Med Chem 48(20):6340–6349

97. K RHanson, E A Havir (1981) Phenylalanine ammonia-lyase In The Biochemistry of Plants, Vol 7: Secondary Plant Metabolites Conn, E E edt; Academic Press, New York; pp 577–625

98. H Griesbach, H (1981) Lignins In The Biochemistry of Plants, Vol 7: Secondary Plant Metabolites Conn, E E edt; Academic Press, New York; pp 457–478

99. C N Sarkissian, Z Shao, F Blain, R Peevers, H S Su, R Heft, T M S Chang, T S Scriver (1999) A different approach to treatment of phenylketonuria: Phenylalanine degradation with recombinant phenylalanine ammonia lyase, Proc Natl Acad Sci USA 96(5):2339–2344

100. B Schuster, J Retey (1994) Serine-202 is the putative precursor of the active site dehydroalanine of phenylalanine ammonia lyase FEBS Lett 349(2):252–254

101. B Langer, D Röther, J Retey (1997) Identification of essential amino acids in phenylalanine ammonia-lyase by site-directed mutagenesis Biochemistry 36(36):10867–10871

102. M Baedeker, G E Schulz (2002) Structures of two histidine ammonia-lyase modifications and implications for the catalytic mechanism Eur J Biochem 269(6):1790–1797

103. T F Schwede, J Retey, G E Schulz (1999) Crystal structure of histidine ammonia-lyase revealing a novel polypeptide modification as the catalytic electrophile Biochemistry 38(17):5355–5361

104. D Röther, L Poppe, G Morlock, S Viergutz, J Retey (2002) An active site homology model of phenylalanine ammonia-lyase from *Petroselinum crispum*, Eur J Biochem 269(12):3065–3075

105. A Skolaut, J Retey (2001) 1,4-Dihydrophenylalanine – its synthesis and behavior in the pheny-lalanine ammonia-lyase reaction Archiv Biochem Biophys 393(2):187–191

106. J Zoń, N Amrhein, R Gancarz (2002) Inhibitors of phenylalanine ammonia-lyase: 1-aminobenzylphosphonic acid substituted in the benzene ring Phytochemistry 59(1):9–21

107. C Appert, J Zoń, N Amrhein (2003) Kinetic analysis of the inhibition of phenylalanine ammonia-lyase by 2-aminoindan-2-phosphonic acid and other phenylalanine analogues Phytochemistry 62(3):415–422

108. J Zoń, P Miziak, N Amrhein, R Gancarz (2005) Inhibitors of Phenylanine Ammonia-Lyase (PAL): Synthesis and Biological Evaluation of 5-Substituted 2-Aminoindane-2-phosphonic Acids Chemistry and Biodiversity 2(9):1187–1194

109. H Ritter, G E Schulz (2004) Structural Basis for the Entrance into the Phenylpropanoid Metabolism Catalyzed by Phenylalanine Ammonia-Lyase The Plant Cell 16(12):3426-3436

110. J C Calabrese, D B Jordan, A Boodhoo, S Sariaslani, T Vannelli (2004) Crystal Structure of Phenylalanine Ammonia-Lyase: Multiple Helix Dipoles Implicated in Catalysis Biochemistry 43(36):11403–11416

111. B Langer, M Langer, J Retey (2001) Methylidene-imidazolone (MIO) from histidine and pheny-lalanine ammonia-lyase, Adv Protein Chem 58:175–214

112. L Maier, P J Diel (1994) Synthesis, physical and biological properties of the phosphorus analogs of phenylalanine and related compounds, Phosphorus Sulfur 90(1–4):259–279
113. L E Chirlian, M M Francl (1987) Atomic charges derived from electrostatic potentials – a detailed study, J Comp Chem 8(6):894–905
114. G Naray-Szabo (1984) Quantum chemical calculation of the enzyme ligand interaction energy for trypsin inhibition by benzamidines, J Am Chem Soc 106(16):4584–4589

CHAPTER 9

THEORETICAL STUDIES OF THE TRANSITION STATES ALONG THE REACTION COORDINATES OF [NIFE] HYDROGENASE

HIROSHI NAKANO, PAWEŁ SZAREK, KENTARO DOI, AKITOMO TACHIBANA

Department of Micro Engineering, Kyoto University, Kyoto 606–8501, Japan

Abstract: [NiFe] hydrogenase has recently received attention as an enzyme for catalyzing hydrogen production. We review the theoretical investigations of the catalysis mechanism. The hydrogen production reaction occurs at the active site of the hydrogenase and the active site has several paramagnetic and several EPR-silent states, the structures of which are still controversial. Moreover, different catalysis mechanisms have been proposed. We review the proposed mechanisms focusing on the reaction paths

Keywords: [NiFe] hydrogenase, hydrogen, fuel cell, *Desulfovibrio gigas*, *Desulfovibrio vulgaris* Miyazaki F, density functional theory

1.   INTRODUCTION

With the growing need to develop alternative energy resources, hydrogen has attracted attention as a candidate fuel and the manufacturing and application technologies of low-cost hydrogen have been developed for the commercialization of fuel cells. Hydrogenases are considered to be a useful material in this application.[1–9] Hydrogenases are enzymes that catalyze under anaerobic conditions by the reaction

$$2H^+ + 2e^- \underset{\longleftarrow}{\overset{\longrightarrow}{}} H_2. \tag{9-1}$$

The hydrogenases, [NiFe] hydrogenase, [NiFeSe] hydrogenase and [Fe] hydrogenase, have been studied. [NiFe] hydrogenase has Ni and Fe atoms at its active site, which play a central role in the catalytic reaction. It is found in anaerobic

W. A. Sokalski (ed.), Molecular Materials with Specific Interactions, 399–432.
© 2007 *Springer.*

bacteria, such as *Desulfovibrio gigas* (Dg) and *Desulfovibrio vulgaris* Miyazaki F (DvMF). The structure of [NiFe] hydrogenase from Dg has been investigated by several groups[10–14] and that from DvMF by Higuchi *et al.*[15–17] Groups[18–20] have investigated [NiFe] hydrogenase from other bacteria by X-ray crystallography and EPR spectroscopy. [NiFeSe] hydrogenase (with a selencysteine in the place of a cysteine residue at the active site) from *Desulfomicrobium baculatum*[21] and [Fe] hydrogenase (with a [Fe_4S_4] cluster and a [2Fe] cluster at the active site, very similar to the active site of [NiFe] hydrogenase) from *Clostridium pasteurianum*[22] and *Desulfovibrio desulfuricans*[23] have also been crystallized and investigated by X-ray crystallography.

We have focused our study on [NiFe] hydrogenase from Dg and DvMF. Figure 9-1 shows the entire structure of [NiFe] hydrogenase from Dg. The active site is shown in the magnified part of Figure 9-1 and its structure is shown in Figure 9-2. These figures show the three diatomic ligands L1, L2, and L3 coordinated to the Fe atom and a bridging atom X between the Ni and Fe atoms. The ligand X in Dg is assigned to μ-O and in DvMF to μ-S.[11–17] The diatomic ligands L1, L2 and L3 are also different between Dg and DvMF; two of the ligands have been identified as CN and one as CO for Dg,[11–14,24–26] while it has been proposed that L1 can be identified as SO, CO, or CN and L2 and L3 as CN or CO for DvMF.[15–17] The bacteria *Allochromatium vinosum* has a similar active center and same ligand pattern as Dg.[27] The SO ligand in DvMF is considered an important factor in characterizing the properties of DvMF as it causes a peculiar function of the enzyme.

Studies of synthetic active sites of [NiFe] hydrogenase and [Fe] hydrogenase have been conducted[28–35] and breakthroughs are expected for the mass production of hydrogen molecules. For this purpose, a more basic study of real hydrogenases

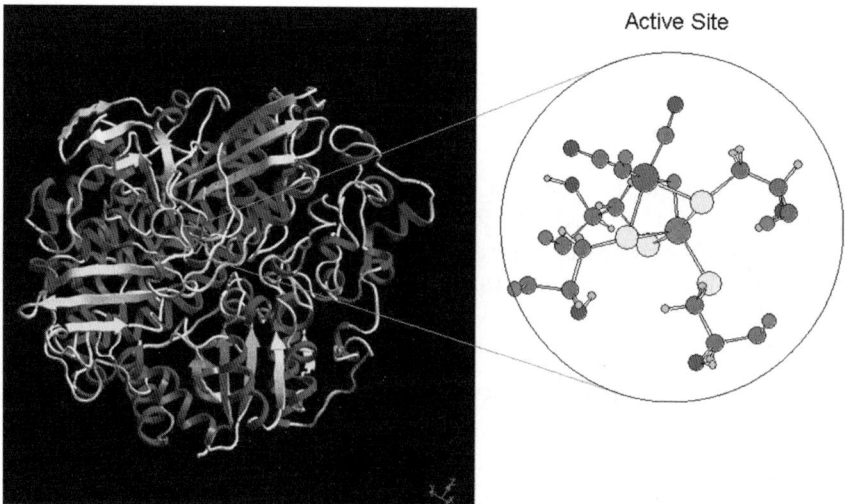

Figure 9-1. Structure of the [NiFe] hydrogenase of *Desulfovibrio gigas*, showing an enlargement of its active site

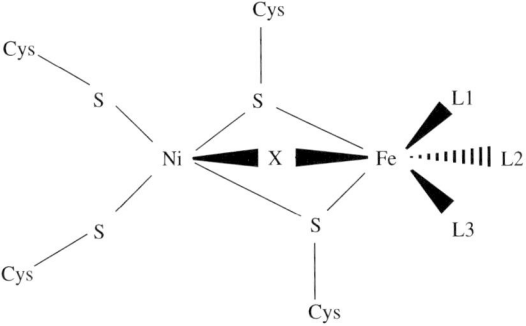

Figure 9-2. Computational model of the active site

in bacteria is required. In addition, theoretical studies have been conducted investigating the catalytic mechanism inside these large proteins by density functional study (DFT), semi empirical study, quantum mechanics / molecular mechanics (QM/MM) study and other techniques. Unfortunately, only a small part of the entire protein can be computed by DFT since current computers are not powerful enough. Hence, a model system to be calculated by DFT must be identified. Typically the active site of [NiFe] hydrogenase is used as the model system.

For [NiFe] hydrogenase, many mechanisms of the catalytic reaction have been suggested.[36,37] One of these schemes is shown in Figure 9-3.[37,38] The oxidized system (Ni-A and Ni-B) is reduced under an atmosphere of H_2 and switches to a reduced system. This scheme has four paramagnetic states: Ni-A, Ni-B, Ni-C and Ni-L and three EPR-silent states: Ni-SU, Ni-SI and Ni-R. Volbeda *et al.* have found that Ni-SI has the two different states, denoted by Ni-SI$_I$ and Ni-SI$_{II}$.[10,11] The Ni-A

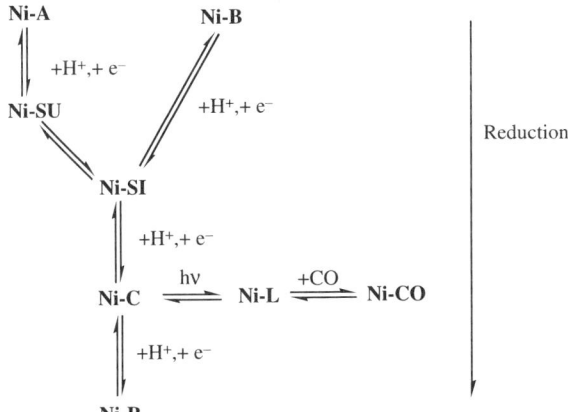

Figure 9-3. Reduction scheme showing the relation of each state[37,38]

and Ni-B states have been obtained under aerobic conditions. The oxidized system is considered to be catalytically inactive and the reduced system is considered to be active. Ni-A can be activated slowly while Ni-B can be activated rapidly and these forms are called the unready and ready forms, respectively.[39,40] [NiFe] hydrogenase in the reduced system catalyzes the H_2 production process effectively. Ni-C has a central role in the catalyzation and is changed into Ni-L under illumination at low temperature ($< 100\,K$). Competing with H_2 production, CO inhibits the activation process to give a Ni-CO state.[41-45] Researchers are divided in their views of the structures of the EPR-silent states, such as Ni-SU, Ni-SI and Ni-R.[36,37] Volbeda *et al.* propose that there are two different states in NI-SI[10-12] and that these EPR-silent states are one part of the catalytic cycle. Lately, DvMF in the oxidized system has been activated by H_2S elimination under an atmosphere of H_2.[15,16] We have found that the reverse cycle of the activation process can be a H_2 production process itself. That is, [NiFe] hydrogenase has two H_2 production processes: one is the normal process in the reduced system as mentioned above and the other is the reverse reaction of the initial stage of H_2S elimination in the oxidized system. This novel mechanism will be discussed in detail in Section 2.2.

Several groups have investigated the catalytic system of [NiFe] hydrogenase and their findings are not yet consistent. Some light has been shed on this complicated puzzle by excellent review articles focusing on experimental investigations,[34,46-50] synthetic studies of hydrogenase[34,46] and theoretical studies.[37,51,52] In this paper we review the published studies, comparing the characteristics of the proposed catalytic mechanisms, focusing especially on the transition states. Identification of the transition state and the activation energy barrier is critical to determining which reaction coordinate plays a crucial role. However, none of the reviews have focused on transition states in relation to the catalytic reaction coordinates.

2.   THEORETICAL INVESTIGATIONS OF [NIFE] HYDROGENASE

2.1.   Active Site of Dg

Dg has been investigated by several groups and the structure of the active site determined by experimental investigations is consistent among these groups. In this section we review the mechanisms proposed by these groups. In Section 2.1.1 we summarize the computational details of each of the studies. In Section 2.1.2 we review the optimized structure of each state, the paramagnetic states, Ni-A, Ni-B, Ni-C, and Ni-L, as shown in Figure 9-4, and the EPR-silent states, Ni-SU, Ni-SI and Ni-R, as shown in Figure 9-5. In the Section 2.1.3 we show each group's reaction mechanism.

2.1.1.   *Computational details*

We detail here the functional and basis set of each groups' density functional calculations.

(i) Pavlov *et al.*[53-55] performed B3LYP[56,57] energy calculations with the large 6-311 + G-(2d, 2p) basis set. They used the LANL2DZ[58-60] set in the B3LYP[56,57] geometry optimizations in Gaussian94.[61]

(ii) Hall *et al.*[62-64] optimized the geometries with the B3LYP[56,57] functional and the double-ζ basis set. They used the modified version of the Hay and Wadt effective core potentials (ECPs)[65] in Gaussian98.[61]

(iii) Gioia *et al.*[66,67] optimized the geometries with the BLYP[56,68] functional and the double-ζ basis set (D95)[69] on the first-row atoms and Los Alamos ECPs[58-60] on the S, Fe and Ni atoms in Gaussian94.[61]

(iv) We optimized the geometries with the B3LYP[56,57] functional. We used the LanL2DZ basis set[56-60] with the Huzinaga polarization function[70] for Fe, Ni and S and with the Dunning[69] function for the other atoms. In the QM/MM study,[71-75] we used the same functional and basis set for the QM region and used the UFF method for the MM region. We used the ONIOM method[76-79] in Gaussian03.[61]

(v) Stein *et al.* in ref. 80 used the BLYP[56,68] functional and the DZVP basis set in DGauss4.0.[81]

(vi) Stein *et al.* in ref. 82 optimized the geometries with the BP86[68,83-85] functional and the non-relativistic *Slater*-type DZP basis set and used a *g*- and hyperfine-tensor study with the relativistic *Slater*-type DZP basis set and TZP basis set. In refs. 38, 52, 82, and 86 Stein *et al.* used the Amsterdam Density Functional package (ADF).[87,88]

(vii) Stein *et al.* in refs. 38 and 86 optimized the geometries with the BP86[68,83-85] functional and the double-ζ *Slater*-type basis set and examined a *g*- and hyperfine-tensor study by the zero-order regular approximation (ZORA)[89-92] implemented in ADF[87,88] with the same basis set as in ref. 80.

(viii) Amara *et al.*[93] performed QM/MM[71-75] calculations. They used the B3LYP[56,57] functional with the double-ζ basis set for the QM region. They used the Hay and Wadt ECPs for the non-metal atoms[58-60] and the Dunning ECPs for the Ni and Fe atoms.[69] For the MM region, they used the potential energy due to covalent interactions accounting for the bonds, bond angles, proper and improper dihedral angles, and the potential energy due to the nonbonding interactions (Coulomb and Lennard-Jones). This calculation was performed using the quantum mechanical CADPAC program.[94]

2.1.2. The structure of each state

Several groups have proposed structures of each state from theoretical analysis based on experimental studies. There is good agreement for the structures of the paramagnetic states Ni-A, Ni-B, Ni-C and Ni-L, while the proposed structures of the EPR-silent states do not agree. In contrast to the other groups, Pavlov *et al.* and Hall *et al.* have proposed that H_2 attacks the Fe atom in the first step of the mechanism. Hence we show their paramagnetic states in Figure 9-4(a) and their EPR-silent states in Figure 9-5(a). The paramagnetic and EPR-silent states of the other groups are shown in Figure 9-4(b) and Figure 9-5(b), respectively.

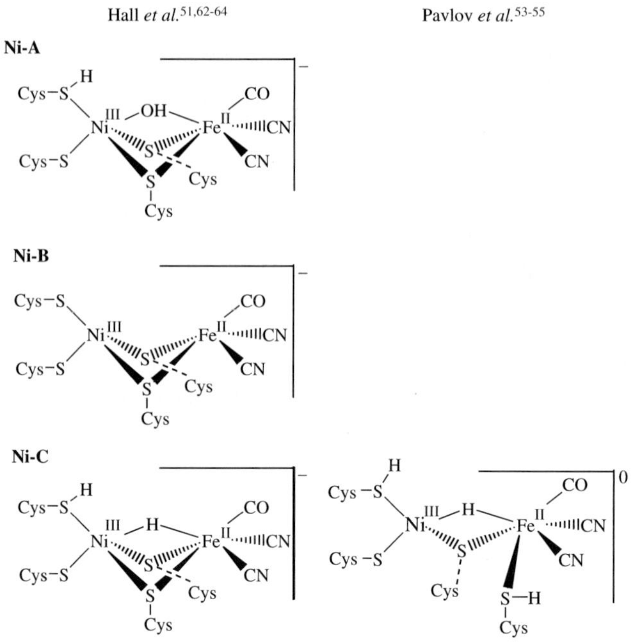

Figure 9-4(a). Structure of the paramagnetic states, such as Ni-A, Ni-B and Ni-C of Pavlov *et al.* and Hall *et al.* from the theoretical investigations

Ni-A is more stable than Ni-B, and Ni-A is reduced slowly into the activated state Ni-SU. Ni-A has been considered to have O^{2-} as the bridging ligand X. Hall *et al.*[51,62–64] compared the optimized structure of each candidate for the structure of Ni-A and determined that X is OH^-, as shown in Figure 9-4(a). Ni-B is reduced faster than Ni-A and is activated into Ni-SI. With theoretical and experimental data for Ni-B, Gioia *et al.* proposed that there is no bridging ligand X in the structure of Ni-B.[66,67] However, Stein *et al.* concluded that the ligand X for Ni-B is OH^- by comparing the experimental structure by X-ray crystallography with the optimized structure of DFT.[52,80,82] They also calculated the structural and spectroscopic data of the Ni-B state by the relativistic DFT study with the ZORA and compared them with the experimental data of the hyperfine- and g-tensors,[38,86] and thus confirmed that X is OH^-. Other groups have also investigated Ni-A and Ni-B with ZORA.[95,96]

The paramagnetic Ni-C state is a part of the catalytic cycle and has the redox state Ni(III)Fe(II). Hall *et al.* proposed that the ligand X in Ni-C is an H atom and one of the cysteine residues is protonated, based on a comparison between their DFT study and experimental data.[51,62–64] Other groups have come to the same conclusion.[51,52,55] Recently, however, Stein *et al.* have stated a slightly different opinion. Based on g- and hyperfine-tensor experimental data and their relativistic DFT calculations, Stein *et al.* conclude that Ni-C has no protonated cysteine residue.[38]

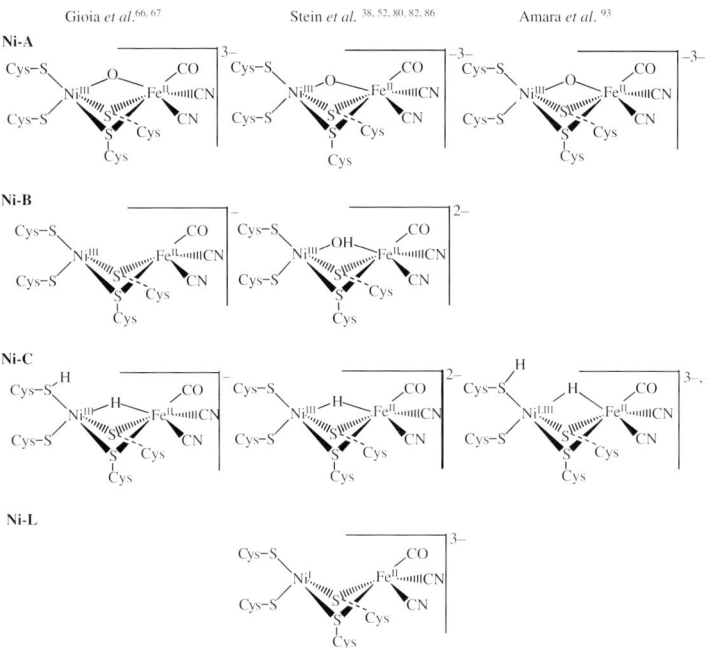

Figure 9-4(b). Structure of the paramagnetic states, such as Ni-A, Ni-B, Ni-C, and Ni-L of Gioia *et al.*, Stein *et al.* and Amara *et al.* from the theoretical investigations

In addition, Pavlov and Siegbahn *et al.* propose that one of the bridging cysteine ligands is liberated from the Ni atom, accompanying a H atom.[53–55] Ni-L has not been examined much because it is not important to the catalytic cycle. Stein *et al.* consider that Ni-L has a vacancy at X and has no protonated ligand.[38,52,80,82,86]

We next review the EPR-silent states Ni-SU, Ni-SI, and Ni-R. The proposed structures of these states by each group are shown in Figure 9-5. The determination of the structures of EPR-silent complexes is harder than for those of the paramagnetic complexes and hence the structures are still controversial. Ni-SU is produced by the reduction of Ni-A. Ni-SI is produced by the reduction of Ni-B or Ni-SU. As mentioned above, Ni-B is reduced and transformed into Ni-SI faster than Ni-A into Ni-SI. Ni-SU and Ni-SI have the same redox state Ni(II)Fe(II), resulting in their EPR-silent properties. Stein has investigated the structure of Ni-SI and proposed that the Fe atom of Ni-SI is coordinated by H_2O.[38,52,80,82,86] Gioia *et al.* and Hall *et al.* propose that there is no atom at the position of the ligand X in Ni-SI.[51,62–64,66,67] Hall *et al.* suggest structures for two Ni-SI states, based on the experimental data by Volbeda *et al.*,[10,11] called Ni-SIa and Ni-SIb. They state that Ni-SIa is transformed into NI-SIb by protonation at the S atom of a cysteine residue.[51] Amara *et al.* suggest a rather unique structure by QM/MM study[93] and called the two Ni-SI states Ni-SI1 and Ni-SI2, again based on Volbeda *et al.*[10,11] They suggest that NI-SI1 and Ni-SI2 have a H atom as the bridge ligand X. The structure of NI-SI2 with a H atom added is in surprisingly good agreement with the experimental structure.[93]

Ni-R is also an EPR-silent state and has the redox state Ni(II)Fe(II). Hall *et al.* have shown that Ni-R has H_2 bound to the Fe atom by DFT calculations and considering the electronic states.[51,62–64] Gioia *et al.* and Stein *et al.* propose similar structures.[66,67] Both groups state that the bridging ligand X is a H atom and Ni-R

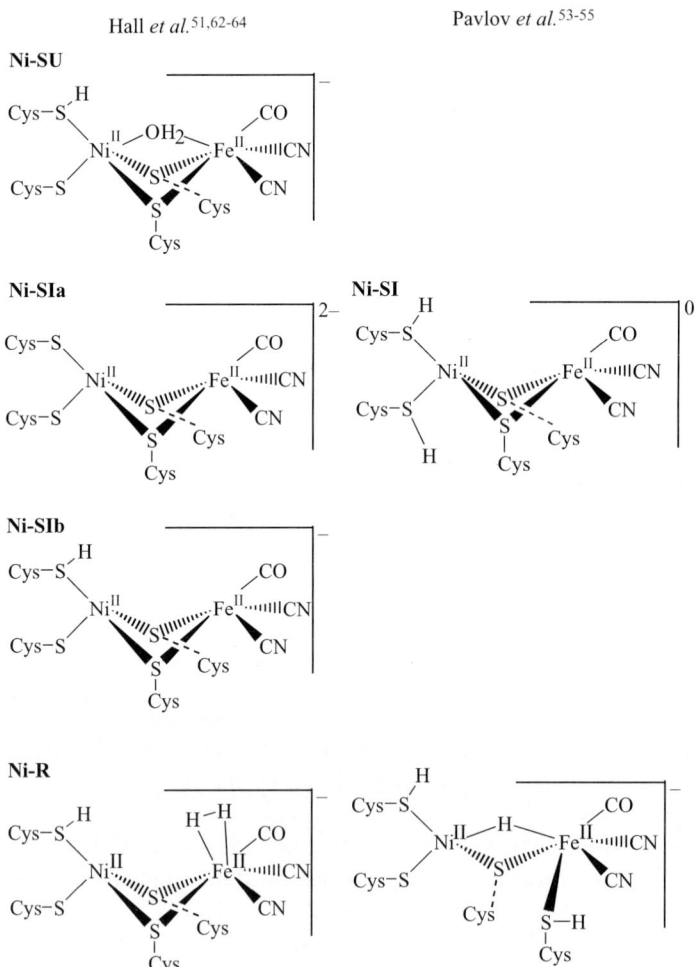

Figure 9-5(a). Structure of the EPR-silent states, such as Ni-SU, Ni-SI, and Ni-R of Pavlov *et al.* and Hall *et al.* from theoretical investigations. Note that Hall *et al.* suggest the existence of two Ni-SI states, Ni-SIa and Ni-SIb

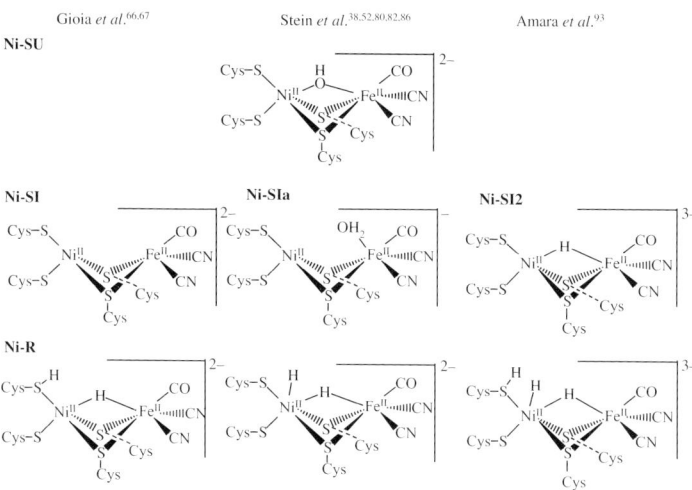

Figure 9-5(b). Structure of the EPR-silent states, such as Ni-SU, Ni-SI, and Ni-R of Gioia *et al.*, Stein *et al.* and Amara *et al.* by theoretical investigations

has another coordinated H atom. Gioia *et al.* propose that the H atom coordinates to one of the cysteine residues bound to the Ni atom and Stein proposes that it coordinates to the Ni atom directly.

We have to admit subtlety for the local spin state of Ni. Hall *et al.* suggest that Ni-SIa and Ni-R are high-spin Ni(II) states based on L-edge XAS data[98] and on this assumption they optimized the structures, giving a distorted tetrahedral coordination of the Ni atom. They found that high-spin structures were in better agreement with the experimental structure than the low-spin structures, confirming their prediction that Ni-SIa and Ni-R are high-spin complexes. However, other groups consider Ni-SIa and Ni-R to be low-spin Ni(II) complexes.[37,98] The low-spin states have square planar coordination. Further theoretical studies are clearly required on this point. Stein *et al.* and Gioia *et al.* have recently investigated the Ni-S$_4$ complexes as a model of the [NiFe] hydrogenase active site together to lead the conclusion that the Ni-SIa and Ni-R is spin-crossover state and the density functional BP86 is the most suited functional to describe the structural features and Ni-SI and Ni-R are spin crossover states.[99] They suggested that high-spin states with B3LYP have too stable energy, compared with the calculations with BP86 or B3LYP*.[100–102]

Recently we calculated the energy of Ni-SIa by QM/MM calculations. We considered the active site as a QM region, the atoms within 7.0 Å of the active site as an MM-free region and the other atoms within 13.0 Å of the active site as an MM-fixed region, as shown in Figure 9-6. Atoms in the MM-free region can move during the optimization process, while atoms in the MM-fixed region cannot move. The total energy of the low-spin state was 1.5 kcal/mol less than that of the high-spin state. The QM region of the low-spin state was 10.2 kcal/lmol less stable

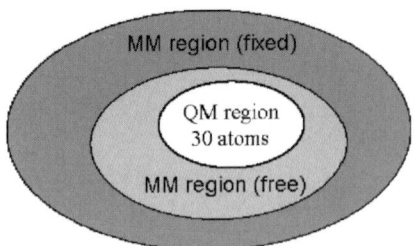

Figure 9-6. Scheme of the QM, MM-free, and MM-fixed regions

than the high-spin state, while the energy of the MM region was 11.7 kcal/mol more stable than the high-spin state. Based on this calculation, we conclude that Ni-SI is a low-spin complex. This difference of 1.5 kcal/mol is small and hence we did not attempt to make a more accurate calculation. A similar calculation for Ni-R is currently underway.

2.1.3.   *Catalytic mechanism of hydrogen production*

Several groups have proposed catalytic cycles,[38,42,53,54,62,66,67,80] each characterized by the atom that coordinates to the Ni atom, Fe atom or cysteine residues and each cycling through the paramagnetic state Ni-C and the EPR-silent states Ni-SIa and Ni-R. The first of these proposed cycles, by Pavlov *et al.*, simulated the catalytic mechanism with a relatively small model of the active site. Figure 9-7(a)[53] shows the mechanism of Pavlov *et al.* The transition states are also calculated and the activation energy is found to be 7.9 cal/mol, as shown in Figure 9-8(a). [53] The mechanism is unique compared with the results of the other groups in that the bond between the bridging cysteine residue and Ni atom is cut and the bimetallic CN ligand of the Fe atom moves toward the vacant position between the Fe and Ni atoms. Pavlov *et al.* later slightly modified the mechanism, as shown in Figure 9-7(b).[54] Again the transition states are calculated and the activation energy is found to be 3.1 kcal/mol, as shown in Figure 9-8(b).[54] Including the solvent effect, the energy is 6.4 kcal/mol. As in the unmodified mechanism,[53] the bond between the bridging cysteine residue and Ni atom is cut. However, it has been commented that this result is achieved because the model is too small. Amara *et al.* claim that the small size of the model allows the bridging cysteine ligands to move too much;[93] the ligands are bonded to amino acid chains and cannot move freely.

The proposed catalytic mechanism of Hall *et al.* is shown in Figure 9-9.[51,62-64] They calculated the transition states using the same model as Pavlov *et al.* The activation energy is 14.2 kcal/mol when the model of the active site has a neutral charge and is 12.4 kcal/mol when the model has a minus charge, as shown in Figure 9-10.[63]

The proposed mechanism of Dole *et al.* is shown in Figure 9-11[42] and has been confirmed by Gioia *et al.*[66,67] They calculated the optimized structure by DFT and performed frontier orbital analysis. From their analysis, they proposed that H_2 reacts with the Ni atom to give two seperate H atoms. Recently we have simulated this mechanism and have found the transition state in the low-spin state. The energy diagram and the structure of the calculation model are shown in Figure 9-12. The Mulliken atomic charge densities of the Fe, Ni, three S atoms and two H atoms are shown in Table 9-1. The three S atoms and two H atoms are identified by suffixes in Figure 9-11. The Mulliken atomic charge is known to be unreliable for the transition metals and hence we pay careful attention to it. At the first step, H2 is bonded to the Ni atom or S3 atom. Next, H1 is trapped between the Fe and Ni atoms, which is indicated by the Mulliken atomic charge of H1 and H2.

Finally, the charge is absorbed by the Fe atom. The electron density on the S1 atom also increases between the transition state and Ni-R, which is consistent with the findings of Gioia *et al.*,[67] and hence it has an active role in the H_2 cleavage. The activation energy is 29.3 kcal/mol, while the reverse reaction, that is, the hydrogen production, requires only 10.4 kcal/mol.

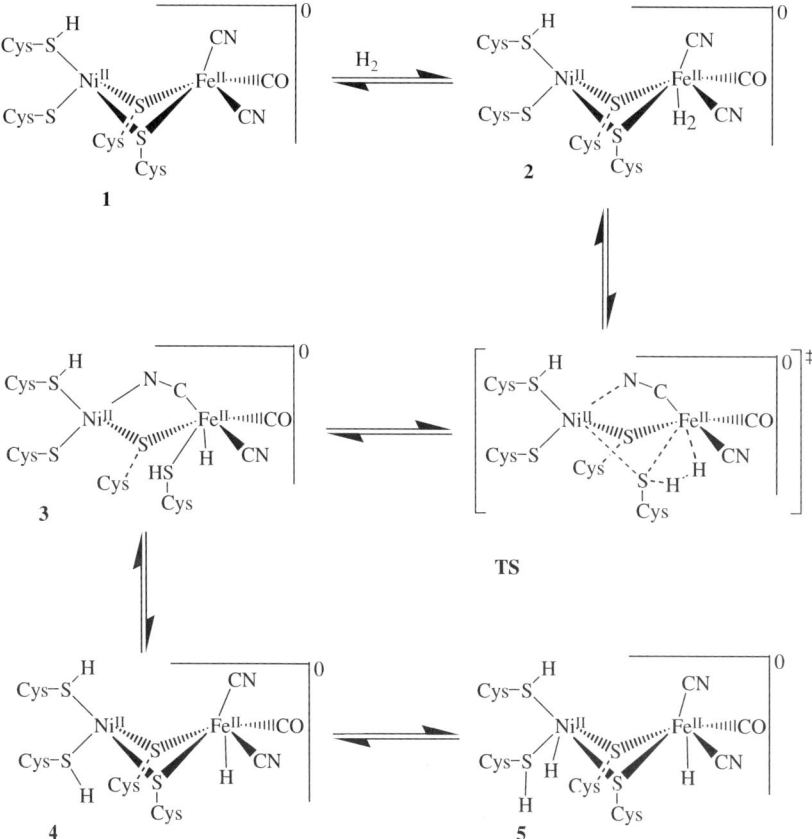

Figure 9-7(a). Catalytic mechanism of Pavlov *et al.* All of the states are calculated as quartet states[53]

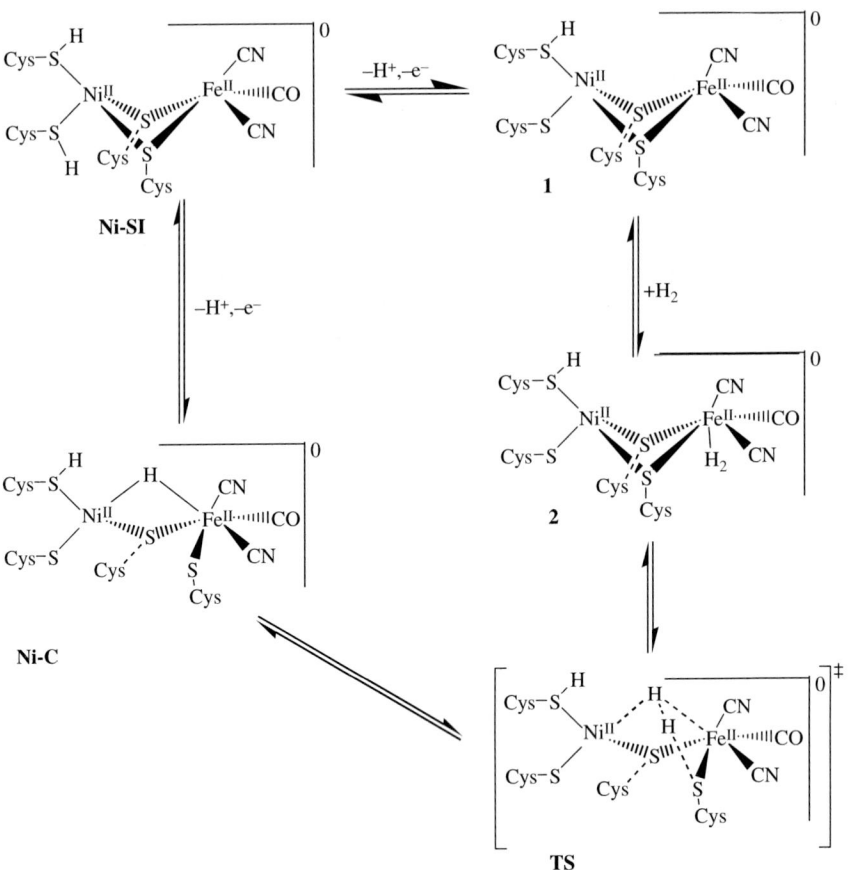

Figure 9-7(b). Modified catalytic mechanism of Pavlov *et al.* All of the states were calculated as quartet states [54,55]

The suggested catalytic mechanism of Stein *et al.* is shown in Figure 9-13. [38,52,86] This mechanism is more complicated than the others. In this reaction, the solvent H_2O has an important role in the heterolytic cleavage of H_2. One of the H atoms of H_2 is attracted to the O atom of H_2O, forming an H_3O^+ ion. This results in the bridging ligand X being a H atom. These considerations are based on relativistic DFT research with ZORA and the experimental g- and hyperfine-tensors. [38,86]

Amara *et al.* suggested a rather unique mechanism, as shown in Figure 9-14. NI-SI2 of this mechanism has a μ-H^-, based on a QM/MM study. The structure of NI-SI2 with μ-H^- is in good agreement with the experimental structure. [93]

To conclude, several controversial mechanisms have been proposed based on theoretical investigations. To characterize a mechanism it is necessary to calculate the reaction path including the transition states. The transition states and activation energies will then specify the catalytic mechanism.

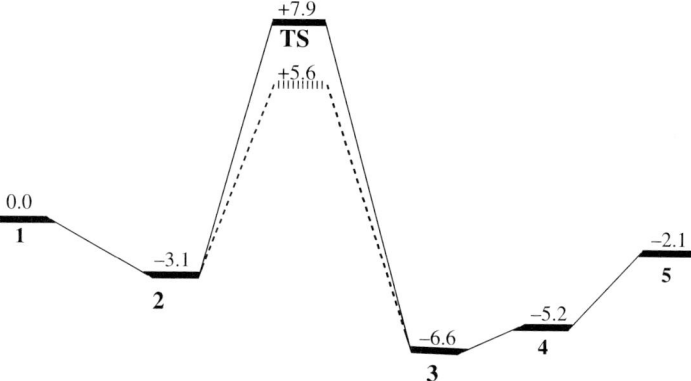

Figure 9-8(a). Energy diagram corresponding to the catalytic mechanism of Pavlov *et al.* shown in Figure 9-7(a).[53] The unit of energy is kcal/mol

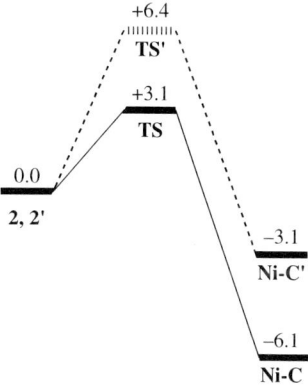

Figure 9-8(b). Energy diagram corresponding to the modified catalytic mechanism of Pavlov *et al.* shown in Figure 9-7(b). Here 2′, TS′ and Ni − C′ include solvent effect.[54] The unit of energy is kcal/mol

2.2. Active Site of DvMF

Unlike Dg, the structure of the active site of DvMF is still controversial. Higuchi *et al.* have investigated the structure by X-ray crystallography[15–17] and Tüker *et al.* have performed semi-empirical calculations.[103,104] Stein *et al.* investigated the structure of [NiFe] hydrogenase from DvMF by DFT calculations and g- and hyperfine-tensor calculations as well as for Dg.[38,80] The active site of DvMF is shown in Figure 9-2. It has been proposed that the L1 ligand is SO, CO or CN and the

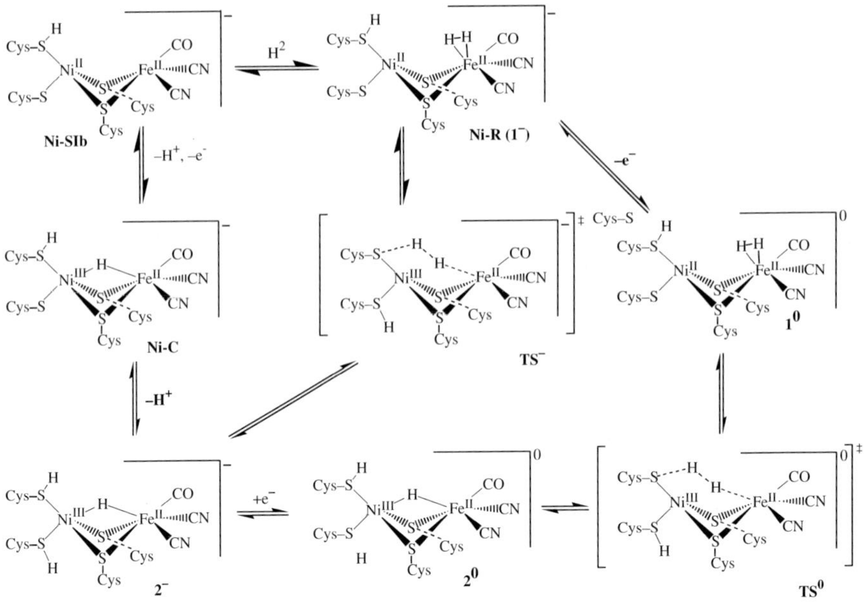

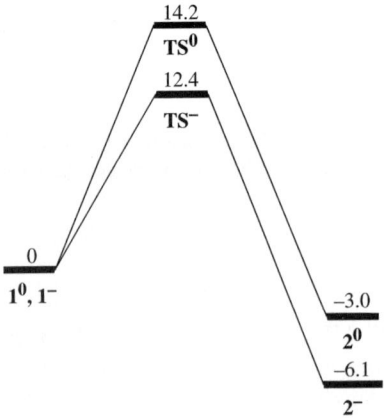

Figure 9-9. Catalytic mechanism of Hall *et al.* The states are calculated as low-spin states. There are two catalytic cycle. One cycles through the neutral states including the transition state, while the other includes only the anion states during the cycle.[51,62–64] The Ni-R state is denoted by **1⁻**, which releases one electron to become **1⁰**. State **2⁰** is an intermediate complex and receives one electron to become **2⁻**

Figure 9-10. Energy diagram of Hall *et al*, shown in Figure 9-9.[63] Each state corresponds to a state in Figure 9-9. The unit of energy is kcal/mol

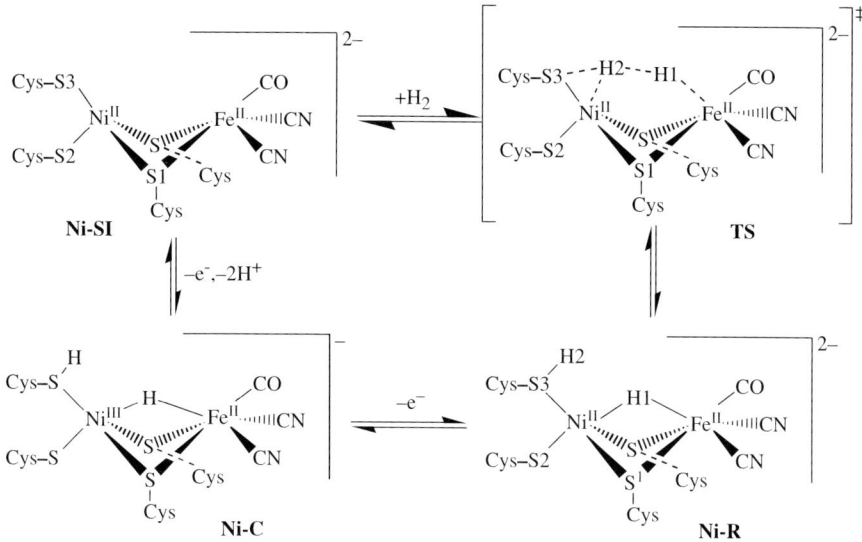

Figure 9-11. Mechanism of Dole *et al.* and Gioia *et al.* Our transition state is inserted into it. All the states are were calculated as low-spin states[42,66,67]

L2 and L3 ligands are CN or CO, but the exact nature of the ligands has not been determined. The widely accepted ligand pattern is SO for L1, CN for L2 and CO for L3. We have made a theoretical investigation of the pattern: CO for L1, CN for L2 and CO for L3. We assumed that there exists a mechanism holding the bridge ligand X during the catalytic reaction. Amara *et al.*[93] suggested a fixed bridge ligand

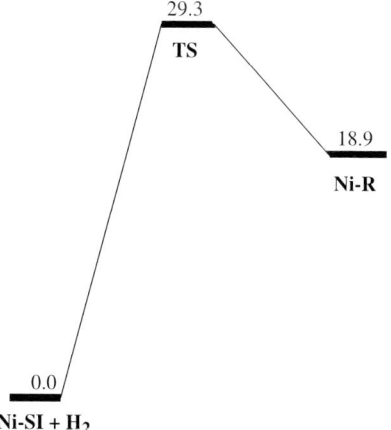

Figure 9-12. Our energy diagram based on the mechanism shown in Figure 9-11. This diagram is only for the low-spin states. The unit of energy is kcal/mol

Table 9-1. Mulliken atomic charge densities in all states in Figure 9-10. The positions of the three S atoms and two H atoms are shown in Figure 9-9, distinguished by the suffixes

	Fe	Ni	S1	S2	S3	H1	H2
$Ni-SI+H_2$	−0.44	−0.01	−0.05	−0.28	−0.28	−0.07	0.09
TS	−0.86	0.02	0.02	−0.31	−0.18	0.00	0.27
Ni-R	−0.89	0.04	−0.19	−0.09	−0.31	0.14	0.19

of a H atom during the reaction. We considered a bridge ligand S atom fixed during the catalytic cycle, as shown in Figure 9-15, because it is experimentally found to be the most probable bridge ligand in DvMF. [15,16] We introduce our preliminary results in this section. Other patterns are being studied, incorporating possible ligands of bridging and nonbridging characters with various redox and spin states under the influence of the environment.

2.2.1.   The optimized structure of the DvMF model

We have devised a DvMF model by substituting a methyl group for cysteine (Cys) in the structural formula shown in Figure 9-15, where S1 is the bridging atom between the Fe atom and Ni atom, S2 is one of the bridging S atomsof the cysteine amino acids and S3 is one of the S atoms of the cysteine amino acids coordinated to the Ni atom only. We then optimized the structure. The optimized atomic distances and

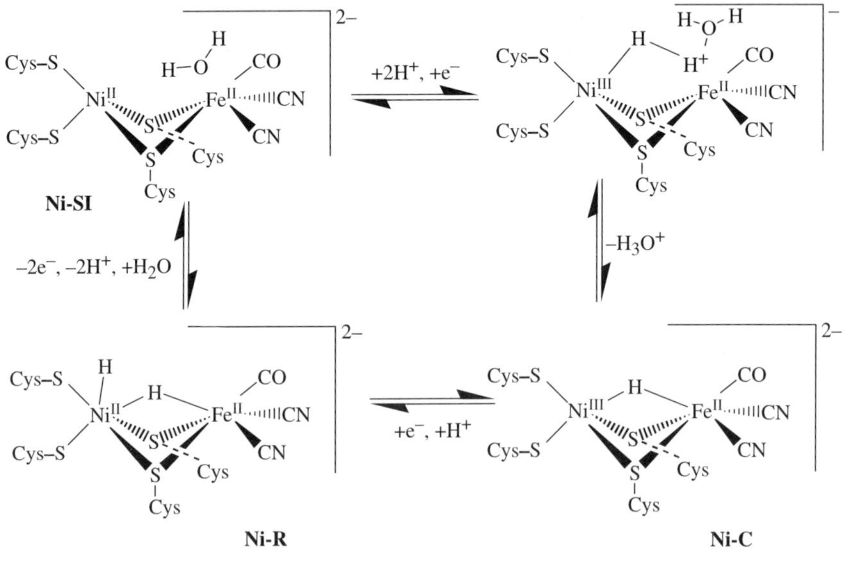

Figure 9-13. Catalytic mechanism of Stein *et al.* characterized by the mediation of H_2O. [38,52,80,82,86] All the complexes are calculated as the low-spin states [80]

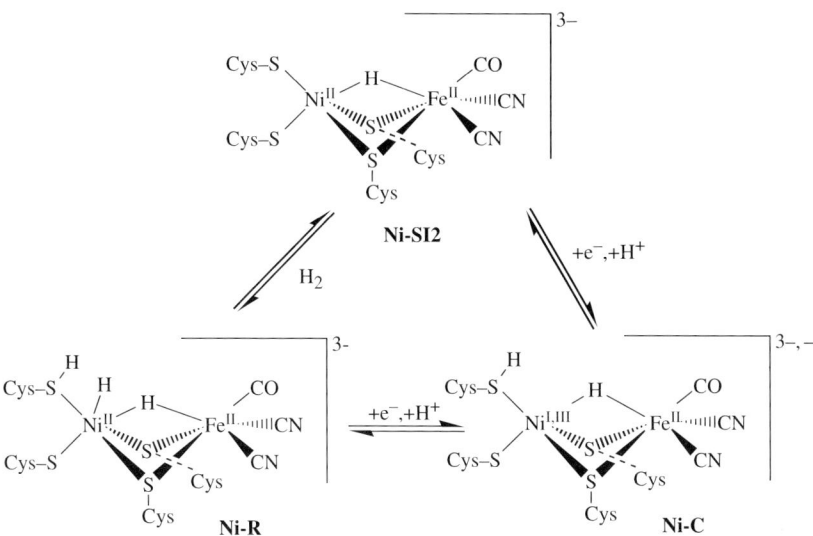

Figure 9-14. Mechanism of Amara *et al.* All the states are calculated as low- and high-spin states[93]

angles of the DvMF model in the low-spin and high-spin oxidized states (**1a** and **1b**, where the suffixes **a** and **b** denote the low-spin state and the high-spin state, respectively) are tabulated in Table 9-2.

Model **1b** is less stable by 19.46 kcal/mol than **1a**, but **1b** closely resembles the structure derived from X-ray crystallography;[15,16] the optimized structural parameters of **1b** agree with the X-ray structural parameters, except for the Ni-Fe distance. On the other hand, the distance between the Ni atom and S2 atom of **1a** is 3.25 Å while the distance measured in the X-ray structure is 2.37 Å. The Ni-Fe distances of both calculated spin states do not agree with the X-ray measured distance; the Ni-Fe distance of the X-ray structure is 2.55 Å while that of **1a** is 3.41 Å and that

Table 9-2. Atomic distances and angles in **1a** and **1b**

distance (Å)	**1a**	**1b**	exp.[15,16]
Ni-Fe	3.409	2.919	2.55
Ni-S1	2.288	2.263	2.16
Ni-S2	3.252	2.432	2.37
Fe-S1	2.418	2.284	2.22
Fe-S2	2.363	2.387	2.37
angle (deg)	**1a**	**1b**	exp.[15,16]
S1-Ni-S2	41.0	88.0	91.1
S1-Fe-S2	53.1	88.6	89.7

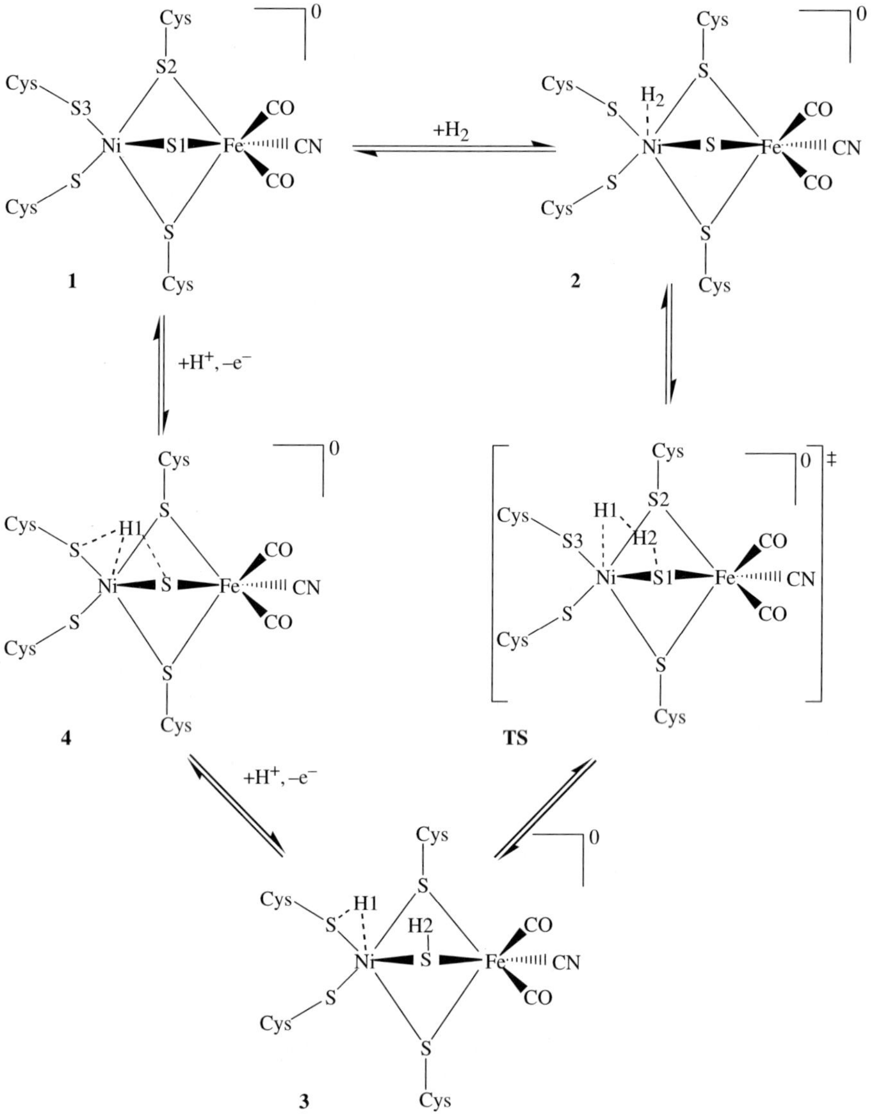

Figure 9-15. Reaction of Eq. (2) (clockwise) and reverse reaction, Eq. (3) (counterclockwise)

of **1b** is 2.92 Å. The differences are too large to be regarded as uncertainties in the calculation. The same differences were reported by Stein *et al.*[80] The differences are due to deviations of the bond angles of S1-Ni-S2 and Si-Fe-S2. In fact, there is not much interaction between the Fe and Ni atoms. Therefore, the differences hardly affect the reaction mechanism.

2.2.2. *Transition states of the activation process in the oxidized system*

We examined the reaction mechanism of the activation process in the oxidized system and performed a more detailed investigated of DvMF[17] (which has a S atom at the bridging ligand of the active site). We defined the following mechanism (shown in Eq. (9-2)) based on the work of Higuchi *et al.*[15–17] As already mentioned, an oxidized system of DvMF can be activated by H_2S elimination under an atmosphere of H_2. The origin of the S atom in H_2S is not yet known, however, Higuchi *et al.* have assumed that it is a bridging S atom.[15,16] In the initial stage of H_2S elimination, DvMF model **1**, which is called the Ni-A state in general, makes a complex with H_2 (complex **2**). Then, the intermediary **3**, with one of the hydrogen atoms abstracted by the S1 atom, is obtained through the transition states **TS** as follows:

$$\mathbf{1a}(d) + H_2 \rightarrow \mathbf{2a}(d) \rightarrow [\mathbf{TSa}](d) \rightarrow \mathbf{3a}(d);$$

$$\mathbf{1b}(q) + H_2 \rightarrow \mathbf{2b}(q) \rightarrow [\mathbf{TSb}](q) \rightarrow \mathbf{3b}(q), \tag{9-2}$$

where d and q in the parentheses denote doublet and quartet states respectively. Each low-spin state is more stable than the corresponding high-spin state. The activation energy in the low-spin states is 34.33 kcal/mol. That of the high-spin states is 21.39 kcal/mol. Therefore, the activation process such as the H_2S elimination is not easy to take place to some extent in the low-spin state, but the reaction become easier to happen after the low-spin state is excited and transferred to the high-spin state.

It is important to also consider the reverse cycle of the reaction shown in Eq. (9-2). This reverse cycle can generate H_2 with a very low activation energy. We describe this reverse cycle in detail, presenting the structures, electron states and energy diagrams of the complexes and transition states in the reaction of Eq. (9-2) in the next section, Section 2.2.3.

2.2.3. H_2 *production in the oxidized system*

Here, we consider the reaction cycle of the H_2 production process in the oxidized system based on the reverse reaction of that given in Eq. (9-2), as shown in Figure 9-15,

$$\mathbf{1a}(d) + 2H \rightarrow \mathbf{4a}(\mathbf{4a'})(s) + H$$

$$\rightarrow \mathbf{3a}(d) \rightarrow [\mathbf{TSa}](d) \rightarrow \mathbf{2a}(d) \rightarrow \mathbf{1a}(d) + H_2;$$

$$\mathbf{1b}(q) + 2H \rightarrow \mathbf{4b}(\mathbf{4b'})(t) + H$$

$$\rightarrow \mathbf{3b}(q) \rightarrow [\mathbf{TSb}](q) \rightarrow \mathbf{2b}(q) \rightarrow \mathbf{1b}(q) + H_2, \tag{9-3}$$

where s and t in the parentheses respectively denote singlet and triplet states. In the reaction of Eq. (9-3), protons and electrons are added to the system from outside the hydrogenase or from ferredoxins along the transport chain of the amino acids. This has been discussed by Pavlov *et al.*[53-55] The optimized species in the low- and high-spin states are shown in Figure 9-16(a) and (b), respectively. The DvMF models **1a** and **1b** can capture H radicals easily. After one H radical is adsorbed on S1(**4a** and **4b**), S3(**4a′**, **4b′**) or Ni(**4b″**) atoms, the complex **2** is formed by capturing another H radical through the complex **3** and **TS** by the counterclockwise cycle in Figure 9-15. Energy diagrams for the reaction of Eq. (9-3) are shown in Figure 9-17. The reaction can generate H_2 with a very low activation energy; the activation energy is 6.74 kcal/mol for the reverse cycle in the high-spin states (**3b → TSb → 2b**) and is 16.94 kcal/mol in the low-spin states (**3a → TSa → 2a**).

For all models, the configuration of the ligands coordinated to the Ni atom is tetrahedral in the high-spin state and square planar in the low-spin state. The Mulliken atomic spin densities and charges of each state are shown in Table 9-3. In the high-spin state, the S1, S3 and Ni atoms have 0.5–0.8 spin densities, while the S1 and Ni atoms in **1a** have hardly any spin. In addition, each of the S1, S3 and Ni atoms in **4b**, **4b′**, and **4b″** have 0.4–0.9 spin densities except the atoms abstracting the first H radical. The second H radical is captured easily on atoms with large spin densit.

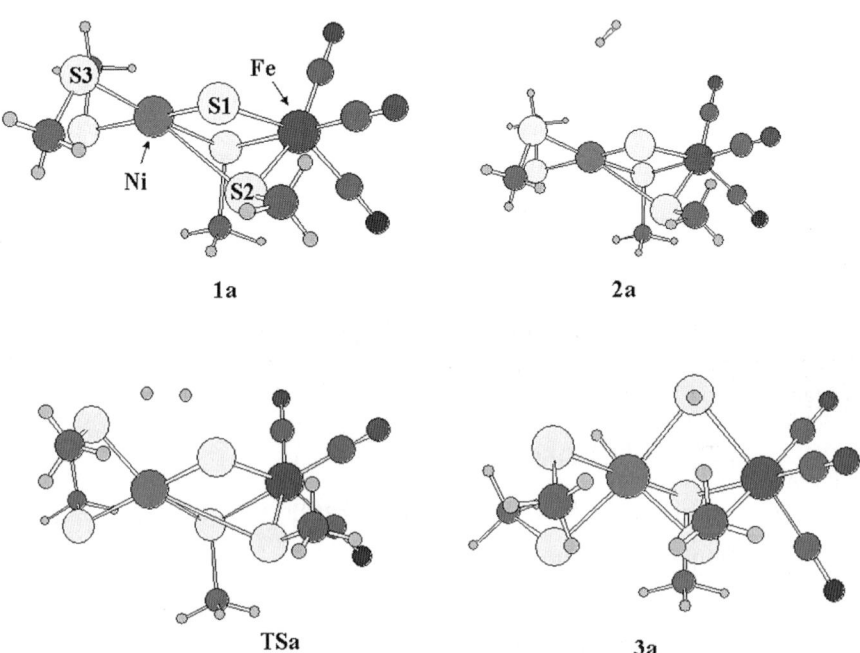

Figure 9-16(a). Optimized species in the low spin state

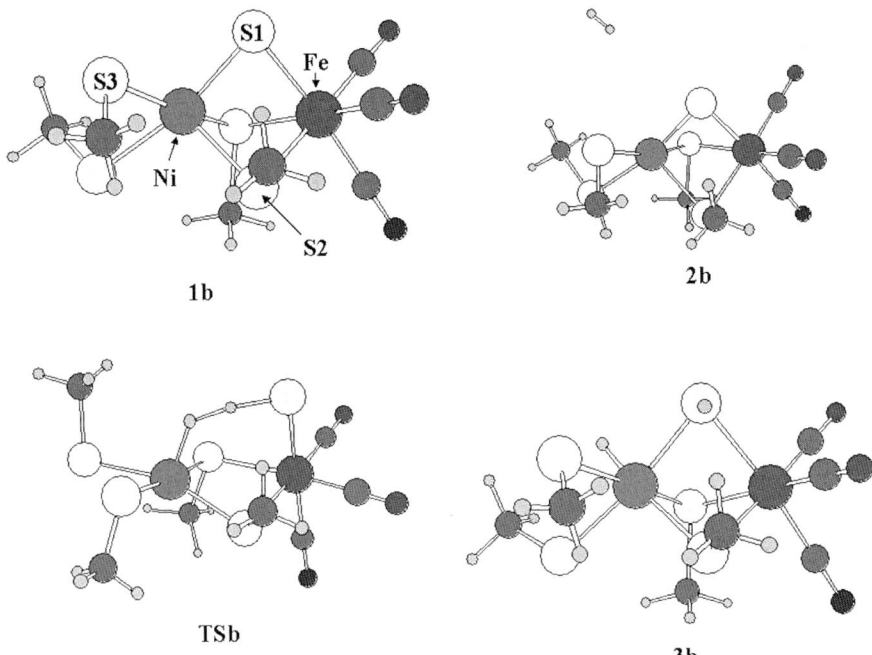

Figure 9-16(b). Optimized species in the high spin state

2.2.4. String model of the catalytic mechanism

In [NiFe] hydrogenase, catalyzation does not necessarily occur in the neutral state of the active site. Therefore we have to consider electron transfer during catalyzation. Hence, we calculated the energy of the anion and dianion complexes in the catalytic reaction, where the structures of each state are assumed to be optimized structures and transition states of the neutral state. This is the starting point in the analysis of the string model.[105–107]

The energy diagram is shown in Figure 9-18. From this diagram, we can see that the reaction occurs easily in the anion and dianion states. In the anion state, especially, the activation energy of the low-spin state is lower than the high-spin state and the complex $3a^-$ is the most stable of the other complexes. In the dianion states the energy of TSb^{2-} is lower than that of $3b^{2-}$, which indicates that the catalytic mechanism proceeds with no potential barrier if $3b$ and $3b^-$ get two or one electron respectively. The charge distributions and spin densities are given in Table 9-4.

Therefore, we suggest two catalytic mechanisms of the reduced system of DvMF:

$$3a^- \rightarrow [TSa^-] \rightarrow 2a^-, \tag{9-4}$$

$$3b^- \xrightarrow{+e^-} 3b^{2-} \longrightarrow [TSb^{2-}] \xrightarrow{-e^-} 2b^-. \tag{9-5}$$

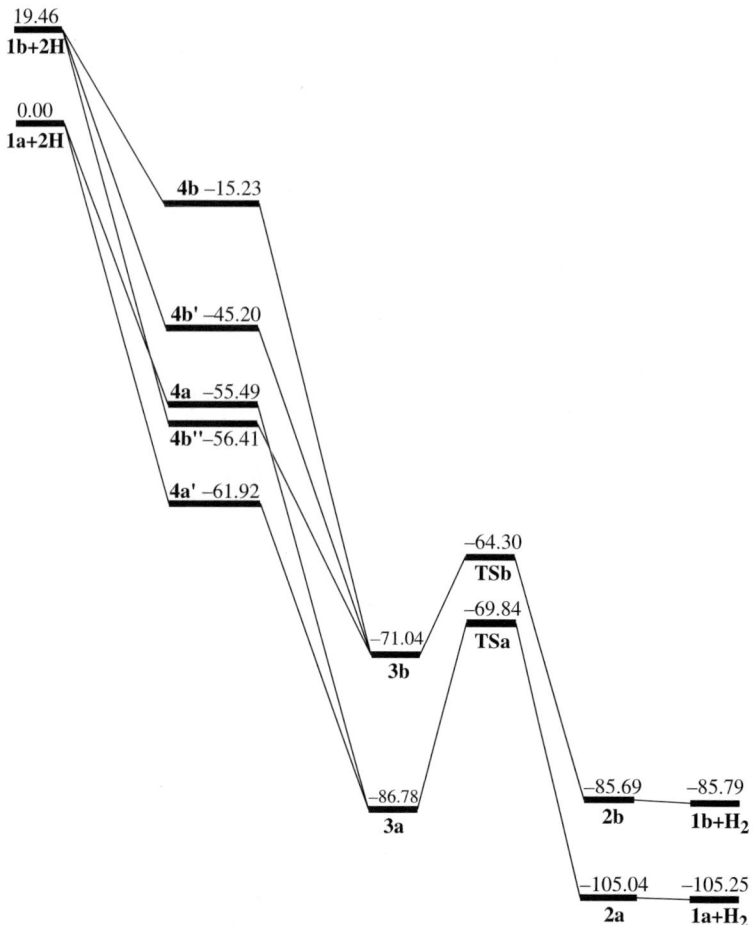

Figure 9-17. Energy diagram of the catalytic cycle. The unit of energy is kcal/mol

These mechanisms need to be optimized for each electric state and the activation energies need to be calculated. The mechanisms proceed more readily than the mechanism of Eq. (9-3) in the neutral state.

2.2.5.   Other patterns of ligands at Fe atom

We have investigated other patterns of ligands, such as L1=CN, L2=CO, L3=CN. This is identical to the active site of Dg with an S atom as the bridge ligand X. Figure 9-19 shows the energy diagram. The low-spin state of this model has a similar catalytic mechanism and energy diagram; the high-spin state is currently under investigation. We are also currently calculating other ligand patterns, such as the most probable pattern, L1=SO, L2=CN, L3=CO.

Table 9-3. Mulliken atomic spin densities and charge densities in all complexes

spin densities	Ni	Fe	S1	S2	S3	H1	H2
1a	−0.04	0.01	0.01	0.00	0.49		
1b	0.67	0.28	0.80	0.35	0.48		
2a	−0.04	0.01	0.01	0.00	0.49	0.00	0.00
2b	0.67	0.29	0.80	0.35	0.48	0.00	0.00
3a	−0.03	0.00	0.04	0.00	0.48	0.00	0.00
3b	0.67	0.82	0.26	0.07	0.53	−0.08	−0.01
4a	0.00	0.00	0.00	0.00	0.00	0.00	
4a′	0.00	0.00	0.00	0.00	0.00	0.00	
4b	0.75	−0.02	0.05	0.34	0.44	0.00	
4b′	0.93	0.28	0.80	0.29	0.04	0.00	
4b″	−0.11	0.04	0.85	0.21	0.49	0.01	
TSa	−0.05	−0.01	0.00	0.22	0.16	0.06	−0.04
TSb	0.61	0.45	0.86	0.06	0.43	−0.03	0.22

charge densities	Ni	Fe	S1	S2	S3	H1	H2
1a	0.09	−0.57	−0.06	0.12	−0.03		
1b	0.16	−0.88	0.00	0.05	−0.02		
2a	0.07	−0.57	−0.06	0.12	−0.03	0.00	0.02
2b	0.14	−0.89	0.00	0.05	−0.02	0.03	−0.01
3a	−0.33	−0.80	−0.10	0.01	0.04	0.23	0.20
3b	−0.06	−0.95	−0.16	0.05	−0.03	0.18	0.22
4a	0.20	−0.79	−0.34	0.01	0.07	0.19	
4a′	0.08	−0.54	−0.11	0.12	−0.08	0.25	
4b	0.15	−0.81	−0.17	0.03	−0.03	0.20	
4b′	0.18	−0.86	−0.06	0.06	−0.10	0.22	
4b″	−0.34	−0.80	0.08	0.04	0.05	0.24	
TSa	0.08	−0.65	−0.13	0.05	−0.03	0.04	0.08
TSb	−0.02	−0.81	−0.08	0.05	0.01	0.11	0.01

2.2.6. *Quantum energy density*

The electronic interaction in the H_2 production process in Section 2.2.3 can be expressed in terms of the quantum energy densities[108–113] based on the regional DFT.[108–114] The electronic kinetic energy density $n_T(\vec{r})$ is defined as

$$n_T(\vec{r}) = \frac{1}{2}\sum_i \nu_i \left[\left\{ -\frac{\hbar^2}{2m}\Delta\psi_i^*(\vec{r}) \right\}\psi_i(\vec{r}) + \psi_i^*(\vec{r})\left\{ -\frac{\hbar^2}{2m}\Delta\psi_i(\vec{r}) \right\} \right]$$

$$(9\text{-}6)$$

where m is the mass of an electron, $\psi_i(\vec{r})$ is the natural orbital and ν_i is the occupation number of $\psi_i(\vec{r})$.[108–111] $n_T(\vec{r})$ is important to the discussion on bond formation because the sign of $n_T(\vec{r})$ has a physical meaning with respect to electronic interaction; in the region $n_T(\vec{r}) > 0$ (electronic drop region, R_D) electrons can move freely in a classical fashion, whereas electrons cannot enter the region

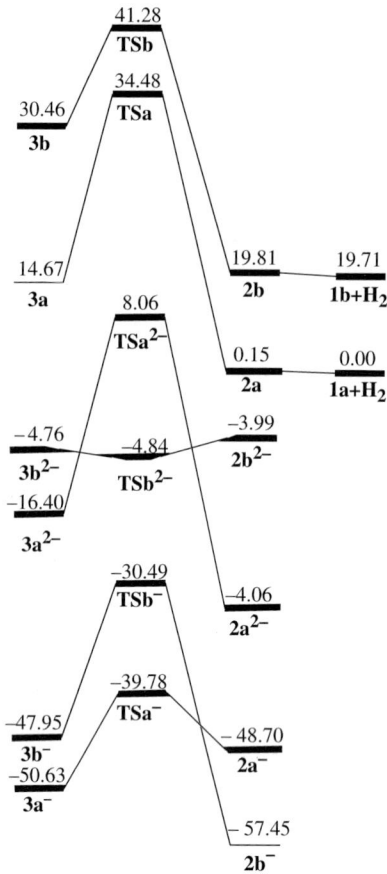

Figure 9-18. String model of the catalytic mechanism in the oxidized system of DvMF. The unit of energy is Kcal/mol. The energies in this diagram do not include the zero-point energy

$n_T(\vec{r}) < 0$ (electronic atmosphere region, R_A) in a classical sense.[108–111] The total electronic force density $\vec{F}^S(\vec{r})$ is given by

$$\vec{F}^S(\vec{r}) = \vec{\tau}^S(\vec{r}) + \vec{X}^S(\vec{r}), \qquad (9\text{-}7)$$

where $\vec{\tau}^S(\vec{r})$ and $\vec{X}^S(\vec{r})$ denote the electronic tension density and electronic external force density, respectively.[108–111] $\vec{\tau}^S(\vec{r})$ has a quantum mechanical origin and is given by $\vec{\tau}^S(\vec{r}) = (\tau^{Sk}(\vec{r}))$ with

$$\tau^{Sk}(\vec{r}) = \frac{\hbar^2}{4m} \sum_i \nu_i \left\{ \psi_i^*(\vec{r}) \frac{\partial \Delta \psi_i(\vec{r})}{\partial x^k} - \frac{\partial \psi_i^*(\vec{r})}{\partial x^k} \Delta \psi_i(\vec{r}) \right.$$

$$\left. + \frac{\partial \Delta \psi_i^*(\vec{r})}{\partial x^k} \psi_i(\vec{r}) - \Delta \psi_i^* \frac{\partial \psi_i(\vec{r})}{\partial x^k} \right\}, \qquad (9\text{-}8)$$

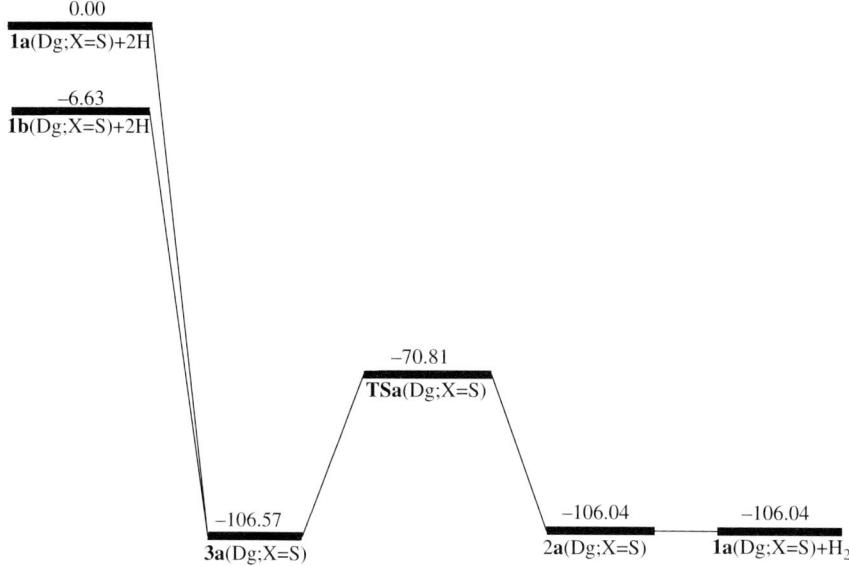

Figure 9-19. Energy diagram of the catalytic cycle for L1=CN, L2=CO, and L3=CN. The mechanism is the same as that in Figure 9-12 and the diagram shows only about low spin states. The unit of energy is kcal/mol

for $k = 1, 2, 3$. In the stationary state, $\vec{\tau}^{S}(\vec{r})$ balances $\vec{E}(\vec{r})$, the electric field acting on an electron.[108–111] The detail of each bond can be expressed in terms of the stress tensor density, which is given by a 3×3 matrix $\overset{\leftrightarrow S}{\tau}(\vec{r}) = (\tau^{Skl}(\vec{r}))$ with

$$
\tau^{Skl}(\vec{r}) = \frac{\hbar^2}{4m} \sum_i \nu_i \left\{ \psi_i^*(\vec{r}) \frac{\partial^2 \psi_i(\vec{r})}{\partial x^k x^l} - \frac{\partial \psi_i^*(\vec{r})}{\partial x^k} \frac{\partial \psi_i(\vec{r})}{\partial x^l} \right.
$$
$$
\left. + \frac{\partial^2 \psi_i^*(\vec{r})}{\partial x^k \partial x^l} \psi_i(\vec{r}) - \frac{\partial \psi_i^*}{\partial x^l} \frac{\partial \psi_i(\vec{r})}{\partial x^k} \right\}, \tag{9-9}
$$

for $k, l = 1, 2, 3$. $n_T(\vec{r})$, $\vec{\tau}^{S}(\vec{r})$, the largest eigenvalues of $\overset{\leftrightarrow S}{\tau}(\vec{r})$ and their eigenvectors in the formation and cleavage of chemical bonds in the TS can be calculated using the MR DFT program,[115] as shown in Figure 9-19.

As shown in Figure 9-20(a), in the low-spin state, the R_D due to the Ni atom is not directly connected to the R_D due to H atoms in **TSa**. This means that electrons cannot transfer classically between the Ni atom and H atoms. The area between S3 and the H atoms is filled with continuous R_D, but the compressive stress, for which the largest eigenvalue of $\overset{\leftrightarrow S}{\tau}(\vec{r})$ is negative,[112,113] is distributed widely in the S3-H area, as shown in Figure 9-20(b), that is, the chemical-bond interaction in the S3-H area has been lost in **TSa**. On the other hand, in the high-spin state, not only

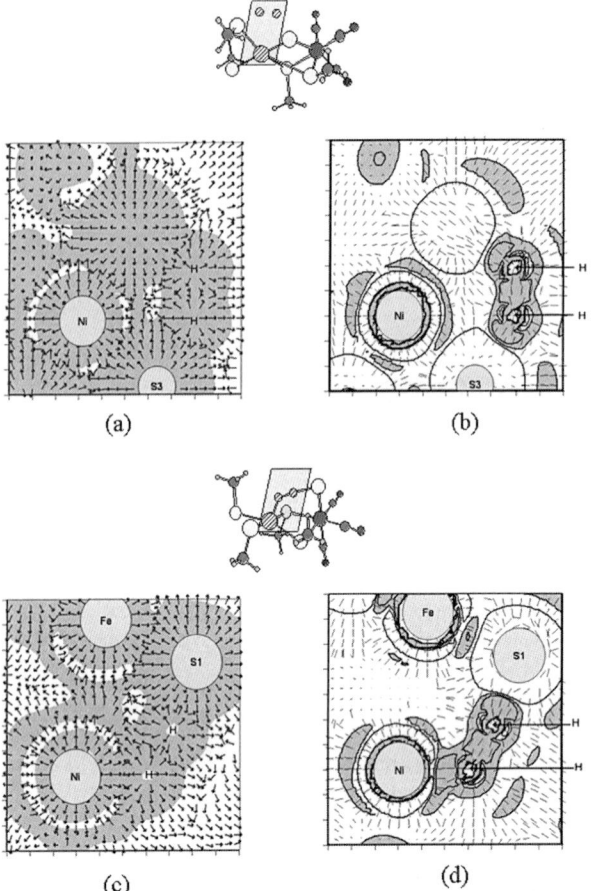

Figure 9-20. Maps of the quantum energy densities:(a) $n_T(\vec{r})$ (background shading), $\vec{\tau}^S(\vec{r})$ (arrows), (b) The largest eigenvalues of $\overset{\leftrightarrow S}{\tau}(\vec{r})$ (background shading) and their eigenvectors (short lines) in **TSa**; The cross section of **TSa** is displayed above (a) and (b), with the centers of the Ni atom and two H atoms are included. (c) $n_T(\vec{r})$ (background shading), $\vec{\tau}^S(\vec{r})$ (arrows), (d) The largest eigenvalues of $\overset{\leftrightarrow S}{\tau}(\vec{r})$ (background shading) and their eigenvectors (short lines) in **TSb**. The cross section of **TSb** is displayed above (c) and (d), with the centers of the Ni atom and two H atoms are included. The gray areas in (a) and (c) denote R_D regions and those in (b) and (d) denote the tensile stresses. The contours in (b) and (d) denote the values of -0.01, 0.0, and 0.01

is classical electron transfer allowed between the Ni and H atoms in **TSb** in terms of the continuous R_D, as shown in Figure 9-20(c), but Figure 9-20(d), also shows that the eigenvectors of the tensile stress, where the largest eigenvalue of $\overset{\leftrightarrow S}{\tau}(\vec{r})$ is positive,[112,113] in the Ni-H area have a "spindle structure."[112,113] Such a spindle structure is observed in typical covalent bonds,[112,113] and therefore it is considered that the character of the Ni-H covalent bond seen in **3b** remains strong even in **TSb**

Table 9-4. Mulliken atomic spin densities and charge densities in anion complexes ($2a^-$, $2b^-$, $3a^-$, $3b^-$, TSa^-, TSb^-) and dianion complexes ($2a^{2-}$, $2b^{2-}$, $3a^{2-}$, $3b^{2-}$, TSa^{2-}, TSb^{2-})

spin densities	Ni	Fe	S1	S2	S3	H1	H2
$2a^-$	0.00	0.00	0.00	0.00	0.00	0.00	0.00
$2b^-$	0.66	−0.02	0.10	0.27	0.51	0.00	0.00
$3a^-$	0.00	0.00	0.00	0.00	0.00	0.00	0.00
$3b^-$	0.62	−0.01	0.11	0.00	0.62	0.00	−0.10
TSa^-	0.00	0.00	0.00	0.00	0.00	0.00	0.00
TSb^-	0.37	0.04	0.28	0.09	0.46	−0.07	0.30
$2a^{2-}$	0.01	0.75	0.12	0.13	−0.01	0.00	0.00
$2b^{2-}$	1.31	0.02	0.08	0.04	0.61	0.21	0.00
$3a^{2-}$	0.60	−0.01	0.06	0.00	0.18	0.00	−0.11
$3b^{2-}$	1.21	0.05	0.10	0.05	0.59	0.00	0.20
TSa^{2-}	0.74	0.02	0.06	0.00	0.06	0.00	0.00
TSb^{2-}	1.22	0.01	0.14	0.13	0.60	−0.02	0.17
charge densities	Ni	Fe	S1	S2	S3	H1	H2
$2a^-$	−0.07	−0.64	−0.06	0.10	−0.18	−0.04	0.06
$2b^-$	0.08	−0.91	−0.21	−0.02	−0.07	0.07	−0.05
$3a^-$	−0.31	−0.80	−0.18	−0.04	−0.12	0.18	0.21
$3b^-$	−0.29	−0.87	−0.20	−0.02	−0.08	0.17	0.21
TSa^-	0.00	−0.71	−0.14	−0.01	−0.06	−0.01	0.07
TSb^-	−0.08	−0.84	−0.34	0.00	−0.04	0.09	0.02
$2a^{2-}$	−0.13	−0.66	−0.17	0.02	−0.22	−0.06	0.00
$2b^{2-}$	−0.19	−1.04	−0.33	−0.08	−0.10	0.01	0.17
$3a^{2-}$	−0.38	−0.87	−0.24	−0.06	−0.23	0.17	0.18
$3b^{2-}$	−0.41	−0.98	−0.22	−0.07	−0.11	0.14	0.06
TSa^{2-}	−0.14	−0.79	−0.19	−0.04	−0.12	−0.04	0.06
TSb^{2-}	−0.08	−0.91	−0.45	−0.07	−0.13	0.12	−0.05

and that the height of the energy barrier from **3b** to **TSb** is greatly suppressed due to the large covalent-bond-like interaction in **TSb**.

3. CONCLUSION

[NiFe] hydrogenase has received much attention as a material for use in hydrogen production. However, the catalytic mechanism is still controversial and the structures of some states have not yet been determined. Therefore further experimental and theoretical investigations are required. It is certain that the active site receives or gives electrons to other parts of the hydrogenase and the environment. There is also no consensus among research groups on a value of the activation energy of hydrogen cleavage. Catalytic mechanisms for Dg have been proposed, shown in Figures 9-7, 9-9, 9-11, 9-13 and 9-14, with activation energies of 3.1–29.3 kcal/mol. [NiFe] hydrogenase of DvMF has catalytic mechanisms in both the oxidized state and the reduced state. Further theoretical investigations are required because this reaction might occur more readily by electron transfer under the well-ordered control of the redox states.

ACKNOWLEDGEMENTS

This work was supported in part by the Center of Excellence (COE) for Research and Education on Complex Functional Mechanical Systems program of the Ministry of Education, Culture, Sports, Science and Technology, Japan. The computations were performed using Fujitsu Primepower HPC2500 at the Academic Center for Computing and Media Studies, Kyoto University.

REFERENCES

1. Cammack R (1999) Hydrogenase sophistication. Nature 397:214–215
2. Albracht SPJ (1994) Nickel hydrogenases: in search of the active site. Biochim. Biophys. Acta 1188:167–204
3. Adams MWW (1990) The structure and mechanism of iron-hydrogenases. Biochim. Biophys. Acta 1020:115–145
4. Graf E-G, Thauer RK (1981) Hydrogenase from methanobacterium thermoautotrophicum, a nickel-containing enzyme. FEBS Lett. 136:165–169
5. Nicolet Y, Lemon BJ, Fontecilla-Camps JC, Peters JW (2000) A novel FeS cluster in Fe-only hydrogenases. Trends Biochem. Sci. 25:138–143
6. Peters JW (1999) Structure and mechanism of iron-only hydrogenases. Curr. Opin. Struct. Biol. 9:670–676
7. Horner DS, Heil B, Happe T, Embley TM (2002) Iron hydrogenases – ancient enzymes in modern eukaryotes. Trends Biochem. Sci. 27:148–153
8. Nicolet Y, Cavazza C, Fontecilla-Camps JC (2002) Fe-only hydrogenases: structure, function and evolution. J. Inorg. Biochem. 91:1–8
9. Lyon EJ, Shima S, Buurman G, Chowdhuri S, Batschauer A, Steinbach K, Thauer RK (2004) UV-A/blue-light inactivation of the "metal-free" hydrogenase (Hmd) from methanogenic archaea. The enzyme contains functional iron after all. Eur. J. Biochem. 271:195–204
10. Volbeda A, Charon M-H, Piras C, Hatchikian EC, Frey M, Fontecilla-Camps J-C (1995) Crystal structure of nickel-iron hydrogenase from *Desulfovibrio gigas*. Nature 373:580–587
11. Volbeda A, Garcin E, Piras C, de Lacey AL, Fernandez VM, Hatchikian EC, Frey M (1996) Fontecilla-Camps JC, Structure of the [NiFe] hydrogenase active site: evidence for biologically uncommon Fe ligands. J. Am. Chem. Soc. 118:12989–12996
12. Hallahan DL, Fernandez VM, Hatchikian EC, Cammack R (1986) Proton-tritium exchange activity of activated and deactivated forms of *Desulfovibrio gigas* hydrogenase. Biochim. Biophys. Acta. 874:72–75
13. Volbeda A, Fontecilla-Camps JC (2003) The active site and catalytic mechanism of NiFe hydrogenases. Dalton Trans. 4030–4038
14. Teixeira M, Moura I, Xavier AV, Moura JJG, LeGall J, DerVartanian DV, Peck HD Jr, Huynh BH (1989) Redox intermediates of *Desulfovibrio gigas* [NiFe] hydrogenase generated under hydrogen. J. Biol. Chem. 264:16435–16450
15. Higuchi Y, Yagi T, Yasuoka N (1997) Unusual ligand structure in Ni–Fe active center and an additional Mg site in hydrogenase revealed by high resolution X-ray structure analysis. Structure 5:1671–1680
16. Higuchi Y, Ogata H, Miki K, Yasuoka N, Yagi T (1999) Removal of the bridging ligand atom at the Ni–Fe active site of [NiFe] hydrogenase upon reduction with H_2, as revealed by X-ray structure analysis at 1.4 Å resolution. Structure 7:549–556

17. Ogata H, Mizoguchi Y, Mizuno N, Miki K, Adachi S, Yasuoka N, Yagi T, Yamauchi O, Hirota S, Higuchi Y (2002) Structural studies of the carbon monoxide complex of [NiFe] hydrogenase from *Desulfovibrio vulgaris* Miyazaki F: suggestion for the initial activation site for dihydrogen. J. Am. Chem. Soc. 124:11628–11635

18. Rousset M, Montet Y, Guigliarelli B, Forget N, Asso M, Bertrand P, Fontecilla-Camps JC, Hatchikian EC (1998) [3Fe-4S] to [4Fe-4S] cluster conversion in *Desulfovibrio fructosovorans* [NiFe] hydrogenase by site-directed mutagenesis. Proc. Natl. Acad. Sci. USA 95:11625–11630

19. Volbeda A, Montet Y, Vernede X, Hatchikian EC, Fontecilla-Camps JC (2002) High-resolution crystallographic analysis of *Desulfovibrio fructosovorans* [NiFe] hydrogenase. Int. J. Hydrogen Energy 27:1449–1461

20. Matias PM, Soares CM, Saraiva LM, Coelho R, Morais J, Le Gall J, Carrondo MA (2001) [NiFe] hydrogenase from *Desulfovibrio desulfuricans* ATCC 27774: gene sequencing, three-dimensional structure determination and refinement at 1.8 Å and modelling studies of its interaction with the tetrahaem cytochrome. J. Biol. Inorg. Chem. 6:63–81

21. Garcin E, Vernede X, Hatchikian EC, Volbeda A, Frey M, Fontecilla-Camps JC (1999) The crystal structure of a reduced [NiFeSe] hydrogenase provides an image of the activated catalytic center. Structure 7:557–566

22. Peters JW, Lanzilotta WN, Lemon BJ, Seefeldt LC (1998) X-ray crystal structure of the Fe-only hydrogenase (CpI) from *Clostridium pasteurianum* to 1.8 angstrom resolution. Science 282: 853–1858

23. Nicolet Y, Piras C, Legrand P, Hatchikian EC, Fontecilla-Camps JC (1999) *Desulfovibrio desulfuricans* iron hydrogenase: the structure shows unusual coordination to an active site Fe binuclear center. Structure 7:13–23

24. Bagley KA, Duin EC, Roseboom W, Albracht SPJ, Woodruff WH (1995) Infrared-detectable group senses changes in charge density on the nickel center in hydrogenase from *Chromatium vinosum*. Biochemistry 34:5527–5535

25. Happe RP, Roseboom W, Pierik AJ, Albracht SP, Bagley KA (1997) Biological activition of hydrogen. Nature 385:126

26. de Lacey AL, Hatchikian EC, Volbeda A, Frey M, Fontecilla-Camps JC, Fernandez VM (1997) Infrared-spectroelectrochemical characterization of the [NiFe] hydrogenase of *Desulfovibrio gigas*. J. Am. Chem. Soc. 119:7181–7189

27. Bleijlevens B, Faber BW, Albracht SP (1997) The [NiFe] hydrogenase from *Allochromatium vinosum* studied in EPR-detectable states: H/D exchange experiments that yield new information about the structure of the active site. J. Bio. Inorg. Chem. 6:763–769

28. Sellmann D, Geipel F, Lauderbach F, Heinemann FW (2002) $[(C_6H_4S_2)Ni(\mu - 'S_3')Fe(CO)(PMe_3)_2]$: a dinuclear [NiFe] complex modeling the $[(RS)_2Ni(\mu-R)_2Fe(CO)(L)_2]$ core of [NiFe] hydrogenase centers. Angew. Chem. Int. Ed. 41:632–634

29. Rauchfuss TB (2004) Research on soluble metal sulfides: from polysulfido complexes to functional models for the hydrogenases. Inorg. Chem. 43:14–26

30. Darensbourg MY, Lyon EJ, Smee JJ (2000) The bio-organometallic chemistry of active site iron in hydrogenases. Coord. Chem. Rev. 206–207:533–561

31. Osterloh F, Saak W, Haase D, Pohl S (1997) Synthesis, X-ray structure and electrochemical characterisation of a binuclear thiolate bridged Ni-Fe-nitrosyl complex, related to the active site of NiFe hydrogenase. Chem. Commun. 10:979–980

32. Davies SC, Evans DJ, Hughes DL, Longhurst S, Sanders JR (1999) Synthesis and structure of a thiolate-bridged nickel-iron complex: towards a mimic of the active site of NiFe-hydrogenase. Chem. Commun. 19:1935–1936

33. Smith MC, Barclay JE, Cramer SP, Davies SC, Gu W-W, Hughes DL, Longhurst S, Evans DJ (2002) Nickel–iron–sulfur complexes: approaching structural analogues of the active sites of NiFe-hydrogenase and carbon monoxide dehydrogenase/acetyl-CoA synthase. Dalton Trans. 2641–2647

34. Marr AC, Spencer DJE, Schröder M (2001) Structural mimics for the active site of [NiFe] hydrogenase. Coord. Chem. Rev. 219–221: 1055–1074

35. George SJ, Cui Z, Razavet M, Pickett CJ (2002) The di-Iron subsite of all-iron hydrogenase: mechanism of cyanation of a synthetic {2Fe3S}-carbonyl assembly. Chem. Eur. J. 8:4037–4046

36. Cammack R, Frey M, Robson R (2001), Hydrogen as Fuel-Learning from Nature. Taylor and Francis, London

37. Bruschi M, Zampella G, Fantucci P, De Gioia L (2005) DFT investigations of models related to the active site of [NiFe] and [Fe] hydrogenases. Coord. Chem. Rev. 249:1620–1640

38. Stein M, Lubitz W (2004) Relativistic DFT calculation of the reaction cycle intermediates of [NiFe] hydrogenase: a contribution to understanding the enzymatic mechanism. J. Inorg. Biochem. 98:862–877

39. Fernandez VM, Hatchikian EC, Patil DS, Cammack R (1986) ESR-detectable nickel and iron-sulphur centres in relation to the reversible activation of *Desulfovibrio gigas* hydrogenase. Biochim. Biophys. Acta 883:145–154

40. Fernandez VM, Hatchikian EC, Cammack R (1985) Properties and reactivation of two different deactivated forms of *Desulfovibrio gigas* hydrogenase. Biochim. Biophys. Acta 832:69–79

41. Roberts LM, Lindahl PA (1994) Analysis of oxidative titrations of *Desulfovibrio gigas* hydrogenase; implications for the catalytic mechanism. Biochemistry 33:14339–14350

42. Dole F, Fournel A, Magro V, Hatchikian EC, Bertrand P, Guigliarelli B (1997) Nature and electronic structure of the Ni-X dinuclear center of *Desulfovibrio gigas* hydrogenase. Implications for the enzymatic mechanism. Biochemistry 36:7847–7854

43. de Lacey AL, Stadler C, Fernandez VM, Hatchikian EC, Fan H-J, Li S, Hall MB (2002) IR spectroelectrochemical study of the binding of carbon monoxide to the active site of *Desulfovibrio fructosovorans* Ni-Fe hydrogenase. J. Biol. Inorg. Chem. 7:318–326

44. Happe RP, Roseboom W, Albracht SPJ (1999) Pre-steady-state kinetics of the reactions of [NiFe]-hydrogenase from *Chromatium vinosum* with H_2 and CO. Eur. J. Biochem. 259:602–608

45. Davidson G, Choudbury SB, Gu Z, Bose K, Roseboom W, Albracht SPJ, Maroney MJ (2000) Structural examination of the nickel site in *Chromatium vinosum* hydrogenase: redox state oscillations and structural changes accompanying reductive activation and CO binding. Biochemistry 39:7468–7479

46. Bouwman E, Reedijk J (2005) Structural and functional models related to the nickel hydrogenases. Coord. Chem. Rev. 249:1555–1581

47. Best SP (2005) Spectroelectrochemistry of hydrogenase enzymes and related compounds. Coord. Chem. Rev. 249:1536–1554

48. de Lacey AL, Fernández VM, Rousset M (2005) Native and mutant nickel–iron hydrogenases: unravelling structure and function. Coord. Chem. Rev. 249:1596–1608

49. Volbeda A, Fontecilla-Camps JC (2005) Structure–function relationships of nickel–iron sites in hydrogenase and a comparison with the active sites of other nickel–iron enzymes. Coord. Chem. Rev. 249:1609–1619

50. Evans DJ, Pickett CJ (2003) Chemistry and the hydrogenases. Chem. Soc. Rev. 32:268–275

51. Fan H-J, Hall MB (2001) Recent theoretical predictions of the active site for the observed forms in the catalytic cycle of Ni-Fe hydrogenase. J. Biol. Inorg. Chem. 6:467–473

52. Stein M, Lubitz W (2002) Quantum chemical calculations of [NiFe] hydrogenase. Curr. Opin. Chem. Biol. 6:243–249

53. Pavlov M, Siegbahn PEM, Blomberg MRA, Crabtree RH (1998) Mechanism of H-H activation by nickel-iron hydrogenase. J. Am. Chem. Soc. 120:548–555

54. Pavlov M, Blomberg MRA, Siegbahn PEM (1999) New aspects of H_2 activation by nickel-iron hydrogenase, Int. J. Quant. Chem. 73:197–207

55. Siegbahn PEM, Blomberg MRA, Pavlov M, Crabtree RH (2001) The mechanism of the Ni-Fe hydrogenases: a quantum chemical perspective. J. Biol. Inorg. Chem 6:460–466

56. Lee C, Yang W, Parr RG (1988) Development of the Colle-Salvetti correlation-energy formula into a functional of the electron density. Phys. Rev. B 37:785–789

57. Becke AD (1993) Density-functional thermochemistry. III. The role of exact exchange. J. Chem. Phys. 98:5648–5652

58. Hay PJ, Wadt WR (1985) Ab initio effective core potentials for molecular calculations. Potentials for the transition metal atoms Sc to Hg. J. Chem. Phys. 82:270–283

59. Wadt WR, Hay PJ (1985) Ab initio effective core potentials for molecular calculations. Potentials for main group elements Na to Bi. J. Chem. Phys. 82:284–298

60. Hay PJ, Wadt WR (1985) Ab initio effective core potentials for molecular calculations. Potentials for K to Au including the outermost core orbitals. J. Chem. Phys. 82:299–310

61. Frisch MJ, Trucks GW, Schlegel HB, Scuseria GE, Robb MA, Cheeseman JR, Montgomery JA Jr, Vreven T, Kudin KN, Burant JC, Millam JM, Iyengar SS, Tomasi J, Barone V, Mennucci B, Cossi M, Scalmani G, Rega N, Petersson GA, Nakatsuji H, Hada M, Ehara M, Toyota K, Fukuda R, Hasegawa J, Ishida M, Nakajima T, Honda Y, Kitao O, Nakai H, Klene M, Li X, Knox JE, Hratchian HP, Cross JB, Adamo C, Jaramillo J, Gomperts R, Stratmann RE, Yazyev O, Austin AJ, Cammi R, Pomelli C, Ochterski JW, Ayala PY, Morokuma K, Voth GA, Salvador P, Dannenberg JJ, Zakrzewski VG, Dapprich S, Daniels AD, Strain MC, Farkas O, Malick DK, Rabuck AD, Raghavachari K, Foresman JB, Ortiz JV, Cui Q, Baboul AG, Clifford S, Cioslowski J, Stefanov BB, Liu G, Liashenko A, Piskorz P, Komaromi I, Martin RL, Fox DJ, Keith T, Al-Laham MA, Peng CY, Nanayakkara A, Challacombe M, Gill PMW, Johnson B, Chen W, Wong MW, Gonzalez C, Pople JA (2004), Gaussian 03, Revision C.02. Gaussian, Inc., Wallingford CT

62. Niu S (1999) Thomson LM, Hall MB, Theoretical characterization of the reaction intermediates in a model of the nickel-iron hydrogenase of *Desulfovibrio gigas*. J. Am. Chem. Soc. 121:4000–4007

63. Niu S, Hall MB (2001) Modeling the active sites in metalloenzymes 5. The heterolytic bond cleavage of H_2 in the [NiFe] hydrogenase of *Desulfovibrio gigas* by a nucleophilic addition mechanism. Inorg. Chem. 40:6201–6203

64. Li S, Hall MB (2001) Modeling the active sites of metalloenzymes. 4. Predictions of the unready states of [NiFe] *Desulfovibrio gigas* hydrogenase from density functional theory. Inorg. Chem. 40:18–24

65. Couty M, Hall MB (1996) Basis sets for transition metals: optimized outer p functions. J. Comput. Chem. 17:1359–1370

66. De Gioia L, Fantucci P, Guigliarelli B, Bertrand P (1999) Ab initio investigation of the structural and electronic differences between active-site models of [NiFe] and [NiFeSe] hydrogenases. Int. J. Quant. Chem. 73:187–195

67. De Gioia L, Fantucci P, Guigliarelli B, Bertrand P (1999) Ni-Fe hydrogenases: a density functional theory study of active site models. Inorg. Chem. 38:2658–2662

68. Becke AD (1988) Density-functional exchange-energy approximation with correct asymptotic behavior. Phys. Rev. A 38:3098–3100

69. Dunning TH Jr, Hay PJ (1976), Gaussian Basis Sets for Molecular Calculations. In: Schaefer HF III, (ed) Modern Theoretical Chemistry, vol. 3, Plenum, New York, pp 1–27

70. Huzinaga S (1984) Gaussian Basis Sets for Molecular Calculations. Elsevier, New York, pp 23–24

71. Warshel A, Levitt M (1976) Theoretical studies of enzymic reactions: Dielectric, electrostatic and steric stabilization of the carbonium ion in the reaction of lysozyme. J. Mol. Biol. 103:227–249

72. Singh UC, Kollman PA (1986) A combined ab initio quantum mechanical and molecular mechanical method for carrying out simulations on complex molecular systems: Applications to the $CH_3Cl +$ Cl^- exchange reaction and gas phase protonation of polyethers. J. Comput. Chem. 7:718–730

73. Field MJ, Bash PA, Karplus M (1990) A combined quantum mechanical and molecular mechanical potential for molecular dynamics simulations. J. Comput. Chem. 11:700–733

74. Gao J (1996) In: Lipkowitz KB, Boyd DB (eds) Reviews in Computational Chemistry, vol. 7, VCH Publishers, New York, pp 119–185

75. Mordasini TZ, Thiel W (1998) Computational chemistry column. Combined quantum mechanical and molecular approaches. Chimia 52:288–291

76. Maseras F, Morokuma K (1995) IMOMM: a new integrated ab initio + molecular mechanics geometry optimization scheme of equilibrium structures and transition states. J. Comput. Chem. 16:1170–1179

77. Svensson M, Humbel S, Froese RDJ, Matsubara T, Sieber S, Morokuma K (1996) ONIOM: a multilayered integrated MO + MM method for geometry optimizations and single point energy predictions. A test for diels-alder reactions and $Pt(P(t-Bu)_3)_2 + H_2$ oxidative addition. J. Phys. Chem. 100:19357–19363

78. Dapprich S, Komáromi I, Byun KS, Morokuma K, Frisch MJ (1999) A new ONIOM implementation in Gaussian98. Part I. The calculation of energies, gradients, vibrational frequencies and electric field derivatives. J. Mol. Struct. (Theochem) 461–462, 1–21

79. Vreven T, Morokuma K (2000) On the application of the IMOMO (integrated molecular orbital + molecular orbital) method. J. Comput. Chem. 16:1419–1432

80. Stein M, Lubitz W (2001) The electronic structure of the catalytic intermediate Ni-C in [NiFe] and [NiFeSe] hydrogenases. Phys. Chem. Chem. Phys. 3:5115–5120

81. Dgauss4.0. (1995) Cray Res. Inc., San Diego

82. Stein M, Lubitz W (2001) DFT calculations of the electronic structure of the paramagnetic states Ni-A, Ni-B and Ni-C of [NiFe] hydrogenase. Phys. Chem. Chem. Phys. 3:2668–2675

83. Becke AD (1986) Density functional calculations of molecular bond energies. J. Chem. Phys. 84:4524–4529

84. Perdew JP (1986) Density-functional approximation for the correlation energy of the inhomogeneous electron gas. Phys. Rev. B 33:8822–8824

85. Perdew JP (1986) Erratum: Density-functional approximation for the correlation energy of the inhomogeneous electron gas. Phys. Rev. B 34:7406–7406

86. Stein M, van Lenthe E, Baerends EJ, Lubitz W (2001) Relativistic DFT calculations of the paramagnetic intermediates of [NiFe] hydrogenase. Implications for the enzymatic mechanism. J. Am. Chem. Soc. 123:5839–5840

87. Theoretical Chemistry Amsterdam Density Functional (ADF) (2000) Rev. 2000.02, Vrije Universiteit De Boelelaan, Amsterdam

88. te Velde G, Bickelhaupt FM, Baerends EJ, Fonseca Guerra C, van Gisbergen SJA, Snijders JG, Ziegler T (2001) Chemistry with ADF. J. Comp. Chem. 22:931–967

89. van Lenthe E, Baerends EJ, Snijders JG (1993) Relativistic regular two-component Hamiltonians. J. Chem. Phys. 99:4597–4610

90. van Leeuwen R, van Lenthe E, Baerends EJ, Snijders JG (1994) Exact solutions of regular approximate relativistic wave equations for hydrogen-like atoms. J. Chem. Phys. 101:1272–1281

91. van Lenthe E, Baerends EJ, Snijders JG (1994) Relativistic total energy using regular approximations. J. Chem. Phys. 101 9783–9792

92. Sadlej AJ, Snijders JG, van Lenthe E, Baerends EJ (1995) Relativistic regular two-component Hamiltonians, four component regular relativistic Hamiltonians and the perturbational treatment of Dirac's equation. J. Chem. Phys. 102:1758–1766

93. Amara P, Volbeda A, Fontecilla-Camps JC, Field MJ (1999) A hybrid density functional theory/molecular mechanics study of nickel-iron hydrogenase: investigation of the active site redox states J. Am. Chem. Soc. 121:4468–4477

94. Amos RD, Alberts IL, Andrews JS, Colwell SM, Handy NC, Jayatilaka D, Knowles PJ, Kobayashi R, Laidig KE, Laming G, Lee AM, Maslen PE, Murray CW, Rice JE, Simandiras ED, Stone AJ, Su M-D, Tozer DJ (1995) The Cambridge Analytic Derivatives Package Issue 6, Cambridge University, Cambridge

95. Stadler C, de Lacey AL, Hernandez B, Fernandez VM, Conesa JC (2002) Density functional calculations for modeling the oxidized states of the active site of nickel-iron hydrogenases. 1. Verification of the method with paramagnetic Ni and Co complexes. Inorg. Chem. 41:4417–4423

96. Stadler C, de Lacey AL, Montet Y, Volbeda A, Fontecilla-Camps JC, Conesa JC, Fernandez VM (2002), Density functional calculations for modeling the active site of nickel-iron hydrogenases. 2. Predictions for the unready and ready states and the corresponding activation processes. Inorg. Chem. 41:4424–4434

97. Fan H-J, Hall MB (2002) High-spin Ni(II), a surprisingly good structural model for [NiFe] hydrogenase. J. Am. Chem. Soc. 124:394–395

98. Wang C-P, Franco R, Moura JJG, Moura I, Day EP (1992) The nickel site in active *Desulfovibrio baculatus* [NiFeSe] hydrogenase is diamagnetic. Multifield saturation magnetization measurement of the spin state of Ni(II). J. Biol. Chem. 267:7378–7380

99. Bruschi M, De Gioia L, Zampella G, Reiher M, Fantucci P, Stein M,(2004). A theoretical study of spin states in Ni-S$_4$ complexes and models of the [NiFe] hydrogenase active site. J. Biol. Inorg. Chem. 9:873–884

100. Reiher M, Salomon O, Hess BA (2001) Reparameterization of hybrid functionals based on energy differences of states of different multiplicity. Theor. Chem. Acc. 107:48–55

101. Reiher M (2002) Theoretical study of the Fe(phen)$_2$(NCS)$_2$ spin-crossover complex with reparametrized density functionals. Inorg. Chem. 41:6928–6935

102. Salomon O, Reiher M, Hess BA (2002) Assertion and validation of the performance of the B3LYP* functional for the first transition metal row and the G2 test set. J. Chem. Phys. 117:4729–4737

103. Tüker L (2003) A theoretical study on [NiFe] hydrogenase core from *Desulfovibrio vulgaris* Miyazaki F. J. Mol. Struct. (Theochem) 664–665: 175–181

104. Tüker L, Eroglu I, Yücel M, Gündüz Y (2004) PM3 (UHF) type quantum chemical treatment of the models for Ni–C state of [NiFe] hydrogenase from *Desulfovibrio vulgaris*. J. Mol. Struct. (Theochem) 672:169–174

105. Tachibana A, Fueno H, Tanaka E, Murashima M, Koizumi M, Yamabe T (1991) String model for the rate constant of nonadiabatic solvation in the hydration reaction of carbon dioxide. Int. J. Quant. Chem. 39:561–583

106. Tachibana A, Fueno H, Yamato M, Yamabe T (1991) Second-order perturbational treatment of normal coordinates in the string model for the hydration reaction of formaldehyde. Int. J. Quant. Chem. 40:435–456

107. Tachibana A (1991) String model of chemical reaction coordinate. J. Math. Chem. 7:95–110

108. Tachibana A (2001) Electronic energy density in chemical reaction systems. J. Chem. Phys. 115:3497–3518

109. Tachibana A (2002) First- Principle Theoretical Study on the Dynamical Electronic Characteristics of Electromigration in the Bulk, Surface and Grain Boundary. In: Baker SP (ed) Stress Induced Phenomena in Metallization. American Institute of Physics, New York, pp 105–116

110. Tachibana A (2002) Field Energy Density In Chemical Reaction Systems. In: Brändas EJ, Kryachko ES (eds) Fundamental Perspectives in Quantum Chemistry, A Tribute to the Memory of Per-Olov Löwdin, Vol. 2, Kluwer Academic Publishers, Dordrecht, pp 211–239

111. Tachibana A (2002) Energy Density in Materials and Chemical Reaction Systems. In: Sen KD (ed) Reviews in Modern Quantum Chemistry: A Celebration of the Contributions of Robert Parr, Vol. 2, World Scientic, Singapore, pp 1327–1366

112. Tachibana A (2004) Spindle structure of the stress tensor of chemical bond. Int. J. Quant. Chem. 100:981–993

113. Tachibana A (2005) A new visualization scheme of chemical energy density and bonds in molecules. J. Mol. Modeling 11:301–311

114. Tachibana A (1999) Chemical potential inequality principle. Theor. Chem. Acc. 102:188–195

115. Nakamura K, Doi K, Tachibana A (2004) Molecular Regional DFT program package, ver. 1 Tachibana Lab., Kyoto University, Kyoto

CHAPTER 10

BACTERIORHODOPSIN ENERGY LANDSCAPE: CURRENT STATUS

V. RENUGOPALAKRISHNAN

Children's Hospital, Harvard Medical School, Boston, MA 02115, USA

Abstract: The folding and stability of bacteriorhodopsin remains of great interest in view of its technological importance. Single molecules of bacteriorhodopsin are unfolded by attaching them to the tip of an AFM probe and then applying force < 50 pico Newtons can be pulled one or more at a time. These experiments provide force profiles of individual chains which exhibit dependence and independence on rest of the helices until all of them are unfolded. Unlike differential scanning calorimetric studies which provide the global thermodynamic profile of proteins, AFM dynamic force probe methods provide a wealth of force profiles of the individual chains at a single molecule level which can then be reconstituted to map the energy landscape of bacteriorhodopsin. Energy landscape of bacteriorhodopsin from dynamic force probe method using atomic force spectroscopy is reviewed in this chapter

Keywords: Atomic Force Microscopy (AFM), Dynamic Force Spectroscopy (DFS), Unfolding Pathway, Bacteriorhodopsin, Protein Energy Landscape

INTRODUCTION

Theoretical analysis of protein conformations has since its beginning considered isolated single protein molecule and attempted to compare the results so derived with time averaged experimental studies which are restricted to the bulk phase. Interestingly the theoretical methods have been extended to large aggregates of protein molecules due to increased computational power in recent times whereas experimental methods to study proteins have reached single molecule level due to technological advances. In the early days while the theoretical studies of protein molecules and their reasonable agreement with experimental studies of proteins in the bulk phase have provided reaffirmation of the validity of theoretical methods, yet a strict one-to-one comparison between them was a matter of contention.

Early attempts to calculate protein energy landscapes have been reported in the literature[1,2] Spectroscopic studies of proteins, especially using atomic

433

W. A. Sokalski (ed.), Molecular Materials with Specific Interactions, 433–451.
© 2007 *Springer*.

force microscopy (AFM) and dynamic force spectroscopy (DFS) on individual protein molecules began to surface since 2000[3–12]. The resolution of AFM and DFS have reached nanometer scale. Therefore AFM in the imaging mode and DFS as a function of temperature have surged to the forefront as important methods in nanotechnology. Zhang and Zhang[13] have reviewed single molecule mechanochemistry.

DFS using AFM has offered for the first time precise measurement of inter-molecular forces arising from non-covalent interactions contributing to the stability of proteins. The historical evolution from single molecules to collective aggre-gates of molecules has several hurdles before we can meaningfully harvest the full potential of these methods. In DFS, single molecule AFM is extended to measure unfolding forces at different pulling speeds exerted on bR at different temperatures. As a result DFS permits resolution of the width of potential barriers crossed and for determining natural transition rates over these barriers, see Figure 10-1, Best et al.[14]

Figure 10-2 shows the typical forces encountered as a function of the stretching distances where applied force less than ∼20 pN trigger unfolding processes, Clausen-Schaumann et al.[5]

Ultimate goal of DFS studies is to create an energy landscape of the protein which would serve as blue print for rational protein engineering studies.[15,16] Reconstituting

No Caption Found

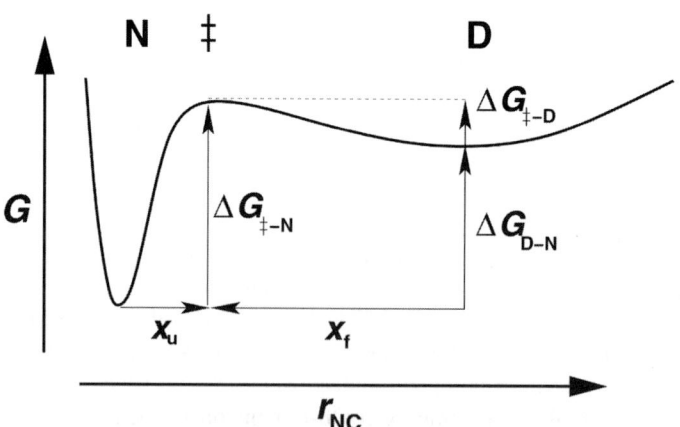

Best, Robert B. et al. (2002) Proc. Natl. Acad. Sci. USA 99, 12143-12148

Figure 10-1. Two-state model for the interpreting mechanical unfolding experiments. The activation energies for folding and unfolding are given by $\Delta\Delta G\ddagger_{-D}$ and $\Delta\Delta G\ddagger_{-N}$ respectively, and x_u and x_f are respectively the distances along the reaction coordinate from the native state (N) and denatured state (D) to the transition state ($\ddagger$). The reaction coordinate is taken here to be the distance between N and C termini, r_{NC}[14]

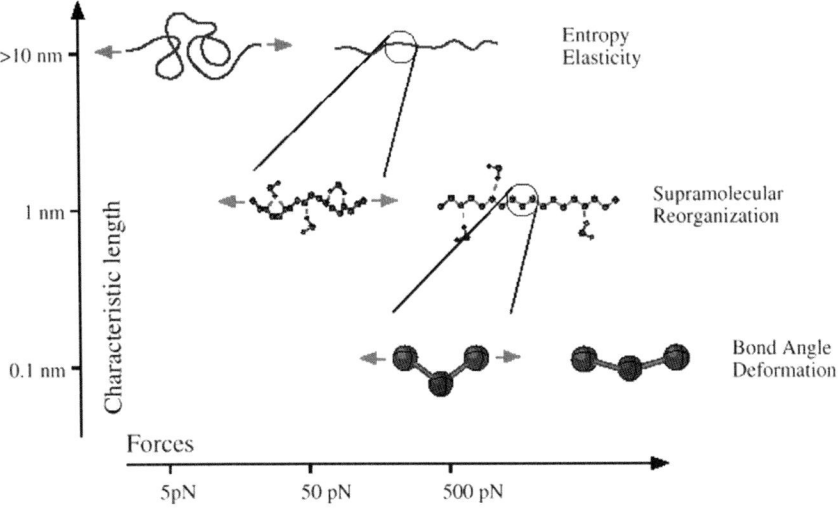

Figure 10-2. Typical forces and length scales in single molecule spectroscopy[5]

the large number of force profiles obtained by the exertion of pN force on the individual chains or domains of a protein molecule to construct the energy landscape of a protein, see a hypothetical energy landscape shown in Figure 10-3, Wise et al. [17] is experimentally and theoretically a daunting challenge.

Calculation of a complete energy landscape is a formidable challenge. [18–22] A hypothetical energy landscape of bR adapted from Wise et al. [17] is shown in Figure 10-3.

Development of time-dependent DFS studies have to await further technological challenges in AFM where the lag phase of a mechanical contraption like a tip of AFM probe riding on a cantilever hovering over the rugged surface of a protein anchored on a platform by its very nature is slow compared to time-scales involved in the conformational dynamics from one peak to another in the energy landscape of a protein. As we encountered similar problems in molecular dynamics (MD), density functional studies (DFT), and 3D multi-nuclear NMR studies, analysis of voluminous data is very tedious.

In this review, we will focus on the energy landscape of a light activated protein, bacteriorhodopsin (bR), and the point mutations we have induced in bR to enhance its thermal, photochemical, and proton-pumping characteristics.

An energy landscape of bR is a complex multi-dimensional potential energy rugged in texture dotted with peaks and valleys. Partial snapshots of this are obtained upon anchoring forces between 100 and 200 pico Newtons (pN) for different helices which were observed to manifest different unfolding characteristics revealing the individual identity of the seven helices. [23–26] The energy landscape of individual TM helices of bR was mapped by monitoring the pulling speed dependence of the unfolding forces and applying Monte Carlo (MC) simulation. [27,28]

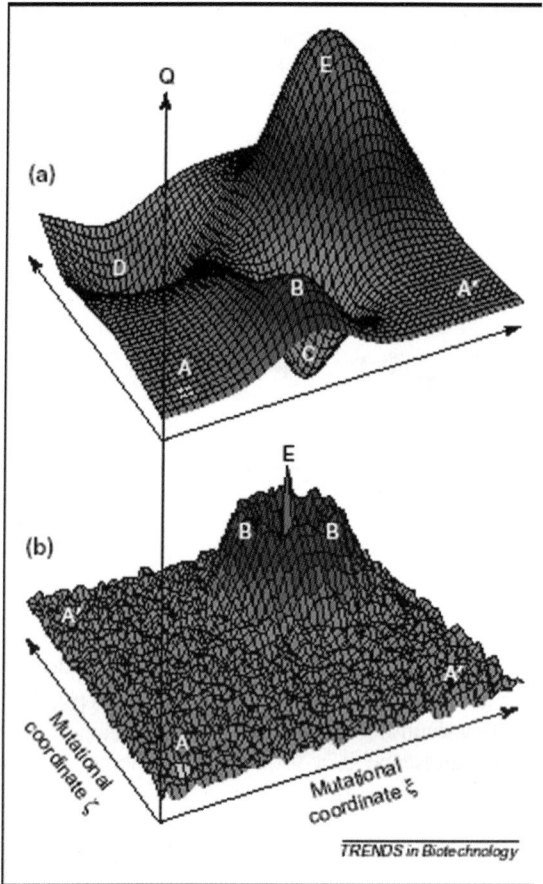

Figure 10-3. A schematic energy landscape of a protein[17]

Single helices formed independently stable units characterized by a single potential barrier. Mechanical unfolding of the helices was triggered by 3.9–7.7 Å extension with unfolding rates were of the order $10^{-3}s^{-1}$. The unfolding pathway or trajectory reflect distinct pulling speed-dependent unfolding routes in their energy landscapes. It was also observed that the unfolding forces of the secondary structure considerably decreased upon increasing the temperature from 8 to 52 °C. The probability of individual unfolding pathways of bR was significantly influenced by temperature. At lower temperatures, the helices manifested significant stability to individually establish potential barriers against unfolding forces whereas they predominantly unfold collectively at elevated temperatures, highlighting co-operativity of biological phenomena.

1.   UNFOLDING PATHWAY OF WILD TYPE BR

Each one of the helices in bR manifested certain intrinsic stability towards the overall stability of the transmembrane (TM) helix. The helices A,B,C,D,E,F,G possess an unique intrinsic potential barrier against unfolding or denaturation. The forces exerted on individual helices vary with temperature and from the force profiles of individual chains, the energy landscape of bR can be mapped. Enhancement of thermal stability of bR[16,17] by site-specific mutations has occupied the attention of our laboratory for many years.

Oesterhelt et al.[23] have reported unfolding pathways of individual wild type bR(w-bR) using AFM and DFS. A selection of typical unfolding profiles are shown in Figure 10-4 A (reproduced with permission from Oesterhelt et al.[23])

Figure 10-4 shows the extension in nm as a function of exerted force in pN. Four peaks located at 10, 30, 50, 70 nm is common to all the unfolding profiles. The relative positions of the second and fourth spikes are similar in contrast to first and third spikes which show variation. Figure 10-4 B shows superposition of 11 force profiles into a composite spike revealing the position of the first spike varies statistically whereas the third spike splits into two spikes. The increasing force at the slope of the second spike thus reflects the stretching of the already unfolded 88 amino acid residues from F and G helices, F-G loop region, and overlapping residues in the E-F loop. On increasing the force to 100 to 200 pN, the remaining membrane anchor is destabilized, see Figure 10-4 C. We can describe the unfolding of the remaining five helices which sequentially follow the F and G helices. Therefore AFM and DFS studies of Oesterhelt et al.[23] of w-bR from purple membrane (PM) patches from *Halobacterium Salinarum (HS)* reveal an individual identity for each one of the 7 helices A through G in which F and G helices at the C-terminal unfold first which then triggers a domino effect of unfolding of E and D helices following which B and C helices unfold one after another like a pack of destabilized cards.

2.   HOW DOES PH INFLUENCE THE UNFOLDING PATHWAY OF W-BR

For thel pH range from 4.2 to 10. the force profiles have similar overall gross features. When we examine them carefully, it is possible to discern subtle differences (Figures 10-6–10-9). Oesterhelt et al.[23] have previously assigned the main peaks to different processes: the peaks below 20 nm include unfolding of helices G and F. At 27, 45, and 65 nm, helices E and D, B and C, and A unfold, respectively. Whereas these main peaks remain more or less unaltered for the pH range, side peaks vary significantly. An unique advantage of single molecule experiments is we can access each molecule at a time. This unique option allows the discrimination between the molecules as individuals as well as between different pathways. Based on the analysis of each unfolding trace the traces may be sorted and grouped according to certain criteria. Muller et al.[24] have performed this task for only one pH value per

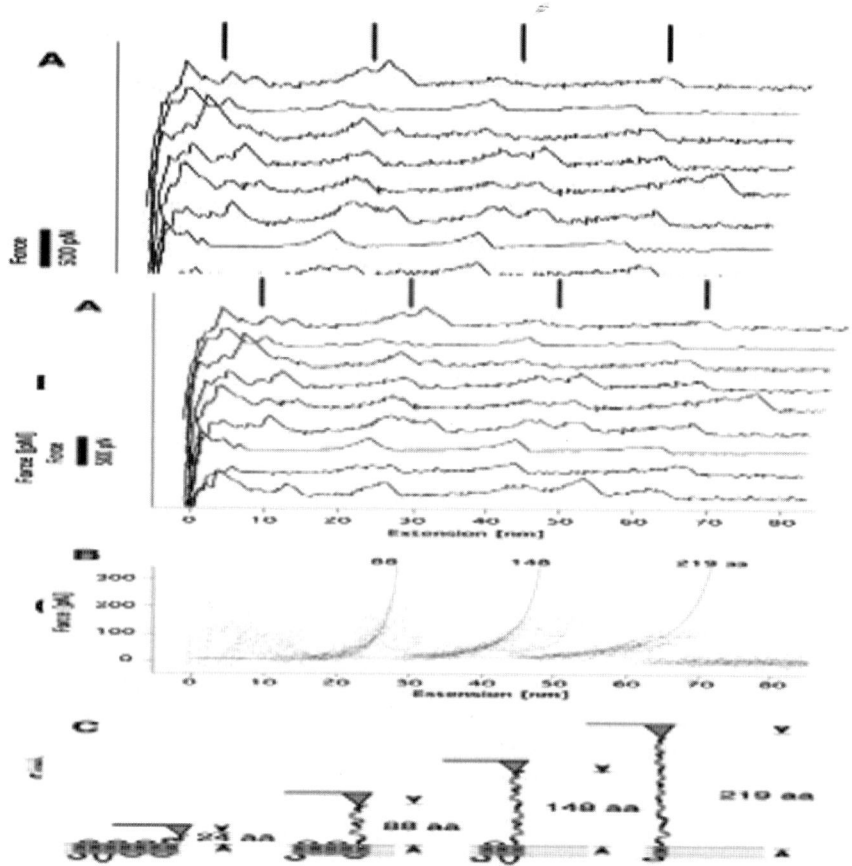

Figure 10-4. Controlled extraction of an individual BR from native purple membrane. (A) Typical high-resolution AFM topograph of the cytoplasmic surface of a wild-type purple membrane. (B) The stylus and protein surface were separated at a velocity of 40 nm/s while the force spectrum was recorded (512 or 4096 pixels). The interaction between tip and surface, which is expressed in the marked discontinuous changes in the force, indicates a molecular bridge between tip and sample. This bridge reaches far out to distances up to 75 nm, which corresponds to the length of one totally unfolded protein. (C) After the adhesive force peaks were recorded, a topograph of the same surface was taken to show structural changes. [23]

block. All blocks were analyzed for pH 4.2 except for the data in Figures 10-6 and 10-11, which were recorded at pH 7.8 to be comparable with the M state data in Figure 10-5 *D*.

3.   UNFOLDING HELICES G AND F

The low extension part, below 30 nm, of all traces superimposed in Figure 10-5 *B* was analyzed individually. Three different main groups became apparent that were superimposed in Figure 10-6, *a–c*. The first group of traces exhibited only the 36-aa

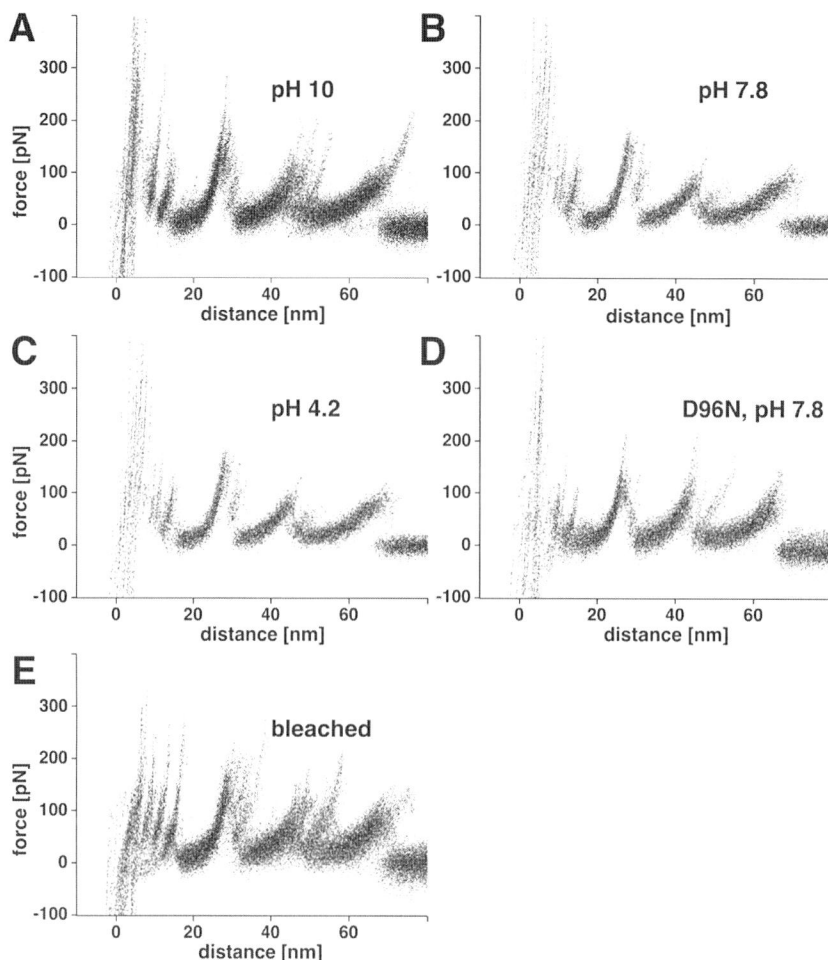

Figure 10-5. Unfolding BR at various conditions. To show common unfolding patterns among single-molecule events, the force spectra were superposed. *(A-C)* BR unfolded at pH 10 (*n* = 31), pH 7.8 (*n* = 32), and pH 4.2 (*n* = 20), respectively. *(D)* BR mutant D96N (*n* = 18) unfolded at pH 7.8. *(E)* exhibited a SD of 12.9 pN (*n* = 20) (Ref. Müller et.al. [24])

peak (Figure 10-6 *a*). In a second group an additional peak at 48 aa occurred with slightly higher probability (Figure 10-6 *b*). The peaks below 5 nm could not be ordered in any systematic way and presumably arise from stretching of the C-terminus. Their variation in position reflects the different attachment sites of the molecule at the tip and thus the length variation of the freely fluctuating segment of the chain. The schematic in Figure 10-6 depicts the model that corresponds to the measured positions of the barriers. According to this model the sequence of the extraction/unfolding process is as follows. First the free C-terminal chain is stretched and then helix G

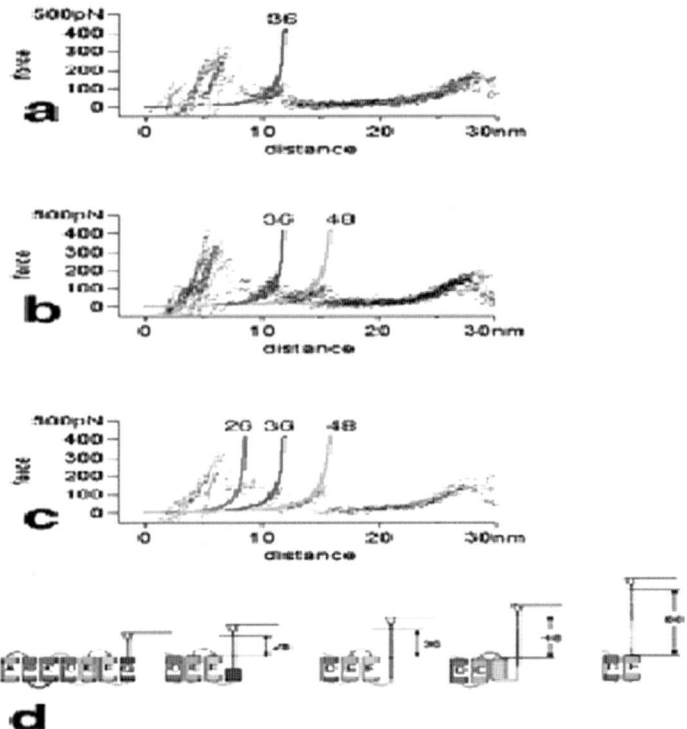

Figure 10-6. Unfolding pathways of transmembrane -helices G and F. (*a*) Unfolding helices G and F in two steps. (*b*) Unfolding of helices G and F and their connecting loop in a three-step process. (*c*) Helices G and F and loop GF unfold in a four-step process. (*d*) Schematic drawing of the unfolding pathways of helices G and F and of loop FG. [24]

unfolds. Then the force acts on the GF loop (peak at 36 aa), and in ∼ 65% of the traces this loop is stretched and pulled through the membrane resulting in the peak at 48 aa (Figure 10-6 *b*). Alternatively, the loop may be extracted together with helix F so that this peak is skipped (Figure 10-6 *a*), and the force starts rising only when it acts directly on helix E. The forces that are required to overcome both barriers are both ∼ 100 pN, the first one slightly higher than the second.

4.   UNFOLDING HELICES E AND D

The trace segments of Figure 10-6 c, showing interactions separated between 15 and 40 nm from the membrane surface, were analyzed accordingly. Janovjak et al. 2002[25] found four distinctly different groups of traces that are depicted in Figure 10-8. In the simplest case, which accounts for ∼ 22% of all traces, one peak at 88 aa is seen (Figure 10-8 *a*). It was found that in ∼ 12% of the cases, an intermittent peak at 94 aa, which reflected a barrier around aa 154 of BR (derived

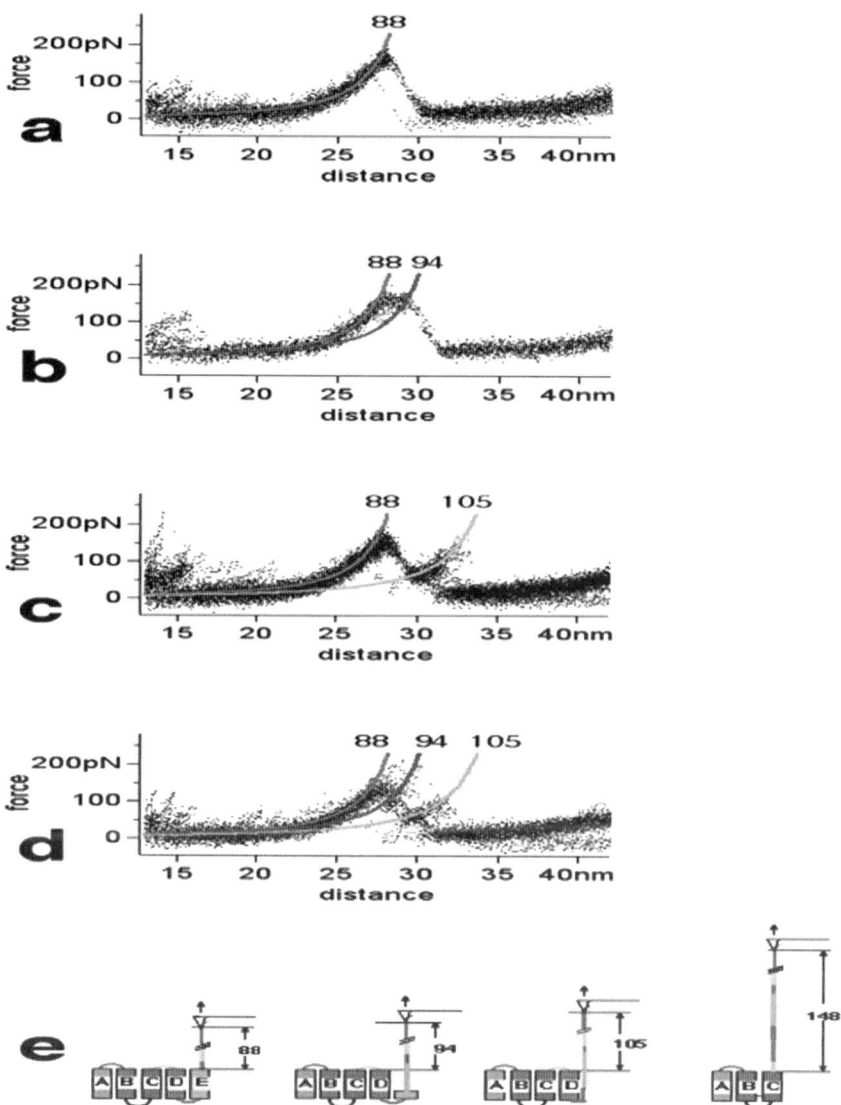

Figure 10-7. Unfolding pathways of transmembrane helices E and D. After unfolding helices F and G, the polypeptide chain, bridging the AFM tip and purple membrane, exhibits a length of 88 aa (red fit). All other helices remain embedded in the purple membrane. (*a*) The structural motif of helices E and D, loop ED, and loop DC unfold in a single step uponexceeding an average pulling force of 167 ± 20 pN ($n = 20$). (*b*) Helices E and D unfold in a two-step process. First, helix E unfolds partly (at 169 ± 22 pN), thereby lengthening the stretched polypeptide to 94 aa (blue fit). After this, the force pulls the remaining part of helix E and, on the hydrophilic loop, connecting helices E and D located on the opposite, extracellular surface. Upon exceeding an average pulling force of 169 ± 21 pN, the remaining part of helix E, the loop ED, helix D, and the cytoplasmic loop CD are unfolded simultaneously ($n = 10$). (*c*) Helices E and D unfold in an alternate two-step process. First, part of helix E and the loop ED

from 248-94 aa; Figure 10-8 *b*) was observed. In 45% of the traces a peak was found at 105 aa (Figure 10-8 *c*), which, based on the model (Figure 10-8 *e*), corresponds to a state where helix E is completely unfolded, but helix D is still intact. Approximately 20% of the traces show all three peaks (Figure 10-8 *d*), which means that the bR molecules measured here went through both intermittent states upon unfolding. The peak heights were ~ 160 pN for the first two barriers and significantly lower for the third (90 pN). The most striking feature of this set is the potential barrier in the proximity of aa 154 of bR.

5.   UNFOLDING HELICES C AND B

In the length window between 35 and 55 nm Muller et al. 2002[24] found again four different groups of traces (Figure 10-8). The majority of the traces exhibited no extra peak between148 and 220 aa, indicating a simultaneous unfolding of helices B and C. A minor fraction of the traces (9%) showed an additional peak at 158 aa (Figure 10-8 *b*) and 35% a second peak at 175 aa (Figure 10-8 *c*). The first case would fit to the extracellular BC loop still untouched, whereas in the second case this loop is completely stretched. In both cases helix B is intact. In 10% of the traces we find all three peaks (Figure 10-8 *d*), indicating that both intermittent states are visited on the unfolding pathway. All peaks are ~ 100 pN in height.

6.   UNFOLDING HELIX A

In 65% of the traces the last peak (Figure 10-9) occurs at 65 nm, corresponding to a stretched unfolded polypeptide of 220 aa in length (Figure 10-9 *a*). In these traces the last helix is pulled out of the membrane in a single step at forces of ~ 100 pN. In the other cases, a second peaks follows (Figure 10-9 *b*). This second peak is smeared out considerably, and the rupture point varies. Drawn in blue is the WLC fit for the fully stretched length of 232 aa from bR. Because this last peak also occurs on multilamellar membrane stacks (see discussion below) it must reflect the destabilization of the N-terminus, its possible interaction with the neighboring proteins, and the pulling through the hydrophobic membrane.

◄───

Figure 10-7. connecting both helices unfold at 161 ± 14 pN, thereby lengthening the stretched polypeptide to 105 aa (green fit). Upon exceeding an average pulling force of 86 ± 23 pN, helix D and loop CD are unfolded (*n* = 39). (*d*) Helices E and D and loop ED unfold in a three-step process. First, part of helix E unfolds at 152 ± 22 pN, thereby lengthening the stretched polypeptide to 94 aa (blue fit). Second, what remains from helix E and loop ED is pulled into the membrane at 135 ± 30 pN, lengthening the polypeptide strand to 105 aa (green fit). Third, helix D and loop CD unfold at a pulling force above 83 ± 23 pN (*n* = 19). (*e*) Schematic drawing of the unfolding pathways found. The total number of force curves shown corresponds to 88

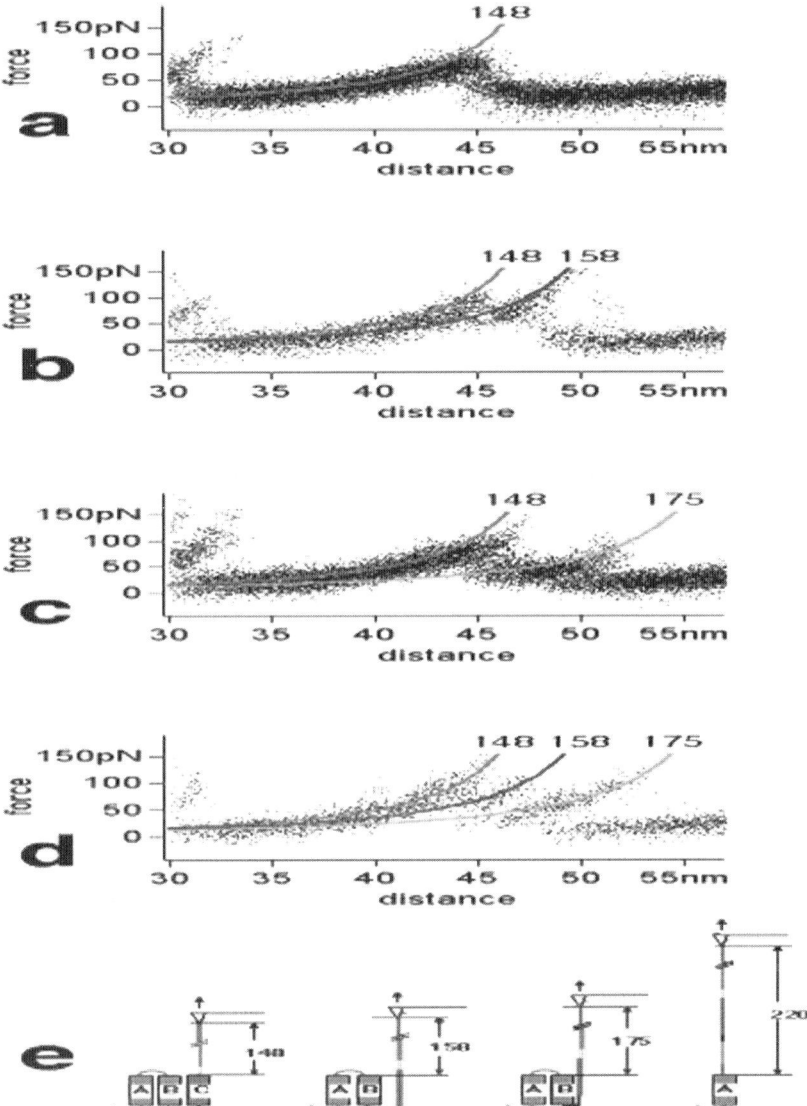

Figure 10-8. Unfolding pathways of transmembrane helices C and B. After unfolding helices E and D, the polypeptide chain, bridging the AFM tip and purple membrane, exhibits a length of 148 aa (red fit). Helices C, B, and A remain embedded in the purple membrane. (*a*) Helices C and B unfold in a single step upon exceeding an average pulling force of 99 ± 16 pN ($n = 40$). (*b*) Helices C and B unfold in a two-step process. First, helix C unfolds at 109 ± 18 pN, thereby lengthening the stretched polypeptide to 158 aa (blue fit). After this, the force pulls on the hydrophilic loop connecting helices C and B located on the opposite, extracellular surface. Upon exceeding an average pulling force of 105 ± 15 pN, the extracellular loop BC, helix B, and the cytoplasmic loop AB are unfolded simultaneously ($n = 8$). (*c*) Helices C and B unfold in an alternate two-step process. First, helix C and the loop connecting these helices unfold at 95 ± 20 pN, thereby lengthening the stretched polypeptide to 175 aa (green fit).

7.   STABILITY OF THE LOOPS

One remarkable finding of this study is the measured potential barrier associated with the N-terminus and the extracellular loops connecting the transmembrane α-helices. To exclude adhesion of the loops to the mica surface as a potential explanation we performed the same experiments on the upper membrane of double-layered purple membrane patches like the ones shown in Figure 10-10 and of purple membrane adsorbed onto hydrophobic graphite (data not shown). In both cases, we did not observe a change in the adhesion peak positions and distributions. Because it would be highly unlikely that a hypothetic adhesive interaction of the loops with mica is the same as with another purple membrane or with graphite, we conclude from these experiments that the loops are stable structural elements. Thus, a potential barrier comparable with the one that is associated with the unfolding of the α-helices needs to be overcome to stretch the loops and to pull them through the membrane. Interestingly, these forces required to overcome the barriers do not depend in an obvious way on the length of the loop (i.e. 102 pN for loop GF, 4 aa; 135 pN for loop ED, 3 aa; and 109 pN for loop CB, 17 aa). This indicates that the process is dominated by an activation barrier. Because these forces are on the order of 100 pN, the width of these barriers must be far less than the thickness of the membrane to be compatible with measured unfolding free energy changes. This again speaks for a breakup of a structure. On the other hand, x-ray and electron diffraction studies on crystallized bR shows these loops to exhibit a well defined structural conformation. The B-factors and temperature factors of the bR structures are similar for all extracellular loops and the transmembrane α-helices, indicating that they exhibit equally high conformational stability.[29-32] This finding was also confirmed by experiments determining the solution structure of truncated bR loops, which showed conformations close to those observed on intact bR.[33]

8.   TEMPERATURE DEPENDENCE OF UNFOLDING PROFILES OF W- BR

Janovjk et al.[25] have investigated the effect of temperature on the unfolding pathway.

Figure 10-10 shows a selection of force-extension traces recorded on single bR molecules. An interpretation of a typical trace exhibiting common features observed among all curves is given at the top of Figure 10-10. After separating AFM tip and purple membrane, the C-terminal polypeptide of bR is extended. Further

---

Figure 10-8. Upon exceeding an average pulling force of 80 ± 17 pN, helix B and loop AB are unfolded ($n = 31$). (*d*) Helices C and B unfold in a three-step process. First, helix C unfolds at 108 ± 26 pN, thereby lengthening the stretched polypeptide to 158 aa (blue fit). Second, loop BC is pulled into the membrane at 116 ± 33 pN (green fit). Third, helix B unfolds at pulling forces above 87 ± 31 pN ($n = 9$). (*e*) Schematic drawing of the unfolding pathways found. The total number of force curves shown corresponds to 88

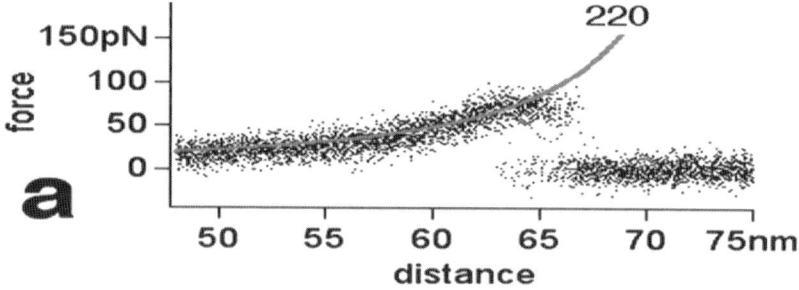

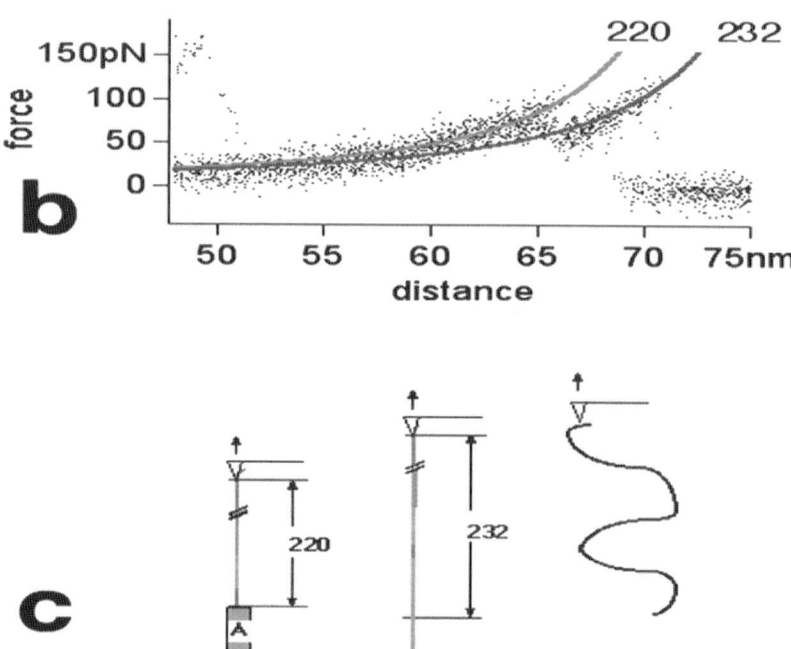

Figure 10-9. Unfolding pathways of the transmembrane helix A. After unfolding helices C and B, the polypeptide chain, bridging the AFM tip and purple membrane, exhibits a length of 220 aa (red fit). Only helix A remains embedded in the membrane. (*a*) Helix A and the N-terminal end are pulled through the membrane within a single step at average pulling force of 87 ± 9 pN ($n = 12$). (*b*) Helix A unfolds at 99 ± 11 pN, and the N-terminal end anchors the polypeptide ($n = 6$). The length of the stretched polypeptide corresponds to 232 aa (blue fit). After this, the force pulls on the hydrophilic N-terminus located on the opposite, extracellular surface. By exceeding a pulling force of 105 ± 11 pN, the polypeptide end is pulled through the membrane. (*c*) Schematic drawing of the unfolding pathways found. The total number of force curves shown corresponds to 18

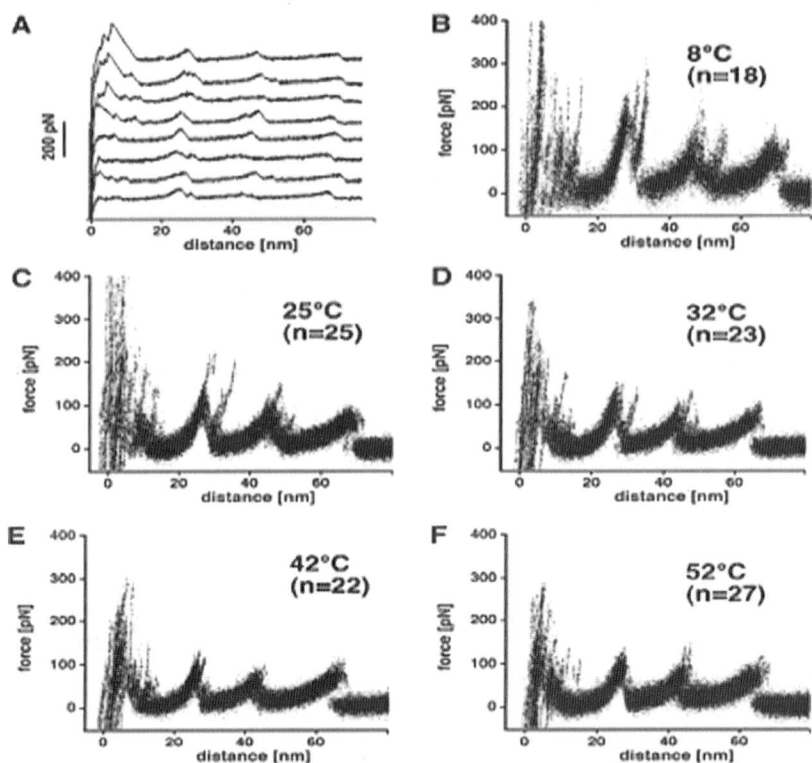

Figure 10-10. Unfolding BR from native purple membrane at various temperatures. (A) Force curves of individual BR molecules recorded at 25<@2;>C. To show common unfolding patterns among single-molecule events, the force spectra recorded at different temperatures were superimposed. (B–F) BR unfolded at 8<@2;>C (B), 25<@2;>C (C), 32<@2;>C (D), 42<@2;>C (E) and 52<@2;>C (F) in 300 mM KCl, 20 mM Tris-HCl with a pH of 7.8 being adjusted for each temperature

separating tip and membrane stretches the C-terminal end and the force builds up in a gradual but non-linear fashion. At a certain force, the first transmembrane helices G and F unfold. This increases the length of the molecular bridge between tip and membrane, the cantilever relaxes and the force drops abruptly. By further separation of the AFM tip and membrane surface, the polypeptide chain of the unfolded structural elements extends. As soon as the polypeptide is stretched again, the force rises. At a certain force, the next secondary element of bR unfolds. As shown previously, the fitted contour length of the force-extension curve and the secondary structure model of bR suggest that helices G and F, D and E, and B and C unfold pairwise.[23] The remaining seventh helix, A, is then pulled from the membrane in a single step. Beyond an extension of 70 nm no interaction is measured.

To see to what extent these unfolding events of secondary structural elements depend on the temperature, force extension curves were recorded at 8 °C

(Figure 10-10 B), room temperature (25 °C; Figure 10-10 C), 32 °C (Figure 10-10 D), 42 °C (Figure 10-10 E), and 52 °C (Figure 10-10 F). Each graph shows a multitude of force extension traces, each one recorded on one single bR (such as shown in Figure 10-10 A). In these figures, 25 traces are superimposed. This kind of graphic representation highlights common features through the accumulation of the measured points and at the same time still represents the individualism of traces. Independent of the temperature adjustment, each curve exhibited a richness of detailed information on the mechanics of this molecule. It becomes clear that the main peaks at 27, 45 and 65 nm remain at their position (Figure 10-10), but that the rupture forces of these unfolding events decrease with increasing temperature. The steepest decrease of the rupture forces was observed between 8 and 32 °C. Above 32 °C, the rupture force decreased only slightly, showing a fluctuation of a similar range as the standard deviation of the mean value.

Similarly to the main peaks, the side peaks did not change their position (contour length) upon variation of the temperature. To see whether the rupture force (Figure 10-11) or the probability (Figure 10-12) of the side peaks change with temperature, they were analyzed from each single force-extension curve. Interestingly, the average rupture force of all side peaks decreased with increasing temperature (Figure 10-12). However, the frequency of the side peaks decreased

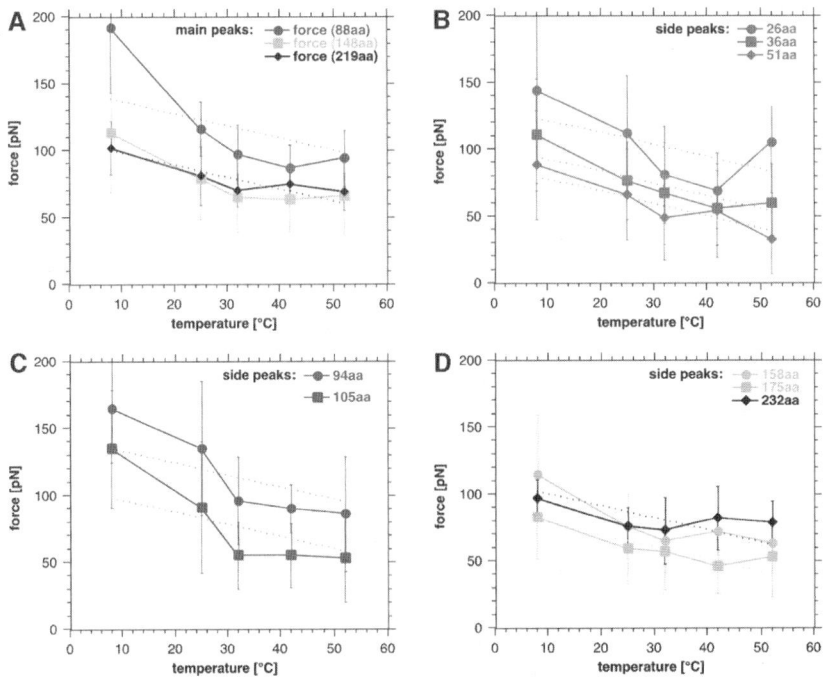

Figure 10-11. Unfolding forces of secondary structural elements depend on temperature. (A) Rupture forces of main peaks, which exhibited no side peaks

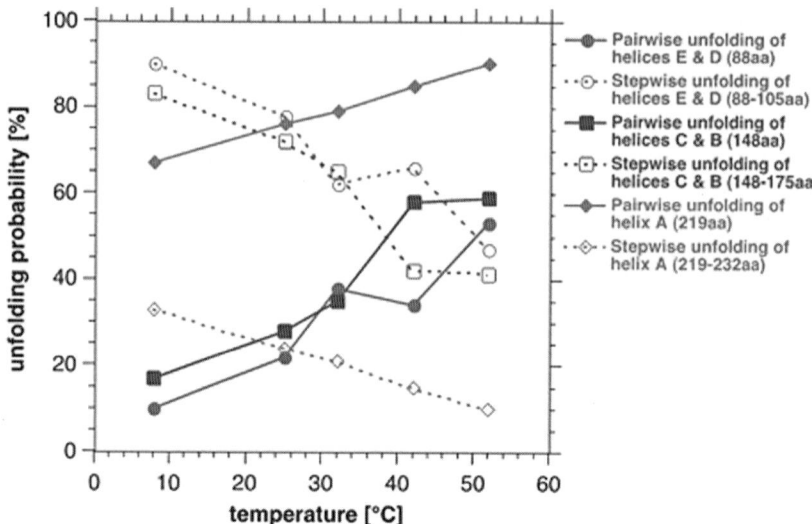

Figure 10-12. Probability of unfolding pathways depends on temperature. The occurrence of main force peaks exhibiting no side peaks (solid lines) increased with increasing temperature. The probability of single structural elements, such as helices or loops, to unfold in a separate event decreases with increasing temperature

with increasing temperature (Figure 10-12). Accordingly, the frequency of the main peaks increased with the temperature. This indicates that the pairwise unfolding of transmembrane -helices is favored with increasing temperature, while with decreasing temperature, the unfolding probability of single secondary structure elements, such as helices and loops, is enhanced.

Probability of unfolding pathways, shown in Figure 10-12, depends on temperature. The occurrence of main force peaks exhibiting no side peaks (solid lines) increased with increasing temperature. As a consequence, the probability of the main peaks exhibiting side peaks (dashed lines) decreased significantly. Solid lines represent probabilities for the pairwise unfolding of transmembrane -helices E and D (88 aa, red), C and B (148 aa, blue) and of helix A (219 aa, green). The probability of their stepwise unfolding is presented by the dotted lines. This indicates -helices of bR unfold preferentially pairwise at elevated temperatures. The probability of single structural elements, such as helices or loops, to unfold in a separate event decreases with increasing temperature.

CONCLUSION

Energy landscape of bacteriorhodopsin from AFM studies has been reviewed in this review. Recently we have embarked on a project to determine the unfolding pathway of bR mutants where mutations in F and G helices were introduced by site-directed mutagenesis, e.g. [E9Q/E194Q/E204Q] bR. This mutant was designed to attenuate proronation-deprotonation kinetics of w-bR. Site specific mutations were introduced in F and G helices to enhance proton pumping activity of bR which can be modulated by mutations of its Asp and Glu residues. Simultaneous mutation of Glu^9, Glu^{74}, Glu^{194}, and Glu^{204} profoundly alters the proton pumping of bR. All the four Glu residues are in the extracellular region. Of these four Glu residues, Glu^{74} is located in the beta sheet linking helices B and C and is not in close proximity to the other Glu residues. Earlier attempts to mutate Asp residues, especially Asp^{85} and Asp^{96} yielded considerable increase in the life time of M state. Basic to all these is the fundamental role of protonation-deprotonation of bR is influenced by Glu and Asp residues. The DFS studies on the triple mutant of bR is in progress in collaboration with Prof. Daniel Mueller, BIOTEC, University of Technology Dresden, Dresden, Germany.

ACKNOWLEDGEMENTS

Author would like to than European Biochemical Society, American Association for the Advancement of Science, Biophysical Society, Elsevier, Current Opinion in Structural Biology, National Academy of Sciences USA for permission to reproduce Figures. VR is supported by grants from NSF.

REFERENCES

1. Troyer JM, Cohen FE (1995a) Protein conformational landscapes: energy minimization and clustering of a long molecular dynamics trajectory. Proteins: Structure, Function, and Genetics 23(1):97–110
2. Troyer JM, Cohen FE (1995b) Protein conformational landscapes: energy minimization and clustering of a long molecular dynamics trajectory. Proteins 23(1):97–110
3. Zhang W, Zhang X (2003) Single molecule mechanochemistry of macromolecules. Progress in Polymer Science 28(8):1271–1295
4. Zlatanova J, Lindsay SM, Leuba SH (2000) Single molecule force spectroscopy in biology using atomic force spectroscopy. Prog. Biophys. Mol. Biol. 74:37–61
5. Clausen-Schaumann H, Seitz M, Krautbauer R, Gaub HE (2000) Force spectroscopy with single bio-molecules. Curr. Opinion Chem. Biol. 4:524–530
6. Janshoff A, Steinem C (2001a) Energy lndscapes of ligand-receptor couples probed by dynamic force spectroscopy. Chem Phys Chem 2:577–579
7. Janshoff A, Steinem C (2001b) Scanning force microscopy of artificial membranes. Chem Bio Chem 2(11):798–808
8. Rief M, Grubmuller H (2002) Force spectroscopy of single biomolecules. Chem Phys Chem 3:255–261
9. Evstigneev M, Reimann P (2003) Dynamic force spectroscopy: Optimized data analysis. Phys. Rev. E 68:045103®
10. Raible M, Evstigneev M, Reimann P, Bartels FW, Ros R (2004) Theoretical analysis of dynamic force spectroscopy experiments on ligand-receptor complex. J. Biotech. 112:13–23
11. Haustein E, Schwille P (2004) Single-molecule spectroscopic methods. Curr. Opinion Struct. Biol. 14:531–540

12. Levy R, Maaloum M (2005) Specific molecular interactions by force spectroscopy: From single bonds to collective properties. Biophys. Chem. 117:233–237
13. Zhang W, Zhang X (2003) Single molecule mechanochemistry of macromolecules. Progress in Polymer Science 28(8):1271–1295
14. Best RB, Fowler SB, Toca-Herrera JL, Clarke J (2002) A simple method for probing the mechanical unfolding pathway of proteins in detail. Proc. Natl. Acad. Sci. USA 99:12143–12148
15. Renugopalakrishnan V, Garduño-Juárez R, Narasimhan G, Verma CS, Wei X, Pingzuo Li (2005a) Rational design of termally sable poteins: Relavence to bionanotechnology. J. Nanosci. Nanotech. 5:1759–1767
16. Renugopalakrishnan V, Wei X, Narasimhan G, Verma CS, Pingzuo Li, Anumanthan A (2005b) Enhancement of protein thermal sability: Towards the dsign of robust proteins for bionanotechnological applications in Bionanotechnology. In: Renugopalakrishnan V, Lewis RV (eds) Dordrecht, The Netherlands, pp 117–140
17. Wise KJ, Gillespie NB, Stuart JA, Krebs MP, Birge RR (2002) Optimization of bacteriorhodopsin for bioelectronic devices. Trends in Biotechnology 20(9):387–394
18. Frauenfelder H., McMahon BH (2000) Energy landscape and fluctuations in proteins. Annalen der Physik (Berlin) 9(9–10):655–667
19. He J, Zhang Z, Shi Y, Liu H (2003) Efficiently explore the energy landscape of proteins in molecular dynamics simulations by amplifying collective motions. J. Chem. Phys. 119(7):4005–4017
20. Levy Y, Papoian GA, Onuchic JN, Wolynes PG (2004) Energy landscape analysis of protein dimers. Israel J. Chemistry 44(1–3):281–297
21. Nevo R, Brumfeld V, Kapon R, Hinterdorfer P, Reich Z (2005) Direct measurement of protein energy landscape roughness. EMBO Reports 6(5):482–486
22. Schug A, Wenzel W, Hansmann UHE (2005) Energy landscape paving simulations of the trp-cage protein. Journal of Chemical Physics 122(19):194711/1–194711/7
23. Oesterhelt F, Oesterhelt D, Pfeiffer M, Engel A, Gaub HE, Muller DJ(2000) Unfolding pathways of individual bacteriorhodopsins. Science 288:143–146
24. Müller DJ, Kessler M, Oesterhelt F, Moller C, Oesterhelt D, Gaub H (2002) Stability of bacteriorhodopsin α-helices and loops analyzed by single-molecule force spectroscopy. Biophys.J. 83(6):3578–3588
25. Janovjak H, Kessler M, Oesterhelt D, Gubb G, Muller DJ (2003) Unfolding pathways of native bacteriorhodopsin depend on temperature. The EMBO J. 22:5220–5229
26. Janovjak H, Struckmeier J, Hubain M, Kedrov A, Kessler M, Müller DJ (2004) Probing the energy landscape of the membrane protein bacteriorhodopsin Structure 12:871–879
27. Rief M, Fernandez JM, Gaub HE (1998a) Elastically coupled two-level-systems as a model for biopolymer extensibility. Phys. Rev. Lett. 81:4764–4767
28. Rief M, Gautel M, Schemmel A, Gaub HE (1998b) The mechanical stability of immunoglobulin and fibronectin III domains in the muscle protein titin measured by atomic force microscopy. Biophys. J. 75:3008–3014
29. Belrhali H, Nollert P, Royant A, Menzel C, Rosenbusch JP, Landau EM, Pebay-Peyroula E (1999) Protein, lipid, and water organization in bacteriorhodopsin crystals: A molecular view of the purple membrane at 1. 9 A. Structure Fold. Des. 7:909–917
30. Essen L., Siegert R, Lehmann WD, Oesterhelt D (1998) Lipid patches in membrane protein oligomers: crystal structure of the bacteriorhodopsin-lipid complex. Proc. Natl. Acad. Sci. USA 95(20):11673–11678
31. Leucke H, Schobert B, Richter HT, Cartailler JP, Lanyi JK (1999) Structure of bacteriorhodopsin at 1.55 A resolution. J. Mol. Biol. 291:899–911

32. Mitsuoka K, Hirai T, Murata K, Miyazawa A, Kidera A, Kimura Y, Fujiyoshi Y (1999) The structure of bacteriorhodopsin at 3.0 A resolution based on electron crystallography: Implication of the charge distribution. J. Mol. Biol. 286:861–862

33. Katragadda M, Alderfer JL, Yeagle PL (2001) Assembly of a Polytopic Membrane Protein Structure from the Solution Structures of Overlapping Peptide Fragments of Bacteriorhodopsin. Biophys J. 81(2):1029–1036

CHAPTER 11

DIMERIZATION AND OLIGOMERIZATION
OF RHODOPSIN AND OTHER G PROTEIN-COUPLED
RECEPTORS

SŁAWOMIR FILIPEK, ANNA MODZELEWSKA
AND KRYSTIANA A. KRZYŚKO

International Institute of Molecular and Cell Biology, 4 Ks. Trojdena St, 02–109 Warsaw, Poland

Abstract: Dimerization, and more generally oligomerization, of G protein-coupled receptors (GPCRs) is experimentally proven and possibly all GPCRs act in oligomeric form. The coupling with G protein, phosphorylation by kinase and binding to arrestin what starts internalization process have also been shown to be influenced by the oligomeric state of the receptors. Cooperative interactions within homo- and heterodimers of GPCRs may be critical for the propagation of an external signal across the cell membrane, activation of a G protein and passing the signal down to effector proteins

Keywords: Rhodopsin; GPCR; membrane proteins; dimerization; oligomerization; G protein; arrestin; signal transduction

1.   INTRODUCTION

G protein-coupled receptors (GPCRs) represent a very large superfamily of receptors essential for signaling across cell membranes.[1,2] In human genome about 1000 genes encode GPCRs with half of them being an odor and taste receptors and the other half are receptors of endogenous ligands and light.[3] Each GPCR responds to a single or a few ligands by activating trimeric (consisting of α, β and γ subunits) G proteins. Then the $G_{\alpha\beta\gamma}$ protein dissociates into G_α and $G_{\beta\gamma}$ and one of them (depending on a specific pathway) modulates enzymes or membrane channels leading to highly amplified signaling cascade. Such processes are responsible for vision, taste, smell, neurotransmission and involve responses to small molecules like opioids, hormones, peptides, but also proteins like proteases and chemokines. For this broad range of ligands GPCRs are important targets for pharmacological intervention and the large fraction of current drugs are directed toward them.[4] Even

453

W. A. Sokalski (ed.), Molecular Materials with Specific Interactions, 453–467.

though GPCRs are so ubiquitous still only a small fraction of their pharmacological potential is being recognized.[2]

Recently it was shown that rhodopsin forms dimers in a shape of long double rows,[5,6] and that dimerization is also an important feature of other GPCRs and can affects their function. The classical view is that receptors (including GPCRs) operate as monomeric proteins. However, a growing amount of experimental pharmacological, biochemical and biophysical data suggests that these receptors form functional homo- and heterodimers as well as higher oligomers.[7] Furthermore, their oligomeric assemblies have important functional roles.[8,9] The concept of dimerization is important not only from scientific point of view, but also technology like pharmacological and new materials industries. The changes in ligand-binding and signaling properties that accompany and influence homo- and hetero-dimerization could give rise to new classes of pharmacologically active compounds. Oligomerization changing properties of single protomers (monomeric units) by modulating of ligand binding may be used for tuning of sensors for specific molecules or their groups.

The current state of research is that GPCRs operate and exist as dimers. However, even a couple of years ago the monomeric state of GPCR was commonly accepted view and a few cases of dimerization were treated as an exception. Gradually, due to growing evidence of experimental data the hypothesis of dimerization became a dominant belief. There are even suggestions that GPCRs its whole life cycle in a cell spend as dimers (both hetero- and homodimers), starting from formation of dimers in endoplasmic reticulum with the help of dimer-probing cytosolic chaperons.[10] Such dimers could by transported by golgi apparatus to plasma membrane. Facing the extracellular space the dimers may be dynamically regulated by ligand binding. In case of heterodimerization different effects like positive or negative cooperation of ligand binding as well as potentiating or attenuating of signaling via G protein may take place. Also heterodimerization may promote internalization of the whole dimer upon binding of agonist of one of the receptors.

The recent explosion of number of papers on GPCR dimerization is summarized in several reviews.[8,11,12] This chapter is devoted to briefly show importance of oligomerization of GPCRs, current state of knowledge in this area and potential applications.

2.   G PROTEIN-COUPLED RECEPTORS

GPCRs constitute the largest family of cell surface proteins involved in signaling across cell membranes. Mammalian GPCRs are commonly divided into three subfamilies: A (the most populated – called rhodopsin like), B and C based on the sequence. More recently, these receptors have been classified into five distinct groups based on phylogenetic analyses of sequences from the human genome. This classification system is called GRAFS, which is an acronym for five different groups: Glutamate (former class C), Rhodopsin (former class A), Adhesion, Frizzled/taste and Secretin (former class B).[13]

This classification was performed by sequence hidden Markov models to identify and classify all GPCRs from 13 eukaryotic genomes. It was observed that the five main families found in mammals were present before the divergence of nematodes and that several other classes of GPCRs were not present within the vertebrate line. The superfamily of GPCRs has a very dynamic gene repertoire, as evidenced by the several expansions of various families of GPCRs that have been found. Further, the number of GPCRs has gradually increased with increased complexity of the organism. [14]

Searches in human, mouse and other genomic databases resulted in the identification of more than 180 protein predictions belonging to the glutamate family of GPCRs. This study also shows that the pheromone-receptor subgroup has undergone independent expansions leaving the human genome without all pheromone receptors. [15]

2.1.   Importance of GPCRs

GPCRs modulate a wide range of physiological processes and are implicated in numerous diseases. Therefore they form the largest class of therapeutic targets. G-protein coupled receptors represent the primary mechanism by which cells feel external environment and different stimuli, and pass the information to the interior of the cell. Abnormalities (usually by mutations but also by risk factors) of delicate balance in signaling mechanism often go to diseases and disorders, e.g. hypertension, hypertrophy, inflammatory diseases, cancer, fibrosis, diabetes, and diseases of central nervous system like Alzheimer disease.

G-protein coupled receptors constitute the largest family of signal transduction membrane proteins. They mediate responses of many bioactive molecules including biogenic amines, amino acids, peptides, lipids, nucleotides and proteins. As a result, GPCRs play a crucial role in many essential physiological processes like neurotransmission, cellular metabolism, secretion, cell growth, immune defense and differentiation.

Differences in structure-function and signal transduction of particular receptor types represent major challenge to understand the role of GPCRs in cells and pharmaceutical treatment. Elucidation of these differences and their mechanisms will greatly advance receptor biology, pharmacology and therapeutics.

2.2.   Rhodopsin as a Template

Despite the broad range of possible actions of GPCRs, they all share a common seven α-helical transmembrane motif of the structure. The ligand binding site is located either at the extracellular region or within the transmembrane α-helical bundle, and the cytoplasmic loops are responsible for coupling to G proteins and activate them (GPCRs act in an enzymatic way) whereas other proteins like arrestins stop enzymatic action of these receptors.

The most extensively studied GPCR is rhodopsin. Rhodopsin is the only GPCR with resolved three dimensional structure at atomic detail.[16–18] The rhodopsin structure may serve as a template for building other GPCRs, since the transmembrane segments of these receptors are highly homologous (especially for the class rhodopsin-like receptors).[19] Other components of the signaling machinery have conserved structures as well: the high-resolution structures of G proteins as well as all arrestins show only small structural variance.[20] Such observations suggest that the mechanistic model of the G protein activation by GPCRs and the arrestin-mediated desensitization and internalization processes must be conserved as well.

Beside of clear advantage of having the rhodopsin structure revealed, there is the dark side of it. The structure was elucidated in its inactive, dark, form. Various experiments suggest some rearrangements of transmembrane helix bundle during activation process of GPCRs. Only an active conformation (so called the Meta II state) represents the rhodopsin structure able to bind to other proteins. This conformation is taken while binding the agonists, therefore, such structure is needed for effective drug design. Experiments of electron cryomicroscopy from 2D crystals revealed that density maps of Meta I state (the state that precede Meta II and is reached in microseconds by rhodopsin) are very similar to inactive rhodopsin structure.[21] New experiments are needed to shed some light on active rhodopsin in Meta II structure (reached by rhodopsin in milliseconds).

2.2.1.   Crystallographic structures

The recently published structure of rhodopsin pushed the resolution limit to 2.2 Å. The new structure completely resolved the polypeptide chain and provided details of ligand (retinal) binding site (PBD accession code 1U19).[17] Because the space group (P4₁) was the same as in the former structures of rhodopsin, the previous trace of the backbone was retained. The other data from Schertler's group (1GZM)[18] provided the data for a crystal belonging to another crystallographic group (P3₁). This structure, resolved to 2.65 Å, shows different orientation of the cytoplasmic loop between transmembrane helices V and VI, whereas the C-terminal region is not seen at all, contrary to the structure from P4₁ space group crystals. Both crystal structures represent dimers of rhodopsin. Unfortunately, this is not a native dimer since protomers are in upside-down position to each other. An interface is formed by helix I in 1U19 and helix V in 1GZM. Such bottom-up arrangements are more stable in the absence of membrane what highlights the great role of phospholipids in forming the appropriate interface in the native rhodopsin dimer. The comparison of the 1U19 and 1GZM dimeric structures is shown on Figure 11-1.

In the structure of rhodopsin one can differentiate several groups of amino acids called microdomains. They were found to be conserved among nearly all GPCRs because they are necessary for these receptors either to take the proper shape or to function properly (to be activated by the ligand, or light in case of rhodopsin, but not spontaneously).[22] The microdomains are shown on Figure 11-2 on the structure of rhodopsin. Microdomain DRY (ERY in case of rhodopsin) located on helix III and NPxxY, located on helix VII, are responsible for activation process of GPCRs.

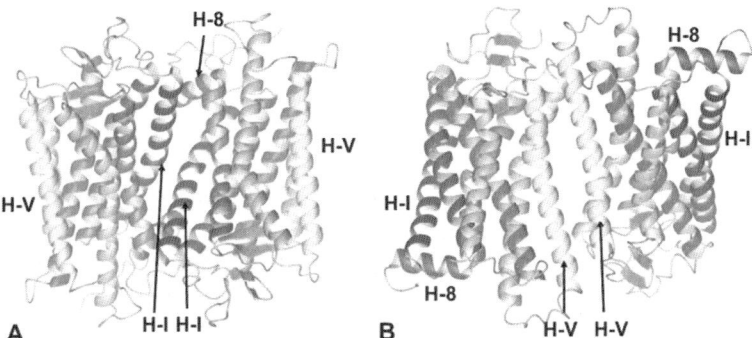

Figure 11-1. The comparison of crystal structures of bottom-up rhodopsin dimers. Transmembrane helices are marked with roman numbers. Cytoplasmic helix 8 is parallel to the membrane. A. The structure 1U19 (P4$_1$ cryst. group). B. The structure 1GZM (P3$_1$ cryst. group)

DRY links helices III and VI together and releases upon activation by agonist what makes it possible to form an active site for G protein on cytoplasmic side of the receptor. NPxxY glues helices VII and 8. Strongly conserved disulfide bridge links extracellular loop between helices IV and V with helix III. It also prevents the receptor against accidental activation. Proline on helix VI introduces a strong kink into this helix what is required for activation of the receptor and taking the proper shape. The ligand (retinal) is covalently bound to rhodopsin. In all other GPCRs its endogenous ligands form nonbonded interactions only.

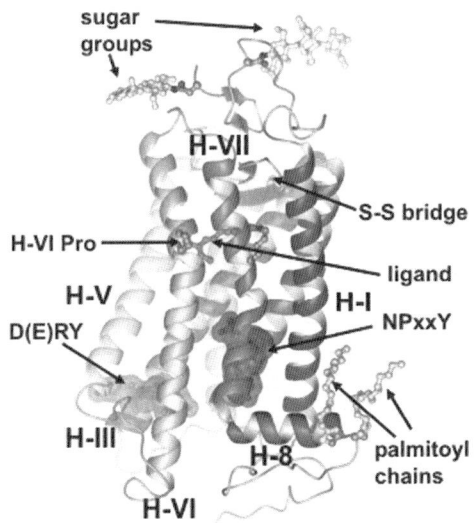

Figure 11-2. Microdomains of rhodopsin

3.   DIMERIZATION AND OLIGOMERIZATION

Oligomerization may generate the novel binding sites for agonists and antagonists or alternate structural response to ligands. Both possibilities can be achieved together. Cooperative effects may also contribute to binding ligands – binding a ligand by one protomer may affect binding the same or another ligand by the second protomer. Furthermore, each protomer may play a different role in oligomeric receptor, one binding a ligand and other(s) binding and activating G protein.

How GPCRs operate is one of the most fundamental problems in the field of transmembrane signal transduction. A growing body of pharmacological, biochemical and biophysical data strongly suggest that these receptors form functional homo- and heterodimers as well as higher-order oligomers.

3.1.   Experimental evidence

In the last years oligomerization of GPCRs became experimentally evident. Biophysical methods based on luminescence and fluorescence energy transfer confirmed the existence of dimeric and even higher-oligomeric structures. It is suggested that dimerization plays a role in various aspects of receptor biogenesis and function. In some cases receptors dimerize spontaneously and in others dimerization is promoted by a ligand.

The issue of dimerization of type C (glutamate receptor class) is clearly set. Type C GPCRs form dimers by covalent linkage of disulfide bond at their extracellular N-terminal domains (e.g. metabotropic glutamate receptor) or by strong non-covalent, intermolecular interactions (e.g. two γ-aminobutyric acid, GABA, receptor subtypes). But is dimerization state also required for their function?

3.1.1.   Passing a signal

The first convincing evidence that the dimerization is necessary for signal transduction came from research on GABA receptors (GbR). Only the coexpression of receptor subtypes GbR_1 and GbR_2 on the cell surface made it possible to pass a signal through the membrane and activate a specific G protein. Using a combination of chimeric receptors of GbR_1 and GbR_2 a transactivation mechanism was proposed in which GbR_1 binds GABA and subsequently GbR_2 activates suitable G protein.[9]

It was confirmed experimentally on α_{1B}-adrenoceptor and G_{11} that GPCRs are able to bind G_α and $G_{\beta\gamma}$ independently and interactions with $G_{\beta\gamma}$ contribute considerably to the ability of agonists to activate alpha subunit of G protein. Studies on leukotriene B4 BLT1 receptor and a G protein consisted of α_{i2}, β_1 and γ_2 subunits also provided evidence for a pentameric complex involving two GPCRs and a trimeric G protein.[23]

Both protomers of metabotropic glutamate (mGlu) receptors are required to be activated to obtain the highest binding to G protein. Extracellular part of mGlu receptors forms so called Venus Flytrap domain (VFT). It is composed of two lobes that stay open without agonist but is closing when agonist is bound to the cleft between lobes. It was reported that closure of one VFT per dimer was necessary to activate that receptor, but the closure of both VFT domains was required for full activation of G protein.[10]

Bioluminescence (BRET) and fluorescence (FRET) resonance energy transfer methods revealed the existence of adenosine (A_{2A}) and dopamine (D_2) heterodimers in living cells. An interface between them involve helix H-V and/or H-VI and a cytoplasmic loop between them from D_2R and H-IV from $A_{2A}R$.[24] Significantly larger size of G proteins relative to the cytoplasmic surface of GPCRs supports the idea that the activating platform for G proteins is the dimeric (or oligomeric) form of the receptor.

3.1.2. Drugs targeted at dimeric state

Opioid receptors belong to family A (rhodopsin like) GPCRs and exist as three types μ, δ and κ. The recent experiments that opioid agonist selectively activates heterodimers but not monomers or homodimers *in vivo*[25,26] are very important because most of other data on dimerization of GPCRs come from *in vitro* experiments. 6'-guanidinonaltrindole (6'-GNTI) is an analgesic and activates δ − κ heterodimers only. Receptor synergy or a combination of κ agonistic effect together with δ antagonism couldn't explain experimental results. δ − κ heterodimers are restricted specifically to spinal cord so using such drugs is tissue specific and provides an approach to design analgesics with reduced side effects. This could lead to new classes of selective pharmaceuticals directed to oligomeric state of GPCRs.

3.1.3. Atomic Force Microscopy (AFM) measurements

AFM measurements revealed the oligomeric structure of rhodopsin in native rod cell disc membranes.[5,27] The functional unit is a dimer and the whole structure is a mosaic of rhodopsin oligomers composed of long chains of dimers linked perpendicular to the axis of a dimer. Based on distances between rhodopsin monomers it was possible to build a model of rhodopsin oligomer.[6] This model, together with crystal structures of G protein, allowed us to build the model of a complex of transducin (rhodopsin G protein) and rhodopsin in oligomeric state.[28] Crystallographic structure of rhodopsin shows that its cytoplasmic surface is too small to bind big trimeric G protein and, what is more important, to accommodate all crosslinking data.

Modeling may provide valuable information about interfaces of interacting proteins and how oligomeric state of GPCRs affects binding of other proteins. This is especially attractive taking into account that direct proof of the structure of complexes involving membrane proteins is very difficult via crystallization or NMR measurements.

3.2. Modeling the Complexes of Oligomeric Rhodopsin

Rhodopsin being the only GPCR with resolved 3D structure is still used as a template for structure building of other GPCRs. Currently, only the ground-state structure of rhodopsin has been described. Hence, there is a great role of modeling in elucidating structural aspects of activating and passing the signal while incorporating the oligomeric state of rhodopsin.

Several methods were used for predicting the interfaces between proteins. Many of the computational methods used for modeling protein-protein complexes are similar to those used to model protein-ligand complexes. Due to the size of the computational task involved in docking of two large protein structures, an approximation is often used to treat them as rigid bodies. For instance program FTDOCK was used in searching for optimum interactions between two rigid proteins.[29] Addition of electrostatic component greatly improved ability for finding final structures. Furthermore, side chain flexibility and the effect of solvation[30] were also implemented for this task. Protein-protein interface can also be modelled by molecular surface fitting with surface flexibility implicitly addressed through liberal intermolecular penetration.[31] Fuzzy logic algorithm was also implemented in shape complementarity problem of interacting proteins.[32,33]

The above methods have been developed based on the observations of complexes formed by globular and soluble proteins. Therefore it is not known whether the interfaces of the membrane proteins, especially GPCRs, have similar features. For the small number of structural data of membrane proteins and especially protein-protein complexes it is difficult to apply the same techniques or to use the same scoring function. Among new methods evolutionary trace (ET)[34] shows high sensitivity for the prediction of the interfaces of protein complexes. Recently, this method was used for GPCRs[35] and it was reported that evolutionary trace residue cluster corresponds to the dimer interface. Dean *et al.*[36] also used ET method for class A, B and C GPCRs and identified clusters of trace residues. Nemoto and Toh[37] improved ET method by introduction of structural information. The procedure involved projection of 3D coordinates on a 2D plane, identification of exposed and inner residues, and identification of candidates for interface residues.

Recently, an extensive review describing computational approaches towards structure-function analysis of GPCRs used so far came in.[38] Among all aspects of modeling of GPCRs authors demonstrate computational experiments on dimerization/oligomerization processes and modeling GPCR - G protein interactions.

3.2.1. Modeling the oligomeric state

Based on the Atomic Force Microscope measurements of distances between rhodopsins in the paracrystals as well as energetic considerations, we constructed a model with helices IV and V forming an interface between rhodopsin molecules. Oligomers in the model were built from separate dimers and linked together by a long cytoplasmic loop between helices V and VI.[6] We assumed that a dimer is a repetitive motif in the oligomer (forming a double row of monomers), therefore tetramers and higher structures are connected in an identical manner. Optimization of the oligomer structure was performed maintaining frozen parts of $C\alpha$ atoms from transmembrane helices.

We enhanced the model by addition of phospholipids and performed molecular dynamics in periodic box. Three types of phospholipids were used with phosphatidylcholine headgroups on the intradiscal (extracellular for other GPCRs) side and phosphatidylethanolamine and phosphatidylserine headgroups (three times more phosphatidylethanolamine headgroups than phosphatidylserine) on the cytoplasmic side to mimic native membranes of rod outer segments. All three types of phospholipids contain the saturated stearoyl chain (18:0) in the *sn*1 position and the polyunsaturated docosahexaenoyl chain (22:6n-3) in the *sn*2 position.

The whole system was soaked in water, counterions were added and molecular dynamics was conducted in periodic box. The final model of rhodopsin oligomer is shown on Figure 11-3. Phospholipids flow on the left and right side of the central double row in channels two phospholipids wide. Thinner channels one phospholipid wide, hidden below the extended cytoplasmic loops of TMH-V and TMH-VI, flow perpendicularly to the wider channels.

3.2.2.   The complex of rhodopsin and G protein

Having the model of rhodopsin oligomer it was possible to start building a much bigger model involving trimeric G protein. It was known that transducin (rhodopsin G protein) binds to activated rhodopsin with its C-terminal helix. Taken together crosslinking data we have found that the long N-terminal helix of transducin binds to a special groove on rhodopsin's surface formed by the cytoplasmic loops and C-terminus. It was not necessary for the second rhodopsin in a dimer to be

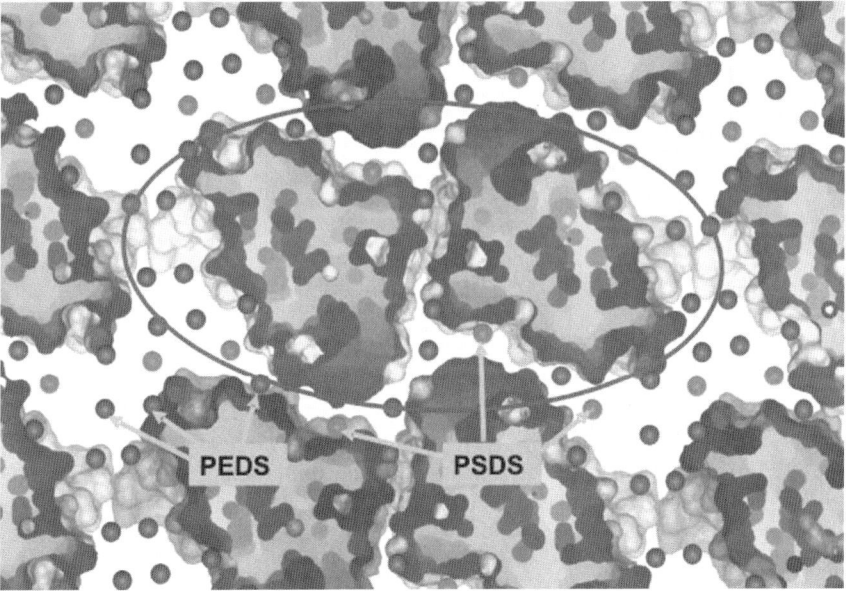

Figure 11-3. The model of rhodopsin oligomer. View from cytoplasmic side while cytoplasmic loops of rhodopsin molecules were removed. A single dimer is marked by ellipse. Positions of phospholipids are indicated by balls. PEDS – phospholipids with ethanolamine heads, PSDS – with serine heads

activated. It anchors Gt_α and facilitates binding to the activated rhodopsin. The whole trimeric transducin is big enough to cover and bind two rhodopsin dimers (Figure 11-4). The adjacent rhodopsin dimer, composed of inactive monomers, gives an additional surface to stabilize the complex. After unbinding the transducin beta-gamma subunit, the remaining alpha part can bind to the second molecule of transducin and facilitate docking to an adjacent rhodopsin tetramer, giving rise to positive cooperativity in binding of transducin.[28]

Both alpha and gamma subunits of transducin are modified by addition of hydrophobic chains and presence of them is required for binding to rhodopsin.[39] Phospholipid channels on both sides of rhodopsin double rows make it possible to soak hydrophobic chains of transducin in the membrane even in maximally crowded rhodopsin oligomer structure (Figure 11-3).

3.2.3.   The complex of rhodopsin and arrestin

The second model of the complex involving oligomeric rhodopsin we created was a complex of arrestin and rhodopsin dimer. Arrestin possesses two large concave binding sites separated by 3.8 nm. The same distance was measured in AFM and it is the distance between rhodopsin monomers forming oligomeric structures. Furthermore, these two concave lobes of arrestin were found by crosslinking experiments to bind rhodopsin. Therefore, we built a model of one arrestin molecule bound to a rhodopsin dimer. Such a structure is shown in Figure 11-5. There is not only complementarity of shapes of binding lobes arrestin and rhodopsin dimer.

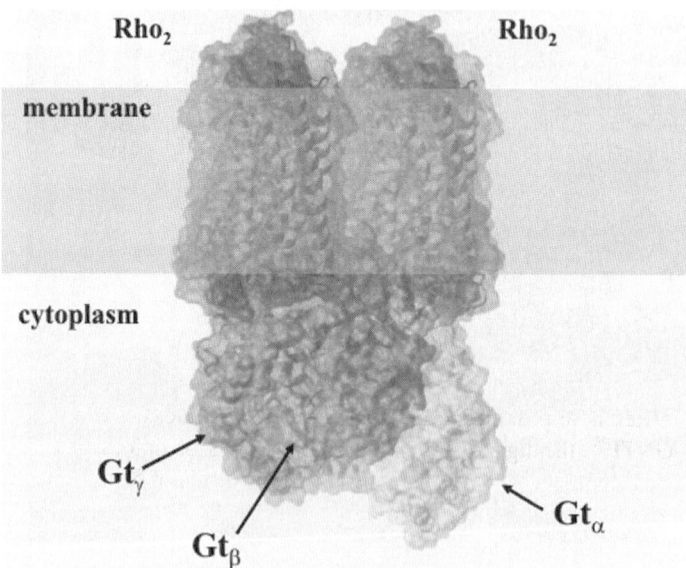

Figure 11-4. The model of the complex of two rhodopsin dimers and trimeric transducin (rhodopsin G protein)

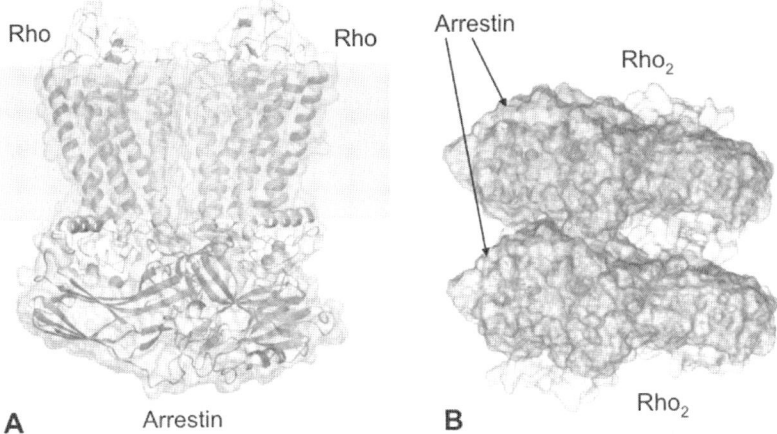

Figure 11-5. The model of arrestin and rhodopsin complex 1:2. A. One arrestin molecule bound to rhodopsin dimer. B. Two arrestin molecules bound to two adjacent rhodopsin dimers. View from cytoplasmic side

During formation of the complex the long range electrostatic forces pull the two interacting parts together since there is a strict complementarity of electrostatic potential of binding surfaces of both arrestin and a rhodopsin dimer. The loose flaps of arrestin (negatively charged) fit themselves into rhodopsin cytoplasmic cavities. C-terminal ends of rhodopsins are also flexible and they fill tightly into both lobes of arrestin (Figure 11-5A). Both lobes of arrestin are built with many positively charged residues what makes it feasible to bind to rhodopsin dimer even in case when only one protomer is activated and phosphorylated. Numerous experiments confirmed that both lobes are required for binding and deactivation of the receptor. The model also shows positive cooperativity between arrestin molecules bound to adjacent rhodopsin dimers in oligomeric double row (Figure 11-5B).

4. CONCLUSIONS AND CHALLENGES

In recent years pharmacological, biochemical and biophysical methods confirmed the existence of dimeric and higher-oligomeric structures involving GPCRs so oligomerization of G protein coupled receptors became experimentally evident. Oligomeric state makes new possibilities in some unexpected areas. For instance the binding of 6'-GNTI[25], the ligand preferring heterodimers of opioid receptors $\delta - \kappa$, may also be possible after some greater oligomeric structure of δ and κ opioid receptors is formed what generates new binding sites for specific drugs. Some cross-linking experiments on GPCRs may be explained only after oligomerization and not dimerization is assumed. Further studies are needed to reveal physiological and pathological consequences of oligomerization *in vivo*. At the molecular level unraveling the structural organization of GPCR oligomers and defining the rules that underlie selection process during homo- and hetero-oligomerization is the most challenging task that could lead to specific drugs directed to this state of GPCRs.

Development of materials that involve GPCRs in close proximity to each other requires recognizing of their properties in oligomeric state. The properties may be totally different from those investigated from single receptors and demonstrate unexpected complexity. Rhodopsin is still the only GPCR with three-dimensional structure known. Therefore, modeling of oligomeric state of G protein-coupled receptors relies on rhodopsin structure. Also interactions of rhodopsin within the oligomer and with other proteins may be valuable for construction of analogous complexes for other GPCRs.

Material science can also benefit from oligomeric state of proteins. New retinal nano-ceramic materials with pillared hybrid micro-structures were fabricated for potential applications in optical holographic data storage. It was observed that the Schiff bases in retinal have substantial effect on optical properties of nano-ceramic films as well as diffraction efficiency for holographic storage. This study indicates feasibility of optimizing optical properties of nano-ceramic clay systems using Schiff bases for a variety of photonic applications.[40]

Bacteriorhodopsin (bR) is investigated for several years for its spectral properties and potential usage in spectral devices. Recently the real time image processing was demonstrated by recording and reconstructing the transient photoisomerizative grating formed in the bR film using Fourier holography. Desired spatial frequencies including both high and low band in the object beam are reconstructed by controlling the reference beam intensity. This technique was used to process mammograms in real-time for identification of microcalcifications buried in the soft tissue for early detection of breast cancer. A feature of the technique is the ability to transient display of selected spatial frequencies in the reconstructing process which enables the radiologists to study the features of interest.[41]

The usage of rhodopsin for technological purposes is far less advanced and new research is needed to fully understand even basic processes that rhodopsin participate in like excitation of chromophore. Recently, quantum mechanics/molecular mechanics strategy was applied to the investigation of the excited state of the visual pigment rhodopsin (Rh). As a consequence, the simulation of the absorption and fluorescence of Rh and its retinal chromophore in solution allows for a nearly quantitative analysis of the factors determining the properties of the protein environment. Authors demonstrated that the Rh environment is more similar to the "gas phase" than to the solution environment and that the so-called "opsin shift" originates from the inability of the solvent to effectively "shield" the chromophore from its counterion. The same strategy was used to investigate three transient structures involved in the photo isomerization of Rh under the assumption that the protein cavity does not change shape during the reaction. Most importantly, it was shown that the mechanism of the approximate 30 kcal/mol photon energy storage observed for Rh is not consistent with a model based exclusively on the change of the electrostatic interaction of the chromophore with the protein/counterion environment.[42]

The energy storage and the molecular rearrangements due to the primary photochemical event in rhodopsin was also investigated in another study by using quantum mechanics/molecular mechanics hybrid methods in conjunction with high-resolution structural data of bovine visual rhodopsin. The analysis of the reactant and product molecular structures revealed the energy storage mechanism as determined by the detailed molecular rearrangements of the retinyl chromophore. These results are particularly relevant to the development of the structure-function relations in prototypical G-protein-coupled receptors.[43]

Copolymer-based membrane technology may enable the development of a versatile class of nanoscale materials in which biomolecules, such as membrane proteins, can be reconstituted. These active materials possess a broad applicability in areas such as the enhancement of existing technologies or production of current-generating films for power sources. For example, these active materials can be integrated with fuel cell ion transport membranes in order to improve its ability to retain leaking protons. Also, the demonstration of protein-driven current production across these membranes represents a possible alternative power source that is both highly efficient and light in weight. Large-area copolymer biomembranes functionalized by bacteriorhodopsin (bR) and cytochrome c oxidase (COX) ion transport proteins was fabricated already. Among their many advantages over conventional lipid-based membrane systems, block copolymers can mimic natural cell biomembrane environments in a single chain, enabling large-area membrane fabrication using methods such as Langmuir-Blodgett (LB) deposition. Following the large-scale insertion of proteins into block copolymer LB films, authors have demonstrated significant pH changes based upon light-actuated proton pumping.[44]

REFERENCES

1. Mirzadegan T, Benko G, Filipek S, Palczewski K (2003) Sequence analyses of G-protein-coupled receptors: similarities to rhodopsin. Biochemistry 42(10):2759–2767

2. Bartfai T, Benovic JL, Bockaert J, Bond RA, Bouvier M, Christopoulos A, Civelli O, Devi LA, George SR, Inui A, Kobilka B, Leurs R, Neubig R, Pin JP, Quirion R, Roques BP, Sakmar TP, Seifert R, Stenkamp RE, Strange PG (2004) The state of GPCR research in 2004. Nat. Rev. Drug Discov. 3(7):574–626

3. Takeda S, Kadowaki S, Haga T, Takaesu H, Mitaku S (2002) Identification of G protein-coupled receptor genes from the human genome sequence. FEBS Lett. 520(1–3):97–101

4. Ballesteros J, Palczewski K (2001) G protein-coupled receptor drug discovery: Implications from the crystal structure of rhodopsin. Curr. Opin. Drug Discov. Devel. 4:561–574

5. Fotiadis D, Liang Y, Filipek S, Saperstein DA, Engel A, Palczewski K (2003) Atomic-force microscopy: Rhodopsin dimers in native disc membranes. Nature 421(6919):127–128

6. Liang Y, Fotiadis D, Filipek S, Saperstein DA, Palczewski K, Engel A (2003) Organization of the G protein-coupled receptors rhodopsin and opsin in native membranes. J. Biol. Chem. 278(24):21655–21662

7. Angers S, Salahpour A, Bouvier M (2002) Dimerization: An emerging concept for G protein-coupled receptor ontogeny and function. Annu. Rev. Pharmacol. Toxicol. 42:409–435

8. Terrillon S, Bouvier M (2004) Roles of G-protein-coupled receptor dimerization – From ontogeny to signalling regulation. EMBO Rep. 5(1):30–34

9. Moepps B, Fagni L (2003) Mont Sainte-Odile: A sanctuary for GPCRs. Confidence on signal transduction of G-protein-couple receptors. EMBO Rep. 4(3):237–243

10. Bulenger S, Marullo S, Bouvier M (2005) Emerging role of homo- and heterodimerization in G-protein-coupled receptor biosynthesis and maturation. Trends Pharmacol. Sci. 26(3):131–137

11. Milligan G (2004) G protein-coupled receptor dimerization: Function and ligand pharmacology. Mol. Pharmacol. 66(1): 1–7

12. Park PSH, Filipek S, Wells JW, Palczewski K (2004) Oligomerization of G protein-coupled receptors: Past, present, and future. Biochemistry 43(50):15643–15656

13. Fredriksson R, Lagerstrom MC, Schioth HB (2005) Expansion of the superfamily of G-protein-coupled receptors in chordates. Ann. N. Y. Acad. Sci. 1040:89–94

14. Schioth HB, Fredriksson R (2005) The GRAFS classification system of G-protein coupled receptors in comparative perspective. Gen. Comp. Endocrinol. 142(1–2):94–101

15. Bjarnadottir TK, Schioth HB, Fredriksson R (2005) The phylogenetic relationship of the glutamate and pheromone G-protein-coupled receptors in different vertebrate species. Ann. N. Y. Acad. Sci. 1040:230–233

16. Palczewski K, Kumasaka T, Hori T, Behnke CA, Motoshima H, Fox BA, Le Trong I, Teller DC, Okada T, Stenkamp RE, Yamamoto M, Miyano M (2000) Crystal structure of rhodopsin: A G protein-coupled receptor. Science 289(5480):739–745

17. Okada T, Sugihara M, Bondar AN, Elstner M, Entel P, Buss V (2004) The retinal conformation and its environment in rhodopsin in light of a new 2.2 angstrom crystal structure. J. Mol. Biol. 342(2):571–583

18. Li J, Edwards PC, Burghammer M, Villa C, Schertler GF (2004) Structure of bovine rhodopsin in a trigonal crystal form. J. Mol. Biol. 343(5):1409–38

19. Filipek S, Teller DC, Palczewski K, Stenkamp R (2003) The crystallographic model of rhodopsin and its use in studies of other G protein-coupled receptors. Annu. Rev. Biophys. Biomol. Struct. 32:375–397

20. Ridge KD, Abdulaev NG, Sousa M, Palczewski K (2003) Phototransduction: crystal clear. Trends Biochem. Sci. 28(9):479–487

21. Ruprecht JJ, Mielke T, Vogel R, Villa C, Schertler GF (2004) Electron crystallography reveals the structure of metarhodopsin I. EMBO J. 23(18):3609–3620

22. Filipek S, Stenkamp RE, Teller DC, Palczewski K (2003) G protein-coupled receptor rhodopsin: A prospectus. Annu. Rev. Physiol. 65:851–879

23. Baneres JL, Parello J (2003) Structure-based analysis of GPCR function: Evidence for a novel pentameric assembly between the dimeric leukotriene B-4 receptor BLT1 and the G-protein. J. Mol. Biol. 329(4):815–829

24. Canals M, Marcellino D, Fanelli F, Ciruela F, de Benedetti P, Goldberg SR, Neve K, Fuxe K, Agnati LF, Woods AS, Ferre S, Lluis C, Bouvier M, Franco R (2003) Adenosine A(2A)-dopamine D2 receptor-receptor heteromerization – Qualitative and quantitative assessment by fluorescence and bioluminescence energy transfer. J. Biol. Chem. 278(47):46741–46749

25. Waldhoer M, Fong J, Jones RM, Lunzer MM, Sharma SK, Kostenis E, Portoghese PS, Whistler JL (2005) A heterodimer-selective agonist shows in vivo relevance of G protein-coupled receptor dimmers. Proc. Natl. Acad. Sci. USA 102(25):9050–9055

26. Park PSH, Palczewski K (2005) Diversifying the repertoire of G protein-coupled receptors through oligomerization. Proc. Natl. Acad. Sci. USA 102(25):8793–8794

27. Fotiadis D, Liang Y, Filipek S, Saperstein DA, Engel A, Palczewski K (2004) The G protein-coupled receptor rhodopsin in the native membrane. FEBS Lett. 564(3):281–288

28. Filipek S, Krzysko KA, Fotiadis D, Liang Y, Saperstein DA, Engel A, Palczewski K (2004) A concept for G protein activation by G protein-coupled receptor dimers: the transducin/rhodopsin interface. Photochemical & Photobiological Sciences 3(6):628–638

29. Gabb HA, Jackson RM, Sternberg MJE (1997) Modelling protein docking using shape complementarity, electrostatics and biochemical information. J Molecular Biology 272(1):106–120

30. Jackson RM, Gabb HA, Sternberg MJE (1998) Rapid refinement of protein interfaces incorporating solvation: Application to the docking problem. J. Mol. Biol. 276(1):265–285

31. Duhovny D, Nussinov R, Wolfson HJ (2002) Efficient unbound docking of rigid molecules, in Algorithms in Bioinformatics, Proceedings. vol. 2452, Springer-Verlag, Berlin, pp 185–200

32. Exner TE, Keil M, Brickmann J (2002) Pattern recognition strategies for molecular surfaces. I. Pattern generation using fuzzy set theory. J. Comput. Chem. 23(12):1176–1187

33. Exner TE, Keil M, Brickmann J (2002) Pattern recognition strategies for molecular surfaces. II. Surface complementarity. J. Comput. Chem. 23(12):1188–1197

34. Lichtarge O, Bourne HR, Cohen FE (1996) Evolutionarily conserved G (alpha beta gamma) binding surfaces support a model of the G protein-receptor complex. Proc. Natl. Acad. Sci. USA 93(15):7507–7511

35. Madabushi S, Gross AK, Philippi A, Meng EC, Wensel TG, Lichtarge O (2004) Evolutionary trace of G protein-coupled receptors reveals clusters of residues that determine global and class-specific functions. J. Biol. Chem. 279(9):8126–8132

36. Dean MK, Higgs C, Smith RE, Bywater RP, Snell CR, Scott PD, Upton GJG, Howe TJ, Reynolds CA (2001) Dimerization of G-protein-coupled receptors. J. Med. Chem. 44(26):4595–4614

37. Nemoto W, Toh H (2005) Prediction of interfaces for oligomerizations of G-protein coupled receptors. Proteins 58(3):644–660

38. Fanelli F, De Benedetti PG (2005) Computational Modeling approaches to structure-function analysis of G protein-coupled receptors. Chem. Rev. 105(9):3297–3351

39. Herrmann R, Heck M, Henklein P, Kleuss C, Hofmann KP, Ernst OP (2004) Sequence of interactions in receptor-G protein coupling. J. Biol. Chem. 279(23):24283–24290

40. Wu PF, Bhamidipati M, Coles M, Rao D (2004) Biological nano-ceramic materials for holographic data storage. Chem. Phys. Lett. 400(4–6):506–510

41. Kothapalli SR, Wu PF, Yelleswarapu CS, Rao D (2004) Medical image processing using transient Fourier holography in bacteriorhodopsin films. Appl. Phys. Lett. 85(24):5836–5838

42. Andruniow T, Ferre N, Olivucci M (2004) Structure, initial excited-state relaxation, and energy storage of rhodopsin resolved at the multiconfigurational perturbation theory level. Proc. Natl. Acad. Sci. USA 101(52):17908–17913

43. Gascon JA, Batista VS (2004) QM/MM study of energy storage and molecular rearrangements due to the primary event in vision. Biophys. J. 87(5):2931–2941

44. Ho D, Chu B, Lee H, Montemagno CD (2004) Protein-driven energy transduction across polymeric biomembranes. Nanotechnology. 15(8):1084–1094

CHAPTER 12

MOLECULAR DYNAMICS SIMULATIONS OF HYDROGEN ADSORPTION IN FINITE AND INFINITE BUNDLES OF SINGLE WALLED CARBON NANOTUBES

HANSONG CHENG[1], ALAN C. COOPER[1], GUIDO P. PEZ[1],
MILEN K. KOSTOV[2], M. TODD KNIPPENBERG[1,3],
PAMELA PIOTROWSKI[3] AND STEVEN J. STUART[3]

[1] *Air Products and Chemicals, Inc. 7201 Hamilton Boulevard, Allentown, PA 18195-1501*
[2] *Department of Physics, Pennsylvania State University, University Park, PA 16802-6300*
[3] *Department of Chemistry, Clemson University, Clemson, SC 29634*

Abstract: Molecular dynamics simulations have been used to systematically study hydrogen storage in single walled carbon nanotubes of various diameters and chiralities using a recently developed curvature-dependent force field. Several fundamental issues related to the effects of nanotube size, chirality and the thickness of nanotube bundles have been examined. A novel methodology for the analysis of effective average adsorption energy and storage capacity was developed. Our simulation results suggest strong dependence of H_2 adsorption energies on the nanotube diameter but less dependence on the chirality. Substantial lattice expansion upon H_2 adsorption was found. The average adsorption energy increases with the lowering of nanotube diameter (higher curvature) and decreases with higher H_2 loading. The calculated H_2 vibrational power spectra and radial distribution functions indicate a strong attractive interaction between H_2 and nanotube walls. The calculated diffusion coefficients are much higher than what has been reported for H_2 in microporous materials such as zeolites, indicating that diffusivity does not present a problem for adsorption energy and effective capacity hydrogen storage in carbon nanotubes. We show that adsorption energy and effective storage capacity can be defined in a distance-dependent manner, providing a more comprehensive understanding of adsorption behavior

Keywords: carbon nanotubes; hydrogen adsorption, molecular dynamics

1.   INTRODUCTION

The sorption and storage of hydrogen by various new structural forms of carbon, which are inherently light weight materials, has recently gained widespread attention as possible enabling technology of a future hydrogen economy. The use of polymer

469

W. A. Sokalski (ed.), Molecular Materials with Specific Interactions, 469–485.
© 2007 *Springer.*

electrolyte membrane (PEM) fuel cells for power generation and for onboard vehicle applications depends on efficient hydrogen storage. It has been known for some time that high-surface-area activated carbons and certain alkali-metal graphite intercalation compounds will reversibly adsorb considerable quantities of hydrogen, but only at cryogenic temperatures, due to the low heat of adsorption of H_2. Therefore such systems do not offer practical or economic advantages over the use of compressed or liquefied hydrogen. The discovery of single walled carbon nanotubes (SWNT) has prompted the investigation of these materials for the separation and adsorption of gases at ambient temperature and moderate pressures. SWNT are composed of a single graphene sheet (one layer of graphite), rolled into a seamless cylinder with a diameter that generally ranges from 0.7 to 2.0 nm. Unfortunately, the reported storage capacity of SWNT varies widely depending on the quality of the nanotubes and the pretreatment of the samples.[1-4] Of fundamental importance is the reported heat of adsorption (~ 4.5 kcal/mol)[1,2], which is substantially higher than what has been found in graphitic carbon, such as graphite and activated carbons (0.9–1.25 kcal/mol)[5-9] and even graphite intercalation compounds (~ 2.3 kcal/mol for KC_{24}).[10-13]

The current consensus appears to be that the unusually high H_2 adsorption energy on SWNT is related to the curvature of the graphitic walls of the nanotubes, which forces the carbon atoms to adopt a quasi-sp^2/quasi-sp^3 hybridization.[14] Curved carbon thus provides an attractive force to adsorb hybridization state intermediate between sp^2 and sp^3.[14] The exterior of a curved carbon surface thus provides an attractive force in the adsorbtion of molecular hydrogen. Nevertheless, many fundamental issues related to the effects of curvature and chirality and how these give rise to the strong H_2 adsorption enthalpy in SWNT remain to be addressed in detail in order to develop a comprehensive understanding of the overall adsorption phenomenon. These issues include:

(1) The distribution of molecular hydrogen throughout a SWNT lattice as a funtion of the nanotube diameter/chirality. The ratio of endohedral (within the nanotube)/exohedral H_2 population in the lattice is largely determined by the energetics associated with adsorption in these sites. To answer this question, careful examination of various H_2 population distributions in the nanotube bundle is required.

(2) The population of H2 that interacts effectively with SWNT a given H_2 loading. Does the diameter and chirality of the tubes affect this quantity? Is the year 2010 gravimetric capacity target of 6.5 wt. % H_2 set by the U.S. Department of Energy feasible?

(3) The range of SWNT diameters and/or a specific chiralities for which H_2 adsorption is energetically most favorable.

In this chapter, we attempt to address these fundamental questions by performing extensive molecular dynamics simulations. In previous publications,[15,16] we have demonstrated using *ab initio* molecular dynamics based on local density functional theory that H_2 adsorption energies in a lattice of (9,9) armchair SWNT at a variety of temperatures with 0.4 wt. % hydrogen loading are significantly higher than in graphite and graphite intercalation compounds. Despite the over-binding expected from local density functional theory calculations, the calculated adsorption energies were in a close proximity to a reported experimental values.[1,2] The simulations showed that H_2 adsorption in SWNT is a highly dynamic phenomenon with very high H_2 mobility in the lattice and substantial SWNT framework deformation at moderate temperatures. Therefore any simulation method must include a way to accurately simulate the deformation in the SWNT walls at a finite temperature.

The computational expense of *ab initio* calculations makes the method impractical for use in performing large-scale or long-time dynamics simulations, such as those required to answer the questions above. Therefore, we developed a novel empirical force field approach that can be used to describe molecular interactions in a curved carbon environment[18]. The goal of this new model is to accurately describe the interactions of H_2 with SWNT in the range between two extreme diameters: SWNT with an infinite radius (i.e. graphene, a single layer of graphite), and SWNT with one of the smallest possible radii, the (2,2) armchair SWNT. In the case of tubes with small radii, the carbon atoms acts as highly reactive radicals. A carbon atom in a highly curved carbon surface readily adopts a pseudo-tetrahedral configuration, resulting in high electron density on the exterior surface of the SWNT.[15,18] The graphene sheet, on the other hand demonstrates only weak van der Waals interactions with H_2. It is therefore anticipated that a nanotube with a radius in between these extremes would respond to H_2 uptake with a stronger interaction than graphite but with a weaker force than the smallest SWNT, and that the intensity of the interaction would vary with the radius of curvature of the nanotube. Using this new potential, the effect of SWNT curvature are investigated to determine if interstitial sites in the bundle are more energetically favorable adsorption sites than the endohedral sites, and whether the weaker van der Waals interactions are sufficient to populate the endohedral sites.

With this new potential, we perform classical molecular dynamics simulations in order to address the fundamental issues of hydrogen storage in single walled carbon nanotubes outlined at the beginning of this chapter. The nanotubes chosen in this study include infinite bundles of three armchair (n,n), three zigzag (n,0), and one chiral (m,n) structure, with a range of diameters in addition to several finite bundles of SWNT. Heats of adsorption for H_2 are calculated in each of these nanotubes at ambient temperature with loadings ranging from 0.4 wt. % to 6.5 wt. % to gauge the feasibility of hydrogen storage in these systems.

2.   COMPUTATIONAL METHODS

The results of molecular dynamics simulations are critically dependent on the quality of the force field. In principle, any potential that accurately describes SWNT should have terms that reflect the curvature of the tube. However, curvature effects are largely ignored in force fields published in the literature.[20–23] Some potentials, such as those based on reactive bond-order formalism, correctly account for the local curvature in describing the covalent bonding interactions.[24–26] But the adsorption

energy is most strongly affected by the non-bonding interaction terms. These have
been parametrized using graphite, completely ignoring the curvature effect.

In order to include curvature-dependence in both the covalent and non-bonding
interactions, we used the adaptive intermolecular reactive bond-order (AIREBO)
potential,[24] with modified van der Waals interactions. This potential uses the same
bonding interactions as Brenner's REBO potential,[25,26] both of which correctly
account for local curvature dependence in the covalent bonding interactions.
Chemisorption is thus treated accurately, but there is no explicit or implicit curvature
dependence in the Lennard-Jones (L-J) parameters used to describe the non-
bonded van der Waals interactions (physisorption). Consequently, we modified the
Lennard-Jones parameters to make them explicitly dependent on the curvature of
the nanotube.

The van der Waals interaction was refit for H_2 adsorption in SWNT by first fitting
the L-J parameters for H-H interactions to the recent high level *ab initio* results on
interactions between H_2 molecules reported by Diep and Johnson,[27] yielding the
center of mass separation of 3.4 Å, and the Lennard-Jones parameters $\sigma_{HH} = 2.65$ Å
and $\varepsilon_{HH} = 17.4$ K. The nonbonding interaction between carbons is not expected
to have a pronounced effect on the H_2 adsorption energy and thus we retain the
standard AIREBO potential parameters of $\sigma_{CC} = 3.40$ Å and $\varepsilon_{CC} = 33.0$ K.

For the interaction potential between hydrogen and carbon, we introduce a new
procedure to derive the Lennard-Jones parameters from existing parameters that
are appropriate for carbon atoms with sp^2 and sp^3 hybridizations. These parameters
may come from existing force fields, and may have been obtained using either
experimental or *ab initio* results. The L-J parameters σ and ε are made explicitly
dependent on the radius of the nanotube, r, using the following equations:

$$\sigma(r) = f(r)\sigma_{sp^2} + [1 - f(r)]\sigma_{sp^3}, \tag{12-1}$$

$$\varepsilon(r) = \begin{cases} f(r)\varepsilon_{sp^2} + [1 - f(r)]\varepsilon_{sp^3}^{exo} & \text{exohedral} \\ f(r)\varepsilon_{sp^2} + [1 - f(r)]\varepsilon_{sp^3}^{end} & \text{endohedral} \end{cases} \tag{12-2}$$

where $f(r) = \left(1 - \frac{r_0}{r}\right)^\lambda$ and $r_0 = 1.356$ Å is the radius of a (2,2) nanotube, which
is assumed to demonstrate purely sp^3 bonding. The non-bonding interactions in
a nanotube with a large radius will thus be similar to those in a graphite sheet
while these interactions can be substantially enhanced in small radius nanotubes,
since $\varepsilon_{sp^3} > \varepsilon_{sp^2}$. For this potential, we chose λ to be 0.62 so that the van der
Waals surface of an endohedral H_2 aligned along the axis of a nanotube touches
the van der Waals surface of the SWNT wall. This procedure can be used to derive
curvature-dependent L-J parameters from curvature-independent values of σ and
ε. The specific parameters used in this study, which are obtained from previous
ab initio calculations,[14] are shown in Table 12-1. Because the σ_{C-H} parameter is
essentially insensitive to the SWNT curvature, varying by only 2.0% from sp^2 to sp^3
carbon, we chose a single value of $\sigma_{C-H} = 2.78$ Å for both sp^2 and sp^3 carbon. The
strength of the interaction varies considerably, however, both with hybridization and

Table 12-1. Lennard-Jones parameters for sp^2 and sp^3 carbons used for nonbonding interactions (units: σ in Å and ε in meV)

	$\varepsilon_{C\ C}$	σ_{C-C}	ε_{C-H} (end)	ε_{C-H} (exo)
sp^2	2.410	3.40	2.24	2.24
sp^3	2.340	3.57	4.00	11.0

direction of curvature, as illustrated by the variation of values of ε_{C-H} in Table 12-1. The detailed fitting procedure can be found in Ref. 18.

3. H$_2$ ADSORPTION IN AN INFINITE BUNDLE OF SWNT

The SWNT systems chosen in the present studies include 3 armchair nanotubes and 3 zigzag nanotubes with diameters ranging from 4 Å to 12 Å, and 1 chiral nanotube with a diameter of 8.28 Å. The nanotubes were carefully chosen to address the fundamental issues of curvature and chirality and the effect of each on the adsorption capacity. First, to understand the curvature effect on hydrogen uptake, we selected nanotubes with diameters varying from about 4 Å to 12 Å. Next, to investigate the effect of nanotube chirality, we intentionally chose the nanotubes of different chiral architectures with similar diameters. Finally, to study the capacity of a given nanotube, we included three different H$_2$ loadings at 0.4 wt. %, 3.0 wt. % and 6.5 wt. %, respectively, in our MD simulations.

In the simulations with the highest H$_2$ loading of 6.5 wt. %, the lattice constants of the periodic lattice were adjusted independently during the equilibration process in order to reduce the diagonal components of the pressure tensor towards ambient pressure. Substantial lattice expansion upon H$_2$ uptake was observed, particularly for the smaller diameter nanotubes. The volume of both exohedral and endohedral sites decreases as the nanotube diameter becomes smaller, so more of the H$_2$ must be accommodated via lattice dilation. In no case was the system volume allowed to surpass twice the unloaded volume, thus providing a limit on the lattice swelling. Consequently, the 6.5 wt. % simulations were performed at higher than ambient pressure, ranging from 7500 bar for the (3,3) tube with 4.1 Å diameter down to ~ 1000 bar for the (15,0) tube with 11.8 Å diameter. These elevated pressures even at near 100% lattice swelling suggest that hydrogen adsorption at the DOE target of 6.5 wt. % presents difficulties.

3.1. H$_2$ Distribution

In order to obtain accurate estimates of adsorption energies, it is important to understand how H$_2$ molecules are distributed in the lattice of each SWNT type. Specifically, for a given nanotube type and H$_2$ loading, we must first determine the ratio of H_2 adsorption at the endohedral sites vs. the exohedral sites. In the simulations of infinitely long (periodic) SWNT, migration of H_2 between endohedral and exohedral sites can not occur. We therefore performed a series of simulations to determine the optimal H_2 distribution across both the endohedral and exohedral sites. It is reasonable to assume that the distribution is dictated by the adsorption thermodynamics and that entropic effects arising from differences in pore geometry and free pore volume are negligible. That is, the free energy can be reasonably approximated by the internal energy. This approximation will be most accurate when the free volume is similar in the endohedral and exohedral pores, as will be the case for high loadings and for near-optimal hydrogen distributions. Thus we assume that the ratio corresponding to the highest average heat of adsorption represents the thermodynamically most favorable distribution. For each nanotube type, we sampled the endohedral/exohedral ratio with distributions ranging from 100% endohedral to 100% exohedral, in steps of 10%, in order to search for the optimal distribution of H_2 in the SWNT bundles. The calculated optimal H_2 distributions and the adsorption energies are shown in Table 12-2.

At 0.4 wt. % loading, H_2 molecules are found to be exclusively adsorbed in the interstitial space in all cases, due to the stronger interactions on the exterior of the curved carbon surface. The recent experiments conducted by Shiraishi *et al.*[2] suggest that the hydrogen that is adsorbed at ambient temperature under high hydrogen pressure (9 MPa) and subsequently desorbed at near-ambient temperatures under vacuum at 0.3 wt. % loading, is physisorbed in the exohedral sites. Our results are consistent with their findings. As the H_2 loading increases, the endohedral sites in nanotubes with larger diameters begin to become populated. This is due to a competition between the more energetically favorable adsorption at the exohedral position and the higher repulsive interactions when the density of H_2 becomes too

Table 12-2. The calculated optimal distribution of H_2 and the average adsorption energy (kcal/mol) per H_2. The statistical uncertainty of the averaged adsorption energy is approximately ±0.5 kcal/mol

Nanotube (n,m)	SWNT diameter (Å)	0.4 wt. % H_2 loading		3.0 wt. % H_2 loading		6.5 wt. % H_2 loading	
		Endo/ exo ratio	ΔE_{ad} (kcal/ mol)	Endo/ exo ratio	ΔE_{ad} (kcal/ mol)	Endo/ exo ratio	ΔE_{ad} (kcal/ mol)
(3,3)	4.068	0:100	5.7	0:100	4.9	0:100	3.7
(5,5)	6.780	0:100	5.1	20:80	2.4	30:70	1.7
(9,9)	12.204	0:100	3.8	40:60	1.4	50:50	1.1
(5,0)	3.914	0:100	5.3	0:100	5.1	0:100	2.8
(10,0)	7.828	0:100	4.8	20:80	2.1	30:70	0.9
(15,0)	11.743	0:100	3.3	30:70	1.1	50:50	0.6
(8,4)	8.285	0:100	4.3	20:80	1.9	30:70	0.6

great in the interstitial sites. Consequently, the exohedral sites will be filled first upon H_2 uptake and subsequently the endohedral sites will be populated when the interstitial space of the nanotube bundles becomes saturated with hydrogen. The endohedral adsorption begins at lower loadings for nanotubes with smaller radii. It is worth noting that the H_2 adsorption in the two smallest diameter nanotubes, (3,3) and (5,0), is exclusively exohedral because the exceedingly small tube diameters result in large repulsive interactions between endohedral hydrogens and the tube walls. The general trend of the H_2 population distribution in the SWNT bundles is that the population at the endohedral sites increases with H_2 loading and the nanotube radius with no apparent dependence on tube chirality.

3.2.   Heat of Adsorption

The average heats of adsorption at room temperature, calculated for the optimal exohedral/endohedral distribution, are shown in Table 12-2. These averages are calculated using Equation (3), with no attempt to correct for any distance-dependent variation in adsorption energy. These results indicate that the adsorption becomes stronger as the diameter of the tube decreases. This is consistent with the observation that the curvature of small nanotubes is greater and thus the carbon hybridization, as determined by C-C-C bond angles, is more consistent with sp^3, leading to a stronger interaction with H_2. This result is expected since the stronger H_2 interaction with smaller nanotubes are built into the force field. Shiraishi *et al.* recently reported an experimentally measured heat of adsorption of 4.84 kcal/mol with a loading of 0.3 wt. % H_2 in SWNT of ca. 1.4 nm diameter.[2] Our calculated heats of adsorption energies for SWNT of similar diameter and H_2 loading are in good agreement with their experimental results. Furthermore, the calculated average heat of adsorption decreases as the H_2 loading increases. In particular, at the DOE gravimetric density target of 6.5 wt. % H_2, the heat of adsorption is lower by several kcal/mol, relative to lower loadings. This trend is consistent with our previous *ab initio* MD studies.[15] Detailed analysis of the MD trajectories at 6.5 wt. % loading indicates that H_2 molecules form a double layered structure at the endohedral sites in the larger tubes, and that the exohedral sites have a high density of H_2, with considerable lattice dilation. As a consequence, the H_2 molecules that reside at short distances from the nanotube walls interact strongly with SWNT bundles and those with large distances from the walls experience only a very small attraction from the nanotube walls, reducing the average heat of adsorption. The large repulsive interactions in these systems at higher than ambient pressure also contribute to the decreased heat of adsorption. Adsorption energies are significant only for the smallest nanotubes, with a diameter close to 4 Å at this loading.. These data clearly demonsrate that the average heat of adsorption decreases when the $H_2 - H_2$ repulsion is strong (at high pressure, i.e. high loading), when the H_2-SWNT distance is large (at high loading and large SWNT radius) and when the H_2-SWNT interaction is weak (at large SWNT radius).

3.3.   Radial Distribution Functions

Figure 12-1 displays the calculated radial distribution functions for H_2 adsorbed in a number of types of SWNT at 6.5 wt. % loading. The feature at *ca.* 0.74 Å in Figure 12-1(a) corresponds to a H-H bond distance that is somewhat broadened compared to gas-phase H_2 at 300 K. H_2 molecules interacting favorably with the SWNT wall have their H—H bond elongated, while shorter intramolecular distances are due to strong repulsions in the high-pressure environment. Figure 12-1(b) represents the intermolecular distance between H_2 molecules. The first peak occurs at H-H distances of between 2.7 Å and 3.0, varying somewhat with different systems. This primarily reflects the variations in H_2 density for systems at different pressures.. The not-insignificant density of intermolecular hydrogen contacts at distances shorter than $\sigma_{HH} = 2.65$ Å in each of these systems indicates the presence of a substantial amount of intermolecular repulsion, causing much of the reduction in H_2 adsorption enthalpy at these high loadings.

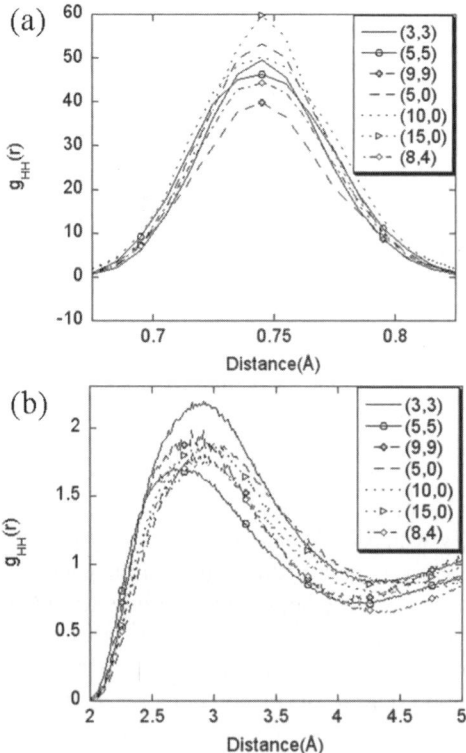

Figure 12-1. Radial distribution function for H-H separations. (a) Intramolecular distance. (b) Intermolecular distance

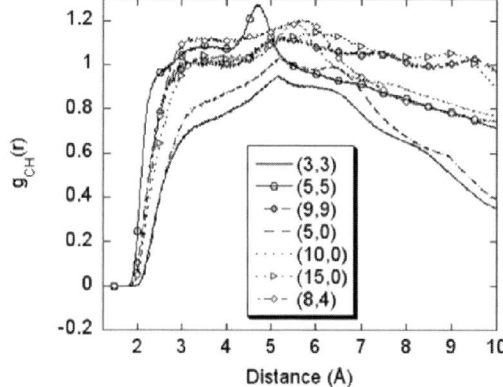

Figure 12-2. Radial distribution function for C-H separations

Figure 12-2 shows the radial distribution function for C-H distances at 6.5 wt. % loading. The sharp rise near 2 Å represents the distance of closest approach of H_2 molecules to the SWNT, demonstrating purely physisorption behavior with no formation of covalent bonds. Once again, there is substantial density below $\sigma_{C-H} = 2.78$ Å, reflecting the high H_2 density at this loading. The variation in distance of closest approach is somewhat surprising, however, and does not directly track the system pressure or tube radius. The H_2 molecules remain the largest distance away from the smaller (3,3) and (5,0) tubes, despite these being at the largest pressures, and they approach most closely to the intermediate-radius tubes. The sharp features in these radial distribution functions, and the decay below a value of 1 at long distances, are a result of evaluating the pair correlation functions at distances longer than half the shortest side of the unit cell.

3.4. H_2 Vibrational Spectrum

Velocity autocorrelation functions were calculated from the MD trajectories. A vibrational power spectrum was then generated by taking the Fourier transform of these autocorrelation functions, in an attempt to further analyze the vibrational dynamics of the adsorbed hydrogen. Figure 12-3 displays the calculated H_2 vibrational power spectrum for various SWNT bundles at 6.5 wt. % H_2 loading. The feature around 4250 cm^{-1} corresponds to the H-H stretching frequency, and is close to the calculated gas-phase value for the H-H stretching frequency with the AIREBO model. However, the H-H stretching frequencies of the hydrogen adsorbed in SWNT are significantly broadened when compared with the gas phase frequencies, which is another indication of the wide variation in H_2 environments – both interacting strongly with the SWNT and compressed by the high densities. The peak position is largely unchanged in different SWNTs. Contrary to previous studies with similar potentials at very low loading,[26] we see no evidence of a systematic red-shift in the

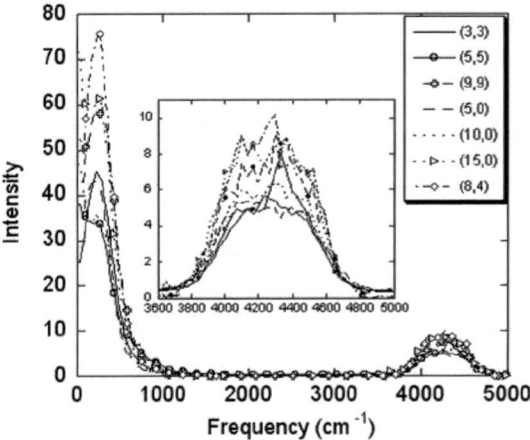

Figure 12-3. H$_2$ vibrational power spectra

H$_2$ vibrational frequency upon adsorption, which is likely due to the large loading of H$_2$ that gives rise to significant H$_2$-H$_2$ repulsion in addition to H$_2$-SWNT attractive interactions.

3.5.   H$_2$ Diffusion Coefficients in SWNT

Hydrogen adsorption is a highly dynamic phenomenon. The adsorption energies are sufficiently weak that the hydrogen molecules are not statically bound, but exchange adsorption sites and diffuse throughout the lattice. In an attempt to further describe this behavior, we calculated the H$_2$ diffusion coefficients at 6.5 wt. % loading in the nanotube bundles from the MD trajectories. The diffusion coefficients were evaluated from the limiting slope of the mean square displacement (MSD) curve with time (excluding both the initial, transient ballistic motion as well as the statistically noisy final region). The mean square displacement was calculated separately in the longitudinal (z) direction and the lateral (x and y) directions, as either 1/2 or 1/4 the slope of the MSD curve, respectively. The results are summarized in Table 12-3. The lateral diffusion coefficients are non-zero, because of the possibility of migration between interstitial pores. Indeed, the lateral diffusion coefficient is quite large, considering that it includes contributions from the endohedral H$_2$ molecules that do not contribute to lateral diffusion. The longitudinal diffusion coefficient is even larger, in all but one of the systems, although there appears to be no correlation with nanotube radius, H$_2$ density, or chirality. A large diffusion coefficient is advantageous, as it assists in the loading and unloading of H$_2$ during the duty cycle of the material. Consequently, it is a strong selling point for SWNT as H$_2$ storage materials that the calculated H$_2$ diffusion coefficients in SWNT appear much larger than what those that are typically found in microporous materials such as zeolites.[28-30] Nevertheless, the calculated diffusion coefficients are still much

Table 12-3. The calculated H2 diffusion coefficients in SWNT bundles (unit: 10^{-5} cm^2/s)

	(3,3)	(5,5)	(9,9)	(5,0)	(10,0)	(15,0)	(8,4)
Diameter (Å)	4.07	6.78	12.20	3.91	7.83	11.74	8.29
Lateral	27.3	35.3	40.9	28.8	62.2	54.8	36.2
Longitudinal	29.2	100.1	31.5	65.5	92.4	61.7	57.5

smaller than those reported in a recent equilibrium molecular dynamics study by Skoulidas et al., which were three to ten times larger.[31] Possible reasons for this discrepancy are that the potential parameters used in the Skoulidas et al. simulations were derived from interactions with planar graphite, and the nanotubes were kept rigid during the entire simulation. Thermal fluctuation in the SWNT walls, already observed to be important in previous *ab initio* molecular dynamics simulations,[15] will act to decrease the diffusion coefficient. In addition, while the interaction of H$_2$ with planar graphitic surfaces is quite weak, and should thus lead to fast diffusion, the stronger interactions with curved SWNT surfaces in the current curvature-dependent force field will also act to reduce the diffusivity. Finally, the elevated pressures in the current study will also decrease the diffusion coefficient.

4. H$_2$ ADSORPTION IN FINITE SWNT BUNDLES

Experimentally, carbon nanotubes are produced as finite bundles or ropes. Although infinite SWNT bundles can serve as a model for bundles consisting of a very large number of tubes, smaller bundles present a substantial fraction of the accessible adsorption volume on their exterior surface, rather than in interior sites. It is therefore important to also investigate H$_2$ adsorption behavior in finite SWNT bundles. To simplify the calculations, we consider H$_2$ adsorption in another extreme case: a thin SWNT bundle made of seven close-packed (9,9) nanotubes. In this case, the distances of H$_2$ from the surface of the SWNT bundle can vary significantly, depending on its loading and its interactions with the curved carbons. Although the total adsorption energy is easily quantified, the normalized, per-particle adsorption energies are more ambiguous if some of these particles are far away from the surface and are not interacting with the substrate. Using the computational procedure outlined in Sec. 2, we can quantitatively characterize the physisorption strength and effective adsorption capacity as a function of adsorption distance. Given a finite distance between H$_2$ and SWNT, the average adsorption energy and effective capacity adsorbed within that distance can be determined simultaneously.

Figure 12-4 displays the H$_2$ density distribution obtained over the course of a simulation for the 1.51 wt. % and 6.5 wt. % H$_2$ loadings. These distributions were obtained by determining the locations of the centers of mass of each H$_2$ molecule at every MD step. In both cases, the calculated MD trajectories show that H$_2$ molecules quickly diffuse around the exterior of the nanotube bundle. The adsorption process is highly dynamic on the exterior walls of the bundle. While molecules at the

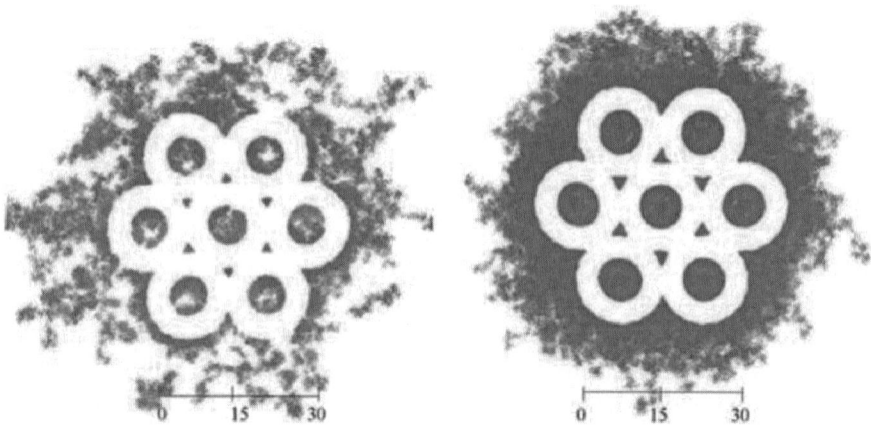

Figure 12-4. Calculated distribution of H_2 molecules in a (9,9) SWNT bundle at 1.51wt% and 6.5wt% loading. Scale bar units: Å

endohedral sites are completely confined, the exohedral molecules are capable of traveling far away from the nanotube bundle.The H_2 density in the interstitial channels and in close proximity to the outer walls of the SWNT bundle is greater than at distances far away from the nanotube walls, reflecting the interactions between the H_2 and the SWNT bundle. There is a high density of H_2 molecules in the twofold groove sites between two nanotubes on the exterior of the bundle. The tight adhesion energies holding the SWNT together in the bundle prevent substantial dilation of the bundle and exchange between interstitial sites, or between these sites and the exterior of the bundle. Consequently, the groove sites have the highest density of H_2. Although the time-averaged density distributions appear to show cylindrically symmetric tubes, the dynamics simulations reveal considerable nanotube deformation. All degrees of freedom were free to move in the simulations, and were thermostatted at 300 K. A close analysis of he density distribution also reveals that the endohedral adsorption is ordered, with evidence of layering.

Figure 12-5 shows the calculated H_2 density, wt.%, the average adsorbate energy for distant particles, and the adsorption energy and their variation with the adsorption distance , calculated using the new approach described in Section 2. The highest H_2 density is around 2.7 Å for both 1.51 wt. % and 6.5 wt. % loadings. At the higher loading, there is an additional peak at around 5.4 Å, reflecting the layered structures apparent in endohedral adsorption. The highest average adsorption energies are obtained for short distances of $r \leq 2.6$ Å, at which the corresponding weight percent of the adsorbed H_2 is extremely small. Subsequently, the average H_2 adsorption energy decays as more H_2 molecules at larger distances are included in the energy average. For 1.51 wt. % H_2 loading, the average adsorption energy drops to 1.5 kcal/mol by a distance of $r = 3.3$ Å, within which the amount of adsorbed H_2 is only 0.8 wt. %. The distance-dependent adsorption energy indicates that beyond that distance there is little interaction between H_2 molecules and the nanotube bundle;

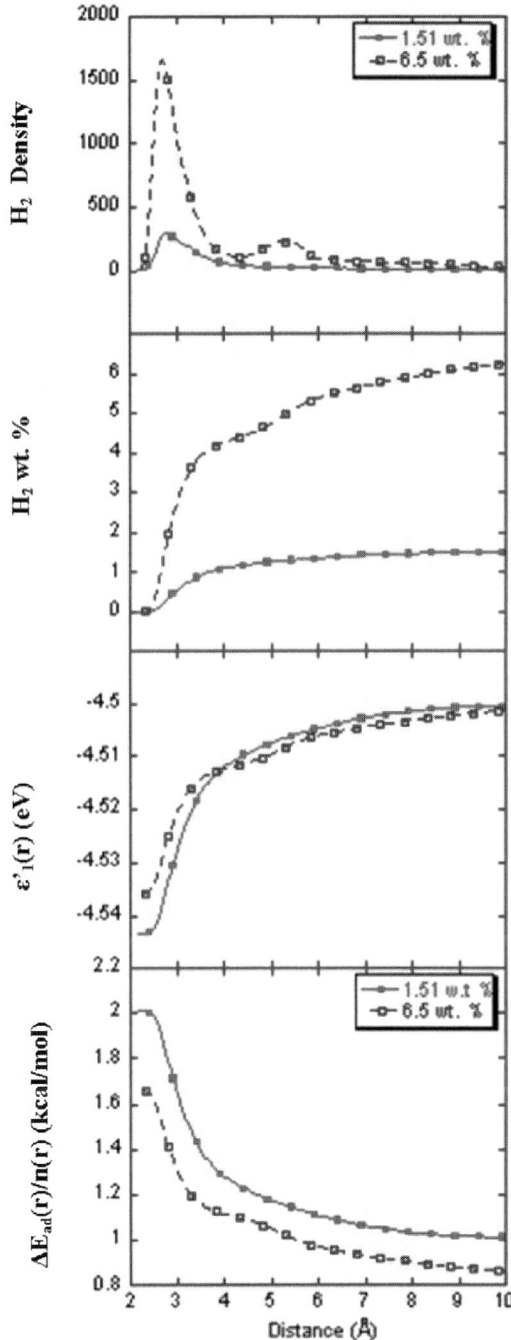

Figure 12-5. Distance dependence of the calculated H$_2$ density, wt.%, distant-particle adsorbate energy energy, and adsorption energy

thus this represents the effective adsorption capacity for this nanotube bundle. For 6.5 wt. % loading, the highest adsorption energies (1.65 kcal/mol) are found for < 0.3 wt. % H_2 within 2.3 Å of the nanotube surface. The effective adsorption energy decreases to 1.5 kcal/mol as H_2 reaches to its highest density at 2.7 Å, at which point the corresponding effective adsorption capacity of SWNT is only 1.6 wt. %. Figure 12-5 thus indicates that of the 6.5 wt. % of H_2 at these loadings, only approximately 5 wt. % interacts with SWNT with an average adsorption energy of at least 1.0 kcal/mol. This 5 wt. % is adsorbed at distances of within ~ 5.4 Å from the surface and there is little attractive force from the SWNT beyond that. Note that different choices for the analytic form of the distance dependence of the energy in Eq. (5) result in only small changes to the numerical results.

5.   CONCLUSIONS

Single walled carbon nanotubes have shown some promise to be a viable adsorbent for hydrogen storage. However, many fundamentally important issues have remained unanswered. In this chapter, we have attempted to address some of these issues by performing a series of molecular dynamics simulations using a recently developed curvature-dependent force field. Our main objective is to explore several important aspects of H_2 adsorption in SWNT with diameters ranging from 4 Å to 12 Å. We have found that the H_2 distribution in the various nanotube lattices varies significantly with the nanotube diameter and H_2 loading. For low H_2 densities, H_2 tends to adsorb in the exohedral sites. As the H_2 loading increases, the endohedral sites become more populated. The larger the nanotube diameter, the higher the relative population in the endohedral sites. However, the population at the endohedral sites never exceeds 50% of the total adsorbed H_2 for tubes up to 12 Å in diameter. The calculated heats of adsorption for H_2 in larger diameter nanotubes at low H_2 loading are in good agreement with the reported experimental data. The H_2 adsorption energy increases as the nanotube diameter decreases. It was found that substantial SWNT lattice dilation could occur upon H_2 adsorption near the DOE gravimetric target of 6.5 wt. %, and that high adsorbed H_2 densities increase repulsion among H_2 molecules, giving rise to lower adsorption energies. For H_2 loading at 6.5 wt. %, the calculated radial distribution functions for H-H distance indicate that the average H-H bond length are slightly broadened around its gas-phase value, corresponding to a significant attractive interaction between H_2 and SWNT, while the bond compression can be attributed mostly to H_2-H_2 repulsions. This conclusion is also clearly supported by the calculated H_2 vibrational power spectra, which show a very broad feature around the gas-phase H_2 stretching frequency. Given the fact that the practically acceptable heat of adsorption is found only for very small carbon nanotubes at 6.5 wt% loading, we conclude that it would be difficult to reach the DOE target with the SWNT with a diameter larger than 6 Å. Finally, the calculated diffusion coefficients for H_2 in SWNT are much larger than those that have been reported in microporous materials such as zeolites, and have been calculated with less limiting assumptions than in previous simulations,[26] showing that the transport properties of SWNT are superb, and make them a promising H_2 storage material. In all cases, the SWNT lattices undergo substantial thermal deformation in the presence of adsorbed hydrogen.

We have developed a powerful methodology for performing detailed analysis on the effective adsorption capacity and average adsorption energy for H_2 adsorption in SWNT. We show that both the average adsorption energy and effective adsorption capacity are strongly dependent on the distance of H_2 molecules from the SWNT surface. The method enables simple but accurate determination of adsorption energy as well as effective capacity simultaneously. The methodology is general and can be applicable to many other physisorption systems.

In the present studies, all SWNT are treated with an infinite or a finite bundle of homogeneous nanotubes. Shi and Johnson recently showed that packing defects due to nanotube inhomogeneity that give rise to large interstitial channels in the lattice can be important for gas adsorption.[32] This effect is not discussed here. In addition, there are a number of issues related to the simulation technique that also need to be addressed in future studies. In Table 12-2, it is apparent that the H_2 adsorption energy of the armchair nanotubes appears to be slightly larger than that of the zigzag nanotubes of similar diameter. Although there are potentially differences in adsorption energy for tubes of different chirality, this specific result is likely an artifact resulting from the force field used in our calculations. The REBO[25,26] and AIREBO[24] force fields both fail to predict the correct relative stability of aromatic molecule analogs of armchair and zigzag nanotubes. For example, contrary to quantum-mechanical and experimental results, the AIREBO force field predicts the energy of phenanthrene (armchair) relative to anthracene (zigzag) to be about 1.9 kcal/mol at 298 K, while the experimental result, using measured heats of formation, is -7.1 kcal/mol[33]. Likewise, for SWNT of similar size, the AIREBO force field predicts that the zigzag architecture is more stable than the armchair. We have tested a number of other force fields available to us, such as CVFF, COMPASS, CFF95, MMFF94, Sybyl, etc. and found that, without exception, these force fields all predict incorrect relative stability of armchair and zigzag aromatic structures. The fact that the zigzag nanotubes are in fact less stable than the armchair structures, as correctly predicted by quantum mechanical calculations, implies that H_2 adsorption should be stronger in the zigzag tubes than in the armchair tubes. We thus expect that the relative H_2 adsorption energies between zigzag and armchair tubes are the reverse of those shown in Table 13.2. We plan to refine the AIREBO force field in our future work to account for the strong chirality effect.

ACKNOWLEDGEMENTS

We are very grateful to Prof. Milton Cole and Dr. Brian Peterson for many stimulating discussions. SJS gratefully acknowledges financial support by the Department of Energy (DE-FG02-01ER45889) and the National Science Foundation (CHE-0239448). This work is supported in part by funding provided by the

U.S. Department of Energy's Office of Energy Efficiency and Renewable Energy within the Center of Excellence on Carbon-based Hydrogen Storage Materials (DE-FC36-05GO15074).

REFERENCES

1. Dillon AC, Jones KM, Bekkedahl TA, Kiang CH, Bethune DS, Heben MJ (1997) Storage of hydrogen in single-walled carbon nanotubes. Nature 386:377–379
2. Shiraishi M, Takenobu T, Ata M (2003) Gas-solid interactions in the hydrogen/single-walled carbon nanotube system. Chem. Phys. Lett. 367:633–636
3. Liu CY, Fan Y, Lu M, Cong HT, Cheng HM, Dresselhaus MS (1999) Hydrogen Storage in Single-Walled Carbon Nanotubes at Room Temperature. Science 286:1127–1129
4. Ye Y, Ahn CC, Witham C, Fultz B, Liu J, Rinzler AG, Colbert D, Smith KA, Smalley RE (1999) Hydrogen adsorption and cohesive energy of single-walled carbon nanotubes. Appl. Phys. Lett. 74:2307–2309
5. Pace EL, Siebert AR (1959) Heat of adsorption of parahydrogen and orthodeuterium on graphon. J. Phys. Chem. 63:1398-1400
6. Dericbourg J (1976) Adsorption de l'hydrogene sur le graphite. Surf. Sci. 59:565–574
7. Constabaris G, Sams JR Jr., Halsey GD Jr., (1961) The interaction of H_2, D_2, CH_4, and CD_4 with graphitized carbon black. J. Phys. Chem. 65:367–369
8. Pez G, Steyert W, (1985) *U.S. Patent* 4:580–404
9. Benard P, Chahine R (2001) Determination of the adsorption isotherms of hydrogen on activated carbons above the critical temperature of the adsorbate over wide temperature and pressure ranges. Langmuir 17:1950–1955
10. Watanabe K, Soma M, Onishi T, Tamaru K (1971) Sorption of molecular hydrogen by potassium graphite. Nature 233:160
11. Watanabe K, Knodow T, Soma M, Onishi T, Tamaru K (1973) Molecular-sieve type sorption on alkali graphite intercalation compounds. Proc. Roy. Soc. Lond. A. A333:51–67
12. Lagrange P, Metrot A, Herold AC (1972) Physisorption of hydrogen on KC_{24}. C. R. Acad. Sci. Ser. C275:765
13. Terai T, Takahashi Y (1989) Formulation of isotherms for low-temperature absorption of H_2 and D_2 on KC_{24} prepared from natural graphite. Synth. Met. 34:329–334
14. Okamoto Y, Miyamoto Y (2001) Ab initio investigation of physisorption of molecular hydrogen on planar and curved graphemes. J. Phys. Chem. B. 105:3470–3474
15. Cheng H, Pez GP, Cooper AC (2001) Mechanism of hydrogen sorption in single-walled carbon nanotubes. J. Am. Chem. Soc. 123:5845–5846
16. Canto G, Ordejón P, Cheng H, Cooper AC, Pez GP (2003) First-principles molecular dynamics study of the stretching frequencies of hydrogen molecules in carbon nanotubes. New J. Phys. 5:124.1–8
17. Cheng H, Pez GP, Kern G, Kresse G, Hafner J (2001) Hydrogen adsorption in potassium-intercalated graphite of second stage: an ab initio molecular dynamics study. J. Phys. Chem. B. 105:736–742
18. Kostov MK, Cheng H, Cooper AC, Pez GP (2002) Influence of carbon curvature on molecular adsorptions in carbon-based materials: a force field approach. Phys. Rev. Lett. 89:146105–1–146105–4
19. Cheng H, Pez GP, Cooper AC (2003) Spontaneous cross linking of small-diameter single-walled carbon nanotubes. Nano. Lett. 3:585–587

20. Frankland SJV, Brenner DW (2001) Hydrogen Raman shifts in carbon nanotubes from molecular dynamics simulation. Chem. Phys. Lett. 334:18–23

21. Wang Q, Johnson JK (1999) Molecular simulation of hydrogen adsorption in single-walled carbon nanotubes and idealized carbon slit pores. J. Chem. Phys. 110:577–586

22. Williams KA, Eklund PC (2000) Monte Carlo simulations of H_2 physisorption in finite-diameter carbon nanotube ropes. Chem. Phys. Lett. 320:352–358

23. Darkrim F, Levesque D (1998) Monte Carlo simulations of hydrogen adsorption in single-walled carbon nanotubes. J. Chem. Phys. 109:4981–4984

24. Stuart SJ, Tutein AB, Harrison JA (2000) A reactive potential for hydrocarbons with intermolecular interactions. J. Chem. Phys. 112:6472–6486

25. Brenner DW, Shenderova OA, Harrison JA, Stuart SJ, Ni B, Sinnott S (2002) A second generation reactive empirical bond order (REBO) potential energy expression for hydrocarbons. J. Phys.: Cond. Matt. 14:783–802

26. Brenner DW (2000) The art and science of an analytic potential. Physica Status Solidi B 217:23–40

27. Diep P, Johnson JK (2000) An accurate H_2–H_2 interaction potential from first principles. J. Chem. Phys. 112:4465–4473

28. Skoulidas AI, Sholl DS (2001) Direct tests of the darken approximation for molecular diffusion in zeolites using equilibrium molecular dynamics. J. Phys. Chem. B. 105:3151–3154

29. Skoulidas AI, Sholl DS (2002) Transport diffusivities of CH_4, CF_4, He, Ne, Ar, Xe, and SF_6 in silicalite from atomistic simulations. J. Phys. Chem. B. 106:5058–5067

30. Maginn EJ, Bell AT, Theodorou DN (1993) Transport diffusivity of methane in silicalite from equilibrium and nonequilibrium simulations. J. Phys. Chem. 97:4173–4181

31. Skoulidas AI, Ackerman DM, Johnson JK, Sholl DS (2002) Rapid transport of gases in carbon nanotubes. Phys. Rev. Lett. 89:185901-1–185901-4

32. Shi W, Johnson JK (2003) Gas adsorption on heterogeneous single-walled carbon nanotube bundles. Phys. Rev. Lett. 91:015504-1–015504-4

33. NIST Chemistry WebBook (October 3, 2005); http://webbook.nist.gov/chemistry

CHAPTER 13

THE REMARKABLE CAPACITIES OF (6,0) CARBON AND CARBON/BORON/NITROGEN MODEL NANOTUBES FOR TRANSMISSION OF ELECTRONIC EFFECTS

PETER POLITZER, JANE S. MURRAY, PAT LANE
AND MONICA C. CONCHA
Department of Chemistry, University of New Orleans, New Orleans, LA 70148, USA

Abstract: We have found that at least some (6,0) carbon and carbon/boron/nitrogen model nanotubes possess a remarkable capability for transmitting electronic effects along their full lengths. This can be triggered by even a rather minor asymmetric perturbation at one or both ends of the system. We have analyzed these quite striking effects as they are manifested in the computed electrostatic potentials and local ionization energies on the tube surfaces and, in one instance, in a reorganization of the framework structure. These observations, and some implications, are presented and discussed

Keywords: (6,0) carbon nanotubes, (6,0) C/B/N nanotubes, electrostatic potentials, local ionization energies, charge delocalization

1.   INTRODUCTION

Fullerene, C_{60}, is a hollow sphere composed of five- and six-membered rings of carbon atoms. Its discovery in 1985,[1] and that of carbon nanotubes a few years later,[2,3] can be viewed as epochal events in the evolution of nanomaterials science. (In fairness, reference should be made to other closely-related work in the 1990–1992 period[4–7] as well as earlier; the latter is summarized by Harris.[8]) The remarkable structural, mechanical and electronic properties of carbon nanotubes, and their analogues involving other elements, have inspired a large number of proposed applications, in a variety of areas: electromechanical tweezers[9] and actuators,[10] conductor junctions,[11–13] quantum wires,[14] transistors,[15] capacitors,[16] optoelectronic devices,[17,18] gas storage systems,[19] pollutant traps,[20,21] catalysts,[22] chemical and biosensors,[23–25] etc. For reviews, see Harris,[8] Saito *et al*,[26] Ajayan[27] and Politzer *et al.*[28]

487

W. A. Sokalski (ed.), Molecular Materials with Specific Interactions, 487–504.
© 2007 *Springer.*

In fullerene, each carbon hexagon is surrounded by alternating pentagons and hexagons. Accordingly a three-fold symmetry axis links the centers of two hexagons on opposite sides of the sphere. If fullerene were to be divided by cutting through the center of this axis, and if the two resulting hemispheres were then connected by a cylindrical network of carbon hexagons, the result would be a particular type of carbon nanotube, the (9,0). This label signifies that there are nine six-membered rings around the circumference of the tube, and that each one has two sides parallel to the tube axis. If the caps at the ends of the tube were not restricted to being these fullerene hemispheres but could assume other forms,[8,29] then there could be some other number n of hexagons around its circumference. For instance, a (7,0) structure is shown in Figure 13-1(a). As long as each six-membered ring has two sides parallel to the axis, the tube is in the $(n, 0)$ category. The diameter would of course be different for each n.

In general, many categories of nanotubes are possible, differing in the orientations of the hexagons relative to the tube axis; some examples are in Figure 13-1. Each category is designated by a label (n, m), using a system that has been explained elsewhere;[8,26–28] n and m are always positive integers. The diameters can be found with the formula

$$\text{Diameter} = (L\sqrt{3}/\pi)(n^2 + m^2 + nm)^{1/2} \tag{13-1}$$

in which L is the length of a side of the six-membered rings. It should be noted that only nanotubes in the $(n, 0)$ category have any bonds parallel to the tube axis.

While we have investigated, computationally, model nanotubes with a variety of structures and chemical compositions,[28–32] our focus in this chapter will be upon the type (6,0). Results will be reported for all-carbon, B_xN_x, and C/B/N systems. Nanotubes can also contain other elements, e.g. aluminum or silicon,[33,34] but the B_xN_x combination is of particular interest because of being isoelectronic with C_{2x}, as well as the notable analogies between carbon and boron/nitrogen systems: (a) Borazine, **1**, is a planar molecule with some similarities to benzene,[35] and (b) solid boron nitride, BN, has two forms, one resembling graphite in structure and the other diamond-like.[36,37] Various B_xN_x[38,39] and $C_xB_yN_z$[40,41] nanotubes have now been prepared.

1

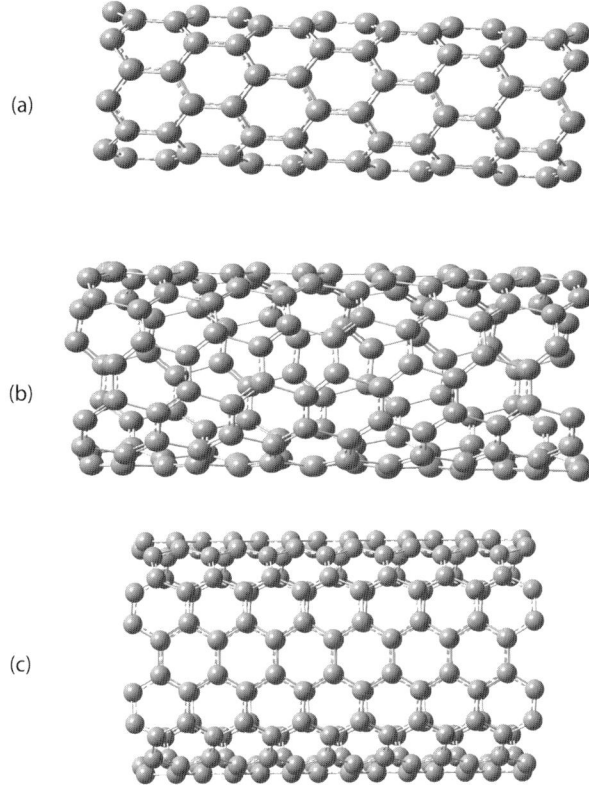

Figure 13-1. Portions of lateral surfaces of carbon nanotubes: (a) (7,0); (b) (7,4); (c) (7,7)

The computed properties that will be discussed in this chapter are the electrostatic potential $V_S(\mathbf{r})$ and the average local ionization energy $\bar{I}_S(\mathbf{r})$. We will begin by reviewing some background for each.

2. PROCEDURES

2.1. Electrostatic Potential

The electrons and nuclei of any system produce an electrostatic potential $V(\mathbf{r})$ in the surrounding space. It is given rigorously by Eq. (2):

$$V(\mathbf{r}) = \sum_A \frac{Z_A}{|\mathbf{R}_A - \mathbf{r}|} - \int \frac{\rho(\mathbf{r}')d\mathbf{r}'}{|\mathbf{r}' - \mathbf{r}|} \tag{13-2}$$

Z_A is the charge on nucleus A, which is located at $\mathbf{R}_A$, and $\rho(\mathbf{r})$ is the system's electronic density function.

V(**r**) is a physical observable, which can be obtained experimentally, by diffraction methods,[42–44] as well as computationally. Its sign in a given region, positive or negative, depends upon whether the effect of the nuclei or the electrons is dominant there.

We have computed V(**r**) on nanotube surfaces, both inner and outer, which we define as the 0.001 electrons/bohr^3 contours of the electronic density, as proposed by Bader *et al.*[45] The resulting surface potentials are labeled $V_S(\mathbf{r})$.

While the patterns of positive and negative regions are themselves of interest, we wish to also characterize $V_S(\mathbf{r})$ quantitatively. We do this by calculating certain statistically-defined properties.[46–50] Those that will be relevant to the present discussion are the most positive and most negative values of $V_S(\mathbf{r})$, $V_{S,max}$ and $V_{S,min}$, and its positive, negative and total variances, σ_+^2, σ_-^2 and σ_{tot}^2:

$$\sigma_{tot}^2 = \sigma_+^2 + \sigma_-^2 = \frac{1}{r}\sum_{i=1}^{r}\left[V_S^+(\mathbf{r}_i) - \bar{V}_S^+\right]^2 + \frac{1}{s}\sum_{j=1}^{s}\left[V_S^-(\mathbf{r}_j) - \bar{V}_S^-\right]^2$$

(13-3)

In Eq. (3), $\bar{V}_S^+$ and $\bar{V}_S^-$ are the average positive and negative values of $V_S(\mathbf{r})$:

$$\bar{V}_S^+ = \frac{1}{r}\sum_{i=1}^{r}V_S^+(\mathbf{r}_i)$$

(13-4)

$$\bar{V}_S^- = \frac{1}{s}\sum_{j=1}^{s}V_S^-(\mathbf{r}_j)$$

(13-5)

The summations in Eqs. (3)–(5) are over the r points where $V_S(\mathbf{r})$ is positive and the s where it is negative, on grids covering the inner and outer nanotube surfaces.

The variances σ_{tot}^2, σ_+^2 and σ_-^2 are measures of the variabilities of $V_S(\mathbf{r})$ and its positive and negative components. They are particularly influenced by the extrema of $V_S(\mathbf{r})$, the $V_{S,max}$ and $V_{S,min}$, due to the terms in Eq. (3) being squared.

In an extended series of studies, we have shown that $V_S(\mathbf{r})$ and the quantities that we use to characterize it provide an effective means for analyzing noncovalent interactions and predicting quantitatively the values of properties that depend upon them, such as boiling points and critical constants, heats of phase transitions, solubilities and solvation energies, partition coefficients, diffusion constants, surface tensions, viscosities, etc. This work has been reviewed on several occasions.[48–50]

2.2. Average Local Ionization Energy

Some time ago, we introduced the concept of an "average local ionization energy" $\bar{I}(\mathbf{r})$, by which we mean the average energy required to remove an electron at the point **r** in the space of a system.[51] Note that we are focusing here upon a point in space, *not* a particular orbital. We defined $\bar{I}(\mathbf{r})$ originally within the framework of Hartree-Fock theory:

$$\bar{I}(\mathbf{r}) = \sum_i \frac{\rho_i(\mathbf{r})\,|\varepsilon_i|}{\rho(\mathbf{r})} \tag{13-6}$$

In Eq. (6), $\rho_i(\mathbf{r})$ is the electronic density function of the **i**th occupied orbital of the system, having energy ε_i. The Hartree-Fock formalism plus Koopmans' theorem[52] provide support for the common interpretation of the $|\varepsilon_i|$ as the electrons' ionization energies; hence our introduction of $\bar{I}(\mathbf{r})$ as the average *local* ionization energy. When computed on the surface of the system, as we normally do, it is denoted by $\bar{I}_S(\mathbf{r})$.

The locations of the lowest values of $\bar{I}_S(\mathbf{r})$, the $\bar{I}_{S,min}$, are of particular interest because these are the sites of the least tightly-held electrons, most reactive toward electrophiles. We have demonstrated that $\bar{I}_S(\mathbf{r})$ correctly predicts the activating/deactivating and directing effects of benzene substituents[51,53] and the reactive sites on other organic molecules,[54] including the nucleotide bases,[55] as well as proton affinities and pK_a.[56] It has been found that the $\bar{I}_{S,min}$ obtained by Kohn-Sham density functional theory are as effective for these purposes as are the Hartree-Fock,[53] even though they do not have the same theoretical support.

The significance of $\bar{I}(\mathbf{r})$ extends well beyond the prediction of reactivity toward electrophiles. It correlates with the shell structures of atoms[57] and their electronegativities,[58] indicates bond strain,[59] and is related to local temperature[60] and local polarizability.[61,62] For a recent review, see Politzer and Murray.[63]

2.3. Computational Approach

We will discuss $V_S(\mathbf{r})$ and $\bar{I}_S(\mathbf{r})$ calculated for (6,0) all-carbon and C/B/N model nanotubes. The computational level, Hartree-Fock (HF) STO-5G//STO-3G, was dictated by the large sizes of the systems. It has been confirmed, however, that minimum basis set STO-5G Hartree-Fock results for both $V(\mathbf{r})$[64,65] and $\bar{I}_S(\mathbf{r})$[65] are quite satisfactory on a relative basis, for seeing trends, which has been our objective. The validity of the HF/STO-3G geometry optimizations can be seen from the good agreement between our C-C and B-N bond lengths and relevant experimental data.[28-31]

When nanotubes are synthesized,[8,26,27] they are typically closed (capped) at both ends. For the (9,0) and the (5,5), the caps can be fullerene hemispheres;[6] in general, however, caps can have various structures, shapes and degrees of curvature.[8] One requirement that they do have to satisfy is imposed by Euler's theorem,[8,26,27] according to which the closure of any hexagonal framework can be achieved only by the introduction of exactly twelve pentagons. Thus, each cap must have six.

For various applications, it may be desired to remove the caps. Several methods are available for doing this, including heating,[66,67] cutting[68] and chemical treatment.[68]

In our computational analyses, we have considered both types of model nanotubes, open and closed. For the former, we have followed the common practice of introducing hydrogens at the ends, to satisfy the unfulfilled terminal valencies.[22,69-73]

3.   RESULTS AND DISCUSSION

3.1.   General

In order to put into perspective the remarkable properties that we have found for $(n, 0)$ model carbon and C/B/N nanotubes, we will first summarize briefly what we have observed in general, for various (n, m). These results have been discussed in detail elsewhere,[28-32] and illustrated with numerous color figures.

Our experience has been that the computed electrostatic potentials $V_S(\mathbf{r})$ on the surfaces of all-carbon tubes are generally weakly positive and rather bland, with little variability. $V_S(\mathbf{r})$ is usually less than 10 kcal/mole. Slightly negative regions are most likely to be found on the outer surfaces of the caps. The total variance, σ_{tot}^2, tends to be less than 20 (kcal/mole)^2; this is typical of alkane and aromatic hydrocarbons, but much less than most other organic molecules, for which σ_{tot}^2 is normally 100–300 (kcal/mole)^2.[46,47]

The carbon nanotubes do, however, demonstrate the interesting consequences of curvature upon $V_S(\mathbf{r})$, which are common to all of the systems that we have studied, regardless of structure (n, m) or composition. As the curvature increases, each point on the inner surface is brought into closer proximity with more nuclei. Thus the inside surfaces are always somewhat more positive (or less negative) than the outside ones. This effect naturally becomes more pronounced as the diameter decreases, and at the caps of the closed tubes. On the other hand, if the tube is unrolled into a flat sheet (graphene), $V_S(\mathbf{r})$ is the same on both sides.[74]

In contrast, increasing curvature makes the outer surfaces more negative. At least two (basically equivalent) explanations of this have been proposed.[75,76] One is that curvature diminishes the $2p\pi - 2p\pi$ overlaps between the atoms, leading to greater electronic localization. Alternately, it can be argued that curvature forces the atoms to deviate from their preferred trigonal planar sp^2 configurations and to assume some degree of sp^3 character, thereby introducing some strain and also electronic localization in the fourth (unfulfilled) valency.

Curvature also affects the computed average local ionization energy $\bar{I}_S(\mathbf{r})$.[28,29,32] On the lateral outer surfaces of carbon tubes, for example, there are minima $\bar{I}_{S,min}$ above the carbons, primarily in the 13–14 ev range, which is considerably less than the 14.8–14.9 ev found for graphene.[74] This difference can be attributed to the same factors that cause the outer surface $V_S(\mathbf{r})$ to be more negative, since this leads to the electrons being less tightly held, more easily removed.

In summary, the most positive potentials (the $V_{S,max}$) will be on nanotube inner surfaces, especially inside caps, while the most negative (the $V_{S,min}$) as well as the lowest ionization energies (the $\bar{I}_{S,min}$) will be on the outsides, again more so on the caps. All of these features are enhanced by increased curvature, e.g. narrower tubes. A somewhat extreme example of this is in Figure 13-2, which shows a closed (6,0)

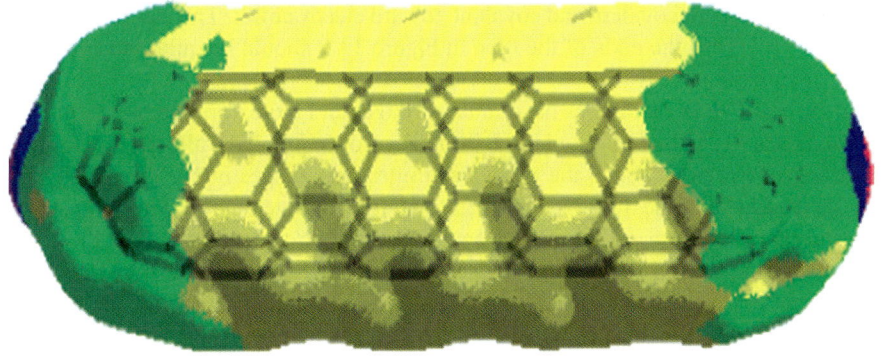

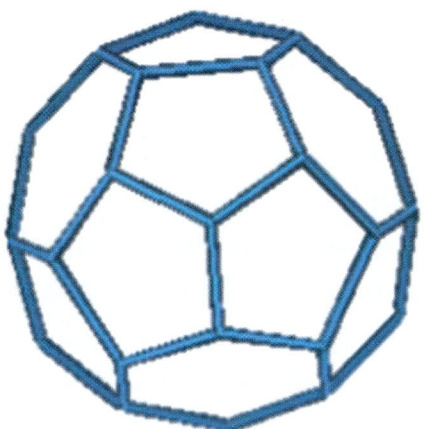

Figure 13-2. Calculated HF/STO-5G//STO-3G electrostatic potential on the 0.001 electrons/bohr³ surface of (6,0) C_{104}. Color ranges, in kcal/mole, are: yellow, between 0 and 15; green, between -10 and 0; blue, between -20 and -10; purple, more negative than -20. The structure of each cap is shown at the bottom of this figure

carbon model nanotube with composition C_{104}. Each cap comes to a point at the center, and thus has relatively high curvature. This causes even the lateral sides to be more positive than is normal for all-carbon tubes, with significant portions having $V_S(\mathbf{r}) > 10$ kcal/mole.[29] The highly curved ends are quite negative and attain $V_{S,min} = -25$ kcal/mole at the tips, with $\bar{I}_{S,min} = 9.0$ ev. (For comparison, the $\bar{I}_{S,min}$ of benzene at this computational level is 14.3 ev.) The inner surface is entirely positive, with $V_{S,max} = 27$ kcal/mole inside the ends. Overall, $\sigma_{tot}^2 = 62.2$ (kcal/mole)². These effects are of course diminished for caps with lesser curvatures.[29]

The more negative $V_S(\mathbf{r})$ and lower $\bar{I}_S(\mathbf{r})$ associated with higher levels of curvature, especially caps, suggest that such outer surfaces should show enhanced reactivity, both noncovalent and covalent, toward electrophiles. This has indeed been found.[27,76,77] Thus the ends of the tube in Figure 13-2 should be quite vulnerable to electrophilic attack.

$B_x N_x$ and C/B/N nanotubes have much stronger and more variable $V_S(\mathbf{r})$ than do carbon ones.[28–32] The $B_x N_x$ and the B/N portions of C/B/N have alternating positive and negative regions above the borons and nitrogens, respectively. The $V_{S,\max}$ and $V_{S,\min}$ tend to be in the 40 to 70 and −25 to −50 kcal/mole ranges, with σ_{tot}^2 usually greater than 200 (kcal/mole)^2. $\bar{I}_S(\mathbf{r})$ again has minima above the atoms, those for borons and nitrogens being higher than those for carbons. For the nitrogens, this simply reflects the higher ionization potential of the nitrogen atom; for the borons, it is due to their positive electrostatic potentials, which make it more difficult to remove an electron. As usual, inner surfaces become more positive and the outer ones more negative, with lower $\bar{I}_{S,\min}$, as curvature increases and at the caps.

3.2.   Functionalized Open Carbon Model Nanotubes

As was mentioned earlier, our open tubes have hydrogens at the ends, simply as a means of avoiding unfulfilled valencies. How are the surface electrostatic potentials and local ionization energies affected by replacing one or more of these hydrogens by functional groups, e.g. OH, NH_2, NO_2, etc.?

Within our experience, the answer is: very little, unless the tube is of type $(n, 0)$. We have investigated (5,5) and (6,1) carbon systems with either an OH or an NH_2 at one end,[28,32] and found the effects upon both $V_S(\mathbf{r})$ and $\bar{I}_S(\mathbf{r})$ to be localized in the vicinity of the substituent. As expected, the oxygen or nitrogen is quite negative ($V_{S,\min} = -31$ to -36 kcal/mole) and their hydrogens are positive ($V_{S,\max} = 29$ to 38 kcal/mole), but the remainder of the tube is little affected, retaining the weak and bland $V_S(\mathbf{r})$ of the nonfunctionalized system. σ_{tot}^2 is between 46 and 56 (kcal/mole)^2.

When we proceed to (6,0) carbon tubes, however, the response to an OH or NH_2 is very different, and involves the entire structure; this can be seen in Figure 13-3(top). $V_S(\mathbf{r})$ changes gradually along the full length of the tube, from strongly positive at the end bearing the substituent to strongly negative at the other.[28,32] The HF/STO-5G $V_{S,\max}$ and $V_{S,\min}$ are about 74 and −44 kcal/mole, and σ_{tot}^2 is 419.1 and 548.7 (kcal/mole)^2 for the OH and NH_2 systems, respectively. (Figure 13-3 was obtained at a higher computational level, HF/6-31G*, than the other results being reported, to verify that the qualitative patterns are the same.)

An even stronger gradation of $V_S(\mathbf{r})$ is produced by an NO_2 group, but with opposite polarity;[28,32] the NO_2 end is negative, $V_{S,\min} = -85$ kcal/mole, with a $V_{S,\max}$ of 59 kcal/mole at the other. $\sigma_{tot}^2 = 933.3$ (kcal/mole)^2, one of the largest values that we have ever obtained. More recently, we have observed analogous $V_S(\mathbf{r})$ patterns with other substituents: F, CN and CH_3. For F and CN, the substituted end is negative; for CH_3, it is the reverse.

The variation of $\bar{I}_S(\mathbf{r})$ in these systems, Figure 13-3(bottom), is consistent with that of $V_S(\mathbf{r})$, although somewhat less dramatic.[28,32] There is an overall gradual

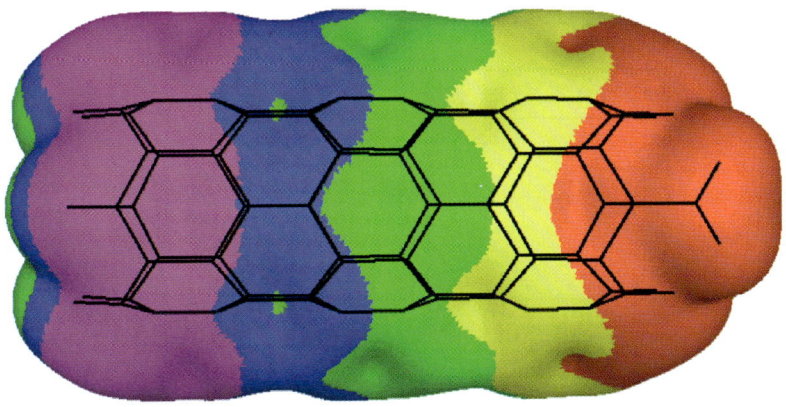

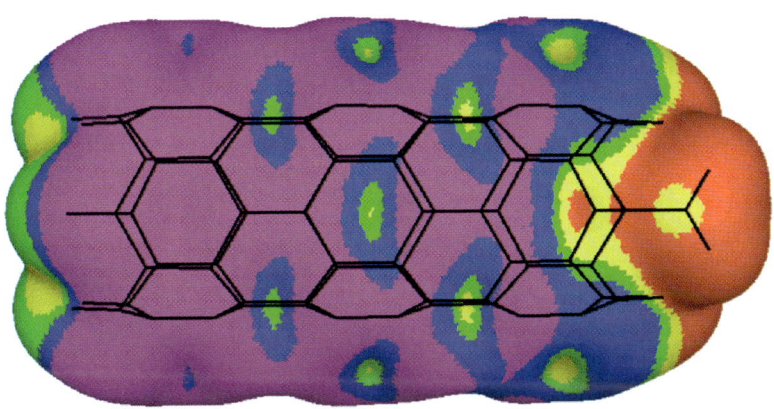

Figure 13-3. Calculated HF/6-31G*//STO-3G electrostatic potential (top) and average local ionization energy (bottom) on 0.001 electrons/bohr^3 surface of (6,0) $C_{72}H_{11}NH_2$ model carbon nanotube. Color ranges (top), in kcal/mole: red, greater than 15; yellow, between 0 and 15; green, between -10 and 0; blue, between -20 and -10; purple, more negative than -20. Color ranges (bottom), in ev: red, greater than 17; yellow, between 15 and 17; green, between 14.5 and 15; blue, between 13 and 14.5; purple, less than 13

decrease in the magnitude of $\bar{I}_S(\mathbf{r})$ in going from the positive to the negative end of the tube.

What is the origin of these striking gradations in $V_S(\mathbf{r})$ and $\bar{I}_S(\mathbf{r})$, extending over the entire lengths of substituted (6,0) carbon tubes? The fact that they do not occur in (5,5) and (6,1) systems draws attention to a key structural difference between these and the (6,0): the presence of a considerable number of C-C bonds parallel to the tube axis, a feature only of $(n,0)$ tubes. These have maximum overlap of $2p\pi$

Table 13-1. Inductive (σ_I) and resonance $(\sigma_R°)$ substituent constants[a]

Substituent	σ_I	$\sigma_R°$
NH_2	0.10	−0.48
OH	0.27	−0.44
CH_3	−0.05	−0.10
F	0.51	−0.34
CN	0.52	0.14
NO_2	0.64	0.19

[a] Reference 81.

atomic orbitals, unimpeded by curvature. We suggest that this structural feature of (6,0) systems, in conjunction with the high polarizabilities of carbon nanotubes,[78-80] facilitates electronic charge delocalization along the tube.

In which direction will this delocalization occur? That depends upon the substituent, as our results show. As can be seen from the inductive and resonance substituent constants in Table 13-1, σ_I and $\sigma_R°$ respectively,[81] the dominant effect of NH_2 and OH groups is donation of electronic charge through resonance, as depicted in **2** and **3**. The donated charge is apparently delocalized toward the other end of the tube, leaving the substituted end positive and producing an increasingly negative potential in the direction away from it.

2   3

The resonance interpretation shown in **2** and **3** is supported by the computed HF/STO-3G bond lengths. The C-NH_2 and C-OH distances in these (6,0) systems are 1.350 and 1.361 A, respectively, both considerably shorter than is typical for the corresponding HF/STO-3G single bond lengths, e.g. 1.491 A in $(H_3C)_2CH-NH_2$ and 1.438 A in $(H_3C)_2CH-OH$. This indicates some double bond character in the NH_2 and OH links to the (6,0) tube, as predicted by **2** and **3**. Also consistent with **2** is the fact that the C$-NH_2$ portion of the (6,0) system is planar, whereas it is pyramidal in the (6,1).[28]

In contrast to NH_2 and OH, Table 13-1 shows that F, CN and NO_2 are primarily inductive withdrawers of electronic charge. Thus, while they also cause a tube-long delocalization, it is toward them, and their ends become the strongly negative ones.

The CH_3 group is an interesting case; it is listed as both a resonance and inductive donor, but a weak one. Nevertheless it also produces a full-length gradation in

$V_S(\mathbf{r})$ and $\bar{I}_S(\mathbf{r})$, although not as pronounced as those due to NH_2 and OH. As with the latter two, the substituted end is the positive one. In all of these instances, it is notable that a single functional group, whether a donor or withdrawer, is enough to affect so drastically the entire tube surface.

3.3.   Closed Model Nanotubes

The charge delocalization that thas been observed in open functionalized (6,0) carbon tubes is manifested as well in closed C/B/N systems. We will consider first a (6,0) tube having the composition $C_{52}B_{26}N_{26}$.[29] Structurally it has two rows of carbons parallel to the axis, then two rows of alternating borons and nitrogens, followed by two more of carbons, etc. An interesting feature of the carbons, which has already been encountered in other $C_{2x}B_xN_x$ tubes,[28,31] is that the C-C distances fall into two groups: roughly half are relatively short, 1.35–1.38 A, suggesting significant double bond character, while the others are 1.42–1.48 A. The short C-C distances are a departure from all-carbon tubes, in which the computed C-C bond lengths are 1.40–1.45 A.[30,31]

  $V_S(\mathbf{r})$ for this $C_{52}B_{26}N_{26}$ shows the usual positive and negative regions above the borons and nitrogens. Above the carbons, however, is a zig-zag pattern of negative potentials, reaching -10 kcal/mole.[29] This is a marked contrast to the relatively weak and bland $V_S(\mathbf{r})$ of all-carbon tubes (see, for example, Figure 13-2), and presumably reflects the presence of the shorter "near-double-bonds." (The C=C bond in ethylene has an HF/STO-5G $V_{S,min}$ of -10.5 kcal/mole.)

  In this $C_{52}B_{26}N_{26}$ tube, the stoichiometry of one cap is C_5B_4N while the other is C_5BN_4. Very interesting is what happens when the lone boron and nitrogen at the ends are replaced by carbons, so that the caps are C_6B_4 and C_6N_4. There develops a tube-long gradation in $V_S(\mathbf{r})$, from strongly positive at the C_6B_4 end to strongly negative at the C_6N_4.[29] The result is very similar to what was described above for functionalized open (6,0) carbon systems. σ_{tot}^2 jumps from 205.7 to 319.2 (kcal/mole)^2, while $V_{S,max}$ and $V_{S,min}$ go from 64 and -31 to 87 and -36 kcal/mole. A change of just one atom on each cap suffices to produce this dramatic effect!

  Our second closed-tube example concerns the (6,0) C_{104} shown in Figure 13-2. $V_S(\mathbf{r})$ has the general features associated with all-carbon tubes, although the lateral sides and the inside are more positive and the outsides of the caps are more negative (as mentioned earlier) due to the perturbation of the system by the pointed, highly-curved caps; each of them has just a single atom at the tip. The C-C bond lengths are normal for an all-carbon tube, mostly 1.43–1.45 A.

  Figure 13-4 shows the unexpected consequences of replacing the two carbons at the tips of the caps, one by a boron, the other by a nitrogen, to form a (6,0) tube with stoichiometry $C_{102}BN$. Six of the twelve rows of C-C bonds parallel to the tube axis develop shortened lengths, 1.35–1.37 A, and above these are the usual negative potentials associated with near-double-bond character. Thus there are three zig-zag rows of negative $V_S(\mathbf{r})$ along the length of the tube. Furthermore, the large negative regions over the caps in C_{104} (Figure 13-2) have been replaced by small

Figure 13-4. Calculated HF/STO-5G//STO-3G electrostatic potential on the 0.001 electrons/bohr³ surface of $C_{102}BN$. Color ranges, in kcal/mole, are: red, greater than 10; yellow, between 0 and 10; green, between −5 and 0; blue, between −10 and −5; purple, more negative than −10. At the tips can be seen the small positive (red, at left) and negative (purple and blue, at right) regions over the boron and the nitrogen, respectively

positive and negative ones over the boron and the nitrogen, respectively. Again, all of this has been brought about by changing just two atoms!

These findings have possible implications for chemical reactivity; for example, will the C=C near-double-bonds in $C_{102}BN$ undergo addition reactions? We have carried out a preliminary HF/STO-3G test of this and found that HCl did indeed add to $C_{102}BN$, with $\Delta E(0 \text{ K}) = -49$ kcal/mole. For HCl + ethylene, the HF/STO-3G $\Delta E(0 \text{ K})$ is −58 kcal/mole. While the computational level was very low, these results are at least encouraging. They further suggest that addition processes may also occur at the short C-C bonds that we have found in $C_{2x}B_xN_x$ systems.

3.4. Applications in Area of Nonlinear Optics

We have not failed to recognize that appropriately designed (6,0) carbon and C/B/N nanotubes may display considerably enhanced nonlinear optical activity. This term refers to the response of the dipole moment of a molecule (or the polarization of bulk material) to the oscillating electric field of electromagnetic radiation.[82-85] The component of the dipole moment along an axis i in the presence of an electric field ε can be represented by a Taylor series:

$$\mu_i(\varepsilon) = \mu_i(0) + \sum_j \alpha_{ij}\varepsilon_j + \frac{1}{2}\sum_{jk} \beta_{ijk}\varepsilon_j\varepsilon_k + \frac{1}{6}\sum_{jkm} \gamma_{ijkm}\varepsilon_j\varepsilon_k\varepsilon_m + \cdots$$

(13-7)

In Eq. (7), the tensors $\boldsymbol{\alpha}, \boldsymbol{\beta}$ and $\boldsymbol{\gamma}$ are labeled the polarizability and the first and second hyperpolarizabilities of the molecule, respectively. In the analogous expression for bulk polarization, the corresponding quantities are the first-, second- and third-order susceptibilities.

Nonlinear optical activity reflects the nonlinear response of $\mu_i(\varepsilon)$ to electromagnetic radiation, which Eq. (7) shows to be governed by the first and second hyperpolarizabilities, β and γ. A high level of such activity can have important applications in a variety of electro-optical devices,[82,86,87] such as frequency converters, modulators, switches, etc.

It is known that nonlinear optical activity is often associated with the presence of electron-donating and – withdrawing groups linked by a conjugated bridge,[82,84,88] e.g. **4**.[89]

$$(N \equiv C)_2 C = CH - C \equiv C - C_6 H_4 - N(CH_3)_2$$
$$\mathbf{4}$$

A common example is *para*-nitroaniline, **5**:

The remarkable charge delocalization capacity that we have found for (6,0) carbon tubes stimulated us to examine, in analogy to **5**, the consequences of introducing an NH_2 group (donating) at one end of an open (6,0) and an NO_2 (withdrawing) at the other.[28,32] Not surprisingly, this led to stronger gradations in $V_S(\mathbf{r})$ and $\bar{I}_S(\mathbf{r})$ than any others that we have observed. $V_{S,max}$ and $V_{S,min}$ were 76 and -87 kcal/mole, with $\sigma_{tot}^2 = 1036\,(\text{kcal/mole})^2$. In order to assess the possible nonlinear optical activity, we used the local density functional SVWN/6-31G* procedure to estimate the first hyperpolarizability, β, for both *para*-nitroaniline and the NH_2/NO_2-substituted tube. The magnitude of β was nine times larger for the latter!

There have already been investigations of fullerene and nonfunctionalized nanotubes for nonlinear optical activity.[85,90–94] Our present results are at least suggestive that this may be particularly enhanced in appropriately functionalized (6,0) systems.

4. SUMMARY AND FUTURE WORK

Open all-carbon (6,0) model nanotubes and, in at least one instance, a closed (6,0) C/B/N system, have been found to have an unexpected capacity for charge delocalization. This can be triggered by even a seemingly minor asymmetric perturbation at one or both ends of the tube. We have pointed out implications for nonlinear optical activity, among other possible applications.

Another aspect of the unusual nature of (6,0) systems is the structural reorganization that accompanied the conversion of C_{104} to $C_{102}BN$, producing chains of C=C near-double-bonds. This could be of considerable significance with respect to chemical reactivity.

We believe that this strikingly facile tube-long transmission of electronic effects is related to the significant number of C-C bonds that are parallel to the tube axis, which is of course common to all in the $(n,0)$ category. However there are many questions to be answered. An obvious one is whether these features of the $(6,0)$ tubes are also present in other $(n,0)$. A preliminary investigation of an OH-substituted open carbon $(8,0)$ tube did show the charge delocalization, but this will have to be studied extensively for various $(n,0)$ with different lengths and functional groups. It is relevant to note, in this context, the enhanced electrical conduction reported for $(n,0)$ carbon nanotubes by Zipper *et al.*[95,96]

Another point that we want to examine is the role of the alignment or nonalignment of the C-H bonds at the ends of the open $(6,0)$ carbon tubes that we functionalize. If the C-H bonds at the two ends are colinear, then we say that they are aligned; if not, then they are nonaligned. In the example in Figure 13-1(a), they are aligned, and this has been the case for all those tubes for which the $V_S(r)$ and $\bar{I}_S(r)$ have been reported in this chapter. In view of the extreme sensitivity of $(6,0)$ systems to any asymmetric perturbation, we believe that the consequences of nonalignment need to be investigated. This can be done by simply increasing or decreasing the length of the tube by one ring of hexagons. All of these studies are in progress.

ACKNOWLEDGEMENT

We are grateful to Professor Tore Brinck for carrying out the computations required for Figure 13-3, and for preparing the figure.

REFERENCES

1. Kroto HW, Heath JR, O'Brien SC, Curl RF, Smalley RE (1985) C_{60}: Buckminsterfullerene. Nature 318:162
2. Iijima S (1991) Helical microtubules of graphitic carbon. Nature 354:56–57
3. Ebbesen TW, Ajayan PM (1992) Large-scale synthesis of carbon nanotubes. Nature 358:220–221
4. Fowler PW (1990) Carbon cylinders: a class of closed-shell clusters. J. Chem. Soc., Faraday Trans. 86:2073–2077
5. Wang S, Buseck PR (1991) Packing of C_{60} molecules and related fullerenes in crystals: a direct view. Chem. Phys. Lett. 182:1–4
6. Dresselhaus MS, Dresselhaus G, Saito R, (1992) Carbon fibers based on C_{60} and their symmetry. Phys. Rev. B 45:6234–6242
7. Mintmire JW, Dunlap BI, White CT (1992) Are fullerene tubules metallic? Phys. Rev. Lett. 68:631–634
8. Harris PJF (1999) Carbon Nanotubes and Related Structures. Cambridge University Press, Cambridge, UK
9. Kim P, Lieber CM (1999) Nanotube nanotweezers. Science 286:2148–2150
10. Baughman RH, Cui C, Zakhidov AA, Iqbal Z, Barisci JN, Spinks GM, Wallace GG, Mazzoldi A, De Rossi D, Rinzler AG, Jaschinski O, Roth S, Kertesz M (1999) Carbon nanotube actuators. Science 284:1340–1344

11. Saito R, Dresselhaus G, Dresselhaus MS (1996) Tunneling conductance of connected carbon nanotubes. Phys. Rev. B 53:2044–2050

12. Treboux G, Lapstun P, Silverbrook K (1999) An intrinsic carbon nanotube heterojunction diode. J. Phys. Chem. B 103:1871–1875

13. Chico L, Crespi VH, Benedict LX, Louie SG, Cohen ML (1996) Pure carbon nanoscale devices: nanotube heterojunctions. Phys. Rev. Lett. 76:971–974

14. Tans SJ, Devoret MH, Dai H, Thess A, Smalley RE, Geerlings LJ, Dekker C (1997) Individual single-wall carbon nanotubes as quantum wires. Nature 386:474–476

15. Tans SJ, Verschueren ARM, Dekker C (1998) Room-temperature transistor based on a single carbon nanotube. Nature 393:49

16. Frackowiak E, Delpeux S, Jurewicz K, Szostak K, Cazorla-Amoros D, Beguin F (2002) Enhanced capacitance of carbon nanotubes through chemical activation. Chem. Phys. Lett. 361:35–41

17. Liu X, Si J, Chang B, Xu G, Yang Q, Pan Z, Xie S, Ye P, Fan J, Wan M (1999) Appl. Phys. Lett. 74:164

18. Freitag M, Martin Y, Misewich JA, Martel R, Avouris Ph (2003) Photoconductivity of single carbon nanotubes. Nano Lett. 3:1067–1071

19. Zhang X, Cao D, Chen J (2003) Hydrogen adsorption storage on single-walled carbon nanotube arrays by a combination of classical potential and density functional theory. J. Phys. Chem. B 107:4942–4950, and papers cited

20. Long RQ, Yang RT (2001) Carbon nanotubes as a superior sorbant for nitrogen oxides. Ind. Eng. Chem. Res. 40:4288–4291

21. Long RQ, Yang RT (2001) Carbon nanotubes as superior sorbent for dioxin removal, J. Am. Chem. Soc. 123:2058–2059

22. Halls MD, Schlegel HB (2002) Chemistry inside carbon nanotubes: the Menshutkin S_N2 reaction. J. Phys. Chem. B 106:1921–1925

23. Kong J, Franklin NR, Zhou CW, Chapline MG, Peng S, Cho KJ, Dai HJ (2000) Nanotube molecular wires as chemical sensors. Science 287:622–625

24. Chen RJ, Bangsaruntip S, Drouvalakis KA, Wong Shi Kam N, Shim M, Li Y, Kim W, Utz PJ, Dai H (2003) Noncovalent functionalization of carbon nanotubes for highly specific electronic biosensors. Proc. Nat. Acad. Sci. 100:4984–4989

25. Chen RJ, Choi HC, Bangsaruntip S, Yenilmez E, Tang X, Wang Q, Chang Y-L, Dai H (2004) An investigation of the mechanism of electronic sensing of protein adsorption on carbon nanotube devices. J. Am. Chem. Soc. 126:1563–1568

26. Saito R., Dresselhaus G, Dresselhaus MS (1998) Physical Properties of Carbon Nanotubes, Imperial College Press, London

27. Ajayan PM (1999) Nanotubes from carbon. Chem. Rev. 99:1787–1800

28. Politzer P, Murray JS, Lane P, Concha MC (2005) Computed electrostatic potentials and local ionization energies on model nanotube surfaces. In: Balandin AA, Wang WL (eds) Handbook of Semiconductor Nanostructures and Devices, American Scientific Publishers, Los Angeles (in press)

29. Politzer P, Lane P, Concha MC, Murray JS (2005) Effect of different caps on model nanotube surface properties. Microelectronic Eng. 81:485–493

30. Peralta-Inga Z, Lane P, Murray JS, Boyd S, Grice V, O'Connor CJ, Politzer P (2003) Characterization of surface electrostatic potentials of some (5,5) and (n,1) carbon and boron/nitrogen nanotubes. NanoLett. 3:21–28

31. Politzer, Lane P, Murray JS, Concha MC (2005) Comparative analysis of surface electrostatic potentials of carbon, boron/nitrogen and carbon/boron/nitrogen model nanotubes. J. Mol. Mod. 11:1–7

32. Politzer P, Murray JS, Lane P, Concha MC, Jin P, Peralta-Inga Z (2005) An unusual feature of end-substituted model carbon (6,0) nanotubes. J. Mol. Mod. 11:258–264

33. Mavrandonakis A, Froudakis GE, Schnell M, Muehlhaeuser M (2003) From pure carbon to silicon-carbon nanotubes: an ab initio study. Nano Lett. 3:1481–1484

34. Zhang D, Zhang RQ (2003) Theoretical prediction on aluminum nitride nanotube. Chem. Phys. Lett. 371:426–432

35. Cotton FA, Wilkinson G (1980) Advanced Inorganic Chemistry, 4th edn. Wiley-Interscience, New York

36. Windholz M (ed) (1983) The Merck Index, 10th edn. Merck, Rahway, NJ

37. Burdett JK (1995) Chemical Bonding in Solids. Oxford University Press, New York

38. Bengu E, Marks LD (2001) Single-walled BN nanostructures. Phys. Rev. Lett. 86:2385–2387

39. Bae SY, Seo HW, Park J, Choi YS, Park JC, Lee SY (2003) Boron nitride synthesized in the temperature range 1000–1200 °C. Chem. Phys. Lett. 374:534–541 and references cited

40. Suenaga K, Colliex C, Demoncy N, Loiseau A, Pascard H, Willaime F (1997) Synthesis of nanoparticles and nanotubes with well-separated layers of boron nitride and carbon. Science 278: 653–655

41. Golberg D, Bando Y, Mitome M, Kurashima K, Grobert N, Reyes-Reyes M, Terrones H, Terrones M (2002) Nanocomposites: synthesis and elemental mapping of aligned B-C-N nanotubes. Chem. Phys. Lett. 360:1–7

42. Stewart RF (1972) Valence structure from X-ray diffraction data: physical properties. J. Chem. Phys. 57:1664–1668

43. Politzer P, Truhlar DG (eds) (1981) Chemical Applications of Atomic and Molecular Electrostatic Potentials, Plenum, New York

44. Naray-Szabo G, Ferenczy GG (1995) Molecular Electrostatics. Chem. Rev. 95:829–847

45. Bader RFW, Carroll MT, Cheeseman JR, Chang C (1987) Properties of atoms in molecules: atomic volumes. J. Am. Chem. Soc. 109:7968–7979

46. Murray JS, Brinck T, Lane P, Paulsen K, Politzer P (1994) Statistically-based interaction indices derived from molecular surface electrostatic potentials; a General Interaction Properties Function (GIPF). J. Mol. Struct. (Theochem) 307:55–64

47. Murray JS, Politzer P (1994) A General Interaction Properties Function (GIPF): An approach to understanding and predicting molecular interactions, In: Murray JS, Politzer P (eds) Quantitative Treatments of Solute/Solvent Interactions. Elsevier, Amsterdam, Ch. 8

48. Murray JS, Politzer P (1998) Statistical analysis of the molecular surface electrostatic potential: An approach to describing noncovalent interactions in condensed phases. J. Mol. Struct. (Theochem) 425:107–114

49. Politzer P, Murray JS (1999) Representation of condensed phase properties in terms of molecular surface electrostatic potentials. Trends Chem. Phys. 7:157

50. Politzer P, Murray JS (2001) Computational prediction of condensed phase properties from statistical characterization of molecular surface electrostatic potentials. Fluid Phase Equilib. 185: 129–137

51. Sjoberg P, Murray JS, Brinck T, Politzer P (1990) Average local ionization energies on the molecular surfaces of aromatic systems as guides to chemical reactivity. Can. J. Chem. 68: 1440–1443

52. Koopmans TA (1933) Über die Zuordnung von Wellenfunktionen und Eigenwerten zu den einzelnen Elektronen eines Atoms. Physica 1:104–113

53. Politzer P, Abu-Awwad F, Murray JS (1998) Comparison of density functional and Hartree-Fock average local ionization energies on molecular surfaces. Int. J. Quantum Chem. 69:607–613

54. Politzer P, Murray JS, Concha MC (2002) The complementary roles of molecular surface electro-static potentials and average local ionization energies with respect to electrophilic processes. Int. J. Quantum Chem. 88:19–27

55. Murray JS, Peralta-Inga Z, Politzer P, Ekanayake K, LeBreton P (2001) Computational character-ization of nucleotide bases: Molecular surfaceelectrostatic potentials and local ionization energies, and local polarization energies, Int. J. Quantum Chem. 83:245–254

56. Brinck T, Murray JS, Politzer P (1993) Molecular surface electrostatic potentials and local ionization energies of groups V–VII hydrides and their anions. Relationships for aqueous and gas phase acidities. Int. J. Quantum Chem. 48:73–88

57. Politzer P, Murray JS, Grice ME, Brinck T, Ranganathan S (1991) Radial behavior of the average local ionization energies of atoms. J. Chem. Phys. 95:6699–6704

58. Politzer P, Grice ME, Murray JS (2005) Electronegativity and average local ionization energy. Coll. Czech. Chem. Comm. 70:550–558

59. Murray JS, Seminario JM, Politzer P, Sjoberg P (1990) Average local ionization energies computed on the surfaces of some strained molecules. Int. J. Quantum Chem., Quantum Chem. Symp. 24:645–653

60. Nagy A, Parr RG, Liu S (1996) Local temperature in an electronic systems. Phys. Rev. A 53: 3117–3121

61. Jin P, Brinck T, Murray JS, Politzer P (2003) Computational prediction of relative group polariz-abilities. Int. J. Quantum Chem. 95:632–637

62. Jin P, Murray JS, Politzer P (2004) Local ionization energy and local polarizability. Int. J. Quantum Chem. 96:394–401

63. Politzer P, Murray JS (2006) The average local ionization energy: Concepts and applications, In: Toro-Labbe A (ed) Theoretical Approaches to Chemical Reactivity. Elsevier, Amsterdam, (In press)

64. Politzer P, Murray JS (1991) Molecular electrostatic potentials and chemical reactivity. In: Lipkowitz KB, Boyd DB (eds) Reviews in Computational Chemistry, vol 2. VCH Publishers, New York, Ch. 7, references cited

65. Murray JS, Brinck T, Politzer P (1992) Applications of calculated local surface ionization energies to chemical reactivity. J. Mol. Struct. (Theochem) 255:271–281

66. Dillon AC, Jones KM, Bekkedahl TA, Kiang CH, Bethune DS, Heben MJ (1997) Storage of hydrogen in single-walled carbon nanotubes. Nature 386:377–378

67. A. Fujiwara, K. Ishii, H. Suematsu, H. Kataura, Y. Maniwa, S. Suzuki Y. Achiba (2001) Gas adsorption in the inside and outside of single-walled carbon nanotubes. Chem. Phys. Lett. 336: 205–211

68. Kuznetsova A, Yates JT Jr., Liu J, Smalley RE (2000) Physical adsorption of xenon in open single walled carbon nanotubes: observations of a quasi-one-dimensional confined Xe phase. J. Chem. Phys. 112:9590–9598

69. Mazzoni MSC, Chacham H, Ordejon P, Sanchez-Portal D, Soler JM, Artacho E (1999) Energetics of the oxidation and opening of a carbon nanotube. Phys. Rev. B 60:R2208–R2211

70. Bauschlicher CW, Jr. (2001) High coverages of hydrogens on a (10,0) carbon nanotube. Nano Lett. 1:223–226

71. S. Irle, A. Mews and K. Morokuma, Theoretical study of structure and Raman spectra for model nanotubes in their pristine and oxidized forms, J. Phys. Chem. A **106**, 11973–11980 (2002).

72. Jaffe RL (2003) Quantum chemistry study of fullerene and carbon nanotube fluorination. J. Phys. Chem. B 107:10378–10388

73. Zhou Z, Steigerwald M, Hybertsen M, Brus L, Friesner RA (2004) Electronic structure of tubular aromatic molecules derived from the metallic (5,5) armchair single wall carbon nanotube. J. Am. Chem. Soc. 126:3597–3607

74. Peralta-Inga Z, Murray JS, Grice ME, Boyd S, O'Connor CJ, Politzer P (2001) Computa-tional characterization of surfaces of model graphene systems. J. Mol. Struct. (Theochem) 549:147–158

75. Chen J, Haddon RC, Fang S, Rao AM, Lee WH, Dickey EC, Grulke EA, Pendergrass JC, Chavan A, Haley BE, Smalley RE (1998) J. Mater. Res. 13:2423

76. Srivastava D, Brenner DW, Schall JD, Ausman KD, Yu M-F, Ruoff RS (1999) Predictions of enhanced chemical reactivity at regions of local conformational strain on carbon nanotubes: kinky chemistry. J. Phys. Chem. B 103:4330–4337

77. Dekker C (1999) Carbon nanotubes as molecular quantum wires. Phys. Today 52(5):22

78. Benedict LX, Louie SG, Cohen ML (1995) Static polarizabilities of single-wall carbon nanotubes. Phys. Rev. B 52:8541–8549

79. Wan X, Dong J, Xing DY (1998) Optical properties of carbon nanotubes. Phys. Rev. B 58:6756–6759

80. Jensen L, Astrand P-O, Mikkelsen KV (2004) The static polarizability and the second hyperpolarizability of fullerenes and carbon nanotubes. J. Phys. Chem. A 108:8795–8800

81. Exner O (1978) In: Chapman NB, Shorter J (eds) Correlation Analysis in Chemistry, Plenum Press, London, ch. 10

82. Williams DJ (1984) Organic polymeric and non-polymeric materials with large optical nonlinearities. Angew. Chem. Int. Ed. Engl. 23:690–703

83. Beratan DN (1991) In: Marder SR, Sohn JE, Stucky GD Marder SR, Sohn JE, Stucky GD (eds) New Materials for Nonlinear Optics. ACS Symposium Series 455, American Chemical Society, Washington, DC., p. 89

84. Kanis DR, Ratner MA, Marks TJ (1994) Design and construction of molecular assemblies with large second-order optical nonlinearities. Quantum mechanical aspects. Chem. Rev. 94:195–242

85. Bredas JL, Adant C, Tackx P, Persoons A, Pierce BM (1994) Third-order nonlinear optical response in organic materials: theoretical and experimental aspects. Chem. Rev. 94:243–278

86. Wilson WL (2001) In: Moore JH, Spencer ND (eds) Encyclopedia of Chemical Physics and Physical Chemistry, vol III. Institute of Physics, London, C2.15

87. Karna SP (2000) Electronic and nonlinear optical materials: the role of theory and modeling. J. Phys. Chem. A 104:4671–4673

88. Matsuzawa N, Dixon DA (1994) Density functional theory predictions of polarizablities and first- and second-order hyperpolarizabilities for molecular systems. J. Phys. Chem. 98:2545–2554

89. May JC, Lim JH, Biaggio I, Moonen NNP, Michinobu T, Diederich F (2005) Highly efficient third-order optical nonlinearities in donor-substituted cyanoethynyl-ethene molecules, Optics Lett. 30 (in press)

90. Liu X, Si J, Chang B, Xu G, Yang Q, Pan Z, Xie S, Ye P, Fan J, Wan M (1999) Third-order optical nonlinearity of the carbon nanotubes. Appl. Phys. Lett. 74:164–166

91. Jiang J, Dong J, Xing DY (1999) Size and helical symmetry effects on the nonlinear optical properties of chiral carbon nanotubes. Phys. Rev. B 59:9838–9841

92. G. Ya Slepyan, S. A. Maksimenko, V. P. Kalosha, J. Herrmann, E. E. B. Campbell and I V. Hertel, Highly efficient high-order harmonic generation by metallic carbon nanotubes, Phys. Rev. A **60**, R777–R780 (1999).

93. Chen P, Wu X, Sun X, Lin J, Ji W, Tan KL (1999) Electronic structure and optical limiting behavior of carbon nanotubes. Phys. Rev. Lett. 82:2548–2551

94. Mishra SR, Rawat HS, Mehendale SC, Rustagi KC, Sood AK, Bandyopadhyay R, Govindaraj A, Rao CNR (2000) Optical limiting in single-walled carbon nanotube suspensions. Chem. Phys. Lett. 317:510–514

95. Szopa M, Marganska M, Zipper E, Lisowski M (2004) Coherence of persistent currents in multiwall carbon nanotubes. Phys. Rev. B 70:75406–75412

96. Zipper E, Szopa M, Marganska M, Lisowski M, Persistent currents in single- and multiwall carbon nanotubes, paper presented by Zipper E. at Nano and Giga Challenges Conference. Jagellonian University, Krakow, Poland, September 2004

CHAPTER 14

ELECTRONIC PROPERTIES AND FRAGMENTATION DYNAMICS OF ORGANIC SPECIES DEPOSITED ON SILICON SURFACES

JIAN-GE ZHOU AND FRANK HAGELBERG

Department of Physics, Atmospheric Sciences, and Geoscience, Jackson State University, USA

Abstract: This contribution summarizes recent progress in the computational treatment of organic species deposited on silicon surfaces, with emphasis on the Si(100) surface. Representative theoretical studies of various organic species in contact with Si surfaces are surveyed, involving unsaturated hydrocarbons, amines, phosphines, and alcohols as adsorbates. The connection of the presented computational results to spectroscopic measurement is outlined in each individual case. The strengths and the limitations of a finite cluster model for simulating the Si substrate are discussed. Further, a comprehensive investigation of one specific system is presented, namely 1-propanol adsorbed on Si(001)-(2 × 1). It is shown by density functional theory within periodic boundary conditions that 1-propanol in contact with Si(001)-(2 × 1) initially occupies a metastable physisorbed state which turns into a stable chemisorbed ground state by dissociative hydrogen transfer. This fragmentation effect is confirmed by *ab initio* molecular dynamics at room temperature. The adsorbed organic layer induces further surface reconstruction. For the first time, the band structure of the 1-propanole/Si(001) film is determined. The tendency of the energy gap as a function of 1-propanole coverage indicates that the surface becomes increasingly insulating as the areal density of the organic adsorbate is enhanced

Keywords: Silicon surface; CVD; Chemisorption; Proton transfer

1.   INTRODUCTION

The interaction of organic molecules with semiconductor surfaces is a highly topical subject of current experimental as well as computational research. This interest is related to the observation that semiconductor surfaces provide templates for highly localized reactions of organic species.[1] An equally strong motivation for treating composites of periodic semiconductor substrates and organic molecules is the prospect of controlled surface modification for various technological applications in the areas of nanolithography, surface coating, chemical sensing, or

W. A. Sokalski (ed.), Molecular Materials with Specific Interactions, 505–532.
© 2007 *Springer.*

molecular electronics. Thus, very recent developmental efforts aim at the design of novel nanoelectronic elements from mixed organic-inorganic units.[2] Complex molecular architectures are fabricated in the laboratory, involving alkene linker chains anchoring in single crystal or porous silicon and employed to immobilize DNA strands on the respective surfaces. Organic adsorbates attached to semiconductors have also been studied extensively as precursors of thin films on surfaces. Unsaturated hydrocarbons[3,4] may be utilized to grow C layer coatings of Si substrates and may be instrumental for chemical vapor deposition of diamond films on Si surfaces. Further, reactions of organic species with semiconductor substrates have been discussed in the context of the technologically essential process of surface oxidation which is applied to create mask patterns of relevance for diffusion doping of Si or for passivation of p-n junctions.[5,6]

In order to optimize these procedures, an in-depth understanding of the interaction between the surface and the organic species is of crucial importance. This has been recognized by a plethora of computational investigations that, focusing mostly on the Si(001) surface, analyze the interface between the extended substrate and the molecular adsorbate by means of quantum chemistry. We attempt to give a condensed summary of present and recent activities directed at the elucidation of these systems. As far as the method of research is involved, one may distinguish approaches that take into account the periodicity of the substrate[4,5] from those that approximate the semiconductor surface by a finite cluster model.[6,7] The latter are necessarily constrained to the study of local effects, i.e. reactions at well-defined attachment sites while the former may include non-local properties such as the dependence of molecular adsorption or dissociation features on the level of coverage. A natural classification of the computational efforts made in the field of organic units attached to semiconductors is given with the type of the considered adsorbate. We will concentrate on unsaturated hydrocarbons, amines, phosphines, and alcohols. The first part of this survey will give an overview of recent computational investigations on these systems. The following segment of this survey, section 3, will present a case study and provide a detailed description of results obtained for 1-propanole deposited on the $Si(001) - (2 \times 1)$ surface.

2. RECENT PROGRESS IN THE COMPUTATIONAL TREATMENT OF ORGANIC SPECIES DEPOSITED ON SILICON SURFACES

2.1. The Si Surface

Silicon forms a diamond lattice. As Si single crystals are cut, some of the sp^3 bonds that stabilize the diamond structure are cleaved, resulting in the creation of unsaturated, or dangling bonds at the surface. To reduce its surface energy, the system tends to minimize the number of dangling bonds. In this process, new bonds between adjacent Si atoms are formed, associated with geometric surface reconstruction. The reactivity of specific Si surface sites strongly depends on their electronic environment and thus on the features of the reconstructed geometry which varies greatly among different Si surface types. For a review on Si surface chemistry see Waltenburg and Yates.[8]

This survey will focus in particular on the Si(001) surface which has been used as a template in the majority of experimental as well as theoretical investigations on the interaction between organic species and Si. As a consequence of the fact that each Si atom on the native Si(001) surface has two dangling bonds, the atoms of this surface reorganize into Si dimer rows. The interaction between the two atoms arranged in a Si dimer is commonly characterized as a strong σ type bond coexisting with a weak π type bond.[1,7] The question whether the dimer is symmetric or asymmetric, being tilted with respect to the Si(001) surface by a *buckling angle*, has given rise to extensive controversy. Recent experimental[9] and theoretical[4] results support the latter alternative. We point out that the computational description of the buckling effect requires the use of a slab model, involving periodic boundary conditions. Careful analysis of finite cluster models representing the Si(001) surface has yielded the symmetric arrangement as most stable geometry.[10] According to these investigations, the bonding operative in the Si dimer is more adequately understood in terms of a singlet diradical structure than as combination of a σ and a π bond. Adopting the buckled geometry, in contrast, one obtains a distinct polar admixture as the lower Si atom of the buckled pair assumes a partial positive, and the higher a partial negative charge. In this way, a zwitterionic charge distribution emerges.[6] The lower Si position has been identified as the preferred attachment location of N and O constituents of organic species, which is plausible in view of the net positive charge on this site. The upper Si atom has been shown to be the target site of hydrogen transfer.

Other reconstruction schemes than the 2×1 pattern have been reported for the Si(001) surface[11], associated with opposite buckling direction of adjacent dimers in a dimer row. More specifically, if the dimers of two neighboring rows buckle in the same (opposite) direction, a 2×2 (4×2) ordering arises.

The Si(111) surface presents a substantially more complex problem than the Si(001) alternative.[12] Two possible reconstruction types have been observed: A (2×1) - pattern that emerges from cleavage without subsequent annealing, and a (7×7) arrangement characteristic of the annealed surface. The former involves a reorganization of the top layer into an *upper* and a *lower* dimer chain, resulting in a periodic modulation of the surface. The (7×7) reconstruction of Si(111) is the most complicated Si surface geometry. The very successful dimer – adatom – stacking fault (DAS) model[13] involves a unit cell consisting of 102 atoms. More specifically, the unreconstructed Si(111) surface relaxes upon annealing into a three – layer architecture, composed of a dimer layer followed by a *rest atom* layer upon which twelve *adatoms* are distributed. The group of the adatoms is subdivided into *corner* and *center* atoms, depending on the spatial relation to their rest atom neighbors. The benefit of this intricate scheme is a reduction of the number of dangling bonds from 49 to 19.

As Si(111), the Si(110) reconstructed surface forms a very large unit cell, stabilizing in a (16×2) pattern. A wide variety of further reconstruction patterns is induced by atomic impurities adsorbed on the Si surface.

2.2.   Unsaturated Hydrocarbons

The adsorption of hydrocarbon molecules on Si surfaces is an interesting topic of study under various viewpoints. For example, a thin hydrocarbon film coating Si may be applied as a low dielectric in microelectronics and may passivate the surface if covalent bonds are formed between Si atoms and the adsorbate species. Further, unsaturated hydrocarbons play an important role as precursor species for chemical vapor deposition (CVD) of diamond – like films on the Si surface, and of silicon carbide (SiC).

Hydrocarbons containing single C – C bonds have been shown to be non-reactive in contact with the clean Si(001) surface.[14] Unsaturated hydrocarbons, however, involving carbon – carbon double bonds, proved to chemisorb with high probability to Si(001) surface sites, as was first demonstrated by Bozack et al. who reported a sticking coefficient close to unity for propylene at 120 K, irrespective of the initial surface coverage.[8,15] From high resolution electron energy loss spectroscopy (HREELS) it was concluded that both C_2H_2 and C_2H_4 adsorb molecularly on Si(001)[16,17], forming an ordered organic layer on the surface. With thermodynamic arguments, Taylor et al.[18] proposed that C_2H_2 attaches to the surface by di-σ bonding, cleaving the Si dimer bond. An analogous interpretation has been given for the experimental results related to the interaction of C_2H_4 with Si(001).[19,20] Further, HREELS studies,[20] where the Si(001)-(2×1) surface was saturated with C_2H_4 and subsequently exposed to atomic hydrogen, demonstrated co-adsorption of hydrogen atoms on dimers already occupied by C_2H_4 molecules. From this finding it was inferred that the Si – Si dimer bond is broken upon ethylene adsorption. This conclusion was based on the assumption that each Si atom has one dangling bond to which an H atom attaches during the post-hydrogenation stage of the experiment.

This assessment was modified by theoretical results. Fisher et al.[21] used the Car – Parrinello technique in combination with the local density approximation (LDA) method to investigate the interaction between C_2H_2 as well as C_2H_4 adsorbates with the Si(001) surface. In both cases, adsorption directly above the dimer proved to be the most stable arrangement, resulting in a lengthened C – C bond distance for both molecules. Correspondingly, the hybridization of C_2H_2 was found to change from sp to sp^2, and that of C_2H_4 from sp^2 to sp^3, with covalent bonds formed between the C atoms and the Si atoms of the dimer. The dimer bond is seen to remain intact, with only a small change of the Si – Si bond distance. This changes, however, as atomic hydrogen is added. Upon geometric relaxation of the substrate – adsorbate system the two Si dimer atoms are found to move apart from each other, and the electron density between them nearly vanishes. At the same time, H atoms attach to the Si atom sites. The experimentally observed bond cleavage therefore occurs as a consequence of hydrogen addition rather than hydrocarbon adsorption. Very similar results are presented by Pan et al.[22] who studied interaction of Si(001)–(2 × 1) with ethylene, including post – hydrogenation, in the framework of a cluster model.

In the language of cycloaddition chemistry, the adsorption of acetylene or ethylene on by a Si(001) – (2 × 1) dimer may be classified as 'suprafacial' [2+2] reaction where two π bonds are broken and two new σ bonds are formed. Symmetry considerations forbid such a reaction between two alkenes or two disilenes. The 'mixed' reaction between an alkene and a Si dimer, in contrast, has been shown to be very facile. An even more stable product is expected to arise from the Si(001) surface analogue of the *Diels-Alder* or [4+2] reaction, involving the formation of a six-membered Si_2C_4 ring which is less strained and therefore more favored than the Si_2C_2 that emerges from a [2+2] process. Hovis et al.[23] used a combination of STM and Fourier transform infrared (FTIR) spectroscopy to study 2,3 – dimethyl – 1,3 butadiene as well as 1,3 cyclohexadiene on the Si(001) – (2 × 1) surface. The [4+2] reaction was confirmed for 80% of the first and 55% of the second adsorbate, for the [2+2] reaction, the corresponding fractions were reported to be 20% and 35%, respectively. This finding gives experimental evidence for a competition between the two reaction types. While the Diels-Alder cycloaddition appears to be the preferred channel, the [2+2] alternative occurs as well, and with considerable probability. Other experimental[24] and computational[3,25] results, however, suggest a strong dominance of the [4+2] pathway. In particular, it has been proposed[3,25] that an initial [2+2] configuration will undergo fast conversion into the [4+2] structure.

In reaction to this controversial situation, Choi et al. have performed highly correlated computations for 1,3 cyclohexadiene on Si(001)-(2 × 1).[26] In this work, a cluster model was employed in combination with the SIMOMM (Surface Integrated Molecular Orbital/Molecular Mechanics)[27] approach that allows to define a quantum substructure and treat the remaining atoms classically. The largest unit used in[26] comprised a $C_6Si_{10}H_{24}$ quantum subsystem in a model of composition $C_6Si_{63}H_{57}$. For the quantum zone, the CASSCF(6,6) (complete active space SCF) method was used for geometry optimization followed by a single point MCQDPT2 (multiconfigurational quasi-degenerate second-order perturbation theory) calculation at the resulting geometry. As the main conclusion from this study, the [2+2] configuration involves only a small reaction barrier in the order of 4–8 [kcal/mol] which makes the formation of the respective product likely while disadvantaging the [2+2] pathway as compared with the [4+2] mechanism which proceeds without barrier. Importantly, the authors showed that the isomerization from the [2+2] to the [4+2] product is hindered by a considerable barrier in excess of 40 [kcal/mol], contradicting the hypothesis of substantial conversion from the former to the latter species. The separation between these two is incompatible with the view that the product distribution is thermodynamically determined. Instead, this distribution reflects the early stage of the reaction. These findings are consistent with the experiments of Hovis et al.[23] as they provide strong arguments for the coexistence of [2+2] and [4+2] reaction products on the Si(001)-(2 × 1) surface.

We turn now to the special case of the benzene molecule. Numerous experimental studies have been performed on benzene adsorbed to Si(001). In terms
of the attachment geometry, however, the measured data do not yield an entirely
unanimous picture. While the benzene molecule has been shown to adsorb consistently on top of the Si(001) surface dimer rows, its spatial extension, being close
to the spacing between two neighboring dimers, allows for a number of structural
variations.

Figure 14-1 shows some geometries that have been suggested as maximally
stable. A regular (*1, 4 cyclohexadiene-like*) butterfly structure, as presented in
Fig. 14-1a, has been identified by thermal desorption and angle-resolved photoelectron spectroscopy,[28] scanning tunneling microscopy,[29] vibrational infrared (IR)
spectroscopy and near-edge X-ray absorption fine-structure techniques,[30] and ab
initio calculations based on a cluster model.[28] In this configuration, benzene attaches
to one dimer of the Si(001) surface and saturates the dangling bonds of this dimer by
forming a di-σ-bond. An alternative tilted (*1, 3 cyclohexadiene-like*) structure (see
Fig. 14-1b) was detected in further STM measurements.[31] Also, a model involving
a tetra-σ-bond between benzene and two surface dimers, as displayed in Figure 14-
1c, was put forward on the basis of STM observation assisted by FTIR spectroscopy
and semiempirical cluster model calculation.[32]

In response to this inconclusive situation, Silvestrelli et al. have carried out
Density Functional Theory (DFT) studies simulating the extended Si(001) substrate
by use of the Car-Parinello approach[33] in conjunction with the local spin density
approximation (LSDA). From total energy calculations for various optimized
attachment geometries, the tight bridge alternative (Figure 14-1c) has been singled
out as the most stable structure. This configuration, however, turned out to correspond to a very narrow potential energy minmum of the combined system of the
benzene molecule and the Si(001) surface. While other adsorption geometries, such
as displayed in Figures 14-1a and b, have been shown to be metastable, a benzene
molecule impinging on the surface is likely to be trapped temporarily in one of
these states, preceding the eventual conversion into the structure of highest stability.
According to experiment as well as computation,[32] di-bonded 'standard butterfly'

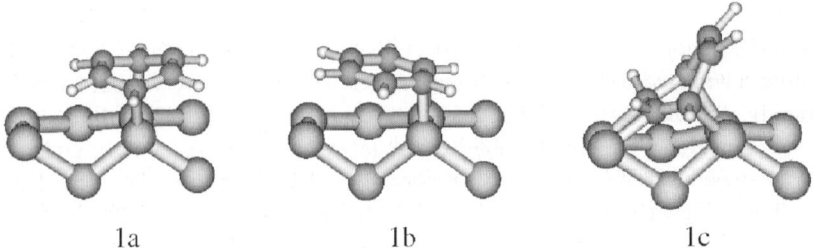

1a   1b   1c

Figure 14-1. Candidate structures for the preferred adsorption geometry of the benzene molecule on
the Si(001)-2 × 1 surface[4]. Shown are the 'Standard butterfly'(a), the tilted (b) and the 'tight bridge'
(c) structures

geometries are separated by sizeable energy barriers from the favored tetra-bonded 'tight bridge' configuration. This observation suggests long conversion times and thus could explain why the energetically favored adsorption structure has not been detected in the majority of the experiments on benzene attached to Si(001).

2.3. Amines

Computational research on the interaction of methyl-, dimethyl-, and trimethylamine with Si surfaces has been performed in response to a variety of experimental studies of these systems. These efforts acknowledge the relevance of Si surface nitridation which gives rise to the formation of silicon nitride films that find applications as insulating and passivating layers in integrated circuits. Thus, photoemission[34,35] and FTIR[1,36] as well as Auger electron (AE), thermal desorption, and low energy electron diffraction (LEED) spectroscopy[37] have been applied to investigate the reactions of amines with Si(001) or Si(111). A common feature of these reactions is that they proceed through the initial formation of a dative bond between the N atom and the lower Si atom in the buckled surface dimer. Basic electronegativity considerations suggest that Si should tend to donate an electron to N since on the Pauling scale it ranks at 1.90 while the corresponding value for N is 3.04. The positive effective charge of the lower Si(001) dimer atom, however, makes it a preferred target for the nucleophilic attack of the nitrogen lone pair. Following this primary step, a physisorbed configuration can stabilize, where the adsorbate remains essentially intact, or further reaction with the host surface may ensue, yielding a chemisorbed situation. Thus, FTIR results[1,36] led to the conclusion that N-H bond cleavage occurs for methyl- and dimethylamine, involving the loss of a H atom to a Si atom site. The rupture of the bond between N and H rather than between N and C has been ascribed to the increase of positive net charge on the N atom as the dative-bonded complex is formed, weakening preferentially the N-H bond.[34,35] From photoemission studies,[34,35] the characteristic N-H bond breaking process does not take place for trimethylamine which forms a stable dative-bonded adduct on the Si(001) surface. This finding has been confirmed by means of AE, LEED and thermal desorption spectroscopy.[37]

We add for completeness that similar observations have been reported for the Si(111) surface[34] where, to our knowledge, the experimental results have not been interpreted by theory. According to core-level and valence-band photoelectron spectroscopy data on the dissociative adsorption of ammonia on Si(111)–7 × 7[38], the surface adatoms associate primarily with NH_2, while the cleaved H radicals connect to rest atoms. This finding has been correlated with the formal charge on the rest atoms and the adatoms, amounting to −1 and about 0.58[34], respectively. Thus, a picture analogous to that proposed for the Si(001) surface emerges. The N constituent of the adsorbate attaches to an electron-deficient component of the Si substrate. For trimethylamine, a physisorbed configuration is identified as the dominant adsorption mode for both, the Si(001) and the Si(111) surface. From X-ray photoelectron spectroscopy (XPS)[34], however, the areal density of trimethylamine molecules on Si(001) exceeds that on Si(111) at a temperature of 300 K approximately by a factor of 2. This result has been interpreted as manifestation of steric effects; it might also be explained by the different reactivities of *corner* and *center* adatoms of Si(111)-7 × 7, as established by STM measurement[39].

Mui et al.[36] report a comparative experimental – theoretical study of amines on both the Si(001)–(2 × 1) and the Ge(001)–(2 × 1) surface. Both substrates were modeled by X_9H_{12} (X = Si, Ge) clusters, utilizing DFT at the BLYP/6-31G(d) level of theory. For both, the Si and the Ge substrate, formation of a X-N dative bond (X = Si, Ge) is the initial step of the reaction between the considered amine species and the semiconductor surface. However, while primary and secondary amines display N-H dissociation when attached to Si(001)–(2 × 1), no such trend is observed for the Ge counterpart of this system. This deviating behavior may be understood in terms of the energy barrier that separates the physisorption from the chemisorption minimum, involving the cleavage of an H atom. For dimethylamine adsorption, this quantity turned out to be about 50% higher for the Ge than for the Si surface. The authors relate this characteristic difference between the two substrates to the different proton affinities of Si and Ge.

2.4.   Phosphines

A large body of experimental data exists on phosphine (PH_3) in contact with Si surfaces. The chief motivation for the interest in composites of Si and phosphine is that the latter is used as a dopant source in the epitaxial growth of thin Si films, involving chemical vapor deposition (CVD) or gas-source molecular beam epitaxy (GSMBE). However, the use of phosphine as a dopant reduces the Si growth rate, diminishing in turn the production efficiency of ultra large scale integration (ULSI) circuits. Most experimental studies on the interaction between phosphine and Si(001) have been carried out in response to this finding. A large range of experimental techniques has been applied in this effort, including LEED,[40] AES,[40–42] secondary ion mass spectrometry (SIMS),[40] XPS,[43,44] thermal desorption spectrometry (TDS),[39,45] temperature programmed desorption,[42,46,47] STM,[41,44,48–50] HREELS,[42] FTIR,[51] and synchrotron core-level photoemission.[49] Summarily, it is observed that PH_3 at room temperature attaches to the electrophilic site of the Si dimer, in analogy to the respective observations made for amines. If the surface coverage does not exceed 0.18 [ML], dissociation of the adsorbate into PH_2 and H is detected, where PH_2 and H stabilize on different sites of the Si dimer. At higher coverage level, the lower number of available dangling Si bonds causes the fraction of dissociated PH_3 molecules to decrease. This tendency, however, is counteracted by temperature. At about 600 [K], PH_3 is entirely converted to PH_2, and around 700 [K] to P. These transition temperatures evidently depend on the coverage level which determines the number of possible attachment sites for the dissociation products.

The interaction of PH_3 with the Si(111)-(7 × 7) surface has been studied by HREELS,[52] demonstrating the onset of dissociative adsorption at about 80 [K].

According to a combination of ultraviolet photoemission spectroscopy (UPS), LEED and ion scattering spectroscopy (ISS),[53] the PH_3 –exposed surface is covered by a P atom layer if annealed to 900 [K]. This observation is also made if PH_3 is adsorbed on the surface at 900 [K].

The experimental findings were interpreted by computational theory in the framework of cluster[51,54] as well as slab[44,55] models. Shan et al.[51] based their respective calculations on a hydrogen terminated Si(001) fragment of composition $Si_{21}H_{20}$. This model contained three surface dimers on the top of eight Si atoms in the second and four and three Si atoms in the third and the fourth layer, respectively. This system was treated by DFT using the hybrid B3LYP method[56] in conjunction with a $6-31+G^*$ basis set. The authors compare various possible PH_3 adsorption geometries, among them a bridge site where P bonds with both dimer atoms, and arrive at the conclusion that the configuration of highest stability involves a 'dangling bond adsorption site' at one of the two dimer atoms. If this position is populated, dimer tilting with respect to the Si(001) is obtained as a consequence of a lowering of the occupied Si atom.

Miotto et al.[55] employed a super-cell geometry to describe the composite of the periodic Si(001) substrate and the finite phosphine adsorbate. Electronic structure calculations were carried out on the basis a modified version[57] of the Car – Parrinello scheme,[33] where the generalized gradient approximation[58] (GGA) was used in combination with pseudopotentials of the Troullier-Martins[59] type. From this work, the physisorbed state associated with the attachment of PH_3 to one of the two Si dimer atoms is metastable, and the decomposition into PH_2 and H, located at the opposite end of the dimer, corresponds to a minimum of lower energy. The activation barrier separating these two structures is distinctly smaller than the thermal energy at the usual growth temperature, implying that PH_3 will decay spontaneously at room temperature, and will be observable in its dissociated form. Similar results have been obtained for NH_3 as well as AsH_3, where the thermal barrier towards chemisorption has been found to be smallest for PH_3 among the three adsorbates considered. The dissociation model gains additional support from the calculation of vibrational frequencies which were found to be in very good agreement with values extracted from measured vibrational spectra. In particular, Si – H stretching and bending modes have been assigned to experimentally detected lines. As pointed out by Shan et al.,[48] H desorbs from the dimer loaded with PH_2 and H at a substantially higher temperature than from the clean Si(001) surface. This is invoked as partial explanation for the observed reactivity degradation in Si(001) CVD processing when P dopants are incorporated.

For Si(111)-(7 × 7), various possible physisorption and chemisorption geometries have been discussed in the framework of a $Si_{25}H_{24}PH_3$ cluster model.[54]

2.5. Alcohols

Zhang et al.[6] performed measurements involving 1 – propanol ($CH_3CH_2CH_2OH$), ethanol (CH_3CH_2OH) and methanol (CH_3OH) adsorbed on the Si(100)-(2 × 1) surface. In this experiment, LEED spectroscopy was combined with AES studies, further, a quadrupole mass spectrometer was utilized for residual gas and thermal desorption analysis. As the primary result of these investigations, the considered alcohol species undergo dissociative chemisorption on Si(100)-(2 × 1) at room temperature. For instance, in the case of 1-propanol, propanal (CH$_3$CH$_2$CHOH) has been identified as a major product of thermal desorption. This species, in all likelihood, results from hydrogen elimination of surface – adsorbed 1-propanole. These findings are in keeping with those obtained from earlier photoemission and infrared spectroscopy experiments on the Si(001) adsorption of ethanol[60] and methanol[61] where ethoxy and methoxy formation has been reported, respectively.

The response of computational theory to these and related observations has, for the most part, been confined to cluster model computations. Thus, Lu et al. report calculated results for CH$_3$OH, CH$_2$O (aldehyde), and HCOOH (formic acid) on Si(001)[62], while Kato et al.[63] investigated CH$_3$OH and CH$_3$NH$_2$ and Zhang et al.[6] refer to the three alcohol species examined in their experiment. All of these efforts are based on the finite Si$_9$H$_{12}$ model of the Si(001) – (2 × 1) surface, and all of them employ the hybrid B3LYP potential. We point out that the computations described in reference[62] involve the use of the ONIOM methodology[64] which allows to subdivide the system into different parts that are treated at different degrees of quantum chemical rigor. Specifically, Lu et al.[62] carry out the optimization of the substrate – adsorbate composite at the B3LYP level and follow up with a single-point CCSD(T) calculation for selected components of the system as a whole, namely the Si dimer in combination with the adsorbed molecule and the passivating H atoms; the B3LYP approach is used for the remaining Si atoms.

In all cases, it was found that the considered alcohol molecule undergoes dissociative adsorption on the Si dimer after having formed initially a dative bond with the electrophilic dimer end. Ref.[6] contains a systematic comparison between the two major dissociation channels involving the –OH component. The two major dissociative reactions consist in the loss of the H atom or that of the OH group. These alternatives are characterized by the dative bond well, by an O–H (O–C) cleavage transition state and a O–H (O–C) cleaved final state. For all adsorbates, the barrier towards O–H dissociation was found to be separated from the energy of the dative bond physisorption minimum by a small difference in the order of a few kcal/mol, while the corresponding quantity for the O–C barrier resulted as 75–85 kcal/mol. For the dissociated product, however, the roles are seen to be reversed, as the loss of the OH group leads to a final chemisorbed state that is 60–80 kcal/mol more stable than that emerging from the loss of the H atom. One concludes that the separation of an H atom proceeds almost spontaneously and is thus kinetically favored, while the detachment of OH yields a deeper energy well in the final state and is therefore thermodynamically preferred. Section 3 of this survey will add more detail to this overall picture for the special case of 1 – propanol adsorption.

Silvestrelli[5] reached a similar conclusion for Si(001) – (2 × 1) interaction with ethanol in the framework of a periodic slab approach, where the Car – Parrinello method was used, instead of a cluster model. Quantitatively, substantial differences between the results emerging from these two different methodologies are noticed. In particular, the energy barrier for H dissociation is found to be higher by an order of magnitude as the cluster treatment is replaced by the slab method.

We emphasize two natural limitations of the finite cluster model. It does not allow to make a statement about the dependence of essential parameters such as adsorption and transition energies on the level of surface coverage, and it does not account adequately for charge delocalization or surface relaxation phenomena. Further, it excludes by definition any information about the modification of the surface band structure as a consequence of the organic molecule adsorption. The following case study of 1-propanol on Si(001) – (2 × 1) is intended to clarify how these elements can be consistently incorporated into the description of the Si surface interaction with organic species.

3. CASE STUDY: ADSORPTION OF 1-PROPANOL ON THE SI(001)-(2 × 1) SURFACE

3.1. Motivation and Methodology

To illustrate the above arguments that speak in favor of a periodic boundary conditions approach to the theme of molecular adsorption on surfaces, a simulation of the interaction between a finite molecular adsorbate and an extended Si(001) substrate is presented in this segment of our survey. More specifically, the adsorption of 1-propanol on the Si(001)-(2 × 1) surface is studied by use of the VASP code,[65] involving a slab geometry. This methodology affords a realistic model of the silicon surface, its reconstruction, and the adsorption of the 1-propanol molecule as well as its fragments. Our first principles approach leads naturally to the distinction between *up* and *down* Si dimer atoms, thus producing the characteristic buckling of the surface silicon dimer. The model allows for an appropriate description of the Si(001) surface with and without the adsorption of the 1-propanol molecule as the reconstruction of the Si surface before and after adsorption can be displayed manifestly. A detailed analysis of the most prominent reactions undergone by 1 - propanol on the Si(001) surface is presented below, including the reaction barriers determining various pathways. Further, the surface band structure within the silicon fundamental gap and the variation of the binding energies, the energy barriers and the energy gap with the degree of coverage are outlined in the following. Four levels of coverage (0.125 ML, 0.25ML, 0.5ML, and 1.0ML) are taken into account. Finally, admission for finite temperature is made, as the characteristic reaction mechanisms at $T = 300\,K$ are compared with those found at $T = 0\,K$.

In all computations, DFT was applied on the level of the generalized gradient approximation (GGA)[66] in conjunction with the PAW potential[67,68]. The wave functions are expanded in a plane wave basis with an energy cutoff of 400 ev, whereas the cutoff for the augmentation charges is 645 eV. The Brillouin zone integrations are performed by use of the Monkhorst-Pack scheme[69] with the origin shifted to the Γ point. A $3 \times 3\,k$ point mesh is utilized for geometry optimization.

The Si(001)-(2 × 1) surface is modeled adopting a supercell geometry with an atomic slab of 5 Si layers where terminating hydrogen atoms passivate the Si atoms. The supercell consists of a 4 × 4 ideal cell, comprising 80 Si atoms and 32 H atoms. The Si atoms in the top four atomic layers are allowed to relax, while the Si atoms in the bottom layer and the adjacent passivating H atoms are fixed to simulate bulk-like termination[70,71]. The vacuum region is about 19 atomic layers, which exceeds the length of the 1-propanol molecule and provides sufficient spacing for the present MD simulation.

The energy barriers characterizing different reaction paths were calculated by the "climbing" Nudged Elastic Band (NEB)[72,73] method with six images, which permits identifying minimum energy paths in complex chemical reactions. Ab initio Molecular Dynamics simulations were performed by use of a Verlet algorithm to integrate Newton's equations of motion. The canonical ensemble was simulated using the Nose' algorithm[74].

We further computed energetic and geometric parameters pertaining to the reconstruction of the clean Si(001) surface. The 2 × 1 reconstructed silicon surface is displayed in this Figure 14-2. This Figure depicts the three silicon top layers, and Figure 14-2b illustrates the buckling angle, i.e. the angle between the dimer row and the horizontal plane. The Si dimers are oriented along the x axis or (110) direction, and the dimer rows are along the y axis or the ($\bar{1}$10) direction.

For the 2 × 1 surface reconstruction with asymmetric Si dimers, the energy gain is 1.6 ev per dimer. The internuclear distance between the two Si centers is 2.32 Å and the distance between the *up* Si atom of one dimer to the *down* Si atom of the next is 5.57 Å. The buckling angle amounts to 18.0°. These results agree with existing experimental[75] data and other calculations[70].

3.2.   The Physisorbed and Chemisorbed Configurations

In this section, the stable physisorbed configurations of the 1-propanol molecule on the Si(001)-(2 × 1) surface are described for a coverage level of 0.125 ML, where three non-dissociative structures are identified. Then seven possible dissociated structures are considered which correspond to chemisorbed configurations. From

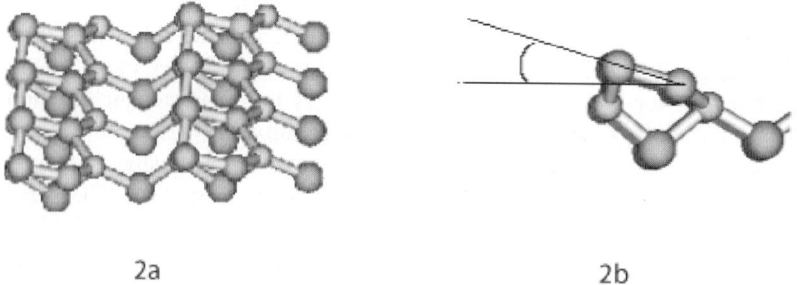

2a   2b

Figure 14-2. The 2 × 1 reconstructed Si(001) surface. The three silicon atom top layers are shown

these seven cases of chemisorption, we select the two most stable ones. Further, the modification of the clean Si(001)-(2 × 1) surface due to the physisorbed and chemisorbed 1-propanol molecules are considered. In an effort to examine which one among the chemisorbed structures is most likely to be observed experimentally, we calculate the energy barriers relevant to the chemisorbed configurations. Three physisorption sites were considered, as shown in Figure 14-3, where only a few surface Si atoms are displayed.

Here we only focus on the adsorbed structures obtained by an exothermic process, i.e. the composite of the surface and the adsorbed species is lower in energy than the free 1-propanol molecule and the clean Si(001)-(2 × 1) surface in separation from each other. Experimentally, it has been demonstrated that the 1-propanol molecule and its fragments are oriented vertically with respect to the surface[6]. This adsorption geometry is therefore adopted for our treatment of the physisorbed and chemisorbed configurations.

The calculated results confirm that the 1-propanol molecule initially interacts with the Si(001)-(2 × 1) surface via the formation of a dative bond between the oxygen atom and the electrophilic *down* Si atom of the surface dimer. Specifically, the O–Si bond may be characterized as a covalent connection arising from the lone pair of the O atom. The 1-propanol molecule remains essentially intact (this motivates the nomenclature I-1, I-2 and I-3 for the physisorbed configurations) at the physisorbed sites, and the O-H bond assumes various orientations with respect to the Si surface. The obtained structures are shown in Figure 14-3 which illustrates that the direction of the O-H bond can be parallel (I-1), antiparallel (I-2) or perpendicular (I-3) to the Si dimer. However, the energies of the three configurations are very close to each other, i.e., the rotation of the 1-propanol molecule around the Si-O bond is quite facile.

The binding energy reported in Zhang et al.,[6] namely 0.39 eV is considerably smaller than the respective value of 0.72 eV found in this work. This discrepancy

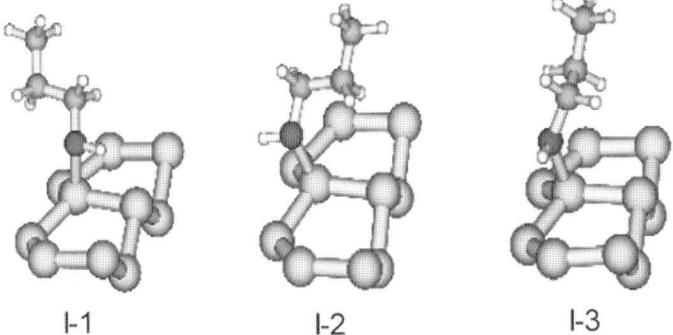

I-1   I-2   I-3

Figure 14-3. Three stable physisorbed configurations of the 1-propanol molecule on the Si(001) surface, where the red, white (smallest), orange and yellow (largest) spheres represent the oxygen, hydrogen, carbon and silicon atoms, respectively

might be attributed to the difference between the single dimer model and the periodic approach followed in the present approach.

For the clean Si(001)-(2 × 1) surface, the buckling angle with the horizontal plane is 18.0°. As a consequence of 1-propanol physisorption, the buckling angles for I-1, I-2 and I-3 become 8.7°, 11.0° and 10.8°, respectively. For the adjacent Si dimer, the corresponding buckling angles are 17.9°, 17.8°, and 17.0°, respectively. As the latter values are close to the angle found for the pure Si(001) surface, 18.0°, the interaction between the 1-propanol molecule and the adjacent Si dimer is quite weak. We have verified that the physisorption in the cases I-1, I-2 or I-3 is a barrierless reaction, which starts from a 1-propanol molecule far from the surface. Once this is physisorbed and attached to the surface through dative bonding, the 1-propanol can proceed to react with the surface via a number of pathways, which break one or more molecular bonds to form dissociated configurations of increased stability. The seven principal dissociated structures arising from H atom loss or O-C bond cleavage are shown in Figure 14-4. For the sake of clarity, we have included only ten Si atoms in this illustration.

The F-1 structure is obtained by breaking the O-H bond and detaching the H atom which attaches subsequently to the *up* Si atom of the same dimer to form a new H-Si bond. This configuration has the second largest binding energy among the chemisorbed structures compared here, namely 2.59 eV. The remaining alkoxy fragment is bonded to a Si surface dimer atom, while the separated H atom forms

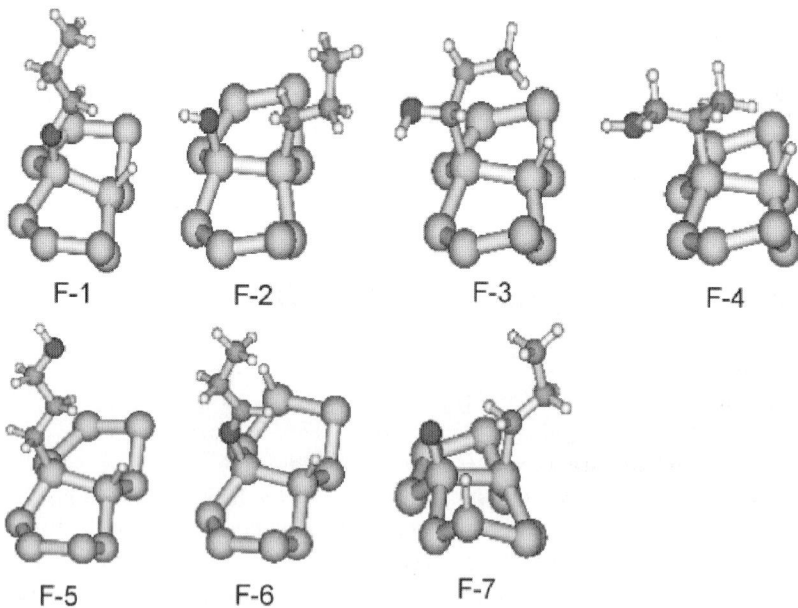

Figure 14-4. The fragmented chemisorption structures of 1-propanol on the Si (001) surface

a bond with the other Si atom of the same dimer. The binding energy decreases as the H atom is attached to a Si atom of an adjacent dimer.

The F-2 configuration is described by C1-O bond cleavage. The OH group and the alkyl fragment are bonded to the same Si dimer. If these two fragments are attached to adjacent dimers, the binding energy decreases. This configuration is thermodynamically most stable, exhibiting the largest binding energy, namely 2.89 eV.

The F-3, F-4 and F-5 configurations are characterized by breaking the C1-H, C2-H C3-H bonds, respectively, where the C1, (or C2, C3) atom is bonded to a Si atom, and the detached H atom forms a new bond with the other Si atom of the same dimer. From the respective binding energy one finds that these C-H cleavage configurations are of lesser stability than both F-1 and F-2.

To examine whether the F-1 and F-2 structures undergo further bond rupture, the configurations F-6 and F-7 were considered. F-6 is described by the cleavage of a C-H bond in F-1, where the H atom is attached to the adjacent dimer. The energy of F-6 is higher than that of F-1 by 0.9 eV. F-7 is obtained from F-2 by further dissociating the O-H bond and attaching the corresponding H atom to the adjacent dimer. This structure is energetically less favored than the original F-2 configuration. From an energetic point of view, F-1 and F-2 are most stable, corresponding to the tendency of 1-propanol to break the C-O bond or the O-H bond. Therefore, the subsequent discussion will be limited to the configurations F-1 and F-2.

Recall that the buckling angle for the clean Si(001)-(2 × 1) reconstructed surface is 18.0°, describing the asymmetric Si dimer, and adopts a modified value upon 1-propanol physisorption, as indicated above. For the chemisorbed configurations F-1 and F-2, however, the buckling angle tends to vanish. In a single dimer model, the accurate description of the Si dimer buckling effect and thus its variation with the adsorption mode is impossible.

3.3. Energy Barriers

To assess which chemisorbed structures are most likely to be observed experimentally, the energy barriers relevant to the chemisorbed configurations were calculated. Table 14-1 shows the energy barriers for the respective reactions.

For the physisorbed structures I-1, I-2 and I-3, the reaction proceeds without barrier. For the cases of chemisorption, we have calculated the energy barriers for the processes that lead from I-1 to F-1, I-2 to F-2 and F-1 to F-6, respectively. The energy barriers have been calculated by the climbing "Nudged Elastic Band" method[72,73] where six equidistant images have been used.

For the transformation to the F-1 configuration, I-1 is the most favorable initial structure since its O-H bond is already oriented parallel to the Si dimer row. This reaction is a proton transfer process from oxygen to the electron-rich, nucleophilic *up* silicon atom of the dimer. The I-1 to F-1 reaction is characterized by an energy barrier of 0.05 eV. The binding energy of I-1 is 0.75 ev which implies that the barrier for the whole process, i.e. adsorption into the I-1 structure followed by

Table 14-1. The energy barriers E_b, and transition state energy levels E_{TS} with respect to the energy of the 1-propanol molecule and the Si(001). surface in separation from each other

Reaction	$E_b(eV)$	$E_{TS}(eV)$
1-propanol + Si(100) → I-1	0	–
I-1 → F-1 (O-H rupture)	0.05	−0.70
I-2 → F-2 (C-O rupture)	1.34	0.62
F-1 → F-6	2.9	0.31

transition to the F-1 structure, is below the initial energy, namely that of the free 1-propanol molecule and a clean Si(001) surface. Since the binding energy of F-1 is 2.59 eV, the I-1 → F-1 process is exothermic.

From the physisorption case I-2 to configuration F-2, involving the breaking of a C-O bond, the energy barrier is 1.34 eV. The binding energy of I-2 is 0.72 eV. Therefore, the transition state energy is higher than the reference energy of the free 1-propanol molecule and the clean Si(001) surface. The binding energy of F-2 is 2.89 ev, making the I-2 → F-2 process exothermic too. Thus the O-C bond cleavage is thermodynamically stable, but the O-H bond cleavage is kinetically favored. In other words, the O-C cleaved final state has the highest binding energy, while the O-H cleavage is hindered by a smaller energy barrier than the O-C bond cleavage. This confirms, on the basis of a more adequate periodic model, the conclusion reached by Zhang et. al.[6] in the framework of a finite cluster approach. The relative energies along the O-C and O-H cleavage reaction paths are schematically illustrated in Figure 14-5.

Zhang et.al.[6] suggested that the initial O-H bond cleavage might be followed by a hydrogen elimination reaction to result in aldehydes and hydrogen. Table 14-1 shows that the energy barrier for the transition from the O-H cleavage configuration F-1 to the configuration F-6 is relatively high. One concludes that the respective reaction is not preferred. Similarly, the transition from F-2 to F-7 configuration is not favored.

The undissociated structures I-1, I-2 and I-3 can be interpreted as metastable precursors for the more stable F-1 configuration. These precursors do not have sufficient binding energy at room temperature to compete as observable reaction products, i.e. the cleavage of H is too fast for any of the physisorbed structures to be detected. The Molecular Dynamics (MD) simulation outlined below gives additional support to this interpretation.

3.4. Band Structure

A sketch of the eight relevant Si dimer units in the top Si layer is shown in Figure 14-6 for unambiguous reference where the horizontal (vertical) corresponds to the [110] ([$\bar{1}$10]) directions, respectively. For 0.125 ML, the 1-propanol molecule

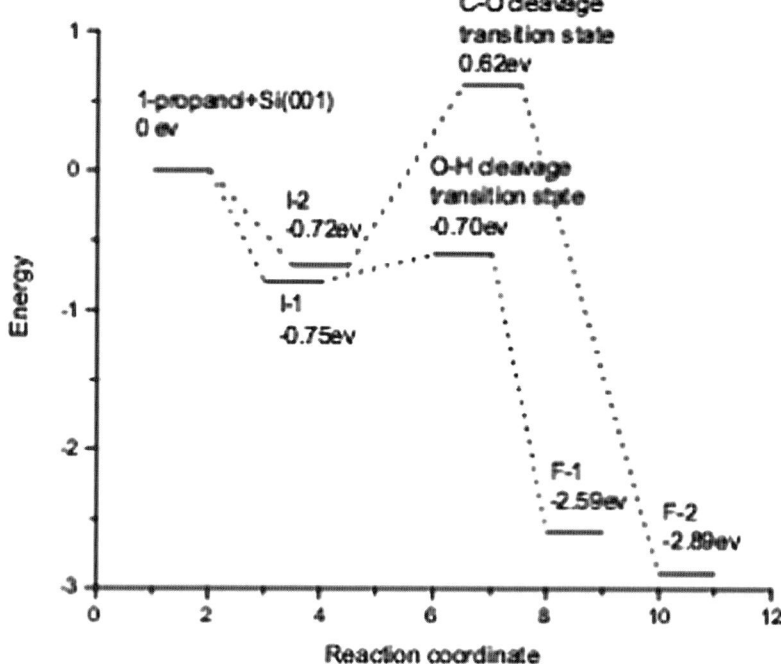

Figure 14-5. The relative energy levels along the O-C and O-H cleavage reaction paths

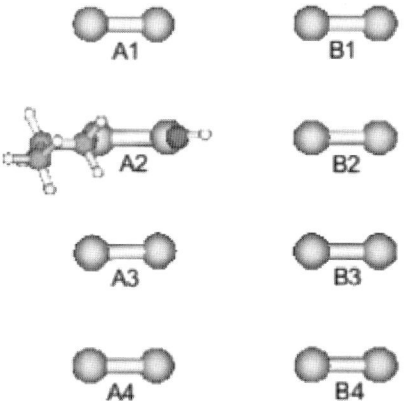

Figure 14-6. The labels and positions of the eight dimers in the top Si layer, where the horizontal direction is along [110] and the vertical one is along [$\bar{1}$10]

or its fragments are adsorbed to the A2 dimer. The *up* Si atoms are located at the left dimer ends, the *down* Si atoms at the right. However, as described above, after the adsorption of the 1-propanol molecule to the *down* Si atom, the latter is raised, i.e., the buckling angle decreases.

The surface band structures within the fundamental band gap of the silicon surface for the configurations I-1, I-2, F-1 and F-2 for 0.125 ML, are depicted in Figure 14-7. The k points Γ, J, K, J′ are four rectangle vertices of the quarter of the surface Brillouin zone (see, e.g., Figure 14-3 in Ref. (Ramstad et al.[76])). Figure 14-7 demonstrates that there are seven, nine, eight and eight surface bands within the fundamental band gap of silicon for the I-1, I-2, F-1 and F-2 configuration, respectively. The remaining valence (conduction) bands lie in the lower (higher) shaded area. In the I-1 configuration, for the valence bands (occupied), the top one is labeled I-1-O1, and the next lower one is I-1-O2, etc. For the conduction band (unoccupied), the bottom one is referred to as I-1-U1, the higher ones are I-1-U2, I-1-U3, I-1-U4, and the highest one is I-1-U5. The same nomenclature is used for the surface bands of the other three configurations.

The two highest surface valence (O1 and O2) and the two lowest surface conduction bands (U1 and U2) contain the information about the adsorption and are thus sensitive to the structural features of the surface. This does not hold for the remaining bands. The conduction bands U4 and U5, for instance, exhibit the same

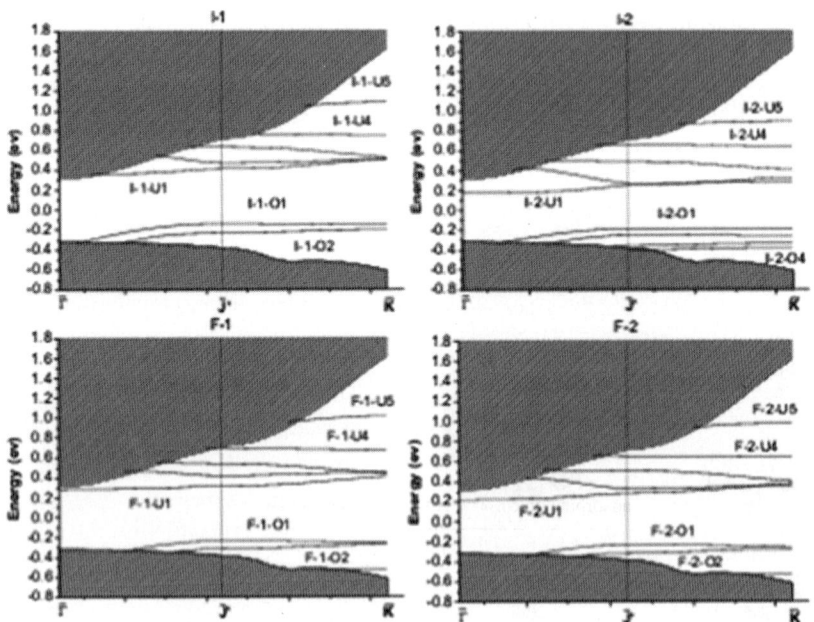

Figure 14-7. Surface band structures for the configurations I-1, I-2, F-1 and F-2 at 0.125 ML. The shaded areas represent the projected bulk band structure, while surface states are shown as solid lines

atomic orbital composition for all four configurations, I-1, I-2, F-1 and F-2. It is therefore sufficient to consider only the top two occupied valence bands (O1,O2) and bottom two unoccupied conduction bands (U1,U2). From our analysis, the band I-1-O1 contains the information about the 1-propanol adsorption. A1 and A3 contributions to the valence band I-1-O1 are found as electronic fingerprints of the adjacent dimers, while the 1-propanol physisorption leaves their geometric structure unaffected, as reflected by their buckling angles.

Generally, the occupied bands within the fundamental band gap of the silicon are composed of the *up* Si atoms, and the unoccupied bands involve the *down* Si atoms, which reflects the finding that the upper silicon atom of the buckled pair is negatively charged (electron rich) and the lower silicon atom positively (electron deficient). Here we note that within the fundamental band gap of silicon, there is no conduction band for a Si-O bonding due to the adsorbate. In case of the acetonitrile adsorption on the silicon surface[77], in contrast, a conduction band with both Si and N contributions is found within this gap, which indicates that the acetonitrile electronic interaction with the silicon substrate might be stronger than that for 1-propanol.

3.5. Dependence on the Level of Coverage

Taking advantage of the slab approach, we discuss the dependence of the binding energy of the four basic configurations (I-1, I-2, F-1, F-2) on the coverage of the 1-propanol molecules. First, we consider the basic configurations I-1, I-2, F-1 and F-2 with the coverage levels 1.00 ML, 0.5 ML, 0.25 ML and 0.125 ML, which corresponds to one 1-propanol molecule attached to one, two, four and eight dimers, respectively. Table 14-2 shows the binding energies of the four configurations of 1-propanol on the Si(001)-(2 × 1) surface.

Table 14-2 shows that the binding energies per 1-propanol molecule for the physisorbed configurations I-1 and I-2 decrease with increasing coverage. This trend appears quite natural since increasing concentration of the adsorbed molecules on the Si(001) surface results in enhanced interaction between the molecules and hence weakens their bond with the substrate. The binding energy for the chemisorbed structure F-1 decreases with increasing coverage too. This may be related to the fact that the alkoxy fragment has similar transverse dimensions as the 1-propanol molecule. However, the binding energy for the chemisorbed configuration F-2 exhibits very little change with the variation of the coverage.

Table 14-2. Binding energies (eV) of the adsorbate on Si(001) at four coverage levels

Coverage	I-1	I-2	F-1	F-2
0.125	0.76	0.72	2.59	2.89
0.250	0.73	0.70	2.57	2.90
0.500	0.68	0.67	2.55	2.91
1.000	0.41	0.37	2.34	2.85

Table 14-3. Energy barriers E_b and transition state energy levels E_{TS} for the dissociation processes I-1 → F-1 and I-2 → F-2 at four levels of surface coverage. The reference for the indicated energy values is the energy of the separated subsystems

Coverage	$E_b(I\text{-}1 \to F\text{-}1)$	$E_{TS}(I\text{-}1 \to F\text{-}1)$	$E_b(I\text{-}2 \to F\text{-}2)$	$E_{TS}(I\text{-}2 \to F\text{-}2)$
0.125	0.05	−0.70	1.34	0.62
0.250	0.02	−0.72	1.34	0.63
0.500	0.02	−0.65	1.32	0.65
1.000	0.05	−0.36	1.21	0.84

This observation is ascribed to the enhanced interaction between the alkyl fragments and OH groups on different dimers with increasing coverage, counteracting the destabilization trend due to the enhanced alkyl density.

The dependence of the energy barriers on the 1-propanol coverage in the interval [0.25 ML, 1.0 ML] is illustrated by Table 14-3, which contains the energy barriers E_b and transition state energy levels E_{TS} with respect to the energy of 1-propanol and Si(001) in isolation from each other. Two processes correspond to O-H bond and C-O bond scission.

From E_b values for the O-H and C-O bond rupture in Table 14-3, we find that the energy barriers for the O-H bond scission are only slightly affected by the level of coverage. However, the energy barriers for the C-O bond breaking E_b decrease with the increasing coverage.

From Table 14-2, the bonding of the chemisorbed structure F-1 (O-H bond scission) is weakened as the coverage increases, and Table 14-3 reveals that the energy barrier with respect to the O-H bond rupture, E_b, increases from 0.5 to 1.00 ML coverage. Thus, the probability of O-H bond scission is somewhat reduced at 1.00 ML. On the other hand, the binding energy for the C-O bond cleavage changes very little as the coverage is varied, while the energy barrier with respect to the rupture of the C-O bond E_b has its minimal value at 1.00 ML. This suggests that at a high coverage level, a small amount of C-O cleavage might occur, as supported by experimental observation[61].

To examine the surface modification upon 1 – propanol adsorption as a function of the coverage level, we analyze the dependence of the energy gap on the coverage with 1 – propanol (or its fragments). Table 14-4 shows their values for the physisorbed configurations I-1 and I-2, and the chemisorbed configurations F-1 and F-2 with 0.125 ML, 0.25 ML, 0.5 ML and 1.0 ML.

The energy gap for the Si(001)-(2 × 1) surface is 0.46 eV, see Table 14-4, which is in keeping with experiment (the corresponding experimental value is about 0.6 eV[78]). The local DFT and GGA procedures tend to underestimate the energy gap of semiconductors by up to 25%[79]. Table 14-4 shows that the energy gaps increase with the level of coverage, which reflects a growing degree of saturation of the silicon dangling bond as induced by the 1-propanol molecules. As a consequence of a higher number of oxygen atoms attached to the surface, and, by the same token, of dative bonds (for the physisorbed configurations I-1 and I-2) or covalent

Table 14-4. Energy gaps ΔE (eV) of the Si(001)-(2 × 1) pure surface compared to those of the 1-propanol adsorption structures I-1, I-2, F-1, and F-2 on Si(001) at four levels of coverage

Coverage	Si surface	I-1	I-2	F-1	F-2
0.125	0.46	0.53	0.40	0.53	0.40
0.250	0.46	0.53	0.47	0.60	0.60
0.500	0.46	0.53	0.53	0.80	0.73
1.000	0.46	0.80	0.73	1.27	1.20

bonds (for the chemisorbed configurations F-1 and F-2) between oxygen atoms and silicon atoms, the surface turns increasingly insulating, i.e., the energy gap widens.

3.6. Room Temperature Molecular Dynamics Calculations

The energy barrier computations have shown that the physisorbed 1-propanol molecule reacts with the Si(001)-(2 × 1) surface by cleavage of the O-H bond. Since the zero temperature transition state analysis may not be able to access all of the relevant phase space volume, a finite temperature *ab initio* MD simulation is performed to include additional possible reactions at T = 300K. The 2 × 1 cell is adopted to carry out the MD simulation (the 2 × 2 cell was used as well, and the results from both approaches were found to agree). In the finite temperature MD calculations all atoms, including the passivating H atoms at the bottom of the slab, are allowed to move. In this manner, a large temperature gradient can be avoided. Lattice parameters are expanded according to the temperature under study using the experimental thermal expansion coefficient in order to prevent the lattice from experiencing internal thermal strain[71].The starting configuration is the physisorbed one, I-1 (see Figure 14-3), the O-H and O-C bond lengths are 1.01 Å and 1.48 Å, respectively. The I-1 physisorbed structure is heated to 300K (room temperature) in 9000 MD steps (9.0 ps, i.e. each step takes 1 fs), followed by another 3000 MD steps at 300 K to evolve the system under conditions of thermal equilibrium. Monitoring the free energy of the system as a function of the evolution time, we assess if the system has reached its equilibrium. From this simulation, it turns out that the free energy fluctuates very little after 10ps, which shows the system is at equilibrium. Every recorded data point represents an average result over an interval of 300 MD steps. In this way, high frequency components due to thermal motion[80] are filtered out.

We consider the time variation of the O-C1 and Si-C1 bond lengths in the MD calculation, which are represented in Figure 14-8. It is seen that the O-C1 bond is not ruptured in the process of the simulation. For times shorter than 6ps, the distance between the C1 and the up Si atom fluctuates around 4.25 Å; between T = 6ps and T = 7ps, it reduces by 1 Å, and after T = 7ps, it oscillates around 4.0 Å,

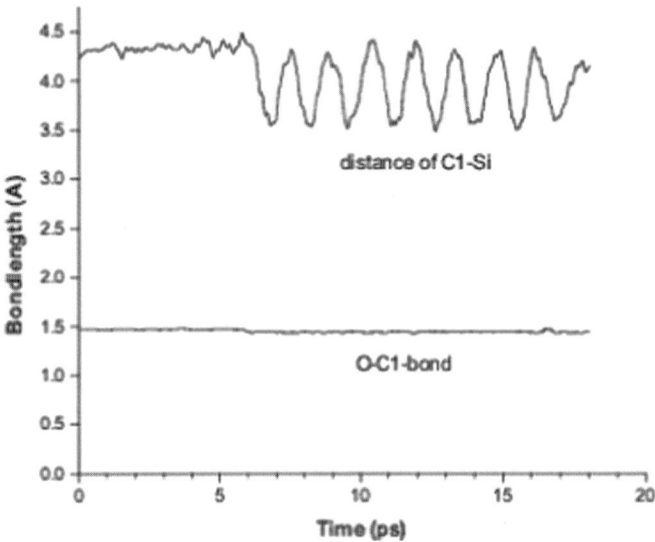

Figure 14-8. Time variation of the Si-C1 distance and O-C1 bond length in the MD evolution. An average over every 300MD steps has been taken to filter out high thermal frequency components

which shows that no bond between C1 and the up Si is formed. This behavior rules out the chemisorbed configuration F-2 as an equilibrium structure.

On the other hand, the time variation of the O-H and Si-H bond lengths in the MD simulation, as displayed in Figure 14-9, illustrates that before T = 6ps, the O-H bond length is about 1.01 Å and the Si-H bond length oscillates around 2.25 Å. In the period of 6ps ~ 7ps, the O-H bond length elongates up to 3.75 Å, and the Si-H bond length shortens to 1.48 Å. This marked change indicates a transition from the metastable physisorbed phase I-1 to the stable chemisorbed phase F-1. Qualifying the F-1 structure as stable makes sense only at room temperature, since the chemisorbed phase F-2 is much more stable than F-1 at still higher temperature. After 7ps, the O-H bond length oscillates with decreasing amplitude, and the Si-H bond length reaches its equilibrium value of 1.48 Å. Figure 14-9 shows that O-H bond scission occurs and the Si-H bond forms between T = 6ps and T = 7ps in the process of heating. The equilibrium structure is the chemisorbed configuration F-1, i.e. the O-H bond is broken (see Figure 14-4), which is consistent with the energy barrier calculation at zero temperature.

Inspection of the MD simulation results shows that the O-H bond is broken, and the H atom reattaches to the *up* Si atom of the same dimer (dimer A2) to form a new H-Si bond. After 7ps, all atoms oscillate around their stable equilibrium positions. Another method of performing the MD simulations consists in setting an initial temperature T = 300 K without any heating and letting the system evolve at this temperature. Following this avenue, we arrive at the same conclusions as reported above.

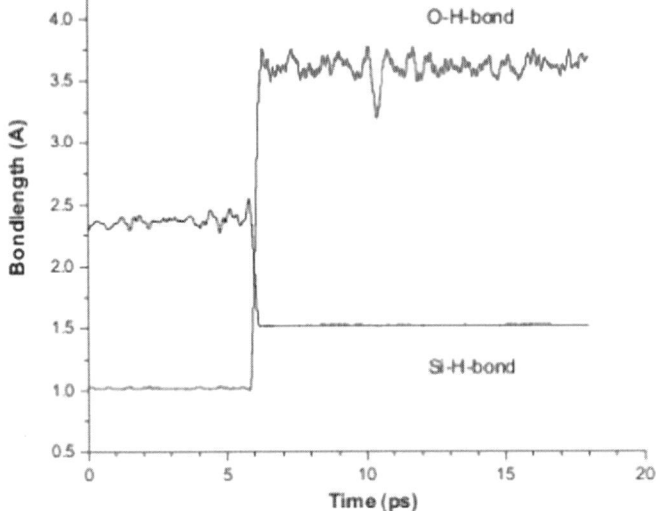

Figure 14-9. Time variation of the O-H and Si-H bond length, demonstrating O-H bond rupture and Si-H bond formation. An average over every 300 MD steps has been taken to filter out high thermal frequency components

4.   SUMMARY

A wide range of organic adsobates on Si surfaces has been investigated by computational theory. This survey has focused on unsaturated hydrocarbons, amines, phosphines, and alcohols. The recent discussion of these systems has centered on several issues deemed as particularly relevant for materials science and its extension into present-day nanotechnology as well as highly challenging from a computational point of view. A common feature of amines, phosphines and alcohols as adsorbed species on the $Si(001) - (2 \times 1)$ surface is the initial formation of a dative bond with the electrophilic end of a Si dimer. Subsequently, the system may undergo dissociative chemisorption involving the loss of a H atom which is preferentially attached to the nucleophilic end of the same substrate dimer. The probability for this process to occur varies sensitively with the nature of the attached species, the temperature as well as the level of coverage. Detailed studies have been devoted to unraveling the complex chemistry of unsaturated hydrocarbons on Si surfaces. As one of the major results of these efforts, not only the analogue of the $[4+2]$ Diels-Alder reaction proceeds on the $Si(001) - (2 \times 1)$ surface, but also a prominent $[2 \times 2]$ pathway is identified which, from *a priori* considerations, was expected to be strongly suppressed.

The theoretical treatment of Si surfaces with organic adsorbates has proceeded along two main avenues: approaches relying on a finite cluster model are to be distinguished from those employing density functional theory in conjunction with periodic boundary conditions. The principal virtue of the first methodology is that it may be readily combined with a broad variety of quantum chemical procedures. Thus, the in-depth understanding of reactions involving unsaturated hydrocarbons on the $Si(001) - (2 \times 1)$ surface was shown to require the use of highly correlated methods. The periodic approach, however, succeeds in reproducing the buckling pattern of the $Si(001) - (2 \times 1)$ reconstruction that has turned out to be of crucial impact on the chemistry of organic species attached to this substrate. Further, the finite cluster model is naturally confined to the description of local properties, while band structure information on the Si/adsorbate interface can be obtained from a periodic treatment. The latter also allows to study the effect of the coverage level on essential parameters such as the adsorption energy of the organic species, or surface relaxation due to the presence of the adsorbate.

The periodic scheme is adopted in our comprehensive case study on $1 -$ propanol in contact with $Si(001) - (2 \times 1)$. It was shown that the 1-propanol molecule interacts with the Si surface through formation of a dative bond, preceding the reaction of the physisorbed 1-propanol molecule with the surface by cleavage of either the O-C or the O-H bond. The O-C bond cleavage is thermodynamically stable, while that of O-H is kinetically preferred. Dative bonding between the $1 -$ propanol species and an Si dimer was characterized in terms of the band structure of the substrate/adsorbate composite. As one of the main results emerging from extensive studies on the coverage dependence of basic energetic properties, the planar 1-propanol density on the $Si(001)$ surface was found to represent a parameter that allows altering the nature of the surface from semiconducting to insulating. Further, recording the time variation of the O-H and Si-H bond lengths by means of ab initio MD simulation demonstrated that the O-H bond length is spontaneously ruptured at room temperature, and the dissociated H atom forms a Si-H bond. The final equilibrium structure at room temperature is the chemisorbed configuration that results from H atom loss to the nucleophilic end of the Si dimer.

ACKNOWLEDGEMENTS

This work is supported by the National Science Foundation through the grants HRD-9805465, NSFESP-0132618 and DMR-0304036, by the National Institute of Health through the grant S06-GM008047, and by the Army High Performance Computing Research Center under the auspices of Department of the Army, Army Research Laboratory under Cooperative Agreement No. DAAD 19-01-2-0014.

REFERENCES

1. Mui C, Wang GT, Bent SF, Musgrave CB (2001) Reactions of methylamines at the $Si(100)$-2×1 surface, J. Chem. Phys. 114: 10170–10180
2. Pike AR, Lie LH, Eagling RA, Ryder LC, Patole SN, Connolly BA, Horrocks BR, Houlton A (2002) DNA on silicon devices: On-chip synthsis, hybridization, and charge transfer, Angew. Chem. Int. Ed., 41: 615–617
3. Konecny R, Doren DJ (1998) Cycloaddition reactions of unsaturated hydrocarbons on the $Si(100)$-(2×1) surface: theoretical predictions, Surf. Sci. 417: 169–188

4. Silvestrelli PL, Ancilotto F, Toigo F (2000) Adsorption of benzene on Si(100) from first principles, Phys. Rev. B 62: 1596–1599

5. Silvestrelli PL (2004) Adsorption of ethanol on Si(100) from first principles calculations, Surf. Sci. 552: 17–26

6. Zhang L, Carman AJ, Casey SM (2003) Adsorption and thermal decomposition chemistry of 1-propanol and other primary alcohols on the Si(100) surface, J. Phys. Chem. B 107: 8424–8432

7. Carman A, Zhang L, Liswood JL, Casey SM (2003) Methylamine Adsorption on and Desorption from Si(100), J. Phys. Chem. B 107: 5491–5502

8. Waltenburg HN, Yates JT Jr (1995) Surface chemistry of silicon, Chem. Rev. 95: 1589–1673

9. Fazleev NG, Fry JL, Weiss AH (2004) Surface states and annihilation characteristics of positrons trapped at the (100) and (111) surfaces of silicon, Phys. Rev. B 70: 165309/1–17

10. Redondo A, Goddard WA III (1982) Electronic correlation and the silicon (100) surface: Buckling versus nonbuckling, J. Vac. Sci. Technol 21: 344–350, Paulus B (1998) Correlation calculations for the reconstruction of the Si(100) surface, Surf. Sci. 408: 195–202

11. Hamers RJ, Tromp RM, Demuth JE (1986) Scanning tunneling microscopy of silicon(001), Phys. Rev. B 34: 5343–5357

12. Monch W (2001) Semiconductor surfaces and interfaces, pp. 219–220 (Third Edition, Springer)

13. Takayanagi K, Tanishiro Y, Takahashi S, Takahashi M (1985) Structure analysis of silicon(111)-7×7 reconstructed surface by transmission electron diffraction, Surf. Sci. 164: 367–392

14. Yates JT Jr (1991) Surface chemistry of silicon-the behaviour of dangling bonds, J. Phys.: Condens. Matter 3: S143–S156

15. Bozack MJ, Taylor PA, Choyke WJ, Yates JT (1986) Chemical activity of the carbon-carbon double bond on silicon surfaces, Surf. Sci. 177: L933–L937

16. Nishijima M, Yoshinobu J, Tsuda H, Onchi M (1987) The adsorption and thermal decomposition of acetylene on silicon(100) and vicinal silicon(100) 0°, Surf. Sci. 192: 383–397

17. Yoshinobu J, Tsuda H, Onchi M, Nishijima M (1987) The adsorbed states of ethylene on silicon(100)c(4×2), silicon(100)(2×1), and vicinal silicon(100) 9°: electron energy loss spectroscopy and low-energy electron diffraction studies, J. Chem. Phys. 87: 7332–7340

18. Taylor PA, Wallace RM, Cheng CC, Weinberg WH, Dresser MJ, Choyke WJ, Yates JT Jr (1992) Adsorption and decomposition of acetylene on silicon(100)-(2×1), J. Am. Chem. Soc. 114: 6754–6760

19. Clemen L, Wallace RM, Taylor PA, Dresser MJ, Cheng CC, Choyke WJ, Weinberg WH, Yates JT Jr (1992) Adsorption and thermal behavior of ethylene on silicon(100)-(2×1), Surf. Sci. 268: 205–216

20. Huang C, Widdra W, Weinberg WH (1994) Adsorption of ethylene on the Si(100)-(2×1) surface, Surf. Sci. 315: L953–L958

21. Fisher AJ, Bloechl PE, Briggs GAD (1997) Hydrocarbon adsorption on Si(001): when does the Si dimer bond break?, Surf. Sci. 374: 298–305

22. Pan W, Zhu T, Yang W (1997) First-principles study of the structural and electronic properties of ethylene adsorption on Si(100)-(2×1) surface, J. Chem. Phys. 107: 3981–3985

23. Hovis JS, Liu H, Hamers RJ (1998) Cycloaddition Chemistry of 1,3-Dienes on the Silicon(001) Surface: Competition between [4+2] and [2+2] Reactions, J. Phys. Chem. B 102: 6873–6879

24. Tepljakov AV, Kong MJ, Bent SF (1997) Vibrational Spectroscopic Studies of Diels-Alder Reactions with the Si(100)-2 × 1 Surface as a Dienophile, J. Am. Chem. Soc. 119: 11100–11101; Tepljakov AV, Kong MJ, Bent SF (1998) Diels-Alder reactions of butadienes with the Si(100)-2 × 1 surface as a dienophile: Vibrational spectroscopy, thermal desorption and near edge x-ray absorption fine structure studies, J. Chem. Phys. 108: 4599–4606

25. Konecny R, Doren DJ (1997) Theoretical Prediction of a Facile Diels-Alder Reaction on the Si(100)-2 × 1 Surface, J. Am. Chem. Soc. 119: 11098–11099

26. Choi CH, Gordon MS (1999) Cycloaddition Reactions of 1, 3-Cyclohexadiene on the Silicon(001) Surface, J. Am. Chem. Soc. 121: 11311–11317

27. Shoemaker JR, Burggraf LW, Gordon MS (1999) SIMOMM: an integrated molecular orbital/molecular mechanics optimization scheme for surfaces, J. Phys. Chem. A 103: 3245–3251

28. Gokhale S, Trischberger P, Menzel D, Widdra W, Droege H, Steinrueck H-P, Birkenheuer U, Gutdeutsch U, Roesch N (1998) Electronic structure of benzene adsorbed on single-domain Si(001)-(2 × 1): A combined experimental and theoretical study, J. Chem. Phys 108: 5554–5564; Birkenheuer U, Gutdeutsch U, Roesch N (1998) Geometrical structure of benzene absorbed on Si(001), Surf. Sci. 409: 213–228

29. Self KW, Pelzel RI, Owen JHG, Yan C, Widdra W, Weinberg WH (1998) Scanning tunneling microscopy study of benzene adsorption on Si(100)-(2 × 1), J. Vac. Sci. Technol A 16: 1031–1036

30. Kong MJ, Teplyakow AV, Lyubovitsky JG, Bent SF (1998) NEXAFS studies of adsorption of benzene on Si(100)-2 × 1, Surf. Sci. 411: 286–293

31. Borovsky B, Krueger M, Ganz E (1998) Metastable adsorption of benzene on the Si(001) surface, Phys. Rev. B 57, R4269–R4272

32. Lopinski GP, Moffat DJ, Wolkow RA (1998) Benzene/Si(100): metastable chemisorption and binding state conversion, Chem Phys Lett 282: 305–312; Lopinski GP, Fortier TM, Moffatt DJ, Wolkow RA, (1998) Multiple bonding geometries and binding state conversion of benzene/Si(100), J. Vac. Sci. Technol. A 16: 1037–1042; Wollow RA, Lopinski GP, Wolkow DJ, Moffat DJ (1998) Resolving organic molecule-silicon scanning tunneling microscopy features with molecular orbital methods, Surf. Sci. 416: L1107–L1113

33. Car R, Parrinello M (1985) Unified approach for molecular dynamics and density-functional theory, Phys. Rev. Lett. 55: 2471–2474

34. Cao X, Hamers RJ (2001) Silicon Surfaces as Electron Acceptors: Dative Bonding of Amines with Si(001) and Si(111) Surfaces J. Am. Chem. Soc 123: 10988–10996

35. Cao X, Hamers RJ (2002) Interactions of alkylamines with the silicon (001) surface, J. Vac. Sci. Tech. B 20, 1614–1619

36. Mui C, Han JH, Wang GT, Musgrave CB, Bent SF (2002) Proton Transfer Reactions on Semiconductor Surfaces, J. Am. Chem. Soc 124: 4027–4038

37. Carman A, Zhang L, Liswood JL, Casey SM (2003) Methylamine Adsorption on and Desorption from Si(100), J. Phys. Chem. B 107: 5491–5502

38. Björkvist M, Göthelid M, Grekh TM, Karlsson UO (1998) NH3 on Si(111)7 × 7: Dissociation and surface reactions, Phys. Rev. B 57: 2327–2333

39. Wolkow R, Avouris P (1988) Atom-resolved surface chemistry using scanning tunneling microscopy, Phys. Rev. Lett 60: 1049–1052

40. Yu ML, Meyerson BS (1984) The adsorption of phosphine on silicon(100) and its effect on the coadsorption of silane, J. Vac. Sci. Technol. A 2: 446–449

41. Wang Y, Chen X, Hamers RJ (1994) Atomic-resolution study of overlayer formation and interfacial mixing in the interaction of phosphorus with Si(001), Phys. Rev. B 50: 4534–4547

42. Colaianni ML, Chen PJ, Yates JT, Jr (1994) Unique hydride chemistry on silicon-PH_3 interaction with Si(100)-(2 × 1), Vac. Sci. Technol. A 12: 2995–2998

43. Yu ML, Vitkavage DJ, Meyerson BS (1986) Doping reaction of PH_3 and B_2H_6 with Si(001), J. Appl. Phys 59: 4032–4038

44. Kipp L, Bringans RD, Biegelsen DK, Northrup JE, Garcia A, Swartz LE (1995) Phosphine adsorption and decomposition on Si(100) 2 × 1 studied by STM, Phys. Rev. B 52: 5843–5850

45. Hirose R, Sakamoto H (1999) Thermal desorption of surface phosphorus on Si(100) surfaces, Surf. Sci. Lett 430: L540–545

46. Maity N, Xia LQ, Engstrom JR (1995) Effect of PH_3 on the dissociative chemisorption of SiH_4 and Si_2H_6 on.Si(100): Implications on the growth on in situ doped Si thin films, Appl Phys Lett 66: 1909–1912

47. Yoo DS, Suemitsu M, Miyamoto N (1995) Hydrogen desorption process of Si(100)/PH_3, J. Appl. Phys. 78: 4988

48. Wang Y, Bronikowski MJ, Hamers RJ (1994) An Atomically Resolved STM Study of the Interaction of Phosphine with the Silicon(001) Surface, J. Phys. Chem. 98: 5966–5973

49. Lin DS, Ku TS, Sheu TJ (1999) Thermal reactions of phosphine with Si(100): a combined photoemission and scanning-tunneling-microscopy study, Surf. Sci. 424: 7–18

50. Lin DS, Ku TS, Chen RP (2000) Interaction of phosphine with Si(100) from core-level photoemission and real-time scanning tunneling microscopy, Phys. Rev. B 61: 2799–2805

51. Shan J, Wang Y, Hamers RJ (1996) Adsorption and Dissociation of Phosphine on Si(001), J. Phys. Chem. 100: 4961–4969

52. Chen PJ, Colaianni ML, Wallace RM, Yates JT Jr (1991) Dissociative adsorption of PH_3 on $Si(111) - (7 \times 7)$: a high resolution electron energy loss spectroscopy study Surf. Sci. 244: 177–184

53. Bozso F, Avouris PH (1991) Adsorption of phosphorus on Si(111): Structure and chemical reactivity, Phys. Rev. B 43: 1847–1850

54. Cao PL, Lee LQ, Dai JJ, Zhou RH (1994) Adsorption and dissociation of PH_3 on Si(100))2×1 and Si(111)7×7: Theoretical study, J. Phys: Condens Matter 6: 6103–6111

55. Miotto R, Srivastava GP, Miwa RH, Ferraz AC (2001) A comparative study of dissociative adsorption of NH_3, PH_3, and AsH_3 on $Si(001) - (2 \times 1)$, J. Chem. Phys. 114: 9549–9556

56. Becke AD (1993) Density-functional thermochemistry. III. The role of exact exchange, J. Chem. Phys. 98: 5648–5652

57. Bockstedte M, Kley A, Neugebaur J, Scheffler M (1997) Density-functional theory calculations for poly-atomic systems: electronic structure, static and elastic properties and ab initio molecular dynamics, Comput Phys. Commun. 107: 187–222

58. Perdew JP, Burke K, Ernzerhof M (1996) Generalized Gradient Approximation Made Simple, Phys. Rev. Lett. 77: 3865–3868

59. Troullier N, Martins JL (1991) Efficient pseudopotentials for plane-wave calculations, Phys. Rev. B 43: 1993–2006

60. Eng J, Raghavachari K, Struck LM, Chabal YJ, Bent BE, Flynn GW, Christman SB, Chaban EE, Williams GP, Radermacher K, Mantl S (1997) A vibrational study of ethanol adsorption on Si(100), J Chem Phys 106: 9889–9898; Casaletto MP, Zanoni R, Carbone M, Piancastelli MN, Aballe L, Weiss K, Horn K (2000) High-resolution photoemission study of ethanol on Si(100)2×1, Surf. Sci. 447: 237–244

61. Casaletto MP, Zanoni R, Carbone M, Piancastelli MN, Aballe L, Weiss K, Horn K (2002) Methanol adsorption on Si(100)2×1 investigated by high-resolution photoemis, Surf. Sci. 505: 251–259

62. Lu X, Zhang Q, Lin MC (2001) Adsorptions of Methanol, Formaldehyde and Formic Acid on the $Si(100) - 2 \times 1$ Surface: A Theoretical Study, Phys. Chem. Chem. Phys. 3: 2156–2166

63. Kato T, Kang SY, Xu X, Yamabe T (2001) Possible Dissociative Adsorption of CH_3OH and CH_3NH_2 on $Si(100) - 2 \times 1$ Surface, J. Phys. Chem. B 105: 10340–10347

64. Maseras F, Morokuma K (1995) A New Ab Initio + Molecular Mechanics Geometry Optimization Scheme of Equilibrium Structures and Transition States, J. Comput. Chem. 16: 1170–1179

65. Kresse G, Furthmueller J (1996) Efficient iterative schemes for *ab initio* total-energy calculations using a plane-wave basis set, Phys. Rev. B54: 11169–11186

66. Perdew JP, Wang Y (1992) Atoms, molecules, solids, and surfaces: Applications of the generalized gradient approximation for exchange and correlation, Phys. Rev. B 46: 6671–6687

67. Kresse G, Joubert J (1999) From ultrasoft pseudopotentials to the projector augmented-wave method, Phys. Rev. B 59: 1758

68. Blöchl PE (1994) Projector augmented-wave method, Phys. Rev. B 50: 17953–17979

69. Monkhorst HJ Pack JD (1976) Special points for Brillouin-zone integrations, Phys. Rev. B 13: 5188–5192

70. Ciani A, Sen P, Batra I (2004) Initial growth of Ba on Si(001), Phys. Rev. B 69: 245308–245319

71. Sen P, Ciraci S, Batra I, Grein C, Sivannthan S (2002) Finite temperature studies of Te adsorption on Si(001), Surf. Sci. 519: 79–89

72. Henkelman G, Uberuaga B, Jonsson H (2000) A climbing image nudged elastic band method for finding saddle points and minimum energy paths, J. Chem. Phys. 113: 9901–9904

73. Jonsson H, Mills G, Jacobsen KW Nudged Elastic Band Method for Finding Minimum Energy Paths of Transitions, in Classical and Quantum Dynamics in Condensed Phase Simulations, ed. Berne BJ, Ciccotti G and Coker DF, 385–405 (World Scientific, 1998)

74. Nosé S (1984) A unified formulation of the constant temperature molecular dynamics methods, J. Chem. Phys. 81: 511–519

75. Over H, Wasserfall J, Ranke W, Ambiatello C, Sawitzki R, Wolf D, Moritz W (1997) Surface atomic geometry of $Si(001) - (2 \times 1)$: A low-energy electron-diffraction structure analysis Phys. Rev. B 55: 4731–4736

76. Ramstad A, Brocks G, Kelly P (1995) Theoretical study of the Si(100) surface reconstruction, Phys. Rev. B 51: 14504–14523

77. Miotto R, Oliveira M, Pinto M, de Leon-Perez F, Ferraz A (2004) Acetonitrile adsorption on Si(001), Phys. 31–2353 Rev. B 69: 235340

78. see [12], p.18

79. Ogitsu T, Schwegler E, Gygi F, Galli G (2003) Melting of Lithium Hydride under Pressure, Phys. Rev. Lett. 91: 175502–175506

80. Lu Z, Wang C, Ho K-M (2000) Structures and dynamical properties of C_n, Si_n, Ge_n, and Sn_n clusters with n up to 13, Phys. Rev. B 61: 2329–2334

CHAPTER 15

RECENT ADVANCES IN FULLERENE DEPOSITION ON SEMICONDUCTOR SURFACES

C. G. ZHOU[1], L. C. NING[1], J. P. WU[1], S. J. YAO[1], Z. B. PI[1], Y. S. JIANG[2], H. CHENG[1,†]

[1] Institute of Theoretical Chemistry and Computational Materials Science, China University of Geosciences, Wuhan, China
[2] Institute of Computational Chemistry, Nanjing University, Nanjing, China

Abstract: Development of novel chemistry on semiconductor surfaces is an area of increasing research interests due to its technological importance. The possibility of depositing fullerenes on semiconductor surfaces via the formation of stable chemical bonds provides an opportunity to design and develop novel materials that meet the increasing stringent technology challenge. In this chapter, we review recent advances in the theoretical modeling of fullerene chemisorption on GaAs and Si surfaces. We show that strong covalent chemical bonds can be formed upon deposition of fullerenes of various sizes on these surfaces, forming well-ordered thin films. The chemical/physical properties of such thin films can be tailored by using different sizes of fullerenes

Keywords: Fullerenes; Semiconductor surfaces; GaAs(001)-c(4 × 4); Si(001)-(2 × 1); Density of States

1.   INTRODUCTION

As the size of microchips evolves from 130nm to 90nm or smaller, the increasing packing density between multilevel interconnects will lead to severe RC delay, power consumption and wire cross talk, which are the major factors limiting device performance. As a consequence, the design of novel semiconductor materials with desired chemical and physical properties has stimulated intense experimental and theoretical efforts.[1,2] For example, there have been ongoing activities to develop materials with low dielectric constants (k) to replace the current silicon dioxide ($k = 4.0$) wire insulator. A particularly active area of research in the past few

---

†Corresponding author, E-mail: chengh2@yahoo.com

W. A. Sokalski (ed.), Molecular Materials with Specific Interactions, 533–563.

years has aimed at developing a robust, well-controlled technology for deposition of well-ordered functional thin films in a pre-determined fashion on surfaces of semiconductor materials, such as silicon and gallium arsenide. The new deposition precursors that go into the thin films include polyarylates,[3] fluorinated and/or siliceous materials,[4,5] and other nanoporous materials.[6] For applications in the next generation of semiconductors, these materials must possess an ultra-low k value to meet more stringent requirements, such as good thermal stability (above 400 °C), good mechanical strength, and high dielectric breakdown fields.[7] The design and development of novel low-k materials that simultaneously satisfy the critical demands on other physicochemical properties have presented a great challenge to the materials community.[7,8]

Many studies have shown that organic precursors containing π-bonds can be deposited on semiconductor surfaces via a cycloaddition process. The organic films so deposited can be very stable due to the formation of strong chemical bonds and well-ordered structures. Recently, there have been several reports on depositing fullerenes on semiconductor surfaces.[9–38] Fullerenes are made of pentagons and hexagons of carbons containing alternative double bonds. In principle, they may also undergo the cycloaddition reactions similar to the organic precursors. Nevertheless, the chemical process may be more complex since several π-bonds of fullerenes may participate in the surface reactions simultaneously, depending on the adsorption sites and the electron delocalization. Fullerene-based precursors may be potentially useful to serve as low-k materials[39] or surface passivation agents upon deposition on semiconductor surfaces due to their unique structural and electronic characteristics. The main advantage of fullerenes lies in their high volumes with ample space within the spheres, their low polarizability and excellent thermal stability.

Most studies on deposition of fullerenes on semiconductor surfaces have focused on silicon[9–28] and, to a much less extent, on GaAs.[29–38] Earlier experimental work using HREELS (high-resolution electron energy loss spectroscopy) on C_{60} adsorption on Si(100)2×1 surface suggested that the molecular attachment to the surface is via physisorption.[13,14] But it turned out in the subsequent experimental studies using HREELS and PES (photoelectron energy spectrum) on Si(111)7×7 and Si(100)2×1 surfaces that the adsorption is in fact of chemisorption in nature.[17] Indeed, recent STM (scanning tunneling microscopy) studies, which have revealed the molecular orientation on the Si(111)-(2×1) surface and the detailed bonding structure, have confirmed the strong interaction between fullerenes and Si surfaces.[16,22] Charge transfer from Si atoms to C_{60} molecules was deemed to occur in the deposition process. Several theoretical studies using the first-principles methods have convincingly demonstrated that strong covalent bonding interactions, rather than van der Waals attractions, between fullerenes and silicon surfaces dominate the adsorption process.[11,17] Unfortunately, to date, detailed adsorption structures and bond strength for fullerenes on GaAs surfaces have not been reported. Very recently, we reported extensive theoretical studies based on first-principles density functional theory on the adsorption of a small fullerene molecule, C_{28}, on the c(4 × 4) reconstructed GaAs(001) surface.[38] It was found that the adsorption is of chemisorption nature dictated by [2 + 2] cycloaddition reaction and/or by simple electron lone-pairs mediated charge transfer reaction from the substrate to the fullerene molecule. The results suggest that the monolayer formed by the C_{28} molecules on the surface is stable and naturally porous.

In this chapter, we summarize our recent theoretical studies on adsorption of fullerenes of various sizes on GaAs and Si surfaces. The fullerenes selected in our studies include C_n, n = 28, 32, 36, 40, 44, 48 and 60. The c(4 × 4) reconstructed GaAs(001) surface and Si(001)-(2 × 1) surface were chosen to be the substrates, both of which have been studied extensively in literature. Our emphasis is on GaAs since it is not only less studied for fullerene adsorption than Si but it is also one of the most important semiconductor materials and has been widely utilized in a broad variety of electronic devices. Our primary objectives are to gain insight into the deposition mechanisms of fullerenes of various sizes on GaAs surfaces and to compare their relative adhesion strengths on the substrates. Understanding their adsorption behavior on semiconductor surfaces should enable us to select the appropriate sizes of molecules for deposition with desired adhesion.

2. COMPUTATIONAL DETAILS

In the present study, we focus on adsorption of a range of fullerenes on the c(4 × 4) reconstructed GaAs(001) surface. For adsorption on the Si(001)-(2 × 1) surface, only C_{60} was considered. Both surfaces were modeled as slabs.

The super cell of the c(4 × 4) reconstructed GaAs(001) surface contains seven layers of GaAs as shown in Figure 15-1(a). The bottom five layers alternate a gallium layer and an arsenic layer, respectively, and the top two layers are constituted only by As atoms. The bottom Ga layer is saturated with hydrogen atoms. The surface cell parameters were taken from the crystal structure of GaAs and the distance between adjacent slabs was chosen to be about 25 Å, large enough to prevent effective interaction between slabs. The unit cell contains 24 Ga atoms, 30 As atoms and 16 H atoms in addition to a fullerene molecule. The closest distance between two nearest neighboring fullerene molecules is in a range between 4.5 Å and 6.0 Å and thus the lateral interaction is relatively small compared with the adsorption energies.

The chosen super cell for the Si(001)-(2 × 1) surface contains six layers of Si, each of which includes eight atoms as shown in Figure 15-1(b). The top surface layer is reconstructed via dimerization to lower down the surface energy and the bottom layer is saturated with hydrogen atoms. The cell parameters were taken from the crystal structure of silicon and the distance between the slabs was chosen to be about 26.0 Å. Upon C_{60} adsorption, the supercell contains 48 Si atoms and 16 hydrogen atoms in addition to 60 C atoms.

All calculations were performed using *ab initio* density functional theory under the generalized gradient approximation (GGA) with the Perdew-Wang

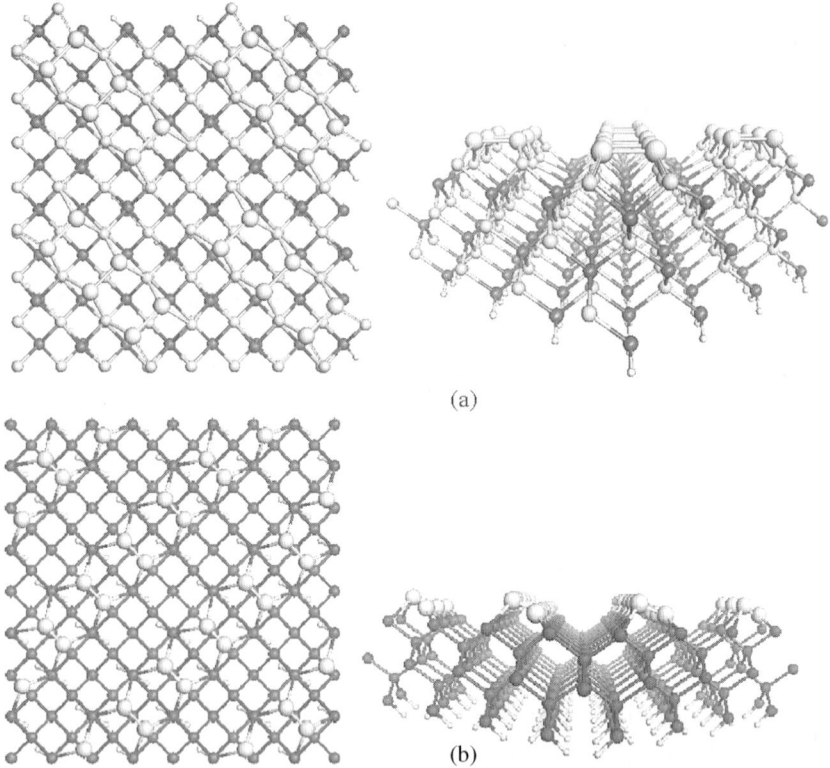

Figure 15-1. (a) Top and side view of reconstructed GaAs(001)-c(4 × 4) surface; The white middle size balls represent As atoms, the dark balls represent Ga atoms; the bottom micro white balls denote the H atoms; and the top layer of As-As dimers are highlighted by larger white balls; (b) Top and side view of reconstructed Si(001)-(2 × 1) surface. The dark balls represent the Si atoms and the top layer Si dimers are highlighted with large white balls, the bottom micro white balls represent the H atoms

exchange-correlation functional.[40–43] The method was implemented in the Siesta simulation code.[44–48] Troullier and J. L. Martins' norm-conserving pseudopotentials were used to describe the core electrons and the localized numerical atomic orbitals of double-ζ augmented with polarization functions as the basis set were employed to describe the valence electrons.[49,50] An energy cutoff of 200 Ry was used to determine a boundary for the wavefunctions of atomic orbitals, beyond which the wavefunctions vanish,[51] with good convergence in the calculated adsorption energies. The Brillouin zone integration was performed using a special k-point approach implemented with the Monkhorst and Pack scheme with $2 \times 2 \times 1$ k-points. The present computational method has been widely utilized in a broad variety of studies on surface phenomena and shown to be highly accurate in providing reliable results on surface structures and energetics.[43,48]

Full structural relaxation for the top 5 layers of the c(4 × 4) reconstructed GaAs(001) surface and the top 4 layers of the Si(001)-(2 × 1) surface together with the adsorbate was performed using the conjugate gradient algorithm without imposing symmetry constraints, while the bottom two layers as well as the hydrogen atoms were kept fixed. The adsorption energy of the fullerene is calculated using Eq. 1:

$$\Delta E_{ads} = -[E(sub+fullerene) - E(sub) - E(fullerene)] \qquad (15\text{-}1)$$

where E(sub + fullerene), E(sub) and E(fullerene) represent the total energies of the surface with a fullerene molecule in the chosen unit cell, the surface itself and the fullerene molecule, respectively. The fullerene energy was calculated by first placing the molecule in a large cubic box with the cell parameters, 20Å × 20Å × 20Å and then fully relaxing the atomic coordinates with the fixed box size using 2 × 2 × 2 k-points.

3. FULLERENE ADSORPTION ON THE C(4 × 4) RECONSTRUCTED GAAS(001) SURFACE

We first performed energy minimization for the GaAs(001)-c(4 × 4) surface. The optimized structure is shown in Figure 15-1(a). The calculated bond distance of the As dimer in the top layer is 2.609Å for the mid-dimer and 2.636Å for the two side dimers. The distance between the top two As layers is 1.23Å, reflecting the quasi-stable nature of the dangling bonds of the As dimers. Figure 15-2(a) shows the calculated density of states (DOS) of the GaAs(001) c(4 × 4) surface. The calculated band gap is about −1.6eV, slightly larger than that of the bulk, which is

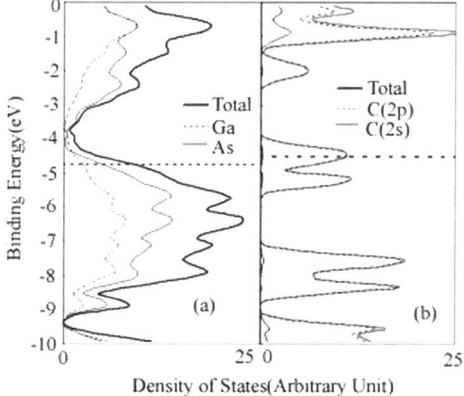

Figure 15-2. The calculated Density of States(DOS) of (a) GaAs(001)-c(4 × 4) surface and (b) gas phase C$_{28}$

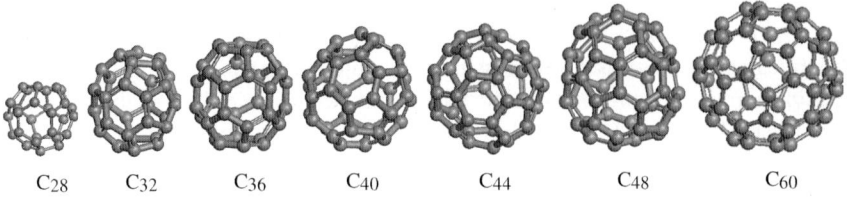

C$_{28}$ C$_{32}$ C$_{36}$ C$_{40}$ C$_{44}$ C$_{48}$ C$_{60}$

Figure 15-3. The fullerene molecules from C$_{28}$ to C$_{60}$

−1.42eV. Both the valence band and the conduction band are essentially dominated by the contribution from the As atoms, especially the As dimers.

3.1.   Structure and Energetics of Fullerenes

To gain insight into the adsorption mechanism of various sizes of fullerenes on the GaAs surface structure shown in Figure 15-1(a), it is important to first understand the structures of these molecules and their relative stability. The fullerenes involved in the present study are C$_{28}$, C$_{32}$, C$_{36}$, C$_{40}$, C$_{44}$, C$_{48}$ and C$_{60}$ (Figure 15-3). The main geometric characteristics of the fully optimized fullerene molecules and the calculated average binding energy per atom are shown in Table 15-1. The calculated HOMO-LUMO gaps are shown in Figure 15-4, which are in good agreement with the available experimental data.[55] It is noteworthy that for C$_{60}$ the calculated bond lengths compare remarkably well with the experimental values (single bond: 1.446 Å; double bond: 1.406 Å),[52-54] suggesting the present computational method is capable of providing accurate results for the fullerene systems.

The adsorption behavior of a fullerene molecule should largely depend on its inherent strains incurred from the curvature. The local stress of a fullerene due to strains near the adsorption site will be relaxed upon bond formation with the substrate as the hybridization of the carbon atoms involved in the bonding changes from sp^2 to sp^3. In general, a fullerene with higher local stress is expected to give a larger adsorption energy. Geometrically, a fullerene with a lower symmetry, such as C$_{36}$, may adsorb on the substrate either vertically or horizontally. The former possesses a larger curvature on the interaction side and is thus expected to be energetically more favorable. On the other hand, a horizontal conformation allows the molecule to interact with more binding sites on surfaces and thus may gain more overall adsorption energy. Furthermore, a smaller fullerene has a larger curvature and a lower binding energy (Table 15-1). In this case, we anticipate that the adsorption energy would be higher.

Table 15-1. The calculated structural parameters of various fullerene molecules

	C$_{28}$	C$_{32}$	C$_{36}$	C$_{40}$	C$_{44}$	C$_{48}$	C$_{60}$
Symmetry	T$_d$	D$_{3d}$	D$_{6h}$	D$_2$	D$_2$	D$_2$	I$_h$
Binding energy (eV)	-6.765	-6.936	-6.994	-7.063	-7.128	-7.169	-7.330

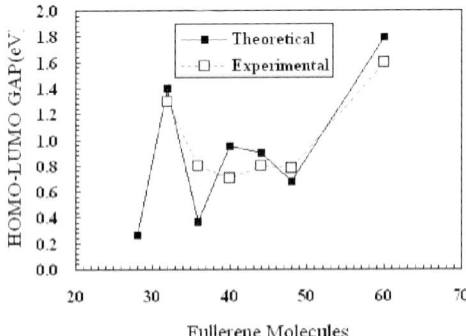

Figure 15-4. The calculated HOMO-LUMO gap of fullerene molecules from C_{28} to C_{60}, comparing with the experimental results

3.2. C_{28} Adsorption on the GaAs(001) c(4 × 4) Surface

There are numerous possible adsorption sites on the surface. We therefore choose the smallest fullerene, C_{28}, to examine energetically most favorable sites for adsorption.

The C_{28} adsorption on GaAs(001) c(4 × 4) surface may exhibit a variety of geometric patterns. The structure and strength are determined by both the surface adsorption sites and the orientation of the molecule. To facilitate the description of the adsorption structure, we first denote the surface single-dimer site, double-dimer site and the trench site with S, D and T and use α, β, γ, δ to represent, in the C_{28} molecule, a single pentagon, a hexagon, a shared double bond of the adjacent pentagon and hexagon, and a shared double bond of the adjacent pentagons, respectively. In the following, we will examine the C_{28} adsorption at the three adsorption sites as outlined above.

3.2.1.   Single-dimer site

C_{28} molecule was first oriented toward a top layer single As dimer with the α, β, γ, δ orientations, respectively, followed by energy minimization of the structures. In all cases, cycloaddition reactions take place and strong covalent bonds are formed. Substantial molecular structural and lattice relaxations occur. The calculated adsorption energies and charge transfer are shown in Table 15-2.

For S^{α}, the structural optimization results in the fullerene molecule falling into the double-dimer site. Four covalent bonds between C_{28} and two arsenic dimers of the top layer are formed; three of the new bonds are with bond distances ranging from 2.126Å to 2.147Å and the fourth one is relatively weak with a bondlength of 2.884Å, reflecting the geometric incommensurability. The arsenic dimer bond distances are subsequently elongated by 0.025Å upon C_{28} adsorption. Likewise, the bond distances of the 4 carbon atoms participated in the surface reaction are also increased slightly compared with their gas-phase values. Part of the fullerene structure near the adsorption site becomes considerably deformed as the molecule forms new bonds with the surface. The adsorption also makes a considerable

Table 15-2. The calculated adsorption energies of C_{28} at different sites of the surface with various orientations

Site	Orientation	Adsorption Energy(eV)	Charge Transfer
Single	S^α	1.88	0.333
	S^β	1.63	0.34
	S^γ	1.33	0.361
	S^δ	1.33	0.216
Double	D^α	2.16	0.296
	D^β	1.95	0.244
Trench	T^α	2.21	0.318
	T^β	3.02	0.359

impact on the top layer structure as the interlayer distance near the binding site between the arsenic layers is changed from 1.23Å to 1.27 Å and the distance between the second As layer and the top Ga layer is increased from 1.25Å to 1.29Å (Figure 15-5(a), (b)).

The attachment of C_{28} at the surface is dictated by two the so-called [2+2] cycloaddition reactions. The highly unsaturated As dimer is reacting with the unoccupied π-orbitals of C_{28}, forming a four-membered ring with two new σ-bonds between As and C atoms. The double-dimer site allows two As dimers and a diene segment of the pentagon in C_{28} to participate the cycloaddition reactions simultaneously. However, one of the two cycloaddition reactions is incomplete due to the structural incommemsurability. Mulliken population analysis indicates that charge transfer from arsenic dimer to C_{28} by 0.333 electron. The reaction gives rise to an adsorption energy of 1.88eV, suggesting that the adsorption structure is highly stable at this site. Figure 15-5(c) displays the calculated density of states spectrum. Compared with the DOS spectrum of the substrate (Figure 15-2(a)), the valence band of the S^α system is broadened slightly across the Fermi level. Much of the contribution comes from the conduction band of C_{28}, which consists of mostly

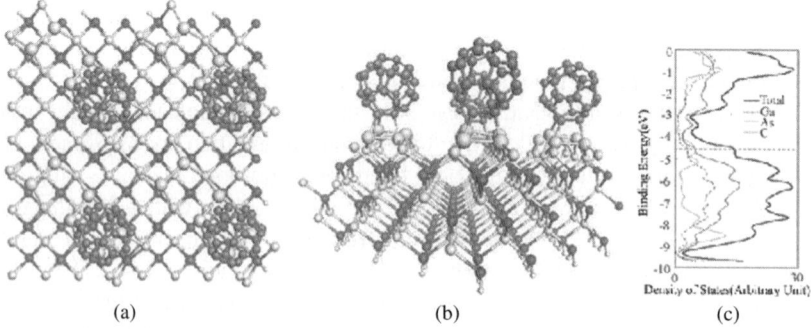

(a)   (b)   (c)

Figure 15-5. Optimized structures with (a) top view and (b) side view of C_{28} on GaAs(001)-c(4 × 4) surface with S^α configuration; and (c) the calculated density of states(DOS) of the absorbed system

the anti-bonding π-orbitals. The DOS spectrum indicates that the overlap between the valence band of the substrate, which is governed primarily by the unsaturated orbitals of As dimers, and the conduction band of C_{28} gives rise to the strong chemisorption, allowing C_{28} to accept electrons from the surface. The change in the valence band and the conduction band is largely dictated by the As dimers and C_{28} and little contribution comes from the rest of the GaAs layers.

For S^β, the fullerene was placed on top of the middle As dimer with a hexagon facing down with the anticipation that a $[2+4]$ cycloaddition reaction might occur. However, energy minimization results in a quite unexpected chemisorption structure: the fullerene molecule inclines to lean to the trench site slightly; the two adjacent pentagons are then reacting with the end atoms of the three As dimers by forming three new covalent bonds, creating two adjacent 6-membered ring structures (Figure 15-6(a), (b)) with a chair conformation. The distance of the C-As σ-bond in the middle is 1.988Å, while the bondlength of the other two on the sides is 2.208 Å. The As dimer looses up with the bondlength increased lightly from 2.636Å to 2.656Å. The two side As dimers bonding with C_{28} tilt up toward the adsorbate and the As dimer in the middle is pushed downward slightly. Considerable relaxation between the top Ga layer and the second As layer upon C_{28} adsorption is observed as the interlayer distance is increased from 1.24Å to 1.32Å. Structural deformation of the fullerene is also noted here as the two pentagons form new chemical bonds with the surface. The calculated adsorption energy is 1.63eV, indicating that C_{28} is stable at this site.

The chemisorption is governed by the charge transfer from the electron lone pairs of the As dimers to the empty π-bands of C_{28} with about 0.34 electron being transferred. Indeed, the calculated DOS spectrum (Figure 15-6(c)) indicates that the empty π-bands of C_{28} and the low lying states of the dimers make most of the contributions to the conduction band near the Fermi level, supporting the partial charge transfer mechanism. For comparison purpose, the calculated density of states of gas phase C_{28} is shown in Figure 15-2(b). Detailed analysis suggests that feature changes observed in the DOS spectrum are attributed mainly to the two top layers

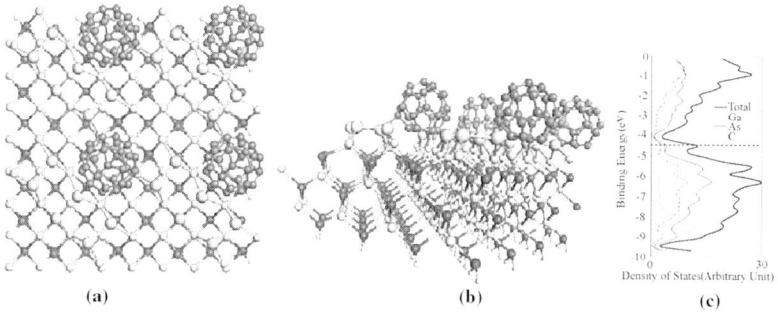

Figure 15-6. Optimized structures with (a) top view and (b) side view of C_{28} on GaAs(001)-c(4 × 4) surface with S^β configuration; (c) the calculated Density of States(DOS) of the absorbed system

of As atoms and C_{28} and the electronic structure of the inner layer atoms remains essentially intact.

Next, we examine the S^γ configuration by aiming the double bond shared by a pentagon and a neighboring hexagon in C_{28} at the single-dimer site. As expected, the $[2+2]$ cycloaddition reaction takes place to form a 4-membered ring structure with the bond distances of the two new σ-bonds being 2.137Å and 2.141Å, respectively (Figure 15-7(a), (b)). However, this configuration is not stable as the standing new ring structure does not adequately support the adsorbate. As a consequence, the fullerene is twisted slightly to form an additional σ-bond with the end atom of a neighboring As dimer with a bondlength of 2.128Å. Compared with S^α and S^β, this chemisorption structure is much less stable. The calculated adsorption energy is only 1.33eV. Like the above two cases, charge transfer from the substrate to the fullerene molecule by 0.361 electron provides the driving force for the surface reaction. The calculated DOS spectrum (Figure 15-7(c)) indeed indicates that the empty π-bands of C_{28} and the electron lone pairs of the As dimers dominates the valence band, consistent with the charge transfer mechanism described in the above.

Finally, we lined up a double bond shared by two adjacent pentagons with the As dimer in the middle and then performed the energy minimization (S^δ). A typical $[2+2]$ cycloaddition reaction takes place with two σ-bonds between carbon atoms and arsenic atoms formed in a 4-memberred ring structure. The optimized structure, shown in Figure 15-8(a) and (b), is symmetric with two bond distances of 2.134Å and 2.137Å, respectively. This structure is not expected to be very stable since the standing 4-ring adsorption structure is very sloppy and can hardly support the fullerene. This is confirmed by the smaller calculated adsorption energy, 1.33eV. The calculated DOS spectrum displayed in Figure 15-8(c) can be interpreted in the same fashion as in S^γ.

It is clear that at the single-dimer site, the most stable adsorption configuration is S^α. The chemisorption process takes advantage of two cycloaddition reactions to support the adhesion of fullerene on the $c(4 \times 4)$ reconstructed GaAs(001) surface. Nevertheless, due to the initial structural alignment prior to the energy minimization,

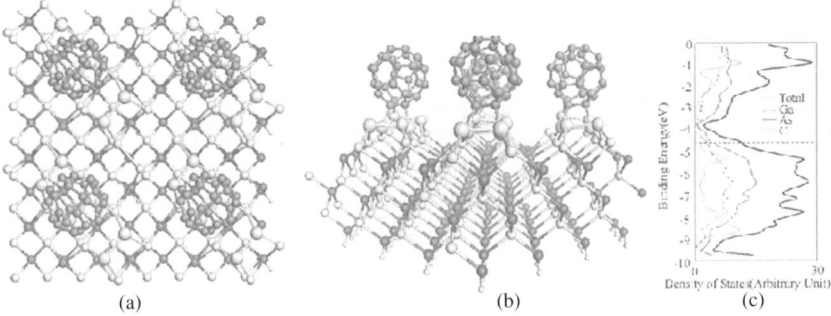

(a)   (b)   (c)

Figure 15-7. Optimized structures with (a). top view and (b). side view of C_{28} on GaAs(001)-c(4 × 4) surface with S^γ configuration; (c) the calculated Density of States(DOS) of the absorbed system

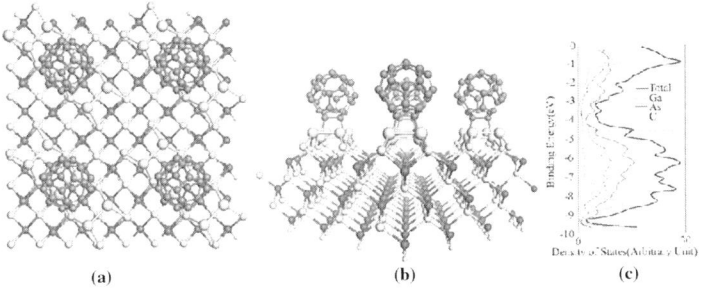

Figure 15-8. Optimized structures with (a). top view and (b). side view of C_{28} on GaAs(001)-c(4 × 4) surface with S^{δ} configuration; (c) the calculated Density of States(DOS) of the absorbed system

these reactions result in only three strong covalent bonds. It is thus expected that with careful alignment of initial configuration of C_{28} toward the As dimers, it is possible to achieve higher adsorption energies at the double-dimer sites.

3.2.2. Double-dimer site

We examined two configurations of C_{28} at the double-dimer site: one with a pentagon lining up with two As dimer bonds (D^{α}) and another with a hexagon facing the dimers (D^{β}). The calculated adsorption energies of the two configurations at this site are shown in Table 15-2.

For D^{α}, the optimized chemisorption structure shown in Figure 15-9(a) and (b) is somewhat similar to the one for S^{α} (Figure 15-5(a), (b)) although the detailed geometric arrangement differs. Like S^{α}, two [2+2] cycloaddition reactions occur; one is complete and another is not due to the incommensurability between the pentagon and the two As dimers. As a consequence, only three genuine σ-bonds are formed with bond distances of 2.162Å, 2.184Å and 2.117Å, respectively. The incomplete cycloaddition yields a weak bond with a bondlength of 3.134Å.

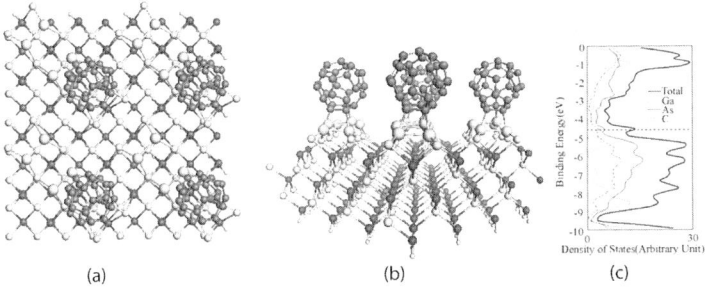

Figure 15-9. Optimized structures with (a) top view and (b) side view of C_{28} on GaAs(001)-c(4 × 4) surface with D^{α} configuration and (c) the calculated density of states(DOS) of the absorbed system

However, unlike S^α, the C=C bond of C_{28} participating the complete cycloaddition is from the shared bond between two adjacent pentagons instead of being between a pentagon and a hexagon as in the case for S^α. Therefore, the geometry is less strained. Indeed, the C-C bond is stretched more significantly upon C_{28} adsorption than in the case of S^α (from 1.452Å to 1.626Å in S^α vs. from 1.448Å to 1.593Å in D^α). The calculated adsorption energy is 2.16eV, slightly higher than the one in S^α. The Mulliken population analysis gives a charge transfer of 0.296 electron from the substrate to the fullerene. The lattice relaxation occurs mainly near the adsorption site. The DOS spectrum shown in Figure 15-9(c) exhibits similar features as in the case of S^α.

Finally, we aligned C_{28} with two parallel C=C bonds of a hexagon well with two As dimers (D^β). Surprisingly, the perfectly well-aligned initial structure did not lead to two [2+2] cycloaddition reactions upon energy minimization. Instead, the two end atoms of the C=C bonds interact directly with the nearby end atoms of the adjacent As dimers to form two C-As σ-bonds with the bond distances of 2.147Å and 2.088Å, respectively. The adsorption yields a 6-membered ring structure with a "boat" conformation. The adsorbate tilts slightly toward the trench site, as shown in Figure 15-10(a) and (b). The calculated adsorption energy is 1.95eV, suggesting this is a stable adsorption site. Once again, the chemisorption is governed by the charge transfer (0.244 electron) from the As dimers to the empty π-bands of fullerene, as is clearly shown in the calculated DOS spectrum (Figure 15-10(c)).

In general, chemisorption of C_{28} at the double-dimer site is energetically more favorable. The fullerene molecule is well supported by the As dimers via formation of C-As σ-bonds.

3.2.3.   Trench site

The trench site is the confined area of the surface surrounded by four 3-dimer unites as shown in Figure 15-1. It is about 5.25Å wide and 16.8Å long. Unlike the As atoms right underneath the dimers, each of the four As atoms inside the trench

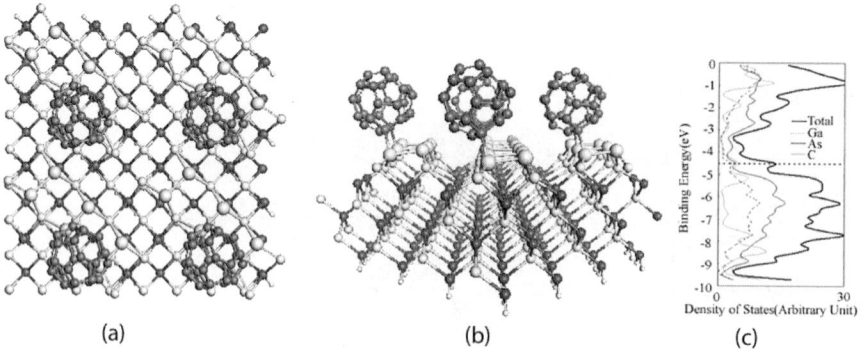

(a)   (b)   (c)

Figure 15-10. Optimized structures with (a) top view and (b) side view of C_{28} on GaAs(001)-c(4 × 4) surface with D^β configuration and (c) the calculated Density of States(DOS) of the absorbed system

area connects with a dimer atom and two Ga atoms in the next layer, leaving one dangling bond. We thus expect that these atoms are also chemically active besides the highly unsaturated dimer bonds.

We place C_{28} on the trench site with a pentagon (T^α) and a hexagon (T^β) oriented toward the surface, respectively. Detailed examination of these structures indicates that C_{28} forms strong covalent bonds with the c(4×4) reconstructed surface.

For C_{28} with the pentagon facing down (T^α), four covalent bonds are formed with the substrate. The calculated bond lengths range from 2.091Å to 2.201 Å. The fullerene molecule interacts with the end atoms of three neighboring As dimers and one atom of the second As layer (Figure 15-11(a) and (b)), forming new σ-bonds. Substantial structural relaxation takes place. First, C_{28} is now partially imbedded in the trench site. Consequently, the surface As dimers are slightly reorganized; the dimer closer to the adsorbate is "squeezed" slightly by C_{28} in the trench, making a shorter bondlength of 2.542Å, while the bond distances of the other dimers become elongated to 2.704Å since one of the end atoms are pulled over toward the fullerene. Second, the chemisorption also results in considerable lattice relaxation as both the second layer As atoms and the top layer Ga atoms shift their positions from their original equilibrium locations; those close to the adsorbate are pulled up and others adjust their positions accordingly, leading to slight misalignment of the top few layers. Consequently, the calculated adsorption energy, which is 2.21eV, is significantly enhanced. Mulliken population analysis suggests that 0.318 electron is transferred from the substrate to the fullerene molecule. Figure 15-11(c) displays the calculated the density of states. Compared with the DOS spectrum of the substrate, the spectrum is broadened, reflecting the top layer structural change.

Finally, we examine the adsorption of C_{28} on the surface with a hexagon oriented to the trench site (T^β). The optimized structure is shown in Figure 15-12(a) and (b). The atoms on the hexagon essentially reside at the same layer of the As dimers. Three distinct σ-bonds between C_{28} and the surface are formed with bond distances of 2.071Å, 2.071Å and 2.201Å, respectively. Two bonds are created via

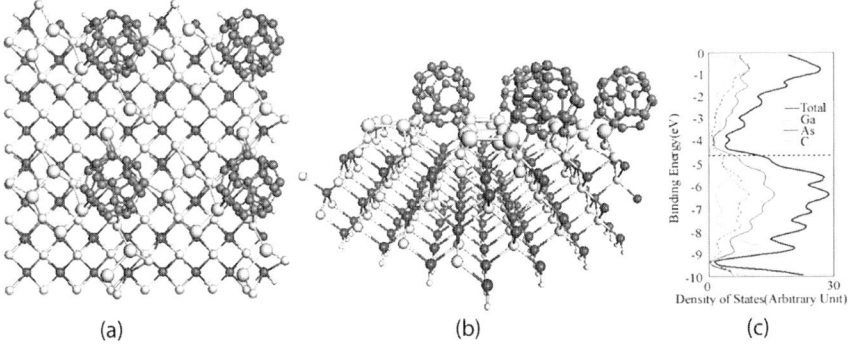

(a)   (b)   (c)

Figure 15-11. Optimized structures with (a) top view and (b) side view of C_{28} on GaAs(001)-c(4×4) surface with T^α configuration and (c) the calculated Density of States(DOS) of the absorbed system

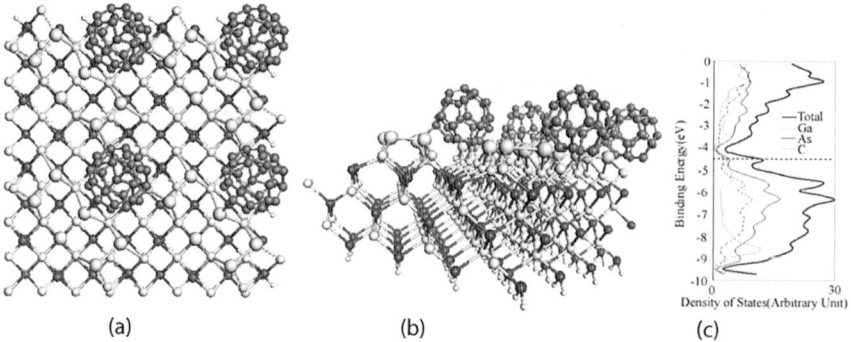

Figure 15-12. Optimized structures with (a) top view and (b) side view of C_{28} on GaAs(001)-c(4×4) surface with T^β configuration and (c) the calculated Density of States(DOS) of the absorbed system

interaction between C_{28} and two atoms in the second As layer; the third one is between an atom in the hexagon of C_{28} and the end atom of the As dimer in the middle of the three dimer unit. The dimer participated in the bond formation is stretched significantly from 2.609Å to 2.792Å. This structural arrangement seems to fit the trench site extremely well. All atoms on the hexagon are within 3.6Å away from the nearby surface atoms to maximize the interactions. Compared with the T^α configuration, this structure appears to cause less structural rearrangement in the lattice and is more stable. Indeed, the calculated adsorption energy, 3.02eV, yields the highest strength at this site among all the adsorption sites we have investigated. The stronger chemisorption arises from the higher charge transfer (0.359 electron) from the surface to the C_{28}. The dangling bonds of As atoms at the trench site as well as the As dimer provide the major driving force for the adsorption. The calculated DOS spectrum (Figure 15-12(c)) has a distinct feature around the Fermi level contributed mostly by the As dimer π-orbitals and the empty π-orbitals of C_{28}.

3.3. $C_n(n = 32, 36, 40, 44, 48, 60)$ Adsorption on the GaAs(001)-c(4×4) Surface

For adsorption of fullerenes larger than C_{28}, we will focus only on adsorption at the trench site, which was shown to be energetically most favorable in the case of C_{28}. We selected several fullerene molecules to test the adsorption strength at other surface sites. But, without exception, the trench site always gave the lowest energy adsorption structures and thus we will discuss adsorption of fullerenes only at this site. With the current range of the fullerenes, the T site is capable of providing one As dimer in the middle of a 3-dimer unit on one side and two opposite ends of two 3-dimer units in the same row to interact with the fullerenes. In addition, the two As atoms with dangling bonds in the second layer at this site also give a strong anchoring force for the molecules.

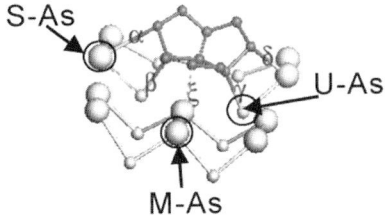

Figure 15-13. The labeling scheme of As functional atoms and bonding situation

To describe the adsorption structures, we first label the atoms of the As dimers as As^S for the two side dimers, As^M for the dimer atoms in the middle and As^U for the unsaturated As atoms in the second layer, as shown in Figure 15-13. The covalent bonds formed between a fullerene and the two As^S atoms of both sides are denoted by α and δ, respectively, and the bond formed between a fullerene and the As^M atom is labeled as ξ. Finally, the covalent bonds between a fullerene and the As^U atoms are labeled as β and γ, respectively. Here, $\alpha, \beta, \gamma, \delta, \xi$ represent not only bonds but also the close contacts. The labeling scheme is shown in Figure 15-13. We now concentrate on the adsorption of C_{32} through C_{60} only at the T site. The main optimized structural parameters and bonding situation are shown in Table 15-3.

Among the fullerene molecules, C_{32} is rather special. It was shown to be much more stable as an individual molecule than other small fullerene molecules.[55] In fact, the time of flight mass spectrum of fullerene anions has shown a great abundance of this species and its ultra-violet photoelectron spectrum exhibits an unusually large HOMO-LUMO gap, 1.30eV.[55] Our calculated HOMO-LUMO gas is 1.40 eV, in good agreement with the experimental result. The most stable adsorption configuration is that the molecule orients itself with two adjacent 5-membered rings facing downward the T-site upon adsorption (Figure 15-14(a), (b)). It forms two strong covalent bonds with two As^U atoms in the second layer with bond lengths of 2.173Å(β) and 2.103Å(γ), respectively. The bonding with the As^S atoms is relatively weak with the bond distances of 3.084Å(α) and 3.214Å(δ), respectively, and the bondlength with the As^M atom is 2.880Å(ξ). This result indicates that

Table 15-3. The selected bond/close contact parameters of the optimized adsorption structures

Bond No.	α	β	γ	δ	ξ
C_{32}	3.214	2.173	2.103	3.084	2.880
C_{36}	2.673	2.129	2.133	2.681	3.501
C_{40}	2.842	2.108	2.106	2.587	3.547
C_{44}	3.705	2.999	2.191	3.142	2.126
C_{48}	2.996	2.945	2.178	2.393	3.107
C_{60}	3.273	3.194	3.227	3.384	3.156

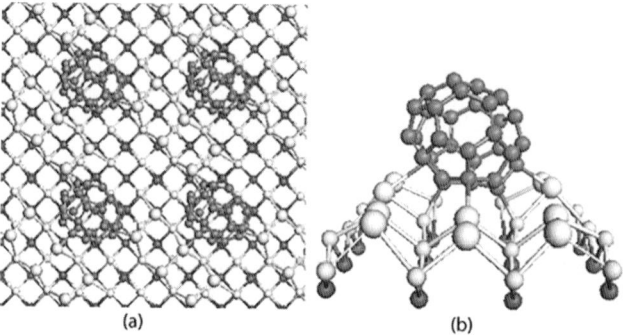

Figure 15-14. Optimized structure of C_{32} on GaAs(001)-c(4 × 4) surface. (a) Top view and (b) local side view

C_{32} is anchored mainly by the unsaturated As atoms in the second layer and the interaction between C_{32} and top layer dimers is relatively weak. As a consequence, the dimers are elongated only slightly upon C_{32} adsorption (approximately 0.05Å) and the substrate relaxation is also very little. The calculated adsorption energy is 2.348eV, suggesting that C_{32} can stick to the substrate quite strongly. Compared with the adsorption energy of C_{28} on the same substrate, the adhesion force for C_{32} is much smaller, mainly due to the exceptional stability of the fullerene. Mulliken population analysis indicates that the adsorption is facilitated by a charge transfer mechanism from the substrate to C_{32} (0.226 electron). The partial charge transfer results mostly from electron-sharing between C_{32} and the dangling bonds of the second layer As atoms to form two covalent bonds.

C_{36} has a geometry of a highly symmetric spheroid. Its equator consists of six fused 6-membered rings and each of its poles is composed of one 6-membered ring. The equator and the poles are connected by twelve 5-membered rings. The height of the spheroid is 5.31Å and its width is 4.34Å. Therefore, C_{36} can interact with the substrate with one of the hexagons either vertically or horizontally. Our calculations suggest that the horizontal adsorption configuration is in fact energetically more favorable (Figure 15-15(a), (b)). This is chiefly because the horizontal configuration allows more bonding sites to interact with the surface. One of the equator hexagons is anchored at the T-site with the C atoms forming two covalent bonds with the As^U atoms. The calculated bond distances are 2.129Å (β) and 2.133Å(γ), respectively, suggesting that the bonding is rather strong. The adsorption results in a slight compression of the anchoring As atoms of the second layer, pushing them downward. In addition, there is one C atom on each of the poles forming a relatively weaker bond with the As^S atom with the bond distances of 2.673Å(α) and 2.681Å(δ), respectively. The relaxation of the side dimers is rather small. The calculated adsorption energy is 2.652eV. The charge transfer from the substrate to C_{36} of 0.118 electron is small, which results mainly from the formation of the covalent bonds between C_{36} and the unsaturated As atoms in the second layer.

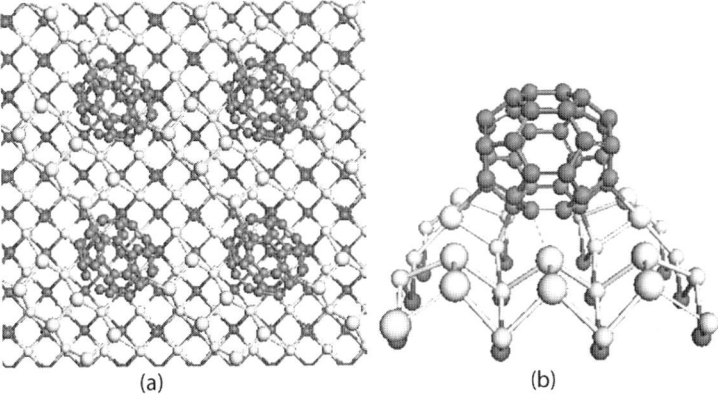

Figure 15-15. Optimized structure of C_{36} on GaAs(001)-c(4 × 4) surface. (a) Top view and (b) local side view

Unlike C_{36}, C_{40} is much less symmetric. Of many optimized stable adsorption configurations, the structure with one of the 6-membered rings, which has most of the sharp corners, is most stable. At these sharp corners, the C atoms adopt a quasi-sp^3 electronic configuration and thus are more reactive than those flat carbons. The optimized structure is shown in Figure 15-16. The bonding of C_{40} with the substrate is similar to that of C_{36} but the bond distances differ slightly. The less symmetric geometry of C_{40} results in significant twist of the 6-membered ring upon adhesion at the T-site; the C atoms forming covalent bonds with the As^U atoms are pulled out of the plane toward the second layer. A small bond relaxation occurs at the side dimers participating the bonding with C_{40}, resulting in a marginal stretch of the dimer bond lengths by approximately 0.012Å. The calculated adsorption energy is 2.423eV with a small charge transfer of 0.086 electron from the substrate to C_{40}.

The structure of C_{44} is also of low symmetry. One of its 6-membered rings with most acute C atoms forms covalent bonds with the substrate, as shown in Figure 15-17. Unlike the other smaller fullerenes, there is only one C atom in C_{44} forming a covalent bond with the As^U atom with a bond length of 2.191Å. The C atom is pulled out from the fullerene considerably. The structural distortion mainly results from the ill-fit of C_{44} at the T-site due to the larger size of the molecule. Another atom in C_{44} also forms a strong covalent bond with the As^M atom with a bondlength of 2.126 Å. Significant deformation of the 6-membered ring participating the surface bonding occurs and the dimer formed by As^M atoms stretches considerably by 0.121Å. Slight bond elongations for other dimers and a marginal lattice relaxation take place upon the fullerene deposition. The calculated adsorption energy is 1.800eV and the electron transfer from the substrate to C_{44} is 0.071.

Like C_{40} and C_{44}, C_{48} is also of low symmetry with a similar adsorption configuration (Figure 15-18). On one side of the fullerene, two C atoms form a $C-As^U$ bond and a $C-As^S$ bond with bond distances of 2.178Å and 2.393Å, respectively. On another side, it also forms additional two relatively weaker bonds with the substrate

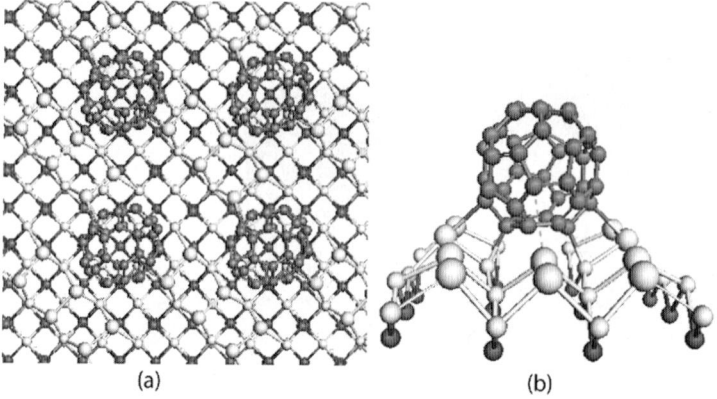

Figure 15-16. Optimized structure of C_{40} on GaAs(001)-c(4 × 4) surface. (a) Top view and (b) local side view

with the bond distances of 2.945Å and 2.996Å. Consequently, the 6-membered ring becomes severely distorted. Very little surface relaxation is observed. The calculated adsorption energy is 1.548eV with a charge transfer of 0.11 from the substrate to C_{48}.

Finally, we calculated the adsorption structure of C_{60} on the surface. Because of its large size and relatively flat sp^2 configuration, its interaction with the substrate is relatively weak with the calculated adsorption energy to be only 1.128 eV. The calculated closest distances between C_{60} and the surface are 3.156Å, 3.194Å, 3.273 Å and 3.227Å, respectively, much longer than the other smaller fullerenes. No appreciable bonding with the top layer dimers was found. Consequently, the dimer structure remains nearly unchanged. We have systematically examined the

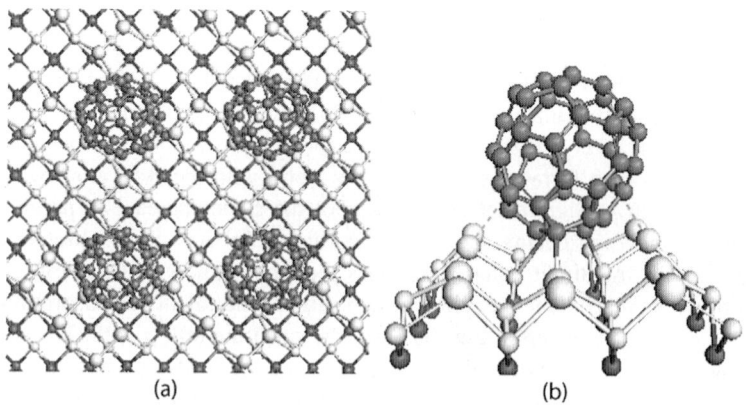

Figure 15-17. Optimized structure of C_{44} on GaAs(001)-c(4 × 4) surface. (a) Top view and (b) local side view

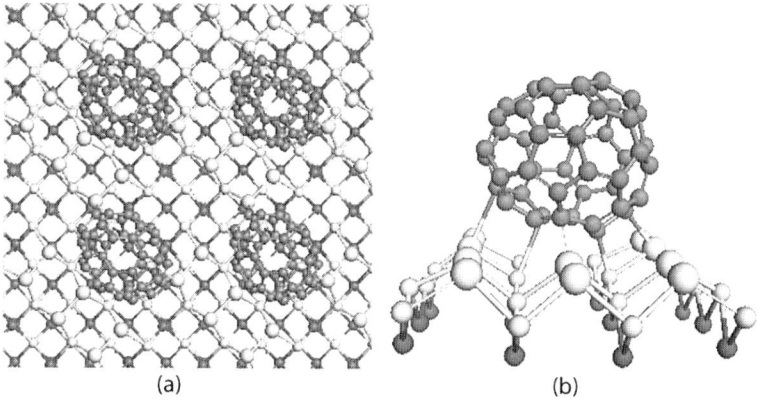

Figure 15-18. Optimized structures with (a). top view and (b). local side view of C_{48} on GaAs(001)-c(4 × 4) surface

adsorption structures at other sites and found that none of the optimized structures is energetically more favorable than the one described here. The optimized adsorption structure at the T-site is shown in Figure 15-19.

3.4. Bonding Analyses

Our extensive search for stable adsorption structures for the above fullerenes indicates that the surface trench site combined with the face-down 6-membered ring of fullerenes with the most acute C atoms gives the energetically most stable adsorption configurations. Covalent bonds can be formed between fullerenes and the substrate. The unsaturated As atoms with a dangling bond in the second layer

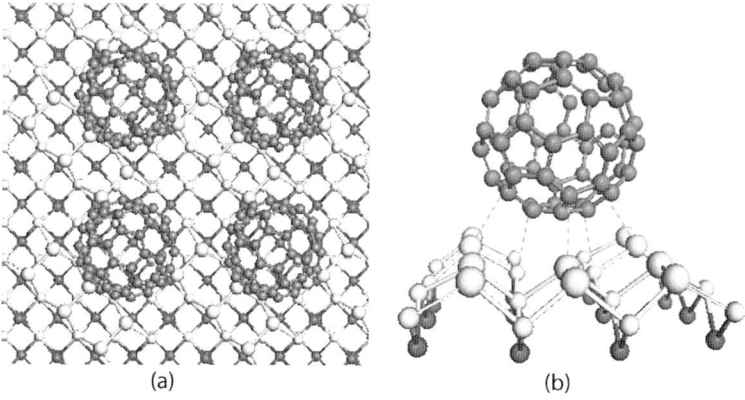

Figure 15-19. Optimized structure of C_{60} on GaAs(001)-c(4 × 4) surface. (a) Top view and (b) local side view

play a key role in anchoring the fullerene molecules. As the size of fullerenes increases, the distances of the covalent bonds increase and it becomes increasingly difficult for the fullerenes to fit in the T-site, suggesting that the bonding in general becomes weaker in the adsorption structures from C_{28} to C_{60}. Indeed, the calculated adsorption energy decreases monotonically, except for C_{32}, as the size of fullerenes increases, as shown in Figure 15-20. The unusually small adsorption energy for C_{32} is attributed to the exceptional stability of the molecule in the gas phase.

Two factors dictate the relative bonding strength of fullerenes on the substrate. The first one is the local curvature of the hexagon in the fullerene participating the bonding with the substrate. The carbon atoms adopt a quasi-sp^2 hybridization forced by curvature. A large curvature gives rise to a higher strain on the atoms; formation of covalent bonds with the surface would relax the stress on the C atoms as their orbital hybridization changes from sp^2 to sp^3. Therefore, the adsorption energy of a fullerene in general increases with curvature. Qualitatively, the relative adsorption strength of fullerenes can also be understood based on their relative gas phase binding energies. It can be seen from Table 15-1 that the calculated average binding energies of the fullerenes increase monotonically from C_{28} to C_{60} as their curvatures decrease. An energetically stable structure tends to be less reactive. Thus one would expect that the adsorption energy decreases as the average binding energy increases. However, the binding energy of C_{32} can not explain the unusually small adsorption energy. This is due to the fact that partial electron transfer from the substrate to C_{32} is needed in order to form stable covalent bonds between C_{32} and the surface. With a large HOMO-LUMO gap of C_{32}, partial electron attachment to the fullerene is difficult, leading to relatively weaker bonding.

The second factor controlling the fullerene adhesion process is whether the molecule can fit well in the narrow size of the T-site to be anchored by the unsaturated As atoms in the second layer. As the fullerene size increases, it becomes

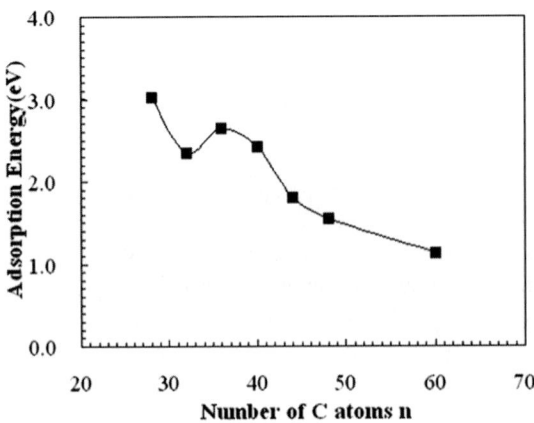

Figure 15-20. The calculated fullerene adsorption energies

increasingly difficult for the molecule to fit in this site. As a result, the distance between a fullerene and the substrate increases with the molecular size, leading to incrementally weaker bonding for large fullerenes. The major driving force for anchoring fullerenes comes from the unsaturated As atoms in the second layer. The As atoms with a dangling bond are in a meta-stable state and ready to form σ-bonds with carbon atoms upon fullerene deposition. This is the primary reason why the surface trenches are the preferred sites for fullerenes adsorption. Bond formation between a large fullerene and the substrate requires substantial structural deformations of the 6-membered ring, leading to a smaller adsorption energy. The top layer As dimers also contribute to the adsorption but its interaction with fullerenes is much weaker. Each atom of the dimers forms two σ-bonds with the second layer As atoms with a bond distance approximately 2.46 Å, while the distance of the dimer bond itself is about 2.60Å. This implies that the two As atoms of the dimer form only a single σ-bond and there is no π-bonding between them. Therefore, there is an electron lone pair on each of the dimer atoms. This electronic structure does not facilitate formation of a covalent σ-bond with a fullerene. As a consequence, surface dimers interact with fullerenes rather weakly. This is remarkably different from the deposition of fullerenes and small unsaturated organic molecules on silicon surfaces, which has been shown extensively to be capable of reacting strongly with the C-C π-bonds via a cycloaddition process with the π-bonds of Si-Si dimers.[20,28,56–76]

Figure 15-2(a) displays the calculated density of states for the c(4 × 4) reconstructed GaAs(001) surface. The DOS spectra for the fullerenes and the adsorption systems are shown in Figure 15-21(a) and (b), respectively. The DOS spectrum of the substrate exhibits a typical semiconductor band gap of about 1.5 eV. Detailed analysis on the spectrum projected to individual atoms suggests that As atoms contribute mostly to both the conduction band and the valence band. The band near the Fermi level comes from the surface states dominated by the As dimers and, in particular, the As atoms with a dangling bond in the second layer. The detailed features of the DOS spectra of fullerenes differ significantly. The band gaps range approximately from 0.3 eV to slightly above 1.7 eV and these materials exhibit considerable metallic or semiconductor characteristics. The dominant contribution to the valence band and conduction band comes from the π-orbitals. The overlap of these orbitals varies with curvature of the molecules. A large curvature forces the π-electrons to localize around nuclei and thus gives rise to a poor overlap. Consequently, these orbitals more likely participate the surface reaction. The conduction band of fullerenes is to accept partial electron transfer from the surface. We also note that the DOS spectrum of C_{32} displays a relatively large band gap, which explains why it is somewhat resistant to the surface reaction. Figure 15-21(b) indicates strong orbital mixing between fullerenes and the substrate and band features of the DOS spectra near the Fermi level change considerably. In all cases, the energy bands of fullerenes move slightly downward, reflecting the partial charge transfer from the substrate to the fullerenes. The downshift results in a shoulder band at the Fermi level, which is contributed by the C and As atoms involved in the bond formation.

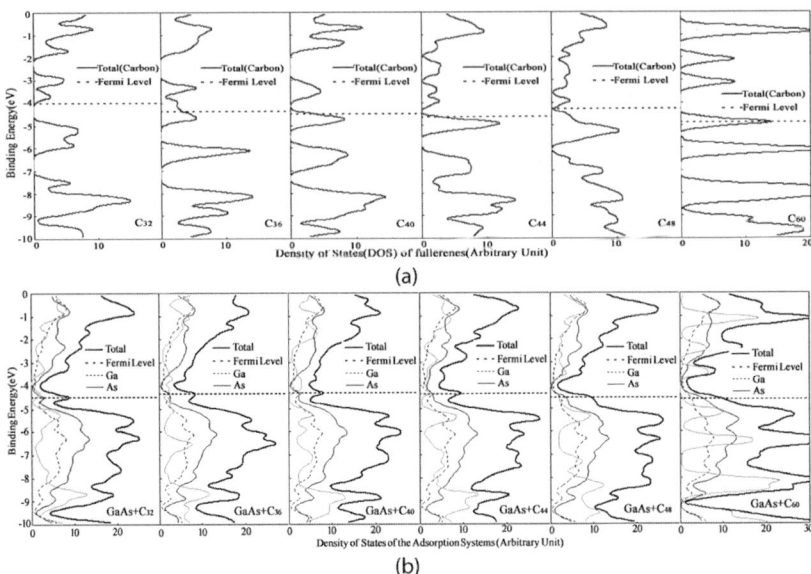

Figure 15-21. The calculated DOS spectra of (a) fullerenes and (b) adsorption systems from C_{32} to C_{60}

As adsorption becomes weaker with molecular size, the band mixing becomes less accordingly. The shoulder band gradually overlaps with the rest of the valence band and the DOS spectrum of an adsorption system more resembles that of the substrate.

4. C_{60} ADSORPTION ON SI(001)-C(2 × 1) SURFACE

There have been many studies on fullerene adsorption on Si surfaces.[9-28] We therefore concentrate only on C_{60} adsorption on Si(001) surface since there has been considerable controversy on the reported adsorption structures and associated adsorption energies on this surface. Essentially, as shown in Figure 15-1(b), there are three surface adsorption sites on Si(001)-(2 × 1) available to interact with C_{60}. The first is the atop site supported by two adjacent Si dimers; the second one is right above the Si dimer and the last one is the trench between two Si dimer rows. Godwin and co-workers performed extensive calculations using local density approximation (LDA) and showed that there are numerous adsorption configurations at each of these sites.[28] But only four of them are energetically most favorable and thus they represent the likely adsorption locations. Accordingly, we will deal only with these sites. We show that in all cases strong chemisorption occurs and covalent bonds between C_{60} and Si(001)-(2 × 1) surface are formed. The overbinding feature observed in the previous LDA calculations by Godwin et al. can be largely corrected.

The four adsorption configurations with favorable energetics selected from the extensive structural samplings are displayed in Figure 15-22. Godwin et al. showed

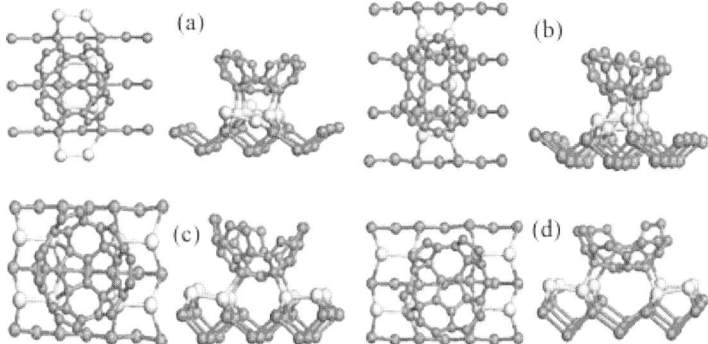

Figure 15-22. The top-view and local side view of C_{60} on Si(001) c(2 × 1) surface. For clarity, only the local bonding arrangements near the adsorption sites are shown here. (a) The C_{60} on top of the two dimers of the same dimer row. (b) C_{60} on top of a dimer. (c) C_{60} at the trench between four dimers. (d) C_{60} at the trench between two dimers

that turning C_{60} slightly around these configurations can also result in stable adsorption structures but the adsorption energies would be higher.

The first configuration concerns that the C_{60} molecule resides on top of two dimers in the same dimer row. As seen in Figure 15-22 (a), two [4+2] independent cycloaddition reactions between two adjacent 6-memberred rings of fullerene and two neighboring Si dimers take place, resulting in formation of two 4-memberred rings between C_{60} and the surface with four new σ-bonds with bond distances of 2.032 Å, 2.065 Å, 2.074 Å, and 2.116 Å, respectively, as shown in Table 15-4. Slight structural distortion of fullerene occurs upon adsorption. The C-C distance of the atoms participating the bond formation is elongated considerably from 1.465 Å to about 1.513 Å as their orbital hybridization changes from sp^2 to sp^3, giving rise to local bonding rearrangement. Consequently, the fullerene undergoes a slight expansion as the height of its cage increases from 7.16 Å to 7.38 Å and the area adsorbed on the Si dimer row becomes rather flat due to the formation of strong covalent bonds between C_{60} and the substrate. The C_{60} adsorption also gives rise to a significant surface relaxation as the Si dimers interacting with the fullerene change from tilting in the same direction to flipping down slightly parallel to the surface. It is remarkable that the dimer bonds are loosen up only slightly upon the

Table 15-4. The calculated adsorption energies and Si-C bond lengths for structures shown in Figure 15-22(a)–(d)

Site	Binding energy (eV)	Si-C bond lengths (Å)
a	2.69	2.032, 2.065, 2.074, 2.116
b	2.47	2.061, 2.039, 3.176, 2.365
c	3.45	2.046, 2.057, 2.317, 2.374
d	3.71	2.055, 2.070, 2.077, 2.081

formation of the C-Si bonds due to the geometric constraint imposed by C_{60}. For the buckled Si atoms participating the bonding, the high end comes down while the low end goes up, leading to a rather flat single bonded structure. The buckled Si dimers not participating the bonding remain tilt, although a certain extent relaxation is observed; the second layer atoms move downward by approximately 0.005 Å; the third layer moves upward by about 0.001 Å and the fourth layer moves downward by 0.008 Å. Mulliken population analysis indicates partial charge transfer from the substrate to C_{60} by 0.148 electron. The chemisorption gives rise to an adsorption energy of 2.69 eV, as shown in Table 15-4, suggesting that the adsorption structure is highly stable at this site. Figure 15-23(a) displays the calculated density of states of the adsorption system at this site. It is clear that the charge transfer occurs from the valence band of Si to the conduction band of C_{60}, consistent with the calculated Mulliken charge.

For the second adsorption configuration, C_{60} molecule was placed on top of a dimer (Figure 15-22(b)). One [2+2] cycloaddition reaction takes place with strong chemical bonding resulting in the formation of two genuine σ-bonds with bond distances of 2.061 Å and 2.039 Å, respectively. Remarkably, two additional weaker σ-bonds between C_{60} and the substrate are also formed from both sides of the C=C bond, yielding the bond lengths of 2.176 Å, 2.365 Å, as shown in Table 15-4. Again, we observe only slight structural distortion of fullerene upon adsorption. For the [2+2] addition, the C-C bond distance is stretched significantly from 1.398 Å to about 1.540 Å. In addition, the fullerene undergoes a slight expansion with the diameter of the cage increased by 0.20 Å. In parallel, substantial surface relaxation occurs upon C_{60} adsorption with Si dimers flipping down slightly to accommodate the C_{60} molecule. The Si-Si dimer bonds are elongated considerably by 0.174 Å, 0.026 Å and 0.048 Å, respectively as C-Si bonds are formed, and the dimers are pushed downward slightly by approximately 0.185 Å. The other top layer atoms not participating the bonding with C_{60} move up slightly, and the second layer atoms also move upward by approximately 0.12 Å; the third layer

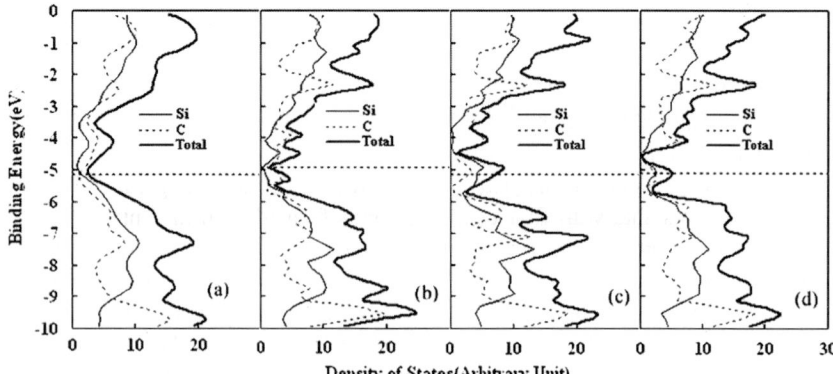

Figure 15-23. The calculated density of states (DOS) of the absorption system

moves downward by about 0.01 Å, the fourth layer moves downward by 0.002 Å. The charge transfer from C_{60} to substrate calculated from Mulliken population analysis is rather small, only 0.046 electron, perhaps resulting from the weaker $[2+2]$ cycloaddition reaction. The calculated adsorption energy is 2.47eV, as shown in Table 15-3, suggesting that the adsorption structure is still very stable at this site but weaker than the first configuration. The calculated density of states spectrum is shown in Figure 15-23(b). Compared with the DOS spectrum for the first configuration (Figure 15-23(a)), the features change significantly. However, the nature of adsorbate-adsorbent interaction remains the same. The change is largely due to the change of local structural rearrangements.

We next place the C_{60} molecule over the dimer trench centered between four dimers, as seen in Figure 15-22(c). C_{60} is now partially imbedded in the trench site, facilitating the interaction with more surface atoms. Four covalent bonds between C_{60} and the substrate are formed; the fullerene molecule interacts with the end atoms of four neighboring Si dimers. The calculated bond lengths are 2.046 Å, 2.057 Å, 2.317 Å and 2.374 Å, respectively, as shown in Table 15-4, reflecting a highly stable bonding environment. Accordingly, the C-C bond distances increases significantly by 0.11 Å, 0.06 Å, 0.05 Å, 0.05 Å, respectively, while the diameter of the cage increases by 0.24 Å. Substantial surface relaxations upon C_{60} chemisorption take place. The surface dimers involved in the chemical bonding flip flat, essentially parallel to the surface, with the bond distance increased by 0.14 Å. The dimers participating the chemisorption come downward considerably by approximately 0.49 Å while those not involved in the bonding move up slightly by about 0.46 Å. Again, the second layer of the substrate is descended by about 0.06 Å upon the chemisorption, and both the third and the fourth layers move downward by about 0.01 Å. The calculated charge transfer from C_{60} to the substrate is 0.148 electron. The chemisorption gives rise to an adsorption energy of 3.45 eV, as shown in Table1, suggesting that the adsorption structure is much more stable at this site than the previous ones formed via $[2+2]$ cycloaddition reactions. The calculated DOS spectrum is shown in Figure 15-23(c). While the basic features of the spectrum remain similar to Figs. 15-23(a) and (b), we observe significant enhancement of DOS around the Fermi level, indicating a considerable metallic character. The change of the physical property is likely due to the strong interaction between C_{60} and the substrate.

Finally, we consider the configuration with the C_{60} molecule residing over the dimer trench centered between two dimers, as seen in Figure 15-22(d). Here, C_{60} is also partially imbedded in the trench site. This structural arrangement appears to be much more stable with four covalent bonds formed with the substrate. The calculated bond lengths are 2.055 Å, 2.070 Å, 2.077 Å and 2.081 Å, respectively, as shown in Table 15-4. Similar to the adsorption pattern in Figure 15-22(c), the fullerene molecule interacts with the end atoms of the four neighboring Si dimers. However, unlike the situation shown in Figure 15-22(c), the local bonding environment allows the molecule to approach the Si dimers more closely, resulting shorter C-Si bonds. All C-C bonds of the atoms involved in the chemisorption are relaxed considerably by 0.11 Å, 0.07 Å, 0.03 Å, 0.04 Å, respectively, and the C_{60} molecule expands by roughly 0.26 Å in diameter. Furthermore, the Si surface relaxation is substantial upon the chemisorption. The buckled dimers involved in the bonding become rather flat and move downward with the bond distances increased by about 0.12Å, while the other dimers remain tilt and are pushed down slightly. The third layer and the fourth layer move downward by approximately 0.02 Å and 0.01 Å, respectively. Mulliken population analysis indicates a rather small charge transfer from C_{60} to substrate by only 0.007 electron. The small charge transfer is largely due to the covalent nature of the chemical bonding. The calculated adsorption energy is 3.71 eV, as shown in Table 15-4, the largest among the four adsorption sites considered in the present work. This result is consistent with what was reported by Godwin et al. Similar to the case of Figure 15-23(c), the calculated DOS spectrum shown in Figure 15-23(d) suggests that this adsorption system is of metallic character with the same charge transfer mechanism.

5. SUMMARY

We have conducted an extensive computational studies using density functional theory on the chemisorption of fullerenes with various sizes on the c(4 × 4) recon-structed GaAs(001) surface and on the adsorption of C_{60} at the Si(001)-c(2 × 1) surface. Adsorption at various adsorption sites coupled with several possible orien-tations of the fullerene molecules was carefully examined. In all cases, we found that the adhesion of fullerenes on the surface is strong and chemically stable covalent bonds between the fullerene and the substrates are formed. The chemisorption results in lattice relaxation and structural distortion of the fullerene molecule to a certain extent, depending on the adsorption site and fullerene orientation. It was found that the strongest chemisorption takes place at the trench site with a hexagon of fullerenes facing down the surface and the weakest adsorption occurs at the single-dimer site. These findings are consistent with the observations of the low temperature experiments on the C_{60} on anisotropic surface,[14] Si(100)-(2 × 1) surface,[16] and on the C_{60} and C_{84} on the Si(110)2 × 1 surface.[12]

The dominant force for the chemisorption is the interaction between the empty π-bands of fullerene, which acts as an electron acceptor, and the electron lone-pairs of the π-orbitals of As, which behaves as an electron donor. The electron lone-pairs reside not only on the As dimers but also on the As atoms confined in the trench area. They allow the surface to undergo either [2+2] cycloaddition reaction or simple charge transfer reactions with the incoming adsorbate with unsaturated bonds. These reactions are competing with each other and the optimal adsorption configuration is highly dependent on the geometric arrangement that gives rise to the maximum bonding. Our results suggest that the [2+2] may not be the only driving force for chemisorption of fullerenes on GaAs(001) surface, a feature that is strikingly different from most of the small organic molecules with unsaturated bonds on silicon surfaces, where the cycloaddition reaction dominates the surface reaction. Instead, the dangling bonds at the trench site can mediate the chemisorption in conjunction with the unsaturated dimer bonds. For the present system, we found this arrangement gives the highest adsorption energy. Finally, in all cases, no $[4+2]$ cycloaddition reaction was found to occur. The results presented in this paper suggest that the adsorption of fullerenes on the $c(4 \times 4)$ reconstructed GaAs(001) surface can be stable and the monolayer is highly porous.

In summary, we have carefully reexamined the chemisorption of C_{60} on Si(001)-c(2×1) surface reported by Godwin et al. as the lowest adsorption energy structures at various surface sites with density functional theory under the generalized gradient approximation with more k-points. As a result, the adsorption overbinding resulting from their LDA calculations is largely corrected and our results should be more accurate. Similar to what was reported by Godwin et al., the trench site was identified as the most stable site for C_{60} chemisorption. However, the relative order of adsorption stability from our studies differs considerably from their results. The configuration with the C_{60} molecule residing over the dimer trench centered between two dimers was found to be much more stable than the one with the C_{60} molecule placed over the dimer trench centered between four dimers which was reported to be more stable by 0.88eV.[28] The relative stability of chemisorption with C_{60} placed over the Si dimers was found to be the least stable sites.

We have attempted to address the chemisorption phenomena of fullerenes on Si surfaces in a different perspective by associating the adsorption with $[2+2]$ cycloaddition reaction commonly seen in organic chemistry. We found that these reactions provide a strong driving force for chemisorption and are primarily responsible for the numerous stable chemisorption configurations at the surfaces. Nevertheless, the cycloaddition reactions alone would not give rise to the most stable structures for the fullerenes on Si surfaces. The highly reactive buckled Si dimers are capable of interacting with the fullerene molecules with both the double bonds and the ends of the double bonds. As a consequence, the trench sites of the Si surfaces are strongly preferred for C_{60} chemisorption.

ACKNOWLEDGEMENT

The Project was supported by the Research Foundation for Outstanding Young Teachers, China University of Geosciences (Wuhan) Grant CUGQNL0519.

REFERENCES

1. Vilan A, Shanzer A, Cahen D (2000) Molecular Control Over Au/GaAs Diodes, Nature 404: 166–168
2. Seker F, Meeker K, Kuech TF, Ellis AB (2000) Surface Chemistry of Prototypical Bulk II-VI and III-V Semiconductors and Implications for Chemical Sensing, Chem Rev 100: 2505–2536
3. Maier G (2001) Low Dielectric Constant Polymers for Microelectronics, Prog Polym Sci 26(1): 3–65
4. Ma Y, Yang H, Guo J, Sathe C, Agui A, Nordgren J (1998) Structural and Electronic Properties of Low Dielectric Constant Fluorinated Amorphous Carbon Films, Appl Phys Lett 72(25): 3353–3355

5. Tsai MH, Whang WT (2002) Low Dielectric Polyimide/Poly(silsesquioxane)-Like Nanocomposite Material, Polymer 42(9): 4197–4207

6. Yang S, Mirau PA, Pai C-S, Nalamasu O, Reichmanis E, Pai JC, Obeng YS, Seputro J, Lin EK, Lee H-J, Sun J, Gidley DW (2002) Nanoporous Ultralow Dielectric Constant Organosilicates Templated by Triblock Copolymers Chem Mater 14(1): 369–374

7. Miller RD (1999) In Search of Low-k Dielectrics, Science 286(5439): 421–423

8. Peters L (1998) Pursuing the Perfect Low-K Dielectric, Semiconductor International 21(10): 64–67

9. Ohno TR, Chen Y, Harvey SE, Kroll GH, Weaver JH, Haufler RE, Smalley RE (1991) C_{60} Bonding and Energy-Level Alignment on Metal and Semiconductor Surfaces, Phys Rev B 44(24): 13747–13755

10. Wang Y (1992) Photoconductivity of Fullerene-Doped Polymers, Nature 356: 585–587

11. Kawazoe Y, Kamiyama H, Maruyama Y, Ohno K (1993) Electronic Structures of Layered C_{60} and C_{70} on Si(100) Surface, Jpn J Appl Phys 32: 1433–1437

12. Wang X-D, Hashizume T, Shinohara H, Saito Y, Nishina Y, Sakurai T (1993) Adsorption of C_{60} and C_{84} on the Si(100)2 × 1 surface studied by using the scanning tunneling microscope, Phys Rev B 47(23): 15923–15930

13. Chen D, Sarid D (1995) An STM Study of C_{60} Adsorption on Si(100)-(2 × 1) Surfaces: from Physisorption to Chemisorption Surf Sci 329(3): 206–218

14. Klyachko D, Chen DM (1995) Ordering of C_{60} on Anisotropic Surfaces, Phys Lett 75(20): 3693–3696

15. Beardmore K, Smith R (1995) C_{60} Film Growth and The Interaction of Fullerenes with Bare and H Terminated Si Surfaces, Studied by Molecular Dynamics, Nucl. Instrum. Methods Phys Res, Sect B 106: 74–79

16. Yao X, Ruskell TG, Workman RK, Sarid D, Chen D (1996) Scanning Tunneling Microscopy and Spectroscopy of Individual C_{60} Molecules on Si(100)-(2 × 1) Surfaces, Surf Sci 366(3): 743–749

17. Suto S, Sakamoto K, Wakita T, Hu C-W, Kasuya A (1997) Vibrational Properties and Charge Transfer of C_{60} Adsorbed on Si(111)-(7 × 7) and Si(100)-(2 × 1) Surfaces, Phys Rev B 56(12): 7439–7445

18. Katircioğlu Ş, Erkoç Ş (1997) Decomposition of C_{60} Molecules on Si(111) Surface, Surf Sci 383: 775–778

19. Suto S, Sakamoto K, Wakita T, Harada M, Kasuya A (1998) Interaction of C_{60} with Silicon Dangling Bonds on the Si(111)-(7 × 7) Surface, Surf Sci 402–404: 523–526

20. Shachal D, Manassen Y (1998) Adsorption of 1,13-tetradecadiene on Si(111)7 × 7 Studied by STM: Possible Electrostatic Interactions Between the Aliphatic Chain and the Surface, Appl Phys A: Mater Sci. & Processing 66: 1229–1231

21. Sakamoto K, Harada M, Kondo D, Kimura A, Kakizaki A, Suto S (1998) Bonding State of the C_{60} Molecule Adsorbed on a Si(111)-(7 × 7) Surface, Phys Rev B 58: 13951–13956

22. Hou JG, Yang J, Wang H, Li Q, Zeng C, Lin H, Bing W, Chen DM, Zhu Q (1999) Identifying Molecular Orientation of Individual C_{60} on a Si(111)-(7 × 7) Surface, Phys Rev Lett 83: 3001–3004

23. Wang H, Zeng C, Li Q, Wang B, Yang J, Hou J, Zhu Q (1999) Scanning Tunneling Spectroscopy of Individual C_{60} Molecules Adsorbed on Si(111)-7 × 7 Surface, Surf Sci 442: 1024–1028

24. Suto S, Sakamoto K, Kondo D, Wakita T, Kimura A, Kakizaki A, Hu CW, Kasuya A (1999) Interaction of C_{60} with Si(111)7 × 7 and Si(100)2 × 1 Surfaces studied by STM, HREELS and PES, Surf Sci 438: 242–247

25. Suto S, Sakamoto K, Kondo D, Wakita T, Kimura A, Kakizaki A (1999) Bonding Nature of C_{60} Adsorbed on Si(111)7 × 7 and Si(100)2 × 1 Surfaces Studied by HREELS and PES, Surf Sci 427–428: 85–90

26. Sakamoto K, Kondo D, Harada M, Kimura A, Kakizaki A, Suto S (1999) Electronic Structures of C_{60} Adsorbed on Si(111)-(7 × 7) and Si(001)-(2 × 1) Surfaces, Surf Sci 433–435: 642–646

27. Iizumi K, Saiki K, Koma A (2002) Investigation of the Interaction Between a C_{60} Epitaxial Film and a Si(111)-7 × 7 Surface by Electron Energy Loss Spectroscopy, Surf Sci 518: 126–132

28. Godwin PD, Kenny SD, Smith R (2003) The Bonding Sites and Structure of C_{60} on the Si(100) Surface, Surf Sci 529: 237–246

29. Kunitsyn AE, Kozyrev SV, Novikov SV, Savel'ev IG, Chaldyshev VV, Sharonova LV (1994) Production of Fullerene Films on GaAs Semiconducting Substrates, Phys Solid State 36: 2573–2579

30. Xue Q, Ling Y, Ogino T, Sakata T, Hasegawa Y, Hashizume T, Shinohara H, Sakurai T (1996) C_{60} Single Crystal Films on GaAs(001) Surfaces, Thin Solid Films 281–282: 618–623

31. Ohno K, Li ZQ, Kamiyama H, Kawazoe Y, Yue Q, Hashizume T, Hasegawa Y, Shinohara H, Sakurai T (1997) A Mechanism of 13% Lattice Expansion in C_{60} FCC(110) Thin Films Grown on the GaAs(001) As-rich Surface, Sci. Reports Research Institutes Tohoku University Series A – Phys Chem & Met 43(1): 61–65

32. Yao JH, Zou YJ, Zhang XW, Chen GH (1997) The (111) Oriented Growth of C_{60} Films on GaAs(100) Substrates, Thin Solid Films 305(1–2): 22–25

33. Dunphy JC, Klyachko D, Xu H, Chen DM (1997) A Novel Bi-directional Step-flow Growth Mode:C_{60} on Ge(100)and GaAs(110) Surf Sci 383: 760–765

34. Sakurai T, Xue QK, Hashizume T, Hasegawa Y (1997) Extraordinary Growth of C_{60} on a GaAs(001) As-rich 2 × 4 Surface, J Vac Sci Tech B 15: 1628–1632

35. Moriarty P, Upward MD, Ma YR, Dunn AW, Beton PH, Teehan D, Woolf DA (1998) Reconstruction Dependent Adsorption of C_{60} on GaAs(111)B, Surf Sci 405: 21–26

36. Colder A, Canut B, Levalois M, Marie P, Portier X, Ramos SMM (2002) Latent Track Formation in GaAs Irradiated with 20, 30, and 40 MeV Fullerenes, J Appl Phys 91(9): 5853–5857

37. Nishinaga J, Ogawa M, Horikoshi Y (2004) Selective Growth of C_{60} Layers on GaAs and Their Crystalline Characteristics, Thin Solid Films 464–465: 323–326

38. Yao SJ, Zhou CG, Ning LC, Wu JP, Pi ZB, Cheng HS, Jiang YS (2005) Chemisorption of C_{28} Fullerene on c(4 × 4) Reconstructed GaAs(001) Surface: A Density Functional Theory Study, Phys Rev B 71: 195316–195322

39. Su JS, Chen YF, Chiu KC (1999) Dielectric Properties of Fullerene Films, Appl Phys Lett 74(3): 439–441

40. Vosko SH, Wilk L, Nusair M (1980) Accurate Spin-Dependent Electron Liquid Correlation Energies for Local Spin Density Calculations: A Critical Analysis, Can J Phys Chem B 58: 1200–1211

41. Perdew JP, Wang Y (1992) Accurate and Simple Analytic Representation of the Electron-gas Correlation Energy, Phys Rev B 45: 13244–13249

42. Perdew JP, Chevary JA, Vosko SH, Jackson KA, Pederson MR, Singh DJ, Fiolhais C (1992) Atoms, Molecules, Solids, and Surfaces: Applications of the Generalized Gradient Approximation for Exchange and Correlation, Phys Rev B 46: 6671–6687

43. Perdew JP, Burke K, Ernzerhof M (1996) Generalized Gradient Approximation Made Simple, Phys Rev Lett 77: 3865–3868

44. Sánchez-Portal D, Ordejón P, Artacho E, Soler JM (1997) Density-Functional Method for Very Large Systems with LCAO Basis Sets, Int J Quant Chem 65: 453–461

45. Artacho E, Sánchez-Portal D, Ordejón P, García A, Soler JM (1999) Linear-Scaling *ab-initio* Calculations for Large and Complex Systems, Phys Status Solidi B 215: 809–817

46. Soler JM, Artacho E, Gale JD, García A, Junquera J, Ordejón P, Sánchez-Portal D (2002) The SIESTA Method for *ab initio* Order-*N* Materials Simulation, J Phys: Condens Matter 14: 2745–2779

47. Caron B, Derome L, Flaminio R, Grave X, Marion F, Mours B, Verkindt D, Cavalier F, Viceré A (1999) SIESTA, a Time Domain, General Purpose Simulation Program for the VIRGO Experiment, Astropart Phys 10: 369–386

48. Kohn W (1999) Nobel Lecture: Electronic Structure of Matter—Wave Functions and Density Functionals, Rev Mod Phys 71: 1253–1266

49. Kleinman L, Bylander DM (1982) Spiral-Vortex Expansion Instability in Type-II Superconductors, Phys Rev Lett 38: 1425–1428

50. Troullier N, Martins JL (1991) Efficient Pseudopotentials for Plane-Wave Calculations, Phys Rev B 43: 1993–2006

51. Ordejón P, Artacho E, Cachau R, Gale J, García A, Junquera J, Kohanoff J, Machado M, Sanchez-Portal D, Soler JM, Weht R (2001) Linear Scaling DFT Calculations with Numerical Atomic Orbitals, Mat Res Soc Symp Proc 677: AA9.6.1–AA9.6.12

52. Hawkins JM, Meyer A, Lewis TA, Loren SD, Hollander FJ (1991) Crystal-Structure of Osmylated C_{60}-Confirmation of the Soccer Framework, Science 252: 312–313

53. Liu S, Lu YJ, Kappes MM, Ibers JA (1991) The Structure of the C_{60} Molecule-X-ray Crystal-Structure Determination of a Twin at 110K, Science 254: 408–410

54. Hedberg K, Hedberg L, Bethune DS, Brown CA, Dorn HC, Johnson RD, Vries MD (1991) Bond Lengths in Free Molecules of Buckminsterfullerene, C_{60} from Gas-Phase Electron Diffraction, Science 254: 410–412

55. Kietzmann H, Rochow R, Ganteför G, Eberhardt W (1998) Electronic Structure of Small Fullerenes: Evidence for the High Stability of C_{32}, Phy Rev Lett 81: 5378–5381

56. Hamers RJ, Hovis JS, Lee S, Liu H, Shan J (1997) Formation of Ordered, Anisotropic Organic Monolayers on the Si(001) Surface, J Phys Chem B 101: 1489–1492

57. Konečný R, Doren DJ (1997) Theoretical Prediction of a Facile Diels-Alder Reaction on the Si(100)-2×1 Surface, J Am Chem Soc 119: 11098–11099

58. John T Yates Jr. (1998) A New Opportunity in Silicon-Based Microelectronics, Science 279: 335–336

59. Hamers RJ, Hovis JS, Greenlief CM, Padowitz DF (1999) Scanning Tunneling Microscopy of Organic Molecules and Monolayers on Silicon (001) Surfaces, Jpn J Appl Phys, Part 1 38: 3879–3887

60. Schwartz MP, Ellison MD, Coulter SK, Hovis JS, Hamers RJ (2000) Interaction of π-conjugated Organic Molecules with π-bonded Semiconductor Surfaces: Structure, Selectivity, and Mechanistic Implication, J Am Chem Soc 122: 8529–8538

61. Hamers RJ, Coulter SK, Ellison MD, Hovis JS, Padowitz DF, Schwartz MP, Greenlief CM, Russell JJN (2000) Cycloaddition Chemistry of Organic Molecules with Semiconductor Surfaces, Acc Chem Res 33: 617–624

62. Bent SF (2002) Attaching Organic Layers to Semiconductor Surfaces, J Phys Chem B 106: 2830–2842

63. Bent SF (2002) Organic Functionalization of Group IV Semiconductor Surfaces: Principle, Examples, Applications, and Prospects, Surf Sci 500: 879–903

64. Konečný R, Doren DJ (1998) Cycloaddition Reactions of Unsaturated Hydrocarbons on the Si(100)-(2×1) Surface- Theoretical Prediction, Surf Sci 417: 169–188

65. Lin JS, Kuo YT, Lee MH, Lee KH, Chen JC (2000) Density Functional Study of Structural and Electronic Properties of Silane Adsorbed Si(100) Surface, J Mol Struct (Theochem) 496: 163–173

66. Linford MR, Fenter P, Eisenberger PM, Chidsey CED (1995) Alkyl Monolayers on Silicon Prepared from 1-Alkenes and Hydrogen-Terminated Silicon, J Am Chem Soc 117: 3145–3155

67. Tao F, Dai YJ, Xu GQ (2002) Tetra-σ Attachment of Allyl Cyanide on Si(111)-7×7, Phys Rev B 66: 035420–035426

68. Cao Y, Yong KS, Wang ZH, Deng JF, Lai YH, Xu GQ (2001) Cycloaddition Chemistry of Thiophene on the Silicon (111)-7 × 7 Surface, J Chem Phys 115: 3287–3296

69. Ohno K, Li ZQ, Kamiyama H, Kawazoe Y, Yue Q, Hashizume T, Hasegawa Y, Shinohara H, Sakurai T (1997) A mechanism of 13% lattice expansion in C-60 FCC(110) thin films grown on the GaAs(001) as-rich surface, Sci. Reports Research Institutes Tohoku University Series A - Phys Chem & Met 43(1): 61–65

70. Yoshinobu J, Fukushi D, Uda M, Nomura E, Aono M (1992) Acetylene adsorption on Si(111)(7 × 7): A scanning-tunneling-microscopy study, Phys Rev B 46: 9520–9524

71. Weiner B, Carmer CS, Frenklach M (1991) Acetylene Reaction With the Si(111) Surface: A Semiempirical Quantum Chemical Study, Phys Rev B 43: 1678–1684

72. Tao F, Wang ZH, Qiao HM, Liu Q, Sim WS, Xu GQ (2001) Covalent Attachment of Acetonitrile on Si(100) Through Si-C and Si-N Linkages, J Chem Phys 115: 8563–8569

73. Carbone M, Piancastelli MN, Casaletto MP, Zanoni R, Comtet G, Dujardin G, Hellner L (2000) Low-Temperature Adsorption States of Benzene on Si(111)7 × 7 Studied by Synchrotron-Radiation Photoemission, Phys Rev B 61: 8531–8536

74. Heringdorf F-JMZ, Reuter MC, Tromp RM (2001) Growth Dynamics of Pentacen Thin Films, Nature 412: 517–520

75. Rochet F, Dufour G, Prieto P, Sirotti F, Stedile FC (1998) Electronic Structure of Acetylene on Si(111)-7 × 7: X-ray Photoelectron and X-ray Absorption Spectroscopy, Phys Rev B 57: 6738–6748

76. Rochet F, Jolly F, Bournel F, Dufour G, Sirotti F, Cantin J-L (1998) Ethylene on Si(001)-2 × 1 and Si(111)-7 × 7: X-ray Photoemission Spectroscopy with Synchrotron Radiation, Phys Rev B 58: 11029–11042

CHAPTER 16

A QUEST FOR EFFICIENT METHODS OF DISINTEGRATION OF ORGANOPHOSPHORUS COMPOUNDS: MODELING ADSORPTION AND DECOMPOSITION PROCESSES

ANDREA MICHALKOVA[1], LEONID GORB[2] AND JERZY LESZCZYNSKI[1,*]

[1] Computational Center of Molecular Structure and Interactions, Department of Chemistry, Jackson State University, 1400 J. R. Lynch Street, P. O. Box 17910, Jackson, MS 39217, USA
[2] U.S. Army Engineer Research and Development Center (ERDC, SpecPro), Vicksburg, MS 39180

Abstract: The problem with a contamination of soil and groundwater by organophosphorus compounds is a widespread environmental concern with environmental deterioration. However, the high cost of remediation becomes evident. Organophosphorus compounds have several applications (agricultural, industrial, and military). Nevertheless, assessments of the hazards from these applications quite often do not take into account chemical processes. The management of contaminants requires considerable knowledge and understanding of contaminant behavior. Unique properties of transition metals and metal oxides such as having high adsorption and catalytic ability have resulted in their applications as natural adsorbents and catalysts in the development of clean-up technologies. An understanding of the physical characteristics of the adsorption sites of selected parts of soil (metal oxides) and transition metals, the physical and chemical characteristics of the contaminant, details of sorption of contaminants on soil, on soil in water solution, and on transition metals, and its distribution within the system is of practical interest. Quantum-chemical calculations provide more insight into the aforementioned characteristics of organophosphorus compounds. This review summarizes experimental studies and the computational techniques and applications which are used to develop theoretical models that explain and predict how transition metals and metal oxides can affect the adsorption and decomposition of selected organophosphorus compounds. The results can contribute to a better knowledge of impact of such processes in existing remedial technologies and in a development of new removal and decomposition techniques

Keywords: organophosphorus compound; nerve agent; adsorption; decomposition; soil; metal oxide; transition metal; cluster approach; surface reactivity; solvent; supermolecular approximation; continuum model; reaction kinetics

Corresponding author, E-mail: jerzy@ccmsi.us

W. A. Sokalski (ed.), Molecular Materials with Specific Interactions, 565–592.

1. INTRODUCTION

1.1. Importance of Catalytic Decomposition of Organophosphorus Compounds

The development of cost effective cleanup technologies for organophosphorus compounds is a high priority for environmental restoration research. Understanding the catalytic decomposition of organophosphorus compounds is of importance for devising new methods for the protection of personnel exposed to chemical warfare agents and other chemically similar industrial compounds.[1] It is vital to have a fundamental knowledge of the chemical-physical behaviors of these organic molecules in soil, groundwater and industrially important materials. One potentially valuable aspect of understanding this fundamental behavior and, we believe, a valuable contribution into characterizing the fate of organophosphorus compounds is the detailed study of their interactions with specific surfaces. Therefore, the fundamental chemistry controlling these interactions should be understood in order to develop the best strategy in the management of contaminated soils. Heterogeneous reactions currently offer one of the most favorable technological routes to the removal of contaminants from air, groundwater and soil. Indeed for example, the application of heterogeneous catalytic methods to automotive emission control represents the most widespread exposure of the public to the benefits of catalytic technology.[2,3] Also, in the case of title compound it can lead to better technologies for their catalytic decomposition since the degradation of organophosphorus compounds is of great importance for health and environmental safety.[1]

Although organophosphorus compounds have agricultural, industrial, and military applications, there is very limited knowledge of their interactions with catalytic surfaces (transition metals and metal oxides). For example, despite the major processes affecting the natural and engineered treatment of organophosphates contaminants have appreciated qualitatively, many questions remain regarding reaction mechanisms. Moreover, despite the existence of some experimental data and the application of several strategies to analyze such finding the distribution of contaminants in the pathways of soil, groundwater and other materials is not clearly characterized yet. Comprehensive study can help to solve these problems and so bring the innovation into the clean-up techniques of contaminants. Quantum-chemical analysis provides a wide array of powerful tools that have been underutilized in deciphering the complex reactions affecting organophosphorus compounds. Since the mid-1980s, computational chemistry has been one of the fastest growing areas of chemistry. This is due to a vast increase in the number of state-of-the-art computing platforms and due to efficient, high-performance computing algorithms. Molecular modeling has become an inseparable part of research activities devoted to gaining an understanding of the molecular basis of chemical processes. The chemical community has already accepted high quality theoretical data not as an addition to the experimental findings but as reliable, independent sources of information concerning molecular structures, properties, and reactivity. It has become a common practice in every field of chemical research including geochemistry and mineralogy to use computational methodologies.

This review provides information related to the methods which are currently most suitable for the remediation of contaminated soils and water. The development of a greater understanding of the physical and chemical processes involved in the degradation of contaminants will facilitate the establishment of more cost-effective and efficient remedial action plans that are protective of human health and the environment.

1.2.   Organophosphorus Compounds, Transition Metals and Metal Oxides

Organophosphates are the most frequently used insecticides worldwide. These compounds cause 80% of the reported toxic exposures to insecticides. Organophosphates produce a clinical syndrome that can be treated effectively only if recognized early. Organophosphates were first discovered more than 150 years ago; however, their widespread use began in Germany in the 1920s, when these compounds were first synthesized as insecticides and chemical warfare agents. Organophosphates form an initially reversible bond with the enzyme cholinesterase. This enzyme is critical for controlling nerve signals in the body since it normally relaxes the activity of acetylcholine (a common neurotransmitter found in the nervous system). Acetylcholine is located at the "neuromuscular junction" where it acts to control muscular contraction. A method of action of normal transmission of acetylcholine is as follows. When a normally functioning nerve is stimulated it releases the neurotransmitter acetylcholine from a terminal to a receptor on the other side. Acetylcholine transmits the impulse to a muscle. When the impulse is sent, the enzyme acetylcholine esterase immediately breaks down the acetylcholine in order to allow the muscle to relax. The transmission upon the organophosphate poisoning is such that these compounds disrupt the nervous system by inhibiting the acetyl-choline esterase enzyme by forming a covalent bond with the site of the enzyme where acetylcholine normally undergoes hydrolysis (breaks down). The result is that acetylcholine builds up and continues to act so that any nerve impulses are continually transmitted, and muscle contractions do not stop. The organophosphate cholinesterase bond can degrade spontaneously re-activating the enzyme or can undergo a process called aging. The process of aging results in irreversible enzyme inactivation.

One class of phosphorus-containing organic chemicals (organophosphates) that disrupt the mechanism of transferring messages to organs by nerves is classified as nerve agents.[4,5] They are also known as nerve gases, though these chemicals are liquid at room temperature. Nerve agents are readily adsorbed by inhalation, ingestion, and dermal contact. Poisoning by a nerve agent leads to contraction of pupils, profuse salivation, convulsions, involuntary urination and defecation, and eventual death by asphyxiation as control is lost over respiratory muscles.[6,7] The number and severity of symptoms which appear vary according to the amount of the agent absorbed and rate of entry into the body. Also, the extent of damage to the victim varies depending on several factors like temperature, wind, air pressure, concentration, type of chemical compound and method of delivery. The effect of inhalational exposure to nerve agent vapor in turn depends on the vapor concentration and the time of exposure.

There are two main classes of nerve agents. The members of the two classes share similar properties.[8,9] The *G-series* is so-named because German scientists first synthesized it. The first nerve agent ever synthesized was GA (Tabun) in 1936. GB (Sarin) was discovered next in 1938, followed by GD (Soman) in 1944 and finally the more obscure GF (Cyclosarin) in 1949. The *V-series* is the second family of nerve agents, and also contains four members: VE, VG, VM, and VX. Tabun or GA (Ethyl N,N-dimethylphosphoramidocyanidate, $C_5H_{11}N_2O_2P$) is a colorless to brown liquid (depending on purity) and volatile (evaporating readily at normal temperatures). Although odorless when pure, Tabun is commonly described as having a faint fruity odor due to common manufacturing impurities. GA is the easiest chemical agent to manufacture in mass quantities to date.[10] Sarin (GB), isopropyl methylphosphonofluoridate ($C_4H_{10}FO_2P$) is similar in structure and biological activity to some commonly used insecticides.[8] Its relatively high vapor pressure means that it evaporates quickly. Soman (GD), 3,3-dimethyl-2-butyl methylphosphonoflouridate ($C_7H_{16}FO_2P$) is both more lethal and more persistent than Sarin or Tabun, but less than Cyclosarin. The VX nerve agent (O-Ethyl-S-[2(di-isopropylamino)ethyl] methylphosphonothioate) is the most well-known of the V-series of nerve agents.[11] VX is odorless and tasteless. VX is an oily liquid that is amber in color and very slow to evaporate. VX is the least volatile of the nerve agents, which means that it is the slowest to evaporate from a liquid into a vapor.[12] Therefore, VX is very persistent in the environment.

Platinum (Pt) and palladium (Pd) with their extraordinary catalytic properties are most commonly catalysts employed in industrial processes. Both Pt [Xe] $4f^{14}5d^96s$ and Pd [Kr] $4d^{10}$ are late transition metals in group VIIIA in the periodic table. Their properties are so unique that platinum may be considered as one of the most versatile, all-purpose, heterogeneous metal catalysts. For this catalytic property, platinum is used in catalytic converters, incorporated in automobile exhaust systems, as well as tips of spark plugs. Both metals can have some applications as catalysts for the decomposition of organophosphorus compounds.[1,13,14] Platinum possesses remarkable resistance to chemical attack, excellent high-temperature characteristics, and stable electrical properties. All these properties have been exploited for industrial applications.[15] Palladium is a soft silver-white metal that resembles platinum. It is the least dense and has the lowest melting point of the platinum group metals. It is soft and ductile when annealed and greatly increases its strength and hardness when it is cold-worked.[16]

Metal oxides, well known for their industrial applications as adsorbents and catalysts, have many potential decontamination applications such as protective filtration systems for vehicles, aircraft, and buildings and the demilitarization of nerve agent munitions and stockpiles.[17,18] An efficient application of metal oxides nanobelts for nerve agent detection has been demonstrated (see for example references 19–23). Metal oxide sensors are commonly used to monitor a variety of toxic and inflammable gases. The sensing mechanism is based on electrical conductance change upon surface reduction-oxidation (redox) reactions with gas species.[19] In recent years, magnesium oxide (MgO)- and calcium oxide (CaO)-like materials have received increased attention as potential adsorbents for the decomposition of chemical warfare agents. Enhancement to the reactivity of nanosize oxides is anticipated due to the increased surface area, the greater amount of highly reactive edge and corner "defect" sites, and unusual, stabilized lattice planes.[17,18] The nanoparticles of MgO are unique. It was concluded that the nanoparticles of MgO exhibit unusual surface morphologies and possess a more reactive surface.[17] It was found that nanocrystalline MgO, being 30% surface, can destroy organophosphorus compounds. MgO in a nanoparticle form has been successfully used as a destructive adsorbent for nerve agents.[24] The term "destructive adsorbent" is intended to describe its ability to efficiently adsorb and at the same time, to chemically destroy incoming adsorbate. Calcium oxide also possesses unique nanoparticle surface properties. X-ray diffraction results have shown the expected cubic internal crystal structure of basis unit of calcium oxide. CaO is highly ionic with a high melting point and large surface area and high surface reactivity. Calcium oxide has some advantages many of which include that it is non-toxic, is easy to handle and store, stable, cheap and exhibits very high capacities for adsorption.[25]

2. COMPUTATIONAL METHODS AND MODELS

Knowledge of molecular structure, transformation mechanisms, and the spectrum of potential intermediates/products of the contaminant is helpful for developing of remediation processes. Understanding and comparing degradation pathways of nerve agents on catalyst (metals and metal oxides) enables theoretical predictions as to the most likely intermediate and final products, thereby shortening the period of expense and experimentation. This review is devoted to explore fundamental capabilities of computational chemistry (CC) techniques including the ab initio methods as tools to characterize properties of nerve agents interacting with such species. Moreover, the nature of the interactions between an adsorbate can be well described by means of quantum-chemical calculations.[26]

The use of CC techniques has many advantages. Accurate CC techniques can be utilized with or without experiments. Thus, compounds that are not even available (including those not yet synthesized) or potentially too hazardous to handle experimentally can be assessed. CC is flexible, environmentally clean and relatively inexpensive. Therefore, the application of current and rapidly developing CC technology will provide environmental scientists/managers the means to expedite risk assessment for the compounds of interest.

2.1.   Quantum-chemical Approximations for the Modeling of Surface Reactivity[27]

Quantum-chemical ab initio calculations have become an alternative to experiments for determining accurately structures, vibrational frequencies and electronic properties as well as intermolecular forces and molecular reactivity.[28–31] Two specific approximations were developed to solve the problems of surface chemistry: periodic approximation, where quantum-chemical method employs a periodic structure of the calculated system and cluster approximation, where a model of solid phase of finite size is created as a cutoff from the system of solid phase (it produces unsaturated dangling bonds at the border of the cluster). Cluster approximation has been widely used for studying interactions of molecules with all types of solids and their surfaces.[32] This approach is powerful in calculating the systems with deviations from the ideal periodic structure like doping and defects.

2.1.1.   *Periodic treatment*

The Vienna *Ab Initio* Simulation Package (VASP) program[33,34] was developed to carry out calculations to obtain the fluctuation trajectory of the selected models. The VASP program uses a rather traditional self-consistency scheme to evaluate the instantaneous electronic ground-state at each molecular dynamics (MD) step so that the wavefunction can be converged to the Born-Oppenheimer surface at each time-step.

A plane-wave basis set with a cut-off energy of 286.744 eV is used for the electron wavefunction which is solved at each MD step by conjugate gradient minimization of the total electronic energy. The Perdew-Wang gradient correction[35] is added to the exchange-correlation functional. For the core region, the optimized Vanderbilt ultrasoft pseudopotentials[36] supplied with the VASP package are used for the C, N, O, Si, Al, Mg and H atoms.

The time-step of 0.5 fs is used to simulate the dynamic system to 4.0 ps. The temperature of 300 K is used throughout the simulations. The MD simulations are performed using the Nosé-Hoover thermostat for temperature control. The Hellmann-Feynman forces acting on the atoms are calculated from the ground-state electronic energies at each time step and are subsequently used in the integration of Newton's equation of motion.

2.1.2.   *Cluster approach*

A physical approach to the electronic structure problems of solids contrasts sharply with the notion that local interactions dominate the structure and properties of molecular systems. Hence, it is very appealing to replace the infinite solid, which is difficult to treat quantum-chemically, by finite sites of interest. Intuitively, cutouts from the bulk or the surface are made and treated like molecules. This type of method is called the cluster approach and the models made as cutouts of the periodic structure are called cluster models.[26]

According to the cluster approach, one can consider a limited number of atoms around the active sites and then can apply the quantum-chemical description to a piece of the solid, referred to as a cluster. This is the easiest and chemically, the best procedure when a local description of the adsorption or catalytic site dominates. To make the cluster neutral, the broken bonds, resulting from an homolitic cut off the chemical bonds in the crystal, are electrically balanced by monovalent atoms. The first approach dealt with monovalent atoms, the parameters of which were specially parameterized (so-called 'pseudoatoms'), but finally the opinion that saturation by hydrogen atoms is sufficient has been widely accepted.[37] In the framework of the cluster approach there has been an enormous number of investigations which include the description of the different properties of the oxide surfaces.[38] The utility of the cluster approach in describing zeolites[32,39−42] an extremely functional class of aluminosilicate minerals exploited for their high capacity for cation exchange has also been demonstrated.

The most important disadvantage of the cluster model is the relatively small size of the considered system. Embedded and dipped cluster models are usually used to overcome this problem. They are considered not only because the cluster models need corrections for neglected interactions with their surrounding but also because they are a promising alternative to the calculation of large cells ("supercells").[32] The extension of the model and the inclusion of the interaction of the cluster with the surroundings into the calculation is realized by different ways depending on the kind of phase (ionic, covalent solids, or metals).[26] For example, in the case of ionic crystal, the surrounding is represented by crystal field with the main long-distance electrostatic part, or by discrete point, charges placed around the cluster instead of rest of crystal. For clusters of molecular crystals, it is possible to perform calculations of large cluster in assumption that all significant interactions are included.

The performance of the cluster approach can be improved dramatically if it is combined with the recently developed ONIOM methodology[43] which is an n-layered integrated molecular orbital and molecular mechanics approach. A three-layered version of the ONIOM approximation allows a quantum–mechanical study of systems which are normally considered with molecular mechanics methods to be performed. The three-layered total energy expression for the ONIOM scheme is defined as

$$E(ONIOM3) = E(High, \ Smodel) + E(Med, \ Imodel) + E(Low, \ Real)$$

$$-E(Med, \ Smodel) - E(Low, \ Imodel) \qquad (16\text{-}1)$$

where High, Med, and Low refer to the levels of approximation, respectively, while Real, Imodel, and Smodel refer to the 'real' system, the 'intermediate model system,' and the 'small model' system, respectively.

In other words the three-layered approach divides a system into active parts treated at a very high level of ab initio molecular orbital theory, a semi-active part that includes important electron contribution and could be treated at a relatively lower level of theory and a non-active part that is handled using the lowest ab initio or even semiempirical approximation. Layer assignments are specified as part of the molecule specification. It can combine any two quantum mechanical (QM) and/or molecular mechanical (MM) computational methods.

2.1.3.   Embedded techniques

Embedding techniques at various levels have been suggested to close the gap between the cluster and the periodic treatment. A physical approach to the electronic structure problems of solids and surfaces contrasts sharply with the intuition of chemists that local interactions dominate the properties of surfaces and adsorption complexes. Hence, it is very desirable to replace the infinite solid, which is very difficult to treat quantum-chemically, by finite models of the sites of interest. In this way it is easy to describe a local site as cluster having a relatively small number of atoms which interact with a potentially infinite number of surrounding atoms through, for example, the Madelung potential which is treated as perturbation. [26,44] These lead to terms such as

$$F_{\mu\nu} = F^0_{\mu\nu} + < \mu(r)|V_{ext}|\nu(r) > \qquad (16\text{-}2)$$

where $F_{\mu\nu}$- matrix elements of Fockian and V_{ext} is an external long-range potential.

Molecular and ionic crystals are the easiest systems to apply to these schemes. The task becomes far more difficult in the case of covalent solids. However, this technique has been recently updated for calculations of covalent solids both at the semiempirical [37] and at the ab initio levels. [38]

2.1.4.   Merits and limits of each approach

Among all of the approaches described above, none is completely satisfactory. A list of their features is described below:

(a) although VASP is very efficient code, the computational burden rises steeply when dealing with large cells; there is also a problem for this type of approximation to describe defects, edge, amorphism, etc. which are of great importance to investigators of surface phenomena;

(b) the embedded technique is also very attractive but requires further development especially for covalent solids for which only a few successful implementations have been reported;

(c) although the cluster approximation is theoretically unsatisfactory, it is very popular because it allows for the use of standard quantum-chemical techniques up to a very high correlated level. The geometrical definition of the cluster is simple, and the cut from the underlying solid usually takes into account the features of the active site which one wishes to model; in particular, it allows one to model such peculiarities as edge, vacancies, defects, etc. [26] The concept of cluster and molecular models reduces the problem of electronic structure and local geometry of the solid to the common problem of determining the geometry and electronic structure of molecules, it reduces the problem of the bonding of molecules or atoms on the surface sites to the problems of molecular reactivity and intermolecular interactions, and it reduces the problem of surface reactions to the problem of potential surfaces for reactions between molecules.

2.2. Quantum-chemical Approximations for the Influence of the Solvent[45]

Current efforts in quantum-chemical modeling of the influence of solvents may be divided into two distinct approaches. The first, the supermolecular approximation, involves the explicit consideration of solvent molecules in quantum-chemical calculations. Another possibility for simulating solvent influence is to replace the explicit solvent molecules with a continuous medium having a bulk dielectric constant. Models of this type are usually referred to as polarized continuum models (PCMs).

2.2.1. Supermolecular approximation

Accurate predictions of solute interactions with a limited number of solvent molecules are possible using the supermolecular approximation. This is an approach based on the consideration of the dissolved molecule together with the limited number of solvent molecules as the unified system. The quantum-chemical calculations are performed on the complex of the solute molecule surrounded by as many solvent molecules as possible. The main advantage of the supermolecular approximation is the ability to take into account such specific effects of solvation as hydrogen bonding between the selected sites of the solvated molecules and the molecules of the solvent. In principle there are only two restrictions for the supermolecular approximation. One of them is the internal limitations of the quantum-chemical methods. The second restriction is the limitation of the current computer technology. Because of such restrictions this approximation coupled with ab initio molecular dynamics is possible only for small model systems.[46–50]

2.2.2. Born-Kirkwood-Onsager continuum model

Probably the simplest quantum-mechanical operators that include interaction with a continuum are the Born and Onsager reaction field models. In the case of neutral solutes, the model could be express as

$$(H_0 - \frac{1}{2}g\mu < \Psi|\mu|\Psi >= E|\Psi >$$
(16-3)

where $g = 2(\varepsilon - 1)/(2\varepsilon + 1)$ is the function of dielectric permittivity and α^3 and α are the solute cavity radii.

This is a generalization of the Onsager reaction field model for a point dipole inside a spherical cavity. For charged solutes, one should also include an ionic Born term, derived by

$$-\frac{1}{2}g_0 < \Psi|\frac{1}{r}|\Psi >$$
(16-4)

where $g_0 = (1 - 1/\varepsilon)\alpha$.

In spite of the Onsager-Born scheme being executed up to the MP2 level in the Gaussian 94–98 packages, there are at least two disadvantages of the scheme. As emphasized above, the Born-Onsager equation includes only the solute's monopole and dipole interaction with the continuum. Such a generalization of the model has appeared in the Kirkwood equation by extending the multipole series to an arbitrary high order. This approach allows for the calculation of the electrostatic part of the free energy of solvation (ΔG_s^0)

$$\Delta G_s^0 = -\frac{1}{2} \sum_{l=0}^{\infty} \sum_{m=-l}^{+l} \sum_{l'=0}^{\infty} \sum_{m=-l}^{l'} M_l^m f_{ll'}^{mm'} M_{l'}^{m'} \tag{16-5}$$

where each component m of every multipole M of order l interacts with all of the reaction field multipole moments induced by the solute multipoles e.g., the $M_{l'}^{m'}$ terms via coupling $f_{ll'}^{mm'}$, called the reaction field factor.

2.2.3.   *Generalized reaction field from surface charge densities*

There is another family of continuum models[51–53] where the influence of the reaction field is modeled by a set of polarized charges distributed on the surface of a molecular cavity. Models of this type are usually referred to as polarized continuum models (PCMs), and they have as their origin the solution of Poisson's equation.

$$\nabla^2 \phi = -\frac{4\pi\rho(r)}{\varepsilon} \tag{16-6}$$

where ε is a dielectric permittivity. Then, the electrostatic part of the free energy of solvation (ΔG_s^0) is defined as

$$\Delta G_s^0 = <\Psi|H^0 + \phi_\sigma|\Psi> -\frac{1}{2}\int_s \phi(r)[\rho_n(r) + \rho_e(r)]d^2r - G_g^0 \tag{16-7}$$

where G_g^0 is a free energy in the gas phase; ρ_n and ρ_e are the potentials created by surface-distributed virtual charges, and the integral represents the cost of polarizing the solvent

$$\sigma(r) = \frac{1-\varepsilon}{4\pi\varepsilon} \frac{\partial}{\partial n}[\phi_\rho(r) + \phi_\sigma(r)]_{s_-} \tag{16-8}$$

where $\phi_\rho(r)$ is the electrostatic potential due to the solute charge distribution, and $\phi_\sigma(r)$ is a potential due to virtual charges $\sigma(r)$.

2.2.4.   *Comparison of continuum models*

Unfortunately, a detailed comparison of the continuum models is available only at the semiempirical level.[54,55] Because the SMx models are specially parametrized to describe free energy of hydration, it is not surprising that they are the best for reproducing this value. A detailed discussion of the advantages and limitations of different types of solvation models with regard to the various types of approximations and different types of organic molecules can be found in references 53–55.

2.3. Computational Approaches for Reaction Kinetics

Two computational approaches are currently available to estimate reaction kinetics (rate constants or half-life time). One is a *canonical variational transition state theory (CVT)* and the second is an *instanton* approach[56,57] which can describe proton transfer in large molecular systems.[58,59] The instanton approach is a quasi-classical method based on the least-action principle. The instanton path is the least-action trajectory for a given temperature which implies that it is the dominant tunneling trajectory at that temperature. Although the exact instanton path is very difficult to calculate for a polyatomic molecule or complex, for rate constant predictions it is sufficient to calculate the corresponding multidimensional instanton action. An approximate expression for this action can be derived from the easily calculated one-dimensional tunneling rate constant through the introduction of appropriate couplings of the tunneling mode to other models, for example the transverse vibrational mode.

CVT approach is particularly attractive due to the limited amount of potential energy and Hessian information that is required to perform the calculations. Direct dynamics with CVT thus offers an efficient and cost-effective methodology. Furthermore, several theoretical reviews[60,61] have indicated that CVT plus multidimensional semi-classical tunneling approximations yield accurate rate constants not only for gas-phase reactions but also for chemisorption and diffusion on metals. Computationally, it is expensive if these Hessians are to be calculated at an accurate level of *ab initio* molecular orbital theory. Several approaches have been proposed to reduce this computational demand. One approach is to estimate rate constants and tunneling contributions by using Interpolated CVT when the available accurate ab initio electronic structure information is very limited.[62] Another way is to carry out CVT calculations with multidimensional semi-classical tunneling approximations.

Both approaches include tunneling corrections and provide approximately the same accuracy. However, the *variational transition state theory* is computationally quite demanding, and at least 40 points on the path of the proton transfer should be available. In contrast, the *instanton* approach uses only vibrational frequencies calculated for local minima and transition states and corresponding values of energy.

3. APPLICATIONS OF TRANSITION METALS AND METAL OXIDES AS CATALYSTS FOR ADSORPTION AND DECOMPOSITION OF ORGANOPHOSPHORUS COMPOUNDS

Within the last several years important basic experimental studies of interactions between organophosphorus compounds and metal oxide surfaces have been carried out. Metal oxides such as MgO, Al_2O_3, FeO, CaO, TiO_2 $\alpha - Fe_2O_3$, ZnO, and WO_3 are currently under consideration as destructive adsorbents for the decontamination of chemical warfare agents.[4,63] For example, several studies have investigated adsorption of dimethyl methylphosphonate (DMMP) (a widely used model compound for the simulation of interactions of phosphate esters with a surface) on the surface of these metal oxides.[64−76] In most of these works the authors have observed that, at first, DMMP is adsorbed molecularly, via hydrogen bonding of the phosphoryl oxygen to a surface at an acid site, followed by stepwise elimination of the methoxy groups which combine with surface hydrogen atoms to yield methanol that evolves from the surface. The final product recorded for the reactions with these oxides is a surface-bound methylphosphonate, with the P-CH$_3$ bond intact. On the basis of observations it was concluded that not only surface defect sites are responsible for the decomposition of DMMP but also that Mg^{2+} and O^{2-} ions in the regular oxide surface also take part in the process. In previous works[77−79] the authors, using solid-state MAS NMR technique, found that Sarin (GB, isopropyl methylphosphonofluoridate (C$_4$H$_{10}$FO$_2$P)), Soman (GD-3,3-dimethyl-2-butyl methylphosphonoflouridate (C$_7$H$_{16}$FO$_2$P), VX and mustard (HD – bis(2-chloroethyl)sulfide) hydrolyze on the surface of the very reactive MgO, CaO and Al$_2$O$_3$ nanoparticles. GD forms both GD-acid and methylphosphonic acid (MPA). VX and GD hydrolyze to yield surface-bound complexes of nontoxic ethyl methylphosphonate and pinacolyl methylphosphonate, respectively. Sarin undergoes initial molecular adsorption on aluminum oxide at unsaturated Al sites followed by slow hydrolysis at room temperature.[80]

This review is mainly devoted to summarize the results of the theoretical studies of organophosphates interacting with catalytic surfaces (transition metal and metal oxide). Therefore, we need to mention that to the best of our knowledge, there were published only a few theoretical works addressing this problem. Some theoretical studies of the interactions of organophosphate compounds with clay minerals and magnesium oxide were also published.

The adsorption of Sarin on the surfaces of magnesium oxide was investigated at the B3LYP/6-31G(d) and MP2/6-31G(d) levels using the representative cluster models.[81] The simplest MgO model contains four oxygen atoms and four magnesium atoms with the chemical formula Mg$_4$O$_4$ prepared using the experimental bulk structure of an MgO crystal (the Mg-O bond length is equal 2.1 Å, and the O-Mg-O bond angles are set to be 90° and 180°). This small model mimics the adsorption and decomposition on the irregular edges, corners, and defect sites. The large (L) model of MgO mimics the adsorption on the regular part of nano-MgO (the size of nanoparticles falls between 2 and 10 nm, or 100–10000 atoms). The L model consists of 16 magnesium (or calcium) atoms and 16 oxygen atoms with the chemical formula Mg$_{16}$O$_{16}$. This work was devoted to studying the interactions of Sarin with hydrated and non-hydrated adsorption sites since the presence of hydroxyl groups can have significant influence on the structure, interactions, and thermodynamic parameters of adsorbed compounds (it was explicitly executed by addition of one or two water molecules to the metal oxide surface).

An analysis of the topological characteristics of electron density was performed for the studied Sarin-magnesium oxide complexes following Bader's Atoms in Molecules Theory (AIM).[82] This analysis can be used to characterize hydrogen bonding solely from the charge density. There have been formulated several effects occurring in the charge density. They include for example, the location of the so-called $(3, -1)$ bond critical points (BCPs, *i.e.*, points where ρ is a minimum along the bond path and a maximum in the other two directions) on the surface of the total charge density, analysis of the electron density (ρ) and analysis of the Laplacian of electron density ($\nabla^2 \rho$) at the BCPs, which are indicative of hydrogen bonding. These effects can be viewed as necessary criteria to conclude that hydrogen bonding is present.[83,84] Based on these criteria one assumes that the C-H...O bond is formed if the value of electron density at the BCP (ρ) ranges between 0.002 and $0.035 \, \text{e/au}^3$ and the value of the Laplacian of electron density $\nabla^2 (\rho)$ is in the range of 0.024–$0.139 \, \text{e/au}^5$.[84]

In addition to these characteristics also the adsorption energy between organophosphorus compounds and the catalytic fragments was estimated in the performed study. The adsorption energy of the organophosphate systems with MgO was corrected by the basis set superposition error (BSSE). The adsorption energy between the M molecule and a surface site S within the surface complex, M-S, is obtained by calculating the three systems involved and by evaluating the difference according to:

$$\Delta E = E(\text{M-S//M-S}) - E(\text{M//M}) - E(\text{S//S}) \tag{16-9}$$

The double slant denotes that all energies are evaluated at the respective equilibrium geometries.[32] The energies obtained at the equilibrium geometry of the complex for each subsystem, e.g. E(M{S}//M-S), are lower than the energies calculated at the same geometry with the basis functions of the respective subsystems alone, e.g. E(M//M-S). The difference is defined as the BSSE

$$\in(\text{M}) = E(\text{M//M-S}) - E(\text{M\{S\}//M-S}) \tag{16-10}$$

$$\in(\text{S}) = E(\text{S//M-S}) - E(\text{S\{M\}//M-S}) \tag{16-11}$$

The BSSE values $\in(\text{M})$ and $\in(\text{S})$ are used to define a counterpoise corrected (CPC) adsorption energy

$$\Delta E^c = \Delta E + \in(\text{M}) + \in(\text{S}) \tag{16-12}$$

A number of locations and orientations of Sarin on the regular nanosurface and on the small fragment of MgO were found. In this study it was revealed that Sarin is physisorbed (the nanosurface and hydroxylated small fragment; this is undestructive adsorption) or chemisorbed (destructive adsorption) on MgO (see Figure 16-1). The physisorption of GB on the surface of MgO occurs due to the formation of hydrogen bonds and ion-dipole and dipole-dipole interactions between adsorbed GB and the surface. The chemisorption occurs due to the formation of covalent bonds between the molecule and the surface. The adsorption results in the polarization and the electron density redistribution of GB. The adsorption energy obtained at the MP2/6-31G(d) level of theory for the most stable chemisorbed system is

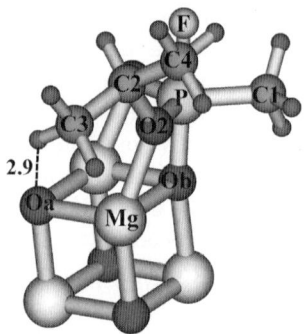

Figure 16-1. The optimized structure of Sarin (GB) adsorbed on the non-hydroxylated Mg_4O_4 fragment obtained at the B3LYP/6-31G(d) level of theory (the A1-GB model)

about $-50\,kcal/mol$. Sarin adsorbed on the nanosurface of hydroxylated small fragment of MgO is much less stabile than Sarin adsorbed on the non-hydroxylated small fragment of MgO. The adsorption energy of Sarin adsorption systems on MgO is proportional to the number and strength of formed covalent bonds between Sarin and the surface.

Also the reaction pathways of Sarin decomposition catalyzed by selected forms of MgO were investigated.[81] Several possible models were considered. In the case of the decomposition on the non-hydroxylated MgO surface the removal of a fluorine from Sarin was modeled. Fluorine was transferred from Sarin into binding distance with the Mg atom of the MgO surface (see Figure 16-2). It was revealed that such a structure provides a reliable model for the reaction mechanism. A two step reaction mechanism was assumed: in the first step Sarin creates a stable adsorbed complex with MgO through three chemical bonds with the MgO surface (the A1-GB model). It is expected that the transfer of the fluorine atom to the surface of MgO is accompanied by a change in the conformation of Sarin. In the second step the bond between P and F is broken (the A1(t)-GB model); the fluorine atom is transferred to the Mg atom of the surface and the remaining part of Sarin adopts the most energetically favorable conformation (the A1(f)-GB model).

The decomposition of Sarin on hydroxylated surface of MgO was also modeled. Following mechanism of the decomposition was suggested: Figure 16-2 the strongly electronegative fluorine affects the hydrogen atom of the hydroxyl group so that this atom is transferred into the binding distance and forms covalent bond with the fluorine atom. Then the bond between the fluorine atom and the phosphorus atom is broken and the HF molecule is formed. The value of the activation energy is about $10\,kcal/mol$. The proposed mechanism is in agreement with experimentally investigated decomposition of DMMP on the hydroxylated MgO surface.[66,69]

An ONIOM study of the adsorption of Sarin on dickite (a 1:1 dioctahedral clay mineral of the kaolinite group)[85] was recently published. For the calculations of the studied systems, the two-layered ONIOM method using combinations of quantum-mechanical methods was applied.[43,86,87] The investigated systems of

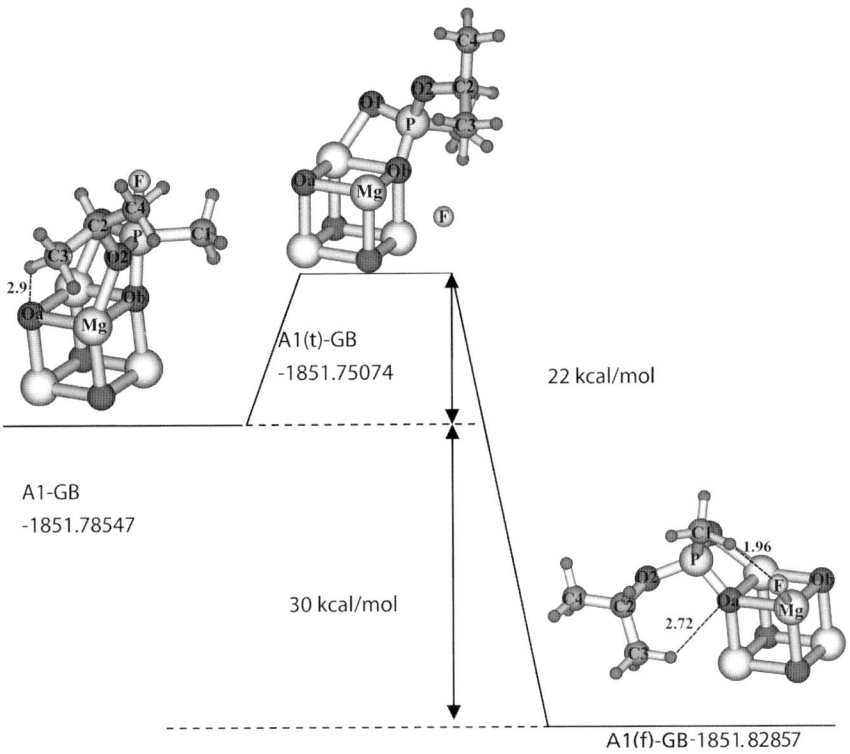

Figure 16-2. The reaction pathway of the decomposition of Sarin (GB) on non-hydroxylated MgO surface from the reactant (the A1-GB model) through transition state (the A1(t)-GB model) to the product (A1(f)-GB model) obtained at the B3LYP/6-31G(d) level of theory

organophosphate with catalytic fragments were divided into two layers which were treated with different computational methods (the B3LYP/6-31G(d) level of theory for the High layer). The layers are conventionally known as the Low (lower layer the mineral fragment) and the High layers (the target molecule with the upper layer of the mineral fragment). The ONIOM energy of the system is then obtained from three independent calculations.

$$E(\text{ONIOM2}) = E\text{model high} - E\text{model low} + E\text{real low} \qquad (16\text{-}13)$$

Real denotes the full system, which only needs to be calculated at the lowest computational level. The additional system is defined as the *model* system. The nature of the interactions was studied using the SCF energy variational-perturbational decomposition scheme proposed by Sokalski et al.[88] In the above scheme ΔE_{HF} is partitioned into the electrostatic exchange $(\varepsilon_{el}^{(10)})$ and the Heitler-London exchange (ε_{ex}^{HL}) first order components and the higher order delocalization (ΔE_{del}^{HF}) term. The delocalization energy accounts for the charge transfer, induction, and other higher order Hartree-Fock terms.[89] The adsorption energy decomposition was performed applying a modified version[90] of the Gamess program.[91] It was found that adsorption on the surface of minerals occurs due to the formation of multiple hydrogen bonds (see Table 16-1) between adsorbed GB and GD and the hydroxyl groups of the octahedral side (see Figure 16-3) and the basal oxygen atoms of the tetrahedral side (see Figure 16-4). The atomic charges of the studied systems were calculated, and the geometrical features were investigated. This type of adsorption results in the polarization and the electron density redistribution of GB and GD on the surface of the mineral. This corresponds to the formation of attractive contacts between GB and GD and the surface of dickite. The polarization of the organic molecule is more significant in the case of adsorption on the octahedral side. The amount of electron density transfers from the mineral

Table 16-1. H...Y and X...Y Hydrogen Bond Distances (Å) and X—H...Y angles (°) Calculated using ONIOM(B3LYP/6-31G(d,p):PM3) (in parentheses) and Electron Density Characteristics ρ (au) and $\nabla^2\rho$ (au) (inner part, B3LYP/6-31G(d,p)) of Hydrogen Bonds in Dickite-GB and Dickite-GD Systems

	H...Y (X...Y)	X-H...Y	ρ	$\nabla^2\rho$
		D7(o)-GB		
HB1	2.003 (2.964)	171.9	0.0207	0.0659
HB2	3.124 (3.591)	111.4	0.0283	0.0857
HB3	2.311 (3.268)	170.3	0.0208	0.0600
HB4	2.207 (3.218)	152.2	0.0060	0.0219
HB5	3.039 (3.966)	142.5	0.0089	0.0247
HB6	2.787 (3.259)	105.9	0.0081	0.0303
HB7	2.578 (3.648)	165.1	0.0054	0.0211
		D7(t)-GB		
HB1	2.85 (3.68)	132.6	0.0035	0.0154
HB2	2.732 (3.785)	161.3	0.0078	0.0266
HB3	2.759 (3.551)	129.0	0.0066	0.0249
HB4	2.726 (3.561)	133.0	0.0041	0.0153
HB5	3.068 (3.65)	113.9	0.0038	0.0149
HB6	2.738 (3.746)	153.0	0.0061	0.0236

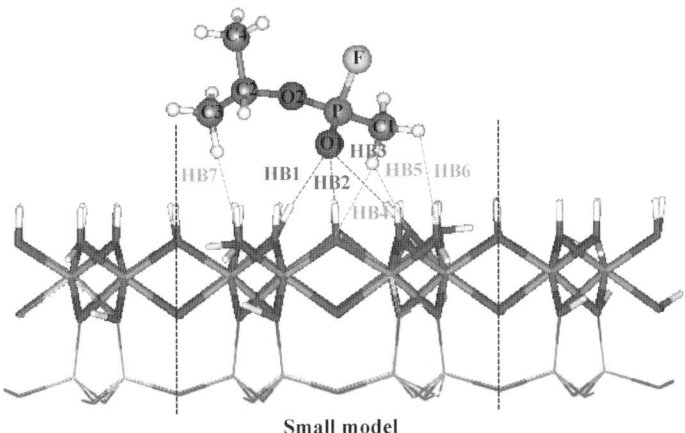

Small model

Figure 16-3. The optimized structure of Sarin adsorbed on the octahedral surface of dickite obtained using the ONIOM(B3LYP/6-31G(d,p):PM3) method

to Sarin and Soman is proportional to the binding strength. The adsorption energies obtained at the ONIOM(B3LYP/6-31G(d,p):PM3) level of theory and using large models of the mineral for the adsorption systems of GB and GD on the octahedral surface of dickite are about -16 and -15 kcal/mol. In the case of adsorption on the tetrahedral surface, the interaction energies of the adsorption systems with GB and GD are approximately -7.0 and -9.0 kcal/mol. GB is adsorbed more preferably on the octahedral aluminum-hydroxide surface than on the tetrahedral silica surface.[85] In the case of adsorption on the octahedral surface, an analysis of the interaction energy components indicates the importance of electrostatic and delocalization binding contributions. In the case of adsorption on the tetrahedral surface, the

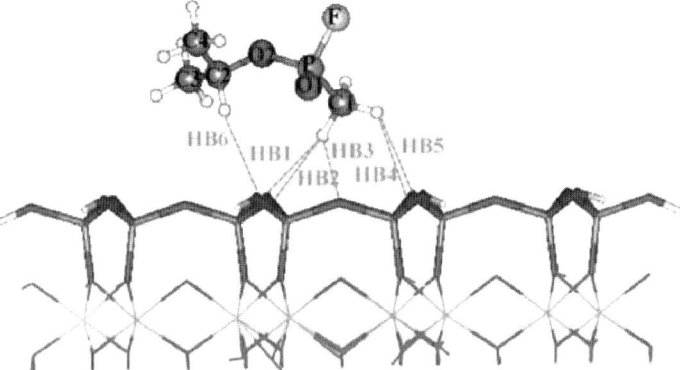

Figure 16-4. The optimized structure of Sarin adsorbed on the tetrahedral surface of dickite obtained using the ONIOM(B3LYP/6-31G(d,p):PM3) method

contribution of the SCF term to the total interaction energy is repulsive. It originates from repulsive interactions between the P=O oxygen atom of Sarin and the basal oxygen atoms of the tetrahedral surface of dickite. The stabilization of Sarin and Soman adsorbed on the tetrahedral surface is secured by correlation forces. The additional local minima of adsorbed Sarin on the tetrahedral and octahedral surfaces of dickite were found. They are characterized by the different involvement of the fluorine atom in intermolecular interactions. H-bonds are weaker in comparison with H-bonds in which the P=O oxygen atom participates.

A study of the structure and interactions of Sarin and Soman with edge tetrahedral fragments of clay minerals has also been performed.[92] The adsorption mechanism of Sarin and Soman on these mineral fragments containing the Si^{4+} and Al^{3+} central cations was investigated. The calculations were performed using the B3LYP and MP2 levels of theory in conjunction with the 6-31G(d) basis set. The number and strength of formed intermolecular interactions have been analyzed using the AIM theory. The charge of the systems and the termination of the mineral fragment are the main contributing factors to the formation of intermolecular interactions in the studied species. In neutral complexes, Sarin and Soman are physisorbed on these mineral fragments due to the formation of C-H...O and O-H...O hydrogen bonds. The chemical bond is formed between a phosphorus atom of Sarin and Soman and an oxygen atom of the -2 charged clusters containing an Al^{3+} central cation (see Figure 16-5) and -1 charged complexes containing a Si^{4+} central cation (chemisorption). Sarin and Soman interact mostly in the similar way with the same terminated edge mineral fragments containing different central cations. Interestingly, the interaction energies of the complexes with an Al^{3+} central cation are larger than corresponding values for the Si^{4+} complexes. The interaction enthalpies of all studied systems corrected for the basis set superposition error were found to be negative. However, based on the Gibbs free energy values only strongly interacting complexes containing a charged edge mineral fragment with an Al^{3+} central cation

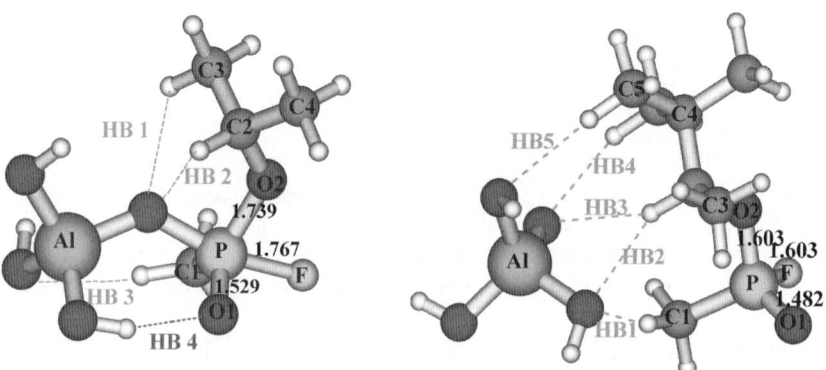

Figure 16-5. The optimized structure of Sarin- and Soman-[AlO(OH)$_3$]$^{2-}$ systems obtained at the B3LYP/6-31G(d) level of theory

are stable at room temperature. It was concluded that Sarin and Soman will be adsorbed preferably on this type of edge mineral surfaces. Moreover, based on the character of these edge surfaces, the tetrahedral edge mineral fragment can provide effective centers for dissociation.

Theoretical studies of the interactions of phosphate with silica and hectorite were published by Murashov and Leszczynski[93] and by Hartzell and co-workers.[94] The ab initio calculations of hydrogen-bonded complexes of dihydrogen and dimethylphosphate anions with ortosilicic acid have shown that the phosphate groups can form strong hydrogen-bonded complexes with the silanols of the silica surface with stabilization energies of ca. -14 kcal/mol per hydrogen bond.[93] A molecular dynamic study of the large tributylphosphate complex of europium provides a test of the sensitivity of force field calculations to predict the behavior of molecules within the interlayer of a trioctahedral smectite clay, hectorite.[94] A cluster quantum-chemical study[95] of DMMP adsorption on magnesium oxide nanoparticles has also been also published. On the basis of the calculation results the possible mechanism of destructive adsorption of DMMP on MgO is discussed. Moreover, theoretical studies of the structures and properties of Sarin and Soman[96] and different conformers of Tabun[97] were also published. The most stable conformers of Sarin and Soman were determined in high-level-correlated calculations.[96] Both molecules are found to have three low-energy conformers each. For both Sarin and Soman two of the lowest energy conformers have almost the same energies with a very small barrier separating the corresponding minima. The third conformer of Sarin is found to lie about 1 kcal/mol above the lowest energy form. For Soman the corresponding value is equal to about 4 kcal/mol. Conformational studies have been carried out on two different enantiomers of Tabun at the DFT and MP2 levels of theory to generate low energy potential energy surfaces in the gas phase as well as in an aqueous environment.[97] The structures of the low energy conformers together with their molecular electrostatical potential surfaces have been compared with those of the non-aged acetylcholinesterase-tabun complex in order to locate the active conformer of the molecule.

In order to understand the origin of the organophosphorus compound interactions with the Pt and Pd surfaces, some effort has been devoted to clarifying the chemical nature of the interaction of DMMP and single-crystal Pt surfaces using vibrational spectroscopy, secondary ion mass spectroscopy, thermal desorption, and Auger spectroscopy.[14,98] The composition of intermediate phosphorus-containing species on a Pt(111) surface have been identified. DMMP adsorbs on this type of surface molecularly through the oxygen of the P=O group. Near 300 K, two possible decomposition mechanisms are proposed that involve PO-C, P-OC, and P-C bond-cleavage. The discovery of exceptionally effective catalysts like Pd and other transition metals[99-102] result in a dramatic improvement in the synthetic application for insertion reactions. Decomposition of DMMP on the Mo(110)[103] and on Ni(111) and Pd(111)[13] surfaces has been studied using Auger spectroscopy. The authors have observed catalytic oxidation of DMMP without the production of a surface phosphate. CO and phosphorus oxide-gas phase species were the products found.

The catalytic decomposition of DMMP with metal catalysts was also studied.[104] The platinum catalyst has shown the best performance for the reaction. It was found that the methoxy groups of DMMP were separated by forming methanol which is decomposed to carbon dioxide and water. Platinum was also found to give the best results as a catalyst of the DMMP decomposition in a study by Graven et al.[105] The product analyses have indicated that the initial reaction was principally hydrolysis with methanol and phosphorus acid as products. In the classical work[106] almost stoichiometric amounts of CO_2 was produced, indicating complete oxidation of DMMP on platinum catalysts. Catalytic oxidation of the organophosphorus compounds is poisoned by P accumulation. It is therefore critical to remove P by forming the oxide before atomic P can be formed. On Pt catalytic oxidation is only possible with high-background pressures of O_2.[14]

The poisoning of Pt oxidation catalysts by phosphorus was also studied by Hegedus and Gumbleton[107] and by Angele and co-workers.[108−110] In a work by Dulcey and co-workers for the first time the thermal desorption of the PO radical was observed in the catalytic decomposition of DMMP on a polycrystalline Pt surface.[111] Catalytic decomposition of Sarin using a Pt catalyst results initially in stoichiometric amounts of the oxidation products: CO_2, HF, H_2O, and H_3PO_4.[106] The application of quantum-chemical methods to platinum and palladium surfaces has been quite limited because of the difficulty to treat large clusters of platinum and palladium atoms.

According to our best knowledge, no theoretical works have been published on the interactions of organophosphates with transition metal surfaces. Our preliminary study on interaction of Sarin with different clusters of Pt and Pd[112] indicates formation of strongly bonded complexes (the interaction energy amounts up to −25 kcal/mol, (see Figure 16-6) that illustrates the optimized geometry of Sarin adsorbed on the Pt_{12} and Pd_{12} clusters).

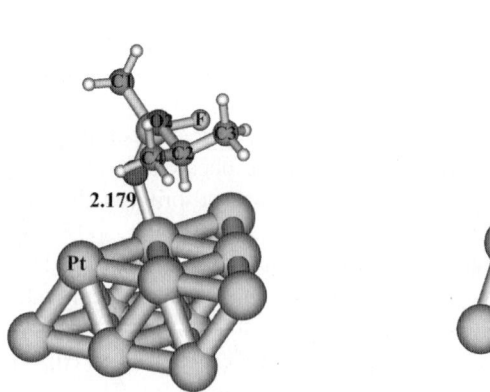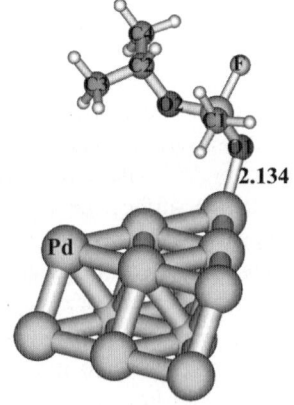

Figure 16-6. The optimized structure of Sarin adsorbed on the Pt_{12} and Pd_{12} clusters obtained at the B3LYP/6-31G(d) level of theory

4.   SUMMARY AND FUTURE RESEARCH AREA

4.1.   Summary

Decomposition of organophosphates is of great importance since it is of relevance to the agricultural, industrial, and military applications of these species. The development of cost effective, cleanup technologies for organophosphorus contaminants possesses a high priority for environmental restoration research. Such development involves the coordination of experimental and theoretical investigations to integrate both technological and fundamental aspects of key processes. Although the major processes affecting the natural and engineered treatment of organophosphates have been appreciated qualitatively, many questions remain regarding reaction mechanisms. Quantum-chemical analysis provides a wide array of powerful tools that have been underutilized in deciphering the complex reactions affecting warfare agents.

The adsorption and reaction of organophosphates on metal oxides have been investigated by a number of groups.[63−79] In most of these works the initial step of the adsorption involves the binding of the phosphoryl oxygen to an acidic site of metal oxide. In most cases, the initial chemisorption leads to the decomposition of the target molecule where different products are formed, depending on the type of organophosphorus compound and metal oxide. For example, decomposition of DMMP on MgO proceeds via a simple path[72] where, in the second step, the loss of two methoxy groups is observed with the presumed evolution of methanol from the surface. Theoretical works of adsorption and decomposition of selected organophosphates were devoted to the study of the dependence of these reactions on several factors and the influence of the nature of the metal oxide and the clay surface.[81,85,92−95] Several different initial positions and orientations of organophosphate molecules on models of metal oxides and clays were found.

The following factors have been identified as having large influence on the surface reactivity of clays and metal oxides as catalysts: the type of the acidic site, the adsorption site, the central cation, the size of the fragment and the presence of water. The organophosphorus compounds can be chemisorbed or physisorbed on selected types of catalytic surfaces. In the case of Sarin it was found that chemisorption on the MgO surface can lead to its decomposition. An energetic diagram that predicts possible intermediates and final products of Sarin decomposition by different adsorption sites of magnesium oxide fragments was theoretically predicted.[81]

Adsorption and decomposition of DMMP and Sarin[14,98−111] on transition metal surfaces such as platinum, palladium, nickel and molybdenum) have been investigated experimentally. Among these transition metals platinum was found to give the best performance in the decomposition of DMMP.[104,105] DMMP undergoes initially molecular adsorption[14,98] through the phosphoryl group. The results of several studies have shown that the main product of DMMP[104,105] and Sarin[106] decomposition is methanol which is in the next step decomposed to carbon dioxide. The biggest problem of catalytic oxidation of organophosphorus compounds on transition metals is poisoning by phosphorus accumulation.[14,107−110]

4.2.   Future Research Area

The knowledge of the interactions of organophosphates with environment is limited. Future work should be devoted to more extensive investigations of the adsorption and interactions of organophosphates (for example nerve agents Tabun and VX) on the surfaces of catalysts (metals and metal oxides) and the degradation pathways of these species. The molecular approach that applies advanced computational chemistry (CC) techniques to develop computational models in order to explain and predict how the selected catalysts (transition metals and metal oxides) affect the adsorption and decomposition provides an efficient way to investigate these phenomena. Using methods and approximations of computational chemistry the following specific tasks should be addressed:

1. Chemical structure of adsorption sites of selected catalysts (transition metals and metal oxides) in the gas phase and under influence of water solution. Since industrial processes affect the surrounding environment including the water and soil, special attention should be paid to the modeling of adsorption of those substances from water solution by metal oxide catalytic surfaces. Research should be devoted to deploying, modifying, and/or assessing existing remediation, characterization and industrial technologies to improve the success in the removing groundwater contamination. As was discussed above, the influence of a water bulk can be modeled using supermolecular approximation or by using another approach in which the explicit solvent molecules are replaced by a continuous medium represented by a dielectric constant (Polarized Continuum Model (PCM)).

2. Chemical and physical characteristics of organophosphate adsorption on catalytic surfaces (metal and metal oxide). The ability of sorption and desorption is one of the primary mechanisms that control or retard the migration of many contaminants and their penetration into the bulk. Therefore, this research area should be devoted to understanding the thermodynamics of organophosphorus compound sorption on the surfaces of interest. Computational techniques are able to predict how thermodynamic characteristics including adsorption enthalpies and Gibbs free energies of organophosphates depend on the structural type of the catalyst (metal and metal oxide) and the type of adsorption site (regular surface, defect, or corner). Therefore, for all studied systems zero-point energies, enthalpies, and Gibbs free energies at room temperature (298 K) could be calculated within the harmonic oscillator-rigid rotor–ideal gas approximation. Also the values of distribution coefficient (K_d) should be calculated since it is the most common method used to describe contaminant adsorption on soil.

3. The degradation pathways of organophosphates on catalysts (thermodynamics and kinetics of its decomposition) should be investigated. This includes evaluation of the products and intermediates of the reduction or decomposition processes. In all possible cases the rate constants and corresponding half-times for organophosphate decomposition products should be also predicted along with the kinetics of their degradation fate.

ACKNOWLEDGEMENTS

This work was facilitated by support from the Department of Defense (DoD) through Engineer Research and Development Center (ERDC) grant No. W912HZ-05-C-0051, the Army High Performance Computing Research Center under the auspices of the Department of the Army grant No. DAAD19-01-2-0014 and the Army Research Laboratory Cooperative agreement No. DAAH04-95-2-0003/contract No. DAAH04-95-C-0008 the content of which does not necessarily reflect the position or policy of the government, and no official endorsement should be inferred. This work was also facilitated by use of the Network Visualization System for Computational Chemistry (NVSCC) (http://www.ccmsi.us/nvscc). We would like to thank the Mississippi Center for Supercomputing Research and the Interdisciplinary Center for Mathematical and Computational Modeling of Warsaw University for a generous allotment of computer time.

REFERENCES

1. Ekerdt JG, Klabunde KJ, Shapley JR, White JM, Yates JT (1988) Surface chemistry of organophosphorus compounds, J Phys Chem 92: 6182–6188
2. Ewing KJ, Lerner B (2001) Infrared detection of the nerve agent sarin (isopropyl methylphosphonofluoridate) in water using magnesium oxide for preconcentration, Applied Spectroscopy 55(4): 407–411
3. Haensel V, Burwell R (1971) Catalysis, Sci Am 225: 46–58
4. Yang Y-Ch, Baker JA, Ward JR (1992) Decontamination of chemical warfare agents, Chem Rev 92: 1729–1743
5. Yang Y-C (1999) Chemical Detoxification of Nerve Agent VX, Acc Chem Res 32: 109–115
6. Nachon F, Asojo OA, Borgstahl G, Masson P, Lockridge O (2005) Role of water in aging of human butyrylcholinesterase inhibited by echothiophate: the crystal structure suggests two alternative mechanisms of aging, Biochemistry 44: 1154–1162
7. Toy A, Walsh E (1987) Phosphorus Chemistry in Everyday Living (Amer. Chem. Soc., Washington, D.C), pp. 319–330
8. Simpson B (2004) Sarin Nerve Gas - Or How I Learned to Stop Worrying and Love Pon-1, BioTech Journal 2: 100–105
9. http://www.dupont.com/safety/en/downloads/NerveAgentChemicalDataSheets.PDF
10. http://en.wikipedia.org/wiki/Tabun_%28nerve_gas%29
11. Yang Y-Ch, Szafraniec LL, Beaudry WT, Rohrbaugh DK, Procell LR, Samuel JB (1996) Autocatalytic Hydrolysis of V-Type Nerve Agents, J Org Chem 61: 8407–8413
12. http://www.dsf.health.state.pa.us/health/cwp/view.asp?a=171&q=233733
13. Guo X, Yoshinobu J, Yates JT, Jr (1990) Decomposition of an organophosphonate compound (dimethylmethylphosphonate) on the nickel(111) and palladium(111) surfaces, J Phys Chem 94(17): 6839–6842
14. Henderson MA, White JM (1988) Adsorption and decomposition of dimethyl methylphosphonate on platinum(111), J Am Chem Soc 110: 6939–6947
15. http://en.wikipedia.org/wiki/Platinum
16. http://en.wikipedia.org/wiki/Palladium

17. Klabunde KJ, Stark J, Koper O, Mohs C, Park DG, Decker S, Jiang Y, Lagadic I, Zhang D (1996) Nanocrystals as Stoichiometric Reagents with Unique Surface Chemistry, J Phys Chem 100: 12142–12153

18. Stark JV, Park DG, Lagadic I, Klabunde KJ (1996) Nanoscale Metal Oxide Particles/Clusters as Chemical Reagents. Unique Surface Chemistry on Magnesium Oxide As Shown by Enhanced Adsorption of Acid Gases (Sulfur Dioxide and Carbon Dioxide) and Pressure Dependence, Chem Mater 8: 1904–1912

19. Yu C, Hao Q, Saha S, Shi L (2005) Integration of metal oxide nanobelts with microsystems for nerve agent detection, Applied Physics Letters 86: 063101/1–063101/3

20. Khaleel A, Kapoor PN, Klabunde KJ (1999) Nanocrystalline metal oxides as new adsorbents for air purification, Nanostructured Materials 11(4): 459–468

21. Ahdjoudj J, Markovits A, Minot C (1999) Hartree-Fock periodic study of the chemisorption of small molecules on TiO2 and MgO surfaces, Catalysis Today 50: 541–551

22. Tomchenko AA, Harmer GP, Marquis BT (2005) Detection of chemical warfare agents using nanostructured metal oxide sensors, Sensors and Actuators B 108: 41–55

23. Lee WS, Lee SC, Lee SJ, Lee DD, Huh JS, Jun HK, Kim JC (2005) The sensing behavior of SnO2-based thick-film gas sensors at a low concentration of chemical agent simulants, Sensors and Actuators B 108: 148–153

24. Koper O, Li YX, Klabunde KJ (1993) Destructive adsorption of chlorinated hydrocarbons on ultrafine (nanoscale) particles of calcium oxide, Chem Mater 5: 500–505

25. Medine GM, Zaikovskii V, Klabunde KJ (2004) Synthesis and adsorption properties of intimately intermingled mixed metal oxide nanoparticles, J Mater Chem 14: 757–763

26. Sauer J (1989) Molecular models in ab initio studies of solids and surfaces: from ionic crystals and semiconductors to catalysts, Chem Rev 89: 199–255

27. Maseras F, Morokuma K (1995) MOMM: a new integrated ab initio + molecular mechanics geometry optimization scheme of equilibrium structures and transition states, J Comp Chem 16: 1170–1179

28. Schaefer HF (1977) Methods of Electronic Structure Theory in: *Modern Theoretical Chemistry* (Plenum Press, New York), Vol. 3

29. Čásky P, Urban M (1980) *Lecture Notes in Chemistry*, No. 16: Ab Initio Calculations. Methods and Applications in Chemistry

30. Hehre WJ, Radom L, Schleyer PvR, People JA (1986) Ab Initio Molecular Orbital Theory (Wiley, New York)

31. Lawley K-P (1987) Ab Initio Methods in Quantum Chemistry (Wiley, New York)

32. Sauer J (1994) Theoretical Study of van der Waals Complexes at Surface Sites in Comparison with the Experiment, Chem Rev 94: 2095–2160

33. Kresse G, Hafner J (1993) Ab initio molecular dynamics for open-shell transition metals, Phys Rev B 48: 13115–13118

34. Kresse G, Furthmülleer J (1996) Efficiency of ab-initio total energy calculations for metals and semiconductors using a plane-wave basis set, Comp Mat Sci 6: 15–50

35. Perdew JP, Chevary JA, Vosko SH, Jackson AK, Pederson RM, Singh DJ, Fiolhais C (1992) Atoms, molecules, solids, and surfaces: applications of the generalized gradient approximation for exchange and correlation, Phys Rev B 46(11): 6671–6687

36. Vanderbilt D, Taole SH, Narasimhan S (1990) Anharmonic elastic and phonon properties of silicon, Phys Rev B 42(17): 11373–11374

37. Mikheikin ID, Abronin IA, Zhidomirov GM, Kazansky VB (1977) Calculations of chemadsorption and elementary events of catalytic reactions within the framework of a cluster model. II. Properties of surface hydroxyl groups of oxides, Kinetics and Catalysis 18: 1580–1583

38. Beran S (1984) Quantum chemical study of the effect of the structural characteristics of zeolites on the properties of their bridging hydroxyl groups, J Mol Catal 26: 31–36

39. Kazansky VB, Serykh AI, Pidko EA (2004) DRIFT study of molecular and dissociative adsorption of light paraffins by HZSM-5 zeolite modified with zinc ions: methane adsorption. Journal of Catalysis 225(2): 369–373

40. Pelmenshchikov AG, Pavlov VI, Zhidomirov GM, Beran S (1987) Effects of structural and chemical characteristics of zeolites on the properties of their bridging hydroxyl groups, J Phys Chem 91: 3325–3327

41. Anchell JL, Hess AC (1996) H2O Dissociation at Low-Coordinated Sites on (MgO)n Clusters, n = 4, 8, J Phys Chem 100: 18317–18321

42. Gorb LG, Rivail JL, Thery V, Rinaldi D (1996) Modification of the local self-consistent field method for modeling surface reactivity of covalent solids, Int J Quant Chem 60: 313–324

43. Svensson M, Humbel S, Froese RDJ, Matsubara T, Sieber S, Morokuma K (1996) ONIOM: A Multi-Layered Integrated MO + MM Method for Geometry Optimizations and Single Point Energy Predictions. A Test for Diels-Alder Reactions and Pt(P(t-Bu)3)2 + H2 Oxidative Addition, J Phys Chem 100: 19357–19363

44. Van Santen RA, Kramer GJ (1995) Reactivity Theory of Zeolitic Broensted Acidic Sites, Chem Rev 95: 637–660

45. Tomasi J, Mennucci B, Cammi R (2005) Quantum Mechanical Continuum Solvation Models, Chem Rev 105: 2999–3093

46. Sponer J, Leszczynski J, Hobza P (1996) Hydrogen bonding and stacking of DNA bases: a review of quantum-chemical ab initio studies, J Biomol Struct and Dynamics 14: 117–135

47. Car R, Parrinello M (1985) Unified approach for molecular dynamics and density-functional theory, Phys Rev Lett 55: 2471–2474

48. Tuckerman M, Laasonen K, Sprik M, Parrinello M (1995) Ab Initio Molecular Dynamics Simulation of the Solvation and Transport of H3O+ and OH− Ions in Water, J Phys Chem 99: 5749–5752

49. Sprik M, Hutter J, Parrinello M (1996) Ab initio molecular dynamics simulation of liquid water: comparison of three gradient-corrected density functionals, J Chem Phys 105: 1142–1152

50. Laasonen K, Sprik M, Parrinello M (1993) "Ab initio" liquid water, J Chem Phys 99: 9080–9089

51. Bianco R, Miertus S, Persico M, Tomasi J (1992) Molecular reactivity in solution. Modeling of the effects of the solvent and of its stochastic fluctuation on an SN2 reaction, Chem Phys 168: 281–292

52. Miertus S, Scrocco E, Tomasi J (1981) Electrostatic Interaction of a Solute with a Continuum. A Direct Utilization of ab initio Molecular Potentials for the Prevision of Solvent Effects, Chem Phys 55: 117–129

53. Cramer J, Truhlar DG (1995) Continuum Solvation Models: Classical and Quantum Mechanical Implementations, in: Computational Chemistry, edited by K. B. Lipkowitz and D. B. Boyd, Vol.7 (VCH Publishers, Inc.)

54. Del Bene J (1985) Molecular orbital theory of the hydrogen bond: XXXII. The effect of H+ and Li+ association on the A—T and G—C pairs J Mol Struct (Theochem) 124: 201–212

55. Tomasi J, Persico M (1994) Molecular Interactions in Solution: An Overview of Methods Based on Continuous Distributions of the Solvent, Chem Rev 94: 2027–2094

56. Smedarchina Z, Zgierski MZ, Siebrand W, Kozlowski PM (1998) Dynamics of tautomerism in porphine: An instanton approach, J Chem Phys 109: 1014–1024

57. Smedarchina Z, Siebrand W, Zgierski MZ, Zerbetto F (1995) Dynamics of molecular inversion: an instanton approach, J Chem Phys 102: 7024–7034

58. Smedarchina Z, Siebrand W, Fernandez-Ramos A, Gorb L, Leszczynski J (2000) A direct-dynamics study of proton transfer through water bridges in guanine and 7-azaindole, J Chem Phys 112: 566–573

59. Smedarchina Z, Fernandez-Ramos A, Siebrand W (2001) DOIT: a program to calculate thermal rate constants and mode-specific tunneling splittings directly from quantum-chemical calculations, J Comp Chem 22: 787–801

60. Truhlar DG, Garrett BC, Klippenstein SJ (1996) Current Status of Transition-State Theory, J Phys Chem 100: 12771–12800

61. Tucker SC, Truhlar DG (1989) Dynamical Formulation of Transition State Theory: Variational Transition States and Semiclassical Tunneling, edited by J. Bertran and I. G. Csizmadia (Advanced Study Institute, Kluwer, Dordrecht), p. 291

62. Gonzalez-Lafont A, Troung TN, Truhlar DG (1991) Interpolated variational transition-state theory: practical methods for estimating variational transition-state properties and tunneling contributions to chemical reaction rates from electronic structure calculations, J Phys Chem 95: 8875–8894

63. Yang Y-C (1995) Chemical reactions for neutralizing chemical warfare agents, Chem Ind 9: 334–337

64. Templeton MK, Weinberg WH (1985) Adsorption and decomposition of dimethyl methylphosphonate on an aluminum oxide surface, J Am Chem Soc 107: 97–108

65. Templeton MK, Weinberg WH (1985) Decomposition of phosphonate esters adsorbed on aluminum oxide, J Am Chem Soc 107: 774–779

66. Li Y-X, Klabunde KJ (1991) Nano-scale metal oxide particles as chemical reagents. Destructive adsorption of a chemical agent simulant, dimethyl methylphosphonate, on heat-treated magnesium oxide, Langmuir 7: 1388–1393

67. Li Y-X, Schlup JR, Klabunde KJ (1991) Fourier transform infrared photoacoustic spectroscopy study of the adsorption of organophosphorus compounds on heat-treated magnesium oxide, Langmuir 7: 1394–1399

68. Atteya M, Klabunde KJ (1991) Nanoscale metal oxide particles as chemical reagents. Heats of adsorption of heteroatom containing organics on heat-treated magnesium oxide samples of varying surface areas, Chem Mater 3: 182–187

69. Li Y-X, Koper O, Atteya M, Klabunde KJ (1992) Adsorption and decomposition of organophosphorus compounds on nanoscale metal oxide particles. In situ GC-MS studies of pulsed microreactions over magnesium oxide, Chem Mater 4: 323–330

70. Henderson MA, Jin T, White JM (1986) A TPD/AES study of the interaction of dimethyl methylphosphonate with iron oxide $(\alpha - Fe_2O_3)$ and silicon dioxide, J Phys Chem 90: 4607–4611

71. Aurian-Blajeni B, Boucher MM (1989) Interaction of dimethyl methylphosphonate with metal oxides, Langmuir 5: 170–174

72. Mitchell MB, Sheinker VN, Mintz EA (1997) Adsorption and Decomposition of Dimethyl Methanephosphonate on Metal Oxides, J Phys Chem B 101: 11192–11203

73. Sheinker VN, Mitchell MB (2002) Quantitative Study of the Decomposition of Dimethyl Methylphosphonate (DMMP) on Metal Oxides at Room Temperature and Above, Chem Matter 14: 1257–1268

74. Mitchell MB, Sheinker VN, Tesfamichael AB, Gatimu EN, Nunley M (2003) Decomposition of Dimethyl Methylphosphonate (DMMP) on Supported Cerium and Iron Co-Impregnated Oxides at Room Temperature, J Phys Chem B 107: 580–586

75. Kanan SM, Tripp CP (2001) An Infrared Study of Adsorbed Organophosphonates on Silica: A Prefiltering Strategy for the Detection of Nerve Agents on Metal Oxide Sensors, Langmuir 17: 2213–2218

76. Kanan SM, Lu Z, Tripp CP (2002) A Comparative Study of the Adsorption of Chloro- and Non-Chloro-Containing Organophosphorus Compounds on WO3, J Phys Chem B 106: 9576–9580

77. Wagner GW, Bartram PW, Koper O, Klabunde KJ (1999) Reactions of VX, GD, and HD with Nanosize MgO, J Phys Chem B 103: 3225–3228

78. Wagner GW, Bartram PW, Koper O, Klabunde KJ (2000) Reactions of VX, GD, and HD with Nanosize CaO: Autocatalytic Dehydrohalogenation of HD, J Phys Chem B 104: 5118–5123

79. Wagner GW, Procell LR, O'Connor RJ, Munavalli S, Carnes CL, Kapoor PN, Klabunde KJ (2001) Reactions of VX, GB, GD, and HD with nanosize Al(2)O(3). Formation of aluminophosphonates, J Am Chem Soc 123: 1636–1644

80. Kuiper AET, van Bokhoven JJG, Medena J (1976) The role of heterogeneity in the kinetics of a surface reaction. I. Infrared characterization of the adsorption structures of organophosphonates and their decomposition, J Catal 43: 154–167

81. Michalkova A, Ilchenko M, Gorb L, Leszczynski J (2004) Theoretical Study of the Adsorption and Decomposition of Sarin on Magnesium Oxide, J Phys Chem B 108: 5294–5303

82. Bader RWF (1990) Atoms in Molecules: A Quantum Theory (Oxford University Press: Oxford)

83. Koch U, Popelier PLA (1995) Characterization of C-H-O Hydrogen Bonds on the Basis of the Charge Density, J Phys Chem 99: 9747–9754

84. Popelier PAL (1998) Characterization of a Dihydrogen Bond on the Basis of the Electron Density, J Phys Chem A 102: 1873–1878

85. Michalkova A, Gorb L, Ilchenko M, Zhikol OA, Shishkin OV, Leszczynski J (2004) Adsorption of Sarin and Soman on Dickite: An Ab Initio ONIOM Study, J Phys Chem B 108: 1918–1930

86. Svensson M, Humbel S, Morokuma K (1996) Energetics using the single point IMOMO (integrated molecular orbital + molecular orbital) calculations: choices of computational levels and model system, J Chem Phys 105: 3654–3661

87. Dapprich S, Komáromi I, Byun KS, Morokuma K, Frisch MJ (1999) A new ONIOM implementation in Gaussian98. Part I. The calculation of energies, gradients, vibrational frequencies and electric field derivatives, J Mol Struct (Theochem) 461–462: 1–21

88. Sokalski WA, Roszak S, Pecul K (1988) An efficient procedure for decomposition of the SCF interaction energy into components with reduced basis set dependence, Chem Phys Lett 153: 153–159

89. Jeziorski B, van Hemert MC (1976) Variation-perturbation treatment of the hydrogen bond between water molecules, Mol Phys 31: 713–730

90. Gora RW, Bartkowiak W, Roszak S, Leszczynski J (2002) New theoretical insight into the nature of intermolecular interactions in the molecular crystal of urea, J Chem Phys 117: 1031–1039

91. Schmidt MS, Baldridge KK, Boatz JA, Elbert ST, Gordon MS, Jensen JH, Koseki S, Matsunaga N, Nguyen KA, Su SJ, Windus TL, Dupuis M, Montgomery JA (1993) General atomic and molecular electronic structure system, J Comp Chem 14: 1347–1363

92. Michalkova A, Martinez J, Zhikol OA, Gorb L, Shishkin OV, Leszczynska D, Leszczynski J (2006) Theoretical Study of Adsorption of Sarin and Soman on Tetrahedral Edge Clay Mineral Fragments, J Phys Chem B, (submitted)

93. Murashov VV, Leszczynski J (1999) Adsorption of the Phosphate Groups on Silica Hydroxyls: An ab Initio Study, J Phys Chem A 103: 1228–1238

94. Hartzell CJ, Cygan RT, Nagy KJ (1998) Molecular Modeling of the Tributyl Phosphate Complex of Europium Nitrate in the Clay Hectorite, J Phys Chem A 102(34): 6722–6729

95. Zhanpeisov NU, Zhidomirov GM, Yudanov IV, Klabunde KJ (1994) Cluster Quantum Chemical Study of the Interaction of Dimethyl Methylphosphonate with Magnesium Oxide, J Phys Chem 98: 10032–10035

96. Kaczmarek A, Gorb L, Sadlej AJ, Leszczynski J (2004) Sarin and Soman: Structure and Properties, Struct Chem 15: 517–525

97. Paukku Y, Michalkova A, Majumdar D, Leszczynski J (2006) Investigation on the low energy conformational surface of tabun to probe the role of its different conformers on biological activity, Chem Phys Lett 422: 317–322

98. Hegde RI, Greenlief CM, White JM (1985) Surface chemistry of dimethyl methylphosphonate on rhodium(100), J Phys Chem 89: 2886–2891

99. Demonceau A, Noels AF, Hubert AJ (1988) Recent Aspects of Transition Metal Catalyzed Reactions of Carbenes in the Realm of Biologically Active Substances in Aspects Homogeneous Catalysis, edited by R. Ugo and D. Reibel, (Publ. Comp., Dordrecht), Vol. 6, pp. 199–232

100. Salomon RG, Kochi JK (1973) Copper(I) catalysis in cyclopropanations with diazo compounds. Role of olefin coordination, J Am Chem Soc 95: 3300–3310

101. R. Paulissen, A. J. Hubert, and Ph. Teyssie (1972) Transition metal-catalyzed cyclopropanation of olefins, Tetrahedron Lett 15: 1465–1466

102. A. J. Hubert, A. F. Noels, A. J. Anciaux, and Ph. Teyssie (1976) Rhodium(II) carboxylates: novel highly efficient catalysts for the cyclopropanation of alkenes with alkyl diazoacetates, Synthesis 9: 600–602

103. V. S. Smentkowski, P. L. Hagans, and J. T. Yates (1988) Study of the catalytic destruction of dimethyl methylphosphonate(DMMP): oxidation over molybdenum(110), J. Phys. Chem 92: 6351–6357

104. S. G. Ryu, J. K. Yang, H. W. Lee, and Y. S. Yang (1995) Decomposition of dimethyl methylphosphonate over alumina-supported precious metal catalysts, Journal of Korean Institute of Chemical Engineers 33: 462–470

105. W. M. Graven, S. W. Weller, and D. L. Peters (1966) Catalytic conversion of an organophosphate vapor over platinum-alumina, Ind. Eng. Chem. Process Des. Dev 5: 183–189 .

106. R. W. Baier, and S. W. Weller (1967) Catalytic and thermal decomposition of sarin, Ind. Eng. Chem. Process Des. Dev 6: 380–385

107. L. L. Hegedus, and K. Baron (1975) Effects of poisoning and sintering on the pore structure and diffusive behavior of platinum/alumina catalysts in automotive converters, Journal of Catalysis 37(1): 127–132

108. B. Angele, and K. Kirchner (1980) The poisoning of noble metal catalysts by phosphorus compounds. I. Chemical processes, mechanisms, and changes in the catalyst, Chem. Eng. Sci 35: 2089–2091

109. B. Angele, K. Kirchner, and E. G. Schlosser (1980) The poisoning of noble metal catalysts by phosphorus compounds. III. The deposition of catalyst poisons in honeycomb catalysts, Chem. Eng. Sci 35: 2101–2105

110. B. Angele, and K. Kirchner (1980) The poisoning of noble metal catalysts by phosphorus compounds. II. The kinetics of poisoning and a mathematical model, Chem. Eng. Sci 35: 2903–2909

111. C. S. Dulcey, M. C. Lin, and C. C. Hsu (1985) Thermal desorption of the phosphoryl (PO) radical from polycrystalline platinum surfaces, Chem. Phys. Lett 115: 481–485

112. A. Michalkova, D. Majumdar, and J. Leszczynski, Adsorption of sarin on platinum and palladium surfaces: An ab initio study, J. Phys. Chem.B 2006 (to be published).

INDEX

CHALLENGES AND ADVANCES
IN COMPUTATIONAL CHEMISTRY AND PHYSICS